TRAFFIC AND GRANULAR FLOW '05

A. Schadschneider   T. Pöschel
R. Kühne   M. Schreckenberg   D. E. Wolf
Editors

TRAFFIC AND GRANULAR FLOW '05

With 402 Figures, 159 in Colour, and 15 Tables

 Springer

Editors

Andreas Schadschneider
Institut für Theoretische Physik
Universität zu Köln
Zülpicher Str. 77
50937 Köln, Germany
e-mail: as@thp.uni-koeln.de

Michael Schreckenberg
Physik von Transport und Verkehr
Universität Duisburg-Essen
Lotharstr. 1
47048 Duisburg, Germany
e-mail: schreckenberg@traffic.uni-duisburg.de

Thorsten Pöschel
Charité
Campus Virchow Klinikum
Centrum für Unfall- und
Wiederherstellungschirurgie
Augustenburger Platz 1
13353 Berlin, Germany
e-mail: thorsten.poeschel@charite.de

Dietrich E. Wolf
Theoretische Physik
Universität Duisburg-Essen
Lotharstr. 1
47048 Duisburg, Germany
e-mail: dietrich.wolf@uni-due.de

Reinhart Kühne
Institut für Verkehrsforschung
Deutsches Zentrum für Luft-
und Raumfahrt (DLR)
Rutherfordstr. 2
12489 Berlin, Germany
e-mail: reinhart.kuehne@dlr.de

Mathematics Subject Classification (2000): 90B20, 90B06, 90B15, 90B18, 60K30, 76T25, 37M05, 37M10, 65C20, 65C35, 68U20, 91A80

Springer is a part of Springer Science+Business Media

springer.com

© Springer-Verlag Berlin Heidelberg 2007

Cover design: WMXDesign GmbH, Heidelberg

Printed on acid-free paper   46/3100/YL - 5 4 3 2 1 0

Preface

The conference series *Traffic and Granular Flow* has been established in 1995 and has since then been held biannually. At that time, the investigation of granular materials and traffic was still somewhat exotic and was just starting to become popular among physicists.

Originally the idea behind this conference series was to facilitate the convergence of the two fields, inspired by the similarities of certain phenomena and the use of similar theoretical methods. However, in recent years it has become clear that probably the differences between the two systems are much more interesting than the similarities. Nevertheless, the importance of various interrelations among these fields is still growing. The workshop continues to offer an opportunity to stimulate this interdisciplinary research.

Over the years the spectrum of topics has become much broader and has included also problems related to topics ranging from social dynamics to biology. The conference manages to bring together people with rather different background, ranging from engineering to physics, mathematics and computer science. Also the full range of scientific tools is represented with presentations of empirical, experimental, theoretical and mathematical work.

The workshop on *Traffic and Granular Flow '05* was the sixth in this series. Previous conferences were held in Jülich (1995), Duisburg (1997), Stuttgart (1999), Nagoya (2001), and Delft (2003). For its 10th anniversary, Berlin was chosen as location, the largest city and capital of Germany. Berlin is also one of the centers for transport related research and hosts many research institutes that have a long history in the fields covered by the workshop.

The *TGF '05* took place from October 10-12, 2005 at the Humboldt University. World-renowned scientists worked here and read famous lectures, such as Max Born, Albert Einstein, Peter Debye, James Franck, Fritz Haber, Otto Hahn, Werner Heisenberg, Gustav Hertz, Jacob van't Hoff, Robert Koch, Max v. Laue, Walter Nernst, Max Planck, Erwin Schroedinger, and Wilhelm Wien, to name only few of the 29 Nobel price laureates of the Humboldt University.

It is one of the most famous venues in the heart of Berlin with locations touching the high-lights and low-lights of German-European and World History. It is located vis-à-vis of the Bebel square where 1933 the Nazi burned books of such famous authors like Karl Marx, Heinrich Heine, Sigmund Freud, Bertolt Brecht, Kurt Tucholsky, and Carl von Ossietzky. The German Reichstag, the house of the parliament, close to the Humboldt University was burned in the same year which was the occasion for the prosecution of dissenters and ended with millions of murdered people in the concentration camps and World War 2.

But also very close to the Humboldt University, at the Brandenburg gate, in November 1989 people were sitting on the Berlin wall celebrating the end of cold war. These pictures went all over the world. They shaped the image of a new, young, open and optimistic Berlin.

We hope that this spirit of openness could also be felt at the conference. Experts from physics, engineering, computer sciences and mathematics experienced a unique forum where current problems and solutions were presented and discussed to deepen the understanding and knowledge of the physics of traffic and the physics of granular media. Both areas have many important applications in society and industry. "Free Flow" is an indispensable prerequisite for acceptable traffic but it is also an existential precondition for mixing powder for production of tablets or packaging in bags and exactly closing. The main goal of the conference was to encourage theorists and practitioners of both areas to a common view on the dynamics of transportation processes for mutual benefit. It attracted nearly 100 participants from all over the world, from almost 20 countries.

The papers presented show the current progress in modelling, computer simulation, experiments and phenomena description as well as the prospectives for application. The importance of the interregulations between both research areas is growing. The conference pays tribute to this development and opens new possibilities for interdisciplinary research. The topics covered are, beside others, vehicular traffic, pedestrian traffic, granular flow, traffic in urban road networks and computer networks and collective phenomena in biological systems.

The conference ignited a broad public interest and the organizers gratefully acknowledge financial support from the German Research Society (Deutsche Forschungsgemeinschaft), from the Technology Foundation Berlin (Technologiestiftung Berlin) and from the German Aerospace Center (Deutsches Zentrum für Luft- und Raumfahrt, DLR).

This conference would not have been possible without many people helping behind the scenes. In particular we like to thank Roberto Aoki, Ute Böttger, Petra Hänssgen, Steffi Lehmann from the DLR and Alireza Namazi from the University of Cologne.

Köln, Berlin, Duisburg
August 2006

Andreas Schadschneider
Thorsten Pöschel
Reinhart Kühne
Michael Schreckenberg
Dietrich E. Wolf

Contents

Part II Transport in Biological Systems

Part III Pedestrians

Part VI Traffic Flow: Empirical Results and Applications

Part I

Granular Flow

Saturn's Rings Seen by Cassini Spacecraft: Discoveries, Questions and New Problems

André Brahic

Université Paris VII Denis Diderot, A.I.M., C.E.A. Saclay, France
Member of the Cassini spacecraft Imaging Team
E-mail: brahic@cea.fr

1 Saturn's Rings: A Natural Laboratory of Granular Flow

The disc around Saturn is a system of colliding particles submitted to the gravitational influence of Saturn and of small nearby satellites. It can be considered as a natural laboratory of granular flow, dynamics, cosmogony, and particle and field physics.

Despite the flood of new information on morphology and optical properties, we have very little evidence about what rings are, how they formed, or how they behave. We can only answer such questions by building theoretical models and comparing their implications with past and future observations.

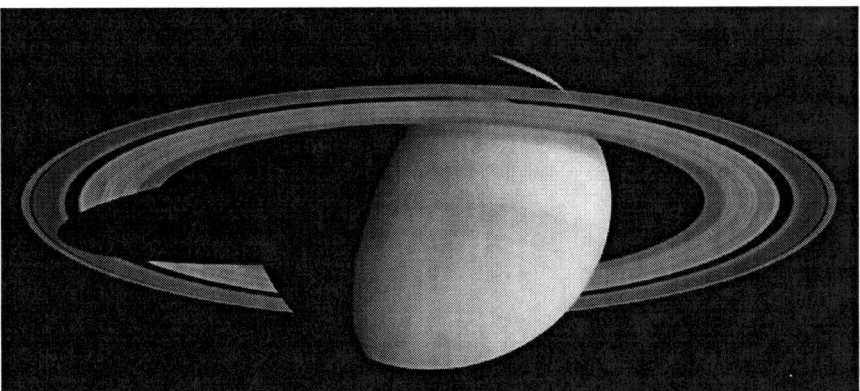

Fig. 1. Saturn's rings. Three main rings (A, B, C) surround Saturn. The D, F, and G rings are too diffuse or too narrow to be observed from the Earth. The external E ring has a maximum density about the orbit of Enceladus. (NASA-JPL/ESA document).

Since they were first discovered by Galileo in 1610, the nature of Saturn's rings has been a continuing challenge to observation and theory. From the beginning, observational resolution seemed to be just short of revealing the essential nature of the rings. The effort to understand rings has always attracted outstanding scientific minds. Galileo's first detection of something strange around Saturn was open to several interpretations. Huygens's revelation of a disc-like structure did not bring any information about rings' nature. Cassini suggested that the rings might consist of a myriad of small particles. Laplace and Maxwell showed that in fact a solid ring would be unstable. Beginning with Poincar, a general picture of collisional flattening and spreading emerged, with structure governed in part by resonances with the satellites. Dynamical theory was adequately consistent with Earth-based observations of seemingly smooth, continuous rings. Optical and radio properties seemed in good agreement with a swarm of small, icy particles of various sizes. Theoretical models seemed in harmony with most observed properties.

Then came the deluge! In a golden decade, our conception of rings underwent a revolution. We learned that rather to be smooth and continuous structures, rings are better characterized as sets of narrow ringlets with sharp edges, sometimes slightly elliptical or kinky in form. In 1977, narrow, black rings were discovered around Uranus as they occulted a star observed from Earth. In 1979, diffuse rings were discovered around Jupiter by the Voyager spacecrafts. In 1984 and 1985, arcs were detected by the author around Neptune as they occulted a star observed respectively from Chile and Hawaii. In 1980 and 1981, the Voyager spacecrafts revealed countless detailed features and structures that had never been imagined.

The structure of the rings is determined by their origin and by dynamical processes which depend upon the sizes and collisional properties of the ring particles, and on the gravitational effects of the satellites. Electromagnetic processes play a role on the motion of charged particles.

The classical ring system consists of three broad rings (A, B, and C from outside towards Saturn) occupying the region between 1.23 and 2.67 Saturnian radii. A faint E ring occupies an extended region outside the main rings and shows a maximum of density about the orbit of Enceladus. The D ring, which fills much of the region between the C ring and the top of Saturn's atmosphere, is too diffuse to be easily observed from Earth. The F and G rings are narrow and faint rings just outside the A ring.

The A ring, which is the outermost of the classical rings, has a typical optical depth of the order of 0.5. Many spiral waves and bending waves can be observed in this ring. Two narrow gaps (the Encke gap and the Keeler gap) are located in the outer portion of the A ring. The B ring is more opaque. Its optical depth varies between 0.7 and more than 2. Its sharp outer boundary coincides with the Mimas 2:1 resonance. It displays much more structure than the A ring. In between the A and B rings, the Cassini division contains broad and diffuse rings separated by narrow gaps. The inner C ring and the Cassini division present a number of similarities. Their optical depth is of the order of 0.2. The C ring is populated by several opaque ringlets.

The E ring is broad and diffuse, composed predominantly of micron-sized particles. It extends between 3 and 8 Saturnian radii. Its optical depth is of the order of 10-7, and it can be observed from Earth only when viewed edge on. The maximum density occurs near Enceladus' orbit, suggesting that this satellite is the source of he E ring.

The rings particles are primarily icy, but there is evidence for albedo, and therefore possibly compositional variations on both local and regional scales. Most of the particles are in the 1 centimetre to 5 metres radius range, but there is reason to suspect the existence of some particles of all sizes up to 10 kilometres in radius.

After the surprises of the Voyager flybys in 1980 and 1981, the Cassini spacecraft, with considerable improvements in resolution and sensitivity, is revealing a system still more complex than foreseen. A collection of never-before-seen phenomena within the rings was seen in the first images that may be evidence of different physical manifestations of particle aggregation, caused by either gravitational instabilities or kinematical effects or both. A small number of structures are described in this article.

The words of Maxwell in his seminal Adam's Prize essay of 1856 are still particularly well adapted to the study of rings: *"I am not aware that any practical use has been made of Saturn's Rings, either in Astronomy or in Navigation . But when we contemplate the Rings from a purely scientific point of view, they become the most remarkable bodies in the heavens... "*.

Saturn's rings system is a so complex laboratory that several fields of physics are needed for the modelling of the structure, behaviour, and evolution of the rings.

2 The Voyager Odyssey

The space exploration has completely changed our understanding of rings. In spite of 370 years of telescopic observations from the Earth, no one imagined before 1980 the wealth and the diversity of structures inside planetary rings. As a prelude to the Voyager flybys in1980 and 1981, the Pioneer 11 spacecraft made the first reconnaissance of Saturn in 1979, providing scientific results on which Voyager could build. Pioneer 11 flew close enough to Jupiter in 1974 to use the pull of the planet to bend its trajectory back on itself. In this way, it could be redirected toward Saturn, on nearly the other side of the solar system. The sling-shot effect of Jupiter's gravity had sent it out of the ecliptic plane into a region of space never before explored. Before the encounter, it was necessary to decide exactly what sort of flyby would be the most productive. Several options were considered, including a plunge directly through the Cassini division, or an aim point in the D ring, about midway between the inner edge of the C ring and the cloud tops, or a flyby well outside the visible

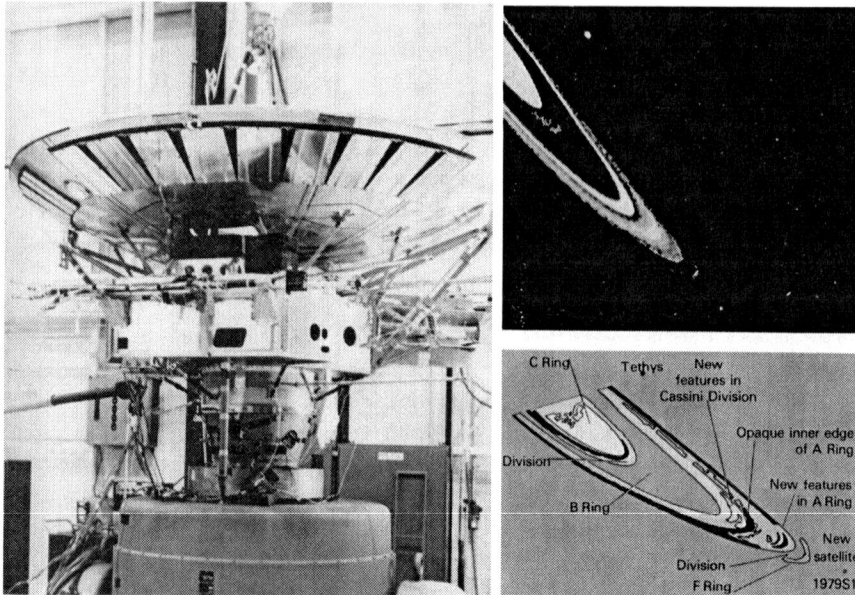

Fig. 2. Saturn's rings photographied by the Pioneer Saturn spacecraft.
The Pioneer 11 spacecraft (left image), which has discovered the F ring, took the
first space image of the rings from space (right image). The rotating spacecraft took
one line of the image at each rotation, this gave a tilted appearance to the rings
which have moved relative to the spacecraft during the exposure time. (NASA-JPL
documents).

rings, at a distance of 2.9 Saturn radii from the centre of the planet. This
last outside option was not only considered much safer, but the flyby distance
was exactly the distance from Saturn at which Voyager 2 would have to cross
the rings in 1981 if it were to continue to Uranus. After considerable debate,
the director of the NASA planetary program overruled the recommendations
of the Pioneer principal investigators to choose the inside option. He selected
the safer outside option. Since the Voyager flybys, the amount of material in
the D ring and the Cassini division has been estimated and we know to day
that Pioneer should have been probably destroyed if an inside option would
have been taken.

On September 1st, 1979, Pioneer 11, renamed Pioneer Saturn, safely
crossed the ring plane and reached, 29 minutes later, the closest approach
to Saturn, just 21000 kilometres above he clouds. The view from the spacecraft would have been truly spectacular, but it was unhappily impossible to
capture it with the on board imaging system. A second ring plane crossing,
at 2.78 Saturn radii, took place two hours after closest approach, without any
indication of damage to the spacecraft. Thanks to this mission, which has
done a large number of varied discoveries including the one of Saturn' F ring,
scientists had learned in addition that a spacecraft can survive inside the in-

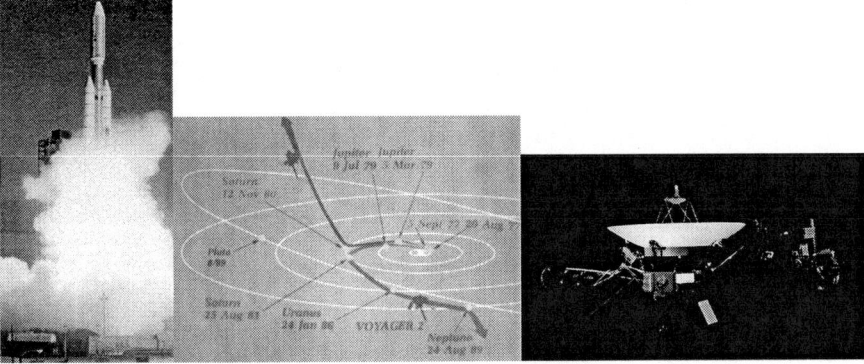

Fig. 3. The Voyager mission. The Voyager spacecrafts (right) were successfully launched in 1977 (left). Voyager 1 visited Jupiter, Saturn and Titan. Voyager 2 flew by Jupiter, Saturn, Uranus, and Neptune successively (centre). (NASA-JPL documents).

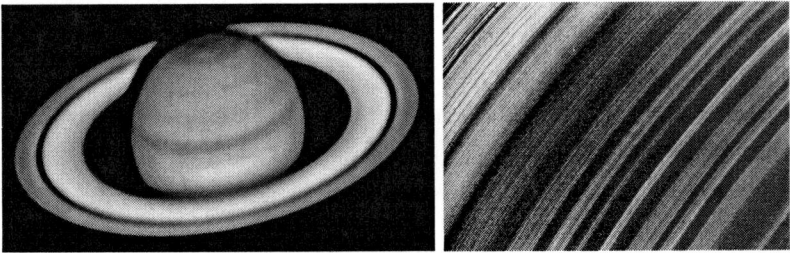

Fig. 4. Saturn's rings. Saturn's ring system is much more dynamic and complex than imagined before the Voyager flybys. Seen from the Earth (left), Saturn's rings look homogeneous with smooth edges. Voyager spacecrafts revealed heterogeneous rings with sharp edges (right). Thousands of structures are visible. (New Mexico University and NASA-JPL document).

tense radiation of Jupiter and Saturn magnetospheres, and that the Saturn's ring plane can be safely crossed outside the A ring.

The six Voyager encounters with the four giant planets were period unparalleled in degree and diversity of discovery, returning far more new information that had been collected in more than three centuries of Earth observations. The closer Voyager came to the planets, the more apparent it became that the scientific richness of the giant planets systems was going to greatly exceed the most optimistic expectations. The study of rings became one of the major goals of the Voyager mission. At the time of the mission design, only Saturn's rings were known and nobody expected the huge amount of surprises unveiled during the two Voyagers Saturn encounters. In 1980 and 1981, rings specialists were really astonished by the thousands of structures observed by the Voyagers inside Saturn's rings. These rings were found to be more complex than previously believed.

3 The Cassini Mission

Cassini Huygens is probably the most ambitious and the most expensive planetary mission ever launched. After a successful launch on October 15, 1997, two Venus flybys on April 26, 1998 and June 24, 1999, a Earth flyby on August 18, 1999, a Jupiter flyby on December 30, 2000, and a Phoebe flyby on June 12, 2004, the spacecraft safely reached Saturn on July1, 2004. At just the right moment, the main engine has been fired for 96 minutes to slow down and the Saturn orbit insertion has been a full success.

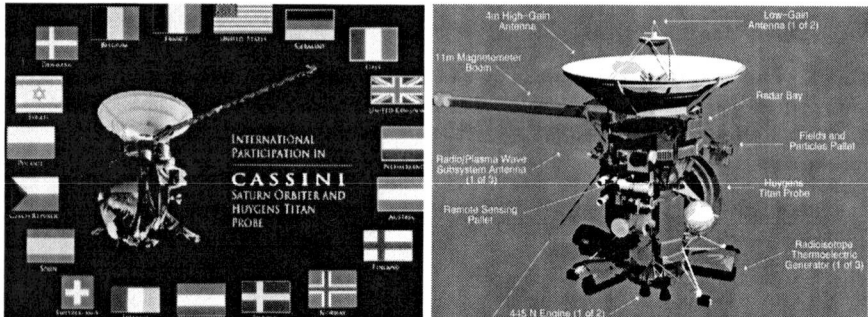

Fig. 5. The Cassini mission. The Cassini Huygens mission is an international collaboration. The flags of the participating countries are represented on the left. The Cassini orbiter (right image) carries 12 instruments. The overall height of the assembled spacecraft is 6.8 metres and its weight is 5.7 tons. (NASA-JPL/ESA documents).

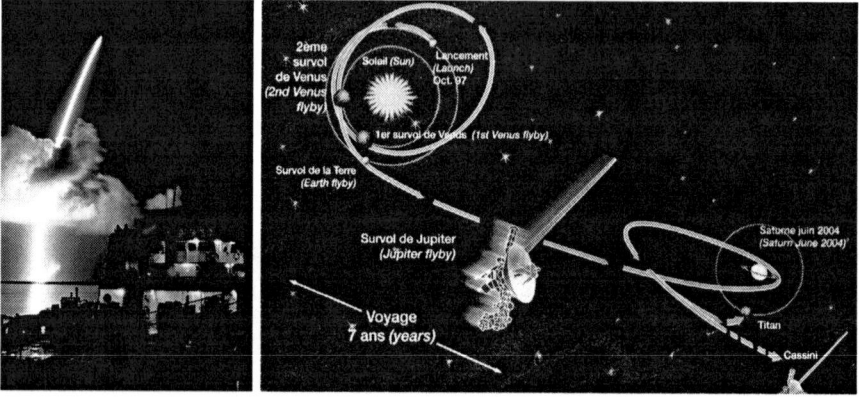

Fig. 6. Cassini's launch and the trajectory to Saturn. The spacecraft has been successfully launched on October 15, 1997. It has been injected into a 6.7 year Venus Venus Earth Jupiter gravity assist trajectory to Saturn. The arrival at Saturn was on July 1, 2004. (NASA-JPL/ESA documents).

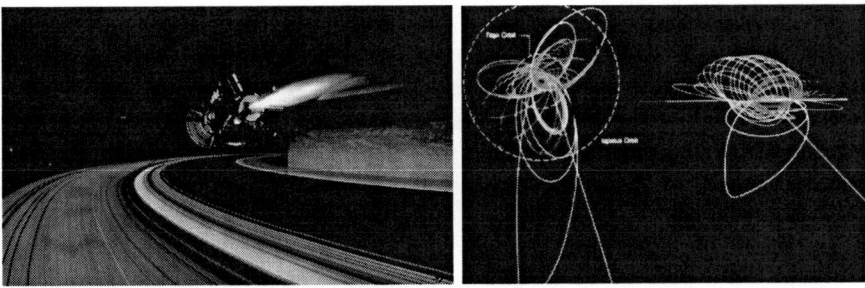

Fig. 7. Revolutions around Saturn. After the Saturn's orbit insertion on July 1, 2004 (left), 73 revolutions around Saturn have been programmed between 2004 and 2008 (right). (NASA-JPL/ESA documents).

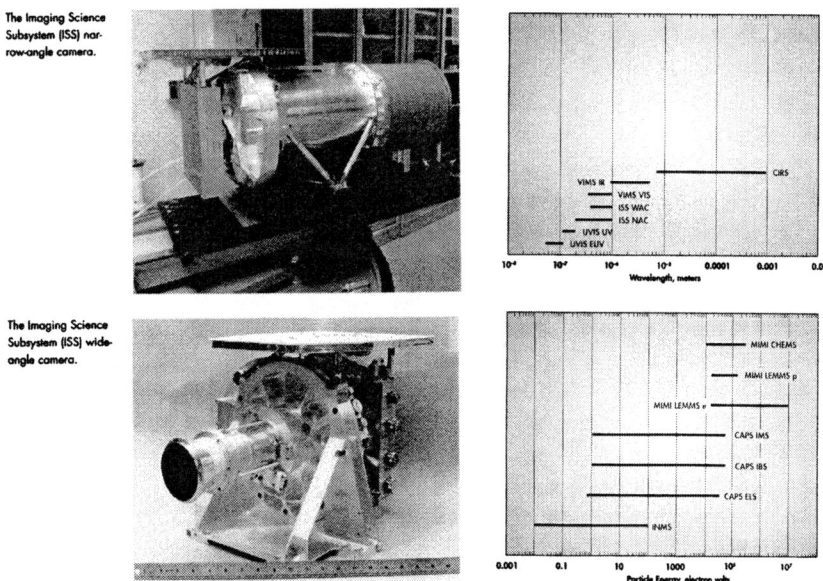

Fig. 8. The Cassini instruments. The Imaging Science Subsystem (ISS) comprises a narrow-angle camera and a wide-angle camera. The narrow-angle camera (upper left), which has a reflective optics with a 2-metre focal length, provides high-resolution images. The wide-angle camera (lower left), which has a refractive optics with a 20-centimetre focal length, provides extended spatial coverage at lower resolution. The bar charts at right show the operating wavelength coverage for the remote-sensing operations (upper right) and energy range for the fields, particles, and waves investigations (lower right). Both cover a range of 1010. (NASA-JPL/ESA documents).

When it reached its target and began orbiting Saturn, it became, at ten astronomical units from the Sun, the farthest robotic orbiter that humankind has ever established in the solar system. The nominal mission duration is four years and the variable orbit design allows an unprecedented exploration of the Saturn system tour over an extended period from a variety of illumination and viewing geometries. Saturn's rings are monitored for temporal changes in a way not previously possible during the Pioneer and Voyager fly by missions.

Until now, the spacecraft is in an excellent state of health and is operating normally. Until July 2008, Cassini will complete 74 orbits of the planet, 44 close flybys of Titan, and many other flybys of icy moons. The mission may last several years longer if there is enough propellant remaining for trajectory corrections. The images of Saturn's atmosphere, rings, and satellites are breathtaking.

4 Saturn's Rings

Since the first observations of Galileo in 1610, Saturn's rings study is one of the oldest scientific subjects. The existence of a ring system around a planet is a natural consequence of collisions between ring particles and of the Roche limit. Theses rings should offer valuable insights into the physics of more exotic and less accessible flat systems such as spiral galaxies, accretion discs, or the primordial solar system.

We do not understand why the ring system is divided in several visually different radial zones at large scales as it can be seen from the Earth even through a modest telescope. But, at a much smaller scale, the rings are still more complex. A number of unexpected structures were discovered in the high resolution images of the Voyager spacecrafts. Spiral waves, bending waves, narrow ringlets, eccentric rings, sharp edges, kinky and braided structures, etc. are common inside the Saturn's ring system. It seems that interactions between rings and nearby small satellites should explain most of the observed features. The possibility of numerous, small satellites occurring within Saturn's ring system was a puzzle the Voyager mission had hoped to solve. Voyager's best-resolution studies of the ring system were aimed at revealing any bodies larger than about 10 kilometres in diameter. Nevertheless, only four moonlets Atlas, Pan, Prometheus, and Pandora - were found in the images. Only one, Pan, was located in the main ring system.

Imaging, stellar occultations and radio occultations from Voyager experiments revealed a remarkable architectural diversity within the rings of all four giant planets. Saturn's rings are representative of all rings in being home to many of the types of features found around Jupiter, Uranus and Neptune. There are eccentric, inclined narrow rings; non-axisymmetric and sharp ring edges; broad, tenuous rings; incomplete arc-like ring segments; small moons shepherding nearby ring material, tightly wound spiral waves; axisymmetric

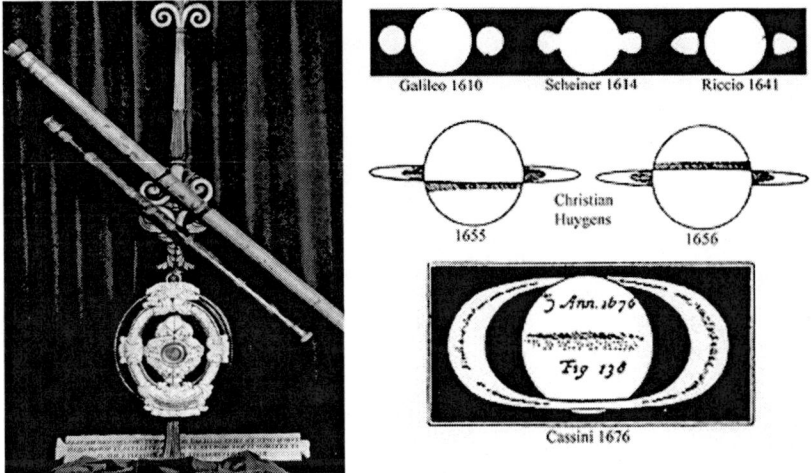

Fig. 9. Galileo discovery. With his refractor (left), Galileo realized during a few nights of July 1610, the most important collection of astronomical discoveries. The changing appearance of the planet Saturn has puzzled 17th century astronomers until Christian Huygens understood Saturn was surrounded by rings and Jean Dominique Cassini discovered the division named after him (drawings on the right). (Florence Observatory and Paris Observatory documents).

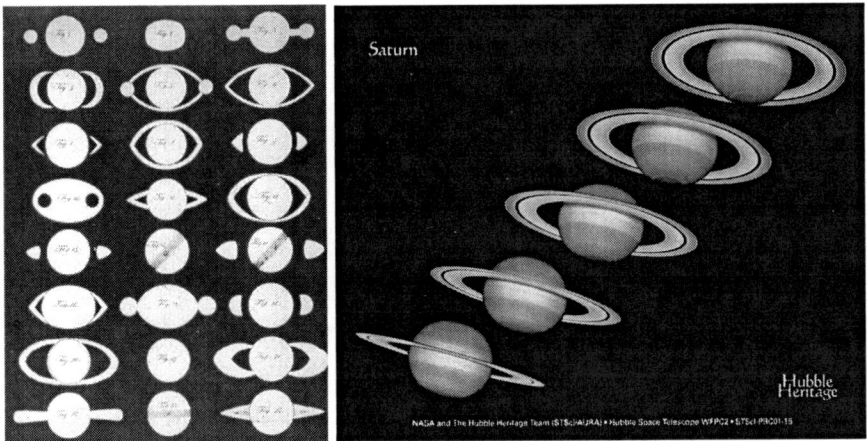

Fig. 10. The changing appearance of the rings. These early drawings of Saturn observed between 1610 and 1654 (left), correspond to the changing appearance of the ring system as seen by the Space Telescope (right). The axis of rotation of Saturn has a fixed direction relative to the stars and is seen more or less open from the Earth during one Saturnian revolution. (right: NASA-JPL/ESA document).

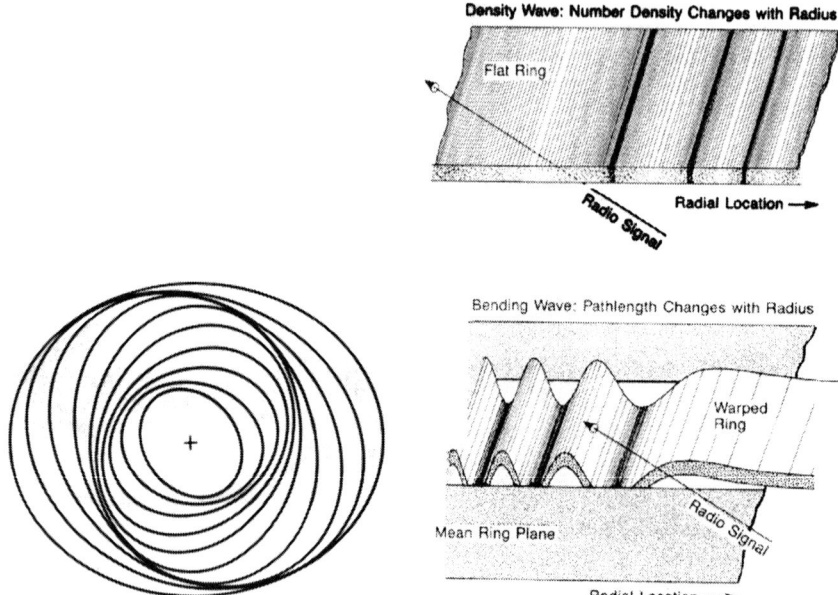

Fig. 11. Spiral waves and bending waves. Schematics of spiral density waves (left). The actual wrapping is in fact much tighter for real density waves in ring systems. Radio occultation is an excellent probe of density and bending waves (right).

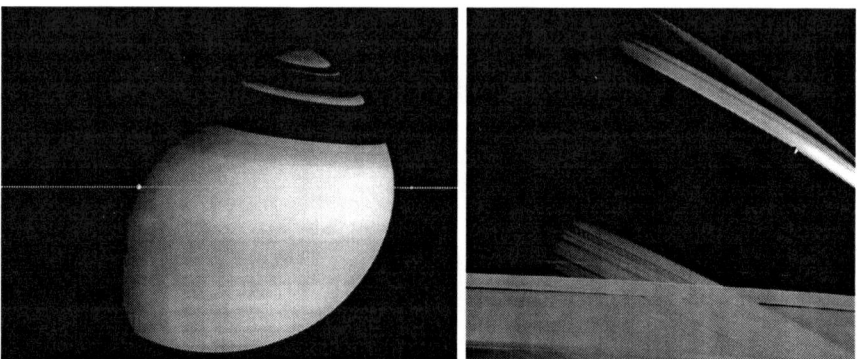

Fig. 12. Saturn's rings. On the left image, the rings are see edge-on. On the right image, the rings are visible in the lower part and their shadow on the northern hemisphere is visible with Mimas and the limb of the planet. (NASA-JPL/ESA document).

but radially irregular features; azimuthally asymmetric ring brightness variations; and a great deal more. A collisional disc if left to itself should spread until it is isolated and featureless, but Saturn's main rings are far from this.

Characterizing ring structure at a spatial scale finer and a spectral range wider than previously possible, determining its root causes, and searching for secular changes in the rings both during the multi-year long Cassini mission and since the Voyager era are prime objectives of the ring investigations at Saturn. Combined studies with visible, ultraviolet, infrared, radar, and radio instruments provide complementary results are now performed.

5 Cassini Observations: New Results on Rings

The Cassini observations around Saturn have produced many new findings, including new rings, new structures, new satellites and refined orbits.

The Cassini spacecraft's instruments have a much better resolution and a much better sensitivity than the Voyager instruments. Orbiting around Saturn, Cassini can observe the evolution of rings as a function of time, with a large variety of phase angles, and at different wavelengths through a number of various filters. Several regions can be seen in reflected light, diffuse light, scattered light, etc. The shadow of the rings on the planet can be observed as well as the planet through the rings.

Fig. 13. Saturn's rings seen under different phase angles. From the Earth, the angle Sun rings observer is limited to 12. From Cassini spacecraft, the rings can be observed under all angles and a large number of features are clearly visible (lower right). The night side of the rings can be observed (upper left), the planet can be seen through the rings (upper right) and transparent and opaque parts of the rings can be distinguished. The opposition effect is visible in Saturn's B ring (lower left). The bright spot occurs where the angle between the spacecraft, the Sun and the rings is near zero. (NASA-JPL/ESA document).

Fig. 14. The Cassini division. Far to be empty, as it was believed before 1979, the Cassini division is populated by many particles of all sizes. A number of structures can be seen in this image. New ringlets have been discovered in Cassini images. Small satellites not yet discovered are probably responsible of the structures. (NASA-JPL/ESA document).

Fig. 15. The mysterious B ring. This detailed view of Saturn's mid-B ring shows intriguing structure, the cause of which has yet to be explained. The image, taken on September 3, 2005, shows a region located between 107200 to 115700 kilometres from Saturn. (NASA-JPL/ESA document).

6 Waves and Wakes

Resonances between small nearby satellites and rings' particles create spiral density waves. The linear increase in wave number with the distance from the resonance gives the surface density. Amplitude's rates of growth and damping give the moon's mass and the ring's viscosity. Non linearity, caused by the self-gravity of density peaks, complicates these relationships.

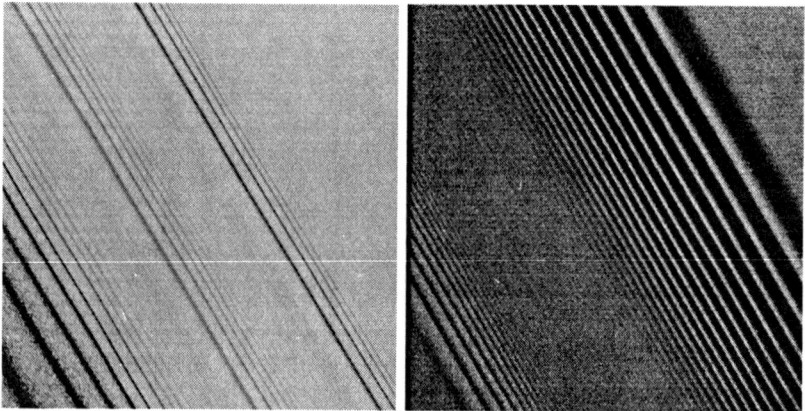

Fig. 16. Rings full of waves. The left image shows three density waves in Saturn's A ring. They are respectively due to Pan, Pandora and Prometheus. The right image shows a bending wave (right) and a density wave (left) in Saturn's A ring. This view shows the dark, unlit side of the rings. (NASA-JPL/ESA document).

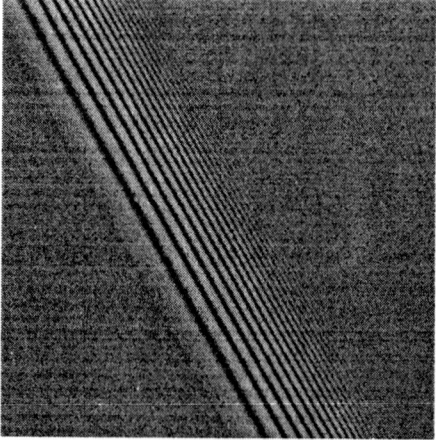

Fig. 17. Prometheus 9:8 resonance. This resonance appears to be a beautiful example of a linear density wave, but attempts to fit the wave reveal some non linearity. The mass derived here is consistent with Voyager and dynamical modelling. A ring surface density of about 39 g/cm2 is consistent with Voyager data. (NASA-JPL/ESA document).

Fig. 18. Atlas 5:4 resonance. Located at this resonance, this wave is on a little peak which truncates it. The Atlas'mass is only 20value. With a value of 1.4 g/cm2, the Cassini Division surface density is only 4height of 6 metres. (NASA-JPL/ESA document).

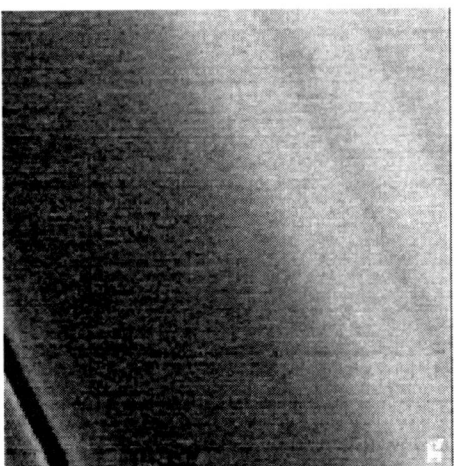

Fig. 19. Pan 7:6 resonance. Pan's mass from this wave is smaller than that derived from Encke Gap edge waves. More waves need to be measured to obtain a better value of the satellite mass. With a value of 2.5 g/cm2 at this location, the Cassini Division surface density is about 6the A ring. A viscosity of 5 cm2/s implies a scale height of 5 metres. (NASA-JPL/ESA document).

7 Ring Edges. Arcs, Clumps and Moonlets

Passing moon excites particle eccentricities. As a result, the edge of a nearby ring shows waves.

The little moon Pan, which is 20 kilometres across, is orbiting within the Encke gap in Saturn's A ring and is responsible for clearing and maintaining this gap. The inner edge of Encke division is shaped by the small satellite Pan and the inner edge wavelength does agree with Pan's position. Pan is responsible for creating stripes, called wakes, in the ring material on either side of it. Since ring particles closer to Saturn than Pan move faster in their orbit, these particles pass the moon and receive a gravitational kick from Pan as they do. This kick causes waves to develop in the gap where the particles have recently interacted with Pan, and also throughout the ring, extending hundreds of kilometres into the rings. These waves intersect downstream to create the wakes, places where ring material has bunched up in orderly manner thanks to Pan's gravitational kick.

Fig. 20. Edge wave in the Encke Gap. The left image reveals two faint, dusty ringlets that occupy the gap along with Pan .The right ringlet occupies nearly the same orbit as Pan, while the other is closer to the gap's inner edge. Not only do the ringlets vary in brightness, but they also appear to move in and out along their length, resulting in notable kinks, which are similar in appearance to those observed in the F ring. One possible explanation for the complex structure of the ringlets is that Pan is not the only moonlet in this gap. The right drawing shows that the particles nearhave most recently interacted with Pan and have just passed the moon. Because of this, the disturbances caused by Pan on the inner gap edge are ahead of the moon. The reverse is true at the outer edge where the particles have just been overtaken by Pan, leaving the wakes behind it. (NASA-JPL/ESA document).

Encke inner edge is significantly non sinusoidal. This is probably a consequence of Pan non-zero eccentricity, but this has to be modelled. Pan creates also waves in the outer edge, but their wavelength is too long to be observed. At the outer edge of the Encke gap and in the region exterior to it, wakes due to Pan are clearly seen. Wakes are simply coherent streamlines, not propagating waves. Farther, streamlines cross and interfere. Wake's density maxima have become very sharp peaks. This can be due to streamline crossing or to self gravity.

Unlike anything seen before, Keeler gap edges show wisps. The spacing between the wisps suggests this phenomenon can be due to a moon located at the centre of the gap.

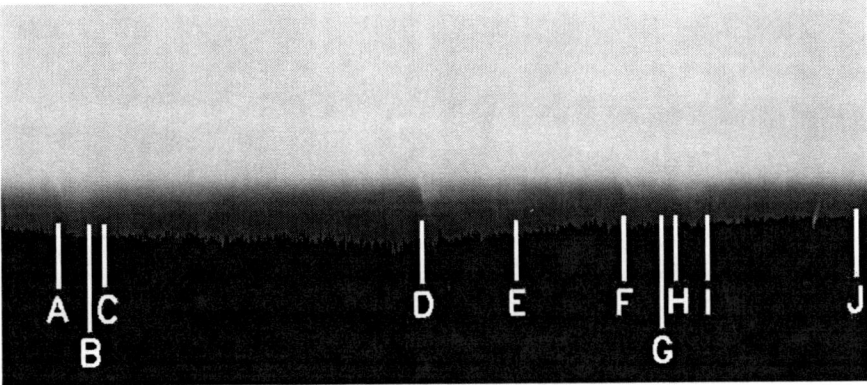

Fig. 21. Fingerprints of an unseen moon? The Keeler gap, a narrow gap 42 kilometres wide, lies approximately 250 kilometres inside the outer edge of the A ring. The above image of the outer gap edge has been stretched by a factor of five and contrast enhanced. Several faint discontinuities or spikes have been discovered. The most easily seen spikes are labelled A through J in this image. These features are similar to the spikes protruding inward from the core of the F ring during Prometheus's passages. These features all move in unison at the orbital speed appropriate for particles at this location. It is likely that the features are caused by the passages of a yet-unseen moonlet on an eccentric orbit within the Keeler gap.

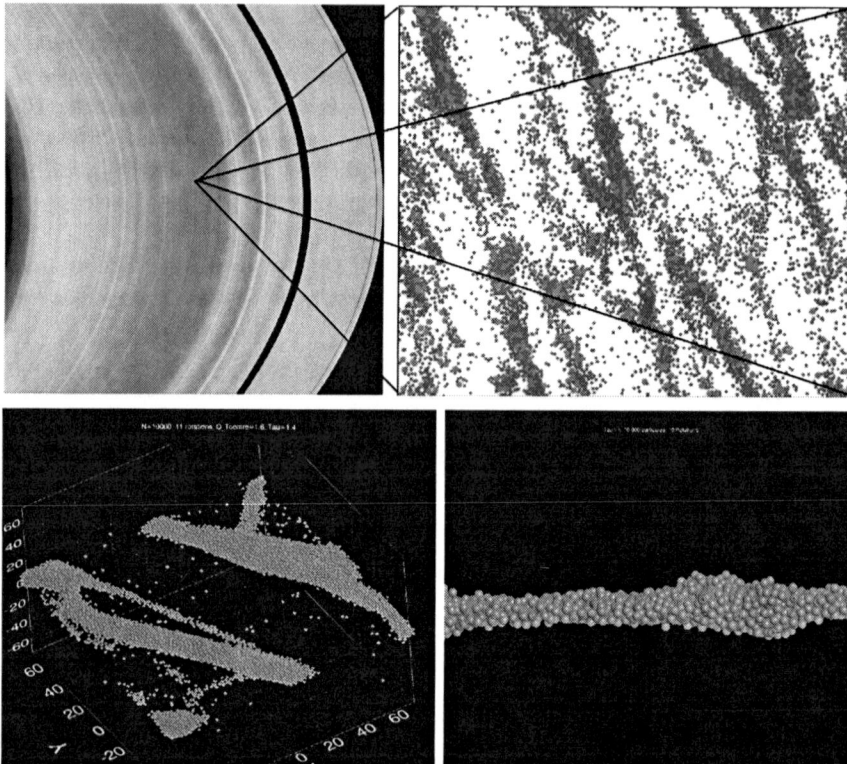

Fig. 22. Wakes and clumps. The upper left image is a false-colour view of Saturn's A ring from the ultraviolet spectrograph instrument aboard the Cassini spacecraft. The ring is the bluest in the centre, where the gravitational clumps are the largest. The thickest black band in the ring is the Encke gap, and the thin black band further to the right is the Keeler gap. The right image is a computer simulation about 150 metres across illustrating a clumpy region of particles in the A ring. The bottom images show the result of a computer simulation by Sbastien Charnoz (face- on on the left and edge-on on the right). Viscosity can enhance density waves induced by gravity and very small axisymmetric structures of about 100 metres large may appear. (NASA-JPL/ESA document).

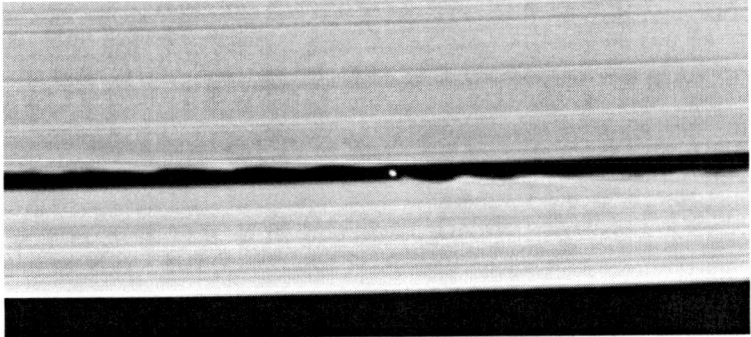

Fig. 23. Wavemaker moon. Cassini's confirmation that a small moon orbits within the Keeler gap in Saturn's rings is made by this image, in which the 7-kilometre-wide body disc is resolved for the first time. The Keeler gap is located about 250 kilometres inside the outer edge of the A ring. (NASA-JPL/ESA document).

Fig. 24. Arcs and clumps in the Encke gap ringlets. Clumps have been observed in three of the four Encke gap ringlets. In the above image, many clumps are clearly visible in the main Encke gap ringlet, as well as waves produced by the tiny moon Pan. (NASA-JPL/ESA document).

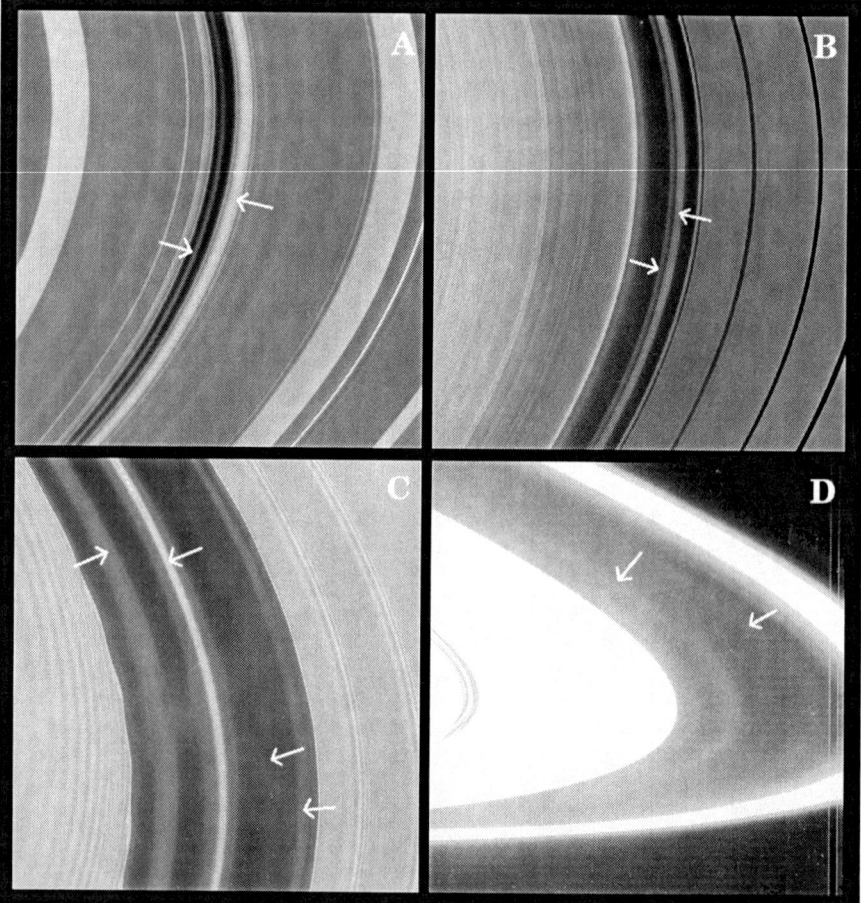

Fig. 25. Clumps in the Pan ringlet. These three images, taken respectively on July 1, 2004, November 15, 2004, and January 30, 2005 show that the clumps in the Pan ringlet move and change as a function of time. Pan is visible at he extreme left of each image. (NASA-JPL/ESA document).

Fig. 26. Diffuse ringlets discovered within Saturn's rings. All images have been processed to bring out low-optical thickness rings. Each arrow points to new rings in the C ring (A image), immediately outside the outer B ring edge (B image), in the Encke gap (C image), and in the F ring region (D image). (NASA-JPL/ESA document).

8 New Ring Phenomena

New features are visible between the A and the F ring. A collection of new ring phenomena, first observed in images taken of the dark side of Saturn's rings immediately after Cassini entered orbit, may be evidence of the clumping and aggregation of ring particles. This phenomenon is caused by the combined gravitational effects of Saturn, orbiting moons, and other ring particles.

Unusual mottled-looking narrow region, with a radial width varying with longitude from 5 to 10 kilometres, have been seen for the first time about 60 kilometres inside the outer edge of Saturn's A ring. The mottled regions are probably caused by particle clumping brought about by gravitational disturbances. The outer A ring edge is sculpted into a seven- lobed pattern called a Lindblad resonance by the co-orbital satellites Janus and Epimetheus. The resonant perturbations in this region are complicated by the presence of these two moons whose orbits are within 50 kilometres of each other.

Other kinds of new features like ropy structures have also been found. For example, at the outer edge of the Encke gap, rope-like features can be seen between the first two wakes nearest the gap edge. Theses ropy features appear to be a product of the enhanced gravitational disturbances that occur when the particles pass through the wakes caused by Pan and consequently are squeezed close together. These disturbances obviously persist even outside

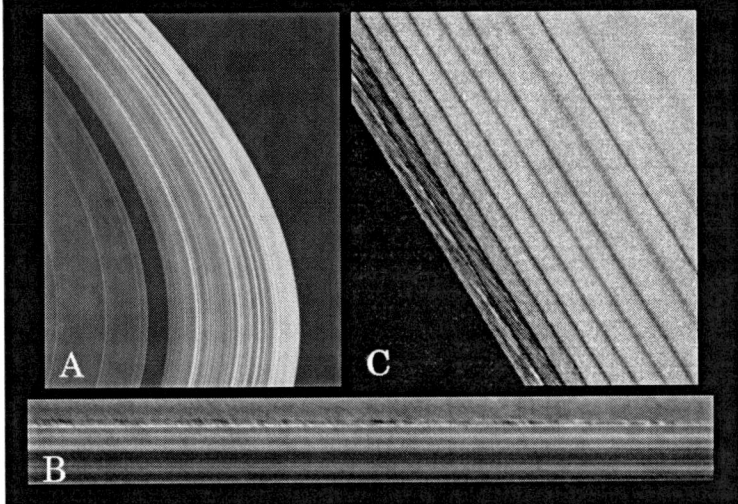

Fig. 27. Examples of new ring phenomena: "mottled" and "ropy" structures. A map projection (B image) of the outer edge of the A ring (A image) shows mottled structures which have never be seen before. The (C) image on the right shows the outer edge of the Encke division and the region exterior to it. The wakes of Pan are clearly seen. A different example of mottled structure is seen in the eight Pan wake from the edge, as well as ropy structure within the first two bands exterior to the gap. (NASA-JPL/ESA document).

the wakes, as is evident in the presence of the ropy structures in the bands in between the wakes.

9 The D Ring

The D ring, which is between the inner C ring and the top of Saturn's clouds, has significantly changed since Voyager. For example, the ringlet located at 72000 kilometres from Saturn's centre, called the D72 ringlet, is now much fainter than it used to be, and its centre of light has shifted of about 200 kilometres inwards and we see new ringlets exterior to the D73 ringlet. There no longer appears to be any wave-like structure in the diffuse material.

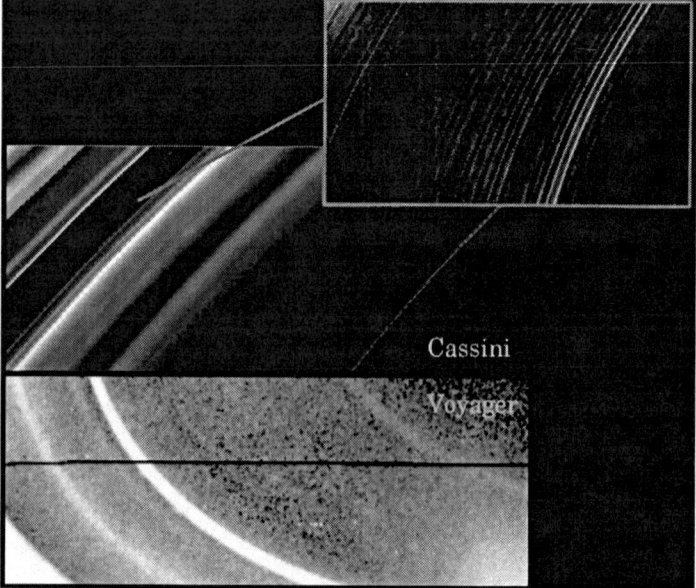

Fig. 28. The D ring. These images from the Cassini and the Voyager missions show that structural evolution has occurred in saturn's D ring during the 25 years separating the two missions. The lower image, taken by Voyager 1 in 1980, shows from left to right, the bright inner edge of the C ring and three discrete ringlets: D 73, D 72, and D 68. The upper image, obtained by Cassini, shows the same region from a similar viewing geometry, but there have been some very significant changes in the appearance of the D ring. The green line marks the inner edge of the C ring. The D 72 ringlet has decreased in brightness by more than an order of magnitude relative to the other ringlets. It has also moved inward bout 200 kilometres relative to the others. With a much higher resolution than was possible for Voyager, Cassini revealed surprising fine-scale structures between the C ring and D 73 (inset). (NASA-JPL/ESA document).

Cassini has observed the D ring at much higher resolution than was possible for Voyager, revealing surprising fine-scale structures with a periodic wave-like structure with a wavelength of 30 kilometres.

10 The F Ring

Saturn's F ring has been one of the most intriguing structures around Saturn since it was first imaged by Voyager 1 spacecraft in 1980 (ref. Voyager). It was variously described as kinks, clumps, strands, and braids. The Voyager images showed features that deviated substantially from those of a simple ellipse. Each particle in the F ring evidently follows its own elliptical orbit around Saturn. The kinks and braids do not represent the paths of individual particles. They are just instantaneous snapshots of a ring containing orbits that vary from place to place. The F ring is surrounded by two shepherding moons, Prometheus (near the inner edge) and Pandora (near the outer edge). They have long been recognized as providing the gravitational pulls that continuously renew these intriguing patterns. A new nearby satellite, S/2004 S6, has been discovered by the Cassini spacecraft. Its orbit can intersect the F ring at high speed.

New narrow diffuse rings have been found in the F-ring region and the evolution of the features has been followed as a function of time. In particular, new Cassini observations show that the faint strands of material, initially interpreted as concentric ring segments, are in fact connected and form a single one-arm trailing spiral winding at least three times around Saturn. The F ring spiral structure has nothing to do with other spiralling structures seen in the main rings of Saturn and contains very little mass. It appears to originate from material somehow episodically ejected from the core of the F ring and then sheared out due to the different orbital speeds followed by the constituent particles. The F ring spiral may be a consequence of moons crossing the F ring and spreading its particle around.

There are many unsolved problems still remaining about the F ring. Perhaps the most fundamental questions concern how the F ring has been formed and why it is so strange. Several factors are important are important. The ring orbits at the edge of the Roche limit, at the boundary inside which tidal forces overcome self-gravity, where accretion is prevented. The rings and the nearby satellites all follow highly eccentric orbits, which means that the ring is highly perturbed when moons approach.

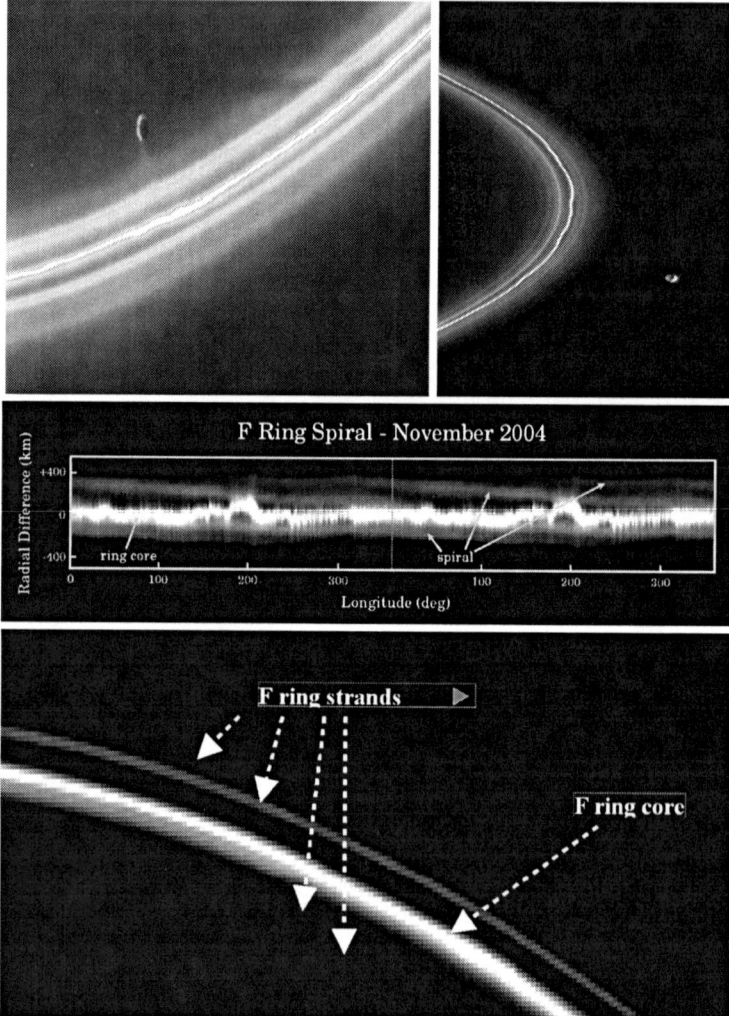

Fig. 29. The F ring spiral. The F ring core is the bright line. The strands appear as dimmer inclined lines below and above the core (bottom image). The map of Saturn's F ring (top image) illustrates how the strands flanking the core of the contorted ring, when examined in detail, actually form a spiral structure wound like a spring around the planet. This spiral may be a consequence of moons crossing the F ring and spreading its particles around. In the middle image, two identical maps of the F ring have been joined, side-by-side, to show the nature of the spiral more clearly. The spiral strand's path across the image begins about 350 kilometres inward the F ring core at about 200 degrees longitude and moves closer to the ring core toward the left. The strand appears to cross the ring core around 100 degrees longitude, after which the distance between the strand and the ring core increases to he left and can be followed, moving even farther outward, wrapping around to the rightmost boundary of the right-hand map and continuing to the left. (NASA-JPL/ESA document).

11 The G Ring

The G ring is a tenuous ring outside the main ring system. An arc is visible in some images of this ring. What makes this part of the G ring brighter than other parts is not clear. However, the existence of this arc might hold clues about how this ring was formed and where the material which makes up this ring comes from.

Fig. 30. The G ring. This image of the G ring shows clearly that this ring has a sharp inner edge and a much smoother outer edge. (NASA-JPL/ ESA document).

12 Colours and Composition

The observation of rings at different wavelengths gives unique information on composition, structure, temperature and nature of the ring particles.

High resolution images in visible wavelengths give detailed information on the structure of the rings and on evolution of features as a function of time.

Ultraviolet observations indicate there is more ice toward the outer part of the rings, than in the inner part, hinting at the origins of the rings and their evolution. The Cassini division and the Encke gap contain thinner and durtier rings than the A ring, indicating a more icy composition.

Infrared observations indicate that the grain sizes in Saturn's rings grade from smaller to larger, related to distance from Saturn. Rings are probably made up of boulder-size snowballs. The size of the ice crystals or grains on the surfaces of those boulders can be determined with infrared data.Rings are made mostly of water in the form of ice. Cassini infrared data are showing that the ring ice is more pure than previously thought, with the most pure ices generally being observed at increasing distances from Saturn. Dirty material is most abundant in the thinnest parts of the rings such as the Cassini division, the Encke gap or other small gaps. This material appears remarkably similar to what Cassini measured on Saturn's moon, Phoebe.

Far infrared data give the temperature of the rings. The data show that the opaque region of the rings, like the outer A ring and the middle B ring, are cooler with temperatures of the order of 70 K, while the most transparent sections, like the Cassini division or the inner C ring, are relatively warmer, with temperatures of the order of 110 K.

Fig. 31. Saturn's rings at different wavelengths. The top image shows the colour of the rings in visible light. The bottom left image in false- colours shows the rings in infrared. The bottom right image shows the rings in ultraviolet. The varying temperatures of Saturn's rings are depicted in the infrared image. Red represents temperatures of about 110 K, and blue 70 K. Green is equivalent to 90 K. The ultraviolet view indicates there is more ice toward the outer part of the rings, than in the inner part. The Cassini division at left contains thinner, dirtier rings than the turquoise A ring, indicating a more icy composition. (NASA-JPL/ESA document).

13 The E Ring and Enceladus

The maximum density of the E ring occurs near Enceladus' orbit, suggesting that this satellite is the source of the E ring. Cassini spacecraft has identified a geologically unique and presently active province at the south pole of Enceladus. This 505-kilometres diameter bright icy moon is active. Cassini imaging, thermal and other data indicate clearly that this satellite is presently heated by some mechanism. Tidal heating associated with the eccentricity of Enceladus' orbit, forced by its 2:1 mean motion resonance with Dione, has long been suspected.

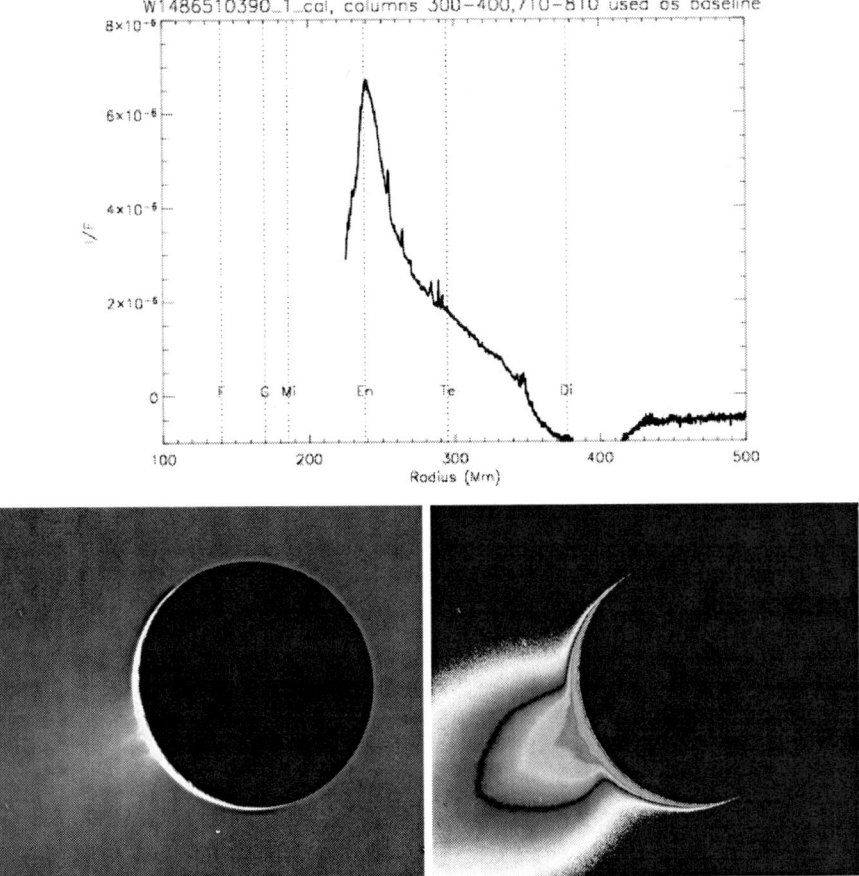

Fig. 32. The E ring and Enceladus. The top graph shows the amount of material as a function of distance to Saturn. At the level of the orbit of Enceladus, there is clearly much more material. The bottom images show fountain-like sources of fine spray of material that towers over the south polar region of Enceladus. The bottom right image is greatly enhanced and colorized in order to make visible the enormous extent of the fainter and larger-scale component of the plume. (NASA-JPL/ESA document).

14 Conclusion

Cassini mission should last at least until July 2008 and may continue several years longer if there is enough propellant remaining for trajectory corrections. Already the first data on Saturn's rings are breathtaking and we can expect the unexpected from future observations. It will take several decades to analyze all the observations and to understand all the observed phenomena. With Saturn's rings, we have a remarkable laboratory of physics and granular flow!

References

Several thousands of articles have been published about Saturn's rings since their discovery. A presentation of the main properties of planetary rings as well as a discussion of dynamical processes can be found in [1], published after the Voyager flybys of Jupiter and Saturn. The first results of the Cassini mission have been presented in [2].

1. *Planetary Rings*, R. Greenberg and A. Brahic, Eds. (Univ. of Arizona Press, Tucson, 1984).
2. C. C. Porco, E. Baker, J. Barbara, K. Beurle, A. Brahic, J. A. Burns, S. Charnoz, N. Cooper, D. D. Dawson, A. D. Del Genio, T. Denk, L. Dones, U. Dyudina, M. W. Evans, B. Giese, K. Grazier, P. Helfenstein, A. P. Ingersoll, R. A. Jacobson, T. V. Johnson, A. McEwen, C. D. Murray, G. Neukum, W. M. Owen, J. Perry, T. Roatsch, J. Spitale, S. Squyres, P. Thomas, M. Tiscareno, E. Turtle, A. R. Vasavada, J. Veverka, R. Wagner, and R. West, Cassini Imaging Science: Initial Results on Saturn's Rings and Small Satellites. Science **307**, 1226 (2005).

Universality Classes for Force Networks in Jammed Matter

Srdjan Ostojic and Bernard Nienhuis

Institute for Theoretical Physics, Universiteit van Amsterdam, Valckenierstraat 65, 1018 XE Amsterdam, The Netherlands

Summary. We study the geometry of forces in some simple models for granular stackings. The information contained in geometry is complementary to that in the distribution of forces in a single inter-particle contact, which is more widely studied. We present a method which focuses on fractal features of the force network and find good evidence of scale invariance of patterns of large forces. The method enables us to distinguish universality classes characterized by critical exponents. Our approach can be applied to force networks in other athermal jammed systems.

1 Jammed Matter and Force Networks

Aggregates of particles can be found in a disordered solid-like state resulting from the phenomenon of jamming [1–5]. Granular materials, colloidal suspensions and molecular liquids are but a few examples of such systems that present a non-zero yield stress while trapped in one of many accessible metastable states. If thermal fluctuations are irrelevant, the forces on each particle must balance. Each stable configuration is thus characterized by a highly irregular network of forces spanning the entire system.

Experimental [6–10] and numerical [11–19] studies have identified two main distinctive features of these force networks. Firstly, strong fluctuations are found in the magnitudes of inter-particle forces. The associated distribution function $P(F)$ displays two characteristic properties: (i) it decays exponentially at large forces and (ii) it exhibits a plateau or small peak at small forces, which has been identified as a signature of jamming. The second experimental observation is that large forces are concentrated along tenuous paths, which have been deemed "force chains". While $P(F)$ has been commonly used also as a characterization of these force chains, strictly speaking it provides no information about the spatial organization of forces. In fact, so far force chains have been identified mainly visually, and a quantitative characterization seems to be lacking.

By drawing an analogy with percolation, in this paper we develop a geometrical description which associates a set of critical exponents with an ensemble of force networks. We apply this approach to three different models of static granular media under uniform pressure. We find that they belong to different geometrical universality classes although $P(F)$ displays similar features in all three of them.

2 Force Clusters

Consider an ensemble of configurations of a fixed number of jammed particles, obtained numerically or experimentally. Each configuration defines a contact graph G, where nodes correspond to particle centers and edges connect particles in contact. Assuming there is no friction, the inter-particle forces are normal to the particle surface, and the underlying force network can be represented by associating with each edge i of G the corresponding force magnitude F_i. To investigate the geometry of forces, rather then the underlying geometry of contacts, we choose a threshold f and look at the subgraph $\bar{G}(f)$ of G obtained by selecting only the edges with $F_i > f$. For f small, $\bar{G}(f)$ consists of a single connected component, but as f increases, $\bar{G}(f)$ breaks up into a number of disconnected clusters. An ensemble of force networks thus induces a family of probability distributions of cluster sizes $\rho(s, f)$ for different thresholds f, the cluster size s being defined as the number of edges in a cluster.

If the forces F_i were distributed independently for each i, e.g. uniformly between 0 and 1, then the force clusters would simply be bond percolation clusters [20]. In that case, in the thermodynamic limit $N \to \infty$, a phase transition occurs at a critical value f_c of f: an infinite cluster exists with probability 1 for $f < f_c$, and with probability 0 for $f > f_c$. At f_c, the cluster sizes are power-law distributed, $\rho(s, f_c) \propto s^{-\tau}$, and the correlation length diverges as $\xi \propto |f - f_c|^{-\nu}$ near the threshold. The scaling exponents τ and ν are universal, they are independent of the underlying geometry, and in fact they do not depend on the local distribution of forces $P(F)$ or even their correlations, as long as these are short-ranged.

In an ensemble of force networks corresponding to a jammed system, force and torque balance on each particle cause dependence and long-range correlations between bonds. Nevertheless, if the average forces are uniform over the extent of the system, we expect to find a critical threshold f_c and an associated set of universal scaling exponents. The analogy with percolation moreover suggests that these exponents are independent of $P(F)$ and thus provide a new, complementary characterization of force networks.

3 Criticality and Finite Size Scaling

An efficient method is necessary to study the existence of scale invariance around the critical threshold from numerical and experimental data. While,

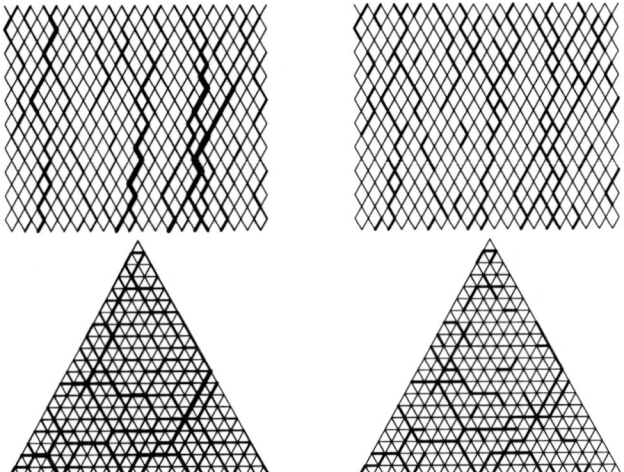

Fig. 1. Examples of force networks (the thickness of the lines is proportional to the force magnitude) and corresponding force clusters close to the critical threshold: packing of 400 grains in Model A (top) and packing of 200 grains in Model B (bottom).

strictly speaking, the system becomes scale-invariant only in the thermodynamic limit, f_c and the associated critical exponents can be extracted from data on systems of finite size using finite size scaling [21]. This describes the scaling of an observable with the system size close to criticality: if a quantity X is expected to diverge as $|f - f_c|^{-\chi}$ near f_c in an infinite system, then in a system of size N, it obeys the scaling law

$$X(N, f) = N^\phi \tilde{X}((f - f_c)N^{1/d\nu}) \tag{1}$$

with d the spatial dimension and $\phi = \chi/d\nu$. The scaling function $\tilde{X}$ depends on a single rescaled variable $x = (f - f_c)N^{1/d\nu}$, and for $x \gg 1$ it behaves as $x^{-\chi}$, while for $x \to 0$ it remains finite.

Using measurements of X in systems of finite sizes, the parameters ϕ, ν and f_c can be obtained from (1) in two steps. Assuming that $X(N, f)$ as function of f displays a maximum $X_m(N)$, from (1) the maxima for different N all correspond to the same maximum of $\tilde{X}$, hence $X_m(N) \propto N^\phi$. Plotting the amplitudes of the maxima versus N, we get the exponent ϕ. The values of f_c and ν can then be obtained by determining the best data collapse in the region around the maximum.

4 Models Studied

Combining the finite-size scaling method with Monte-Carlo simulations, we studied force-cluster criticality in three two-dimensional models of static granular matter under uniform pressure [22, 23]. As all three models – which we will call A, B, and C for further reference – have been introduced earlier in other contexts, here we only define them briefly, without motivating in detail their relevance to granular matter. In our view, they are are the simplest implementations of two fundamental ingredients of force networks, namely force balance on each grain and force randomness.

4.1 Snooker Model

To start with, we consider the "snooker-triangle packing" studied in [26, 27]. It consists of a hexagonal packing of frictionless spherical grains confined within a triangular domain, with the same confining pressure applied on all sides of the triangle. A force network on this packing consists of repulsive forces in vectorial balance on each grain and consistent with the applied pressure. These constraints however do not define a single configuration of forces, but a whole set. Following Edwards' prescription [28], all such force networks are taken to be equally likely, similarly to a micro-canonical ensemble. We sample this ensemble with a Metropolis algorithm, using the parametrization of force networks developed in Ref. [29]. In Fig. 1 we show an example of a force network in this model and the corresponding force clusters for a threshold $f = 0.94$.

4.2 Independent q-Model

We next consider consider the scalar q-model [24], one of the first models introduced to account for the fluctuations of forces and appearance of force chains in a granular packing. Here we consider the massless q-model on a periodic tilted square lattice, which can be interpreted as a packing of rectangular bricks [25]. A uniform pressure is applied on the top of the packing and on each site a brick supports a weight W_{ij}. Each brick transfers *vertical* forces $F_l^{(ij)}$ and $F_r^{(ij)}$ to its bottom left and right neighbors respectively. Vertical force balance is automatically satisfied by considering $F_l^{(ij)}$ and $F_r^{(ij)}$ respectively as fractions q_{ij} and $1 - q_{ij}$ of W_{ij}, and randomness in force transfer is implemented by taking the q_{ij} uniformly distributed between 0 and 1, independently for each site. Fig. 1 shows a force network in this model and the corresponding force clusters for a threshold $f = 0.7$ (for unit external pressure).

4.3 Microcanonic q-Model

Our third model is a variation on the q-model. We consider the same packing as in Sec. 4.2, but now, following Edwards' prescription, all allowed force networks – consisting of sets of vertical forces $\{(F_l^{(ij)}, F_r^{(ij)})\}$ – are equally likely. As shown in Ref. [30], this is equivalent to having the q_{ij} distributed with the joint probability distribution $\prod_{ij} W_{i,j}$. The aim is to examine the influence of the form of the probability distribution by comparing independent and microcanonic q-models, and the difference between scalar and vectorial conservation laws by comparison with the snooker model.

5 Results

For the statistical characterization of clusters, we use the standard methods of percolation theory [20]. A convenient observable to study is the second moment of the distribution of cluster sizes, $\langle s^2(N, f) \rangle = \int s^2 \rho(s, f_c)$, where the system size N is defined as the total number of edges. The probability distribution of cluster sizes $\rho(s, f_c)$ does not usually take into account the percolating cluster, as in the thermodynamic limit this would only add a single cluster of infinite size. Therefore the percolating cluster is also omitted in the calculation of the moments of the distribution. For $f < f_c$, excluding the percolating cluster is equivalent to removing the largest cluster, while for $f > f_c$ removing the largest cluster does not change the scaling properties. To avoid the need to define what "percolating" means in the triangular geometry of the snooker model, we leave out the largest cluster rather then the percolating one. We determined the cluster sizes by a standard depth-first search algorithm [31].

In all three models defined above, we find that $\langle s^2(N, f) \rangle$ displays a maximum as function of f for fixed N. The amplitudes of the maxima as functions of N follow sharp power-laws shown in Fig. 2 (a), thus confirming the existence of a critical threshold in each model. The corresponding critical exponent ϕ is related via the hyper-scaling relation [20] to τ, the exponent of the cluster-size distribution at criticality, and D, the fractal dimension of the incipient cluster as $\phi = \frac{3-\tau}{\tau-1} = D - 1$. Higher moments $\langle s^n(N, f) \rangle$ display a similar scaling with exponents $\phi_n = \frac{n+1-\tau}{\tau-1}$, implying that the full distribution $\rho(s, f_c)$ approaches a scaling form around the critical threshold.

The value of the critical threshold f_c depends on the scale set by the external pressure. Under unit pressure, we found a different f_c for each model. Fig. 3 displays the scaling functions obtained by collapse of the data. The estimated values of the critical thresholds and exponents are summarized in Table 1, where the two-dimensional percolation exponents are also included for reference.

In Fig. 2(b) we show the probability distributions $P(F)$ of force magnitudes. In the independent q-model, $P(F)$ is exactly exponential [32], while in the other two models it is exponential for large forces, and displays a peak at small forces.

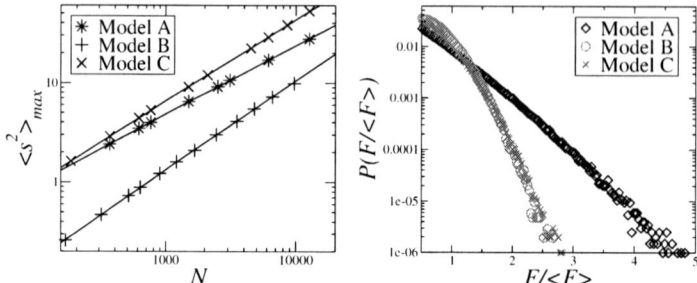

Fig. 2. Results of Monte-Carlo simulations for the three models defined in the text: (a) scaling of the maxima of the second moment $\langle s^2(N, f)\rangle$ of the distribution of cluster sizes (omitting the largest cluster in every configuration), as function of the total system size N; (b) probability distributions $P(F)$ of force magnitudes, obtained from 100 samples of systems of 10^4 particles.

Table 1. Values of the critical threshold f_c and the critical exponents ϕ and ν obtained from Fig.2 and the data collapse shown in Fig.3. For two-dimensional percolation, exact values are shown inside brackets.

	f_c	$\phi = D - 1$	ν
Independent q-model	0.7 ± 0.01	0.69 ± 0.01	3.1 ± 0.1
Snooker model	0.93 ± 0.01	0.89 ± 0.01	1.65 ± 0.1
Microcanonic q-model	0.585 ± 0.05	0.81 ± 0.01	1.65 ± 0.1
Percolation		$0.895(43/48)$	$1.33(4/3)$

6 Discussion

We have introduced a new approach to investigate the geometry of force networks, based on statistics of clusters created by forces larger then a given threshold. The existence of a critical threshold uncovers a scale-invariance of patterns of large forces, which we characterized by the critical exponents ν and ϕ for the correlation length and the second moment of the cluster size distribution. In particular, in each network we identify a fractal object of dimension $D = \phi + 1$, given by the incipient force cluster at the critical threshold. As shown in Table 1, we found three different sets of critical exponents for the three models we studied, implying that they belong to distinct geometrical universality classes, although their $P(F)$ display similar features. Interestingly, for the snooker model, ϕ is very close to the percolation value, but as the values of ν are further apart and the scaling functions are different, it does not belong to the percolation universality class.

Two distinct universality classes could have been expected a priori for the q-models on one hand and the snooker packing on the other. Indeed, the q-models are both directed and include only scalar conservation laws, while the snooker model is isotropic with vectorial conservation laws. The reason for the segregation of the independent and microcanonic in two different universality

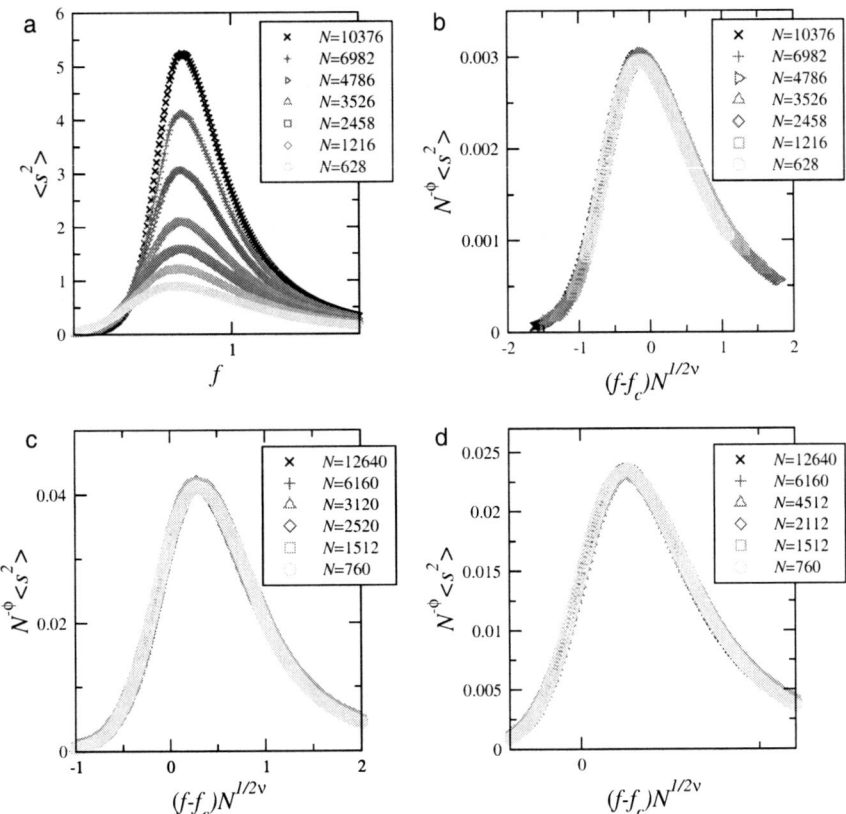

Fig. 3. Scaling of the second moment (omitting the largest cluster in each configuration) as function of the threshold f and the system size N. (a)$\langle s^2\rangle(f)$ for different system sizes in the snooker model. (b-d) Data collapse obtained by expressing the rescaled second moment of cluster sizes $N^{-\phi}\langle s^2\rangle$ as function of the rescaled variable $(f - f_c)N^{1/2\nu}$ for the three models defined in the text: (b) snooker model, (c) independent q-model and (d) microcanonic q-model The values of the corresponding parameters f_c, ϕ and ν are summarized in Table 1. We do not show the data for very small system sizes where the collapse takes place only in a small region around the maximum. For the q-models, the systems studied had the same vertical and horizontal linear sizes.

classes is more subtle. They differ only by the form of the probability distribution of forces, but in the independent case the distribution is Markovian from top towards bottom, while in the microcanonic case no such preferred direction of propagation exists.

While in jammed matter the disorder in the underlying contact geometry plays an important role, we considered here only lattice models with fixed contact geometry. The force-cluster method can be applied in a straightforward fashion to ensembles of forces networks resulting from disordered contact networks. By analogy with other critical phenomena, we however do not expect such randomness to modify the universality class. Indeed, the values we have found for ν in combination with the Harris' criterion [33] suggest that geometric disorder is irrelevant. A further study indeed shows that introducing quenched disorder in the q-models does not modify the universality class, which in turn confirms that the scale-invariance found here is in all aspects similar to equilibrium critical phenomena.

While the results presented here clearly show that our method is able to discriminate between different scaling behaviors, a crucial question is whether any of the models belongs to the same universality class as a realistic two-dimensional system of grains under isotropic pressure. A recent study [34] of packings generated by molecular dynamics simulations showed that packings under isotropic pressure lead to the same scaling behaviour irrespectively of the applied pressure, the polydispersity of the grains, the coefficient of friction and the force law. Remarkably, the corresponding scaling exponents and scaling function appear to be the same as those obtained from the snooker packing.

7 Outlook

The existence of universality classes for force networks raises a number of new questions. First of all, what properties of a jammed system determine the universality class of its force network? Our results suggest that that the isotropy of the applied force and the vector nature of the force balance are essential. On the other hand, packings under static shear might lead to another universality class. Another relevant parameter could be the temperature in thermal systems which exhibit jamming, such as colloids.

Our method based on force cluster criticality is clearly able to discriminate between the many models proposed for force networks [15, 35–39]. In particular, it shows that the Edwards' hypothesis, which proposes to consider all metastable states of a jammed system equally likely, leads quantitatively to the same scaling properties as found in force networks generated by molecular dynamics simulations.

Finally, the method developed here for force networks in jammed matter is clearly more general. It applies in principle to any ensemble of graphs with continuous variables on the edges, such as flux, transport or metabolic networks [40, 41]. The corresponding universality classes could complement the topological characterizations of networks developed in the recent years [42].

References

1. A. Liu and S. Nagel, *Jamming and rheology* (Taylor & Francis, 2001).
2. A. J. Liu and S. R. Nagel, Nature **396**, 21 (1998).
3. V. Trappe, V. Prasad, L. Cipelletti, P. Segre, and D. A. Weitz, Nature **411**, 772 (2001).
4. H. A. Makse and J. Kurchan, Nature **415**, 614 (2002).
5. M. Cates, J. P. Wittmer, J.-P. Bouchaud, and P. Claudin, Phys. Rev. Lett. **81**, 1841 (1998).
6. D. M. Mueth, H. M. Jaeger, and S. R. Nagel, Phys. Rev. E **57**, 3164 (1998).
7. G. Løvoll, K. J. Maløy, and E. G. Flekkøy, Phys. Rev. E **60**, 5872 (1999).
8. D. L. Blair, N. W. Mueggenburg, A. H. Marshall, H. M. Jaeger, and S. R. Nagel, Phys. Rev. E **63**, 041304 (2001).
9. J. M. Erikson, N. W. Mueggenburg, H. M. Jaeger, and S. R. Nagel, Phys. Rev. E **66**, 040301 (R) (2002).
10. J. Brujic, S. F. Edwards, D. V. Grinev, I. Hopkinson, D. Brujic, and H. A. Makse, Faraday Disc. **123**, 207 (2003).
11. F. Radjai, M. Jean, J.-J. Moreau, and S. Roux, Phys. Rev. Lett. **77**, 274 (1996).
12. S. Luding, Phys. Rev. E **55**, 4720 (1997).
13. F. Radjai, D. E. Wolf, M. Jean, and J.-J. Moreau, Phys. Rev. Lett. **80**, 61 (1998).
14. H. A. Makse, D. L. Johnson, and L. M. Schwartz, Phys. Rev. Lett. **84**, 4160 (2000).
15. M. L. Nguyen and S. N. Coppersmith, Phys. Rev. E **62**, 5248 (2000).
16. S. J. Antony, Phys. Rev. E **63**, 011302 (2000).
17. C. S. O'Hern, S. A. Langer, A. J. Liu, and S. R. Nagel, Phys. Rev. Lett. **86**, 111 (2001).
18. C. S. O'Hern, S. A. Langer, A. J. Liu, and S. R. Nagel, Phys. Rev. Lett. **88**, 075507 (2002).
19. L. E. Silbert, G. S. Grest, and J. W. Landry, Phys. Rev. E **66**, 061303 (2002).
20. D. Stauffer and A. Aharony, *Introduction to Percolation Theory* (Taylor & Francis, 1991).
21. V. Privman, *Finite Size Scaling and Numerical Simulations of Statistical Physics* (World Scientific, 1990).
22. H. M. Jaeger and S. R. Nagel, Rev. Mod. Phys. **68**, 1259 (1996).
23. J.-P. Bouchaud, in *Les Houches, Session LXXVII*, edited by J. Barrat (EDP Sciences, 2003).
24. S. N. Coppersmith, C.-h. Liu, S. Majumdar, O. Narayan, and T. A. Witten, Phys. Rev. E **53**, 4673 (1996).
25. M. da Silva and J. Rajchenbach, Nature **406**, 708 (2000).
26. J. H. Snoeijer, T. J. H. Vlugt, M. van Hecke, and W. van Saarloos, Phys. Rev. Lett. **92**, 054302 (2004a).
27. J. H. Snoeijer, T. J. H. Vlugt, W. G. Ellenbroek, M. van Hecke, and J. M. J. van Leeuwen, Phys. Rev. E **70**, 061306 (2004b).
28. S. F. Edwards and R. Oakeshott, Physica A **157**, 1080 (1989).
29. S. Ostojic and D. Panja, in *Powders and Grains 2005* (Barkema, 2005a).
30. S. Ostojic and D. Panja, J. Stat. Mech. p. P01011 (2005b), S. Ostojic and D. Panja, `cond-mat/0403321` (2005b).
31. M. E. J. Newman and G. T. Barkema, *Monte Carlo Methods in Statistical Physics* (Oxford University Press, 1999).

32. J. H. Snoeijer, M. van Hecke, E. Somfai, and W. van Saarloos, Phys. Rev. E **70**, 011301 (2004c).
33. A. Harris and T. Lubensky, Phys. Rev. Lett. **33**, 1540 (1974).
34. S. Ostojic, E. Somfai and B. Nienhuis, Nature, in press.
35. M. Nicodemi, Phys. Rev. Lett. **80**, 1340 (1998).
36. M. L. Nguyen and S. N. Coppersmith, Phys. Rev. E **59**, 5870 (1999).
37. O. Narayan, Phys. Rev. E **63**, 010301 (R) (2000).
38. J.-P. Bouchaud, P. Claudin, D. Levine, and M. Otto, Eur. Phys. J. E **4**, 451 (2001).
39. C. Goldenberg and I. Goldhirsch, Granular Matter **6**, 87 (2004).
40. M. A. de Menezes and A.-L. Barabasi, Phys. Rev. Lett. **92**, 028701 (2004).
41. E. Almaas, B. Kovacs, T. Vicsek, Z. N. Oltvai, and A.-L. Barabasi, Nature **427**, 839 (2004).
42. R. Albert and A.-L. Barabasi, Rev. Mod. Phys. **74**, 47 (2002).

Species Segregation and Dynamical Instability of Horizontally Vibrated Granular Mixtures

Massimo Pica Ciamarra, Alessandro Sarracino, Mario Nicodemi, and Antonio Coniglio

Dip.to di Scienze Fisiche, Università di Napoli "Federico II"
INFM-Coherentia, INFN and AMRA, Napoli, Italy

Summary. We review recent results about the segregation process of a granular mixture of disks on an horizontally oscillating tray. In this condition an initially disordered mixture first segregates via the formation of stripes perpendicular to the driving direction; then stripes merge in a coarsening process. We discuss quantitatively both the short-time and the long-time dynamics of the system, and the dependence of the observed phenomenology on the frequency and amplitude of oscillation of the tray. The same system is also investigated when, instead of being disordered, is initially prepared in two stripes parallel to the driving direction. In this condition the interface between the two stripes manifests an instability which again leads to the formation of stripes perpendicular to the driving direction. Finally, we shortly review the mechanism which have been proposed in order to explain the observed segregation process.

1 Introduction

A granular medium consisting of a collection of dry, cohesionless, identical particles exhibits a wide range of complex behaviours. Despite the simplicity of the constituent particles no reliable mathematical model exist for most of these collective phenomena. Of particular interest is the counter-intuitive phenomenon of species segregation [1]. When subject to an external perturbation, such as vertical or horizontal oscillations, an initially disordered binary mixture of grains (which may differ in size, mass, frictional properties) often segregates their components. Depending on the driving conditions different mechanism, such as percolation [2–4], inertia [5], convection [6], or even purely thermodynamical effects [7], have been shown to be responsible for the segregation process.

Recently segregation of a binary mixture subject to horizontal oscillations has been observed by T. Mullin and co-workers [8–11]. In their experiment a granular disordered monolayer, composed of a mixture of steel spheres and poppy seeds, where placed on a horizontal tray oscillating along the x-axis.

They observed the system to segregate via the formation of a pattern of alternating stripes of particles of the same kind, parallel to the y-axis.

In order to understand the physical mechanism which is responsible for this segregation process we have recently investigated this experiment via soft-core molecular dynamics simulations. Here we review our work [12–14] and present new results relative to the coarsening process of the stripes.

2 Model

We perform soft core Molecular Dynamics simulations of a two-dimensional granular media taking into account grain-grain and grain-tray interactions [15]. Two grains with diameters D_i and D_j in positions $\mathbf{r}_i$ and $\mathbf{r}_j$ interact if overlapping, i.e., if $\delta_{ij} = [(D_i + D_j)/2 - |\mathbf{r}_i - \mathbf{r}_j|] > 0$. The interaction is given by a normal repulsive force with viscous dissipation [16, 17]. In two dimensions this reduces to the linear spring-dashpot model,

$$\mathbf{f}_n = k_n \delta_{ij} \mathbf{n}_{ij} - \gamma_n m_{\text{eff}} \mathbf{v}_{nij}, \tag{1}$$

where k_n and γ_n are the elastic and viscoelastic constants, and $m_{\text{eff}} = m_i m_j/(m_i + m_j)$ is the effective mass. We follow the realistic simulations of [11, 15] and model the interaction with the tray via a viscous force

$$\mathbf{f}_t = -\mu_i (\mathbf{v}_i - \mathbf{v}_{\text{tray}}), \tag{2}$$

where $\mathbf{v}_{\text{tray}}(t) = 2\pi A\nu \sin(\nu t)\mathbf{x}$ is the velocity of the tray and $\mathbf{v}_i$ the velocity of the disk i, plus a white noise force $\xi(t)$ with

$$\langle \xi(t)\xi(t')\rangle = 2\Gamma\delta(t - t'). \tag{3}$$

For the grain-grain interaction, we use the value $k_n = 2\ 10^5$ g cm^{2}s^{-2} and γ_n chosen, for each kind of grains, such that the restitution coefficient is given: $e = 0.8$ [17]. The two components of our mixture have mass $M_h = 1$ g and $M_l = 0.03$ g, and viscous coefficient $\mu_h = 0.28$ g s^{-1} and $\mu_l = 0.34$ g s^{-1}. The white noise has $\Gamma = 0.2g^2cm^2s^{-3}$. Apart from a simple rescaling of masses and lengths, these values are those of reference [11] (and given in private communications), and are taken from direct measurements on the experimental system. We solve the equations of motion by the Verlet algorithm with an integration time-step $dt = 6\mu s$, which is limited by the value of K_n and not, as usual in numerical simulations of the Langevin equation, by the relaxation time m/μ (in our case $dt \ll M_l/\mu_l \ll M_h/\mu_h$). The numerical resolution of stochastic force may be difficult [18]. We have validated the algorithm and the integration time step considering that smaller value of dt reproduces the same results, and that the expected properties of a thermal binary mixture are recovered if the grain-grain interaction is made elastic (i.e. γ_n is set to zero).

The heavier grains of our mixture have diameter $D_h = 1$ cm. We consider lighter grains diameter D_l to be Gaussian distributed with mean value 0.7 cm and 17% polydispersity, or to have the same size of the heavier grains, $D_l = D_h = 1$ cm. We use a tray of width $d_y = 20$ cm and length $d_x = 40$ cm or $d_x = 320$ cm. Our simulations refer to oscillations with amplitude $A = 1.2$ cm and frequency $\nu = 12$ Hz. The qualitative picture we discuss does not change if these values are changed.

The dynamics of the system is determined by the amplitude and the frequency of oscillation, and by the grain properties: size, mass and area fractions. These are defined as

$$\Phi_l = N_l \frac{\pi}{4 L_x L_y} \overline{D_l^2}, \quad \Phi_h = N_h \frac{\pi}{4 L_x L_y} D_h^2, \tag{4}$$

where $\overline{D_l^2}$ is the mean value of the square of small grains diameter $(\overline{D_l^2} \neq D_l^2$ when we considering polydisperse light grains), and N_h (N_l) is the number of heavy (light) grains.

3 Dynamics

The dynamics of a disordered mixture subject to horizontal oscillations can be schematically divided in two steps, stripes formation and stripes coarsening:

- **Stripe formation** After few oscillations of the tray particles of the same species organize in cluster, which rapidly merge and orientate giving rise to a pattern of stripes perpendicular to the driving direction, as shown in Fig. 3.
- **Coarsening** On a much longer timescale a coarsening process takes place: stripes of particles of the same kind merge. Consequently the number of stripes decreases and the mean stripe width increases.

This qualitative explanations of the dynamics is formally described by the temporal evolution of the quantity a characteristic length of the system in the x direction, ξ_x. As usual in coarsening processes this length is defined as:

$$2\pi \xi_x = \int dk_x S_x(k_x) dk_x, \tag{5}$$

where $S_x(k_x) = S(k_x, 0)$ is the structure factor for wave vectors $\vec{k} = (k_x, 0)$ with null y component. Fig. 2 shows the time evolution of ξ_x for a mixture of grains of equal size, and area fraction $\Phi_h = 0.20$, $\Phi_l = 0.31$. This is well described by the following functional form

$$\xi_x(t) = a + b \exp(-t/\tau) + ct^\alpha \tag{6}$$

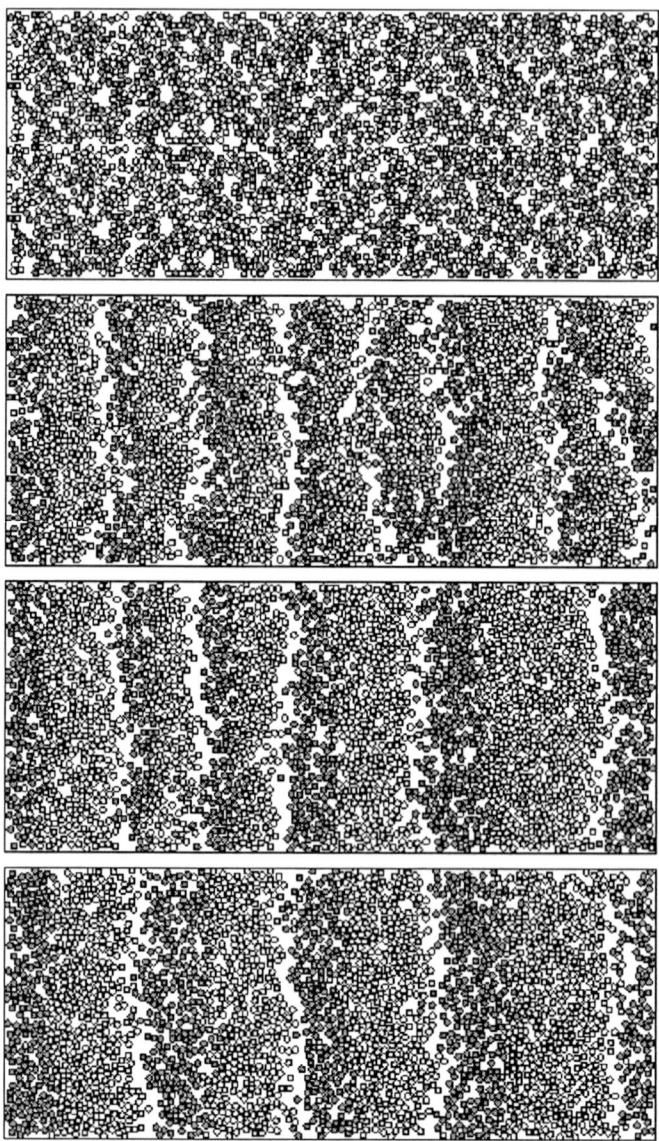

Fig. 1. Evolution of a mixture of equal-sized grains subject to horizontal vibrations. The plots show the state of the system (from top to bottom) after 0, 400, 9800 and 20500 oscillations. Periodic boundary conditions are used in both directions. Note that about 7 stripes are observed after 400 taps, which becomes 4 after 20500 taps.

which combines an exponential relaxation with lasts approximately $\tau \simeq 175$ oscillations, with a subsequent coarsening process. In this last stage the characteristic length grows with a power law with an exponent $\alpha \simeq 0.25$. A similar growth exponent has been observed in [8]. We discuss now in some more detail the short dynamics (stripes formation) and the long time dynamics, coarsening.

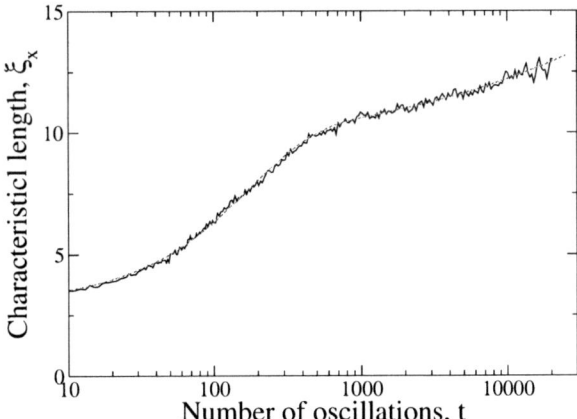

Fig. 2. Temporal evolution of the characteristic length of the pattern. This is well described by eq. 6 (smooth line), which combines an exponential relaxation with a coarsening process.

3.1 Short-Time Dynamics

In a previous work we have examined the short time dynamics of the system [12]. The goal was to understand under which conditions stripes form, an what is the dependence of the initial wavelength of the striped pattern on the properties (amplitude and frequency) of the drive. In order to make direct comparisons with the experiments this analysis has been conducted with monodisperse heavy grains, and polydisperse small grains. Under these conditions, which are those studied by T. Mullin and coworkers, the system can be mixed or segregated in stripes. Moreover, if segregation occurs, stripes of the monodisperse species can be either "fluid" or "crystalline".

   This behaviour, in the (ϕ_h, ϕ_l) plane, is summarized in the diagram of Fig. 3(a) showing the system "fluid" and "crystal" regions along with their segregation properties, for $\nu = 12$ Hz and $A = 1.2$ cm. Large grains are considered to be in a "fluid" configuration when their radial density distribution function, $g(r)$, shows a first peak at $r = D_h$ and a second one at $r = 2D_h$, and to be in a "crystal" configuration when a new peak at $r = \sqrt{3}D_h$ appears [10].

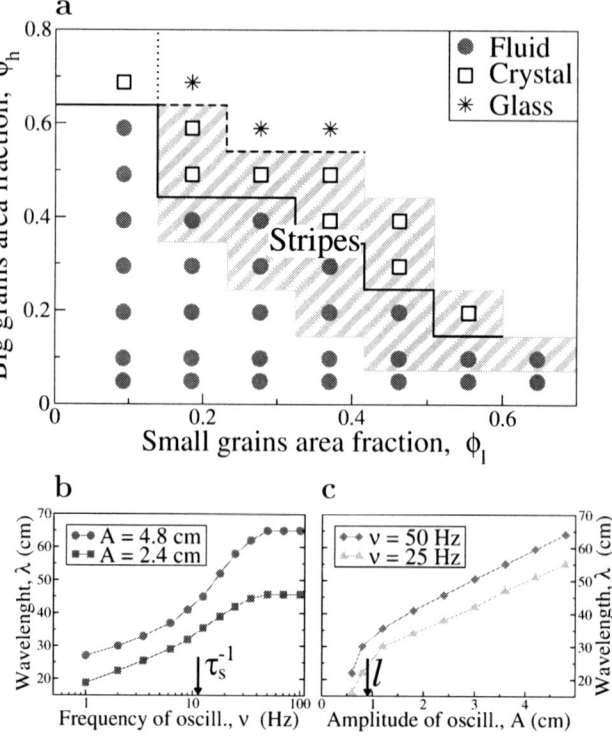

Fig. 3. Panel a Ordering properties of the late stage configurations of the mixture as a function of the area fractions of the two components. The *shaded area* covers the region where segregation via stripes formation occurs. **circles**: large grains are in a fluid state. **squares**: large grains form a crystal. **stars**: the system appears blocked in a "glassy" disordered configuration (see text). When stripes form their characteristic short time length scale λ is a function of the frequency, ν, and of the amplitude, A, of the driving oscillations. This is shown, in the case $\phi_h = 0.30$ and $\phi_l = 0.28$, in **panels b** and **c**.

The system is in a "glassy" state [19] when on the longest of our observation time scales, the system is still far from stationarity.

Fig. 3(a) shows that grains at small concentrations are mixed and in a fluid state. Segregation via stripes formation appears at higher concentrations. At even higher concentrations, large grains form stripes with a crystalline order, as smaller grains are always fluid for their polydispersity. Finally, at very high area fractions, the system is blocked in its starting disordered configuration ("glassy" region). For instance, by increasing ϕ_l at a fixed value of ϕ_h (say $\phi_h \simeq 0.174$), we observe first a transition from a mixed fluid state to a segregated striped fluid and then a transition where the the monodisperse phase crystallize. The experiments of [10], where $\phi_h \simeq 0.174$, show the very same transitions found here at locations differing by a 10%.

In the case $\phi_h = 0.30$ and $\phi_l = 0.28$, where stripes form, we describe their dependence on the dynamics control parameters in Figs. 3(b) and 3(c), showing that the length scale $\lambda = d_x/n$, with d_x later dimension of the tray and n number of stripes, increases as a function of the shaking frequency, ν, and of the amplitude A. These results are to be compared, for instance, with those found in liquid-sand mixtures under oscillating flow: as we will discuss in the next section, in fact, is it possible that in the investigated system stripes form as a result of a dynamical instability of Kelvin-Helmholtz type, the same instability responsible for ripples formation in liquid-sand mixtures. While our results show that the wavelength depends both on the amplitude and the frequency of oscillations, in liquid sand mixture the wavelength depends on the amplitude of oscillation, but not on its frequency [20].

In our system the dependence on ν can be schematically understood by comparison with the characteristic time scales $\tau_h = M_h/\mu_b$ and $\tau_l = M_l/\mu_s$ of the two species (here $\tau_h^{-1} = 0.28$ Hz and $\tau_l^{-1} = 11.3$ Hz): in the limit $\nu \gg \tau_h^{-1}, \tau_l^{-1}$ grains are not able to follow the tray motion and no sensitivity to ν is expected, as well as when $\nu \ll \tau_l^{-1}$ since the grains move with the tray. Analogously, the dependence on A is expected to be substantial when A is at least of the order of the mean grains separation length, $l = (4\phi_h/\pi D_h^2 + 4\phi_l/\pi D_l^2)^{-1/2}$, since under this condition grains strongly interact.

3.2 Long-Time Dynamics

The long time dynamics of the system is characterized by a coarsening process in which the characteristic length of the system increases as a power law, $\xi_x(t) \propto t^\alpha$, with an exponent $\alpha \simeq 1/4$, as Fig. 2 shows.

The term 'coarsening' is usually referred to describe the out-of equilibrium dynamics of a binary mixture (or of a magnetic system) whose temperature is quenched from a high value to a value T_q which is below the coexistence curve. At T_q the stable state is made of two coexisting regions: one region is rich in one component of the binary mixture (has positive magnetization), while the other region is rich in the second component of the mixture (has negative magnetization). Therefore the system, which shortly after the quench is still in a mixed state, spontaneously segregate forming growing domains rich in one species or the other. These characteristic size of these domains grow in time with a power law with an exponent which is $1/4$ for the case of conserved order parameter (the binary mixture case), $1/3$ if the order parameter is not conserved (magnetic system) (for a comprehensive review see [21]).

The long-time dynamics of our system exhibits a phenomenology which closely resemble that of a thermal system undergoing coarsening. However it is worth noting that (at the moment) this coarsening process cannot be interpreted like the phase separation of a binary mixture quenched below the critical point. For instance, it is difficult to introduce a temperature in our system. If we try to define the temperature as the velocity fluctuations, in fact, we end up in a confusing situations: for each species the fluctuations of the x and of the y component of the velocities are different (as if the temperature was a vector), the temperature of the two components are different, and the velocity distributions are not Maxwellian. At the present stage of understanding the notion of temperature and that of stable phase for the investigated system appear to be meaningless, and the coarsening process must be seen as induced by the forcing.

In this respect one may try to devise a simple model to understand the origin of the growth exponent α. The idea is to model the fluctuations of the width of the stripes, and to consider that two stripes merge if in contact [8, 22]. Assuming that all of the trips have the same width in order for two stripes to merge the need to fluctuate of a distance of order ξ_x. Therefore we have:

$$\frac{\partial \xi_x}{\partial t} \propto n \cdot d_y \cdot P(\xi_x) \tag{7}$$

where $n = d_x/(2\xi_x)$ is the number of stripes, $P(x)$ is the probability that a point of a stripe fluctuates of a distance x, and d_y the y length of the tray. In order to estimate $P(x)$ we assume that each point of a stripe makes a random walk in the horizontal direction as a consequence of the various collisions; $P(x)$ is proportional to the time $\tau(x)$ we have to wait for a point to be displaced by x: $P(x) \propto \tau(x) \propto 1/x^2$. With this assumption:

$$\frac{\partial \xi_x}{\partial t} \propto \frac{1}{2} d_x d_y \frac{1}{\xi_x^3}, \tag{8}$$

and therefore $\xi_x(t) \propto t^{1/4}$.

4 Dynamical Instability

We have seen so far that a disordered granular mixture subject to horizontal oscillations segregates via the formation of stripes. Here we discuss the evolution of the same system when the initial state is not disordered. On the contrary the two species are placed in two stripes *parallel* to the driving direction, as show in 4. The solution of the equation of motion of a grain of mass M interacting with the oscillating tray via a viscous force regulated by the coefficient of friction μ is

$$x(t) = -\frac{A}{1 + \tau^2 \nu^2} [\cos(\nu t) + \tau \nu \sin(\nu t)], \tag{9}$$

where $\tau = M/\mu$. In our system the two species, having different relaxation times $\tau_h = M_h/\mu_h = 3.57$ s and $\tau_l = M_l/\mu_l = 0.09$ s, are thus forced to oscillate with different amplitudes and different phases. In the configuration shown in 4 (upper panel) one one may expect the two species to oscillate independently (following equation 9 with different relaxation times), and the initial configuration to be a stable one. But this is not the case. The oscillatory

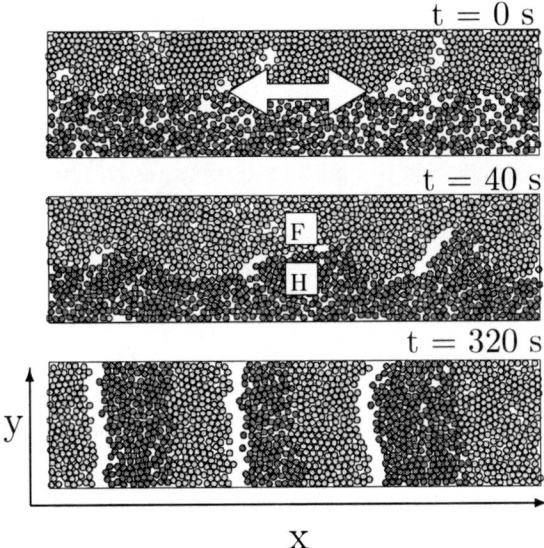

Fig. 4. Evolution of a binary mixture of disks placed on a tray oscillating along the x direction. Here we consider the case in which the diameters of particles of different species are equal. The pictures shows only 1/4 of the system length, which is of 320 D_b. The initial state (t= 0 s) is made of two stripes of particles of different species parallel to the driving direction. As times goes on the flat interface between the two species evolves via the formation of a sine-like modulation (t=40 s). Finally, the wavy interface between grains of different species breaks leading to the formation of a striped pattern as seen before (t=320 s).

motion of the tray induces an oscillating shear velocity at the interface between the two species which causes the interface to evolve via the formation of a modulation with a sine-like shape. As times goes on the amplitude of the modulation grows until it breaks giving rise to the striped pattern seen before.

The mechanism responsible for the evolution of the pattern is understood by considering Fig. 4 at t=40 s, and by making use of fluid-dynamics considerations. Here 'F' and 'H' mark regions in which the horizontal motion of grains of a given species is free, or hindered by the presence of grains of the other species. By virtue of Bernoulli's law the pressure in 'F' is smaller than the pressure in 'H', implying a growth of the perturbed interface. This sets-up a mechanism with a positive feedback, which leads to the formation of the striped pattern. At the moment a closer connection between the investigated system and instability in fluids appears difficult. For instance one is tempted to study the instability via a generalization of the Kelvin-Helmholtz instability (which is a well-known fluid mechanical instability observed when there is a constant shear velocity between two fluid flowing one past the the other [23, 24]) to the case where the shear velocity between the two fluids oscillates in time. However the Kelvin-Helmholtz instability is investigated introducing a typical lengthscale of the system, the capillary length, a combination of gravity and surface tension: in the investigated system gravity plays no role, and there is no surface tension.

A dynamical instability similar to the one of Fig. 4 has been observed both in two fluid systems [25] and in liquid sand mixtures [26]. In these cases, however, gravity stabilizes the interface in a wavy like configuration, and stripes perpendicular to the driving direction are not observed.

5 Conclusion

We conclude by shortly discussing the mechanisms which have been proposed in order to explain the observed segregation process. Originally the phenomenology was attributed to the depletion potential [9], a form of interaction well known in colloidal systems [27]. Two big spheres immersed in bath of smaller ones are subject to an effective potential, due to an entropic effect (the clustering of big spheres increases the free space available to the smaller ones, and consequently the entropy of the system), which is attractive at small distances. This attractive interaction is used to explain the phase separation of the system. The anisotropy of the drive, in turn, is used to explain why the phase separation manifests via the formation of stripes [9]. However in order for the depletion potential to exists it is necessary that the mixture is made of particles of different size. Since we have observed segregation also in the case of equal-sized particles, we can rule out the depletion potential as a possible explanation of the observed phenomenology (in ref. [13] we show that the depletion potential does not explain segregation even when the two species have different size).

Another mechanism responsible for segregation, the 'differential drag', has been proposed in [10, 28, 29]. Shortly, the authors suggest that since the two species are forced to oscillate with different amplitudes and phases (see eq. 9) there is an effective repulsion between particles of the different species, which is responsible for the observed phenomenology. While it is certainly true that such a repulsion exists and could possibly play a significant role in order to explain the observed phenomenology, we note here that the 'equilibrium' configuration of the system, that is the one that minimize this repulsion, is made of stripes *parallel* to driving direction, and not of *perpendicular* stripes as observed. When the stripes are parallel to the driving direction, in fact, particles of different species never interact, while they interact if the stripes are perpendicular to the driving direction.

Finally, in Ref. [12] we have suggested that the segregation process can be related to instability process previously discussed. However there is no direct evidence that this is instability is responsible for the segregation, and it could be that the segregation process and the instability share a common yet unknown microscopic origin.

In conclusion, even though the overall phenomenology of the segregation process of a granular mixture subject to horizontal oscillations is clear, the microscopic origin of the observed phenomenology is still obscure.

References

1. A. Kudrolli, Reports on Progress in Physics **67**, 209 (2004).
2. A. Rosato, K.J. Strandburg, F. Prinz and R.H. Swendsen, Phys. Rev. Lett. **58**, 1038 (1987).
3. J. Bridgewater, Powder Technol. **15**, 215 (1976).
4. J.C. Williams, Powder Technol. **15**, 245 (1976).
5. T. Shinbrot and Fernando J. Muzzio, Phys. Rev. Lett. **81**, 4365 (1998).
6. J.B. Knight, H.M. Jaeger, and S.R. Nagel, Phys. Rev. Lett. **70**, 3728 (1993).
7. M. Tarzia, A. Fierro, M. Nicodemi, and A. Coniglio Phys. Rev. Lett. **93**, 198002 (2004).
8. T. Mullin, Phys. Rev. Lett. **84**, 4741 (2000).
9. P.M. Reis and T. Mullin, Phys. Rev. Lett. **89**, 244301 (2002).
10. P.M. Reis, G. Ehrhardt, A. Stephenson and T. Mullin, Europhys. Lett. **66**, 357 (2004).
11. G. Ehrhardt, A. Stephenson and P.M. Reis, Phys. Rev. E **71**, 041301 (2005).
12. M. Pica Ciamarra, and M. Nicodemi and A. Coniglio, Phys. Rev. Lett. **94**, 188001 (2005).
13. M. Pica Ciamarra, and M. Nicodemi and A. Coniglio, J. Phys.: Condens. Matter **17**, S2549 (2005).
14. M. Pica Ciamarra, and M. Nicodemi and A. Coniglio, in Powders and Grains 2005, edited by R. Gracía-Rojo, H.J. Herrmann and S. McNamara.
15. H.J. Herrmann and S. Luding, Cont. Mech. Thermod. **10**, 189-231 (1998).
16. P.A. Cundall and O.D.L. Strack, Geotechnique **29**, 47-65 (1979).
17. L.E. Silbert, D. Ertas, G.S. Grest, T.C. Halsey, D. Levine and S.J. Plimpton, Phys Rev E **64**, 051302 (2001).
18. R. Mannella, in *Stochastic Processes in Physics, Chemistry, and Biology*, J.A. Freund and T. Pöshel (ed.s) Springer-Verlag (2000); T. Pöshel and T. Schwager, *Computational granular dynamics*, Springer (2004).
19. *"Unifying concepts in granular media and glasses"*, (Elsevier Amsterdam, 2004), Edt.s A. Coniglio, A. Fierro, H.J. Herrmann, M. Nicodemi.
20. A. Stegner, J.E. Wesfreid, Phys. Rev. E **60**, R3487 (1999).
21. N. Goldenfeld, Lecture on phase transitions and the renormalization group, Addison-Wesley, 1992.
22. P.A. Mulheran, J. Phys. I France **4**, 1 (1994).
23. H. Lamb, Hydrodynamics (Cambridge Univ. Press, Cambridge, 1932).
24. S. Chandrasekhar, Hydrodynamic and Hydromagnetic stability. (Clarendon, Oxford, 1961).
25. A.A. Ivanova, V.G. Kozlov and P. Evesque, Fluid Dynamics **36** 362, (2002).
26. M.A. Scherer, F. Melo, M. Marder, Phys. of Fluids **11**, 58-67 (1999).
27. S. Asakura and F.Oosawa, J. Chem. Phys. **22**, 1255 (1954).
28. C.M. Pooley, and J.M. Yeomans, Phys. Rev. Lett. **93**, 118001 (2004).
29. P. Sánchez, M.R. Swift, and P.J. King, Phys. Rev. Lett. **93**, 184302 (2004).

Lattice Versus Lennard-Jones Models with a Net Particle Flow

Manuel Díez-Minguito, Pedro L. Garrido, and Joaquín Marro

Institute 'Carlos I' for Theoretical and Computational Physics, and
Departamento de Electromagnetismo y Física de la Materia,
Universidad de Granada, E-18071 - Granada, Spain

Summary. We present and study lattice and off-lattice microscopic models in which particles interact via a local anisotropic rule. The rule induces preferential hopping along one direction, so that a net current sets in if allowed by boundary conditions. This may be viewed as an oversimplification of the situation concerning certain traffic and flow problems. The emphasis in our study is on the influence of dynamic details on the resulting (non-equilibrium) steady state. In particular, we shall discuss on the similarities and differences between a lattice model and its continuous counterpart, namely, a Lennard–Jones analogue in which the particles' coordinates vary continuously. Our study, which involves a large series of computer simulations, in particular reveals that spatial discretization will often modify the resulting morphological properties and even induce a different phase diagram and criticality.

1 Introduction

Many systems out of equilibrium [1, 2] exhibit spatial striped patterns on macroscopic scales. These are often caused by transport of matter or charge induced by a drive which leads to heterogeneous ordering. Such phenomenology occurs in flowing fluids [3], and during phase separation in colloidal [4], granular [5, 6], and liquid–liquid [7] mixtures. Further examples are wind ripples in sand [8], trails by animals and pedestrians [9], and the anisotropies observed in high temperature superconductors [10, 11] and in two–dimensional electron gases [12, 13].

Studies of these situations, often described as nonequilibrium phase transitions, have generally focused on lattice systems [14–18], i.e., models based on a discretization of space and in considering interacting particles that move according to simple local rules. Such simplicity sometimes allows for exact calculations and is easy to be implemented in a computer. Moreover, some powerful techniques have been developed to deal with these situations, including nonequilibrium statistical field theory. However, lattice models are perhaps a too crude oversimplification of fluid systems so that the robustness of such an approach merits a detailed study.

The present paper describes Monte Carlo (MC) simulations and field theoretical calculations that aim at illustrating how slight modifications of dynamics at the microscopic level may influence, even quantitatively, the resulting (nonequilibrium) steady state. We are also, in particular concerned with the influence of dynamics on criticality. With this objective, we take as a reference the *driven lattice gas* (DLG), namely, a kinetic nonequilibrium Ising model with conserved dynamics. This system has become a prototype for anisotropic behavior, and it has been useful to model, for instance, ionic currents [17] and traffic flows [19]. In fact, in certain aspects, this model is more realistic for traffic flows than the standard *asymmetric simple exclusion process* [14, 15]. Here we compare the transport and critical properties of the DLG with those for apparently close lattice and off–lattice models. There is some related previous work addressing the issue of how minor variations in the dynamics may induce dramatic morphological changes both in the early time kinetics and in the stationary state [20–22]. However, these papers do not focus on transport nor on critical properties. We here in particular investigate the question of how the lattice itself may condition transport, structural and critical properties and, with this aim, we consider nearest–neighbor (NN) and next–nearest–neighbor (NNN) interactions. We also compare with a microscopically off–lattice representation of the *driven lattice gas* in which the particles' spatial coordinates vary continuously. A principal conclusion is that spatial discretization may change significantly not only morphological and early–time kinetics properties, but also critical properties. This is in contrast with the concept of universality in equilibrium systems, where critical properties are independent of dynamic details.

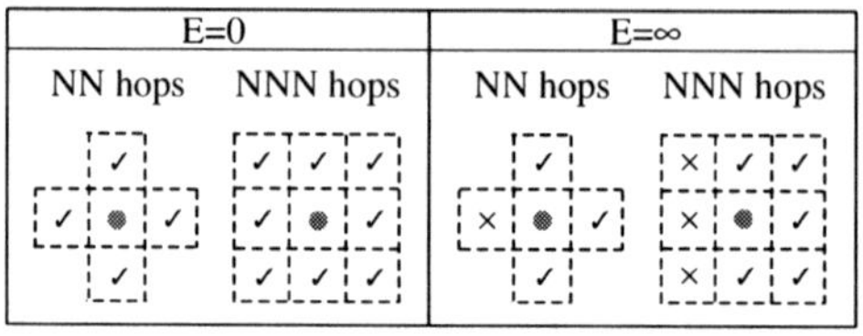

Fig. 1. Schematic comparison of the sites a particle (at the center, marked with a dot) may occupy (if the corresponding site is empty) for nearest–neighbor (NN) and next–nearest–neighbor (NNN) hops at equilibrium (left) and in the presence of an "infinite" horizontal field (right). The particle–hole exchange between neighbors is either forbidden ($\times$) or allowed ($\checkmark$), depending on the field value.

2 Driven Lattice Gases

The *driven lattice gas*, initially proposed by Katz, Lebowitz, and Spohn [23], is a nonequilibrium extension of the Ising model with conserved dynamics. The DLG consists of a d-dimensional square lattice gas in which pair of particles interact via an attractive and short–range Ising–like Hamiltonian,

$$H = -4 \sum_{\langle j,k \rangle} \sigma_j \sigma_k . \tag{1}$$

Here $\sigma_k = 0(1)$ is the lattice occupation number at site k for an empty (occupied) state and the sum runs over all the NN sites (the accessible sites are depicted in Fig. 1). Dynamics is induced by the competion between a heat bath at temperature T and an external driving field E which favors particle hops along one of the principal lattice directions, say horizontally ($\hat{x}$), as if the particles were positively charged. Consequently, for periodic boundary conditions, a nontrivial nonequilibrium steady state is set in asymptotically. MC simulations by a biased *Metropolis* rate reveal that, as in equilibrium, the DLG undergoes a second order phase transition. At high enough temperature, the system is in a disordered state while, below a critical point (at $T \leq T_E$) it orders displaying anisotropic phase segregation. That is, an anisotropic (striped for $d = 2$) rich–particle phase then coexists with its gas. It is also found that the critical temperature T_E monotonically increases with E. More specifically, for $d = 2$, assuming a half filled square lattice in the large field limit (in order to maximize the nonequilibrium effect), one has a *nonequilibrium* critical point at $T_\infty \simeq 1.4 T_0$, where the equilibrium value is $T_0 = 2.269 J k_B^{-1}$. It was numerically shown that this belongs to a universality class other than the Onsager one, e.g., MC data indicates that the order parameter critical exponent is $\beta_{\mathrm{DLG}} \simeq 1/3$ [17, 24] (instead of the Onsager value 1/8).

   Other key features concern the two–particle correlation function $C(x,y)$ and its Fourier transform $S(k_x, k_y)$, i.e., the structure factor. As depicted in the left graph of Fig. 2, correlations are favored (inhibited) along (against) the field direction. In fact, the DLG shows a slow decay of the two–point correlations due to the spatial anisotropy associated with the dynamics [25]. This long range behavior translates into a characteristic discontinuity singularity at the origin ($\lim_{k_x \to 0} S_\| \neq \lim_{k_y \to 0} S_\perp$) in the structure factor [16], which is confirmed in Fig. 2.

   How do all these features depend on the number of neighbor sites to which a particle can hop? Or in other words, how robust is the behavior when extending interactions and accessible sites to the NNN?

   Previous work has shown that extending hopping in the DLG to NNN leads to an inversion of triangular anisotropies during the formation of clusters [21], and also that dramatic changes occur in the steady state, including the fact that, contrary to the DLG with NN interactions, the critical temperature decreases with increasing E [22]. However, other important features such as correlations and criticality seem to remain invariant. Analysis of the parallel $(C_\parallel)$ and transverse $(C_\perp)$ components reveals that correlations are quantitatively similar for the DLG and for the DLG with NNN interactions (henceforth NDLG) —although somehow weaker for the latter case. Also persists a slow decay of correlations which yield to the discontinuity at the origin of $S(k_x, k_y)$. These facts are shown in Fig. 2.

On the other hand, recent MC simulations of the NDLG indicate that the order parameter critical exponent is $\beta_{NDLG} \approx 1/3$ [26], as for the DLG. The *anisotropic diffusive system* approach [28], which is a Langevin–type (mesoscopic) description, predicts this critical behavior. In both cases, DLG and NDLG, the Langevin equations, as derived by coarse graining the master equation, lead to $\beta = 1/3$. These two Langevin equations are identical, except for new entropic terms in the NDLG due to the presence of additional neighbors [27].

The fact that extending particle hops and interaction to the diagonal sites leaves invariant both correlations and criticality seems to indicate that the two systems, DLG and NDLG, belong to the same universality class.

3 A Driven Off-Lattice Gas

In order to deep further on this interesting issue, we studied to what extent the DLG behavior depends on the lattice itself. With this aim, we considered a driven system with continuous variation of the particles' spatial coordinates —instead of the discrete variations in the DLG— which follows as close as possible the DLG strategy. In particular, we analyzed an off–lattice, microscopically–continuum analog of the DLG with the symmetries and short–range interaction of this model.

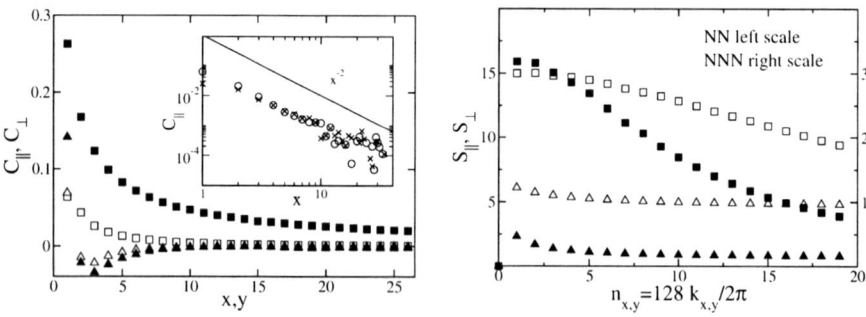

Fig. 2. Parallel (squares) and transverse (triangles) components of the two–point correlation function (left) and the structure factor (right) above criticality with NN (filled symbols) and NNN (empty symbols) interactions for a 128×128 half filled lattice. The inset shows the x^{-2} power law decay in $C_\parallel$ for both discrete cases: DLG (○) and NDLG (×).

3.1 The Model

Consider a *fluid* consisting of N interacting particles of mass m confined in a two–dimensional box of size $L \times L$ with periodic (toroidal) boundary conditions. The particles interact via a truncated and shifted Lennard–Jones (LJ) pair potential [30]:

$$\phi(r) \equiv \begin{cases} \phi_{LJ}(r) - \phi_{LJ}(r_c), & \text{if } r < r_c \\ 0, & \text{if } r \geq r_c, \end{cases} \qquad (2)$$

where $\phi_{LJ}(r) = 4\epsilon \left[(\sigma/r)^{12} - (\sigma/r)^6 \right]$ is the LJ potential, r is the interparticle distance, and r_c is the *cut-off* which we shall set at $r_c = 2.5\sigma$. The parameters σ and ϵ are, respectively, the characteristic length and energy. For simulations, all the quantities were reduced according to ϵ and σ, and k_B and m are set to unity.

The uniform (in space and time) external driving field E is implemented by assuming a preferential hopping in the horizontal direction. This favors particle jumps along the field, as it the particles were positively charged; see dynamic details in Fig. 3. As in the lattice counterpart, we consider the large field limit $E \to \infty$. This is the most interesting case because, as the strength of the field is increased, one eventually reaches saturation, i.e., particles cannot jump against the field. This situation may be formalized by defining the transition probability per unit time (*rate*) as

$$w(\eta \to \eta'; E, T) = \frac{1}{2} \left[1 + \tanh(E \cdot \delta) \right] \cdot \min \left\{ 1, exp(-\Delta\Phi/T) \right\}. \qquad (3)$$

Here, any configuration is specified by $\eta \equiv \{\mathbf{r}_1, \cdots, \mathbf{r}_N\}$, where $\mathbf{r}_i$ is the position of the particle i, that can move anywhere in the torus, $\Phi(\eta) = \sum_{i<j} \phi(|\mathbf{r}_i - \mathbf{r}_j|)$ stands for the energy of η, and $\delta = (x_i' - x_i)$ is the displacement corresponding to a single MC trial move along the field direction, which generates an increment of energy $\Delta\Phi = \Phi(\eta') - \Phi(\eta)$. The biased hopping which enters in the first term of Eq. (3) makes the *rate* asymmetric under $\eta \leftrightarrow \eta'$. Consequently, Eq. (3), in the presence of toroidal boundary conditions, violates detailed balance. This condition is only recovered in the absence of the driving field. In this limit the *rate* reduces to the Metropolis one, and the system corresponds to the familiar *truncated and shifted two–dimensional LJ fluid* [29, 30]. Note that each trial move concerning any particle will satisfy that $0 < |\mathbf{r}_i' - \mathbf{r}_i| < \delta_{max}$, where δ_{max} is the maximum displacement in the radial direction (fixed at $\delta_{max} = 0.5$ in our simulations).

MC simulations using the *rate* defined in Eq. (3) show highly anisotropic states (see Fig. 3) below a critical point which is determined by the pair of values (ρ_∞, T_∞). A linear interface forms between a high density phase and its vapor: a single strip with high density extending horizontally along $\hat{x}$ throughout the system separates from a lower density phase (vapor). The local structure of the anisotropic condensate changes from a strictly hexagonal packing of particles at low temperature (below $T = 0.10$), to a polycrystalline–like structure with groups of defects and vacancies which show a varied morphology (e.g., at $T = 0.12$), to a fluid–like structure (e.g., at $T = 0.30$,) and, finally, to a disordered state as the temperature is increased further. This phenomenology makes our model useful for interpreting structural and phase properties of nonequilibrium fluids, in contrast with lattice models, which are unsuitable for this purpose. Skipping the microscopic structural details, the stationary striped state is similar to the one in lattice models, however.

3.2 Transport Properties

Regarding the comparison between off–lattice and lattice transport properties, the left graph in Fig. 4 shows the net current j as a function of temperature. Saturation is only reached at $j_{max} = 4\delta_{max}/3\pi$ when $T \to \infty$. The current approaches its maximal value logarithmically, i.e., slower than the exponential behavior predicted by the Arrhenius law. The sudden rising of the current as T is increased can be interpreted as a transition from a poor–conductor (low–temperature) phase to a rich–conductor (high–temperature) phase, which is reminiscent of ionic currents [17]. This behavior of the current also occurs in the DLG. Revealing the persistence of correlations, the current is nonzero for any low T, though very small in the solid–like phase. From the temperature dependence of j one may estimate the transitions points between the different phases, in particular, as the condensed strip changes from solid to liquid ($T \approx 0.15$) and finally changes to a fully disordered state ($T \approx 0.31$).

The current is highly sensitive to the anisotropy. The most relevant information is carried by the transverse–to–the–field current profile $j_\perp$, which shows the differences between the two coexisting phases (right graph in Fig. 4).

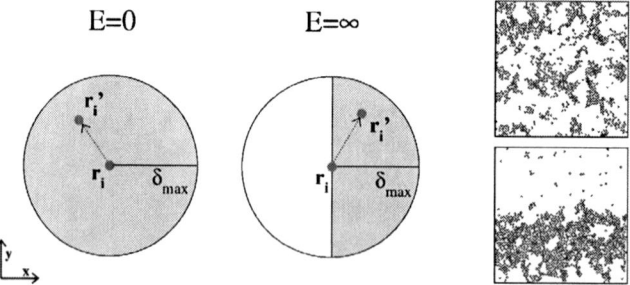

Fig. 3. Schematic representation of the accessible (shaded) region for a particle trial move at equilibrium (left) and out-of-equilibrium (right), assuming the field points along the horizontal direction ($\hat{x}$). The right hand side shows typical steady state configurations above (upper snapshot) and below (lower snapshot) criticality in the large field limit.

Above criticality, where the system is homogeneous, the current profile is flat on the average. Otherwise, the condensed phase shows up a higher current (lower mean velocity) than its mirror phase, which shows up a lower current (higher mean velocity). Both the transversal current and velocity profiles are shown in Fig. 4. The current and the density vary in a strongly correlated manner: the high current phase corresponds to the condensed (high density) phase, whereas the low current phase corresponds to the vapor (low density) phase. This is expectable due to the fact that there are many carriers in the condensed phase which allow for higher current than in the vapor phase. However, the mobility of the carriers is much larger in the vapor phase. The maximal current occurs in the interface, where there is still a considerable amount of carriers but they are less bounded than in the particles well inside the *bulk* and, therefore, the field drives easily those particles. This enhanced current effect along the interface is more prominent in the lattice models (notice the large peak in the current profile in Fig. 4). Moreover in both lattice cases, DLG and NDLG, there is no difference between the current displayed by the coexisting phases because of the particle–hole symmetry. Such a symmetry is derived from the Ising–like Hamiltonian in Eq. (1) and it is absent in the off–lattice model.

3.3 Critical Properties

A main issue is the (nonequilibrium) liquid–vapor coexistence curve and the associated critical behavior. The coexistence curve may be determined from the density profile transverse to the field $\rho_\perp$. This is illustrated in Fig. 5. At high enough temperature above the critical temperature the local density is roughly constant around the mean system density ($\rho = 0.35$ in Fig. 5). As

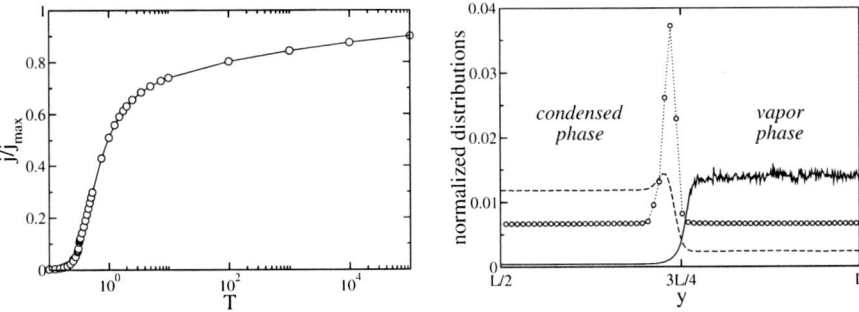

Fig. 4. Left graph: Temperature dependence of the net current for the driven LJ fluid. Right graph: Transverse–to–the–field current profiles below criticality. The shaded (full) line corresponds to the current (velocity) profile of the off–lattice model. For comparison we also show the current profile of the DLG with NN interactions (circle–dotted line). Since each distribution is symmetric with respect to the system center of mass (located here at $L/2$) we only show their right half parts.

T is lowered, the profile accurately describes the striped phase of density ρ_+ which coexists with its vapor of density ρ_- ($\rho_- \leq \rho_+$). The interface becomes thinner and less rough, and ρ_+ increases while ρ_- decreases, as T is decreased. As an order parameter for the second order phase transition one may use the difference between the coexisting densities $\rho_+ - \rho_-$. The result of plotting ρ_+ and ρ_- at each temperature is shown in Fig. 5. The same behavior is obtained from the transversal current profiles (Fig. 4). It is worth noticing that the estimate of the coexisting densities $\rho_\pm$ is favored by the existence of a linear interface, which is simpler here than in equilibrium. This is remarkable because we can therefore get closer to the critical point than in equilibrium.

Lacking a *thermodynamic* theory for "phase transitions" in non–equilibrium liquids, other approaches have to be considered in order to estimate the critical parameters. Consider to the rectilinear diameter law $(\rho_+ + \rho_-)/2 = \rho_\infty + b_0(T_\infty - T)$ which is a empirical fit extensively used for fluids in equilibrium. This, in principle, has no justification out of equilibrium. However, we found that our MC data nicely fit the diameters equation. We use this fact together with a universal scaling law $\rho_+ - \rho_- = a_0(T_\infty - T)^\beta$ to accurately estimate the critical parameters. The simulation data in Fig. 5 thus yields $\rho_\infty = 0.321(5)$, $T_\infty = 0.314(1)$, and $\beta = 0.10(8)$, where the estimated errors in the last digit are shown in parentheses. These values are confirmed by the familiar log–log plots. Compared to the equilibrium case [29], one has that $T_0/T_\infty \approx 1.46$. This confirms the intuitive observation above that the field acts in this system favoring disorder. On the other hand, our estimate for the order–parameter critical exponent is fully consistent with both the extremely flat coexistence curve which characterizes the equilibrium two–dimensional LJ fluids and the equilibrium Ising value, $\beta_{\text{Ising}} = 1/8$ (non–mean–field value).

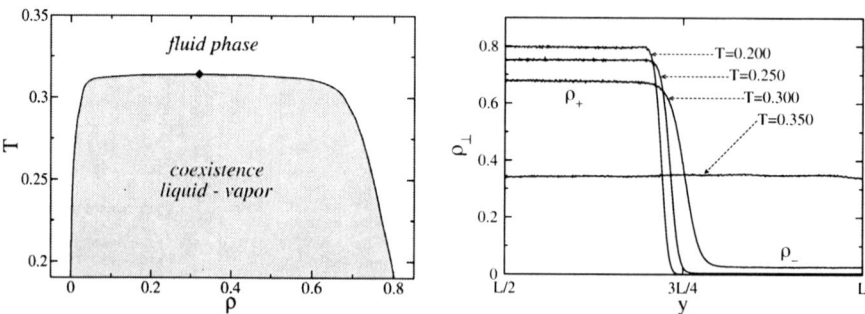

Fig. 5. The temperature–density phase diagram (left graph)was obtained from the transversal density profile (right graph) for $N = 7000$, $\rho = 0.35$, and different temperatures. The coexistence curve separates the liquid–vapor region (shaded area) and the liquid phase (unshaded area). The diamond represents the critical point, which has been estimated using the scaling law and the rectilinear diameter law (as defined in the main text).

Although the error bar is large, one may discard with confidence the DLG value $\beta_{\mathrm{DLG}} \approx 1/3$ as well as the mean field value. This result is striking because our model *seems* to have the symmetries and short–range interactions of the DLG. Further understanding for this difference will perhaps come from the statistical field theory.

4 Final comments

In summary, we reported MC simulations and field theoretical calculations to study the effect of discretization in *driven diffusive systems* In particular, we studied structural, transport, and critical properties on the *driven lattice gas* and related non–equilibrium lattice and off–lattice models. Interestingly, the present *Lennard–Jones* model in which particles are subject to a constant driving field is a computationally convenient prototypical model for anisotropic behavior, and reduces to the familiar LJ case for zero field. Otherwise, it exhibits some arresting behavior, including currents and striped patterns, as many systems in nature. We have shown that the additional spatial freedom that our fluid model possesses, compared with its lattice counterpart, is likely to matter more than suggested by some naive intuition. In fact, it is surprising that its critical behavior is consistent with the one for the Ising equilibrium model but not with the one for the *driven lattice gas*. The main reason for this disagreement might be the particle–hole symmetry violation in the driven *Lennard–Jones* fluid. However, to determine exactly this statement will require further study. It also seems to be implied that neither the current nor the inherent anisotropy are the most relevant feature (at least regarding criticality) in these driven systems. Indeed, the question of what are the most relevant ingredients and symmetries which determine unambiguously the universal properties in driven diffusive systems is still open. In any case, the above important difference between the lattice and the off–lattice cases results most interesting as an unquestionable nonequilibrium effect; as it is well known, such microscopic detail is irrelevant to universality concerning equilibrium critical phenomena.

We acknowledge very useful discussions with F. de los Santos and M. A. Muñoz, and financial support from MEyC and FEDER (project FIS2005-00791).

References

1. H. Haken: Rev. Mod. Phys. **47**, 67 (1975)
2. M.C. Cross, P.C. Hohenberg: Rev. Mod. Phys. **65**, 851 (1993)
3. R.G. Larson: *The Structure and Rheology of Complex Fluids* (Oxford University Press, New York 1999)
4. J. Dzubiella, G.P. Hoffmann, H. Löwen: Phys. Rev. E **65**, 021402 (2002)
5. P.M. Reis, T. Mullin: Phys. Rev. Lett. **89**, 244301 (2002)
6. P. Sánchez, M.R. Swift, P.J. King: Phys. Rev. Lett. **93**, 184302 (2004)
7. C.K. Chan: Phys. Rev. Lett. **72**, 2915 (1994)
8. Z. Csahók, C. Misbah, F. Rioual, A. Valance: Eur. Phys. J. E **3**, 71 (2000)
9. D. Helbing: Rev. Mod. Phys. **73**, 1067 (2001)
10. J. Hoffman, E.W. Hudson, et al.: Science **295**, 466 (2002)
11. J. Strempfer, I. Zegkinoglou, et al.: Phys. Rev. Lett. **93**, 157007 (2004)
12. U. Zeitler, H.W. Schumacher, et al.: Phys. Rev. Lett. **86**, 866 (2001)
13. B. Spivak: Phys. Rev. B **67**, 125205 (2003)
14. T.M. Liggett: *Interacting Particle Systems* (Springer Verlag, Heidelberg 1985)
15. V. Privman: *Nonequilibrium Statistical Mechanics in One Dimension* (Cambridge University Press, Cambridge 1996)
16. B. Schmittmann, R.K.P. Zia: 'Statistical Mechanics of Driven Diffusive Systems'. In: *Phase Transitions and Critical Phenomena, Vol. 17*, ed. by C. Domb and J.L. Lebowitz (Academic, London 1996)
17. J. Marro, R. Dickman, *Nonequilibrium Phase Transitions in Lattice Models* (Cambridge University Press, Cambridge 1999)
18. G. Ódor: Rev. Mod. Phys. **76**, 663 (2004)
19. T. Antal, G.M. Schütz: Phys. Rev. E **62**, 83 (2000)
20. J.L. Vallés and J. Marro: J. Stat. Phys. **43**, 441 (1986)
21. A.D. Rutenberg, C. Yeung: Phys. Rev. E **60**, 2710 (1999)
22. M. Díez–Minguito, P.L. Garrido, J. Marro: Phys. Rev. E **72**, 026103 (2005)
23. S.Katz, J.L. Lebowitz, H. Spohn: Phys. Rev. B **28**, 1655 (1983); J. Stat. Phys. **34**, 497 (1984)
24. A. Achahbar, P.L. Garrido, J. Marro, M. A. Muñoz: Phys. Rev. Lett. **87**, 195702 (2001); E.V. Albano, G. Saracco: Phys. Rev. Lett. **88**, 145701 (2002); *ibid.* Phys. Rev. Lett. **92**, 029602 (2004)
25. P.L. Garrido, J.L. Lebowitz, C.Maes, H.Spohn: Phys. Rev. A **42**, 1954 (1990)
26. A. Achahbar et al. (unpublished)
27. M. Díez–Minguito et al. (unpublished)
28. F. de los Santos, P.L. Garrido, M.A. Muñoz: Physica A **296**, 364 (2001)
29. B. Smit, D. Frenkel, J. Chem. Phys. **94**, 5663 (1991)
30. M. Allen, D. Tidlesley: *Computer Simulations of Liquids* (Oxford University Press, Oxford 1987)

Dune Formation

Hans J. Herrmann

Departamento de Física, Universidade Federal do Ceará
Campus do Pici, 60451-970, CE, Brazil
(on sabbatical leave from ICP, University of Stuttgart)
hans@icp.uni-stuttgart.de

Summary. Dunes are ubiquitous and exist in many forms in deserts and along coasts. They are a consequence of the wind moving sand grains by a mechanism called "saltation". In order to describe the formation and evolution of dunes one must understand the surface flux of sand. Using the equation of motion of turbulent air in the approximation of Jackson and Hunt for gentle hills one obtains a set of equations for dune motion. These equations reproduce very well field measurements. They also allow to study in detail the collision of dunes and the stability of dune fields since their solution is many orders of magnitude faster that real time observations.

1 Introduction

We all know the beautiful landscapes formed by dunes as for instance seen in Figure 1. They are a consequence of the forces exerted on the grains by the wind and the resulting particle flux. The first to systematically study airborne sand transport was the British brigadier R. Bagnold who, during the time of World War II did experiments in wind channels and field measurements in the Sahara. He presented the first expression for the sand flux as function of the wind velocity. Since then more refined expressions have been proposed. Bagnold also described for the first time the two basic mechanisms of sand transport: saltation and creep, and wrote the classic book on the subject which still is consulted very much [1].

If the ground is covered by sand and has no vegetation the sand flux on the surface modifies the shape of the landscape and spontaneously creates patterns on different scales: ripples in the range of ten to twenty centimeters and dunes in the range of two to two hundred meters. The change of the topography can be described by a set of coupled equations of motion which contain as variable fields the shear stress of the wind and the sand flux. These equations allow to explain among others the different dune morphologies, their velocity and their formation.

Fig. 1. Typical desert landscape showing the characteristic sharp edges separating the slip faces from wind driven regions.

In this article we will first introduce the properties of the turbulent wind field, then present the mechanisms of sand transport and then we will discuss dune formation.

2 The Wind

Air is a Newtonian fluid of density $\rho = 1,225 \mathrm{kgm}^{-3}$ and has a dynamic viscosity $\mu = 1.78 \times 10^{-5} \mathrm{kgm}^{-1}\mathrm{s}^{-1}$ which is defined as

$$\tau = \mu \frac{dv}{dz}$$

where τ is a small applied shear stress and $\frac{dv}{dz}$ the resulting velocity gradient. Its state is fully described by the velocity field $\mathbf{v}(\mathbf{r})$ and the pressure field $p(\mathbf{r})$ when we assume constant temperature and density. Its time evolution is given by the Navier-Stokes equations and the incompressibility condition. The solution of this equation is mainly characterized by the dimensionless Reynolds number defined through

$$Re = \frac{Lv}{\nu}$$

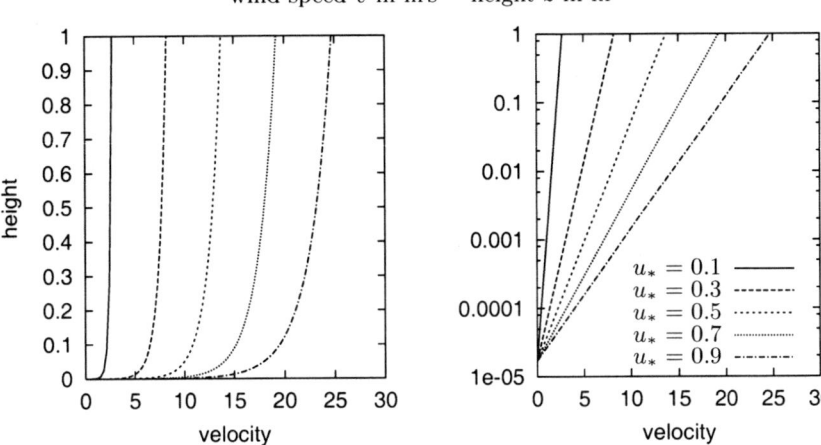

Fig. 2. Velocity profile of the atmospheric boundary layer above a surface with a roughness length $z_0 = 1.7 \ 10^{-5}$ m; left: linear scale, right: semi-log plot.

where $\nu = \frac{\mu}{\rho}$ is the kinematic viscosity. L and v are a characteristic length and a characteristic velocity of the problem as it could be given by the boundary conditions. Re represents the ratio of inertial forces to viscous forces. For low Reynolds numbers the flow is laminar. For high Reynolds numbers the flow is turbulent which means that there are strong spatial and temporal fluctuations on different scales all the time. This situation is typical outdoor even at moderate wind velocities due to the enormous size of the atmosphere. This complex behaviour arises from the fact that for large Re the Navier-Stokes equation is dominated by the non-linear inertia term.

The critical Reynolds number at which the atmospheric boundary layer becomes turbulent is in the order of 6000 [2]. Using the mixing length theory of Prandtl [4] one obtains the well known logarithmic profile of the atmospheric boundary layer illustrated in Figure 2.

$$v(z) = \frac{u_*}{\kappa} \ln \frac{z}{z_0}, \tag{1}$$

where z_0 denotes the roughness length of the surface and $u_* = \sqrt{\tau/\rho}$ the shear velocity. The shear velocity u_* characterizes the flow and has the dimensions of a velocity, although it is actually a measure of the shear stress. The roughness length z_0 is either defined by the thickness of the laminar sublayer for aerodynamically smooth surfaces or by the size of surface perturbations for aerodynamically rough surfaces.

The spatial and temporal fluctuations can be of small scale and high frequency and therefore it is generally too expensive to simulate them directly

Fig. 3. Velocity field in a longitudinal cut along the central slice of a Barchan dune. One clearly sees the flow separation and a large eddy that forms in the wake of the dune.

in practical applications. Instead, the Navier Stokes equations can be time–averaged or ensemble–averaged, or otherwise manipulated to remove the small scale dynamics, which results in a modified set of equations that are computationally more accessible.

One of these approaches for turbulence is the semi–empirical standard k-ϵ model [6] which is based on transport equations for the turbulent kinetic energy k and its dissipation rate ϵ. In the derivation of the k-ϵ model one assumes that the flow is fully turbulent, and the effects of molecular viscosity are negligible.

There are many programs, packages, and libraries available that have been developed to solve the turbulent Navier Stokes equation with different boundary conditions using the k-ϵ model. Nevertheless, three–dimensional turbulent flow on large scales is still a challenge and limited by the performance of processors and memory. We have chosen here the commercial code FLUENT V5.0 [7].

We show in Figure 3 the velocity field of the wind over a crescent-shaped obstacle which is in fact the topography of a real Barchan dune (see Figure 6) measured in Marocco. We see from the cut in wind direction Figure 3 that behind the dune an eddy of relatively low velocity is formed while the strong wind seems to follow above an imaginary continuation of the initial hill following the line $s(x)$ that delimits the eddy (separation line).

The three dimensional calculations using FLUENT are very time consuming from a computational point of view. It is not possible to use it in an iterative calculation as needed to follow the evolution of a dune where the surface and thus the boundary evolves in time. Furthermore, the theoretical understanding is limited by using such a "black–box" model.

A dune or a smooth hill can be considered as a weak modification of the surface that causes a perturbation of the air flow. An analytical calculation of the shear stress perturbation due to a two dimensional hill has been performed first by Jackson and Hunt [8]. Later, the work has been extended to three dimensional hills and further refined [9–13]. After a rather lengthy calculation they obtain for the Fourier transformation of the shear stress perturbation $\hat{\tau}_x$ in wind direction,

$$\hat{\tau}_x(k_x, k_y) = A \frac{h(k_x, k_y)k_x}{\sqrt{k_x^2 + k_y^2}} (k_x + iB|k_x|), \tag{2}$$

where $|k| = \sqrt{k_x^2 + k_y^2}$, $\gamma = 0.577216$ (Euler's constant) and A and B depend logarithmically on $\ln L/z_0$.

The non-local convolution integral term is a direct consequence of the pressure perturbation over the hill. The second local term is a correction that comes from the non–linearity of the Navier Stokes equation and represents the effect of inertia. The calculation of the shear stress of the air onto a smooth surface using eq. (2) is computationally very efficient. The limitation of this analytical formula is that it can only be used for surfaces with slopes having less than 30 degrees.

One way to treat this problem of flow separation is to divide the flow into two parts by the separating streamline $s(x)$ that reaches from the brink at which one has the flow separation to the ground.

The area enclosed by the separating streamline and the surface, called the *separation bubble*, a re–circulating flow develops,whereas the (averaged) flow outside is laminar as shown in Figure 3. The general idea, suggested by ref. [10], is that the air shear stress $\tau(x)$ on the windward side can be calculated using the envelope that comprises the dune and the separation bubble Figure 4. In Figure 4 we see the streamline calculated using FLUENT for a test dune that is modelled by a circle segment and a brink position ten meters before the maximum of this circle segment [14]. The dotted line is a fit using an ellipse segment.

3 Sand Transport

Sand consists of grains with diameters d which range from $d \approx 2\,\text{mm}$ for very coarse sand to $d \approx 0.05\,\text{mm}$ for very fine sand. The sand itself is mostly composed of quartz (SiO_2) which has a density ρ_{quartz} of $2650\,\text{kg}\,\text{m}^{-3}$. Dune sand

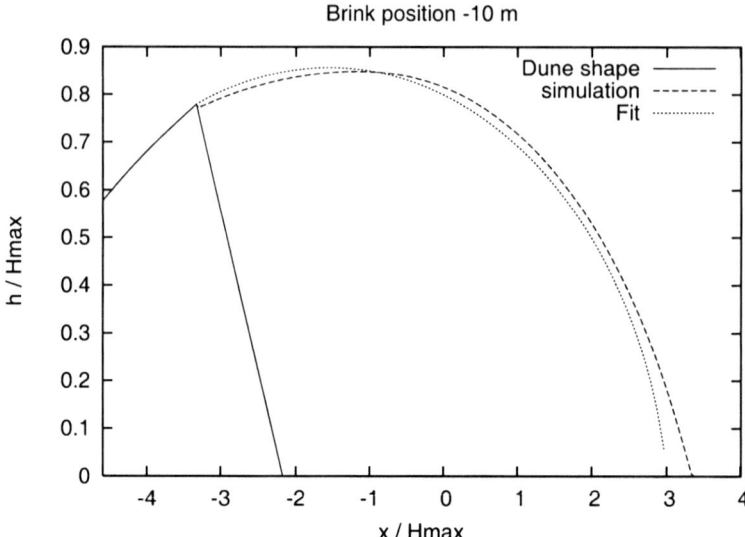

Fig. 4. Separation streamline as calculated with FLUENT (dashed line) and elliptic fit (dotted line) for a dune with a brink position ten meters before the maximum (from ref. [14]).

has a quite sharply peaked distribution of diameters because the transport produces a natural mechanism of size segregation.

One can also distinguish sand grains with respect to their shape [15]. An moving fluid such as air exerts a *drag force* F_d which acts in the direction of the flow. For turbulent flow it scales quadratically with the velocity due to Newton's drag law:

$$F_d = \beta \rho u_*^2 \frac{\pi d^2}{4},$$ (3)

where β is a phenomenological parameter. Gravity and inertia oppose the aerodynamic forces. Be $\rho' = \rho_{\text{quartz}} - \rho_{\text{air}}$ the reduced density of the sand grains in the air and the minimal shear stress required to move a grain called *aerodynamic entrainment threshold* $\tau_{ta} = \rho_{\text{air}} u_{*ta}^2$. This aerodynamic entrainment fluid threshold shear stress τ_{ta} on a flat surface is directly proportional to the reduced density ρ' and the diameter d of the grains. Shields [17] introduced a dimensionless coefficient Θ that expresses the ratio of the applied tangential force to the inertial force of the grain,

$$\Theta(Re_*) \equiv \frac{\tau_{ta}}{\rho' g d}.$$ (4)

Ref. [1] used the dimensionless Shields parameter Θ to define the fluid threshold shear velocity u_{*ta},

$$u_{*ta} = \sqrt{\Theta \frac{\rho' g d}{\rho_{\text{air}}}}. \tag{5}$$

This expression is only valid as long as cohesive and adhesive forces can be neglected and thus for grain diameters larger than 0.2 mm. The typical value for the fluid threshold shear velocity $u_{*ta} = 0.25\,\mathrm{m\,s^{-1}}$ is obtained for $d = 250\,\mu\mathrm{m}$ using $\Theta = 0.012$.

During sediment transport when sand grains are flying in the air, they impact onto the bed. The momentum transfer from an impacting grain to grains resting on the ground lowers the threshold for entrainment. This has already been observed by Bagnold [18] who called this lowered threshold *impact threshold* u_{*t}. The impact threshold shear velocity u_{*t} can be calculated in an analogous way and expressed by eq. (5) replacing the Shields parameter by an effective value $\Theta_{eff} = 0.0064$.

Different mechanisms of aeolian sand transport such as suspension and bed–load can be distinguished according to the degree of detachment of the grains from the ground. Bed–load transport can further be divided into saltation, reptation, and creep. Small grains are suspended in air and can travel long distances on irregular trajectories before reaching again the ground.

Saltation is the most relevant bed-load mechanism transport mechanism. To initiate saltation some grains have to be entrained directly by the air. This is called *direct aerodynamic entrainment*. The entrained grains are accelerated by the wind along their trajectory mainly by the drag force before they impact onto the bed again. The interaction between an impacting grain and the bed is called *splash process* and can produce a jet of grains that are ejected into the air. It is currently the subject of theoretical and experimental investigations [19, 20]. Finally, the momentum transferred from the air to the grains gives rise to a deceleration of the air. Due to this negative feedback mechanism saltation reaches a constant transport rate after some transient time.

Using again FLUENT it is possible to calculate the saltation layer on the grain level and obtain the loss of velocity of the wind due to the negative feedback for different heights as done in ref. [27]. An example is given in Figure 5 were we see that the velocity loss occurs mostly around a specific height, namely the one which is typical for the grain trajectories. One also notices that the loss is proportional to the amount of transported grains given through the particle flux q as illustrated in the inset.

When already flying in the air, aerodynamic forces and gravity act on the grain and determine its trajectory [1, 22–24]. The trajectory is close to that of a simple ballistic trajectory. More elaborate calculations [23, 26] have shown that the simple approximation using the ballistic formula gives values which are overstimated by 10–20%.

Measurements performed in wind tunnels [28, 29] show that sand flux starts at a threshold u_{*t} and scales with the cube of the shear velocity ($q \propto u_*^3$) for high shear velocities ($u_* \gg u_{*t}$). In the vicinity of the threshold the functional dependence is not well understood and empirical and theoretical flux

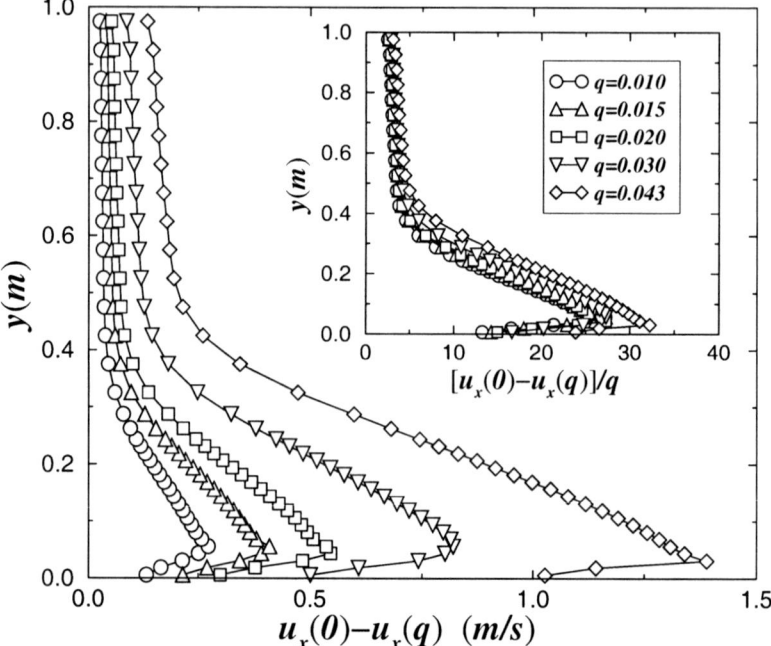

Fig. 5. Profile of the difference of the wind velocity without and with grain transport as function of the height for different fluxes q. The inset shows that this velocity loss scales with q (from ref. [27]).

relations differ considerably. In fact recent calculations using FLUENT yield a quadratic dependence of the form $q \propto (u_* - u_{*t})^2$ [27]. The simplest flux relation that predicts a cubic relation between sand flux q and shear velocity u_* was proposed by Bagnold [1],

$$q_B = C_B \frac{\rho_{\text{air}}}{g} \sqrt{\frac{d}{D}} u_*^3, \tag{6}$$

being d the real grain diameter and $D = 250\,\mu m$ a reference grain diameter. Later the threshold u_{*t} was incorporated into the sand flux relations in order to account for the fact that sand transport cannot be maintained below a certain shear velocity. Many phenomenological sand flux relations have been proposed and have been summarized for instance in ref. [15]. A sand flux relation that is widely used is the one by Lettau and Lettau [30],

$$q_L = C_L \frac{\rho_{\text{air}}}{g} u_*^2 (u_* - u_{*t}) \tag{7}$$

C_L being a fit parameter. Analytical calculations that predict the sand flux by averaging over the microscopic processes have deepened very much the understanding of aeolian sediment transport [22, 23, 31–33].

The relations of the form $q(u_*, ...)$ discused up to now assume that the sediment transport is in steady state, i.e. the sand flux is saturated. In order to overcome this limitation and to get information about the dynamics of the aeolian sand transport, numerical simulations based on the grain scale have been performed [21, 24, 34]. They showed that on a flat surface the typical time to reach the equilibrium state in saltation is approximately two seconds, which was later confirmed by wind tunnel measurements [28].

Assuming that each splash event produces on average the same number of ejected new particles the number of saltating grains would increase exponentially in time. Each accelerated grain, however, removes momentum from the wind field. Therefore after a saturation time T_s the flux must saturate to a value q_s. From this microscopic picture Sauermann et al. [35–37] have derived an equation describing this evolution of the flux towards saturation

$$\frac{\partial q}{\partial x} = \frac{1}{l_s} q \left(1 - \frac{q}{q_s} \right), \tag{8}$$

l_s being the "saturation length".

Let us emphasize that $T_s(\tau, u)$ and $l_s(\tau, u) = T_s u$ are not constant, but depend on the external shear stress τ of the wind and on the mean grain velocity u. We can relate the characteristic time T_s and length l_s of the saturation transients to the saltation time T and the saltation length l of the average trajectory of a saltating grain,

$$T_s = T \frac{\tau_t}{\gamma(\tau - \tau_t)}, \qquad l_s = l \frac{\tau_t}{\gamma(\tau - \tau_t)} \tag{9}$$

τ_t being the entrainment threshold shear stress and γ a constant. For typical wind speeds, the time to reach saturation is in the order of $2\,\mathrm{s}$ [21, 24, 34]. Assuming a grain velocity of 3–$5\,\mathrm{m\,s^{-1}}$ [25] we obtain a length scale of the order of $10\,\mathrm{m}$ for saturation. This length scale is large enough to play an important role in the formation of dunes.

4 The Complete Model

The lee side of a hill has the tendency to steepen. If the wind blows long enough from the same direction, the lee side will reach the angle of repose $\Theta \approx 34°$, which is the steepest stable angle of a free sand surface. If this angle is exceeded, avalanches start to slide down the hill until the surface has relaxed to a slope equal or below the angle of repose. In that case the responsible for sand flux is not the wind but gravity.

Without having to take into account the individual avalanches this effect can be implemented by redistributing the sand in such a way that the slip

Fig. 6. Field of Barchan dunes near Laâyoune, Morocco.

face is always a straight line with a slope corresponding to the angle of repose Θ. In two dimensions this is easy. In three dimensions this process, however is not straightforward. Bouchaud et al. (BCRE) [38] proposed a set of equations to describe avalanches which allows implementing locally and iteratively even in three dimensions the formation of surfaces having the angle of repose as their steepest inclination. Therefore these BCRE equations seem adequate to describe the dynamics of the slip face.

The complete model is defined by the three variable fields $h(x, y)$, $q(x, y)$ and $\tau(x, y)$. τ is calculated from h through the Fourier transformation of eq. (2). Then q is obtained from τ through eq. (8) using q_s from eq. (7) and l_s from eq. (8). The new topography h is then obtained from q using mass conservation:

$$\frac{\partial h}{\partial t} = \frac{1}{\rho_{\text{sand}}} \nabla_s q \tag{10}$$

In regions where $\nabla_s h > \tan\Theta$ slip occurs and the just mentioned BCRE equations are applied. ∇_s denotes the spatial derivative in direction of the strongest gradient. Once $h(x, y)$ is obtained one goes back to calculate again $\tau(x, y)$ etc. In this way one iteratively obtains the time evolution of the three fields.

The above system of iteratively solved coupled equations describes fully the motion of the free granular surface under the action of wind and gravity and can be used to calculate formation, evolution and shape of dunes. A natural consequence of the two different driving mechanisms, wind and gravity, is that the solution will separate in two regions: those for which the slope was larger than Θ and where therefore the BCRE equation was applied, ie the slip faces, and those where this was not the case. Theses two regions are separated by characteristic sharp edges which are the typical feature of sandy landscapes as seen in Figure 1.

5 The Motion of Dunes

Dunes are land formations of sand of heights, ranging typically from 1 to 500 meters which have been shaped by the wind. These topographical structures are found typically where large masses of sand have accumulated, which can be in the desert or along the beach. Correspondingly one distinguishes desert dunes and coastal dunes. Dunes can be mobile or fixed. Fixed dunes are older and are either "fossilized" which means transformed into a cohesive material, precurser to sand stone, or fixed because the average wind at their site over some period is zero. Otherwise the sand moves if the winds are strong enough, that means typically stronger than 4 meters a second.

As we all know, the beautiful landscapes (Fig. 6) formed by dunes are characterized by very gentle hills interruped by sharp edges called brink lines, delimiting regions of steeper slope, called slip-faces, lying in the wind shadow. Depending on the amount of available sand and the variation of the wind direction, one distinguishes different typical dune morphologies that have been classified by geographers into over 100 categories. The most well-known are longitudinal, transverse and Barchan dunes; other common dunes are star dunes, ergs, parabolic dunes and draas.

If the wind always comes from the same direction, one obtains transverse dunes if there is much sand and crescent shaped Barchan dunes (from a Turkish word) if little sand is available. Barchans exist in large fields in Marocco, Peru, Namibia etc. as seen in Figure 6). Their velocity ranges from 5 to 50 m per year and is inversely proportional to their height.

An interesting question about Barchans is their shape. In Figure 7 we see a longitudinal cut through the highest point of Barchans of different size normalized in such a way that they all have the same maximum [39].

On the windward side all curves fall on top of each other, while the crest lies more inwards for increasing dune height. The numerical solution of the equations of the last section reproduced precisely these profiles [40] but on top they do not have the uncontrollable fluctuations that come always from field data. Similarly also the transverse cuts scale with height and the numerical calculation also agrees with the observation.

One consequence of the above similarity relations is a well-known linear dependence between dune height, length and width as has been already reported by Bagnold [1]. Viewing the dune from the top, the brink has the shape of a parabola [39]. Due to the competition between the saturation length and the size of the separation bubble behind the dune one can calculate for the minimal height of a stable dune to be about 1,5 meters. The shear stress of the wind and the sand flux on a Barchan dune have also been measured and very favourably been compared to the numerical results of the equations [41].

With these computer dunes it has lately been shown [42] that when a small Barchan bumps into a larger one it can either be swallowed (if it is too small), or it can coalesce but produce at each horn a new baby Barchan (breeding), or it can, if the two initial dunes are of similar size separate again after some

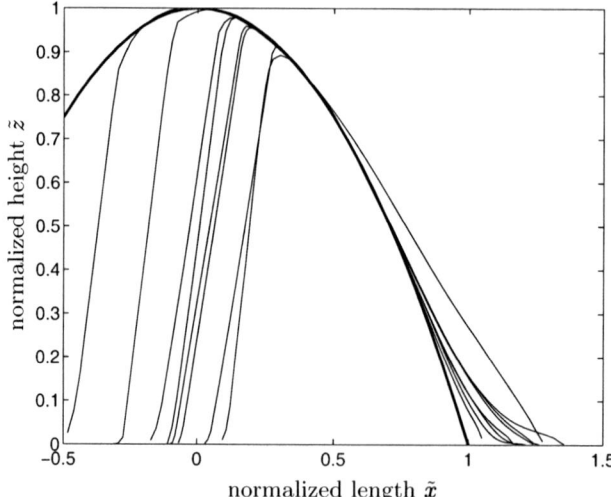

Fig. 7. Longitudinal profiles of eight dunes along their symmetry plane normalizing the length scales such that the shapes collapse on top of each other (from ref. [39]).

exchange of sand (solitary behaviour). In this last case, it looks as if the dunes do cross each other unaltered except for an eventual change in their size.

The system of equations of motion for dunes has also been used to calculate entire systems of dunes and virtual landscapes. An example is shown in Figure 8 where a constant influx of sand is used as boundary condition on the left while on the right one has a free boundary.

6 Conclusion

In this article we have shown that using known expressions for turbulent flow and using the transport mechanism of saltation, it is possible to formulate up a set equations of motion for a wind driven free granular surface. These three coupled equations containing as variable fields the shear stress of the wind, the sand flux at the surface and the profile of the landscape must be complemented by the BCRE equations in regions where the slope exceeds the angle of repose in order to correctly describe the slip faces. The resulting system of equations can be solved iteratively using appropiate boundary conditions and initializations. The solutions produce patterns that not only ressemble those observed in nature but also agree very well quantitatively with field measurements of shapes, sand fluxes and dune velocities.

The simulation of dune motion on the computer allows make predictions over long time scales since in the real world dune motion is very slow. One can also predict the effect of protective measures like the BOFIX-technique

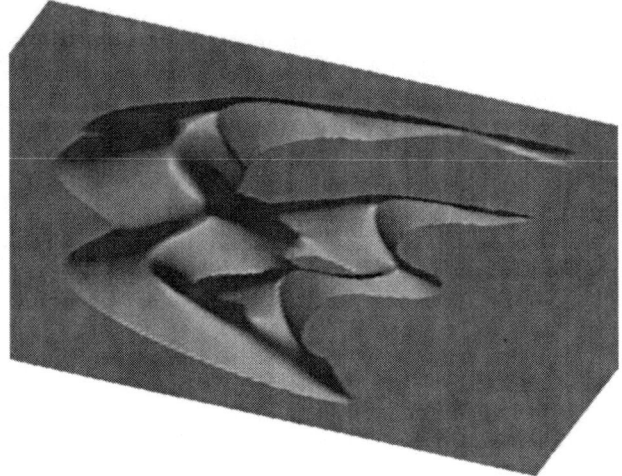

Fig. 8. Complex dune pattern, calculated with the full three dimensional model. Wind is blowing from the left to the right. When Barchan dunes are too close they interact, get eventually connected, and form complex dune structures. The large dunes are shielding the small dunes from the arriving sand flux which then constantly loose volume.

of Meunier [43] and even calculate the dunes on Mars [44]. Recently also the interaction with vegetation has been implemented [45]

References

1. Bagnold, R. A. (1941). *The physics of blown sand and desert dunes*. London: Methuen.
2. Houghton, J. T. (1986). *The physics of atmospheres*, Volume 2nd edn. Cambridge: Cambridge Univ. Press.
3. Kármán, T. (1935). Some aspects of the turbulence problem. *Proc. 4th Int. Congr. Appl. Mech. Cambridge*, 54–91.
4. Prandtl, L. (1935). The mechanics of viscous fluids. In W. F. Durand (Ed.), *Aerodynamic theory*, Volume Vol. III, pp. 34–208. Berlin: Springer.
5. Sutton, O. G. (1953). *Micrometeorology*. New York: McGraw–Hill.
6. Launder, B. E. and Spalding, D. B. (1972). *Lectures in Mathematical Models of Turbulence*. London, England: Academic Press.
7. Fluent Inc. (1999). Fluent 5. Finite Volume Solver.
8. Jackson, P. S. and Hunt, J. C. R. (1975). Turbulent wind flow over a low hill. *Q. J. R. Meteorol. Soc. 101*, 929.
9. Sykes, R. I. (1980). An asymptotic theory of incompressible turbulent boundary layer flow over a small hump. *J. Fluid Mech. 101*, 647–670.
10. Zeman, O. and Jensen, N. O. (1988). Progress report on modeling permanent form sand dunes. *Risø National Laboratory M-2738*.

11. Carruthers, D. J. and Hunt J. C. R. (1990). *Atmospheric Processes over Complex Terrain*, Volume 23, Chapter Fluid Mechanics of Airflow over Hills: Turbulence, Fluxes, and Waves in the Boundary Layer. Am. Meteorological. Soc.
12. Weng, W. S., Hunt, J. C. R., Carruthers, D. J., Warren, A., Wiggs, G. F. S., Livingstone, I. and Castro, I. (1991). Air flow and sand transport over sand dunes. *Acta Mechanica (Suppl.) 2*, 1-22.
13. Hunt, J. C. R., Leibovich, S. and Richards, K. J. (1988). Turbulent wind flow over smooth hills. *Q. J. R. Meteorol. Soc. 114*, 1435-1470.
14. Schatz, V. and Herrmann H.J. (2005). Numerical investigation of flow separation in the lee side of transverse dunes. preprint for Geomorphology.
15. Pye, K. and Tsoar, H. (1990). *Aeolian sand and sand dunes*. London: Unwin Hyman.
16. Chepil, W. S. (1958). The use of evenly spaced hemispheres to evaluate aerodynamic forces on a soil surface. *Trans. Am. Geophys. Union 39*, 397-403.
17. Shields, A. (1936). Applications of similarity principles and turbulence research to bed–load movement. Technical Report Publ. No. 167, California Inst. Technol. Hydrodynamics Lab. Translation of: Mitteilungen der preussischen Versuchsanstalt für Wasserbau und Schiffsbau. W. P. Ott and J. C. van Wehelen (translators).
18. Bagnold, R. A. (1937). The size–grading of sand by wind. *Proc. R. Soc. London 163*(Ser. A), 250–264.
19. Nalpanis, P., Hunt, J. C. R. and Barrett, C. F. (1993). Saltating particles over flat beds. *J. Fluid Mech. 251*, 661–685.
20. Rioual, F., Valance, A. and Bideau, C. (2000). Experimental study of the collision process of a grain on a two–dimensional granular bed. *Phys. Rev. E 62*, 2450–2459.
21. Anderson, R. S. (1991). Wind modification and bed response during saltation of sand in air. *Acta Mechanica (Suppl.) 1*, 21–51.
22. Owen, P. R. (1964). Saltation of uniformed sand grains in air. *J. Fluid. Mech. 20*, 225–242.
23. Sørensen, M. (1991). An analytic model of wind-blown sand transport. *Acta Mechanica (Suppl.) 1*, 67–81.
24. McEwan, I. K. and Willetts, B. B. (1991). Numerical model of the saltation cloud. *Acta Mechanica (Suppl.) 1*, 53–66.
25. Willetts, B. B. and Rice, M. A. (1985). Inter-saltation collisions. In O. E. Barndorff-Nielsen (Ed.), *Proceedings of International Workshop on Physics of Blown Sand*, Volume 8, pp. 83–100. Memoirs.
26. Anderson, R. S. and Hallet, B. (1986). Sediment transport by wind: toward a general model. *Geol. Soc. Am. Bull. 97*, 523–535.
27. Almeida, M.P., Andrade Jr, J.S. and Herrmann, H.J. (2005). Aeolian transport layer. *Phys.Rev.Lett.* in print, cond-mat/0505626.
28. Butterfield, G. R. (1993). Sand transport response to fluctuating wind velocity. In N. J. Clifford, J. R. French, and J. Hardisty (Eds.), *Turbulence: Perspectives on Flow and Sediment Transport*, Chapter 13, pp. 305–335. John Wiley & Sons Ltd.
29. Rasmussen, K. R. and Mikkelsen, H. E. (1991). Wind tunnel observations of aeolian transport rates. *Acta Mechanica Suppl 1*, 135–144.
30. Lettau, K. and Lettau, H. (1978). Experimental and micrometeorological field studies of dune migration. In H. H. Lettau and K. Lettau (Eds.), *Exploring the world's driest climate*. Center for Climatic Research, Univ. Wisconsin: Madison.

31. Ungar, J. E. and Haff, P. K. (1987). Steady state saltation in air. *Sedimentology 34*, 289–299.
32. Sørensen, M. (1985). Estimation of some eolian saltation transport parameters from transport rate profiles. In O. E. B.-N. et al. (Ed.), *Proc. Int. Wkshp. Physics of Blown Sand.*, Volume 1, Denmark, pp. 141–190. University of Aarhus.
33. Werner, B. T. (1990). A steady-state model of wind blown sand transport. *J. Geol. 98*, 1–17.
34. Anderson, R. S. and Haff, P. K. (1988). Simulation of eolian saltation. *Science 241*, 820.
35. Sauermann, G., Kroy K. and Herrmann H. (2001), A continuum saltation model for sand dunes. *Phys. Rev. E 64*, 31305.
36. Kroy, K., Sauermann G. and Herrmann H. J. (2002), A minimal model for sand dunes. *Phys. Rev. Lett. 88*, 054301.
37. Kroy K., Sauermann G. and Herrmann H. J. (2002), Minimal model for aeolian sand dunes *Phys. Rev. E 66*, 31302
38. Bouchaud, J. P., Cates, M. E., Ravi Prakash J., and Edwards S. F. (1994). Hysteresis and metastability in a continuum sandpile model. *J. Phys. France I 4*, 1383.
39. Sauermann, G., Poliakov, A., Rognon, P. and Herrmann, H. J. (2000), The shape of the Barchan dunes of southern Marocco, *Geomorphology 36*, 47-62.
40. Schwämmle, V. and Herrmann, H. J. (2003). A model of Barchan dunes including lateral shear stress, *EPJE 16*, 591-594.
41. Sauermann G., Andrade J. S., Maia L. P. Costa U. M. S., Araújo A. D. and Herrmann H. J. (2003), Wind velocity and sand transport on a Barchan dune, *Geomorphology 1325*, 1-11.
42. Schwämmle V. and Herrmann H. J. (2003), Budding and solitary wave behaviour of dunes *Nature 426*, 619-620.
43. Meunier J. and Rognon P. (2000), Une méthode écologique pour détruire les dunes mobiles, *Secheresse 11*, 309-316.
44. Ribeiro Parteli E.J., Schatz V. and Herrmann H.J. (2005), Barchan dunes on Mars and on Earth, *Powders and Grains 2005*,eds. R. Garcia-Rojo, H.J. Herrmann and S. McNamara (Balkema, Leiden, 2005), p.959-962.
45. Duran O. and Herrmann H.J. (2005) Dune mobility competing with vegetation, submitted to Nature.

Dynamics of Aeolian Sand Heaps and Dunes: The Influence of the Wind Strength

Sebastian Fischer[1,3] and Klaus Kroy[2,3]

[1] Physik Department, TU München, 85748 Garching, Germany
[2] Institut für Theoretische Physik, Universität Leipzig, Augustusplatz 10/11, 04109 Leipzig, Germany
[3] Hahn–Meitner Institut, Glienicker Straße 100, 14109 Berlin, Germany

Summary. The so-called *minimal model* provides an efficient minimum mathematical description of aeolian sand dune formation based on turbulent boundary layer calculations and a mean-field like saltation model. It allows us to analyze the effect of environmental conditions – uncontrollable in the field – on the characteristic shapes of dunes systematically. While the previously studied stationary solutions obtained under periodic boundary conditions are "unphysical" in the sense that they correspond to unstable fixed points of the equations, the solutions for open boundary conditions are shown to be strongly constrained by the unstable manifolds of these fixed points. For morphological evolution under periodically (e.g. seasonally) changing wind conditions a rule of thumb emerges, saying that the shapes of comparatively small/large dunes are slaved by the unstable manifolds pertaining to the actual/time-averaged environmental conditions, respectively.

From the interplay of turbulent air flow and erodible soil a plenitude of different shapes emerges. Among these, we will focus our attention on barchan dunes, which form wherever unidirectional winds prevail and sand supply is sparse. These dunes are crescent shaped with downwind pointing horns, have gentle upwind slopes and a slip-face dominating the downwind side, as sketched in Fig. 1(a). Whereas sand transport is of aeolian nature throughout the upwind face and the horns, it is due to gravity on the slip-face, where avalanches maintain the surface at the angle of repose. Barchan dunes commonly organize in large fields and present the most mobile dunes found on earth with migration velocities in the range of $20 - 70$m/yr and as such are a hazard in arid regions.

The turbulent flow field above a dune is essentially scale free, which leaves the dimensions of the sand grains as the only elementary length scale in the phenomenon of dune formation. With roughly six orders of magnitude separating typical dune lengths and sand grain diameters one would naturally expect dunes to be scale invariant. This is actually not the case. Careful measurements of barchan dunes in Morocco [1] (cf. Fig. 2) have provided clear

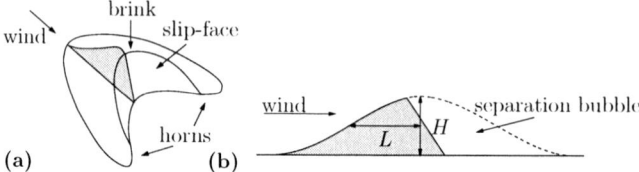

Fig. 1. (a) Sketch of a barchan dune. (b) Central slice of a barchan dune. Beyond the brink, inside of the so-called separation bubble the air flow near the ground is stagnant or even reverse. As a consequence, there is no wind erosion on the slip face. The height profile can be parameterized by height H and windward length L at half height of the common envelope of the dune profile and its separation bubble

evidence that longitudinal cuts along the symmetry line (shaded in Fig. 1) of different barchan dunes are not scale invariant. The solution to this puzzle is provided by a peculiarity of aeolian sand transport, which conveys the information on the grain size to the much larger scales of saltation trajectories. In particular, the so-called *saturation length* ℓ_s [2] is proportional to the grain size, but moreover dependent on the immersed density and on wind strength. The presence of this characteristic length, which is typically of the order of $10^{-2} - 10^{-1}$ dune lengths, is responsible for the observed scale dependent shape. Nevertheless, the windward profiles can be superimposed onto a master curve once lengths and heights are rescaled *independently*, i.e. instead of being self-similar, they are merely self-affine [1]. Accordingly, all windward profiles can be obtained from a universal scaling function, the shape function $f(x/L) \equiv h(x)/H$ (centered at the origin), where h denotes the envelope profile of the dune and its separation bubble as illustrated by the sketch in Fig. 1(b). In Fig. 2 we reproduce from [1] bare and rescaled data for dune profiles measured in Morocco. The characteristic height H and length L of each dune can be extracted by scaling the data onto the universal shape function. In Fig. 2(b), we compare the scaled data with the semi-empirical shape function $f(\xi) = \cos^{2.3}(0.74\xi)$ proposed in [3] (on the basis of numerical solutions of the minimal model; see below). The collapse of the upwind parts of the profiles is most convincing, while differences become pronounced on the downwind side where the position of the brink is shifted upwind with increasing size of the dunes. The fact that a more or less universal parameterization of longitudinal windward dune profiles by only two parameters works, justifies the convenient characterization of barchans in terms of e.g. height-length relations common in the geomorphological literature.

It is inevitable in field studies that environmental conditions such as the average wind strength and sand supply have to be accepted as they are, often with added difficulty to collect reliable data about them. Apart from that, the slow migration speeds render systematic measurements under fixed environmental conditions impossible. This is where a mathematical model of

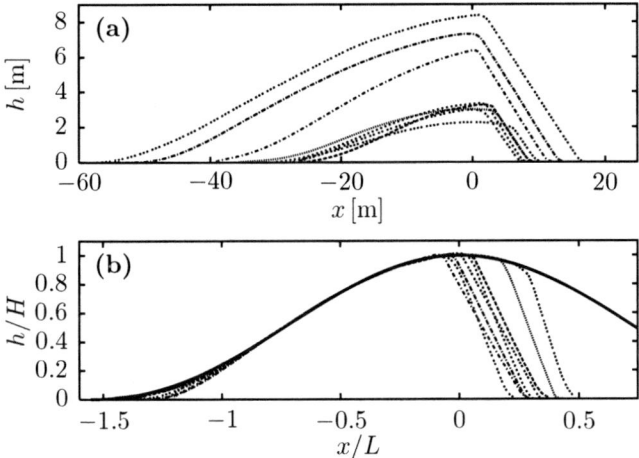

Fig. 2. (a) Longitudinal cuts along the symmetry line of barchan dunes measured in Morocco (adapted from [1]). Heights are in the range of $3 - 9$m and lengths $40 - 80$m. (b) Although dunes are not scale invariant, data collapse of the windward profiles can be achieved when heights and lengths are rescaled independently. The rescaling parameters H and L were obtained by fitting the height profiles to the shape function (*solid line*) mentioned in the main text

dune formation and dune migration can provide further insights. Historically, the interest in modeling aeolian sand transport and structure formation goes back to R. A. Bagnold, who published his seminal book *The Physics of Blown Sand and Desert Dunes* in 1941 [4]. Approximately 60 years later, sound understanding of the two major aspects of dune formation and propagation, namely turbulent air flow and aeolian sand transport, was achieved, so that a minimal model for the formation of barchan dunes could be formulated [5, 6]. In this development, the perturbative treatment of the stationary flow field above small hills due to Hunt and coworkers [7, 8], and the non-equilibrium sand transport model proposed by Sauermann *et al.* [2] play a key role. On the basis of the 2D minimal model the basic trend of the afore-mentioned morphological differences between barchans measured in the field could be explained.

In the following we shortly review the constituents entering the minimal model and investigate the stability of its stationary solutions with respect to perturbations of the wind strength. For constant sand supply, we present a qualitative discussion of the influence of a periodically changing wind strength on dune dynamics.

1 The Model

Two important variables govern the surface evolution of a dune (and of any erodible surface); the shear stress τ the wind exerts onto the surface and the flux q of sand grains carried along by the wind.[4] From the latter, changes of a 2D height profile h follow simply from mass conservation,

$$\partial_t h = -\varrho_{sand}^{-1} \partial_x q \,, \tag{1}$$

where ϱ_{sand} is the average density of the dune. On the time scale of surface evolution, the air flow as well as the sand flux adapt instantaneously to the height profile h. Hence stationary formulae for shear stress τ and sand flux q can be used in the model, leaving the height profile h as the only dynamic variable.

The shear stress τ follows from a perturbative calculation of the turbulent boundary layer above smooth hills, completed by a heuristic model for the separation bubble downwind of the brink, which smoothly extends the height profile over the slip-face [5–9] (Fig. 1). The stationary shear stress perturbation $\hat{\tau}[h'] := \tau/\tau_0 - 1$ caused by the apparent obstacle consisting of the dune and its separation zone, is a functional of the slope h' of the envelope,

$$\hat{\tau} = A \left[\int_{-\infty}^{+\infty} d\xi \, h'(\xi)/(\pi(x - \xi)) - Bh'(x) \right] \,. \tag{2}$$

Here τ_0 denotes the (unperturbed) reference shear stress, the parameters are $A = 3.2$ and $B = 0.25$ for the central slice of a barchan dune [6]. The shear stress perturbation $\hat{\tau}$ is essentially independent of the actual size of the dune, except for negligible logarithmic dependencies hidden in the prefactors A and B, and scales with its aspect ratio $\varepsilon = H/L$. Due to turbulent fluctuations the air flow over a symmetric hill is asymmetric, and the maximum shear stress is slightly displaced in the upwind direction with respect to the apex of the hill. This symmetry breaking by the turbulent flow is a crucial prerequisite for structure formation [6, 10].

Sand transport over a dune mainly takes place in saltation, i.e. sand grains move in consecutive jumps if the shear stress τ exceeds some threshold value τ_t. During their flight the grains extract momentum from the wind which they impart to the sand bed upon impact. The number of grains in the saltation cloud adjusts itself according to a feedback mechanism between acceleration of the grains by the wind and flow deceleration by the grains such that in equilibrium each impacting grain is replaced by one dislodged grain [11, 12]. The characteristic length scale for the equilibration of this feedback process is the aforementioned saturation length ℓ_s.

Since the wind strength above a dune varies locally, the sand flux incessantly has to adapt to changing wind conditions, and an equilibrium between

---

[4] Here we follow the notation found in the literature. In 2D, the sand flux q actually has dimensions of a sand transport rate per unit width [2].

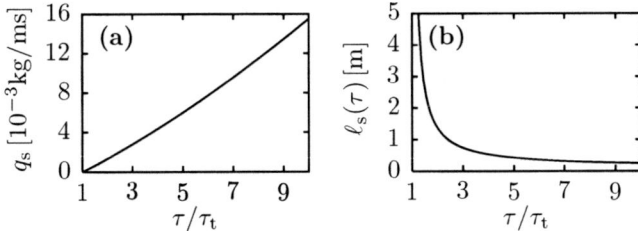

Fig. 3. Shear stress dependence of the kinetic coefficients q_s and ℓ_s of the sand flux equation (3). **(a)** The saturated flux q_s increases nearly linearly with the shear stress τ. **(b)** The saturation length ℓ_s sets the scale for saturation transients. It diverges when the shear stress τ approaches the threshold τ_t, indicating a slowing down of the equilibration near the threshold

the wind and the sand flux can never be established on the dune. This feature is captured by a continuum model, which calculates the stationary nonequilibrium sand flux $q[\tau]$ from the stationary shear stress τ according to

$$\ell_s \partial_x q = q(1 - q/q_s) .\tag{3}$$

The equilibrium or saturated sand flux $q_s(\tau)$ and the saturation length $\ell_s(\tau)$ are local functions of the shear stress $\tau(x)$, whereas the actual flux $q(x)$ is a retarded functional of $\tau(x)$ [2]. This results in a spatial lag of q with respect to q_s, which in competition with the symmetry breaking of the shear stress τ determines the morphology of stationary dunes [5, 6]. Since the shear stress τ drops below the threshold τ_t inside of the separation bubble, aeolian sand transport breaks down there. Sand crossing the brink settles over a short distance $l_{dep} \ll \ell_s$ and eventually avalanches down the slip-face. In the minimal model, those avalanches are considered to relax the surface instantaneously, the actual implementation follows the so-called BCRE model [13, 14].

2 Solutions

2.1 Stationary Solutions

Stationary solutions $h(x, t; \tau_0, q_{in})$ of (1) for a given pair of external parameters τ_0 and q_{in} correspond to shape invariantly moving dunes. Depending on whether the influx q_{in} is finite or vanishes, we distinguish two types of stationary solutions: *heaps* with continuous slopes throughout the whole profile form at finite influx, whereas *dunes* with a slip-face form at zero influx. The shape transition is a consequence of the saturation length ℓ_s fixing the length of stationary heaps. For any given reference shear stress τ_0, stationary heaps are enforced to steepen with increasing volume, and eventually develop a slip-face [5, 6]. That stationary dunes are obtained at vanishing influx only, is due to the minor technical simplification that transverse sand losses are dismissed as "second order" effects. (If desired, they could easily be accounted for by an analytical expression [3].) Rescaling heights and lengths by H and L, respectively, leads to a collapse of the windward profiles onto the universal shape function $f(\xi, t)$, which is independent of the reference shear stress τ_0 and the influx q_{in}. The profiles $h(x, t; \tau_0, q_{in})$ thus depend on the parameters τ_0 and q_{in} only implicitly through their characteristic heights H and lengths L. As anticipated in the introduction, this implies that a morphological analysis may be restricted to a reduced phase space.

2.2 Instability of Stationary Solutions with Respect to Wind Changes

To better understand the nature of the transient solutions of the minimal model, we start by studying the response of the equations of motion to perturbations of the stationary solutions of equation (1). For definiteness, we consider a stationary solution for external parameters $\tau_0/\tau_t = 2.0$ and $q_{in}/q_{s,0} = 0.4$, and perturb it by switching the reference shear stress to a higher/lower value, respectively, cf. Fig. 4. Here we introduced the abbreviation $q_{s,0} = q_s(\tau_0)$ for the saturated flux over an unperturbed sand bed. The numerical analysis shows that an augmentation of the shear stress to $\tau_0/\tau_t = 2.2$ at fixed influx ratio, which is equivalent to a shortening of the saturation length ℓ_s by roughly 15 percent, leads to enhanced erosion of the windward side of the heap. The length L of the heap decreases towards the stationary value corresponding to the new shear stress value $\tau_0/\tau_t = 2.2$. But since flux saturation is already reached somewhat upwind of the crest, sand is deposited on the top of the heap and the height H increases away from its reference value for the stationary solution. For a decreased reference shear stress $\tau_0/\tau_t = 1.8$ we observe reciprocal behavior; the saturation length ℓ_s increases, which results in decreased erosion of the windward foot but increased erosion of the top (since the flux q now takes longer to saturate). This results in the length L changing towards the stationary solution corresponding to $\tau_0/\tau_t = 1.8$ whereas the height H shrinks away from it. In summary, the stationary solutions — or fixed points — of the model equations are unstable with respect to changes of the wind strength; attractive with respect to length L but repulsive with respect to height H. As seen from Fig. 4(c) this translates into shape attraction, i.e. attraction with respect to the aspect ratio $\varepsilon = H/L$, while it leads to repulsion with respect to volume $V \propto HL$.

2.3 Growing and Shrinking Solutions

In the preceding paragraph we showed that stationary solutions are unstable with respect to perturbations of the wind strength. Augmentation of the

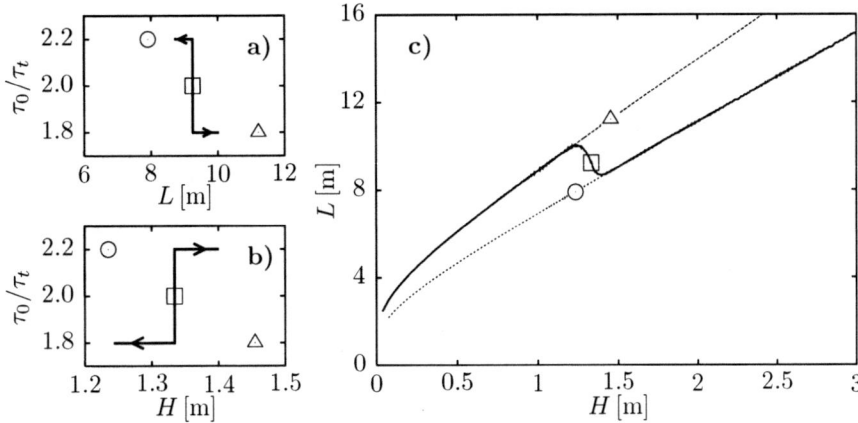

Fig. 4. Response of a stationary solution to a change of the reference shear stress. Open symbols denote stationary solutions with influx ratio $q_{in}/q_{s,0} = 0.4$ and reference shear stress $\tau_0/\tau_t = 1.8$ (*triangle*), $\tau_0/\tau_t = 2.0$ (*square*), and $\tau_0/\tau_t = 2.2$ (*circle*), broken lines the corresponding unstable manifolds. The solid lines are the numerically calculated trajectories with arrows indicating the direction of the dynamic evolution. (**a**) Initial response of the length L: upon changing the reference shear stress from $\tau_0/\tau_t = 2.0$ to $\tau_0/\tau_t = 2.2$ ($\tau_0/\tau_t = 1.8$) the length L is attracted towards the fixed point corresponding to the new shear stress value. (**b**) Initial response of the height H: a change of the reference shear stress leads to repulsion of the height H away from the new fixed point. (**c**) Except for an initial transient, both trajectories (*solid lines*) are attracted by the unstable manifolds of the new fixed points

reference shear stress leads to indefinitely growing heaps, whereas diminution results in heaps that shrink until they vanish. Figures 5 and 6 show snapshots illustrating the evolution of two such solutions obtained from a stationary heap for external parameters $\tau_0/\tau_t = 2.0$ and $q_{in}/q_{s,0} = 0.4$ by a quench of the reference shear stress to $\tau_0/\tau_t = 2.2$ and $\tau_0/\tau_t = 1.8$, respectively. A complete track record of the growth history is provided in Fig. 4(c) for the scale variables H and L. Interestingly, growth and shrinkage take place in a rather organized fashion: for any given reference shear stress τ_0 and influx q_{in}, trajectories starting from arbitrary initial shapes and volumes converge towards the unstable manifold of the corresponding fixed point [15]. In other words, growing or shrinking profiles run through a series of preferred shapes selected by the prevailing wind strength and sand supply. The dynamics is asymptotically governed by the unstable manifold of the fixed point, which considerably reduces the a priori high dimensional state space. In particular, any morphological evolution under fixed wind conditions is bound to follow these unstable manifolds.

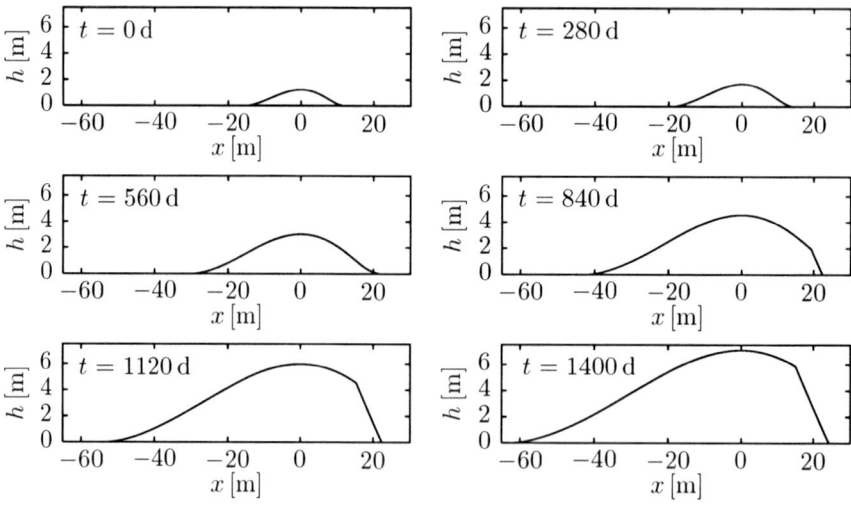

Fig. 5. Snapshots of a growing heap for reference shear stress $\tau_0/\tau_t = 2.2$ and influx $q_{in}/q_{s,0} = 0.4$. With increasing volume, the heap steepens and eventually becomes a dune with a slip-face

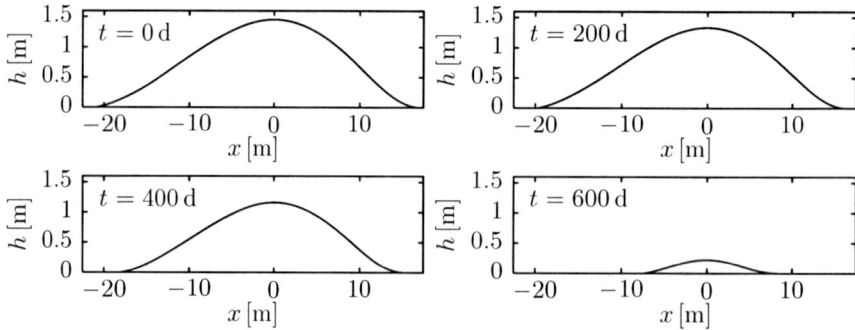

Fig. 6. Snapshots of a shrinking heap for reference shear stress $\tau_0/\tau_t = 1.8$ and influx $q_{in}/q_{s,0} = 0.4$. With increasing distance from the fixed point the volume decreases faster until the heap is finally completely eroded

2.4 The Influence of Periodic Wind Variations

In the field, one often encounters periodically changing wind conditions which might occur for example seasonally or daily. In this paragraph we provide a qualitative discussion of both scenarios. While changing the reference shear stress we keep the influx ratio $q_{in}/q_{s,0}$ fixed. Figure 7 shows numerical results obtained for fixed influx ratio $q_{in}/q_{s,0} = 0.4$ and reference shear stress periodically switching every 180 days between $\tau_0/\tau_t = 2.2$ and $\tau_0/\tau_t = 1.8$. The volume of the initial heap is chosen such that it grows in time. The unstable manifold of the fixed points corresponding to the two wind speeds alternately attract the trajectory. For small volume, the trajectory switches rapidly from one unstable manifold to the other, i.e. the dune profile can adapt rapidly to the new "shape attractor". With growing volume, the relaxation process takes increasingly longer due to the increasing amount of sand that has to be moved to achieve the required shape change. An indicative reference time scale is provided by the time needed to completely rebuild the dune, which is of the order of $\varrho_{sand} V / q_{s,0}$. Large dunes cannot adiabatically adapt their shapes to the shape attractor corresponding to the new environmental conditions. Their shapes are rather the result of an effective time averaged environmental condition. Thus their trajectories in the reduced phase space of Fig. 7, follow asymptotically the unstable manifold of the average reference shear stress, in our example given by $\tau_0/\tau_t = 2.0$.

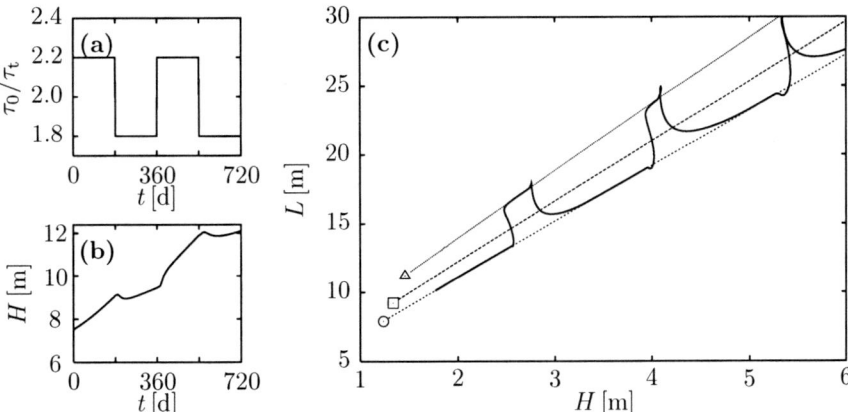

Fig. 7. Evolution of a growing dune subject to periodically changing wind strength. (a) Shear stress protocol. The reference shear stress is changed every 180 days from $\tau_0/\tau_t = 2.2$ to $\tau_0/\tau_t = 1.8$ and vice versa while the influx ratio $q_{in}/q_{s,0} = 0.4$ is kept fix. (b) Evolution of the height H. (c) Reduced phase space trajectory of a growing heap recorded for three cycles (*solid line*). Broken lines represent the unstable manifolds of the fixed points corresponding to the alternating environmental conditions $\tau_0/\tau_t = 1.8$ (*triangle*) and $\tau_0/\tau_t = 2.2$ (*circle*), and to the time averaged wind shear stress $\tau_0/\tau_t = 2.0$ (*square*), respectively. As long as the dune volume is small, the trajectory switches rapidly between the unstable manifolds

The asymptotic regime is more conveniently studied for a small heap subject to a more frequent change of the reference shear stress, e.g. every 12 hours, as shown in Fig. 8. The reaction of the shape to the rapidly alternating environmental conditions has diminished to tiny oscillations around the unstable manifold corresponding to the average reference shear stress $\tau_0/\tau_t = 2.0$.

One may certainly expect deviations from this idealized behavior for larger shear stress differences and in particular if the lower shear stress falls close to (or locally even below) the threshold for sand transport, so that the nonlinear dependence of the saturation length ℓ_s on the reference shear stress becomes important. However, the two idealized scenarios provide a powerful rule of thumb giving a rough classification for the shape evolution of dunes in the field and are helpful in understanding some of the apparently less systematic shape variations of measured dune shapes with dune size (cf. Fig. 2): The shapes of small dunes follow adiabatically the unstable manifold corresponding to the actual environmental conditions, while the shapes of large dunes reflect essentially the long time averaged environmental conditions.

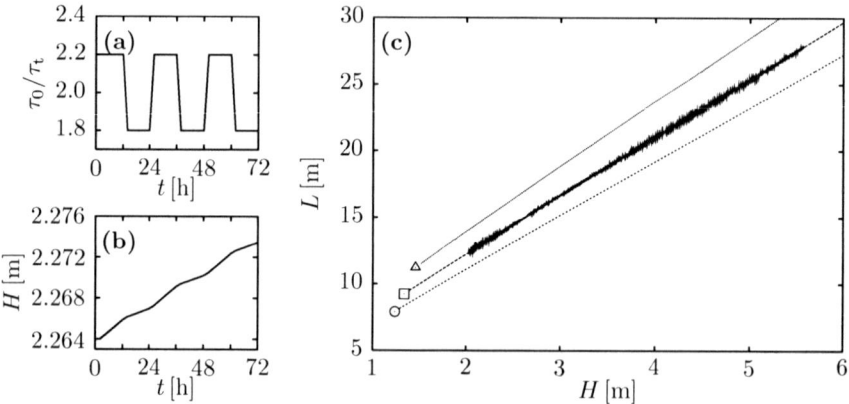

Fig. 8. The same as in Fig. 7 but for a periodic change of the reference shear stress every 12 hours (**a**). The periods of alternating winds are so short that the trajectory averages over both unstable manifolds rather than relaxing onto them (**c**). Asymptotically, it follows the course of the unstable manifold corresponding to the average reference shear stress $\tau_0/\tau_t = 2.0$

3 Conclusions

Previous systematic studies of the minimal model of aeolian sand dunes were mostly restricted to the simple but unstable stationary solutions obtained under constant wind strength and periodic flux conditions, i.e. constant and equal in- and outflux. Allowing for open boundary conditions and varying environmental conditions may at first sight seem tantamount to opening a Pandora box of complex non-equilibrium behavior. However, as we have demonstrated above, due to a tight organization of the infinite dimensional phase space of non-equilibrium dune shapes by a low dimensional set of shape attractors, the possible *generic* non-equilibrium shape evolutions are highly constrained and the actual complexity is much lower than one might have expected. As a paradigmatic example pertinent to field studies, we have numerically analyzed the growth of small heaps and large dunes subject to a daily and seasonally alternating wind strength, respectively. The emerging picture can be summarized by a rule of thumb saying that the shapes of small dunes are slaved by the unstable manifolds of the fixed-point solution for the actual environmental conditions, while the shapes of large dunes are basically determined by corresponding shape attractors pertaining to the averaged environmental conditions.

References

1. G. Sauermann, P. Rognon, A. Poliakov, H. J. Herrmann: Geomorphology **36**, 47 (2000)
2. G. Sauermann, K. Kroy, H. J. Herrmann: Phys. Rev. E **64**, 031305 (2001)
3. K. Kroy, S. Fischer, B. Obermayer: J. Phys.: Condens. Matter **17**, 1229 (2005)
4. R. A. Bagnold: *The physics of blown sand and desert dunes*, (Methuen, London 1941)
5. K. Kroy, G. Sauermann, H. J. Herrmann: Phys. Rev. Lett. **88**, 054301 (2002)
6. K. Kroy, G. Sauermann, H. J. Herrmann: Phys. Rev. E **66**, 031302 (2002)
7. P. S. Jackson, J. C. R. Hunt: Q. J. R. Meteorol. Soc. **101**, 929 (1975)
8. J. C. R. Hunt, S. Leibovich, K. J. Richards: Q. J. R. Meteorol. Soc. **114**, 1435 (1988)
9. O. Zeman, N. O. Jensen: *Progress report on modeling permanent form sand dunes*, (Technical Report M-2738, Risø National Laboratory 1988)
10. B. Andreotti, P. Claudin, S. Douady: Eur. Phys. J. B **28**, 341 (2002)
11. *Aeolian Grain Transport 1 & 2*, ed. by O. E. Barndorff-Nielsen, B. B. Willetts, (Acta Mechanica Suppl., Springer, Wien, New York, 1991)
12. R. S. Anderson, P. K. Haff: Science **241**, 820 (1988)
13. J. P. Bouchaud, M. E. Cates, J. Ravi Prakash, S. F. Edwards: J. Phys. France I **4**, 1383 (1994)
14. J. P. Bouchaud: 'A phenomenological model for avalanches and surface flows'. In: *Physics of dry granular media*. ed. by H. J. Herrmann, J.-P. Hovi, S. Luding, (NATO-ASI Series, Kluwer academic publishers, Dordrecht 1998)
15. S. Fischer, K. Kroy, in preparation

Granular Shearing and Barkhausen Noise

Andrea Baldassarri[1], Fergal Dalton[2], Alberto Petri[2], Luciano Pietronero[1,3], Giorgio Pontuale[2], and Stefano Zapperi[1,3]

[1] Dipartimento di Fisica, Università "La Sapienza", P.le A. Moro 2, 00185 Roma, Italy
[2] Istituto dei Sistemi Complessi - CNR, via del Fosso del Cavaliere 100, 00133 Roma, Italy
[3] Istituto dei Sistemi Complessi - CNR, Via dei Taurini 19, 00185 Roma, Italy

Summary. We report the result of new experiments on a granular medium sheared in a Couette geometry. We investigate in particular the dependence of the resulting stick-slip patterns on the imposed shear rate. A model equation based on a stochastic description of the internal forces of the granular medium allows us to recover the experimental results and unveils similarities between the stick slip motion and the Barkhausen noise emitted by a ferromagnet during a hysteresis cycle. We show that, because of the stochastic nature of forces in the granular medium, there is indeed a correspondence in the statistical properties between shear rate fluctuations in granular media and displacement of magnetic domain walls under a varying external field. The main difference between the two systems consists of a characteristic behavior on the part of the granular medium, but which is not exhibited by the Barkhausen effect. The stochastic model proposed allows us to ascribe this behavior to the moment of inertia of the plate and, on the basis of available data from the statistical properties of friction in solid-on-solid systems and in polymer mono-layers, suggests the possibility of describing a larger class of driven instabilities in terms of similar general mechanisms.

1 Introduction

A remarkably broad variety of fundamental and applied problems in the physical, life, and materials sciences originate from an interplay between disorder and non-equilibrium dynamics. Given the significance of fluctuations and noise in practically every branch of science and technology, an interdisciplinary approach based on the cross-fertilization of materials science with techniques and ideas drawn from the physics of fluctuations would have an immense potential in the understanding and development of new and exciting areas of modern physics. As an example, fluctuations play a role in material synthesis since deposition and growth are stochastic processes and typically involve the nucleation of defects and other disordered structures (clusters, inclusions, rough surfaces). Similarly disorder affects the material during device operation and typically gives rise to noise, leading eventually to failure. While noise certainly represents an experimental nuisance in many cases, it also provides a well established method for material characterization. Examples include the Barkhausen noise in ferromagnetic materials, voltage noise in conductors and acoustic emission in deformed solids.

Granular matter is utilized at some stage in the production of almost all products which exist today, and yet the complexity they exhibit is understood only in the most general conditions, despite its acknowledged importance in the scientific community. In fact, the shear response of granular media has been investigated for more than a century in different scientific frameworks [1], ranging from soil mechanics [2] and earth sciences [3, 4] to physics [5, 6]. A large variety of experimental settings and theoretical methodologies, or direct particle simulations [7, 8], have been applied to the problem in the search of general relationships such as rate and state friction laws [4, 9]. However, it has been observed that under a slow loading rate, the shear response of granular media typically displays large fluctuations [10–12], with regimes in which motion is intermittent and erratic, the so-called stick-slip phase. The necessity of obtaining a mechanical description to be used in application has pushed most of the past theoretical activity to the analysis of averaged properties, in the search for constitutive macroscopic laws. On the contrary, a detailed analysis of slip statistics is of primary importance if one wishes to understand the mechanisms which are at the base of the wide fluctuations observed. They are, for example, the distinguishing feature of "crackling noise", which denotes an intermittent activity with widely fluctuating amplitudes [13]. This phenomenon is observed in different contexts such as magnetic materials [14, 15], ferroelectrics, type II superconductors [16], fracture [17, 18] and plasticity [19, 20] to name just a few. In addition it shares analogies with earthquake phenomenology, where the statistics typically displays wide fluctuations and recorded sizes of seismic events span several decades [21, 22].

Common aspects observed in such apparently different systems have suggested the existence of a deeper correspondence in the underlying physical properties [13], and one may hope that any general insight or understanding gained in these phenomena may guide research in other fields. In this article, starting from a laboratory experiment on sheared granular matter, we discuss a model which quantitatively reproduces the observations and, at the same time, is based on a simple but intuitive hypothesis which sheds light on the similarities between different phenomena. Indeed, we show that, because of the stochastic nature of forces in the granular medium, there is a correspondence in the statistical properties between shear rate fluctuations in granular media and domain wall velocities in ferromagnets (Barkhausen effect).

Fig. 1. A photograph of the experimental set-up

2 Granular Experiment and Dynamic Statistics

The adopted experimental setup consists of a circular Couette cell filled with mono-disperse glass beads. An annular plate is driven over the surface of the cell by a motor *via* a torsion spring, exerting a shear stress on the granular medium. A picture of the experimental apparatus is shown in Fig. 1 and a detailed description of the setup is provided in [23]. As the spring winds up, the torque on the plate increases until the plate slips. We focus our analysis on the stick-slip regime, observed at low driving angular velocities when the system cannot support a steady sliding regime. The apparatus allows us to measure the deflection of the torsion spring and the angular position θ of the plate as functions of time, from which we obtain the instantaneous position of the plate and its derivatives. The reaction torque F of the medium can be derived by the equation of motion:

$$I\ddot{\theta} = k(\omega t - \theta) - F, \tag{1}$$

where I is the moment of inertia of the disk, k is the stiffness of the torsional spring, and ω is the angular velocity of the driving motor. The first term on the right hand side is simply the force exerted by the spring. The second term describes the "friction" F exerted by the medium and counteracting the motion of the plate. This force being very irregular, the typical instantaneous

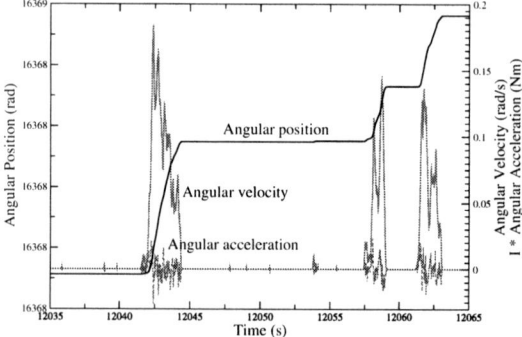

Fig. 2. A sample time series of angular position, velocity and acceleration. The portion of signal between two consecutive zeros of the velocity defines a slip event, with the associate time duration and size

velocity signal displays a very erratic behavior with pulses of widely fluctuating magnitudes as those reported in Fig. 2. From similar previous experiments its probability distribution is expected to obey a power-law [11, 12].

The present experiment allows to observe that also other physical quantities exhibit asymmetric statistical distribution with long tail. For example the static and dynamic torque [24]. But in order to get a more complete characterization of the system from the statistical point of view more quantities can be investigated. Typical quantities employed in crackling noise are the size S and duration T of the events. For the granular medium a slip event is defined as the portion of signal between two consecutive zeros of the plate velocity (see Fig. 2). As an example some distributions of duration T are shown in Fig. 3. They are found to display a monotonic decrease at small scales for different values of the driving velocity ω, but followed sometimes by a characteristic peak at larger scales. It will be seen below that the presence and the position of the peak are related to the moment of inertia of the driving plate. Such a feature is absent in the aforementioned Barkhausen noise since in that case the motion of the domain walls is over-damped.

3 Proposed Model Equation

The instantaneous frictional torque F of (1), which acts as a friction term, may depend in principle on the angular position, its derivatives, and on a variety of state variables [3], including memory terms. This complexity arises from the disordered arrangement of grains, where forces propagate via a complicated network with both elastic and frictional interactions playing fundamental roles [25]. The network changes in time in a pseudo-plastic way [26],

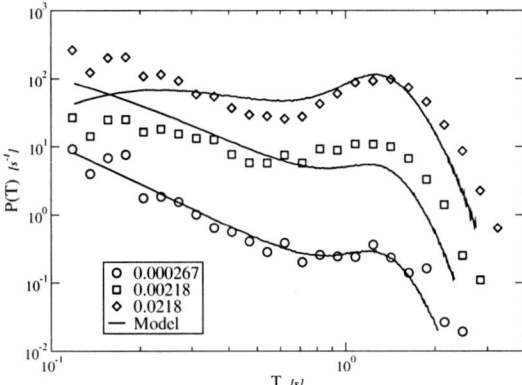

Fig. 3. Some distributions of the slip duration T from the experimental data (points) and from the stochastic model equation discussed in the text (lines). The legend reports the drive angular velocity (rad/s)

making the detailed description of the resulting torque a formidable problem with a huge number of dependent degrees of freedom.

The above considerations suggest to considering F itself as a random quantity and to implement a Brownian scheme for tackling the problem [27]. In the Brownian motion the medium reaction to the particle motion is contained into two terms: one term is of the deterministic type, and represents the medium average viscosity; the second term is completely stochastic and just depends on time. This simplified approach which yields a very effective description of the diffusion of a particle in a liquid, is found to yield as well a simplified but highly effective description of the non-equilibrium system under consideration. In the present case the adoption of a Brownian force hypothesis is supported from the observation that the average behavior of F as function of the instantaneous plate velocity v behaves very smoothly and, worth noticing, has a shape similar to that found in the steady sliding regime in some dry solid-on-solid friction experiments [28]. In the present case it is well fitted with the function (see e.g. Fig. 5 in Ref. [27]):

$$\bar{F}(v) = F_0 + \gamma(v - 2v_0 \ln(1 + v/v_0)),\qquad(2)$$

where F_0 is the average static friction torque, v_0 corresponds to the minimum in the average torque and γ is the velocity damping coefficient. The fact that similar functions serve for both systems constitutes a confirmation that similar macroscopic general behaviors can be shared by systems that are quite different microscopically.

Thus the force is split F into two additive terms $F = \bar{F}(\dot{\theta}) + F_f$: In the first term the dependence on velocity is confined to a deterministic law, analogous to the viscous reaction of a fluid, while the second term is independent of the velocity and introduces the fluctuating part. The properties of F_f being un-known, we proceed to investigate its power spectrum, which bears information on the correlation properties. Since the motion of the plate is intermittent, it seems more natural to choose the angular position θ instead of the time as parameter, in order to obtain a description as simple as possible. The power spectrum of the fluctuating part is shown in Fig. 4 and its shape indicates that the random force is not a white noise, but is correlated. This can be in-terpreted as a direct consequence of the disordered structure of the force chain network [29, 30] present in the granular medium: As the disk slips by a small angle $\delta\theta$, the friction torque changes by a random amount δF_f which repre-sents a fraction of the total torque. This is because for small displacement the rearrangement in the granular structure is limited. On the other hand, under subsequent, or large, slips, the rearrangement of the structure will be more complete, and the fluctuations, increasing linearly for small angles, will decay according to some cut-off function g: $\langle (F(0) - F(\theta))^2 \rangle \propto \theta \cdot g(\theta/\theta_0)$, with $\lim_{\theta \to 0} g = 1$ and $\lim_{\theta \to \infty} g = 0$ In order to decide which kind of func-tion best fits the experimental spectrum a more detailed analysis would be necessary. In any case the above facts can be expressed mathematically by assuming that the force itself performs a bounded random walk

$$\frac{dF_f}{d\theta} = \eta(\theta) - f(F_f), \tag{3}$$

where $f(F_f)$ is some bounding function: $f(F_f) > 0$. The simplest choice is a linear force: $f(F_f) = aF_f$ ($F_f > 0$). Thus a is the inverse correlation length and η is an uncorrelated stochastic process with variance D, that is

$$\langle \eta(\theta)\eta(\theta') \rangle = D\delta(\theta - \theta').$$

By choosing a Gaussian distribution for η, the above equation describes an Ornstein and Uhlenbeck process with Lorentzian power spectrum, which seems to approximate quite well the experimental one:

$$S(k) = \left\langle \left| \int d\theta F_f(\theta) \exp(-i\theta k) \right|^2 \right\rangle = \frac{2D}{a^2 + k^2}. \tag{4}$$

4 Parameter Dependence of the Distribution Peak

Despite some crude approximations, the model described in the previous sec-tion seems successful in capturing the main statistical features of the stick-slip motion observed in the experiments and in reproducing the broad distribution characterizing the slip events [27].

Figure 3 displays the distributions of the event duration for different driv-ing velocities and the corresponding distributions as obtained by the model.

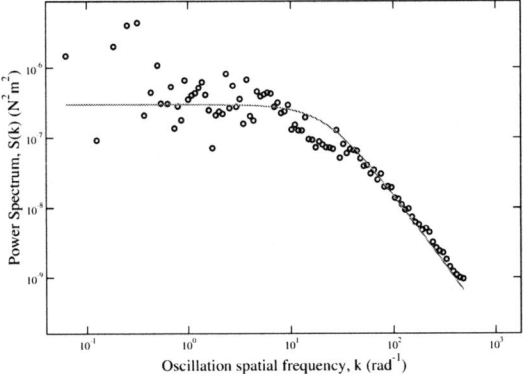

Fig. 4. The power spectrum of the fluctuating part of the force $F_f(\theta)$

It is worth noticing that the latter are not curve fit to the distributions. They are probability distributions of the data generated by the model, (1), whose parameters I, k, ω are fixed by the experiment, the parameters F_0, γ, v_0 are derived from the fit of the friction law (2), and a is the torque correlation length determined experimentally (Fig. 4).

One notable feature of the model is that it allows us to clarify the peak displayed by these and by the size distributions [27] and which is not present in the case of the Barkhausen noise. In fact by dropping out the fluctuating

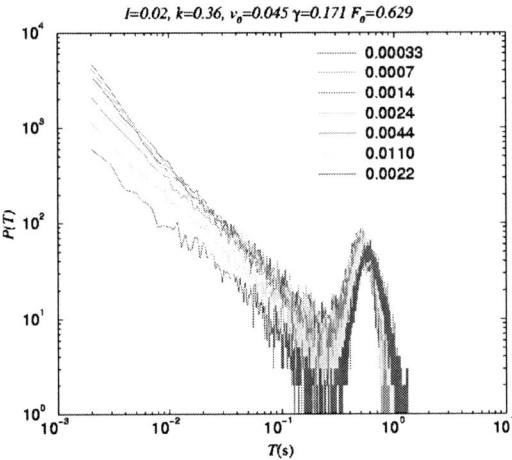

Fig. 5. The probability distributions for the slip duration as obtained by the stochastic model for different values of the driving velocity

part of the force $F_f(\theta)$ in

$$I\ddot{\theta} = k(\omega t - \theta) - \bar{F}(\dot{\theta}) - F_f(\theta) , \tag{5}$$

an almost periodic motion of the plate is obtained, the probability distribution of event duration becoming delta function whose location is proportional to $\sqrt{(I/k)}$. On the other hand if the inertial term is neglected the motion is over-damped and power laws are obtained, reproducing the leftmost part of the observed distributions and thus showing that the presence of the peak is related to the inertia of the plate. Actually, the peak is enhanced also by the instability in the friction law. Figure 5 shows the behavior of the duration distribution $p(T)$ from the simulation of the model with different driving velocities. As expected from the experiment the dependence of the peak position on the driving velocity is very weak. On the contrary, it changes according with the square root of the inverse spring constant (Fig. 6). One task of forthcoming experiments will be to change the inertia of the plate to check how peak location varies[4].

5 Analogies with Different Systems

As stated in the introduction the proposed approach not only allows to explain some of the observed behavior, but strengthens the possibility of finding common description for phenomena observed in unrelated fields, e.g. as unrelated as the granular from magnetic materials. In fact an equation of motion similar to (5) describes a different "crackling noise", the Barkhausen noise. The Barkhausen noise is the (magnetic) noise emitted by ferromagnets during an hysteresis cycle due to the magnetic domain motion. As the external magnetic field changes, domains with the direction of internal field in agreement with the external fields grow, while opposite domain shrink. Fluctuations in the domain wall motion generate the Barkhausen noise which, although known since almost a century [14], received a quantitative description only in the last decades [15] by means of a stochastic equation similar to (5). In fact the generic cartesian coordinate x for the motion of the domain wall position can be described by [15]:

$$\Gamma\frac{dx}{dt} = H_a + H_d + H_p . \tag{6}$$

Here the left hand side is the damping, H_a is the applied external field, H_d is the demagnetizing field of the material and $H_p(x)$ describes the pinning of the domain walls by defects at random positions. If the inertial term is included, this equation generalizes to the same form as (5), with a correspondence among physical quantities given in Table 1:

---

[4] Actually the value of I which yields results closer to the experimental ones is larger than the true value of about 50%. We believe this to be due to some layers of grains moving together with the plate during its motion and corresponding to three or four layers of grains.

Granular	$\bar{F}(\dot\theta)$	ωt	$-k\theta$	$F_f(\theta)$
Ferromagnet	$\Gamma\dot{x}$	H_a	H_d	$H_p(x)$

Table 1. Correspondence between the sheared granular and the Barkhausen noise stochastic equations

A noticeable point is that even in the case of the Barkhausen noise the fluctuating term H_p undergoes a bounded random walk, analogous to (3).

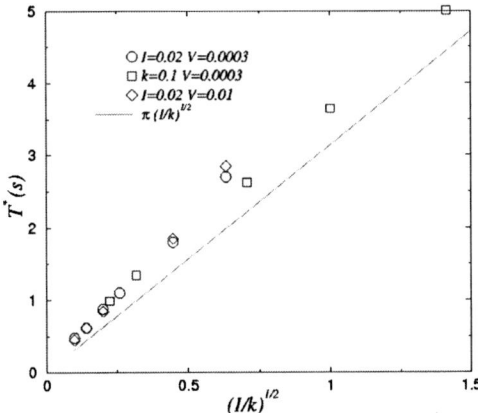

Fig. 6. The peak position in the probability distributions of the slip duration at different driving velocity and spring constant

6 Summary and Perspectives

In this work we have addressed the statistical behavior of some quantities characterizing the stick-slip regime of a granular medium subject to shear stress in a Couette cell, showing that experimental data can be interpreted to a good quantitative extent in terms of a stochastic equation of motion. The noise term in the equation represents the fluctuating part of the force, which undergoes a bounded Brownian motion and is therefore exponentially correlated. The motion is damped by a viscous term which only depends on the instantaneous velocity of the plate applying the shearing. Of course in perspective it is desirable that these terms will be derived from first principles, starting from the microscopic structure of the medium.

The stochastic equation desribing the granular system is of the same type describing the Barkhausen noise in ferromagnetic materials. Moreover, the similarities of some statistical properties of the investigated granular system

[24] with those of some solid-on-solid [28, 31], and of PET filaments systems [32] strongly suggest that a wide class of different phenomena can be described by the same type of stochastic equation.

References

1. O. Reynolds: Phil. Mag. **20**, 469 (1885)
2. D. W. Taylor: *Fundamentals of soil mechanics.* (Wiley, New York 1948)
3. C. Marone: Annu. Rev. Earth. Planet **26**, 643 (1998)
4. P. Segall, J.R. Rice: J. Geoph.Res. **100**, 22155 (1995)
5. R. A. Bagnold: Proc. Roy. Soc. **195**, 219 (1966)
6. S. Nasuno: Phys. Rev. Lett. **79**, 949 (1997)
7. P. A. Thomson, G. S. Grest: Phys. Rev. Lett. **67**, 1751 (1991)
8. H. J. Tillemans, H. J. Hermann: Physica A **217**, 261 (1995)
9. F. Lacombe, H. J. Hermann, S. Zapperi: E. Phys. J. E. **2**, 181 (2000)
10. B. Miller, C. O'Hern, R. P. Behringer: Phys. Rev. Lett. **77**, 3110 (1996)
11. F. Dalton, D. Corcoran: Phys. Rev. E **63**, 61312 (2001)
12. F. Dalton, D. Corcoran: Phys. Rev. E **65**, 31310 (2002)
13. J. Sethna, K. A. Dahmen, C. R. Myers: Nature **410**, 242 (2001)
14. H. Barkhausen: Physik Z. **20**, 401 (1919)
15. G. Bertotti: *Hysteresis in Magnetism.* (Academic Press, San Diego 1998)
16. S. Field, J. Witt, F. Nori, X. Ling: Phys. Rev. Lett. **74**, 1206 (1995)
17. A. Petri, G. Paparo, A. Vespignani, A. Alippi, M. Costantini: Phys. Rev. Lett. **73**, 3423 (1994)
18. A. Garcimartín, A. Guarino, L. Bellon, S. Ciliberto: Phys. Rev. Lett. **79**, 3202 (1997)
19. G. Ananthakrishina, S. J. Noronha, C. Fressengeas, L. P. Kubin: Phys. Rev. E **60**, 5455 (1999)
20. M. C. Miguel, A. Vespignani, S. Zapperi, J. Weiss, J. R. Grasso: Nature **410**, 667 (2001)
21. B. Gutenberg, C. F. Richter: Bull. Seismol. Soc. Amer. **34**, 185 (1944)
22. I. Main: Rev. Geoph. **34**, 433 (1996)
23. F. Dalton, A, Petri, G. Pontuale, L. Pietronero, in these Proceedings.
24. F. Dalton, F. Farrelly, A. Petri, L. Pietronero, L. Pitolli, G. Pontuale: Phys. Rev. Lett. **95**, 138001 (2005)
25. C. Goldenberg, I. Goldhrish: Nature **433**, 188 (2005)
26. T. S. Majmudar, R. P. Behringer: Nature **435**, 1079 (2005)
27. A. Baldassarri, F. Dalton, A. Petri, S. Zapperi, G. Pontuale, L. Pietronero, preprint cond-mat/0507533
28. F. Heslot, T. Baumberger, B. Perrin, B. Caroli, C. Caroli: Phys. Rev. E **49**, 4973 (1994)
29. R. Albert, M. A. Pfeifer, A. L. Barabasi, P. Shiffer: Phys. Rev Lett. **82**, 205 (1999)
30. J. Geng, R. P. Behringer: Phys. Rev. E **71**, 11302 (2005)
31. A. Johansen and P. Dimon and C. Ellegaard and J. S. Larsen, H. H. Rugh: Phys. Rev. E **48**, 4779 (1993)
32. B. Briscoe, A. Winkler, M. J. Adams: J. Phys. D: Appl. Phys. **18**, 2143 (1985)

Component Analysis of Granular Friction

Fergal Dalton[1], Alberto Petri[1], Giorgio Pontuale[1], and Luciano Pietronero[1,2]

[1] Istituto dei Sistemi Complessi, CNR Area Ricerca Tor Vergata, Via del Fosso del Cavaliere 100, 00133 Roma, ITALY
[2] Dipartimento di Fisica, Università "La Sapienza", P.le A. Moro 2, 00185 Roma, Italy

Summary. We perform an analysis of the stick-slip properties of a granular bed. The granulate is confined to a circular channel and sheared by an overhead top plate with a stick-slip motion. We attempt to decompose the frictional torque F_f subtended by the medium into its independent components by graphical and phenomenological analyses. We find clear functional dependence on the position, velocity and acceleration of the plate and the residual torque signal shows some dependence on the properties of the stick events. This article is related to that of Baldassarri et al. in these proceedings.

1 Introduction

In recent times, the study of granular materials (GM) has become something of a paradigm for complex systems in general [1]. Even though GMs constitute one of the most widely used materials in industrial processes, and also to a large extent in daily life, our ability to understand and predict their behaviour remains essentially at a phenomenological level [2]. Fluctuations are at the very core of GM behaviour, and it is only recently that the scientific community has come to appreciate the importance of these fluctuations [3].

GMs exhibit a very diverse behaviour, constituting solids, liquids or even gases under different conditions, or even a mixture of two or all three simultaneously [4]. To make the problem even more intractable, GMs are macroscopic particles which interact frictionally and/or collisionally, dissipating energy in either case through a highly non-linear interaction [5,6]. Therefore, to maintain a steady dynamic state, it is necessary to apply energy to the system which, of course, pushes the system out of equilibrium, resulting in an energy gradient within the granulate and hence, to the extent that a temperature can be defined [7–9], a temperature gradient.

Furthermore, granular materials can exist in a glassy state, in which the particles do not condense to a crystalline solid state, but instead follow a dynamic which progressively slows due to frustration between particles [10–12]. The transition to this state has been extensively studied for colloids, gels and spin glasses and, for GM, has become known as the jamming or rigidity transition [13–15]. The glassy state of GMs is generally accentuated by the presence of disorder, for example heterogeneous particles, which inhibit the emergence of the crystalline state.

Previous work on the imaging of GMs with photoelastic grains [16, 17] or confocal microscopy [18] has clearly demonstrated that GMs are highly disordered systems, where stress is carried along highly directional chains of grains [16, 19]. This is true in the static and slowly sheared states in which frictional interactions between grains are the dominant effect [20](p. 78). On the other hand, our research in this article on the transition to the fluid state of sheared GMs, where collisional contacts become the dominant interaction [21]. In particular, we wish to identify which state parameters of the system contribute to friction, and where possible to identify the functional dependence. Common sense dictates that the shear velocity should play an important role [5, 6, 22] though current research is not yet conclusive. Jaeger et al. [23] have also proposed a microscopic model which generates static and dynamic friction, with an increasing friction for high shear velocity.

In previous work, we have identified a solid/fluid transition for the system under study here [24]; below the transition, the GM exhibits a constant resistance to shear which, above the transition, increases rapidly. A similar experimental setup has also been utilised to demonstrate that the system exhibits criticality, though not universally [25, 26] and found some evidence for a second-order rigidity phase transition in the system [27].

More recently, using data from the present experiments, Baldassarri et al. [3, 28] have used experimental results as parameters for a simple macroscopic dynamic model which broadly reproduces the characteristics of the motion observed. The model essentially consists of a Langevin equation in which the frictional force is the sum of a deterministic function of the velocity and a random fluctuating component (1) which follows a confined random walk. Interestingly, the results indicate that there is an almost complete analogy between stick-slip granular shear and magnetic domain wall motion under an applied magnetic field. The comparison between experiment and model, however, is not perfect: the power-spectrum experimentally obtained, for example, is not a perfect Lorentzian, and the event distributions obtained often differ from the model values by up to a factor of 10.

In this preliminary study, at the risk of complicating the model with potentially many other parameters, we wish to improve this analysis to see if other dependencies can be identified, in particular, on the angular acceleration $\ddot{\theta}$ and a *non-fluctuating* component dependent on the angular displacement θ.

$$I\ddot{\theta} = \kappa(\omega_D t - \theta) - F_f(\theta, \dot{\theta}, \ddot{\theta}, ...)$$
$$F_f = \overline{F_v}(\dot{\theta}) + F_r(\theta) \qquad (1)$$
$$\frac{\mathrm{d}F_r}{\mathrm{d}\theta} = \eta(\theta) - aF_r$$

2 Experiments

The experiments discussed in this article were carried out on 2 mm glass beads in an opaque circular Couette channel of mean radius 140 mm and width 80 mm (see figure 1). The depth of beads in the channel is typically 80 mm and the medium is sheared from overhead by an annular aluminium top plate weighing 1 kg which has a layer of beads glued to its lower surface. The layer of beads does not extend the full width of the channel, but leaves a gap of approximately 5 mm on either side to avoid individual grains jamming the system at the boundary. A photograph of the apparatus is shown elsewhere in these proceedings [3].

The top plate is driven by a variable speed motor *via* a gearing mechanism and a torsion spring. The gearing mechanism and variable speed allows us to investigate for driving speed $3 \times 10^{-4} < \omega_D < 0.3$ rad/s while the torsion spring allows us to amplify the resulting fluctuations.

The experimental system is similar to that presented in [25], though here the dilation mechanism is much improved and allows free expansion of the granular material. Additionally, spatial and temporal resolution are improved by a factor of approximately 10.

Though we are unable to investigate the interior of the granulate, we assume that the lower layers of GM will crystallise while the upper layers remain more heterogeneous due to the intermittent shearing. Therefore, before each

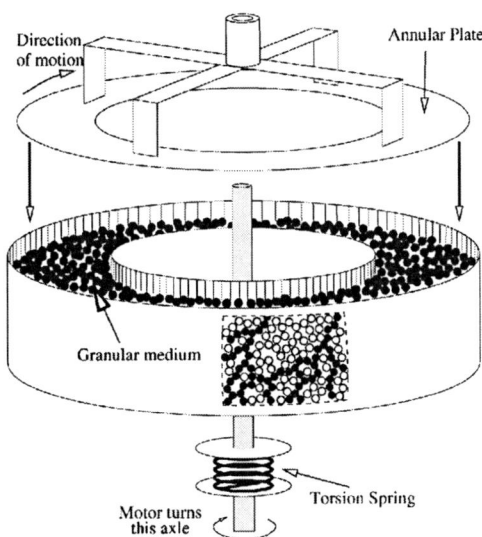

Fig. 1. The experimental apparatus. The annular top plate is forced to rotate over a granular medium in a stick-slip fashion.

series of experiments, the system is run for a long time (appx. 1000 revs) in order to approach a steady state. Experiments were conducted at ambient temperature ($18 < T < 23°C$) and relative humidity (R<50%).

The experiment is constructed in such a way as to facilitate the accurate measurement of the position of the annular top plate, and the torque acting thereon. This is achieved by means of an angular encoder with a resolution of 3.5×10^{-5} rad on both sides of the torsion spring, sampled at high frequency ($\sim$ 10 kHz). The difference between the two signals yields the instantaneous torque, while the differentiation of the signal from the top plate yields the angular velocity and acceleration.

The graphs presented in this article are chosen to depict what is, at this time, a preliminary analysis involving a subset of the entire ensemble of experiments. While the results give a "feeling" for the system's behaviour, we must state that they are not averaged over the ensemble though this will naturally be the ultimate objective of these analyses. Nevertheless, the majority of experiments yield qualitatively similar results in which the parameters and quality of the various curvefits can change. The driving velocity w_D for the data presented here, unless specified otherwise, is below the critical driving velocity at which the system fluidises.

3 Results

To give an idea of the behaviour observed, in figure 2 we show the motion of a high and a low driving velocity experiment. At high driving velocity, $w_D > 0.1$ rad/s, the system exhibits a continuous fluctuating motion, which is intermittently broken by stick events. On the contrary, the stick-slip phase is, on average, still, but intermittently broken by slip events. The driving velocity is shown in both cases by the dashed lines and so, it is evident that in the stick-slip phase, the fluctuations in velocity greatly outweigh the mean w_D, during some events by a factor of 20.

In figure 3 the typical distribution of the friction for low driving velocity is shown. In previous work, we have attempted to provide a "best-fit" to the distribution, and concluded that either a Lognormal (2), Gumbel (3) or Gamma (4) distribution will adequately describe the curve [29]. At higher driving velocity however, the distribution changes to a symmetric Gaussian. Similar results have also been obtained for solid-on-solid friction [30] and polymer films [31]. We have argued [24] that this feature could be used to infer the presence of force chains in the stick-slip state, and a fluidised stratum of 3 or 4 particle layers at high driving [24].

$$p(T) = \frac{1}{\sqrt{2\pi}(T - T_0)\sigma} \exp\left(-\frac{\ln^2 \frac{T-T_0}{T-T_0}}{2\sigma^2}\right) \tag{2}$$

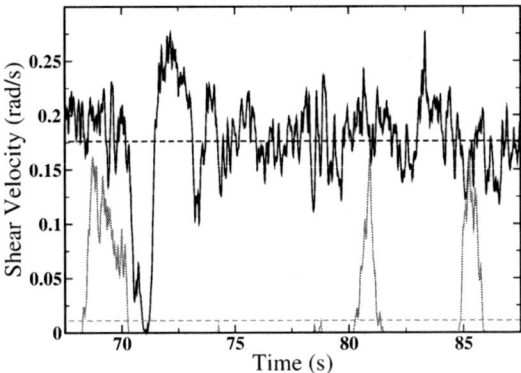

Fig. 2. The high velocity regime, characterised by steady sliding (black) is contrasted by the low velocity regime, with stick-slip events (red) of the experiment. The relevant driving velocities are shown as dashed lines.

$$p(T) = \frac{1}{\alpha} \left[\exp\left(-\frac{T - T_0}{a} - A \exp\frac{T - T_0}{a} \right) \right]^B \tag{3}$$

$$p(T) = \left(\frac{T - T_0}{a} \right)^\alpha \exp\left(-\frac{T - T_0}{a} \right) \tag{4}$$

In figure 4 we demonstrate the correlation between the instantaneous frictional torque F_f (eqs. (1)), and the instantaneous velocity, which is well fitted by a curve of the form (5). This curve, and others in this article, are relatively insensitive to changes in the driving velocity ω_D *as long as it remains below*

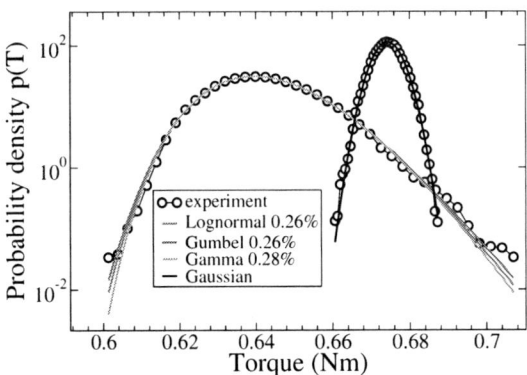

Fig. 3. The distribution of torque in the experiment for both low (left) and high (right) ω_D.

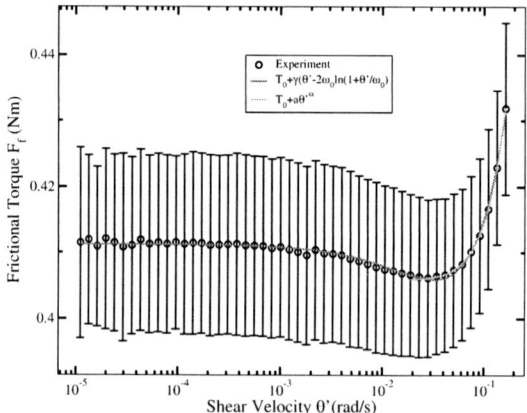

Fig. 4. The correlation between the frictional torque F_f and the instantaneous shear velocity $\dot{\theta}$. The curvefits are to (5) (red) and a simple quadratic increase (green) above the minimum.

the fluidisation threshold, this example being driven at $\omega_D = 0.01$ rad/s. As ω_D increases this curve loses definition at low values of $\dot{\theta}$, progressively decaying until fluidisation (even above fluidisation some vestiges of the original form may still be seen, for example the rising tail, though we do not present any similar analysis of these curves here). The friction is a linearly decreasing function of velocity at low velocity, leading to a static and dynamic friction; a minimum in the friction indicates that there is a preferred angular velocity ω_0 where dissipation is minimised:

$$F_v(\dot{\theta}) = F_0 + \gamma[\dot{\theta} - 2\omega_0 \ln(1 + \dot{\theta}/\omega_0)] \tag{5}$$

where γ is a high velocity damping coefficient. Additionally, we observe that above the minimum, the friction grows approximately as $\dot{\theta}^\alpha$ with $\alpha \simeq 2$, suggesting fluid friction.

When we subtract this velocity dependence from F_f, we can proceed to observe the residual dependence on other parameters. In figure 5, we plot this residual torque as a function of angular acceleration $\ddot{\theta}$. There is a clear functional dependence which for many experiments can be fitted by (6), while others show two linear segments with the same general trend.

$$F_a(\ddot{\theta}) = A(e^{-a\ddot{\theta}} - 1) \tag{6}$$

The existence of this curve suggests that there is an additional inertia I_{GM} due to the mobilisation of grains, though it is not intuitive why an additional inertia should have the form observed. Baldassarri et al [3, 28] have shown that the additional inertia corresponds to the mobilisation of approximately three or four layers of grains.

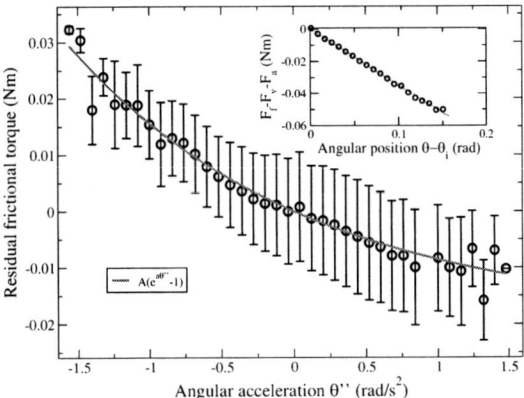

Fig. 5. The correlation between the residual frictional torque $F_f - F_v(\dot{\theta})$ and the instantaneous shear acceleration $\ddot{\theta}$. The curvefit is of the form (6). The inset shows the average dependence of the residual torque on the plate's angular displacement within each event: $F_p(\theta - \theta_i)$ where θ_i is the angular position at the start of event i (see text).

Nonetheless, we can still plot the remaining residual frictional torque: $(F_f - F_v(\dot{\theta}) - F_a(\ddot{\theta}))$ which, according to [3] we would expect to be a function of angular displacement, fluctuating as a random walker in space, encoding the noise characteristics of the system. We show this plot on the left of figure 6 together with the velocity of the system. It is clear that this residual friction is not actually such a random walk, but instead decreases during events as the plate spins and the spring winds down.

However, our analysis may still proceed. We wish to obtain only the fluctuating component of the friction and must therefore eliminate the decrease during each event, due to the unwinding of the spring, which may be considered as a deterministic component of the frictional torque depending on the displacement of the plate relative to the start of the event (shown in the inset to figure 5). In the same way that we have subtracted the components $F_v(\dot{\theta})$ (shown in fig. 4) and $F_a(\ddot{\theta})$ (fig. 5), we can also subtract this dependence on $\theta - \theta_i$ (inset to fig. 5), which is given by $F_p(\theta - \theta_i) = -\kappa(\theta - \theta_i)$. The resultant signal is shown on the right of figure 6 together with the velocity of the top plate as a function of the angular displacement θ. This signal now seems to fluctuate randomly though more energetically for the largest events. We plot, therefore, the standard deviation of this noise signal for each event as a function of the event size and the event maximum velocity. The resulting curves, shown in figure 7 are generally a power-law function of the maximum velocity or size, before increasing exponentially above the transition velocity of ~0.1 rad/s, corresponding to the point at which the grains fluidise.

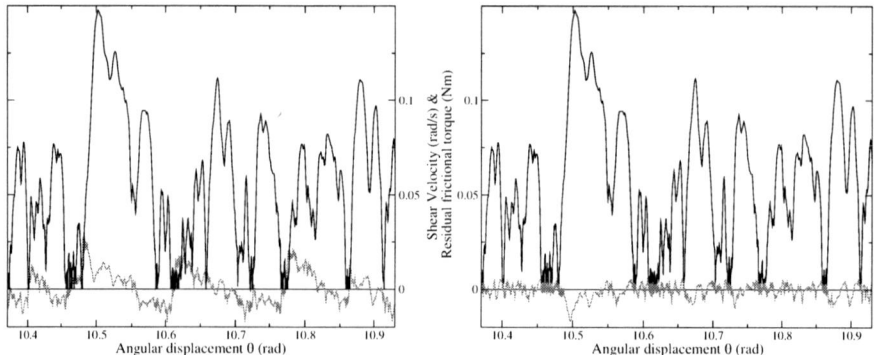

Fig. 6. The left graph shows the residual torque $F_f - F_v(\dot{\theta}) - F_a(\ddot{\theta})$ (red) and the plate velocity (black). Events are clearly identifiable in the velocity signal. The right graph shows the residual torque $F_f - F_v(\dot{\theta}) - F_a(\ddot{\theta}) - F_p(\theta - \theta_i)$ (where $F_p(\theta - \theta_i)$ is the drift due to the macroscopic motion of the torsion spring), and the plate velocity.

4 Discussion

In this preliminary study, we have extended the analysis performed by Baldassarri et al. [3, 28], and have found that the frictional response of a granular medium may be expressed, in addition to the deterministic and fluctuating components already revealed, as deterministic functions of the system acceleration and displacement.

Therefore, we conclude that the original model (1), while broadly capturing the motion, is unable to accurately reproduce all the details due to its deliberate simplicity. The present analysis, though complicating the model, gives the possibility of refining the model and may form the basis for further speculation.

Although below a certain size, the noise characteristics are related to the size of the event in a scale-free manner, above this critical size, the noise signal behaves differently. It seems that this transition may mark the transition from solid to fluid within the granulate and so any future developments to the model must necessarily include a second-phase in which the response of the medium is qualitatively different.

Certainly, the analysis does not finish here. We cannot exclude the possibility that there may be other state variables which govern the frictional response. Indeed, the overhead pressure, and even ambient conditions will influence the outcome of an experiment, and there will almost certainly be memory effects which will vastly increase the search-space in which correlations may be found.

In future work, we wish to quantitatively improve the reliability of this analysis using "independent component analysis" in which any arbitrary set of signals are compared and an orthogonal set of relations determined. Our current understanding indicates that this method will enable the identification of both cross-terms (involving, for example, both velocity and acceleration) and non-linear terms. However, it is yet to be ascertained if this type of analysis is applicable to the data at hand.

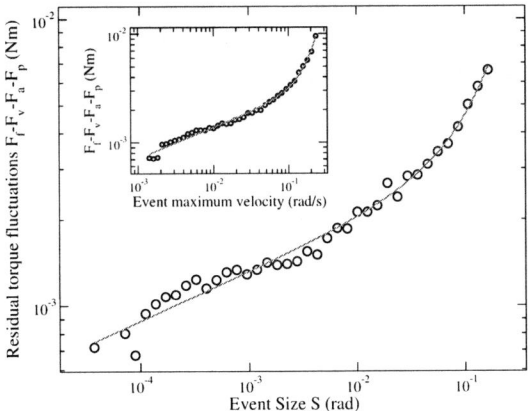

Fig. 7. Fluctuations in the adjusted residual torque as a function of the event size. Above a certain size, the fluctuations change in properties. The inset shows the same data against the event maximum velocity. It appears that the fluidisation transition causes the alteration to the noise properties.

References

1. Hernán A. Makse, Jasna Brujić, and Sam F. Edwards. Statistical mechanics of jammed matter. In Heye Henrichson and Dietrich E. Wolf, editors, *The Physics of Granular Media*. Wiley, 2004.
2. R. P. Behringer, D. Howell, L. Kondic, S. Tennakoon, and C. Veje. Predictability and granular materials. *Physica D*, 133:1–17, 1999.
3. A. Baldassarri, F. Dalton, A. Petri, L. Pietronero, G. Pontuale, and S. Zapperi. Granular shearing and barkhausen noise. 2005. Proceedings of the Traffic and Granluar Flow '05 conference held at Berlin, October 2005.
4. H. M. Jaeger, S. R. Nagel, and R. P. Behringer. Granular solids, liquis and gases. *Rev. Mod. Phys.*, 68:1259, 1996.
5. F. Lacombe, S. Zapperi, and H. J. Herrmann. *Eur. Phys. J. E*, 2:181–189, 2000.
6. Mark O. Robbins. Jamming, friction and unsteady rheology. In A. J. Liu and S. R. Nagel, editors, *Jamming and Rheology: Constrained Dynamics on Microscopic and Macroscopic Scales*. Taylor and Francis, London, 2000.
7. Bob Behringer. Taking the temperature. *Nature*, 415:594–595, 2002.
8. Alain Barrat, Jorge Kurchan, Vittorio Loreto, and Mauro Sellitto. Edwards' measures for powders and glasses. *Phys. Rev. Lett.*, 85:5034–5037, 2000.

9. H. A. Makse and J. Kurchan. *Nature*, 415:614–617, 2002.
10. Leonardo E. Silbert, Deniz Ertaş, Gary S. Grest, Thomas C-Halsey, and Dov Levine. Analogies between granular jamming and the liquid-glass transition. *Phys. Rev. E*, 65:051307, 2002.
11. L. Berthier, L. F. Cugliandolo, and J. L. Iguain. Glassy systems under time-dependent driving forces: Application to slow granular rheology. *Phys. Rev. E*, 63:051302, 2001.
12. Anita Mehta and G. C. Barker. Glassy dynamics in granular compaction. *J. Phys.: Condens. Matter*, 12:6619–6628, 2000.
13. Antonio Coniglio and Mario Nicodemi. The jamming transition of granular matter. *J. Phys.: Condens. Matter*, 12:6601–6610, 2000.
14. Einat Aharonov and David Sparks. Rigidity phase transition in granular packings. *Phys. Rev. E*, 60:6890–6896, 1999.
15. Patrick Mayor, Gianfranco D'Anna, and Gérard Gremaud. Jamming in a weakly perturbed granular media. *Materials Sc. and Eng. A*, 370:307–310, 2004.
16. D. Howell, R. P. Behringer, C. Veje. Stress fluctuations in a 2d granular couette experiment: a continuous transition. *Phys. Rev. Lett.*, 82:5241, 1999.
17. B. Miller, C. O'Hern, and R. P. Behringer. Stress fluctuations for continuously sheared granular materials. *Phys. Rev. Lett.*, 77:3110–3113, 1996.
18. Jasna Brujić, Sam F. Edwards, Dmitri V. Grinev, Ian Hopkinson, Djordje Brujić, and Hernán A. Makse. 3d bulk measurements of the force distributions in a compressed emulsion system. *Faraday Discuss.*, 123:207–220, 2003.
19. A. Drescher and G. de Josselin de Jong. Photoelastic verification of a mechanical model for the flow of a granular material. *J. Mech. Phys. Solids*, 20:337–351, 1972.
20. C.H. Scholz. *The mechanics of earthquakes and faulting*. Cambridge University Press, England, 1990.
21. T. G. Drake. Structural features in granular flows. *J. Geophys. Res.*, 95:8681–8696, 1990.
22. P. A. Thompson and G. S. Grest. *Phys. Rev. Lett.*, 67:1751–1754, 1991.
23. H. M. Jaeger, Chu-Heng Liu, S. R. Nagel, and T. A. Witten. *Europhys. Lett.*, 11:619–624, 1990.
24. Fergal Dalton, Francis Farrelly, Alberto Petri, Luciano Pietronero, Luca Pitolli, and Giorgio Pontuale. Shear stress fluctuations in the granular liquid and solid phases. *Phys. Rev. Lett.*, 95:138001, 2005.
25. F. Dalton and D. Corcoran. *Phys. Rev. E.*, 65:31310–31315, 2002.
26. F. Dalton and D. Corcoran. *Phys. Rev. E.*, 63:61312–61314, 2001.
27. R. Lynch, D. Corcoran, and F. Dalton. The onset to criticality in a sheared granular medium. In A. Méndez-Vilas (Ed.), *Recent Advances in Multidisciplinary Applied Physics*, pages 369–374, 2003. Proceedings of the First International Conference on Applied Physics (APHYS-2003).
28. A. Baldassarri, F. Dalton, A. Petri, S. Zapperi, G. Pontuale, and L. Pietronero. Brownian forces in sheared granular matter. 2005. cond-mat/0507533.
29. F. Dalton, A. Petri, G. Pontuale, and L. Pietronero. Stress fluctuations and the solid/fluid transition in a sheared granular bed. In R. García-Rojo, H. J. Herrmann, and S. McNamara (Eds.), *Powders and Grains 2005*, page 353, 2005.
30. A. Johansen, P. Dimon, C. Ellegaard, J. S. Larsen, and H. H. Rugh. *Phys. Rev. E*, 48:4779–4790, 1993.
31. B. Briscoe, A. Winkler, and M. J. Adams. *J. Phys. D: Appl. Phys.*, 18:2143–2143, 1985.

Granular Flow and Pattern Formation on a Vibratory Conveyor

Christof A. Krülle[1], Andreas Götzendorfer[1], Rafał Grochowski[2], Ingo Rehberg[1], Mustapha Rouijaa[1], and Peter Walzel[2]

[1] Experimentalphysik V, Universität Bayreuth, D–95440 Bayreuth, Germany
[2] Mechanische Verfahrenstechnik, Universität Dortmund, D–44227 Dortmund, Germany

Summary. Vibratory conveyors are well established in routine industrial production for controlled transport of bulk solids. Because of the complicated interactions between the vibrating trough and the particles both glide and throw movements frequently appear within one oscillation cycle. Apart from the amplitude and frequency, the form of the trajectory of the conveyor's motion also exerts an influence. The goal of our project is a systematic investigation of the dependence of the transport behavior on the three principle oscillation forms: linear, circular and elliptic. For circular oscillations of the shaking trough a non-monotonous dependence of the transport velocity on the normalized acceleration is observed. Two maxima are separated by a regime, where the granular flow is much slower and, in a certain driving range, even reverses its direction. In addition, standing waves oscillating at half the forcing frequency are observed within a certain range of the driving acceleration. The dominant wavelength of the pattern is measured for various forcing frequencies at constant amplitude. These waves are not stationary, but drift with a velocity equal to the transport velocity of the granular material, determined by means of a tracer particle. Finally, the fluidization of a monolayer of glass beads is studied. At peak forcing accelerations between $1.1\,\mathrm{g}$ and $1.5\,\mathrm{g}$ a solid-like and a gas-like domain coexist. It is found that the number density in the solid phase is several times that in the gas, while its granular temperature is orders of magnitude lower.

1 Introduction

Vibratory conveyors are highly used for discharging, conveying, feeding, dosing and distributing bulk materials in many branches of industry, for example in the chemical and synthetic materials industries, food processing (Fig. 1(a)), sand, gravel, and stone quarries, for small-parts assembly mechanics (Fig. 1(b)), the paper-making industry, sugar or oil refineries, and foundries [1–3]. In addition to transport, vibration can be utilized to screen, separate, compact or loosen product. Open troughs are used for conveying bulk materials, closed tubes for dust-sealed goods, and work piece-specific rails for conveying oriented parts.

Fig. 1. (a) Linear vibratory conveyors in the food industry, (b) Vibratory bowl feeder in an automated assembly chain at SUSPA company, Sulzbach-Rosenberg

Some of the main advantages of vibratory conveyors are their simple construction and their suitability to handle hot and abrasive materials. In addition, they are readily used in the food industry, since they can easily be kept complying to hygienic standards by using stainless steel troughs. Some disadvantages of vibratory conveyors are their noisy operation, the induced vibrations on their surroundings and their limited transport distance. Furthermore, the granular material may be damaged when it is subjected to extreme accelerations normal to the trough.

Important properties to be considered regarding the granular material (or bulk solid for engineers) are: particle size distribution and shape, friction between particles and trough and inter-particle friction, modulus of elasticity of the particles and/or the bulk, cohesion, layer thickness, and the permeability of air.

Considering the many parameters involved, the performance of a vibratory conveyor is difficult to predict theoretically. Obvious design parameters are: vibration mode (linear with or without vibration angle α, circular, elliptic), amplitude A and frequency f of the oscillations, combined as dimensionless *throw number* $\Gamma = A\sin(\alpha)(2\pi f)^2/g$, inclination or declination of the conveyor, smoothness of the trough surface, modulus of elasticity of the trough's inner surface, which can be coated with rubber, plastic, etc., and possible electrostatic charges.

2 Conveying Principles

Three different principles of conveying have to be distinguished [3, 4, 7] (see Fig. 2):

- *Sliding*: Here the deck is moved by a crankshaft mechanism only horizontally with asymmetric forward and backward motions. The material remains always in contact with the trough surface and is transported forward relative to the deck by a stick-slip drag.
- *Throwing*: If the vertical component of the acceleration exceeds gravity, the material loses contact during part of the conveying cycle and is repeatedly

forced to perform ballistic flights. Complicated sequences of a rest phase, a positive (or negative) sliding phase, and a flight phase have to be considered. The net transport in the forward (or even backward) direction depends sensitively on the coefficient of friction between the particles and the trough and on the coefficient of restitution for the collision with the deck.

- *Ratcheting*: Motivated by advances in the investigations of fluctuation-driven ratchets a new transport mechanism has been proposed recently [5, 6]: a horizontal transport of granular particles can be achieved in a purely vertically vibrated system, if the symmetry is broken, instead of the direction of the vibration mode, by an asymmetric periodic sawtooth-shaped profile of the base.

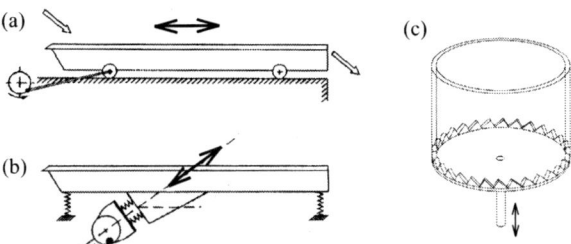

Fig. 2. Three different conveying principles: (a) 'Sliding' by asymmetric horizontal back-and-forth movements, (b) 'Throwing' by linear vibration with throw number $\Gamma > 1$, both images taken from [4]. (c) 'Ratcheting' by vertical vibration on a sawtooth-shaped profile of the base. From [5, 6]

3 Granular Transport

Since the transport phenomena on vibratory conveyors involve the nonlinear interaction of many-particle systems with complex behavior, the investigation of their dynamical properties has become a challenging subject to physicists, too. In the past, most studies dealing with vibrated granular media were based on purely vertical or purely horizontal vibration. Only recently a few experimental explorations of the dynamics of granular beds subject to *simultaneous* horizontal and vertical vibration have been reported [3, 8–14]. The observed phenomena include the spontaneous formation of a static heap, convective flow, reversal of transport, and self-organized spatiotemporal patterns like granular surface waves.

The most important questions currently under investigation are:

- How does the granular transport velocity depend on (i) *external* parameters of the drive like amplitude and frequency of the oscillation, the vibration mode, or the inclination of the trough, and (ii) *internal* bulk parameters like coefficient of restitution, friction coefficients, and the filling height?
- Is it possible to optimize the transport effectivity by suitable modifications of the surface of the trough implying ratchet like profiles?
- What kind of self-organized structures can be expected? Are there clearly characterized instabilities? Which physical mechanism underlie these structures? How is the granular transport effected?
- Are there segregation effects in bidisperse or polydisperse systems? What are the analogies to vertical vibration?
- Can these results eventually lead to optimized industrial devices like conveyor systems, metering devices, sieves, mixers, dryers, or coolers?

4 Experimental Setup

For this purpose, a prototype annular conveyor system has been constructed (Fig. 3) for systematic studies of the transport properties for different oscillation modes, i.e. linear, elliptical, and circular (see Fig. 4) for a long running

Fig. 3. Annular vibratory conveyor with a toroidal trough of radius $R = 22.5$ cm and width $w = 5$ cm, suspended on adjustable columns via elastic bands: (1) Torus-shaped vibration channel, (2) Adjustable support, (3) Elastic band, (4) Vibration module with unbalanced masses, (5) Electric motor with integrated frequency inverter

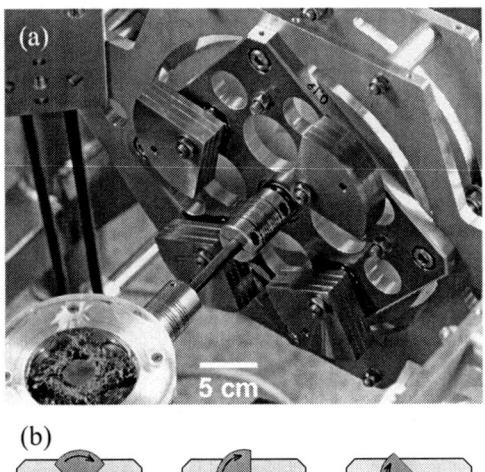

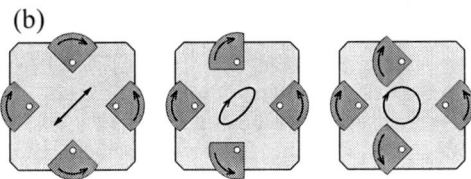

Fig. 4. (a) Driving module with four unbalanced masses, (b) Side views of a driving unit with four rotating unbalanced masses, for three principal modes of oscillation: linear ($\varphi = 0$), elliptical ($\varphi = \frac{\pi}{4}$), and circular ($\varphi = \frac{\pi}{2}$)

time, without disturbing boundary conditions [10–12, 14]. In principle, this conveyor can be excited in all six degrees of freedom individually, or by a combination of two of them. For the first experiments, a vibration mode has been chosen consisting of a torsional vibration $\phi(t) = A/R \cos(2\pi ft)$ around the symmetry axis of the apparatus, superposed with a vertical oscillation $y(t) = A \cos(2\pi ft + \varphi)$ where φ is the fixed phase shift between the two oscillations. If, for example, this phase shift φ is chosen to be $\pi/2$, then each point on the trough traces a circular path in a vertical plane tangent to the trough at that point. In short, the support agitates the granules via a *vertical circular vibration*.

For being able to adjust different vibration modes special driving units have been developed, equipped with four rotating unbalanced masses, as combinations of two unbalanced-mass linear vibrators oriented perpendicularly to each other [10–12].

Characteristic of the unbalanced-mass agitated system is the frequency dependence of the vibration amplitude:

$$A(f) = A_t \frac{f^2}{\sqrt{(f_0^2 - f^2)^2 + (2\zeta f_0 f)^2}} \ . \tag{1}$$

The resonance frequency $f_0 = \frac{1}{2\pi} \sqrt{\frac{k_{\text{eff}}}{M_0 + N \cdot m_u}}$ can be limited to a small value (through appropriate choice of spring constant k_{eff}), so that in the over critical range, $f > 3f_0$, an almost constant terminal amplitude $A_t = r_u \frac{N/2 \cdot m_u}{M_0 + N \cdot m_u}$

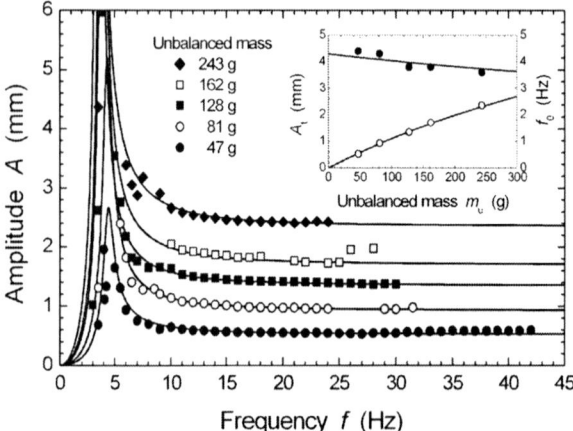

Fig. 5. Amplitude vs. frequency $A(f)$ for different unbalanced mass m_u. The inset shows the dependence of the resonance frequency f_0 and the terminal amplitude A_t on the mass m_u of the $N = 32$ unbalanced masses (net mass of the device $M_0 = 24.3\,\text{kg}$, eccentricity of the unbalanced masses $r_u = 19\,\text{mm}$, effective spring constant $k_{\text{eff}} = 177\,\text{N/cm}$

arises which can be adjusted for fixed eccentricity r_u by the out of balance mass m_u. The frequency response (see Fig. 5) was measured experimentally before every measurement by determination of the position of an affixed LED on the vibrating device recorded with a CCD camera. The vibration amplitude is found by aligning a circular path with radius $A(f)$ to the image data. For excitation frequency $f > 15\,\text{Hz}$, the vibration amplitude is nearly constant and therefore the dimensionless acceleration, i.e. the machine number $K = A(f) \cdot 4\pi^2 f^2/g$ is approximately proportional to f^2. From the measured data, the damping constant ζ can be determined to be 0.08 ± 0.01

An important parameter of the bulk solid that should be chosen carefully for obtaining reproducible results is the layer height. For too thin layers the particles will dilate mutually and start to perform irregular motions across the trough bottom. In practice, as a rule of thumb, the layer of material should at least be 10 particles high. Under these conditions, the bulk solid moves more or less like a solid block, and the transport velocity is not very sensitive to the exact height.

During the transport experiments, the average flow velocity is determined by tracking a colored tracer particle that is carried along with the bulk. This is done automatically with a PC based image processing system, which detects the passage of the tracer through a line perpendicular to the trough and stores each passage time on hard disk.

5 Results

5.1 Flow Reversal

The result for such an experiment is shown in Fig. 6. Below a critical value $\Gamma_c \approx 0.45$ the grains follow the agitation of the tray without being transported. The onset of particle movements is hindered by frictional forces between grains and the substrate. By increasing the acceleration above this threshold a net granular flow with constant velocity is observed. For circular vibration of the trough, surprisingly, the transport velocity is not a monotonous function of Γ, but has two maxima at $\Gamma = 1.2$ and $\Gamma = 4.2$. In between, the granular flow is slower and even *reverses* its direction in the regime $2.6 < \Gamma < 3.8$, whereas an individual glass bead is propagated in the same direction for all accelerations.

The critical Γ values, at which the transport behavior changes qualitatively, are independent of the oscillation amplitude. In the frequency-amplitude parameter space (Fig. 7) the threshold values lie on f^{-2} hyperbola of constant acceleration ('isoepitachs'). However, for $\Gamma > 4$, i.e. beyond the second reversal of flow direction, this scaling behavior is not observed anymore. Depending on the vibration amplitude and the filling height, the third reversal occurs in the range $5 < \Gamma < 7$ [11].

5.2 Linear Vibration Mode

Complementary studies [9] applying linear vibrations have shown, that the flow reversal is a special property of the circular vibration. For linear vibrations, also a non-monotonous dependence of the transport velocity is seen (Fig. 8), with a dip at $\Gamma \approx 5$, but no reversal of the flow. Note that the maximum transport velocity in this case arrives at about 90% of the oscillation velocity $A\omega$, while in the circular case only about 50% can be attained.

5.3 Sand Bag Test

Worth mentioning is the so-called 'sand bag test' [7] routinely performed by the manufacturers of vibratory conveyors. Since an individual grain, such as a single glass bead, dropped onto the substrate will rebound with a high coefficient of restitution and therefore gives an unrealistic transport characteristic, a better approach for modelling the bulk transport is achieved by use of a small fabric bag, filled with the same kind of beads. The strikingly different collective behavior arises from the large number of rapid inelastic collisions of neighboring grains. However, it takes some experience to find a suitable single object for producing reliable data for comparison with an effective one-particle model.

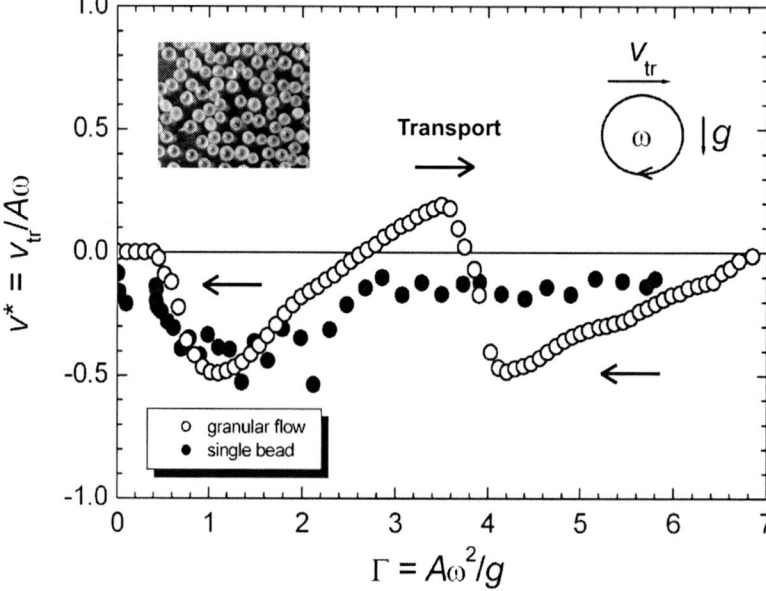

Fig. 6. Normalized transport velocity v^* of a granular flow ($\approx 300\,000$ glass beads with $1\,$mm diameter, see inset) on the vibratory conveyor with circular vibration, compared with the mean velocity of one single glass bead

5.4 Theoretical Description

Such a theoretical approach is based on the following initial assumptions [3, 15, 16]:

- The granular material behaves like a solid body and can be represented by a point mass.
- When the layer of granular material hits the trough after a flight phase a fully non-elastic collision is assumed.
- Rotations of the particle and interactions with the side walls of the trough are neglected.
- The kinetic and static coefficients of friction are set equal or the distinction between them is neglected.
- The air resistance during the flight phase is negligible.

According to these assumptions, Sloot and Kruyt [3] obtained fairly good agreement between theory and experiment for slide conveyors but observed large deviations for linear throw conveyors with vibration angle α. Hongler *et al.* [17] described the dynamics of a vibratory feeder by a set of coupled, nonlinear and strongly dissipative mappings and identified the transport behavior to be determined by periodic and chaotic solutions. First simulations

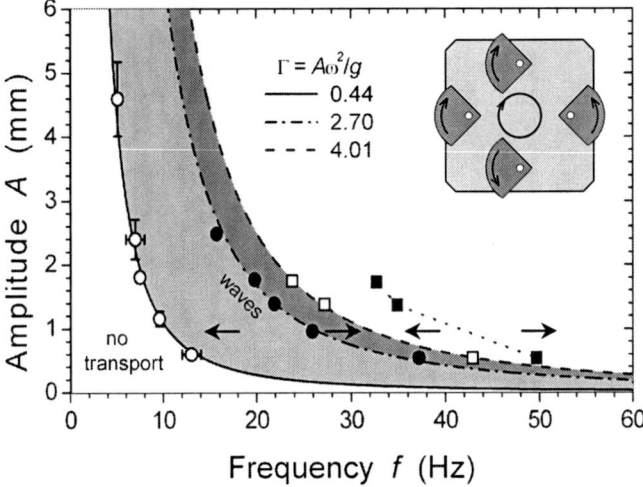

Fig. 7. Phase diagram of the transport behavior at clockwise circular forcing. The alternating arrows correspond to the transport direction of the bulk solid

for conveyors with variable vibration mode (linear, circular, and elliptic) by Landwehr, Lange, and Walzel [18, 19] applied various, more complex collision models taking also a finite coefficient of restitution into account. More recently, an effective single-particle model by El hor and Linz [13, 15, 16] that includes only dynamic friction forces and collisions with complete dissipation of the vertical velocity component in order to understand the theoretical basics of the transport process led to rather good agreement with the experimental results shown in Figs. 6 and 8.

5.5 Onset of Particle Motion

Despite the complex interactions between the particles and the vibrating trough during the transport process, which up to now can only be handled via numerical simulations, it is interesting that the *onset* of motion for a single block subject to static friction can be derived analytically.

A *linear* harmonic motion of the conveyor with amplitude A and vibration angle α (see inset of Fig. 9(a)) can be expressed as $x(t) = A \cos(\alpha) \cos(2\pi ft)$ for its horizontal and $y(t) = A \sin(\alpha) \cos(2\pi ft)$ for the vertical component, respectively. Due to the periodic acceleration a particle with mass m lying on the trough experiences a horizontal force $F_h(t) = m\ddot{x}(t)$ and a modulated effective weight $N(t) = m(g + \ddot{y}(t))$. The mass is hindered from sliding if the resulting frictional force $F(t) = \mu_s N(t)$ is larger than $|F_h(t)|$, where μ_s is the static coefficient of friction. At the onset of particle motion both forces are

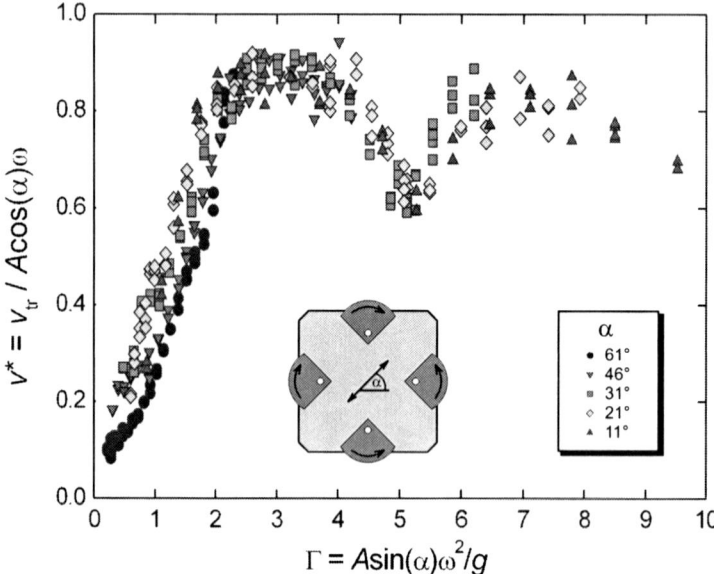

Fig. 8. Normalized transport velocity v^* of a granular flow for linear vibration at different vibration angles α

equal, which leads to the balance equation

$$|m\ddot{x}(t)| = \mu_s m(g + \ddot{y}(t)) .$$

(2)

This condition is met first at the highest point of the conveyor's trajectory where the horizontal acceleration is maximal while the friction force is minimal, yielding a critical throw number Γ_{onset} for which the particle starts to move:

$$\Gamma_{\mathrm{onset}} = \frac{\mu_s \tan(\alpha)}{1 + \mu_s \tan(\alpha)} = \left[1 + \frac{1}{\mu_s \tan(\alpha)}\right]^{-1} .$$

(3)

Figure 9(a) shows the monotonous but nonlinear dependence of Γ_{onset} as a function of both the vibration angle α and the static friction coefficient μ_s.

A similar calculation for the *circular* motion of the conveyor, where both vibration amplitudes are set equal at a fixed phase shift $\varphi = \pi/2$, i.e. $x(t) = A\cos(2\pi f t)$ and $y(t) = A\cos(2\pi f t + \pi/2)$, leads to the expression

$$\Gamma_{\mathrm{onset}} = \frac{\mu_s}{\sqrt{1 + \mu_s^2}}$$

(4)

for the slipping threshold. A comparison of both vibration modes, linear and circular, is made in Fig. 9(b), which shows that it is easier to overcome static

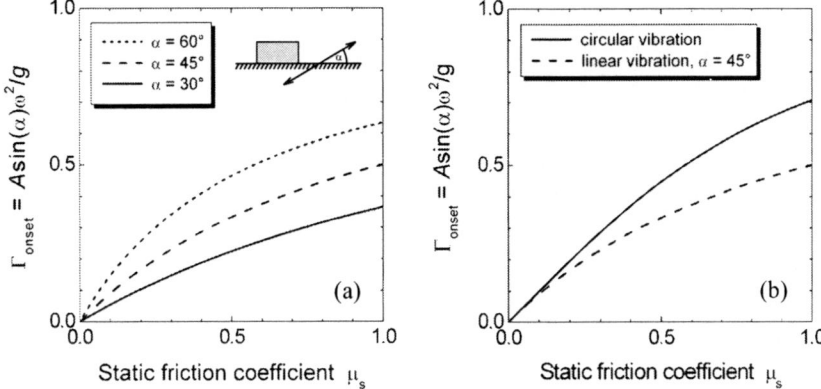

Fig. 9. (a) Critical throw number Γ_{onset} for the onset of particle motion on a linearly vibrating conveyor at various vibration angles α. (b) Comparison of the critical throw number Γ_{onset} for the onset of particle motion for the circular and the ($\alpha = 45°$) linear vibration modes

friction by linear vibrations of the trough. The circular mode seems to be 'softer', i.e. less effective to set particles in motion.

The most general, i.e. *elliptic* case is characterized by both parameters, inclination angle α and phase shift φ, with the corresponding horizontal and vertical orbital components $x(t) = A\cos(\alpha)\cos(2\pi ft)$ and $y(t) = A\sin(\alpha)\cos(2\pi ft + \varphi)$, respectively, which yields the general solution of this problem as

$$\Gamma_{\text{onset}} = \frac{\mu_s \tan(\alpha)}{\sqrt{1 + 2\mu_s \tan(\alpha)\cos(\varphi) + \mu_s^2 \tan^2(\alpha)}}. \qquad (5)$$

These considerations may be seen as a starting point for the systematic study of the transport behavior for all parameters of a vibratory conveyor.

6 Surface Waves

If Γ exceeds 1, the vertical component of the circular acceleration will cause the grains to detach from the bulk followed by a flight on a ballistic parabola. This results in a much less dense-packed, 'fluidized', state with highly mobile constituents. In a certain driving range between $\Gamma_1 = \sqrt{\pi^2 + 1} \approx 3.3$ and $\Gamma_2 \approx 3.7$ a locking of the time-of-flight between successive bounces and the period of the circular vibration occurs [20]. The initially flat bed becomes destabilized, and undulations of the granular surface occur in the range $2.4 < \Gamma < 4.5$, see Fig. 10 [21, 22]. High-speed CCD imaging showed that they oscillate with half

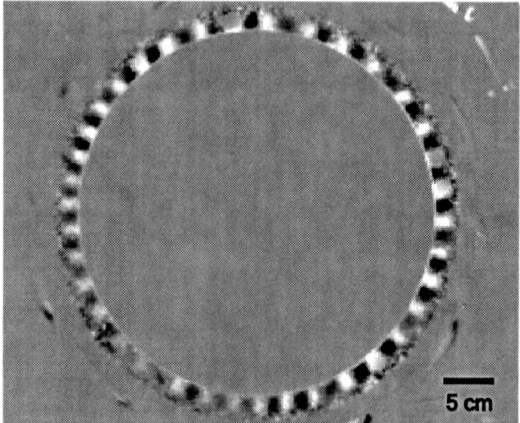

Fig. 10. Granular surface waves ($\lambda = 3.8$ cm) inside a vibratory conveyor with vertical circular motion of the annular trough oscillating with $\Gamma = 3.0$ at a frequency $f = 22.4$ Hz

the excitation frequency ('f/2 waves'). In contrast to previous work [23–27] done at purely vertical vibration the present waves are not stationary but are transported along the annular trough. The drift velocity of the wave pattern can be measured by applying a phase-locked imaging technique with fixed time delay $\Delta t = 2T$. This is done with a camera which is triggered by a pick-up signal taken from the rotating unbalanced masses. From these images space-time diagrams as shown in Fig. 11 can be constructed. The deviation of the striped pattern from the vertical is taken as a measure for the wave speed.

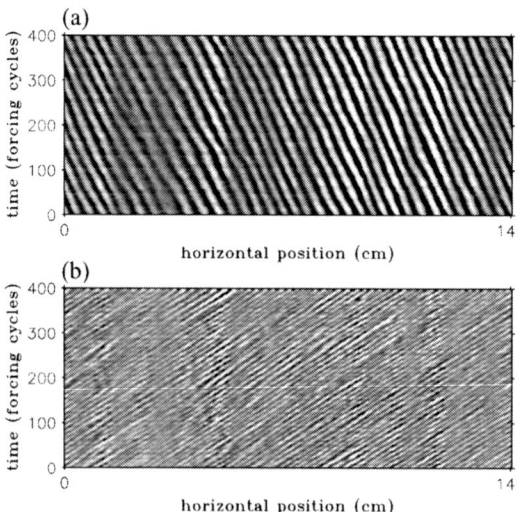

Fig. 11. Space-time diagrams of amplitude $h(\varphi, t)$ of surface waves at maximum wave amplitude every other forcing cycle, for $\Gamma = 2.84$ (a) and $\Gamma = 4.22$ (b). High amplitude appears bright

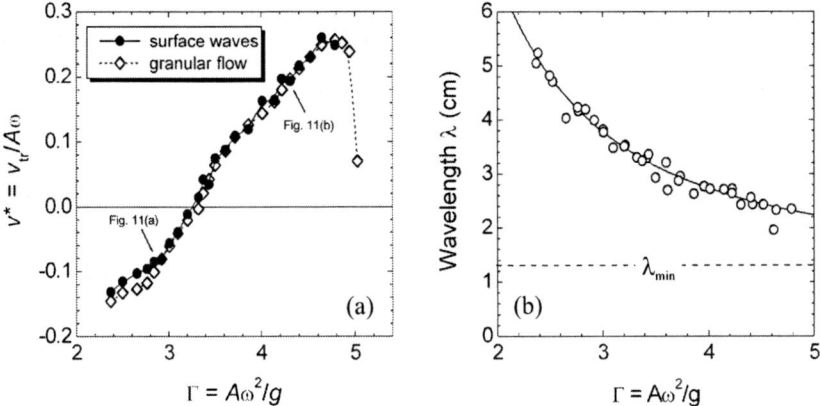

Fig. 12. (a) Scaled velocity v^* of surface waves ($\bullet$) and granular flow ($\diamond$). (b) Circles represent the measured wavelength λ plotted over the normalized acceleration Γ. Throughout the measurement the driving amplitude was kept constant at 1.47 mm. The solid line is the graph of the function $\lambda = 1.3\,\text{cm} + 20\,\text{cm} \cdot \Gamma^{-1.9}$, the best fit to the data

A comparison of the wave speed with the bulk velocity of the transported particles shows that both velocities are identical (Fig. 12(a)). In particular, a reversal of the wave velocity is also possible by adjusting the vibration frequency alone. In the range $\Gamma \approx 4.5$, when the granular flow is reversed a second time, surface waves disappear. The granular surface becomes flat but still oscillates at twice the vibration frequency. Above $\Gamma \approx 5.5$ standing waves are observed again but this time repeating their patterns after four shaker periods ('f/4 waves'). This scenario is reminiscent to the wave patterns found by Bizon *et al.* [28] for vertical vibration of a laterally extended system.

A more detailed analysis of the Γ dependence of the wave length λ shows an algebraic decay as

$$\lambda(\Gamma) = \lambda_{\min} + \Lambda \cdot \Gamma^{\alpha} , \qquad (6)$$

see Fig. 12(b). For the minimum wavelength $\lambda_{\min}$ we obtain a value of 1.3 ± 0.3 cm, which is approximately the depth of the granular layer. The values for the other parameters are $\Lambda = 20 \pm 4$ cm and $\alpha = -1.9 \pm 0.3$. This is consistent with the results of Metcalf *et al.* [26] who examined the dependence of the wavelength on the peak acceleration Γ at constant frequency.

For comparison with theoretical models it is necessary to determine the dynamic surface profile $h(\varphi, t)$ during the transport process with high spatial and temporal resolution. This task has been solved by Pak and Behringer [24] only for a small section of an annular trough. Our container consists of a 2 cm wide annular channel with open top, 7 cm high Plexiglas walls, and a radius of $R = 22.5$ cm giving a circumference of $L_0 = 141$ cm (see Fig. 13). The granular

Fig. 13. Experimental setup with transparent trough and conical mirror placed in the center of the ring. The reflected image of the surface profile is captured with a high-speed CCD camera on top of the mirror

system is observed from the top via a conical mirror placed in the center of the ring, similar to Ref. [29]. Thus a side view of the whole channel is captured with a single high-speed digital camera (resolution: 1280×1024 pixels at rates up to 500 images per second). Figure 14(a) shows an anamorphotic image reflected from the conical mirror. The wavy granular surface is seen as a jagged ring around the tip of the cone. For reconstructing the true shape of the profile $h(\varphi, t)$ digital image processing is performed which delivers $360°$ panoramic side views of the granular profile in the channel as presented in Figs. 14(b) and 15. The spatial resolution is sufficient for detecting single particles of 2 mm size. The channel is lit from outside through diffusive parchment paper wrapped around the outer wall, hence particles appear dark in front of a bright background.

(a)

(b)

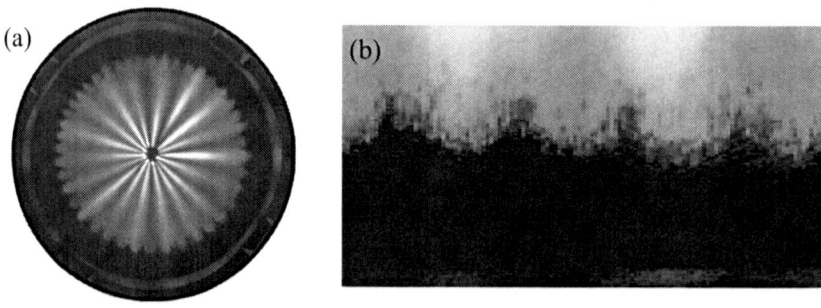

Fig. 14. (a) Anamorphotic image reflected from the conical mirror. (b) Section of the reconstructed granular surface

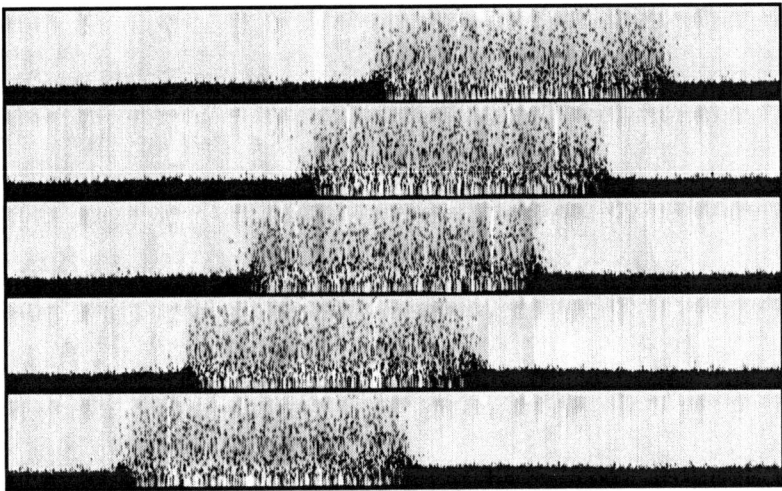

Fig. 15. Snapshots through the inner side wall of the channel covering 360° taken at $y(t) = 0$ during the downwards motion of the container. Time increases from top to bottom by 1.72 seconds (20 cycles) between consecutive snapshots. For clarity all images are stretched in the vertical direction by a factor of four ($f = 11.6\,\text{Hz}$, $\Gamma = 1.23$)

7 Coexistence of Condensed and Fluidized Phases

Finally, a rather surprising pattern has been observed in a *single* layer of monodisperse beads (see Fig. 15). At peak forcing accelerations between 1.1 g and 1.5 g a solid-like and a gas-like domain coexist. The solid fraction L_s/L_0 decreases with increasing acceleration and shows hysteresis (Fig. 16). The sharp boundaries between the two regions travel around the channel faster than the particles are transported. Complementary to our experimental studies a molecular dynamics simulation is used to extract local granular temperature and number density [30]. It is found that the number density in the solid phase is several times that in the gas, while the granular temperature is orders of magnitude lower. This system shows that equipartition of energy can be violated by coexisting gaseous and solid domains, even though particle motion is fully three-dimensional and not restricted by guiding partitions. The rotation of the solid phase in the annular conveyor demonstrates that the coexistence of solid and fluid regions is not caused by small potential inhomogeneities in the forcing, particle container interactions or a tilt of the apparatus.

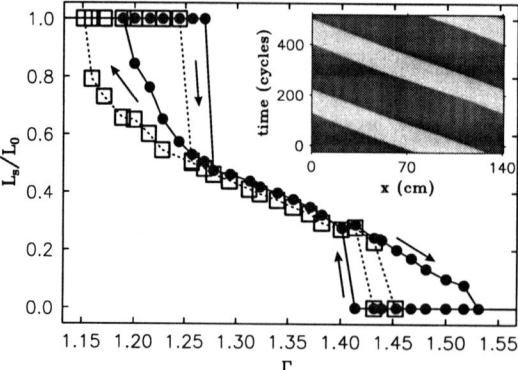

Fig. 16. Solid fraction L_s/L_0 as a function of the peak container acceleration Γ in experiment (*filled circles*) and simulation (*open squares*). Arrows indicate how the system evolves in the hysteresis loops. The inset is a space-time diagram of the granular density. Solid regions appear dark

8 Concluding Remarks

To summarize, the vibratory conveyor system presented here has demonstrated its potential for the systematic investigation of the transport properties of granular materials in a systematic way. Considering the granular pattern formation, the described annular apparatus is unrivaled for its capability to study solid-liquid transitions since, due to the permanent transport of all particles at periodic boundary conditions, the influence of spatial inhomogeneities is excluded: (i) propagating patterns which persist along the whole system cannot be caused by local inhomogeneities of the container and (ii) one can wait until any transient due to coarsening processes of the developing patterns have disappeared.

For industrial applications, the observed reversal effect is relevant as the direction of a granular flow is selected through the frequency of the excitation alone. One can employ such two-way conveyors for example in larger cascading transport systems as control elements to convey the material to different processes as needed.

For a further understanding of the general behavior of granular material on vibratory conveyors the next steps are (i) the development of a better understanding of *segregation phenomena* in multidisperse systems, (ii) the investigation of *clustering patterns* in submonolayers, and (iii) the qualitative modelling of the *spatiotemporal surface structure*. In particular, regarding the travelling oscillon patterns, the model by Eggers and Riecke [31] for pure vertical vibration seems to be a suitable starting point for generalization.

Acknowledgements

We would like to thank H. El hor, F. Landwehr, S.J. Linz, J. Kreft, T. Schnautz, and S. Strugholtz, for valuable discussions. Support by Deutsche Forschungsgemeinschaft (DFG-Sonderprogramm 'Verhalten granularer Medien') is gratefully acknowledged.

References

1. G. Pajer, H. Kuhnt, F. Kuhnt: *Fördertechnik – Stetigförderer*, 5th edn. (VEB Verlag Technik, Berlin 1988)
2. F.J.C. Rademacher, L. Ter Borg: Eng. Res. **60**, 261 (1994)
3. E.M. Sloot, N.P. Kruyt: Powder Technol. **87**, 203 (1996)
4. A.W. Gerstel, J.G.R. Scheublin: Bulk Solids Handling **14**, 573 (1994)
5. I. Derényi, P. Tegzes, T. Vicsek: Chaos **8**, 657 (1998)
6. Z. Farkas, P. Tegzes, A. Vukics, T. Vicsek: Phys. Rev. E **60**, 7022 (1999)
7. F.J.C. Rademacher: Bulk Solids Handling **15**, 41 (1995)
8. S.G.K. Tennakoon, R.P. Behringer: Phys. Rev. Lett. **81**, 794 (1998)
9. R. Grochowski, S. Strugholtz, P. Walzel, C.A. Krülle: Chemie Ingenieur Technik **75**, 1103 (2003)
10. M. Rouijaa, C. Krülle, I. Rehberg, R. Grochowski, P. Walzel: Chemie Ingenieur Technik **76**, 62 (2004)
11. M. Rouijaa, C. Krülle, I. Rehberg, R. Grochowski, P. Walzel: Chem. Eng. Tech. **28**, 41 (2005)
12. R. Grochowski, P. Walzel, M. Rouijaa, C. A. Kruelle, I. Rehberg: Appl. Phys. Lett. **84**, 1019 (2004)
13. R. Grochowski, S. Strugholtz, H. El hor, S.J. Linz, P. Walzel: 'Transport Properties of Granular Matter on Vibratory Conveyors'. In: *Proceedings of International Congress for Particle Technology (PARTEC 2004) at Nuremberg, March 16–18, 2004*
14. C.A. Kruelle, M. Rouijaa, A. Götzendorfer, I. Rehberg, R.Grochowski, P. Walzel, H. El hor, S.J. Linz: 'Reversal of a Granular Flow on a Vibratory Conveyor'. In: *Powders and Grains 2005*, ed. by R. Garca-Rojo, H.J. Herrmann, S. McNamara (Balkema, Leiden 2005) pp. 1185-1189
15. H. El hor, S.J. Linz: J. Stat. Mech. L02005 (2005)
16. H. El hor, S.J. Linz, R. Grochowski, P. Walzel, C.A. Kruelle, M. Rouijaa, A. Götzendorfer, I. Rehberg: 'Model for Transport of Granular Matter on Vibratory Conveyors'. In: *Powders and Grains 2005*, ed. by R. Garca-Rojo, H.J. Herrmann, S. McNamara (Balkema, Leiden 2005) pp. 1191-1195
17. M.-O. Hongler, P. Cartier, P. Flury: Phys. Lett. **135**, 106 (1989)
18. F. Landwehr, R. Lange, P. Walzel: Chemie Ingenieur Technik **69**, 1422 (1997)
19. F. Landwehr, P. Walzel: Chemie Ingenieur Technik **71**, 1167 (1999)
20. F. Melo, P.B. Umbanhower, H.L. Swinney: Phys. Rev. Lett. **75**, 3838 (1995)
21. C.A. Kruelle, S. Aumaître, A.P.J. Breu, A. Goetzendorfer, T. Schnautz, R. Grochowski, P. Walzel: 'Phase Transitions and Segregation Phenomena in Vibrated Granular Systems'. In: *Advances in Solid State Physics 44*, ed. by B. Kramer (Springer, Berlin 2004) pp. 401-414
22. A. Götzendorfer, C.A. Kruelle, I. Rehberg: 'Granular Surface Waves in a Vibratory Conveyor'. In: *Powders and Grains 2005*, ed. by R. Garca-Rojo, H.J. Herrmann, S. McNamara (Balkema, Leiden 2005) pp. 1181-1184
23. S. Douady, S. Fauve, C. Laroche: Europhys. Lett. **8**, 621 (1989)
24. H.K. Pak, R.P. Behringer: Phys. Rev. Lett. **71**, 1832 (1993)
25. F. Melo, P. Umbanhower, H.L. Swinney: Phys. Rev. Lett. **72**, 172 (1994)
26. T. Metcalf, J.B. Knight, H.M. Jaeger: Physica A **236**, 202 (1997)
27. P.B. Umbanhowar, F. Melo, H.L. Swinney: Physica A **249**, 1 (1998)
28. C. Bizon, M.D. Shattuck, J.B. Swift, W.D. McCormick, H.L. Swinney: Phys. Rev. Lett. **80**, 57 (1998)

29. E. van Doorn, R.P. Behringer: 'Wavy Instability in Shaken Sand'. In: *Powders and Grains 1997*, ed. by R.P. Behringer, J. Jenkins (Balkema, Leiden 1997) pp. 397-400
30. A. Götzendorfer, J. Kreft, C.A. Kruelle, I. Rehberg: Phys. Rev. Lett. **95**, 135704 (2005)
31. J. Eggers, H. Riecke: Phys. Rev. E **59**, 4476 (1999)

Erosion Waves: When a Model Experiment Meets a Theory

Eric Clement[1], Florent Malloggi[1], Bruno Andreotti[1], and Igor S. Aranson[2]

[1] ESPCI-Universités Paris 6 and 7, Laboratoire de Physique et Mécanique des Milieux Hétérogénes, UMR7636, 10, rue Vauquelin 75005 Paris, France
[2] Materials Science Division, Argonne National Laboratory, 9700 South Cass Avenue, Argonne, IL 60439

Summary. We present recent results on two laboratory scale avalanches experiments taking place both in the air and under-water. In both cases, a family of solitary erosion/deposition waves are triggered. At higher inclination angles, we show the existence of a linear long wavelength transverse instability followed by a coarsening dynamics and finally, the onset of a fingering pattern. Both experiments strongly differ by the spatial and time scales involved, nevertheless, the quantitative agreement between the stability diagram, the wavelengths selection and the avalanche morphology suggest a common erosion/deposition scenario. These experiments are studied theoretically in the framework of the "partial fluidization" model that was developed earlier to describe dense granular flows. This model identifies a family of propagating solitary waves displaying a behavior similar to the experimental observation. A primary cause for the transverse instability is directly related to the dependence of the solitary wave velocity on the granular mass trapped in the avalanche, a results recovered experimentally.

1 Introduction

Avalanching processes leading to catastrophic transport of various natural materials do not only occur in the air as we know of snow avalanches, mud flows and their catastrophic human and economical toll. Such events frequently happen below the see level as they take many forms from turbidity currents to thick sediment waves sliding down the continental shelf. This is a fundamental feature shaping the submarine morphology. From the modelling of risks point of view, important questions still remain such as to evaluate to which extend an initial triggering event (an earth quake, an eruption..) would be responsible for a subsequent process that might propagate or amplify over large distances as an unstable matter wave. Unfortunately, the dynamics of such catastrophic events remains an issue so far lacking of conceptual clarity [1, 2] since (i) the rheology of the flows involved in an avalanche is complex and still not unravelled, (ii) the physics of erosion/deposition mechanisms is essentially limited to empirical descriptions based on dimensional analysis and semi-empirical formulations. There were several theoretical attempts to describe from a phenomenological point of view the dynamics of erosion waves as an interplay between a rolling phase and a static phase [3, 4]. While extensive laboratory-scale experiments on dry and submerged granular materials flowing on rough inclined plane [5–7] have brought new perspectives for the elaboration of reliable constitutive relations, many open questions still remain such as to understand and model avalanches propagation on erodible substrates [8–10]. It has been shown experimentally that families of localized triangular shape avalanches can be triggered in the metastability domain, between the stoppage angle and maximal avalanche angle. [8]. Also, the shape of other localized droplet-like waves was recently shown to depend strongly on the intimate nature of the granular material used [9]. All these questions are closely related to the compelling need for reliable description of the fluid/solid transition for particulate assemblies in the vicinity of the flow arrest. Here, we present experimental results concerning avalanche fronts developing over an erodible granular substrate, both in the air and under water. We know that avalanche fronts flowing on solid rough substrates (non-erodible) are transversally stable, the transverse coupling due to gravity being essentially a stabilizing mechanism [5, 11]. But, when segregation occurs, an avalanche front on a rough substrate may exhibit a fingering pattern explained by a pinning mechanism [12, 13]. Although the rough grains we use have a narrow polydispersity (25%), we investigate here a quite different mechanism. We demonstrate the existence of a linear transverse instability of the solitary front occurring at higher inclination angles.

Recently, a model of "partially fluidized" dense granular flows was developed to couple a phenomenological description of a solid/fluid transition with hydrodynamic transport equations. It reproduces many features found experimentally such as metastability of a granular deposits, triangular down-hill and balloon-type up-hill avalanches and variety of shear flow instabilities [14, 15]. The model was later calibrated with molecular dynamics simulations [16]. Here the partial fluidization model is applied to the previous situation of solitary avalanches flowing over a thin erodible sediment layer. A set of equations describing the dynamics of fully eroding waves is derived and a family of solitary wave solutions propagating downhill is obtained. The velocity and shape selection of these waves is investigated as well as the existence of a linear transverse instability. The primary cause for the transverse instability is associated with the dependence of the soliton velocity with the mass trapped in the flow. A numerical study is conducted to follow the nonlinear evolution of the avalanche front. All these features are discussed in the context of the previous experimental findings [10]. New perspectives for quantitative contact between modelling and experiments are then underlined.

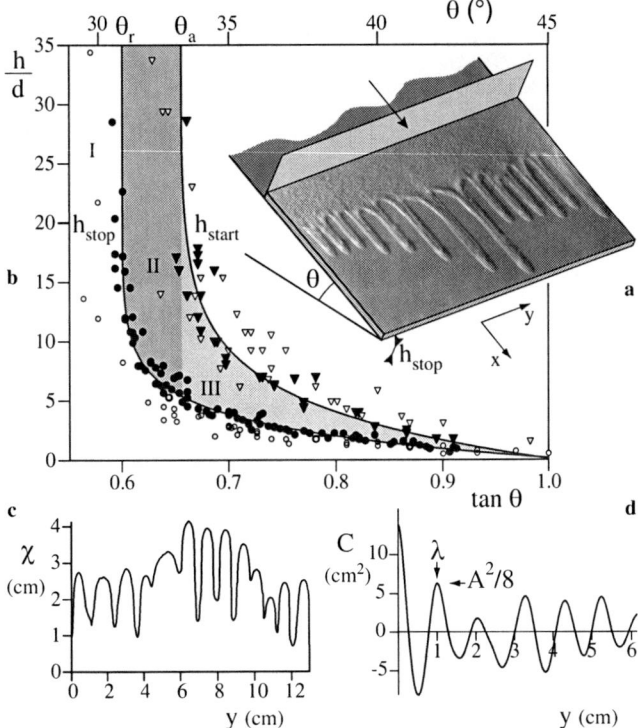

Fig. 1. a Experimental set-up. **b** Stability diagram: h_{stop} is the thickness of the sediment left after an avalanche for a given angle θ, in air (●) and in water (○); $h_{start}(\theta)$ is the maximum stable height of sediment, in air (▼) and in water (▽). $h_{start}(\theta)$ and $h_{stop}(\theta)$ are fitted by the form $h = b\ln((\tan\theta - \mu)/\delta\mu)$ (solid lines). In region I, an avalanche front cannot propagate autonomously down the slope: the perturbation is bound to fade away when the driving stops. Avalanches triggered in region II are stable while they exhibit a transverse instability in region III. In particular, solitary erosion waves are evidenced when starting from the stable height h_{stop}. **c** Front profile $\chi(y)$ obtained after image processing by a correlation technique. **d** The corresponding correlation function $C(y)$ allows to define the average wavelength λ and amplitude A.

2 Experiments on Erosion Waves

2.1 Description of the Set-Ups

The avalanching set-ups consist of a thin layer of grains deposited on a substrate that can be tilted at a value θ (fig. 1a). The dry granular set-up is similar to the one of Daerr et al. [8, 17]. The avalanche track is 70 cm wide and 120 cm long. The granular medium is Fontainebleau sand of a medium size $d = 300 \pm 60$ μm and the track bottom is made of black velvet. For under

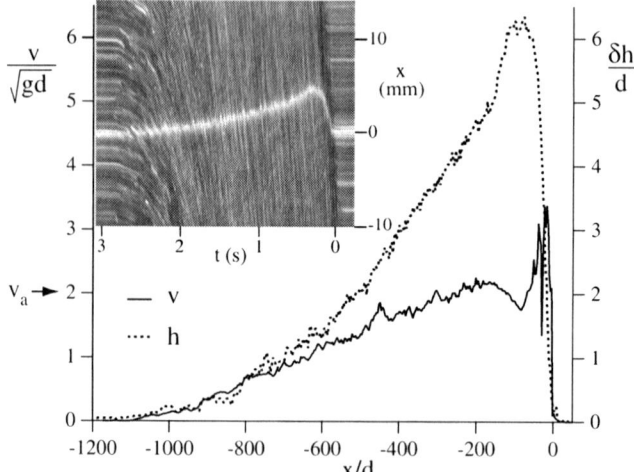

Fig. 2. Solitary erosion wave profile $\delta h = h - h_{\text{stop}}$ rescaled by d (dotted line) and surface velocity profile v rescaled by $\sqrt{gd}$ (solid line) (dry, $\theta = 32\,\text{deg}$, $h_{\text{stop}} = 2.3\text{ mm} = 7.8\ d$, region II). Inset: spatio-temporal diagram done with a fast camera (125 Hz), showing the particle motion as well as the profile height (deflection of the laser sheet). It can be observed that the surface grain velocity tends at the front towards the solitary wave velocity v_a

water avalanches, the set up size is quite smaller. The avalanche track is the bottom of a plexiglass tank that can be tilted up to an angle θ from an horizontal position. The avalanche track width is 15 cm and so is the track length. The granular sediment is an aluminum oxide powder of size either $d = 30\ \mu\text{m}$ or $40 \pm 11\ \mu\text{m}$. To avoid inter-particle cohesion, it is sufficient to maintain the pH value close to 4 by adequate addition of hydrochloric acid [18]. The substrate is initially set at an horizontal position and a fixed mass of powder is poured and suspended by vigorous stirring. A uniform sediment layer of height h then forms within 10 min. The bottom is an abraded but transparent plexiglass plate which offers the possibility to monitor the avalanche dynamics by transparency when illuminated from below. The profile of the avalanche front $h(x,t)$ is obtained with a laser slicing technique and is resolved within $30\ \mu\text{m}$ (0.1 d) in the dry case. The front dynamics is quantitatively monitored by image processing of the avalanche front pictures. The front line equation $\chi(y,t)$ is then extracted (fig. 1c) and the front line auto correlation function $C(y,t) = < \chi(y+y',t)\chi(y',t) >_{y'}$ is computed. Then, the correlation function first maximum is identified from which we define the average wavelength λ and the amplitude $A = 2\sqrt{2C(\lambda)}$ (fig. 1d). In addition, for dry avalanches, we measure the surface velocity field using a Particle Image Velocimetry technique.

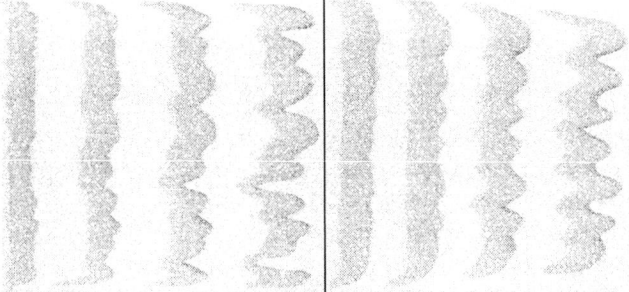

Fig. 3. Flowing part of solitary waves visualized by image difference (air, $d = 300$ μm, $\theta = 35$ deg, time interval 1.1 s), starting from a flat bed (left) or from an initial bed presenting a forced wavelength $\lambda = 6.5$ cm.

2.2 The Sand Layer Stability Diagrams

It has been shown that the stability of dry granular layers of depth h lying on a substrate inclined at an angle θ can be simply apprehended by a diagram with two branches [5] (fig. 1b) $h_{\text{start}}(\theta)$ and $h_{\text{stop}}(\theta)$ with the following interpretation: a uniform deposit of height h will globally loose stability if tilted above the angle θ defined by $h = h_{\text{start}}(\theta)$ and the avalanching process will leave at rest a deposit of height $h_{\text{stop}}(\theta)$. The h_{start} and h_{stop} curves diverge at an asymptotic angle limit, respectively equal to the avalanche angle of the granular pile θ_a and to the repose angle θ_r. Between the two, a domain of metastability for the granular deposit is present. Interestingly, the stability curves obtained for dry and underwater layers bear the same features and fall on the same curve when the deposited height is rescaled d (fig. 1b).

2.3 Solitary Erosion/Deposition Waves

To initiate avalanche fronts both in air and under water, we designed a 'bulldozer' technique where a plate perpendicular to the avalanche track scrapes the sediment at a constant velocity (fig. 1a). Although our results on avalanche stability are valid in the whole metastable region (fig. 1b), we will limit ourselves here to experiments started from a stable sediment layer of height $h_{\text{stop}}(\theta)$. Once an autonomous avalanche front separates from the plate, the bulldozer driving stops. For $\theta_r < \theta < \theta_a$, we always obtain transversely stable autonomous avalanche fronts, both in the wet and dry cases. We observed that the avalanche quickly converges toward a form which then remains constant. Furthermore, this solitary wave is found to be quite insensitive to the avalanche preparation details within a range of scraping velocities or initial masses set into motion. For this systematic study, we have kept a constant scraping velocity at about one-third of the typical avalanche velocity v_a. For each value of the – unique – control parameter θ, there is thus a single possible solitary erosion wave. In water, v_a is of the order of the Stokes velocity $\dfrac{\Delta\rho}{\rho_w}\dfrac{gd^2}{18\,\nu} \simeq 2$ mm/s where $\dfrac{\Delta\rho}{\rho_w} = 3$ is the density contrast between grains and water, ν the water kinematic viscosity and g the gravity acceleration. In the air, the propagation velocity is of the order of $\sqrt{gd} \simeq 5$ cm/s. In figure 2 we show local sediment height h and local surface velocity v profiles for such an avalanche. Independently, we measured the flow rule on homogeneous steady flows and found $v/\sqrt{gh} = \beta(h/h_{\text{stop}} - 1)$, with $\beta = 0.8$, as previously found for sand [6]. Here, this equilibrium relation remains verified in the tail of the avalanche.

2.4 The Transverse Instability

For $\theta > \theta_a$ the neutral wave fronts are transversally unstable. It is worth noticing that for the same angles, avalanches down a solid rough plate are stable (at least in the dry case). After the initial instability, we have identified a sequence of fusion processes increasing the spatial modulation lengths (coarsening scenario). Finally, the transverse destabilization ends up as a fingering pattern. In this final stage, the flowing zones are disconnected one from the others so that the wavelength does not evolve anymore. On figure 4, we display a typical time evolution of the dominant wavelength extracted from the correlation function. In inset, a typical fusion event is displayed to illustrate the coarsening scenario. Because of the competition between unstable modes and the coarsening process taking place, the identification of a generic scenario for the transverse instability is problematic.

This is the reason why, in addition to the experiments started from a flat bed we just described, we performed series of experiments starting from a modulated initial condition. The modulation at a given wavelength is simply produced by imprinting on the sediment surface regularly spaced thin scarification (shallow scratches). We find that the forced modes always fade away in region II, but on the other hand, in region III, the front modulations amplifies exponentially for a wide band of modes. The linear regime is clearly evidenced over one decade in amplitude. Non-linear effects start being visible when the amplitude becomes of the order of 1 cm. The inset of figure 5 shows the dispersion relationship deduced from these experiments, which demonstrates the existence of an initial long wavelength linear instability.

For experiments both in the air and under water performed in the unstable regime, we extract the two characteristic wavelengths. The initial wave length λ_0 would correspond, to the best of our experimental possibilities, to the fastest growing mode of the linear regime. Then, the wave length λ_∞ is taken at the onset of the fingering instability. In fig. 6, we display both wavelengths rescaled by the grains sizes : λ_0/d and λ_∞/d, as a function of the inclination angle θ. The selected wavelengths are typically larger than a grain size by at least two orders of magnitude. Note that the largest wavelengths

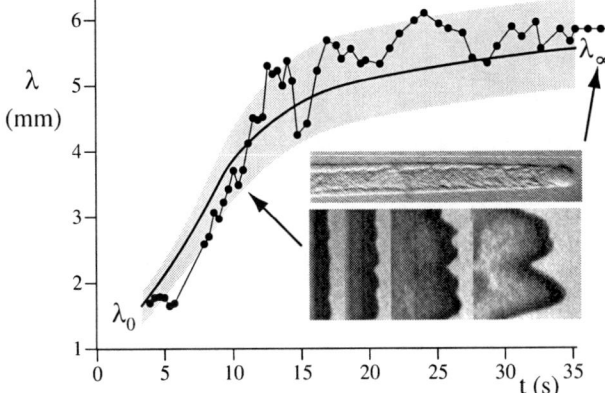

Fig. 4. Time evolution of the wavelength λ (water, $d = 40\ \mu m$, $\theta = 37.1$ deg) in a single typical realization ($\bullet$) and averaged over realizations (solid line) – the shadow zone indicates the standard deviation. After a small plateau at the initial wavelength λ_0, λ increases due to merging processes (lower photograph) until the value λ_∞ which corresponds to the formation of non-interacting fingers (upper photograph).

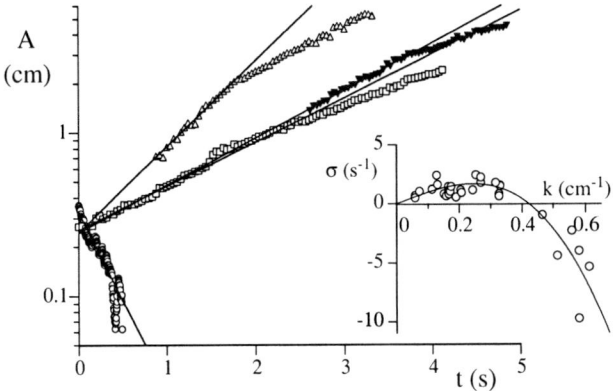

Fig. 5. Time evolution of the amplitude (air, $d = 300\ \mu m$, $\theta = 35$ deg) when the initial condition is forced at a given wavelength $\lambda = 12$ mm ($\bigcirc$), $\lambda = 30$ mm ($\square$), $\lambda = 90$ mm ($\triangle$) and $\lambda = 178$ mm ($\blacktriangledown$). Inset: linear growth rate σ as a function of the wave number k. The solid line is the best fit by $\sigma = \sigma_m |k| \lambda_0 (1 - (k\lambda_0)^2/3)$, with a maximum growth rate $\sigma_m \simeq 2.5\ s^{-1}$ for $\lambda_0 \simeq 4$ cm. Measurement of λ_0 from an undisturbed solitary wave (fig. 3) gives 3.3 cm.

measured are of the order of the track width (1800 d in water and 750 d in air). Furthermore, in the limit of finite size effects and measurements uncertainties, we find that a value $\theta \cong \theta_a$ corresponds to a diverging boundary for the initial

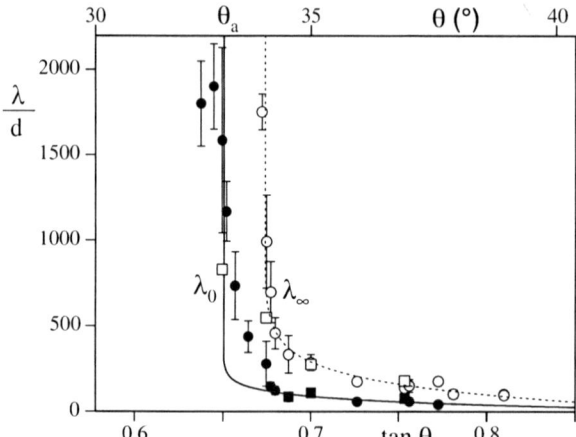

Fig. 6. Initial ($\bullet$) and final ($\bigcirc$) wavelengths rescaled by d as a function of θ ($d =$ 40 μm, in water). The initial wavelength data in air ($\blacksquare$) and water ($\bullet$) coincide, as well as the final wavelength data in air ($\square$) and water ($\bigcirc$). The error bars correspond to the dispersion of the data from a realization to the other. As λ_0 diverges at θ_a, we have superimposed the curve 10 $h_{\text{start}}(\theta)$ (solid line), which is a good approximation of λ_0 to the first order. The dotted line is the best fit of the final wavelength λ_∞ by the same logarithmic form as $h_{\text{start}}(\theta)$ or $h_{\text{stop}}(\theta)$.

wavelength λ_0/d. Hence this is a signature of a zero wave-number instability with a threshold close to θ_a. Another striking feature is the collapse, *on the same curve*, of data obtained in the air and underwater, once rescaled by the grain size. In the range of parameters where the fingering regime is reached before the end of the track, the ratio of the final to the initial wavelength is approximately constant and equal to $\lambda_\infty/\lambda_0 \simeq 3.5$. The presence of a fingering instability is a quite fascinating feature of this avalanching process. Here, the fingering front stems from the onset of localized propagating waves following the transverse instability regime. These fingers are localized matter droplets with levees on the side and propagating in a quasi solitary mode and when they are fully developed, their selected width is found to be quite sensitive to the slope ($\simeq \lambda_0$ for both wet and dry cases).

3 The Partial Fluidization Model

3.1 The Model Presentation

In this Section we apply the partial fluidization model to investigate the avalanche dynamics on a thin erodible sediment layer. A set of equations describing the dynamics of fully eroding waves is derived and a family of solitary solutions propagating downhill is obtained. The velocity and shape selection of these solitary waves is investigated as well as the existence of a linear transverse instability. According to the partial fluidization theory [14], the ratio of the static part of shear stress to the fluid part of the full stress tensor is controlled by an order parameter (OP) ρ, which is scaled in such a way that in granular solid $\rho = 1$ and in the fully developed flow (granular liquid) $\rho \to 0$. At the "microscopic level" OP is defined as a fraction of the number of persistent particle contacts to the total number of contacts. Due to a strong dissipation in dense granular flows, ρ is assumed to obey purely relaxational dynamics controlled by the Ginzburg-Landau equation for generic first order phase transition,

$$\tau_\rho \frac{D\rho}{Dt} = l_\rho^2 \nabla^2 \rho - \frac{\partial F(\rho, \delta)}{\partial \rho}. \tag{1}$$

Here $\tau_\rho, l_\rho \approx d$ are the OP characteristic time and length scales, d is the grain size. $F(\rho, \delta)$ is a free energy density which is postulated to have two local minima at $\rho = 1$ (solid phase) and $\rho = 0$ (fluid phase) to account for the bistability near the solid-fluid transition. The relative stability of the two phases is controlled by the parameter δ which in turn is determined by the stress tensor. The simplest assumption consistent with the Mohr-Coulomb yield criterion is to take it as a function of $\phi = \max|\sigma_{mn}/\sigma_{nn}|$, where the maximum is sought over all possible orthogonal directions m and n.

For thin layers on inclined plane Eq. (1) can be simplified by fixing the structure of OP in z-direction (z perpendicular to the bottom, x is directed down the chute and y in the vorticity direction) $\rho = 1 - A(x, y)\sin(\pi z/2h)$, h is the local layer thickness, A is slowly-varying function. This approximation valid for thin layers when there is no formation of static layer beneath the avalanche. Then one obtains equations governing the evolution h and A, coordinates x, y, height h, and time t are normalized by l_ρ, τ_ρ correspondingly [14, 15],

$$\frac{\partial h}{\partial t} = -\alpha \frac{\partial h^3 A}{\partial x} + \frac{\alpha}{\phi} \nabla \left(h^3 A \nabla h \right) \tag{2}$$

$$\frac{\partial A}{\partial t} = \lambda_0 A + \nabla^2 A + \frac{8(2 - \delta)}{3\pi} A^2 - \frac{3}{4} A^3 \tag{3}$$

where $\nabla^2 = \partial_x^2 + \partial_y^2$, $\lambda_0 = \delta - 1 - \pi^2/4h^2$, dimensionless transport coefficient:

$$\alpha \approx \frac{2(\pi^2 - 8)}{\pi^3 \nu} g \tau_\rho l_\rho \sin \theta, \tag{4}$$

ν is the shear kinematic viscosity, θ is the chute inclination, $\phi = \tan \theta$. Control parameter δ assumes the form $\delta(\tilde\theta) = ((\tan \tilde\theta)^2 - \phi_0^2)/(\phi_1^2 - \phi_0^2)$, $\phi_{0,1}$ are tangents of dynamic and static repose angles correspondingly, $\tan \tilde\theta$ is the local slope of granular layer. Assuming that the slope of the layers $\tan \tilde\theta$ is close to the chute slope $\tan\theta$, we expand the control parameter $\delta \approx \delta_0 + \beta h_x, \delta_0 = \delta(\theta)$, $\beta \approx 1.5 - 3$ depending on the value of θ, see for detail [14, 15]. The last term in Eq. (2) is also due to change of local slope and is obtained from expansion $\tilde{\theta} = \theta + h_x$. This term is responsible for the saturation of the slope of avalanche front (without it the front can be arbitrary steep) [15].

3.2 Solitary Wave Shape Selection

In the coordinate system co-moving with the velocity V Eqs. (2),(3) assume the form

$$\frac{\partial h}{\partial t} = V\partial_x h - \alpha\frac{\partial h^3 A}{\partial x} + \frac{\alpha}{\phi}\nabla\left(h^3 A\nabla h\right) \tag{5}$$

$$\frac{\partial A}{\partial t} = V\partial_x A + \lambda_0 A + \nabla^2 A + \frac{8(2-\delta)}{3\pi}A^2 - \frac{3}{4}A^3 \tag{6}$$

Numerical studies revealed that the one-dimensional Eqs. (5),(6) possess a one-parametric family of localized (solitary) solutions, see Fig 7:

$$A(x,t) = A(x - Vt), h(x,t) = h(x - Vt) \tag{7}$$

Here the boundary conditions take a form $h \to h_0, A \to 0$ for $x \to \pm\infty$, where h_0 is the asymptotic height. The one-dimensional steady state solitary wave solution (7) satisfy

$$V(h - h_0) = \alpha h^3 A\left(1 - \frac{\partial_x h}{\phi}\right) \tag{8}$$

$$-V\frac{\partial A}{\partial x} = \lambda_0 A + \partial_x^2 A + \frac{8(2-\delta)}{3\pi}A^2 - \frac{3}{4}A^3 \tag{9}$$

The solutions can be parameterized by the "trapped mass" m carried by the solitary wave i.e. the area above h_0,

$$m = \int_{-\infty}^{\infty}(h - h_0)dx \tag{10}$$

The velocity V is increasing function of m, see inset Fig. 7a. The family of admissible solutions for a propagative solitary wave terminates at $m = m_c$ and $V = V_c = V(m_c)$. The critical mass m_c decreases with the increase in α. The dependence of V vs m is consistent with experimental data, see inset to Fig. 7b for sandy avalanches in the air. Note that this experimental curve was obtained by collecting the falling sand at the end of the avalanche plane when erosion waves of different sizes were triggered. We also notice that below a mass threshold, no propagation of an erosive wave is possible. The structure of the solutions is sensitive to the value of α: for large α the solution has a well-pronounced shock-wave shape, Fig. 7a, with the height of the crest h_{max} several times larger than the asymptotic depth h_0. For $\alpha \to 0$ the solution assumes more rectangular form, see Fig. 7b, and $h_{max} - h_0 \ll h_0$. The results are consistent with the shape of sand (compare with large α) and glass bead ($\alpha \to 0$) avalanches, see inset of Fig. 7a

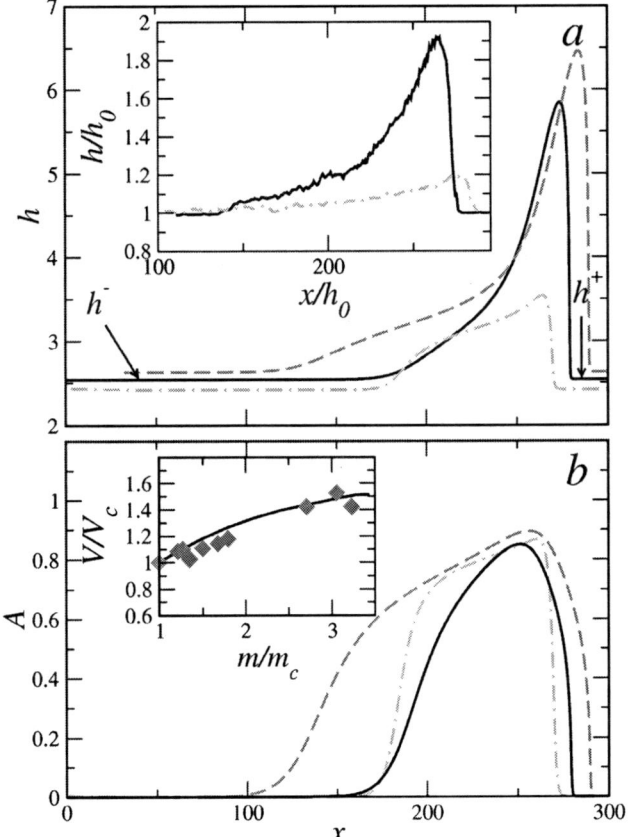

Fig. 7. h (a) and A (b) for various values of m and α. Solid line is for $m = 147.7$, $V = 2.72$, dashed line is for $m = 211$, $V = 3.12$, for $\delta = 1, \alpha = 0.08, \beta = 2$; point-dashed line is for $\alpha = 0.025, \delta = 1.15$, $m = 62$, $V = 0.86$. Inset to Fig. 7a: Representative height profiles for avalanches in sand (solid line) and glass beads (dashed line). Inset to Fig. 7b: V vs m (solid line), diamonds depict data for sand avalanches.

3.3 The Front Linear Transverse Instability

To understand *transverse instability* we focus on the solitary solution with slowly varying position $x_0(y, t)$

$$A(x, t) = \bar{A}(x - x_0(t, y)), \; h(x, t) = \bar{h}(x - x_0(t, y)) \tag{11}$$

Substituting Eq. (11) in Eq. (5) and integrating over x, one obtains

$$\partial_t m = V(m)(h^+ - h^-(m)) - \zeta_1 \partial_y^2 x_0 + \zeta_2 \partial_y^2 m \tag{12}$$

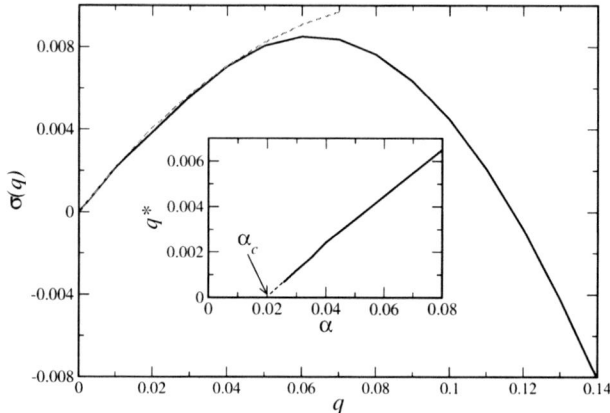

Fig. 8. Growthrate $\sigma(q)$ vs q for $\delta = 1.15$ and $\alpha = 0.08$ for the mass of the avalanche $m = 102$. Solid line: $\sigma(q)$ obtained by numerical stability analysis of one-dimensional solution Eq. (11). Dashed line is solution of Eq. (15). Inset: optimal wavenumber of q^* vs α for $\delta = 1.15$

where $\zeta_{1,2} = const$ is defined as

$$\zeta_1 = \frac{\alpha}{\phi} \int_{-\infty}^{\infty} \left(\bar{A}\bar{h}^3 \partial_x \bar{h} \right) dx, \quad \zeta_2 = \frac{\alpha}{\phi} \int_{-\infty}^{\infty} \left(\bar{A}\bar{h}^3 \partial_m \bar{h} \right) dx$$

Here $h^+ = h(x \to \infty)$ is the height of the deposit layer ahead of the front and $h^- = h(x \to -\infty)$ is the height behind the front, see Fig. 7a. While the value of h^+ is prescribed by the initial sediment height, the value of h^- behind the front is determined by the velocity (or mass) of the front. For steady-state solution $h^+ = h^- = h_0$. For the slowly-evolving solution the difference between h^+ and h^- can be small, however it is important for the stability analysis. These terms are also necessary to describe experimentally observed initial acceleration/slowdown of the avalanches. Substituting Eqs. (11) into Eq. (3) and performing orthogonality conditions one obtains

$$\partial_t x_0 = V(m) + \partial_y^2 x_0 \tag{13}$$

There are also higher order terms in Eq. (13) which we neglect for simplicity. To see the onset of the instability we keep only the leading terms in Eq.(12),(13), using $V(m) \approx V(m_0) + V_m(m - m_0)$, and $\tilde{m} = m - m_0 \ll m_0$:

$$\partial_t \tilde{m} = -\tau \tilde{m} - \zeta_1 \partial_y^2 x_0 + \zeta_2 \partial_y^2 \tilde{m}$$
$$\partial_t x_0 = V_m \tilde{m} + \partial_y^2 x_0 \tag{14}$$

where $m_0 = const$ is the steady-state mass of the solitary wave, and $\tau = V(m_0)\partial_m h^-$. Seeking solution in the form $m, x_0 \sim \exp[\sigma t + iqy]$, q is the transverse modulation wavenumber, for the most unstable mode we obtain from Eq. (14) the growthrate λ

$$\sigma = \frac{-q^2(1 + \zeta_2) - \tau + \sqrt{(q^2(1 - \zeta_2) - \tau)^2 + 4V_m\zeta_1 q^2}}{2} \qquad (15)$$

Expanding Eq. (15) for $q \to 0$ we obtain $\sigma \approx \frac{1}{2}(2V_m\zeta_1/\tau - 1)q^2 + O(q^4)$. The instability occurs if $V_m\zeta_1/\tau - 1/2 > 0$. Substituting τ and using $V_m/h_m = V_h$, we obtain a simple instability criterion:

$$2V_h\zeta_1/V > 1 \qquad (16)$$

Eq. (16) gives a value of threshold α since $\zeta_1 \sim \alpha$. For $\alpha < \alpha_c$ no instability occurs, and the modulation wavelength diverges for $\alpha \to \alpha_c$. Far away from the threshold we neglect τ and then obtain for $\lambda(q)$:

$$\sigma = |q|\sqrt{\zeta_1 V_m} - (1 + \zeta_2)q^2/2 + O(q^3) \qquad (17)$$

The optimal wavenumber q^* is given

$$q^* \sim \sqrt{\zeta_1 V_m} \sim \alpha \qquad (18)$$

Fig. 8 shows $\sigma(q)$ obtained by numerical stability analysis of linearized Eqs. (2), (3) near the one-dimensional solution Eq. (7). For comparison is shown the solution to Eq. (15), with the parameters extracted from the corresponding one-dimensional steady-state problem Eqs. (8),(9). One sees that Eq. (15) gives correct description for small q, however fails to predict $\sigma(q)$ in the whole range of q. For this purpose one needs to include higher order terms. Thus, Eq. (15) gives correct description of the onset of instability and qualitative estimate for the selected wavenumber q^*. Inset to Fig. 8 shows the dependence of optimal wavenumber q^* vs α, obtained by numerical linear stability analysis of the solitary solution. It shows almost linear decrease of q^* with α consistent with Eq. (18). For very small α the plot indicates that $q^* \to 0$ at $\alpha \to \alpha_c$, consistent with Eq. (16). From the qualitative point of view, the transverse instability of planar front is caused by the following mechanism: local increase of solitary wave mass results in the increase of its velocity and, consequently, "bulging" of the front. Since the bulge "rolls" forward, i.e below the level of the avalanche, the granular fluid flows towards the bulge, further draining the trailing regions.

3.4 Beyond the Linear Instability: Coarsening and Fingering

To study the evolution of the avalanche front beyond the initial linear instability regime a fully two-dimensional numerical analysis of Eqs. (2), (3) was performed. Integration was performed in a rectangular domain with periodic

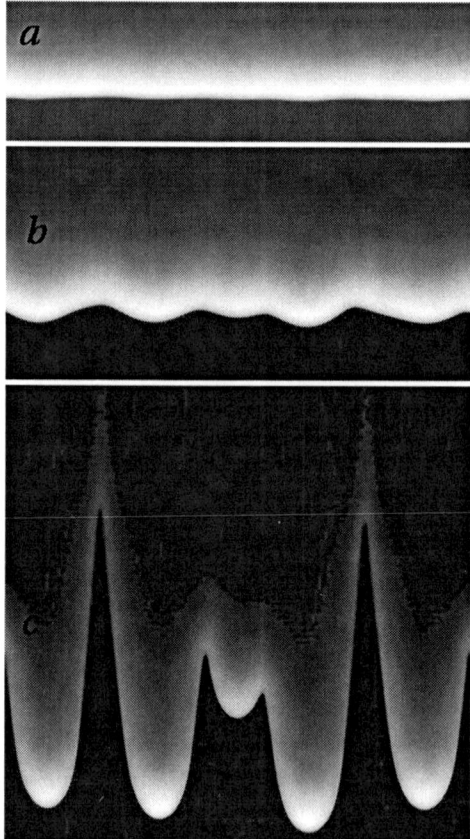

Fig. 9. Grey-coded images of height profile $h(x, y)$ (white corresponds to larger height) for three different moments of time, a) $t = 170$, b) $t = 300$ and c) $t = 500$ units of time. Domain size is 600 units in x direction and 450 units in y direction, only part of domain in x direction is shown. Parameters: $\delta = 1.16$, $\alpha = 0.14$, $\beta = 2$ and initial height $h_0 = 2.285$.

boundary conditions in x and y directions. Number of mesh points was up to 1200×600 or higher. As an initial condition we used a flat state $h = h_0$ with a narrow stipe $h = h_0 + 2$ deposited along the y-direction. To trigger the transverse instability small noise was added to the initial conditions. The initial conditions rapidly developed into a quasi-one-dimensional solution described by Eq. (7). Due to the periodicity in the x-direction, the solitary solution could pass through the integration domain several times. It allowed us to perform analysis in a relatively small domain in the x-direction. The transverse modulation of the solitary wave leading front was observed after about 100 units of time for the parameters of Fig. 9. We observe that modulation initially grows in amplitude, eventually coarsens and leads to the formation of large-scale finger structures.

4 Conclusions

Experimentally, we have investigated the dynamics of underwater and dry granular avalanches taking place on a erodible substrate. We have identified the domain of existence for solitary waves going down the slope without changing form. For angles larger than the avalanche angle, we proved the existence of a linear transverse instability which further develops via a coarsening fusion process and finally ends up as a fingering pattern. The existence of solitary waves provides a new important test to models. For instance, it may be shown that they cannot be recovered in Saint-Venant models that do not include a static erodible layer below the avalanche. The mechanism responsible for the instability remains yet to be identified but the scenario is a standard zero wave number instability of threshold θ_a. The inhibition of this instability on a solid bottom suggests that erosion/deposition processes in the avalanche depth could play a determinant role. Further studies with other materials on different substrates are needed to determine to which extent, the instability is generic. A further challenging experimental issue is to get a more focused vision on the interface separating the jammed and the rolling phases, and its relation to the instability onset. In the final stage of the instability, fingers appear as droplet like solutions of the erosion/deposition process and thus look essentially different from the segregation fingers reported on a rough substrate [12]. Note that their shape is reminiscent of many natural patterns obtained in debris or mud flows [13] which also display surprisingly well selected widths at values about hundreds of a typical rock size.

These experimental findings were put to test in the context of a phase field modelling developed to describe dense granular flows. At the qualitative level the agreement between theory and experiments is impressive. (i) Existence of steady-state solitary avalanches propagating downhill with a shape similar to experiment. (ii) Generic zero wave number (long-wave) transverse instability compatible with the experimental divergence of the selected wavelength close to the instability threshold. Far from the threshold, linear growth rate dependence with q compatible with measurements. (iii) Coarsening in the later development of the instability. (iv) Fingering instability with localized droplet-like avalanches (also similar to those described in [9]). The analysis predicts that the transverse instability ceases to exist when the rescaled transport coefficient α decreases (see Fig. 8). However, the model does not provide explicit expression for α due to the dependence of granular viscosity ν on other external parameters (e.g. local pressure, see [14]). Rough estimates of α can be extracted from the flow rules in Ref. [6] which gives the relation between depth-average velocity $\langle V \rangle$ and height h: $\langle V \rangle / \sqrt{hg} \approx \bar{\beta} h / h_{stop}(\theta) + const$, where dimensionless material constant $\bar{\beta} \approx 1$ for sand and $\bar{\beta} \approx 0.2$ for glass beads. Since flux of grains $J = h\langle V \rangle \approx \bar{\beta}\sqrt{g}h^{5/2}/h_{stop}(\theta)$, to compare with flux expression in Eq. (2), we write for the fully-fluidized state ($A \approx 1$): $J \approx \bar{\beta}\sqrt{g}h^3/h_{stop}^{3/2}(\theta)$. Since the typical time in the problem τ_ρ is of the order collision time $\sqrt{d/g}$, after rescaling $x \to x/d$, $h \to h/d$, $t \to t/\tau_\rho$, we obtain in the dimensionless form the estimate for $\alpha \approx \bar{\beta}(d/h_{stop}(\theta))^{3/2}$. Since $h_{stop} \to \infty$ with the decrease of angle θ, the instability should disappear for smaller angles, which is verified experimentally. The analysis also predicts that the instability could be suppressed for the case of small rheological parameter $\bar{\beta}$ corresponding to smooth glass beads. Thus, the model provides a crucial prediction on the transverse instability mechanism which lies in the dependence of the solitary wave velocity with the flowing mass trapped in the avalanche. This result is recovered experimentally. Still, important questions remains on how to bring more quantitative comparison between the theory and the experimental measurements. In this perspective, a challenging question is a deeper understanding of spatial and temporal parameters involved in the phase field equation (1). Along those lines an important question would be to clarify the qualitative differences observed between smooth glass bead and rough sandy materials as far as the effective flow rules and avalanches shapes are concerned. This is centered around the central and challenging question of erosion and jamming of a dense granular assembly. Note that the fingering patterns found for these granular avalanches, bear remarkable similarities with those existing in thin films flowing down inclined surfaces, both with clear and particle-laden fluids [19]. However, the physical mechanisms leading to this fingering are likely dissimilar: in fluid films, it is driven (and stabilized) by the surface tension, whereas in the granular flow case, the surface tension plays no role.

Acknowledgements

We thank Olivier Pouliquen, Philippe Claudin, Stephane Douady, Lev Tsimring, Tamas Börzsönyi and Robert Ecke for discussions and help. IA was supported by US DOE, Office of Science, contract W-31-109-ENG-38; BA, EC and FM are supported by ANR project "Catastrophes telluriques et Tsunami":PIGE.

References

1. R.M. Iverson, Reviews of Geophysics **35**, 245296 (1997).
2. M.A. Hampton, H.J. Lee and J. Locat, Reviews of Geophysics **34**, 3359 (1996).
3. J-P. Bouchaud, M. Cates, J.R. Prakash and S.F. Edwards, J. Phys. France I **4**, 1383 (1994).
4. A. Aradian, E. Raphaël, and P.-G. de Gennes, C. R. Physique **3**, 187 (2002)
5. O. Pouliquen, Phys. Fluids **11**, 542 and 1956 (1999).
6. G.D.R. Midi (collective work), Eur. Phys. J. E **14**, 341 (2004).
7. O. Pouliquen et al, *Powders & Grains 2005*, p. 850,ed. by R. Garcia-Rojo, H.J. Herrmann, and S. McNamaca; Balkema, Rotterdam.
8. A. Daerr et S. Douady, Nature **399**, 241 (1999).
9. T. Borzsonyi, T.C. Halsey, and R. E. Ecke, Phys. Rev. Lett. **94**, 208001 (2005).
10. F. Malloggi et al. *Powders & Grains*, p. 997, ed. by R. Garcia-Rojo, H.J. Herrmann, and S. McNamaca (Balkema Rotterdam, 2005); cond-mat/0507163, submitted to Phys. Rev. Lett. (2005)
11. O. Pouliquen, J.W. Valance, Chaos **9**, 621 (1999).
12. O. Pouliquen, J. Delour and S.B. Savage, Nature **386**, 816 (1997).
13. G. Felix and N. Thomas, Earth and Planetary Science Letters **221**, 197 (2004).
14. I.S. Aranson and L.S. Tsimring, Phys. Rev. E **64**, 020301 (2001); Phys. Rev. E **65**, 061303 (2002);
15. I.S. Aranson and L.S. Tsimring, submitted to Rev. Mod. Phys. (2005), cond-mat/0507419
16. D. Volfson and L.S. Tsimring and I.S. Aranson, Phys. Rev. E **68**, 021301 (2003).
17. A. Daerr, Phys. Fluids **143**, 2115 (2001).
18. A. Daerr, P. Lee, J. Lanuza and E. Clément , Phys.Rev.E, **67** 065201 (2003).
19. S.M. Troian, E. Herbolzheimer, S.A. Safran, and J.F. Joanny, Europhys. Lett. **10**, 25 (1989); J. Zhou, B. Dupuy, A.L. Bertozzi, and A.E. Hosoi, Phys. Rev. Lett. **94**, 117803 (2005).

Bidisperse Granular Flow on Inclined Rough Planes

Céline Goujon[1,2], Blanche Dalloz-Dubrujeaud[1], and Nathalie Thomas[2]

[1] IUSTI, Université de Provence, 5 rue Enrico Fermi,
   F- 13453 Marseille Cedex 13, France
[2] Present address : Instituto de Investigaciones en Materiales
   Universidad Nacional Autónoma de México,
   Apdo. Postal 70-360, Cd. Universitaria, México D.F. 04510, México

Summary. Experiments on bidisperse dry granular flows on an inclined rough plane were made in order to better understand the rheology of this kind of material. Flows, created by a localised input of a granular matter onto the plane, propagate and spread laterally, being unconfined by the experimental set-up. Because of size segregation, the lateral and vertical inhomogeneous repartitions of particles lead to several effects causing the main difference between bidisperse and monodisperse flow: the outline effect and the interfaces effects. The outline effect is due to the large beads at the front, and at the borders of the flow. It can be interpreted using the relative frictions of the two types of beads on the rough plane. The relative friction has been quantified with experiments on monodisperse granular flows. The interfaces effects deal with the interaction between the layers of large and small beads and with the interaction between the small beads and the plane. It combines friction, dragging and longitudinal separation of th etwo superposed layers of beads.

1 Introduction

The flow of dense granular matter on inclined plane is often encountered in engeineering applications involving the transport of material such as minerals and cereals. It is also common in geophysical situations where rocks avalanches, landslides and pyroclastic flows are natural events consisting in large-scale flow of grains [1, 2]. Most of these granular flows are composed of polydisperse grains. Many chute flow experiments have been carried out and different configurations have been used, changing the boundary conditions from smooth [3, 4] to rough [5] and using several kind of materials [6–9].
This paper presents results of bidisperse granular flow on inclined rough plane. These results have been compared to the one previously observed for monodisperse flows. To understand the modifications observed, it has been necessary to quantify the basal friciton of the beads on the plane (changing the diameter

of flowing beads, the relative roughness change) and with this quantification to identify the mechanisms able to explain the behaviour of bidisperse flows.

2 Experimental Setup

The experimental setup consists in an inclined plane which is made rough by gluing beads of diameter λ with a compactness C. The plane is 1.5 m long and 0.6 m width. We control precisely the angle of inclination θ. The experiments consist in a flow of glass beads of diameter d. Different experiments have been carried out : the measurement of the length L of a deposit left after the release of a finite volume contained in a cap, the measurement of the thickness deposit (h_{stop}) left on the plane by a steady state flow and the measurement of the mean velocity $\bar{u}$ of an unconfined steady state flow, (fig. 1).

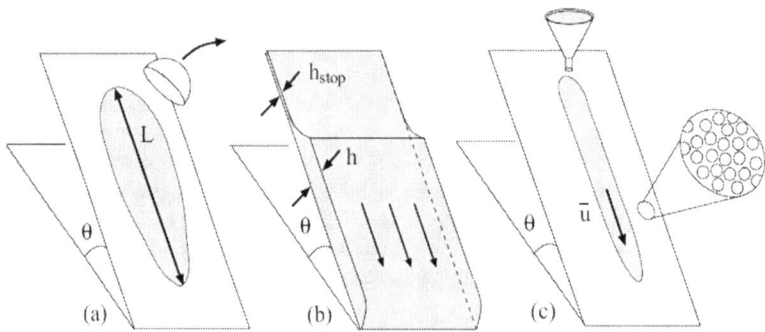

Fig. 1. Experiment configurations: (**a**) Release of a volume contained in a cap: deposit length measurement (**b**) Deposit left by a uniform steady state flow: deposit thickness measurement (**c**) Unconfined steady state flow: mean velocity measurement

3 Bidiperse Granular Flow

First obsevations on bidisperse granular flows show a great variety of morphologies (fig. 2). The morphology of a monodisperse flow presents always the same tear shape. On contrary, bidiperse flows present several morphologies like rapid stop of the flow, tear, fingering, streching. We also see a separation (segregation) between the large and the small beads.

In order to understand this varierty of morphologies, it is necessary to interest to the segregation phenomenum which is observed for each experiment. Because of the presence of two bead sizes, during the flow there is a separation

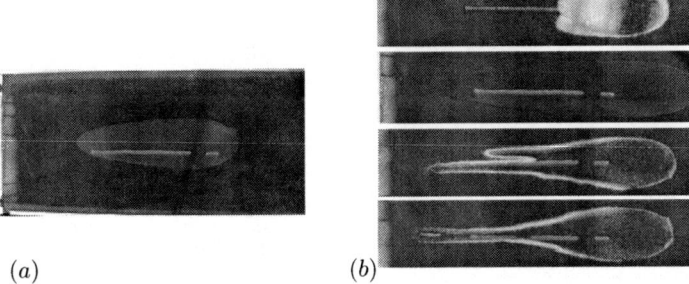

(a)   (b)

Fig. 2. Morphology of deposits (the largest beads are white and the smallest dark grey, the central white line is the laser sheet): (a) General morphology of a monodisperse deposit (b) Morphologies of bidisperse granular flows

by size, with the largest beads at the top of the flow. Because their velocity is higher, the largest beads are also at the front and at the periphery of the flow. This segregation is very rapid compare to the spreading time of the flow. We will consider it is instantaneous in our study. Because of this segregation, we can schematicly represent the flow as two superposed layers of beads (smallest one at the bottom and largest ones at the free surface) (fig. 3a) surrounded by the largest at the periphery of the flow (fig. 3b).

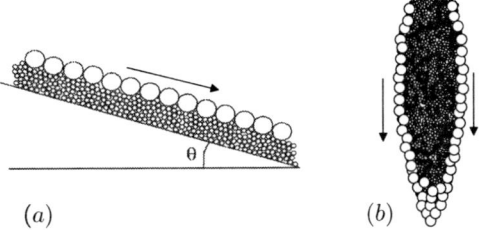

(a)   (b)

Fig. 3. Schematic representation of the flow during the release of a volume containes in a cap. (a) Segregation in the thickness, larger beads come at the top of the flow (b) Segregation at the free surface, larger beads come at the front and all around the flow

Because of this segregation, large and small beads are touching the same plane experiencing different relative roughness. We understand the influence of the relative roughness on the flow by studying monodisperse flow and changing the relative roughness (first part of this work). Understanding this, we will be able to explain the change in morphology of bidisperse flows in the second part.

3.1 Monodisperse Study: Influence of the Relative Roughness

We present results of monodisperse flows on a plane with the following characteristics : $\lambda = 425\ \mu m$ and $C = 0.56$. The diameter of the flowing beads varies
from $d = 150\ \mu m$ to $580\ \mu m$. In all experiments we observe a singularity for a
diameter of flowing bead $d_c = 275\ \mu m$. This singularity corresponds to a minimum length L of the deposit, a maximum thickness h_{stop} and a minimum in
the mean velocity $\bar{u}$ (fig. 4). As the volume is contant and the maximum width
of the deposit imposed by the cap [10], an increase in length is equivalent to
a decrease in thickness.

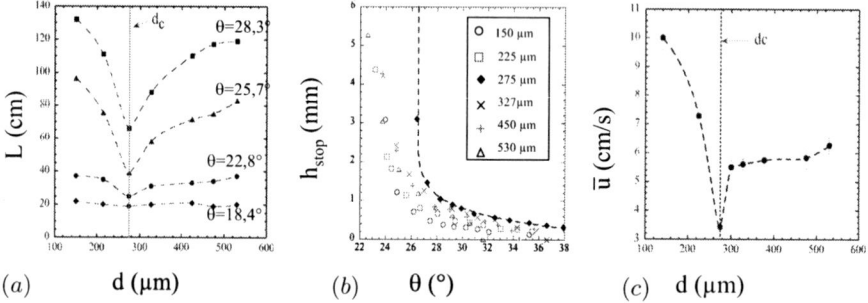

(a)   d (μm)   (b)   θ (°)   (c)   d (μm)

Fig. 4. Experimental results of monodispers flows on plane $\lambda = 425\ \mu m$ and $C =$
0.56 for several diameters of flowing beads: (a) Length of monodisperse deposit vs
the diameter of the flowing beads for several angles of inclination, (b) Thickness
h_{stop} of monodisperse deposit vs the angle of inclination for several diameters of
flowing beads (c) Mean velocity $\bar{u}$ vs the diameter of the flowing beads

On the h_{stop} versus θ curves, two characteristic angles are defined: θ_1 is the
inclination angle under which there is no flow and θ_2 the angle above which
there is no more beads on the plane (fig. 5a). The h_{stop} curve can be fitted
by the expression proposed by Pouliquen [11]:

$$\tan \theta = \tan \theta_1 + (\tan \theta_2 - \tan \theta_1)\ \exp \left(- \frac{h_{stop}}{Cd} \right)$$

where c is an adjustable parameter.

If we look at the evolution of θ_2 with the diameter of the flowing beads,
there is still the singularity at d_c (fig. 5b).

A geometrical model has been proposed to interpret this singularity [10].
It is based on the calculation of the angle required to make the last bead flow
down on the plane. This angle has been compared to the angle θ_2 experi-
mentaly measured and a good correlation has been found. The value of d_c is

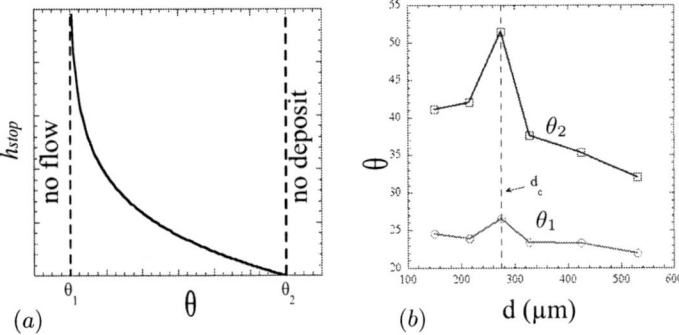

Fig. 5. (a) Evolution of h_{stop} vs the angle of inclination, definition of θ_1 and θ_2, (b) Evolution of θ_1 and θ_2 vs the diameter of the flowing beads on the plane $\lambda = 425 \ \mu m$ and $C = 0.56$

a function of the diameter of the glued beads and their compactness. Some more experiments were carried out on other planes, changing the roughness and the compactness to confirm the model [10].

As proposed by GDR Midi [12], the angle of inclination (and moreover the maximum angle θ_2) can be interpreted in terms of a friction coefficient μ, which is the ratio between the tangential and the normal forces $\mu = \rho g sin\theta / \rho g cos\theta = \tan\theta$. We present this friction coefficient versus an inertial parameter I define as : $I = \gamma d / \sqrt{ghcos\theta}$ where γ is the velocity gradient in the flow (steady state flow), and h the thickness of the flow. This evolution represents the basal friction on the plane. Figure 6 represents the friction coefficient versus the inertial paramter I for the plane $\lambda = 425 \mu m$ and $C = 0.56$ for several flowing beads and shows the influence of the relative roughness. We see that the curve for $d = d_c$ is above all the other ones, which means that, for this diameter, the basal friction is the highest. For this reason, the d_c diameter was interpreted as a maximum of friction. The two limits of the $\mu(I)$ curve correspond to the θ_1 and θ_2 angles. As the curves are classified, we will use the value of $\mu_2 = \tan\theta_2$ given by our model to compare the relative friction of the different flowing beads on the same plane.

So, for each mixture used in bidisperse flows, we know which beads have the highest friction (μ_2) on the plane.

3.2 Bidisperse Flows: Outline and Interface Interactions

Results on bidisperse flows show a rich variety of morphologies. We will discuss in this section the possible mechanisms explaining these morphologies in relation with th values of the relative friction of the different beads. All experiments presented in this section correspond to flows on the plane $\lambda = 1400 \ \mu m$ and $C = 0.56$ giving $d_c = 755 \ \mu m$.

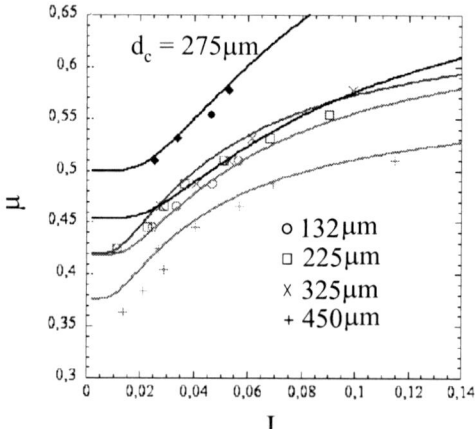

Fig. 6. Rheology of monodisperse flows for different relative roughness, plane $\lambda = 425\ \mu m$ and $C = 0.56$

One strong feature is, for some mixtures, the diminution of the width of the deposit (fig. 2b). This diminution has been quatify by measuring the width along the longitudinal direction (fig. 7).

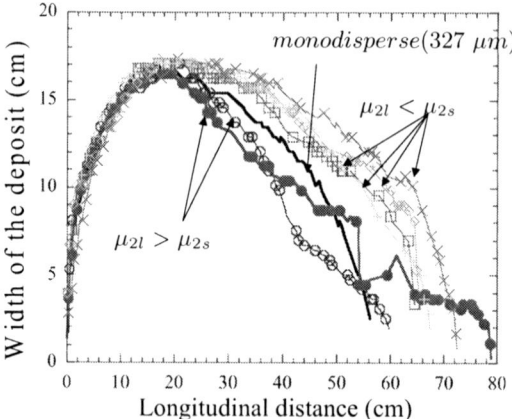

Fig. 7. Evolution of the longitudinal width of a monodisperse and bidisperse flows deposit on the plane $\lambda = 1400\ \mu m$ and $C = 0.56$. Diameter of the small beads $d_s = 327\ \mu m$, percentage of the large beads is $\%l = 50\%$, diameters of the large beads d_l respectively :(○) 670 μm, (●) 755 μm, (□) 1125 μm, (+) 1325 μm, (◇) 1750 μm, (×) 2150 μm

This figure shows that the maximum width does not change but that the decrease in width change compare to the monodisperse flow. If $\mu_{2l} > \mu_{2s}$ (large beads have a greater friction than the small ones) the width of the deposit is smaller than the one of monodisperse flow. If the friction coefficient of the large beads (μ_{2l}) is greater than the one of the small beads (μ_{2s}) the large beads at the periphery confine the flow and avoid the spreading of the front. This mechanism is also a function of the percentage of large beads ($\%l$). If there is too much large beads, the flow is stopped after a very short trajectory ($\%l \geq 80\%$), if the percentage is not sufficient ($\%l \leq 20\%$), the flow is confined on a short distance and the morphology does not change a lot, and for intermediate percentages, there is not enough large beads to stop the flow, but enough to confine the flow on all the longitudinal distance and we observe fingering (fig. 8).

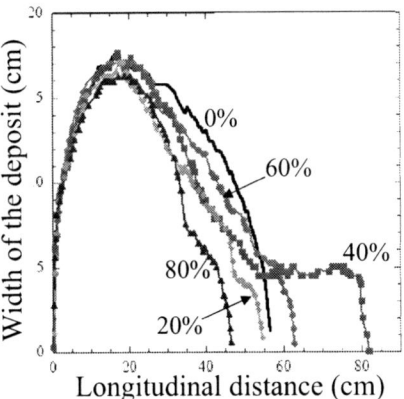 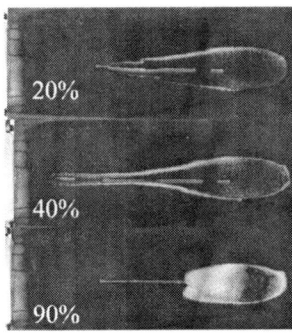

Fig. 8. Evolution of the longitudinal width and shape of a bidisperse flow deposit on the plane $\lambda = 1400 \ \mu m$ and $C = 0.56$, for several percentage of large beads. Diameter of the small beads is $d_s = 327 \ \mu m$, diameter of the large beads is $d_l = 755 \ \mu m$

In conclusion, we can say that the variation of the width of the deposit is sensible to the relative friction of the two species. If the friction of the larger beads μ_{2l} is greater than the friction of the small beads μ_{2s}, the flow of small beads is confined by the large beads, a finger can form or the flow can be stopeed, depending on the volumic fraction of the large one.

When we interest to the interfaces effects, the interface plane/flowing beads has been studied in the first part with the monodisperse flows, the second interface is the one between the large and the small beads layer. The interaction of the large beads and the small beads have been studied, measuring the deposit left by the mixture and comparing the thickness to those of a monodipserse flow. In order to measure only the influence of the large beads on the small beads, we limit the experiments to specific cases for which there is no confining effect (because the confining increases the thickness of the deposit) and for which the deposit is only composed of small beads. We can then compare this thickness to the one left by a monodiperse flow of small beads. To match these two conditions, the choice of the mixtures is done such that : $\mu_{2l} < \mu_{2s}$ (no confining) and $d_l \gg d_s$ (no large beads are trapped in the deposit).

Figure 9 presents the ratio of the deposit thickness left by a bi-disperse flow ($h_{stop,bi}$) on the one left by monodisperse flow of small beads ($h_{stop,mono}$) for several mixtures corresponding to the conditions described previously. These measurements have been done for several diameter ratios and volumic fractions of large beads. The presence of large beads modifies the deposit thickness h_{stop}. The deposit thickness decreases with the fraction of large beads %l, but the size ration of beads does not seem to have a great influence.

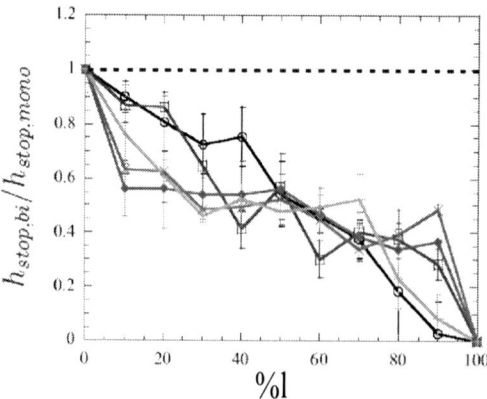

Fig. 9. Variation of the thickness ratio $h_{stop,bi}/h_{stop,mono}$ vs the volumic fraction of large beads %l, on the plane $\lambda = 1400\mu m$ and $C = 0.56$. Diameter of the small beads is $d_s = 327\mu m$, diameters of the large beads are respectively ($\circ$) $1125\mu m$, ($\square$) $1325\mu m$, ($\blacklozenge$) $1750\mu m$, ($+$) $2150\mu m$, ($\times$) $2925\mu m$

A proposed mechanism, for the deposit thickness decrease, is that the large beads modify the velocity gradient in the small beads underlayer. Because of the greater velocity of the large beads, the gradient is increased. This modification of the velocity gradient would lower the deposit thickness h_{stop} according to [11, 12]. To confirm this hypothesis, some steady state flow experiments have been carried out. For all experiments we see a monolayer (one bead thick) of large beads flowing on the underlaying layer of small beads. The velocity of the small beads front ($\bar{u}_s$) and the velocity of the monolayer of the large beads (u_l) have been measured. Assuming that the velocity gradient is linear, the surface velocity of the small beads u_s is twice their mean velocity $\bar{u}_s$. Comparing u_s and u_l, we see that for almost all experiments they have the same value (fig. 10). This means there is continuity at the interface between the velocities of the small and the large beads and that the velocity at the interface results both from the dynamics of small and large beads. This confirm that the velocity gradient in the layer of small beads is modify by the presence of large beads. For few values for which the velocity of the large beads is greater, there is no continuity. A possible explaination is that the large beads may have a rotationel motion on the layer of small beads. These points correspond to high angles of inclination and high percentage of large beads. In these cases, the velocity of the large beads only partially influences the layer of small beads.

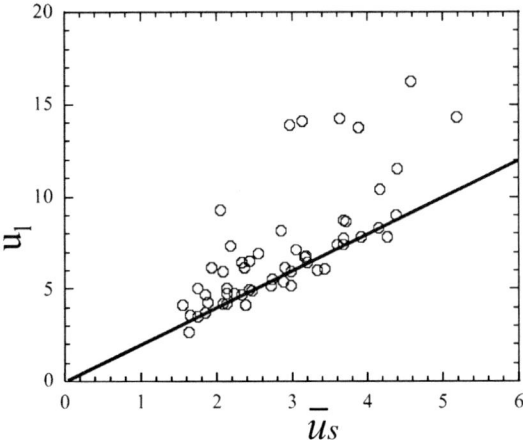

Fig. 10. Comparaison between the mean velocity of the small beads $\bar{u}$ and the velocity of large beads for steady state flows. Diameter of the small is beads $d_s = 327~\mu m$, diameter of the large beads $d_l = 1325~\mu m$, $\%l = 50\%$, plane $\lambda = 1400~\mu m$ and $C = 0.56$, $22° < \theta < 31°$. The line represents the value for which $2\bar{u}_s = u_l$ (velocity continuity)

The large beads modify the velocity gradient of the small beads, and then the rheology of the small beads, which consequently implies a change in the thickness of the deposit of small beads. In fact, for unsteady flows (fig. 9), this thickness differs from $h_{stop,mono}$. The small values of $h_{stop,bi}$ indicates that the velocity gradient of small beads have been increased. We call this mechanism dragging. This explain also why the width of the deposit is increased (fig. 7).

4 Conclusion

This work presents new results on bidiperse granular flows on inclined rough plane. Two types of intercations have been pointed out : outline effect and interface effects. With these two types of interaction, the deposit morphology modifications can be explained qualitatively (narrowing or fingering of the deposit, increase length of the deposit, decrease of the thickness deposit). The amplitude of these modifications deals with the fraction of large beads and the relative friction of each category of beads more than with the size ratio between the beads. To understand these effects, it has been necessary to study the influence of the relative roughness of the plane. This was done with a work on monodisperse flows. The relative friction of the beads on a same plane have been calculated. Thanks this model, it has been possible to interpret the modifications of the deposit morphologies and to propose two mechanisms governing bidisperse flows: the confining and the dragging.

References

1. T. Takahashi, Annu. Rev. Fluid Mech. **13**, 57, (1981)
2. C.S. Campbell et al., J. Geophys. Res. **100**, 8267 (1995)
3. J.S. Patton et al., ASME, J. Appl. Mech. **52**, 172 (1987)
4. H. Ahn et al., J. Appl. Mech. **59**, 119 (1991)
5. D.A. Augenstein, R. Hogg, Powder Technol. **10**, 43 (1974)
6. K. Ridgway, R. Rupp, Chem. Process. Eng., May issue, 82 (1970)
7. D. A. Robinson, S.P. Friedman, Physica A **311**, 97 (2002)
8. S.B. Savage, J. Fluid Mech. **92**, 53 (1979)
9. T.G. Drake, J. Fluid Mech. **225**, 121 (1991)
10. C. Goujon and al., Eur. Phys. J. E **11**, 147-157 (2003)
11. O. Pouliquen et al., Phys. Fluids, **11**, 542-548 (1999)
12. GDR MiDi (collective article) Eur. Phys. J. E **14**,341-365(2004)

Sheared and Vibrated Granular Gas in Microgravity

Yan Grasselli[1], Georges Bossis[2], and André Audoly[2]

[1] EAI Tech-CERAM, rue A. Einstein, BP 085, 06902 Sophia Antipolis Cedex, France
[2] L.P.M.C - Université de Nice, Parc Valrose, 06108 Nice Cedex 2, France

Summary. We present two experimental studies on the flow of model granular media made of nearly elastic spherical particles. Experiments are performed in microgravity conditions inside an airplane undergoing parabolic flights. We first investigate the rheological behaviour of the medium in a cylindrical Couette geometry. The curves, shear stress versus shear rate, are presented and the quadratic dependence on the shear rate is clearly shown. The second series of experiments investigate the behaviour of a vibrated monolayer of spherical particles. A high speed camera is used to record the motion of particles. With an image analysis tracking technique we determine the velocities of the particles from which we retrieve the temperature. The density profiles show the persistence of clusters of particles at the centre of the cell which coexist with a gas phase at the edges of the cell. We conclude by comparing experiments performed in microgravity and in normal gravity.

1 Introduction

The increase in interest in granular matter in recent years has been motivated by the existence of such materials in industrial and geological situations. Granular media exhibit amazing phenomena in both static and dynamic regimes. The shapes of heaps or the formation of arches in a static pile are still intriguing [1, 2]. On the other hand, the rheological properties of a granular have shown several different behaviors depending on the flow range (quasistatic to rapid flows). A non exhaustive list includes non linear waves, inelastic collapse, segregation, convection rolls [3–5]. All these features suggest that the local structure of the granular medium greatly influences its dynamical behavior. These aspects of granular flows have been investigated theoretically, by numerical simulations and experimentally [6–8].

Experiments made in normal gravity do not allow to obtain the rheology of a system of spheres at low or intermediate volume fraction. The use of a vibrating plate combined with a shearing device can help to work with a disordered system but still at packing fraction above 0.64 [9]. Performing experiments in microgravity allows to obtain results at much lower volume fractions where kinetic theories, taking into account the inelastic nature of collisions between particles, are more likely to apply. In this theory [10], the temperature is found to be proportional to the square of the shear rate, the square of a mean free path of the particles and inversely proportional to $1-e^2$, where e is the restitution coefficient of the spheres of the granular medium which drives the energy dissipation phenomenon. Jenkins and Richman [11] have extended this theory by explicitly taking into account the anisotropy of the second moment of the velocity fluctuations. The boundaries, and specifically their roughness properties, also influence the rheology with the existence of a slip velocity occurring at the moving walls of the cell. A comparison of the predictions of the theory by Jenkins [13] and Richmann [14], to our experiments has shown a fair agreement only for the lowest volume fraction ($\nu = 12\%$) but an overestimation of the viscosity at higher volume fraction [15].

An other currently used experiment, which also suffers from the presence of the gravity since the volume fraction remains usually higher than 0.6 [16], consists in vibrating a granular heap. In this case we are interested in the density and the temperature profile, the velocity distribution-usually not Gaussian-and the pair correlation function. These quantities can be obtained thanks to high speed camera which records the trajectory of each particle in a monolayer configuration. For vibrated spheres, the parameter characterizing the vibration is usually the reduced acceleration: $\Gamma^* = a/g$, where a is the acceleration of the container and g the one of the gravity. Nevertheless S. Mc Namara and E. Falcon have recently pointed out [17] that the velocity of the cell, V, was a more pertinent parameter to interpret the evolution of the temperature or pressure with the motion of the shaker when $\Gamma^* \gg 1$. Furthermore these authors have proved by numerical simulation that the usual scaling of temperature and pressure proportional to V^2 was no longer verified, even in zero gravity, if we take into account that the restitution coefficient depends on the velocity of the particles.

In this paper we shall first present and discuss some results obtained in a cylindrical Couette geometry with iron spheres; then in a second part we present a new experiment done in microgravity with a flat cell mounted on a shaker and the trajectories of the particles being recorded by a high speed video camera. Preliminary results concerning the average kinetic energy and the density profiles will be presented in this section. We discuss our results and conclude in the last section.

2 Stress Versus Shear Rate in Cylindrical Couette Geometry

Experiments have been performed in a cylindrical Couette geometry filled with iron spherical particles with diameter σ of $1mm$ and $2mm$ (fig. 1).

The density of the particles ρ_p is about $7500kg/m^3$ and the associated restitution coefficient e is close to 0.9. The cell has a height of $35.8mm$, the

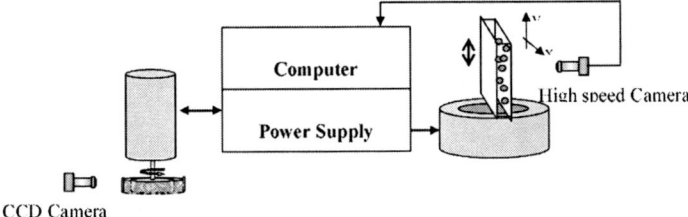

CCD Camera

Fig. 1. Experimental setup used to record the behaviour of a model granular medium submitted to cylindrical Couette shear flow or to periodic oscillations. The CCD camera allows us to obtain a direct visualisation of the structures formed under shear while the high speed camera records at a rate of 950 frames per second the vibrated granular medium. The computer controls the devices through the power supply and stores the images delivered by the high speed camera.

inner cylinder a radius of $R_i = 12.15mm$ and the outer one, a radius of $R_e = 16.6mm$. The number of layers of particles across the shearing gap varies from 1 to 5 in our experimental situations. In this geometry, one can assume that the shear rate is almost constant. By changing the number of particles inside the cell, we are able to study the effect of the solid volume fraction ν (or density) on the rheology of the granular medium. Iron particles of $1mm$ in diameter are glued to the inner cylinder to create a rough surface. The gap between these glued particles over the inner wall of the cell varies between 0.3 and 1 diameter. The inner cylinder is mounted on a rheometer which is computer controlled. A given angular velocity ω is applied to the inner cylinder and the rheometer measures the corresponding torque. This torque is then converted to a shear stress using the geometrical characteristics of the cell. The outer cylinder of the cell is made of glass to have a direct visualisation of the shearing process and to avoid electrostatic effects occurring with plastics.

To cancel the effect of gravity, experiments are performed inside an airplane undergoing parabolic flights. By parabola, we mean, in the following, the time interval during which the granular medium is no longer submitted to gravity. Each flight includes 30 successive parabolas, each lasting about $30s$. The experimental determination of the rheological behavior of the granular medium is performed by measuring the torque exerted on the inner cylinder while changing the angular velocity over a given range. More details on the experimental procedure followed in microgravity and on reproducibility tests are given in [15]. Experimental curves of the shear stress τ versus the square of the shear rate γ^2 for different solid volume fractions ν of particles are presented in figure 2 for the particles of $1mm$ in diameter. As expected, the behavior is clearly quadratic in $\dot\gamma$ and one can also notice an increase with the volume fraction of particles. A parallel with the theory of Jenkins and Richman [13, 14] allows us to compare the experimental slope of these curves to their theoretical values [15]. For instance for a volume fraction $\nu = 42\%$

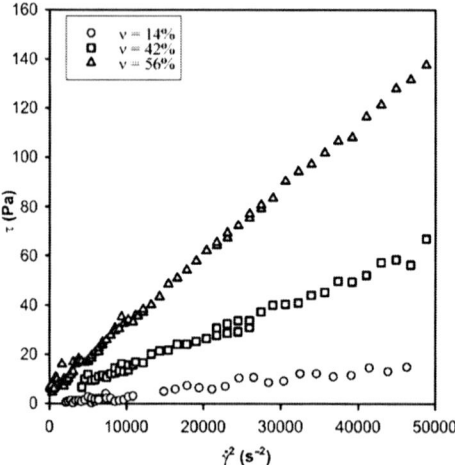

Fig. 2. Shear stress τ as a function of the square of the shear rate $\dot{\gamma}^2$ for different solid volume fractions ν. Iron spherical particles of $1mm$ in diameter. The linear behaviours observed are a clear signature of the quadratic dependence versus $\dot{\gamma}$.

the experimental slope of the stress versus the square of shear rate is 0.0012 for $1mm$ spheres and 0.0026 for $2mm$ spheres whereas the theoretical slopes are respectively 0.0030 and 0.0053. Recent numerical simulations [18] predict a shear stress $\tau = aV^n$ with $n = 2.2 \pm 0.2$ and still a higher slope: $a = 0.015$ for $\nu = 40\%$ [19], but for a flow between two annular plates.

The exponent is compatible with the Bagnold scaling and our experiments in microgravity also prove the validity of this scaling. On the other hand, the value of the prefactor is still to be confirmed. We believe that our experiments are quite well defined, but as this coefficient depends on the slipping velocity which itself depends on the roughness of the wall and on the ratio $(R_e - R_i)/\sigma$ (4.5 with the $1mm$ spheres), the comparison with existing data with numerical simulations is not very easy. Moreover some other phenomena could be specific to our experiment. In particular the outer wall was a glass wall in order to check visually the homogeneity of the density through the cell (small perturbations of gravity along the axis of rotation during the parabola can dramatically change the repartition of spheres through the cell) and also in order to have some information on the structuration of particles.

We can see in figure 3 that when the volume fraction increases we observe clearly the formation of rings of particles. These rings are likely related to the clustering transition and they certainly contribute to reduce the transverse momentum transfer, so it could explain partly why our slope is smaller than the theoretical one. Other experiments with rough spheres also glued on the external cylinder would be worse doing. The existence of clusters when the dissipation rate becomes higher than the energy input can strongly mod-

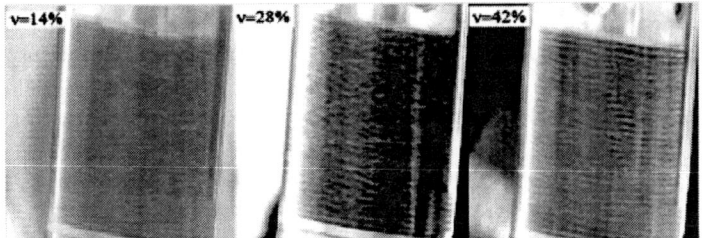

Fig. 3. Experimental pictures of the sheared granular medium in the presence of shear and microgravity. As the volume fraction ν increases, structures made of rings of particles appear in the shear gap.

ify the rheology of the system. This is well observable in vibrated flat cell. In the following section we describe this kind of experiment also realized in microgravity.

3 Vibrated Granular Gas

In order to experimentally determine the granular temperature, a similar model granular medium made of iron spheres with diameter $\sigma = 2mm$ has been studied. The cell containing the medium is mounted vertically on an electromagnetic actuator allowing to apply sinusoidal excitations to the medium with varying amplitude A and frequency f. An accelerometer monitors the induced acceleration a (figs. 1 and 4).

The cell thickness ($2.5mm$) is chosen in such a way that the medium can be treated like a monolayer of granular gas. The experimental cell is of circular shape (diameter of $6.5cm$) enclosed in between of two glass plates to reduce undesired electrostatic effects. We chose a circular shape for the cell so that

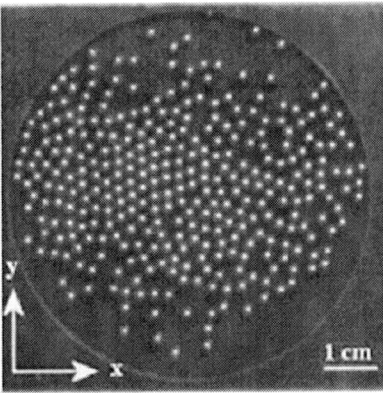

Fig. 4. Typical experimental picture of the vibrated monolayer. A direct image analysis processing on such picture gives the instantaneous position of all particles inside the cell. The imposed oscillation is along the vertical direction y.

during shaking, the energy supplied to the medium through the collisions in between the particles and the walls is better redistributed in the perpendicular direction of the imposed vibration. By changing the number of particles inside the cell, we have the possibility to study the effects of concentration on the temperature from a dilute up to a concentrated situation. At this time, only two different volume fractions have been used: $\nu = 7.9\%$ and $\nu = 16.7\%$. We present in the following preliminary results on this study.

As soon as microgravity occurs, the vibrated particles quickly distribute throughout the entire cell. The motion is recorded with the help of a high speed camera at a frame rate of 950 images per second. In these conditions, we had the possibility to record four full seconds of oscillations giving rise to about 4000 different pictures and so 4000 different times during oscillations. Each picture is then treated with a specific image analysis technique described in reference [20]. This technique allows us to track and to obtain the x and y positions of each single particle in the cell as a function of time (fig. 5). One can notice that the trajectory greatly depends on the location of the particle inside the cell. The motion is of Brownian type for a particle located at the center (particle 1) while for particle 2 one can observe the rebounds between moving walls of the cell on top and with the central cluster of particles at the bottom of its trajectory. The large circle shows the position of the cell at rest. From this experimental determination, we can retrieved the instantaneous velocity $v = \sqrt{v_x^2 + v_y^2}$ where v_x^2 and v_y^2 are, respectively, the x and y component of the velocity of each particle throughout the granular gas and then obtain the granular temperature of the system. Here, we focused on the average value of the temperature as a function of the vertical position y in the cell and not

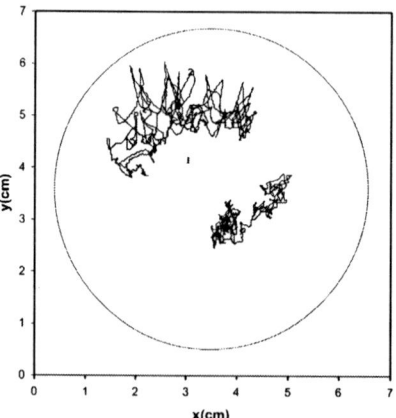

Fig. 5. Experimental trajectory of two test particles in the cell in presence of the oscillation (the trajectories of all other particles are not shown). The y-direction is along the direction of the applied vibration. The type of motion greatly depends on the location of the particle in the cell.

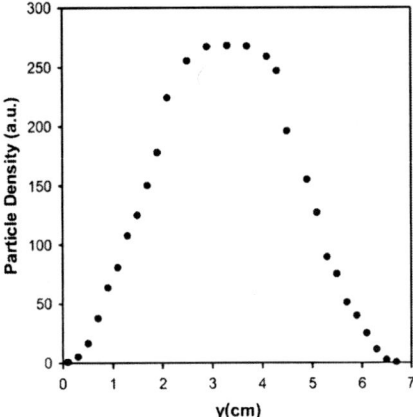

Fig. 6. Averaged particle density (expressed in arbitrary units) as a function of the vertical position y. The horizontal flat region appearing at the center of the cell clearly shows the presence of the cluster of particles. The energy injected in the cell on top and bottom quickly dissipates through the inelastic collisions between the particles.

on the temporal fluctuations. The averaged values are calculated over all the frames of one experiment.

Direct evident observations can be made from an instantaneous view of the cell (fig. 4). We can define "hot" regions corresponding to the top and bottom of the cell where the particles collide with the moving walls of the cell. While the central area corresponds to a cold high density region where the energy introduced into the cell has been dissipated during the collisions in between the particles. We have observed that in our experimental situations, only binary collisions occurred in between particles. The particle distributions through the cell are shown if we plot the average density inside the cell as a function of the vertical position y (fig. 6). The average is calculated over the whole experimental recording time and over all particles.

Moreover, the important parameter on the behaviour of the medium is the amplitude A of vibration rather than the relative acceleration Γ. At large amplitudes and over a period of oscillation, we have more collisions of particles on the wall of the cell and then more energy supplied to the medium. The injected energy in this way is related to the kinetic energy of the particles which is then dissipated through the collisions. The presence of a flat density profile at the center of the cell on figure 6 confirms the presence of the central cluster. Note that the width of this flat region increases for decreasing vibration amplitude. We have averaged the experimental velocities $< v >$ of the particles over a whole experiment as a function of the vertical position y in the cell. Typical results obtained for the volume fraction of $\nu = 16.7\%$ and for two different amplitudes are presented on figure 7.

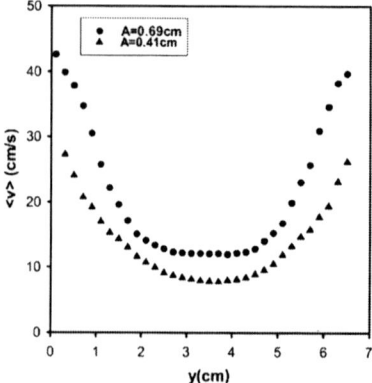

Fig. 7. Averaged particle velocity $< v >$ versus the position y in the cell aligned with the direction of oscillations. Two different amplitudes are reported. As stated in the text, the important parameter is the amplitude of oscillation rather than the acceleration of the system.

One can see, on top and bottom locations, the "hot" regions corresponding to high velocities while at the center of the cell, small velocities are found.

From this experimental determination, we can retrieve the average granular temperature as a function of the velocity V of the system, defined by $A\omega$ where A is the amplitude and ω the angular frequency of the oscillation (fig. 8). The behaviour seems to be almost linear suggesting that the temperature is proportional to the velocity. This can be the case when the restitution coefficient of the particles depends on their velocity as pointed out in reference [17]. But additional experiments are needed to conclude on this point.

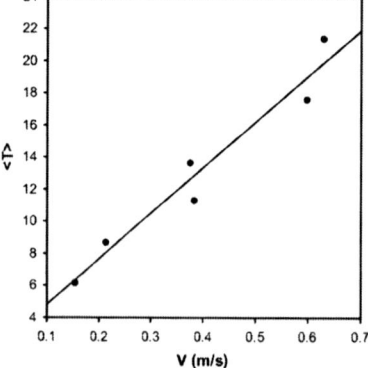

Fig. 8. Average temperature T retrieved from the particle's velocities of the shaken medium as a function of the velocity of the cell V. The behaviour seems to be linear suggesting that the restitution coefficient of the particles depends on the velocity.

Finally, in order to point out the advantages of microgravity for this kind
of study, identical experiments have been realized in the laboratory (so in
presence of gravity) with the cell now placed horizontally. The same experimental conditions concerning volume fraction, amplitude and frequency have
been reproduced in order to compare the two situations. Despite the difficulty to precisely perform the horizontality of the cell, important differences
have been observed. By performing the same processing on the experimental
recording, we can yet bring two conclusions. First, the fact that the particles lie on the bottom plate induces a collective motion of particles because
of rolling motion. This general motion is superposed to the oscillations and
gives rise to undesired effects on the dynamics of the medium. Second, the
differences observed in the density distributions of particles and the average
velocities of the particles through the cell (in the direction of the vibrations)
become critical for the regime of small oscillation amplitudes (fig. 9). Nevertheless, for large amplitudes, the observed deviations are quite small. We
can then conclude that for strong shaking regimes, there is no real profit of
performing such experiments in microgravity while for weak vibrations, the
zero-g condition is quite necessary.

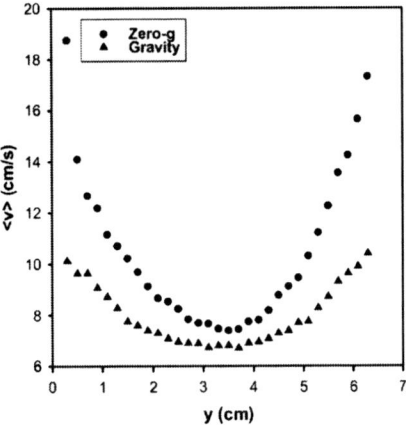

Fig. 9. Compared average velocity versus the vertical position in the cell y, without
and with gravity, for the same oscillation amplitude. Clear deviations are found due
principally to the friction of the particles over the walls of the cell. These differences
increase while decreasing the oscillation amplitude.

4 Conclusion

We have performed an experimental investigation of a model granular medium
composed of nearly elastic particles (iron beads) submitted to a Couette shear
flow or to a periodic oscillation. The experiments have been performed in microgravity, in order to cancel the effects related to the weight of the particles

and to be in an "ideal" situation to test the models and response of rapid shearing flows. We well recover the quadratic dependence of the shear stress versus the shear rate predicted by Bagnold. The experimental slope is found to be smaller than the theoretical one especially at the highest volume fraction; an observation that could be related to the presence of a shear induced ordering.

Some results obtained on a vibrated monolayer of iron spherical particles have been presented. Our processing allows us to track individually each particle as a function of time through the cell. From the calculations of the average particle's velocities, we have determined both the density profile and the average granular temperature which seems to be proportional to the velocity of the system. Processing of experimental data are still underway especially on the study of particle diffusion through the vibrated medium.

Acknowledgements

This work has been supported by the National Center of Spatial Study (CNES) and we acknowledge for the possibility to use specially equipped aircraft for microgravity flights.

References

1. Y. Grasselli, H. Herrmann, Physica A, **246**, 301 (1997)
2. D. Mueth, H. Jaeger, S. Nagel, Phys. Rev. E, **57**, 3164 (1998)
3. R. Behringer, Nonlinear Sci. Today, **3**(3), 1 (1993)
4. C. Campbell, Annu. Rev. Fluid. Mech., **22**, 57 (1990)
5. Y. Grasselli, H. Herrmann, J. Granular Matter, **1**, 1, 43 (1998)
6. M. Hopkins, M. Louge, Phys. Fluid. A, **3**(1), 47 (1991)
7. L. Bocquet, W. Losert, D. Schalk, T. Lubenski, J. Gollub, Phys. Rev. E, **65**, 11307 (2001)
8. P. Haff, J. Fluid. Mech., **134**,401 (1983)
9. S. Savage, J. Jeffrey, J. Fluid. Mech., **110**, 255 (1981)
10. J. Jenkins, M. Richman, J. Fluid. Mech., **192**, 313 (1988)
11. J. Cao, G. Ahmadi, Part. Sci. and Tech., **13**, 133 (1995)
12. R. Bagnold, Proc. R. Soc. London Ser., A **225**, 49 (1954)
13. J. Jenkins, Physics of dry granular media, H. Herrmann et al. eds, Kluwer Academic Publishers, 353 (1998)
14. M. Richman, Acta Mechanica, **75**, 227 (1988)
15. G. Bossis, Y. Grasselli , O. Volkova, J. Phys. Cond. Matter **16**, 3279 (2004)
16. E.R. Novak, J.B. Knight, E. Ben-Naim, H.M. Jaeger, S.R. Nagel, Phys. Rev. E, **17**, 1972 (1998)
17. S. Mcnamara, E.Falcon, Phys.Rev E, 031302 (2005)
18. O Baran, L Kondic - Arxiv preprint cond-mat/0411459, 2004 - arxiv.org
19. Note that in the geometry used by the authors of ref. 18, the angular velocity is equivalent to the shear rate.
20. Y. Grasselli, G. Bossis, Surface Characterization methods, **87**, A. Milling Ed, Marcel Dekker Inc, 269 (1999)

A Domino Model for Granular Surface Flow

Andreas Hoffmann and Stefan J. Linz

Westfälische Wilhelms-Universität Münster, Institut für Theoretische Physik, Wilhelm-Klemm-Str. 9, D-48149 Münster, Germany

Summary. We present and analyze a cellular automaton model for granular surface flow along piles and inclines. It is based on the intuitive idea that granular surface flow happens via successive excitation of small-scale avalanches. We show that this model reproduces several essential experimental results for granular surface flow and give a continuum approximation of its spatio-temporal evolution.

1 Introduction

Since the pioneering work by Jaeger et al. [1] on avalanching in 1989, granular matter [2, 3] experiences considerable scientific interest as one of the prime paradigms of a complex system. Besides other aspects of granular dynamics, the theoretical understanding of surface flow along granular piles and inclines and in drums still presents a major challenge. This is because a variety of properties such as avalanching and continuous surface flow dynamics as well as local properties such as the spatio-temporal surface patterns and front propagation need to be understood in a unifying way (for an overview cf. [5]). For low shear, granular flow is governed by Reynolds' dilatancy, and, therefore, only the few uppermost granular layers close to the surface contribute to the dynamics. For that reason, there are, besides micromechanical simulations of the full many-grain problem (cf. [4] for a thorough overview), two directions of modeling granular surface flow: (i) continuum approaches modeling the boundary layer flow including its interaction with the interface to non-moving bulk [6–8] and (ii) mean-field models that only model global features of that flow such as the dynamics of mean surface inclination and mean velocity [9–15].

In this contribution, we present a cellular automaton model for granular surface flow that is based on the intuitive picture that the flow down a granular pile or incline occurs via successive excitation of small-scale avalanches propagating downhill along the surface. In this picture, the flowing granular matter is modeled in analogy to a sequence of coupled avalanches where each of them follows the rules given by a mean-field model for avalanches [11] and can destabilize granular material below it and even above it. This material can also set off avalanches according to the same dynamics. By this, the surface flow being triggered by successive avalanching processes runs down the pile in a domino-like manner.

In chapter 2 we sketch the essence of the mean-field model, describe the local mechanisms leading to small-scale avalanching and give the mathematical formulation of our cellular automaton model. Using numerical simulations, we show in chapter 3 that the model reproduces essential experimental results, such as the evolution of locally triggered avalanches and fronts [17, 18], the S-shaping of the granular surface in rotated drums [19], and the power spectrum of avalanches found by Jaeger et al. [1]. In chapter 4 we briefly outline a continuum approximation of the domino model.

2 Domino Model

Brief review of the mean-field model
The mean-field model for granular surface flow along piles, heaps, inclines or in drums [11–13] is based on the experimental observation that (1) a flow of granular particles close to the surface sets in if the inclination angle $\phi(t)$ of the pile exceeds the maximum rest angle ϕ_s and (2) stops again if the angle of repose ϕ_r is reached. To model this behavior in the spirit of a mean-field model, only two global variables are taken into account: (1) the (spatially averaged) time dependent angle of inclination $\phi(t)$ of the heap and (2) the mean velocity $v(t)$ of the moving particles that is determined by the kinetic energy of the flow, $E_{kin}(t)$ via $v(t) = \sqrt{E_{kin}(t)/2m}$ where m denotes the total mass of particles in motion. Since an ab-initio derivation of the dynamic equations for $\phi(t)$ and $v(t)$ does not seem to be easily feasible, they are modeled by generalizing the motion of a solid block slipping down on an inclined surface and given by

$$\dot{v} = g[\sin\phi - k_d(v)\cos\phi]\chi(v,\phi) \tag{1}$$
$$\dot{\phi} = -a_{mf}v + \omega. \tag{2}$$

Here, $k_d(v) = b_0 + b_2v^2$ denotes a generalized friction coefficient that combines the frictional forces of the Coulomb type (b_0-term) and the Bagnold type (b_2v^2-term). Since the inclination angle decreases with the speed of the avalanche v, the simplest possible description is expressed by a linear feedback of $\dot{\phi}$ and the velocity v with $a_{mf} > 0$. Including the external rotation rate ω the model is also able to describe the global motion in a rotating drum. The cut-off function $\chi(v,\phi) = \Theta(v) + \Theta(\phi - \phi_s) - \Theta(v)\Theta(\phi - \phi_s)$ with Θ representing the Heaviside function captures the fact that a dynamics starts at $\phi > \phi_s$ and can only go on as long as $v(t) > 0$. Although this model (supplied with some stochastics [14, 15]) can successfully describe many dynamical

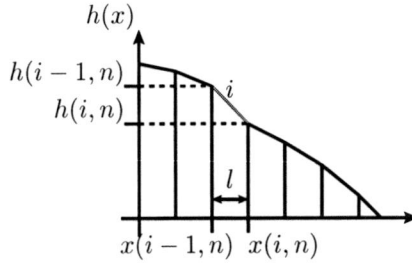

Fig. 1. Subdivision of a heap of granular material into small equidistant cells. Each small cell possesses a individual spatially constant angle $\phi(i)$ with respect to the horizontal; the whole rugged surface profile of the heap, however, is assumed to be continuous.

aspects of granular surface flow such as the spectral behavior of successive avalanches [1] or the hysteretic transition from avalanches to continuous flow in rotated drums [19] as function of the rotation rate w, it fails by construction to explain any spatio-temporal properties of the surface flow.

General idea of the domino model

The aim of the domino model is to incorporate *local* properties of the granular surface flow, i.e. the *spatio-temporal dynamics* of the surface, the transport velocity at the surface and the local inclination angle or the slope. For the following, it turns out to be useful to discretize time in an Eulerian way by subdividing the time axis t in small, equally sized time segments $t_0, ..., t_{N-1}, t_N, t_{N+1}, ...$ with stepsize Δt and labeling the successive time steps by $n = N\Delta t$. The next important step is to subdivide the granular heap of horizontal length L in an array of individual cells of equal width l labeled by the index $i = 1, .., I$, as depicted in Fig. 1. This width should be still at least about an order of magnitude larger than micromechanical scale for grains to allow for the applicability of the mean-field model for the individual cell dynamics. At a fixed time n, the individual cells are specified by the horizontal and vertical positions $x(i, n)$ and $h(i, n)$ measured with respect to the horizontal at the foot of the pile and by the corresponding inclination angles $\phi(i, n)$ with $h(i-1, n) - h(i, n)) = l \tan \phi(i, n)$ of the locally straight cell surface, cf. Fig. 1. The surface elements of the cells are assumed to be able to perform an avalanche-like motion modeled as in the mean-field model [11]; their individual dynamics, however, must be subsequently cross-coupled to that of their nearest neighboring cells by additional constraints. Were all cells uncoupled from their neighbors, the dynamical rules for avalanching of each individual cell i would read

$$v(i, n + 1) - v(i, n) = g[\sin \phi(i, n) - k_{di}[v(i, n)] \cos \phi(i, n)]\Delta t\, \chi(v, \phi) \quad (3)$$
$$\phi(i, n + 1) - \phi(i, n) = [-a_i v(i, n) + w]\, \Delta t \quad (4)$$

with $k_{di}[v(i)] = b_{0i} + b_{2i}v(i)^2$ being the (local) nonlinear friction coefficient including Bagnold friction for larger velocities and w being the external rotation rate in the case of rotating drum flow. The $\chi[v(i, n), \phi(i, n)]$-function is a natural extension of the fact that a local avalanche starts slipping if the inclination angle of cell i exceeds the maximum local rest angle $\phi_s(i)$ and stops if the velocity $v(i, n)$ reaches zero,

$$\chi(v, \phi) = \Theta[v(i, n)] + \Theta[\phi(i, n) - \phi_s(i)] - \Theta[v(i, n)]\Theta[\phi(i, n) - \phi_s(i)]. \quad (5)$$

The angle $\phi_s(i)$ is a specific property of the granular material and must be taken from experiments. The constants b_{0i}, b_{2i} and a_i and their relation to the parameters of the global model are discussed below. Throughout this paper, we assume that these parameters possess the same value for each cell, i.e. $a_i = a$, $b_{0i} = b_0$, $b_{2i} = b_2$, $\phi_s(i) = \phi_s$. Since, however, all individual cells are connected to the neighboring cells, the constraints of (1) mass/area conservation of flowing granular material and (2) continuity of the whole surface profile must be incorporated.

Coupling mechanisms

The domino model is based on the idea that an avalanche starts at some cell i of the pile (e.g. the uppermost one 1) due to addition of grains to that cell implying a local increase of the angle $\phi(i, n)$ or due to (possible) external rotation where all angles $\phi(i, n)$ are simultaneously increased. A local avalanche starts slipping if the inclination angle of the cell i exceeds ϕ_s and then deposits material on the cell below it to the right, see left panel of Fig. 2. Then, the lower cell can also be destabilized (provided $\phi(i + 1) > \phi_s$) and develop a subsequent local avalanche that again deposits material further below and so on. This mechanism characterizes the *flowing-down* process. On the other hand, the slipping of an avalanche at the cell i also induces a transport of the grains from above, i.e. at least from cell $i - 1$, which was previously supported by the resting grains at cell i. Consequently, the upper cell $i - 1$ to the left will also be destabilized. This mechanism is called the *pulling-from-below* process, see right panel of Fig. 2. In the model, it is assumed

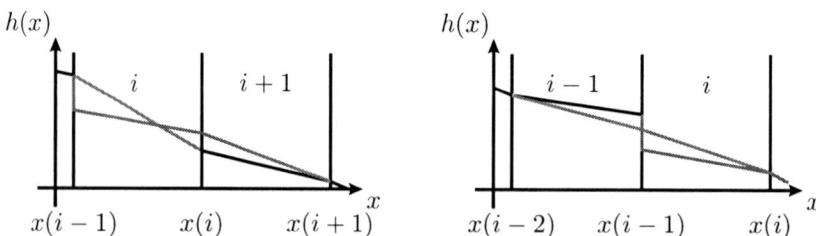

Fig. 2. Left panel: flowing-down process - first step of a small avalanche: The avalanche of cell i runs down and deposits material on cell $i + 1$; the height $h(i)$ of the foot of avalanche i is determined by the conservation of mass criterion, i.e. by area conservation below the cell surfaces. Right panel: pulling-from-below process - second step after the slipping of avalanche at cell i which left a discontinuity at $x(i - 1)$, see left panel. The pulling of i on $i - 1$ is described again by mass conservation.

that local avalanches slip independently of their neighbors and interact with the neighboring cells only via transporting mass to the next cell or pulling mass from the previous one. The cell i performs avalanching according to the mean field model resulting in a change of angle from $\phi(i, n)$ to $\phi(i, n + 1)$. During the transport of grains from cell i to cell $i + 1$ no granular material is lost. Therefore, mass conservation, or equivalently, conservation of the area below the surface at any time n and continuity of the surface determine the local height $h(i, n + 1)$ of the avalanching cell as shown in the left panel of Fig. 2. This process couples the evolution of local height $h(i, n)$ to the velocity $v(i, n)$ of cell i, its angle $\phi(i, n)$ and the angle $\phi(i + 1, n)$ of cell $i + 1$. The height $h(i, n + 1)$ follows from the relation

$$h(i, n + 1) = h(i, n) + \frac{1}{3} l \tan \phi(i, n) - \frac{1}{3} l \tan \phi(i, n + 1). \qquad (6)$$

For all avalanches except the top one, there is the second process that avalanche i destabilizes cell $i - 1$ and grains flow from $i - 1$ to i. This in turn modifies $h(i - 1, n + 1)$ which is now also coupled to $v(i, n)$ and the angles $\phi(i - 1, n)$, $\phi(i, n)$ and $\phi(i, n + 1)$ which follows from the relation

$$h(i - 1, n + 1) = h(i - 1, n) - \frac{2}{3} l \tan \phi(i, n) + \frac{1}{3} l \tan \phi(i, n + 1). \qquad (7)$$

The flowing-down and pulling-from-below processes happen simultaneously so that the surface of the pile stays always continuous.

Cellular automaton formulation

Next, we combine the two afore-mentioned basic coupling mechanisms with the avalanching process for individual cells in form of a spatio-temporally discrete formulation of the entire surface dynamics. Although our basic mechanisms are local, it is important to note that they are triggered by the entire downflow of material, or expressed differently, the starting conditions for an avalanche at cell i at time n generally depends on the entire downflow that reached cell i from all cells $i = 1, ..., i - 1$. In this respect, our model exhibits a long-range dynamics. We consider the dynamics of the system at the cell position i and the time n. The system at time n starts at cell 1 at the top developing an avalanche process that successively moves down to the lowest cell n_{max} if the cells i sequentially fulfill the flow conditions. The initial conditions for the automaton at cell i are $h(i - 1, n)$, $h(i, n)$, $h(i + 1, n)$ together with $\phi(i, n)$ and $v(i, n)$ in the case that cell $(i - 1)$ does not perform an avalanching process at time n. The more interesting case is that an avalanche flows down from cell $i - 1$ and, therefore, deposits material onto cell i. Then, cell i is instantaneously in an intermediate state which we denote by $\phi^+(i, n)$ and the height of $i - 1$ is in an intermediate state $h(i - 1, n')$. This state is intermediate n' because the downflow along cell i at time n will drag away further material from cell $i - 1$. In other words, the automaton at position i and time n finishes the transition of avalanche $i - 1$ from its state at time step n to time step

$n + 1$, excites an avalanche at cell i into a intermediate state n' and gives a tilt $(+)$ to cell $i + 1$. With this notation and using n' for intermediate values, the discrete formulation of one time step of the avalanching process at cell i is described in an algorithmic formulation by:

$$\phi(i, n') = \phi^+(i, n) - av(i, n)\Delta t \tag{8}$$
$$v(i, n + 1) = v(i, n)$$
$$+ g \left(\sin \phi^+(i, n) - \left[b_0 + b_2 v^2(i, n)\right] \cos \phi^+(i, n)\right) \Delta t \, \chi(v, \phi) \tag{9}$$

with $\chi(v, \phi)$ given by (5).Introducing the definition

$$\Delta h(i - 1, i, n, n') = \frac{1}{3}\left[h(i - 1, n') - h(i, n) - l \tan \phi(i, n')\right],$$

the resulting heights after one dynamical step can be calculated as follows

$$h(i, n') = h(i, n) + \Delta h(i - 1, i, n, n') \tag{10}$$
$$h(i - 1, n + 1) = h(i - 1, n') - \Delta h(i - 1, i, n, n') . \tag{11}$$

The entering angles are determined by the local heights and given by

$$\phi^+(i + 1, n) = \arctan\left(\{h(i, n') - h(i + 1, n)\}/l\right) . \tag{12}$$

Given the heights of all cells, all angles are fixed and can be computed from the height values. As input for the cellular automaton we only need to know the angle $\phi(i, n)$ or $\phi^+(i, n)$. The other angles follow passively:

$$\phi(i - 1, n + 1) = \arctan\left(\{h(i - 2, n + 1) - h(i - 1, n + 1)\}/l\right) \tag{13}$$
$$\phi(i, n') = \arctan\left(\{h(i - 1, n + 1) - h(i, n')\}/l\right) . \tag{14}$$

The avalanching step of cell $i - 1$ is then finished. However, the cell i will be further destabilized by developing an avalanche at cell $i + 1$. It will deposit material onto $i + 1$ and is, therefore, in the intermediate state characterized by $h(i, n')$. Now the cell $i + 1$ is in the tilted state $\phi^+(i + 1, n)$ and by this the automaton starts again at position $i + 1$. The effect of external rotation ω can be simply included by generalizing (8) via the substitution

$$-av(i, n) \rightarrow -av(i, n) + \omega . \tag{15}$$

Parameters and experiments

The mean-field model described above possesses four parameters, namely b_0, b_2, a and ϕ_s. In the domino model, these parameters might be, at least in general, dependent on the individual cell of the domino model. Whereas the maximum angle of rest of a pile $\phi_{s,mf}$ in the mean field model can be directly measured in experiments, its local counterpart $\phi_s(i)$ of the domino model might slightly exceed $\phi_{s,mf}$. The parameter b_2 mimics the Bagnold friction and is related to the density and other specific properties of the material under concern. It does not depend on the length scale of the cell. From extrapolations of experiments [9], it follows that b_2 is typically of order 0.001. Also the parameter b_0 can be estimated from avalanching experiments along heaps being basically proportional to the difference between maximum angle of rest ϕ_s and the angle of repose ϕ_r, or in detail $b_0 = \tan\phi_d$ with $\phi_d = \frac{1}{2}(\phi_s - \phi_r)$ at $\dot{v} = 0$. The parameter a_{mf} in the mean-field model can be extrapolated from experiments performed with slowly rotated drums. In Ref. [10] it has been argued that a_{mf} scales with the length of the granular surface L, $a_{mf} = a_{mf}(L) \propto 1/L$. Since we introduce the cell size l as an artificial length scale in the domino model, the parameter a entering in the domino model has to be appropriately adjusted. Specifically, the requirement that e.g. avalanches propagating through the whole system must possess a duration time being *independent* of the cell size l has to be invoked. Extensive numerical experiments performed with our model suggest a scaling relation

$$a = a(l) \propto a_{mf}\frac{L^2}{l^2} \; . \tag{16}$$

3 Selected Results

Dynamics of a chute flow
As a first example to investigate dynamical properties of granular surface flow in our domino model, we study the dynamics of granular media on an incline that can be tilted to arbitrary chute angles ϕ_c. Experimentally this can be realized by a chute setup that allows for the study of stick-slip avalanches, avalanches with restricted size (velocity of avalanche fronts, growth behavior in time and interaction with initially imposed surface structures like small heaps or slope roughness) and even some continuous flow states when periodic boundary conditions are assumed.

Here, we investigate the chute flow under the condition that the chute is subject to periodic boundary conditions at its top and its bottom. This can be realized by putting the granular material on a transport belt running upwards [16] or by realizing a constant inflow/outflow along the chute. From our simulations, we infer that under these conditions a hysteretic transition should be observable if one slowly ramps up and down the inclination angle ϕ_c of the chute bed. Specifically, the important observation is that (1) increasing the inclination angle triggers the flow to start at $\phi_c = \phi_s$ while (2) slowly decreasing the inclination angle again leads to a halt of the chute flow not before $\phi_c = \phi_d$ is reached. The numerical simulation of the mean velocity $\overline{v} = 1/N \sum_{i=1}^{N} v(i,n)$ being constant for a fixed angle ϕ_c are shown in Fig. 3. The slope has been subdivided into 200 small cells with dynamical parameters chosen to reproduce the experiment of Jaeger et al. [1]. They correspond to

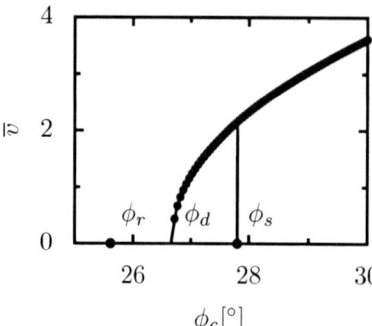

Fig. 3. Mean velocity as function of the chute angle ϕ_c. The chute angle is ramped up and down slowly in steps of 0.1 degrees every 800 seconds. The slope was divided into 200 parts with initially uniform inclination. Dotted line agrees with a behavior $v \propto \sqrt{\phi_c - \phi_d}$.

$\phi_s = 27.8°$, $\phi_r = 25.6°$, $a = 0.5$, $b_0 = \tan{(\phi_s + \phi_r)}/2$, $b_2 = 0.01$. The dotted line in Fig. 3 can be fitted to the analytic form $v \propto \sqrt{\phi_c - \phi_d}$.

Front propagation along a chute

In recent experiments, Daerr and Douady [17] have experimentally shown that single avalanches created by perturbing a static layer of granular material on a rough incline can exhibit two distinct types of dynamics: (1) an avalanche that just propagates downhill or (2) an avalanching front that propagates downwards and upwards. Here we show that both types of avalanching also can be recovered in the domino model. Following the experiments of Daerr and Douady [17], we have performed simulations on chutes with a rough surface profile of the granular material. We add to the angle ϕ_r a normally distributed part between $-(\phi_s - \phi_r)/2$ and $(\phi_s - \phi_r)/2$ being multiplied with a scaling factor r. Then we ramp up the chute by an additional angle $\Delta\phi$ into the

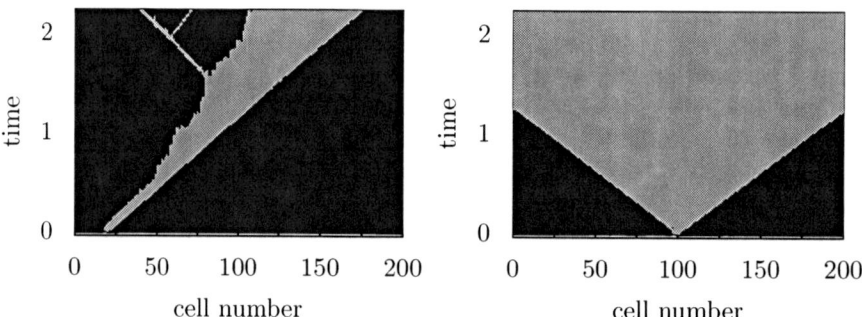

Fig. 4. Space-time plots of two different scenarios of avalanching fronts on chutes triggered in the metastable region. Grey (black) areas denote where the granular surface is moving (at rest). In the left panel the chute is tilted from $\phi_r = 25.6°$ by $\Delta\phi = 1.02°$, whereas in the right panel $\Delta\phi = 1.3°$ exceeds $\phi_d = (\phi_s + \phi_r)/2$ and leads to an avalanche with upwards moving rear front.

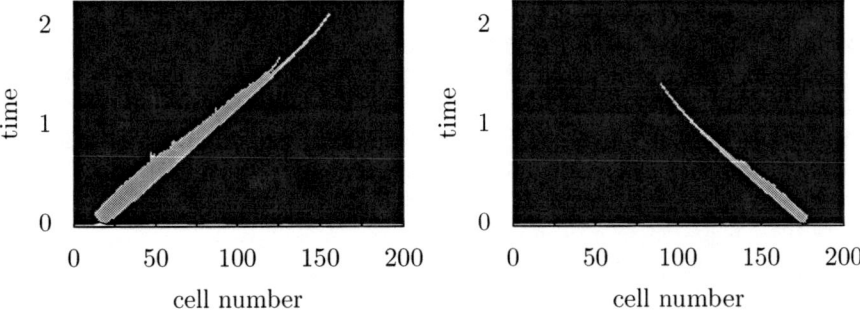

Fig. 5. Space-time plots describing avalanches on chutes inclined at $\phi_r = 25.6°$ and tilted by $\Delta\phi = 0.95°$. The roughness is $r = 0.1$ times $\phi_s - \phi_r = 2.2°$ with ϕ_s and ϕ_r taken from the Jaeger experiments [1]. In the left panel the flow is triggered by a small heap, leading to an avalanche running down the slope that eventually starves out. In the right panel, we created a dip as initial condition. The tilt of the chute is too small to destabilize enough material from above, the flow stops without the material running over the dip.

metastable region. We disturb the surface at one point by adding additional grains, so that locally ϕ_s is exceeded. Depending on $\Delta\phi$ and the scale of roughness r we numerically observe various scenarios of avalanches flowing down the chute with rear fronts either moving downwards or upwards or even combinations of both as function of time, cf. left panel of Fig. 4. Besides a surface flow that propagates from its creation point down to the foot of the pile, also other avalanching structures can be obtained by varying the initial profile of the heap. In Fig. 5 we show localized structures with forward and rear fronts moving downhill or uphill that peter out after some time at some point on the pile. It is essential to note that such backward propagating fronts do not imply upflow of the granular material. The flow direction is always downwards, however the starting point of the surface flow can move uphill in time because of the pulling-from-below mechanism. What are the conditions for the rear front to move downwards or upwards? Our numerical simulations indicate the following. If one starts with a pile surface with averaged angle ϕ in a no-flow configuration and puts locally a small pile with larger inclination angle as a perturbation on it, then there is:

(1) no flow if $\phi < \phi_r$

(2) a global flow if $\phi > \phi_s$

(3) a localized flow with rear front moving downhill that might peter out at some time or not, if $\phi_r < \phi < \phi_d = (\phi_s + \phi_r)/2$ and

(4) a localized flow with a rear front moving uphill if $\phi_d < \phi < \phi_s$.

This seems, at least qualitatively, to agree with the experimental findings of Daerr and Douady [17]. A quantitative comparison between our results and those of Daerr and Douady is difficult, because their results are given with respect to the static height of granular material on their chute. In our model that focusses on the dynamical variations of the surface profile, such a static height is not present by construction. So, the comparison remains qualitative.

One can directly infer from the space-time plots in Fig. 4, that the forward front velocities are constant whereas the rear front velocities differ depending on whether they are moving also downwards or upwards. In the first case, the rear front of the avalanche exhibits as rather rugged dynamics. In the latter case, the rear front is moving with approximately the same speed as the head front. This is in contrast to the arguments of Rajchenbach [20] who supposed a head front moving twice as fast as the rear front; however, it seems to be in accordance with the observations of Daerr and Douady [17].

Power spectrum of avalanches
As mentioned above, an important feature of the mean-field model (when supplied with additional stochastic forces) is that it can correctly reproduce the power spectrum of sequences of avalanches detected by Jaeger et al. [1]. Using the *deterministic* domino model, it is challenging to see whether the influence of multiply coupled small scale avalanches can also generate the spectrum and act similarly as the Gaussian white noise term in the extended mean field model [14]. This is especially interesting since so far there exists no physical explanation of the statistical properties of the noise in the global model. Within our model, there are actually two different ways to determine the power spectrum. First, one can determine the time evolution of the spatially averaged velocity of the avalanching process, $\overline{v}(n) = (1/I)\sum_{i=1}^{I} v(i,n)$, perform a Fourier transform of that signal and compute the spectrum $S_{av}(f) \propto |F[\overline{v}(n)]|^2$ Second, one can fix a specific cell i and measure the time evolution of the velocity signal at this position, $S_{loc}(f) \propto |F[v(i,n)]|^2$. The latter is close to the method experimentally used in Ref. [1]. In Fig. 6 we compare the experimental result [1], i.e. the curve that ends at $(\log_{10}(f) = 1, \log_{10}(S(f)) = -4)$ due to their experimental resolution, with numerical simulations of $S_{av}(f)$ (lower curve for large f) and $S_{loc}(f)$ from the domino model for a specific set of appropriate parameters. The typical shape of the power spectrum, a peak at low frequencies that reflects the time between two avalanching events, followed by broad shoulder and decay behavior that lies between $1/f^3$ and $1/f^4$ is present in all three spectra. Specifically, we find that the decay behavior of $S_{loc}(f)$ agrees very well with the experiments. This is, by the way, to our knowledge the first theoretical explanation of the power spectrum of the avalanching process based on a purely deterministic dynamics.

S-shaped surface distortion
In half-filled drums rotated with a sufficiently large rotation rate ω, a (hysteretic) transition to a continuous surface flow develops. Rajchenbach [19] has shown that the corresponding surface profile is not flat, but possesses an S-shaped distortion (e.g. for a rotation rate of 4.5 rpm). This effect can also be found in our model, see left panel of Fig. 7, indicating that it is an intrinsic property of the granular flow and not primarily an effect of strong centrifugal forces that are so far neglegted in our model. To analyze the origin of the

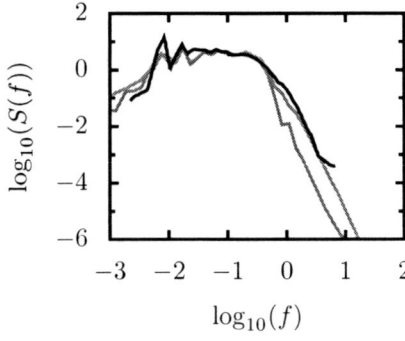

Fig. 6. The two computed power spectra of a sequence of 60 avalanches versus the power spectrum obtained from experimental data by Jaeger et al. [1]. Parameters are $\phi_s = 28.8$, $b_0 = \arctan(\phi_d) = \arctan(26.7°)$, $b_2 = 0.0001$, $\omega = 1.2$. The numerical curves have been smoothed over small frequency scale fluctuations.

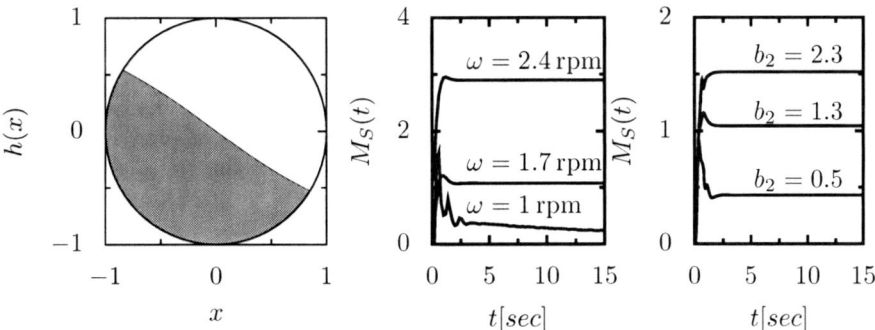

Fig. 7. Left: S-shaped surface distortion under rotation. Middle: Strength of the surface distortion measured by the mean square deviation M_S from the linear regression line. For increasing rotation rates there is a saturation of the distortion, for low values, the surface distortion vanishes after 4 sec of rotation time. Right: Dependence of M_S on the time t for three distinct values of b_2 showing that the Bagnold friction triggers the surface distortion in the model.

surface distortion, we set the Bagnold friction term $\propto b_2$ to small values of $b_2 = 0.003$ or 0.03 and use 4.5 rpm as a fixed rotation rate. Starting from a flat surface and turning on the rotation abruptly, the surface typically deforms initially in an S-shaped way, but becomes flat again after a bigger avalanche has flowed down. Subsequently, the surface stays at a mean angle of about ϕ_d. Increasing Bagnold friction to a value of $b_2 = 1.3$ the average surface angle increases with increasing rotation rates and, even more important for the domino model, the surface profile does not loose its initial S-deformation with ongoing rotation.

To quantify the surface deformation, we introduce the characteristic measure M_S being the sum of the square deviations from the linear regression line of their surface profile. In the middle panel of Fig. 7, we show that low rotation rates (1 rpm) lead to an initial build-up of a S-surface that breaks down again to very low values of M_S, whereas higher rotation rates (1.7 rpm and 2.4

rpm) lead to significantly non-flat surface profiles. In the right panel of Fig. 7, we present the dependence of M_S on time for three different values of b_2. As mentioned before, the initial strong increase and subsequent decrease of M_S vanishes for large values of b_2. The surface stays S-shaped and the saturation value of M_S increases with b_2. Consequently, the effect of the S-shaped surface deformation is a ramification of the Bagnold friction in our domino model.

4 Continuum Approximation

The domino model presented so far possesses the interesting property that it can be continuized in space and time and, therefore, be recast in the form of a system of coupled nonlinear field equations for the spatio-temporal evolution of the height $h(x,t)$ and the local velocity $v(x,t)$ at the surface on a still coarse-grained mesoscopic level. Assuming here for simplicity zero external rotation $\omega = 0$, taking into account that the coupling $a(l)$ scales like $1/l^2$ with the cell size, the continuum limit $l \to 0$ and $\Delta n \to 0$ can be performed by replacing

$$g(i,n) \to g(x,t) \tag{17}$$
$$(1/\Delta t)[g(i,n+1) - g(i,n)] \to \partial_t g(x,t) \tag{18}$$
$$(1/l)[g(i+1,n) - g(i,n)] \to \partial_x g(x,t) \tag{19}$$

with g representing h or v. The non-trivial point of such a continuum limit consists of an adequate treatment of flow down and pulling-from-below mechanisms entering in the domino model. A detailed discussion of that will be presented elsewhere [21]). Here, we just quote the final result given by

$$\partial_t h = q_1 \left\{ \left(1 + (\partial_x h)^2\right) \partial_x v + 2v \left(\partial_x h\right) \left(\partial_x^2 h\right) \right\} \tag{20}$$
$$\partial_t v = q_2[\partial_x h + b_0 + b_2 v^2]\chi(h,v) \tag{21}$$
$$\chi(h,v) = \Theta(v) - \Theta(\partial_x h - s_s) - \Theta(v)\Theta(\partial_x h - s_s). \tag{22}$$

with $q_1 = -(1/3)a_{\mathrm{mf}}L^2$, being independent of the cell size l, $q_2 = -g\cos\phi_d$, and the local maximum slope of rest $s_s = \tan\phi_s$. This system of equations has to be supplied with initial conditions for the velocity $v(x,0)$ (often equals zero) and the height profile $h(x,0)$ at the beginning and boundary conditions for the specific experimental setup. Comparative studies of the domino model and its continuized version show that many properties of the cellular automaton model such as e.g. the power spectrum of avalanches and the front propagation in case of localized perturbations along chutes survive the continuum limit.

5 Conclusion

A cellular automaton model for granular surface flow [11] has been presented that takes spatio-temporal properties of the surface evolution into account by coupling small scale avalanches. As discussed in this contribution, the model provides explanations for several experimentally observed local *and* dynamical effects in a unifying way. A more detailed account of our investigation will be given elsewhere [21].

References

1. H. M. Jaeger, C. Liu, S. R. Nagel : Phys. Rev. Lett. **62**, 40 (1989)
2. H. M. Jaeger, S. R. Nagel, R. P. Behringer : Rev. Mod. Phys. **68**, 1259 (1996)
3. L. P. Kadanoff : Rev. Mod. Phys. **71**, 435 (1999)
4. T. Pöschel, T. Schwager : *Computational Granular Dynamics* (Springer, Berlin Heidelberg New York 2005)
5. GDR MiDi : Eur. Phys. J. E **14**, 341 (2004)
6. J.-P. Bouchaud, M. E. Cates, J. Ravi Prakash, S. F. Edwards : Phys. Rev. Lett. **74**, 1982 (1995)
7. S. Douady, B. Andreotti, S. Douady : Eur. Phys. J. B **11**, 131 (1999)
8. I. S. Aranson, L. S. Tsimring : Phys. Rev. E **64**, 020301 (2001)
9. H. M. Jaeger, C. Liu, S. R. Nagel, T. A. Witten : Europhys. Lett. **11**, 619 (1990)
10. V. G. Benza, F. Nori, O. Pla : Phys. Rev. E **48**, 4095 (1993)
11. S. J. Linz, P. Hänggi : Phys. Rev. E **51**, 2538 (1995)
12. S. J. Linz, P. Hänggi : Phys. Rev. E **50**, 3464 (1994)
13. S. J. Linz, P. Hänggi : Physica D **97**, 577 (1996)
14. W. Hager, S. J. Linz, P. Hänggi : Europhys. Lett. **40**, 393 (1997)
15. S. J. Linz, W. Hager, P. Hänggi : CHAOS **9**, 649 (1999)
16. T. G. Drake : J. Geophys. Res. **95**, 8681 (1990)
17. A. Daerr, S. Douady : Nature **399**, 241 (1999)
18. A. Daerr : Phys. Fluids **11**, 2115 (2001)
19. J. Rajchenbach : Phys. Rev. Lett. **65**, 2221 (1990)
20. J. Rajchenbach : Phys. Rev. Lett. **88**, 014301 (2002)
21. A. Hoffmann, S. J. Linz : in preparation

Morphological Change of Crack Patterns Induced by Memory Effect of Drying Paste

Akio Nakahara and Yousuke Matsuo

Laboratory of Physics, College of Science and Technology, Nihon University, Funabashi, Chiba 274-8501, Japan

Summary. We have experimentally found a method to imprint into paste the direction of future crack propagation in the drying process of paste. The rheological measurement and the drying experiment making morphological phase diagram reveal that the plasticity of paste plays an important role in the memory effect.

1 Introduction

In this work, we have two purposes to do our experiments. First, we are interested in the complex rheology of soft matter, especially the memory effect of paste, from the scientific point of view. Second, we want to find a method to control crack pattern formation in order to avoid serious damages in the field of industry.

To achieve both purposes, we perform drying experiment of paste. We prepare paste by mixing powder with water, pour the mixture into the acryl container, and dry it at 25°C and 30% humidity. Usually isotropic and cellular crack patterns appear as the paste is dried [1]. Here, we propose a method to control future crack propagation which appears in the drying process.

2 Emergence of Anisotropic Crack Patterns

As our first experimental result, we obtain the anisotropic crack pattern as is shown in Fig. 1. Here, we use Calcium Carbonate as powder, and prepare paste by mixing 3000g of powder with 1500g of distilled water. The diameter of the circular container is 500mm. As soon as we pour the mixture into the container, we oscillate the container horizontally in angular direction for 60sec to spread the mixture homogeneously inside the container. After drying the mixture for 3 days, we get the radial crack pattern as is shown in Fig. 1(a).

At first, we thought that the radial crack pattern is induced by the boundary effect of the circular container, but soon we realized that our idea was

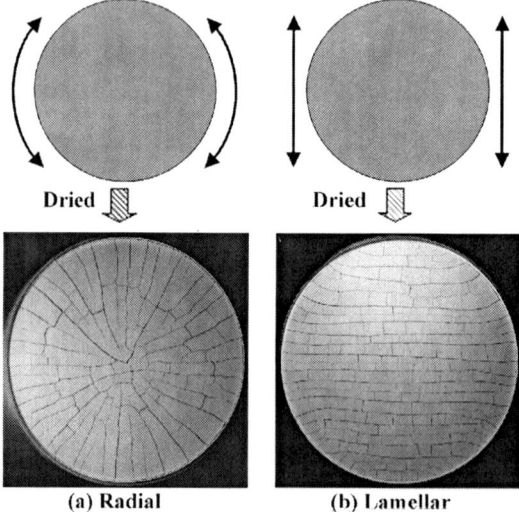

(a) Radial   (b) Lamellar

Fig. 1. Anisotropic crack patterns [2] (a) Radial crack pattern which appears when the container is initially oscillated in an angular direction (b) Lamellar crack pattern which appears when the container is initially vibrated in one direction

wrong. In the next experiment where we vibrated the container horizontally in one direction, we obtain the lamellar crack pattern as is shown in Fig. 1(b). Here, the direction of these anistropic crack is perpendicular to the direction of the initial external vibration. That is, these anistropic crack patterns are induced by the memory of paste on the direction of the initial external vibration, and not by the boundary effect [2].

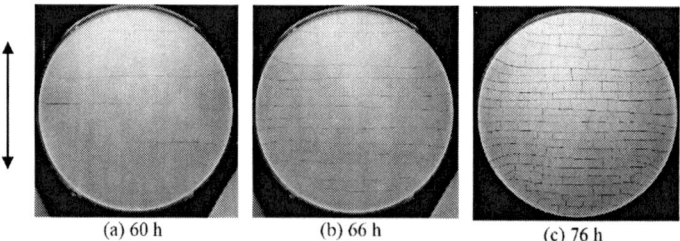

(a) 60 h   (b) 66 h   (c) 76 h

Fig. 2. Time evolution of lamellar crack pattern formation [3]

Figure 2 shows the time evolution of the lamellar crack formation. At the first stage of the crack formation, straight cracks propagate along the direction perpendicular to the direction of the initial external vibration. As time goes on, new cracks are formed between lamellar cracks until the spacing between lamellar cracks becomes about the thickness of the mixture. At the final stage, long rectangular fragments break themselves into shorter pieces and we get the brick structure [3].

3 Rheological Measurement

To investigate the feature of powder of Calcium Carbonate, we perform microscopic observation using Scanning Electron Microscope (SEM), and find that the shape of particles is isotropic and rough. Since the sizes of particles range between 0.5 and 5μm, the rheology of the mixture of powder and water changes drastically as a function of the solid volume fraction, i.e., the volume fraction of powder in the mixture, denoted by ρ.

Thus, we perform rheological measurement of the mixture. Figure 3 shows the value of the yield stress σ_Y as a function of the solid volume fraction ρ. There are two vertical lines in the figure. Below $\rho=25\%$ which corresponds to the Liquid-Limit (LL) line, the value of the yield stress vanishes and the mixture can be regarded as a Newtonian viscous fluid. Above $\rho=54\%$ which corresponds to the Plastic-Limit (PL) line, the mixture is called as semi-solid, and we cannot mix powder with water homogeneously due to the lack of enough water.

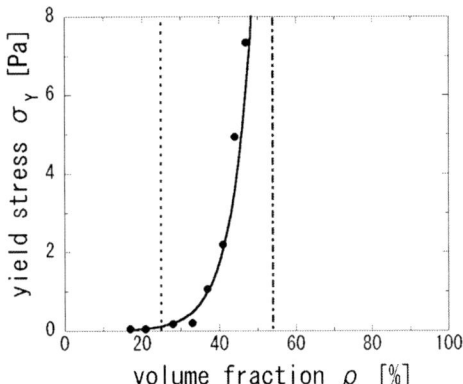

Fig. 3. Yield stress σ_Y as function of solid volume fraction ρ [2]. The dotted and the dashed-and-dotted vertical lines correspond to the Liquid-Limit (LL) line ($\rho=25\%$) and the Plastic-Limit (PL) line ($\rho=54\%$), respectively.

Between LL and PL lines, the mixture has a finite yield stress with plasticity and the value of the yield stress increases drastically as the value of the solid volume fraction increases. Thus, in the following experiments, we systematically change the value of the solid volume fraction ρ.

4 Morphological Phase Diagram of Crack Patterns

Now, we investigate the condition when paste can remember the direction of the initial external vibration. We consider that there are two important parameters in our experiments, one is the solid volume fraction ρ, and the other is the strength of the initial external vibration. Here, the strength of the initial external vibration is expressed as $4\pi^2 r f^2$, where r represents the amplitude and f denotes the frequency of the initial external vibration. In the following experiments, we set the amplitude r of the initial vibration as 15mm, and change the frequency f from 20 to 60rpm. As containers, we use square acryl boxes with sides of 200mm. We fix the mass of powder in the mixture as 360g in each container, so that we can equalize the final thickness of mixture with different solid volume fraction when they dry up and thus we can equalize the characteristic sizes of final crack patterns. Here, the final thickness of the mixture becomes about 7mm.

First, we present the dependence on the solid volume fraction ρ in Fig. 4 by setting the value of the frequency f as 40rpm and changing the solid volume fraction ρ. When the value of ρ is low (ρ=28%), we only get the isotropic and cellular crack pattern. As the value of ρ increases, the pattern changes into lamellar crack pattern (ρ=41%). Note that the direction of the lamellar crack pattern is perpendicular to the direction of the initial external vibration, denoted by the arrow in the figure. As we increases the value of ρ further, however, we again get the cellular crack pattern, but this change can be understood in the morphological phase diagram which will be shown below.

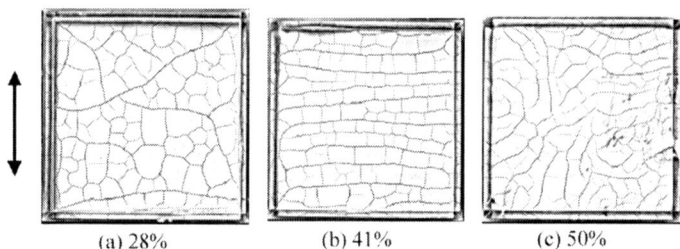

(a) 28%   (b) 41%   (c) 50%

Fig. 4. Dependence on the solid volume fraction ρ. Here, we set the value of the frequency f as 40rpm, and the corresponding strength of the initial external vibration becomes 0.26m/s^2.

Next, we present the dependence on the frequency f of the initial external vibration in Fig. 5 by setting the value of the solid volume fraction as ρ=41%, and changing the frequency f. When the value of f is low (f=20rpm, i.e., the strength of the initial external vibration is 0.07m/s^2), we only get the isotropic and cellular crack pattern, but, as the value of f increases, the pattern changes into lamellar crack pattern (f=40rpm, i.e., strength of 0.26m/s^2).

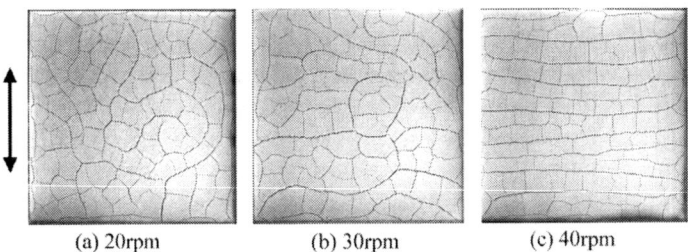

(a) 20rpm   (b) 30rpm   (c) 40rpm

Fig. 5. Dependence on the frequency f of the initial externalvibration. Here, we set the value of the solid volume fraction ρ as 41%.

All these results described above are summarized by the morphological phase diagram of crack patterns, shown in Fig. 6, as a function of the solid volume fraction ρ and the strength $4\pi^2 r f^2$ of the initial external vibration. In the left region of LL line, we only obtain isotropic and cellular crack pattern. The region between LL and PL lines is divided, by the solid and the dashed curves, into three regions, A, B, and C.

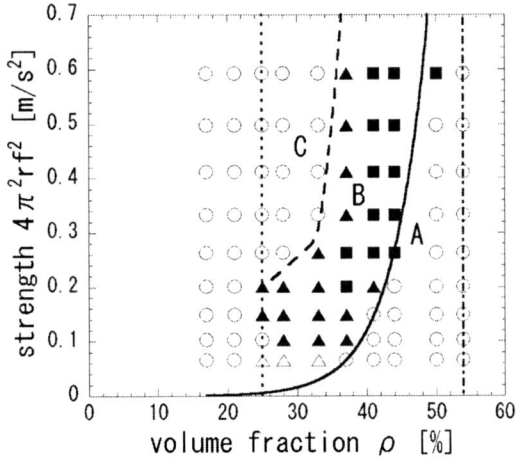

Fig. 6. Morphological phase diagram of crack patterns as a function of the solid volume fraction ρ and the strength $4\pi^2 r f^2$ of the initial external vibration [2]. Open circles denote isotropic cellular crack patterns and solid squares denote lamellar crack patterns. Solid and open triangles denote combinations of isotropic cellular crack patterns and lamellar crack patterns. The region between the LL dotted line and the PL dashed-and-dotted line is divided, by the solid and the dashed curves, into three regions, A, B, and C. Only in region B between the solid and the dashed curves, lamellar crack patterns can appear.

The solid curve in Fig. 6 which divides regions A and B is drawn by equalizing the value of the shearing stress σ induced by the external vibration to that of the yield stress σ_Y shown in Fig. 3. In region A below the solid curve only isotropic and cellular crack patterns appear, while in region B above the solid curve we obtain lamellar crack pattern. That is, when the strength of the external vibration is larger than that of the yield stress of the paste, the initial external vibration causes shear movement of paste and the memory is kept inside the paste due to its plasticity.

However, in region C above the dashed curve we again get only isotropic and cellular crack patterns. This change is explained by checking the fluidity of the mixture at the initial vibration. In region C, some surface waves or turbulent flows appear during the initial vibration. This turbulent flow destroys the microscopic memory inside the mixture, and thus, only isotropic cellular crack patterns appear in region C.

Recently, theoretical approaches based on the elastic-plastic deformation of paste with a nonzero yield stress are proposed to explain the memory effect of a paste to an external mechanical force [4] and the formation of resultant anisotropic crack patterns in the drying process of a paste [5].

5 Concluding Remarks

We experimentally found that we can imprint the direction of an external vibration into paste by applying shear movement to the paste, and the memory in the paste is visualized as the morphology of anisotropic crack patterns which appear in the drying process. We consider that our experimental results can be applied to industry, because, if we can control the direction in which cracks will propagate in the future, we make plans to avoid accidental serious damage.

Acknowledgements

We like to acknowledge H. Uematsu, M. Otsuki, S. Sasa, T. S. Komatsu, and Y. Nakahara for valuable discussions. We thank M. Sugimoto, T. Taniguchi, and K. Koyama for support with rheological measurements at the Venture Business Laboratory of Yamagata University. We also thank Y. Aoyagi, A. Taguchi, K. Nakagawa, and A. Itoh for support with microscopic observations using a SEM at the Advanced Materials Science Center of Nihon University.

References

1. G. Groisman, E. Kaplan: Europhys. Lett. **25**, 415 (1994)
2. A. Nakahara, Y. Matsuo: J. Phys. Soc. Jpn. **74**, 1362 (2005)
3. A. Nakahara, Y. Matsuo: 'Imprinting memory into paste to control crack patterns in the drying process'. In: *Powders and Grains 2005*. ed. by R. Garcia-Rojo, H.J. Herrmann and S. McNamara (A.A.Balkema, Rotterdam 2005) pp. 1081
4. Ooshida T., K. Sekimoto: Phys. Rev. Lett. **95**, 108301 (2005)
5. M. Otsuki: Phys. Rev. E **72**, 046115 (2005)

Hydrodynamic Interactions Between Electrically Charged Grains in Suspensions

Jochen H. Werth[1], Henning Arendt Knudsen[2], and Dietrich E. Wolf[1]

[1] University of Duisburg-Essen, Campus Duisburg, Lotharstr. 1, D-47048 Duisburg, Germany
[2] Dept. of Physics, University of Oslo, P.O.Box 1048 Blindern, NO-0316 Oslo, Norway

Summary. In suspensions with charged particles, electrostatic forces and hydrodynamic interactions are both important to describe the system. We study different models of hydrodynamic interaction for monopolarly charged particles in a nonpolar liquid. In this case, there is no screening of the Coulomb repulsion, so the repulsion between all pairs must be taken into account. For some selected systems we examine different models of hydrodynamic interaction, and we show that anomalies and unphysical behaviour in many cases result. We propose a way to model the interaction which does not suffer from these problems.

1 Introduction

Suspensions appear in many different forms and settings, and the particles as well as the solvent may have different properties. In order to simulate such suspensions, the particles as well as the fluid must be handled in some way. Depending on the system in question one among several methods may be the preferred one. For systems with rather few particles, one might choose an Euler-Lagrange method, meaning that the Navier-Stokes equations are simulated on a grid in the Euler-picture, whereas the particles are followed explicitly - a lagrangian picture. There exist many varieties of these rather detailed methods. A different approach is to follow only the particles explicitly. The effect of the solvent can be incorporated by various models of hydrodynamic interaction. This allows for simulation of systems with many more particles than for the Euler-Lagrange methods.

In this study we examine how various models of hydrodynamic interaction work for a charged suspension. The motivation for studying this system stems from methods in coating processes [1]. Those system contain in reality so many particles that a too detailed hydrodynamic modelling is neither wanted nor possible. Interactions are often modelled as superposition of two-particle in-

teractions, possibly also of three-particle and more-particle interactions. However, it turns out that for systems of many particles, these methods may result
in unphysical behaviour, which we demonstrate for certain test systems, employing different models of hydrodynamic interaction. In particular for denser
system these problems are evident.

Based on our observations we conclude that extreme care must be taken
for denser systems. We propose a way to mend the problem, and tests on the
same systems show that our method does not suffer from the same problems.

2 Models of Hydrodynamic Interaction

Every suspended particle experiences forces and interacts with the solvent.
Assuming that the motion of the fluid is governed by the linear Stokes equation
[2] and denoting the external force and torque on a single particle with $\mathbf{F}$
and $\mathbf{T}$, respectively, the motion of the particle is governed by the following
equation

$$\begin{pmatrix} \mathbf{v} \\ \boldsymbol{\omega} \end{pmatrix} = \begin{pmatrix} \boldsymbol{\mu}^{tt} & \boldsymbol{\mu}^{tr} \\ \boldsymbol{\mu}^{rt} & \boldsymbol{\mu}^{rr} \end{pmatrix} \begin{pmatrix} \mathbf{F} \\ \mathbf{T} \end{pmatrix} . \tag{1}$$

Here, the particle's velocity and angular velocity are denoted by $\mathbf{v}$ and $\boldsymbol{\omega}$,
respectively. The matrix is called the mobility matrix, which contains the
effect of the fluid on the particle, whereby one assumes that the overdamped
limit is valid. For a single particle in an infinitely extended, resting fluid, the
mobility matrix is diagonal: $\boldsymbol{\mu}^{tt} = \frac{1}{6\pi\eta a}\mathbf{I}$ for the translation, and $\boldsymbol{\mu}^{rr} = \frac{1}{8\pi\eta a^3}\mathbf{I}$
for the rotation. Here, $\mathbf{I}$ denotes the unity matrix. The radius of the particle
is a and the viscosity of the fluid is η. As soon as more than one particle is
considered, the motion of particles is coupled, and the mobility matrix grows
in complexity. The structure of the equation of motion is maintained in the
form of Eq. (1) when the velocity vector $\mathbf{v}$ is understood to contain the velocity
of all particles sequentially, and similar for angular velocity, force, and torque.
For two particles, that would mean that the coupling of their translational
motion and the forces acting on them takes the form

$$\begin{pmatrix} \mathbf{v}_1 \\ \mathbf{v}_2 \end{pmatrix} = \begin{pmatrix} \boldsymbol{\mu}_{11}^{tt} & \boldsymbol{\mu}_{12}^{tt} \\ \boldsymbol{\mu}_{21}^{tt} & \boldsymbol{\mu}_{22}^{tt} \end{pmatrix} \begin{pmatrix} \mathbf{F}_1 \\ \mathbf{F}_2 \end{pmatrix} . \tag{2}$$

When it comes to actual values for the entries in the mobility matrix, there
exists a whole range of models of increasing complexity. In the following we
present some of these, which we subsequently use in test simulations.

2.1 Oseen Tensor

For simplicity, we present the effect on translational coupling between two
particles first in the simplest possible setting. Assuming that there is a single
point particle present, upon which the force $\mathbf{F}$ is acting, the resulting flow
field $\mathbf{u}$ in distance $\mathbf{r}$ to the particle is

$$\mathbf{u}(\mathbf{r}) = \frac{1}{8\pi\eta} \left(\frac{1}{r}\mathbf{I} + \frac{\mathbf{rr}}{r^3} \right) \cdot \mathbf{F} = \mathbf{G}(\mathbf{r}) \cdot \mathbf{F} , \tag{3}$$

where $\mathbf{G}$ is the Oseen tensor [2]. The simplest possible interaction model is to calculate the velocity field of each particle individually and to sum up all contributions. Thus, the velocity of particle 1 in a system of N particles is given by

$$\mathbf{v}_1 = \frac{\mathbf{F}_1}{6\pi\eta a} + \sum_{i=2}^{N} \mathbf{u}_i = \frac{\mathbf{F}_1}{6\pi\eta a} + \sum_{i=2}^{N} \mathbf{G}(\mathbf{r}_1 - \mathbf{r}_i) \cdot \mathbf{F}_i . \tag{4}$$

This model was introduced by Kirkwood and Riseman [3]. The resulting mobility matrix is given by

$$\mu_{ii}^{tt} = \frac{1}{6\pi\eta a}\mathbf{I} \quad \text{and} \quad \mu_{ij}^{tt} = \mathbf{G}(\mathbf{r}_i - \mathbf{r}_j), \quad i \neq j . \tag{5}$$

2.2 More Accurate Two-Particle and Many-Particle Models

The simple model of Kirkwood and Riseman fails in reproducing accurate mobilities for suspensions. This can be seen e.g. by the fact that the resulting mobility matrix is not positive definite for particles close to each other [4]. However, for two particles, a solution of the interaction problem exists up to arbitrary order and is presented by R.B. Jones and R. Schmitz [5]. In their paper, the mobility matrices μ_{ii} and μ_{ij} are presented in form of a series expansion in the particle distance $\mathbf{r}$. For identical particles, explicit results up to an order of r^{-20} are given.

Although the Jones-Schmitz calculations are accurate for two particles and the Stokes equation describing the fluid motion is linear, the superposition of two-particle interactions does not lead to a valid mobility matrix. The reason is, that the superposition of two-particle solutions does not fulfill the no-slip boundary condition on the particles' surface in a many-particle system.

Mazur and van Saarloos calculated in principle many-particle interactions [6]. Expanding the mobility matrix again in a series of particle distances r, the terms up to r^{-3} only contain two particle interactions identical to those of Jones and Schmitz. Explicit results are given up to r^{-7}, where three particle interactions (of order r^{-4} and r^{-6}) and four-particle-interactions (of order r^{-7}) occure.

Another widely used model including many-particle hydrodynamic interactions is proposed in [7]. However, in case of particles beeing close to each other, this model again employs two-particle interactions.

3 Simulation of Repulsive Particle Systems

In the following we consider systems of identical particles carrying all the same electric charge and beeing suspended in a nonpolar liquid. There is no

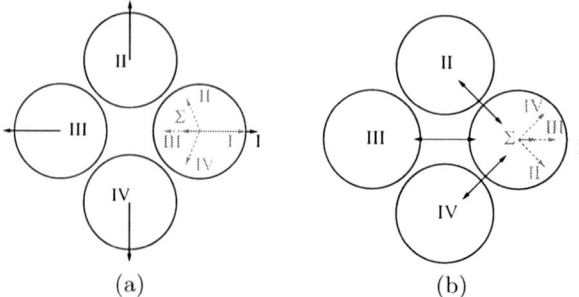

Fig. 1. (a) Four equally charged particles in symmetric positions are shown. On each particle the only(II-IV) or longer arrow(I) represents the total electrostatic force acting on the particles. Each of these forces gives a component to the velocity of particle I, which are shown as dotted lines on I. Schematically, the sum of the four velocities is shown to point in the opposite direction of the actual force of the particle. (b) By considering each pair interaction separately, only the electrostatic force from one neighbour at a time is used when calculating the hydrodynamic interaction. This ensures a resultant velocity in the same direction as the total electrostatic force.

screening of the Coulomb repulsion, so the repulsion between all pairs must be taken into account. Evidently the particles repel each-other. However, hydrodynamic interaction between particles tends to slow this repulsion down. For a few selected systems we examine the different models of hydrodynamic interaction described above, and we show that anomalies and unphysical behaviour in many cases result. We propose a way to model the interaction which do not suffer from these problems.

The idea of the test setups is illustrated in Fig. 1. The four particles in Fig. 1(a) all repel each-other. Due to symmetry one sees directly that the total repulsive force on each particle is of equal size and pointing radially away from the centre. This setup is chosen for two reasons. Firstly, all particles experience equivalent surroundings, meaning that it suffices to show results for one particle. Secondly, it is intuitively clear that the hydrodynamic interaction must act to slow down the expansion as compared to the case without interaction. However, if the interaction is over-estimated the particles do not only slow down, they may effectively attract each-other, which is clearly unphysical.

We show results in three dimensions(3D). The numbers of particles that we have chosen are: 4, 6, 8, 12, and 20, placed in the corners of a tetrahedron, hexahedron, octahedron, dodecahedron, and icosahedron, respectively. Again, this is for simplicity in the presentation, since these forms are perfectly symmetric. Results are shown in Fig. 2, where the x-axis is the distance r between the centres of two neighbouring particles in units of the particle radius a. The y-axis is the calculated velocity of the particles radially outwards normalized with respect to the velocity the particle would have if there were no hydrodynamic interaction (i.e. the total force over stokes friction, $\mathbf{v}_0 = \frac{\mathbf{F}}{6\pi\eta a}$).

All the examples in Fig. 2 show normalized velocities that are smaller than unity. This means that in a sense the hydrodynamic interaction is shown to act in correct direction, that is to say the interaction slows the particles down. For particle distances over some threshold, roughly at least one particle radius between the particles that are closest together, the methods of Kirkwood-Riseman, Jones-Schmitz, and Mazur-van Saarloos show the same results: a substantial slowing down of the particles. For smaller distances, the interaction is clearly over-estimated by the methods since they all lead to attraction of the particles. One might at this point argue that this is only a result of the somehow constructed particle setup. To check this we have performed simulations of dense system with many particles which are placed randomly in space. These systems we also have followed in time, and the particles have indeed attracted each-other in some cases: they have imploded.

The failure of the hydrodynamic modelling which we have demonstrated is severe and leads to very unphysical events. It seems that the superposition

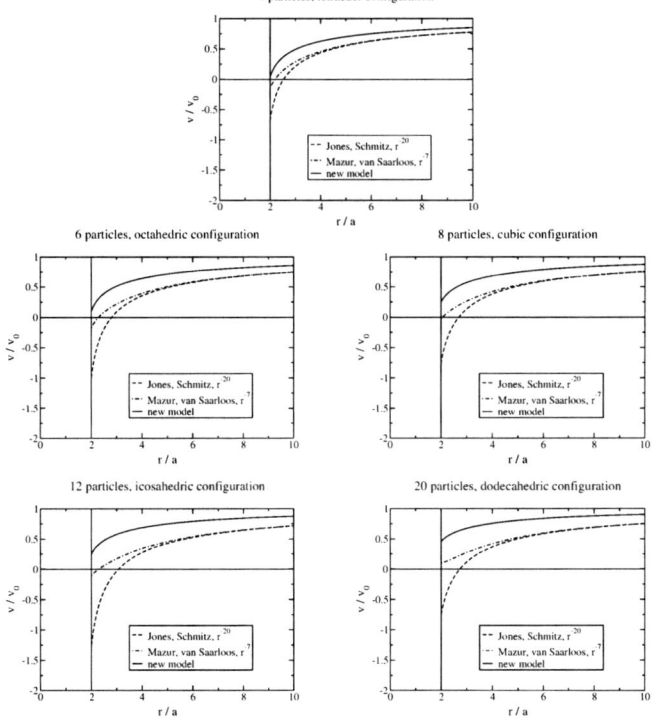

Fig. 2. As a function of particle distance (centre-centre) the resulting velocity of the particles with hydrodynamic interaction is shown. The velocity is normalized as described in the text. Negative values mean that the particles attract each other. This can be observed in all cases for short inter-particle distances when standard hydrodynamic interaction is included. Our proposed method (solid lines) does not show this unwanted feature.

of two-particle interactions, or few-particle interactions is not permitted for many particles. One way to get around this problem may be to only consider the electrostatic repulsive force between each pair of particles, and then calculate the contribution to the velocity based on this force and hydrodynamic interaction for this pair separately. In all the setups shown in Fig.2, we tested out this idea. By using the pairwise interaction of Jones and Schmitz we se that no artificial attraction results(solid curves). We note that whether this method gives the physically correct values is not hereby proved, but we find the idea is promising.

4 Conclusion

We have studied aspects of suspensions with monopolarly charged particles. The effect of hydrodynamic interactions is generally recognized to be important, but its modelling in the case of many particle is highly non-trivial. We considered the situation were the solvent itself is nonpolar so that the coloumb interaction is long-ranged. This leads to rather strong electrostatic interaction, but the motion of the particles are expected to be slowed down due to the hydrodynamic interaction between the particles. We presented some standard interaction models and tested their effect on some particular symmetric particle setup. These setups were chosen because one can more easily, based on physical intuition, distinguish between physical and unphysical behaviour.

It turns out that all the interaction models that are commonly used lead to attraction between the (equally charged!) particles whenever the distances between the particles are too small. We conclude that in particular for dense systems this method cannot be used. Whether they can be used for less dense system for many particles is hereby not proved correct or wrong, but we have observed similar extreme anomalies for simulations of systems with many particles, and we claim that extreme care must be taken if one choses to do so. We do not give the ultimate solution to the problem, but we propose an idea that we find very promising: by only taking the pairwise electrostatic repulsion into account when calculating the hydrodynamic interaction between the particles, the anomalies do no longer occur for the test systems in this study.

References

1. M. Linsenbühler, J.H. Werth, S.M. Dammer, H.A. Knudsen, H. Hinrichsen, K.-E. Wirth, D.E. Wolf, *Cluster Size Distribution of Charged Nanopowders in Suspensions*, submitted to Powder Technology.
2. S. Kim, S.J. Karrila: *Microhydrodynamics; Principles and Selected Applications* (Butterworth - Heinemann, Boston 1991)
3. J.G. Kirkwood, J. Riseman, J. Chem. Phys. **16**, 565 (1948)
4. R.E. De Wames, W.F. Holland, M.C. Shen: J. Chem. Phys. **46**, 2782 (1967)
5. R.B. Jones, R. Schmitz: Physica A **149**, 373 (1988)
6. P. Mazur, W. van Saarloos: Physica A **115**, 21 (1982)
7. B. Cichocki, B.U. Felderhof, K. Hinsen, E. Wajnryb, J. Blawzdziewicz: J. Chem. Phys. **100**, 3780 (1994)

Particle Discharge Process from a Capillary Pipe

Qing-Song Wu[1] *, Mao-Bin Hu[1], Xiang-Zhao Kong[1], and Yong-Hong Wu[2]

[1] School of Engineering Science, University of Science and Technology of China, Hefei 230026, P.R.China
[2] Department of Mathematics and Statistics, Curtin University of Technology, Perth WA6845, Australia

Summary. The particle discharge process from a vertical open-top pipe with a capillary outlet reveals some exceptions to the common belief that the outflux oscillation results solely from dynamic arching of beads at the orifice and that the outflux is not sensitive to the filling height. With beads of a particular size range, the outflux fluctuates greatly with time and the bulk dense granular flow in the pipe shows stop-and-go motion when the filling height is above a threshold. When the filling height falls to the threshold, led by a transitional stage, the outflux and the bulk movement become stable. The dropping velocity variation of the upper surface is measured to study the bulk motion in the pipe. With a heuristic theory, we find that the granular compaction and interstitial air pressure effect are responsible for the stop-and-go oscillation and the transitional behavior.

PACS number(s): 45.70.Mg, 81.05.Rm, 45.70.Vn

1 Introduction

Due to its complex dynamic behaviors, the physics of granular matter has attracted the attention of scientists and engineers in various fields [1–4]. Granular systems consist of particles interacting by inter-particle contacts and they exhibit many interesting phenomena, including density waves, surface pattern, segregation etc. Among them, the problem of guided flows in tubes, pipes, or chutes is of crucial importance for many industrial processes. Experiments, cellular-automata and molecular-dynamic simulations have been carried out to study the outflux behavior of granular material from pipes or silos. But some of the underlying mechanisms are still not clear.

Here, two previously unreported observations in an open-top pipe with capillary outlet are communicated: (i) When the filling height is far above a

---

* Corresponding author, Email:qswu@ustc.edu.cn

threshold, the outflux fluctuates with time and the bulk condensed granular flow above the bottleneck shows uniform stop-and-go motion. (ii) When the filling height falls below the threshold, led by a transitional stage, the outflux and the bulk movement become smooth. We also develop a heuristic model taking account of the granular compaction and interstitial air pressure effect to interpret the above behavior.

This paper is organized as follows: In Sec. II, we briefly review the recent results on granular pipe or silo flow. Sec. III presents the results of our experiment. In Sec. IV, the heuristic model for analyzing the experimental results is presented. Finally, conclusions are given in Sec. V.

2 Granular Pipe Flow and Outflux Oscillation

The outflux Q from a hopper or silo is related to many factors. The outlet width d and the inclination angle θ for the converging hopper wall (as shown in Fig.1a) play an important role. The experiment of Beverloo et al. [5] revealed that $Q \sim d^{5/2}$. When the size of particles is small ($< 400 \mu m$), the flux is far below Beverloo's prediction value due to the interstitial air effect. Computer simulations of Khelil [6] show that $\theta \approx 35°$ is a typical transition point (θ is the angle of the recline wall to the horizontal). The outflux Q does not change significantly when θ is smaller than $35°$, whereas it increases dramatically when θ exceeds $35°$.

Experiments and simulations [7–9] show that the particles usually do not flow uniformly. Instead, they flow in the form of density waves. Experiments of Wu [10] and Veje [11] found that the counter flow of air can induce the

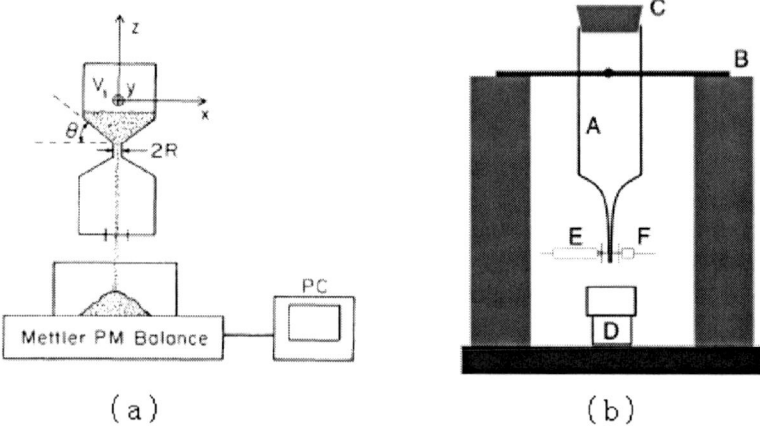

Fig. 1. The outflux ticking experimental setups of Wu et al [10] (a) and Veje et al [11] (b).

ticking of outflows in a closed-top hourglass. The oscillation motion results from the coupling between the flow of sand and convection of air through the sand matrix. When there is a counterflow, they observe ticking depending sensitively on the grain size and distribution. But the ticking disappears when the chamber is open to the air. Their experimental setups are as shown in Fig.1.

It is usually thought that the outflux of a silo or hopper fluctuates due to the dynamic arching at the orifice when the effect of interstitial air is neglectable. Because the arching effect is not sensitive to the granular pressure [15], it is usually thought that the outflux is not sensitive to the filling height [17].

In 1998, Moriyama et al [12] proposed an experiment with a flask attached to the bottom of the granular pipe and a flow meter attached to the outlet of the flask. When the outlet of the flask was fully open, granules could fall rather freely and there are no visible density waves. As the bottom was gradually closed, the pressure in the flask increased, and density waves appeared.

It is clear that the appearance of density waves in the above mentioned experiments is related to the effect of interstitial air. However, density waves can emerge without air. The friction between the particles and the pipe, the dissipative collisions between particles and an environmental electromagnetic field can also induce density waves in the pipe. In general, when the granular particle's diameter is smaller than $400\mu m$, the air effect can not be ignored in granular pipe flow. Chen et al [3] studied the flow of granular nickel particles moving down vertical pipes from a hopper in the presence of a local, horizontal AC electric field. For low V ($< V_c = 2.0kV$), a downward-moving interface exists between the hopper and the electrodes separating a high-density particle region in the lower part from a low-density region in the upper part. For high V($> V_c$), no interface exists and the whole region between the hopper and the electrodes are densely filled.

In 1996, Raafat et al [13] studied the granular density waves in a vertical glass tube. They discovered that particles were in bulk freely falling when the granular density is low. When the granular density is high, the particles were in slow flow. In between there are density waves. In 1999, Aider et al [14] studied the density waves and stop-and-go effect in detail. They investigated the granular flow with different humidity. In Raafat or Aider's experiments, the upper part (hopper) and the lower part (valve) of the pipe have great influence to the the flow structure and dynamic behavior of the granular flows in the pipe.

3 Experimental Observations

Our experiment is carried out in a vertical glass tube without any hopper on the top. The particle flow in the tube and the outflux show some peculiar behaviors different from the experimental observations described above.

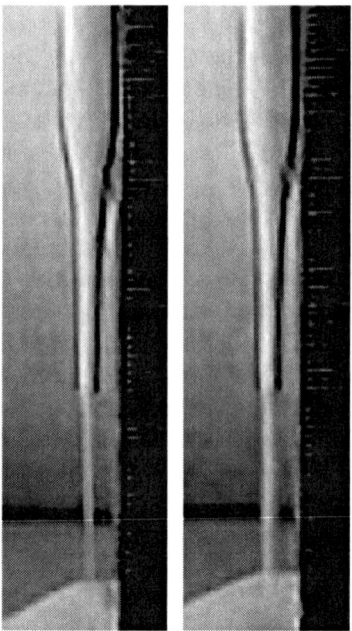

Fig. 2. Outflux oscillation, (left) thin flow and (right) thick flow.

The glass tube is of internal radius $R = 2.5mm$ and length $L = 500mm$ with a radius $r = 0.75mm$ capillary outlet at the bottom. The typical length of the capillary part is $h_c \approx 40mm$. Several granular materials were used including glass beads and sand with diameter $d_g = 0.1 - 0.28mm$, $d_g = 0.28 - 0.40mm$, and $d_g = 0.1 - 0.40mm$. With these granular materials, we found qualitatively similar behavior. For convenience, we concentrate on $d_g = 0.1 - 0.4mm$ glass beads throughout this paper. The outlet size is chosen to be greater than four times of the beads' diameter, so that dynamic arching is not easy to form at the outlet [15]. The granular flow is not stuck even though outflux oscillation is found in the experiment with the above granular species. However, when d_g is greater than 0.45mm, the flow is stuck; and when d_g is smaller than 0.1mm, the flow is smooth.

The flow is initiated by opening the bottom outlet of the pipe after the pipe is fully filled with particles. The outflux from the capillary outlet fluctuated with time when the filling height h_0 was far above a threshold of about 50mm. As shown in Fig.2, a periodic flow phenomenon is observed: a thin flow is always followed by a thick flow. The maximum size difference could be 80%, which is much greater than the outflux oscillation due to dynamic arching (10% − 15%). The fluctuation was quite rhythmic with a frequency of about $3Hz$. This is also different from random oscillation in dynamic arching case.

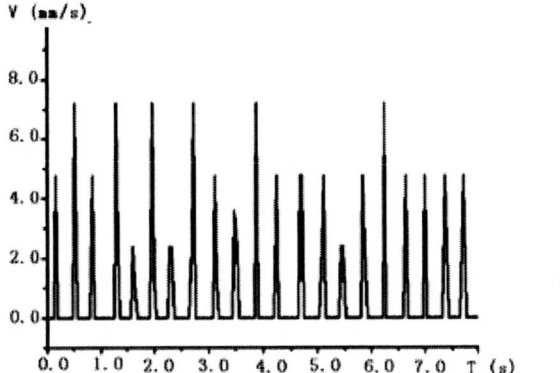

Fig. 3. Upper surface dropping velocity variation at the first stage $(0 \sim 8s)$.

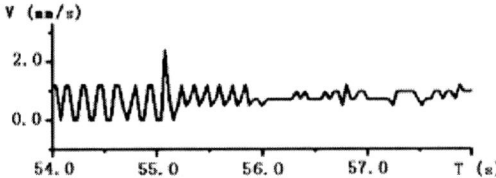

Fig. 4. Transition from the stop-and-go motion to a stable motion.

Simultaneously, the dense bulk movement above the bottleneck showed uniform stop-and-go motion. The outflux was thick when particles flowed, while a thin flow corresponded to the stopping of the bulk above the bottleneck. It is suitable to use the dropping speed of the upper surface as a representation of the bulk motion. We placed a ruler beside the pipe with the minimum scale of millimeter (See Fig.2). A digital camera was used to capture the movement of the granular upper surface with the frequency of 24 frames per second. The position of the surface was read afterwards frame by frame. With the position data, we determined the moving velocity.

As shown in Fig.3, the velocity varies periodically with a well-defined period T. Each T can be separated into two different phases, the active phase T_a (when the particles flow) and the inactive phase T_i (when the particles stop flowing). The peak velocity is $7.2mm/s$ in T_a, corresponding to a maximum flux of $F_{max} \approx 140mm^3/s$. Taking T_i into account, the mean mass flux is $F \approx 35mm^3/s$.

This stop-and-go motion persisted until the filling height of particles dropped to a threshold $(h_0 \approx 50mm)$. The movement gradually became smooth with a transition as shown in Fig.4. Firstly, the speed began to fluctuate more frequently at a lower peak velocity $1.2mm/s$. And the outflux oscillates with the same rhythm. Close inspection of the variation revealed that T_i gradually decreased to zero. At $h_0 \approx 50mm$, the velocity began to oscillate around $V \approx 0.9mm/s$ with no stagnant period ($T_i = 0$). Finally, the motion became stable at $V \approx 0.6mm/s$ with only small variations, corresponding to a mean outflux $F \approx 12mm^3/s$, which was much smaller than that in the stop-and-go phase.

It must be mentioned that no visible air bubble was observed. However, the effect of interstitial air cannot be ignored, since the typical grain size (fraction in millimeter) corresponds to a situation where the pressure fluctuations of the interstitial air are likely to be comparable with the hydrostatic pressure of the grains. So, we must take into account the effect of interstitial air pressure variation in order to develop a sound theory to understand the above behavior.

4 Theory and Analysis

In this part, we propose a heuristic theory considering the granular compaction and interstitial air pressure effect. Unlike fluids, the static pressure within granular material will not increase linearly with depth h as described in Archimedes' law: $P(h) = \rho g h$. It will saturate to a maximum when h is above a threshold because of the container wall's friction.

Using the arguments put forward by Janssen [16], the horizontal stress σ_r is proportional to the vertical stress σ_z: $\sigma_r = K\sigma_z = Kp$, where K is the parameter characterizing the conversion of the vertical stress into the horizontal stress due to the imbricate nature of the particles. So the frictional stress at the wall is $\sigma_{rz} = \mu_w \sigma_r = \mu_w K p$ where μ_w is the friction coefficient for particle-wall contacts. The force equilibrium equation for the slice dh at depth h is:

$$\rho g \pi R^2 \partial h = \pi R^2 \partial P + 2\pi R \mu_w K P \partial h \tag{1}$$

Here $\rho_g (\approx 2.5g/cm^3)$ is the density of the glass. So we get an equation for the static pressure $P(h)$ acting on a slice dh at depth h starting from the upper surface:

$$P(h) = \rho_g g \lambda [1 - exp(-h/\lambda)], \tag{2}$$

with a characteristic height $\lambda = \frac{R}{(2\mu_w K)}$. For rolling friction, $\mu_w = 0.1$. And K is usually set to 0.3. So Eq.(2) shows that $P(h) \approx \rho_g g h$ when $h \ll \lambda$, while it saturates to $\rho_g g \lambda$ when $h \gg \lambda$. There is a transitional region near $h = \lambda$.

We note that the pressure difference on the upper and lower surface of each slice can be written as

$$\pi R^2 dP = \frac{\partial P}{\partial h} \pi R^2 dh = \rho_g g \pi R^2 e^{-h/\lambda} dh \tag{3}$$

The friction by the lateral wall is

$$dF_{frict} = 2\pi R\mu_w KP dh = \rho_g g\pi R^2 [1 - e^{-h/\lambda}] dh \qquad (4)$$

Comparison of these two actions shows that the pressure difference is very small when $h \gg \lambda$. We note $\lambda \approx 4cm$. So it is suitable to use the hydrostatic pressure as the dynamic pressure to calculate the friction, while we ignore the pressure difference in the flowing phase. So the acceleration of the slice can be written as

$$a(h) = g \cdot \Gamma(h) = g - \frac{1}{\rho_g \pi R^2} \frac{\partial F_{frict}}{\partial h} = g - \frac{P(h)}{\rho_g \lambda} \qquad (5)$$

Substituting Eq.(2) into Eq.(5), we get the reduced acceleration

$$\Gamma(h) = exp(-h/\lambda), h \in [0, h_0] \qquad (6)$$

Strictly speaking, the above equation is true only at the very moment of the beginning of the downward motion. We suppose that the dynamic coefficient of friction is identical to the static one. Eq.(6) shows that $\Gamma(h)$ decreases with depth h so that the lower part in the pipe is less accelerated than the upper part. This makes the bulk material in the pipe more compact during the flowing stage [18]. This compact effect is most significant at the capillary part. Due to the radius change in the pipe $(R \rightarrow r)$ and the upward part of the wall's normal force, $\Gamma(h)$ is greater for the slice at the position S_1 where the pipe begins to shrink in contrast with that of the slice at the outlet S_2. This leads to a net accumulative flux f into the capillary part.

4.1 Stop-and-Go Motion and the Outflux Oscillation

When an accumulation of ΔV occurs, the air at the capillary part is compressed slightly, resulting in a small change ΔP_1 in interstitial air pressure. For an isothermal process,

$$\Delta P_1 \approx P_0 \cdot \frac{\Delta V}{\phi V_1} \qquad (7)$$

where P_0 is the environmental air pressure, $V_1 \approx 1.5 mL$ is the volume of the capillary part and a small section above S_1, ϕ is the volume fraction of air. Assume that the accumulative flux f is constant. ΔV increases linearly with time. When the pressure at S_1 equals the maximum granular packing pressure, the flow is hindered, ΔP_1 stops increasing and consequently T_a ends. In the T_i phase, ΔP_1 decreases due to the air current q passing through the granular packing:

$$\frac{d\Delta P_1}{dt} = P_0 \cdot \frac{q}{\phi V_1}. \qquad (8)$$

Assuming that the pressure profile is linearly distributed between the P_{max} site S_1 and the upper/bottom surfaces, we can use Darcy's law:

$$q = -\frac{\kappa\pi}{\eta} \cdot \Delta P_1 \cdot (\frac{r^2}{h_c} + \frac{R^2}{h_0 - h_c}) \tag{9}$$

where κ is the permeability of the bead packing, η is the viscosity of the air and $h_c (\approx 40mm)$ is the typical length of the capillary part. Substituting Eq.(9) into the pressure-decreasing equation and making integration, we have

$$\Delta P_1(t) = \Delta P_1^{max} \cdot exp(-\frac{t}{\tau}) \tag{10}$$

with τ denoting the characteristic time for the pressure attenuation

$$\tau = \frac{\eta\phi V_1}{P_0\pi\kappa(\frac{r^2}{h_c} + \frac{R^2}{h_0 - h_c})} \tag{11}$$

Thus, in the inactive phase T_i, the pressure difference in the capillary part vanishes exponentially, and $T_i \approx \tau$. The permeability $\kappa \approx 4.0 \times 10^{-8} cm^2$ for the bead size we used [19]. The viscosity of air is $2 \times 10^{-4}P$ and P_0 is 1 atm. This leads to $\tau \sim 10^{-1}s$, in agreement with the experiment observations.

ΔP_1^{max} is the maximum of ΔP_1 occurring at the point when the particles are hindered from falling. It can be estimated from Eq.(2) that

$$\Delta P_1^{max} = \rho_g g\lambda \approx 10^{-2} atm \tag{12}$$

From Eq.(7), we can also get

$$\Delta P_1^{max} \approx P_0 \cdot \frac{\Delta V_{max}}{\phi V_1} \tag{13}$$

so $\Delta V_{max} \approx 1.8mm^3$. It is about 1/7 of the total outflux in an oscillation period T. That is, in an oscillation period, 6/7 of the total outflux is discharged in the active phase T_a, whereas the other 1/7 rests at the capillary part and then is discharged in the inactive phase T_i. This results in the thick flow in T_a and a thin flow for T_i.

4.2 Transitional Behavior

To understand the appearance of transition in the ending period of the discharge process, we calculate the characteristic height in Eq.(2): $\lambda = \frac{R}{2\mu_w K} \approx$ 4cm for the granular packing above h_c. When the filling height h_0 falls below $h_c + \lambda$, the required ΔP_1^{max} decreases with h_0. So ΔV_{max} and time T_a decreases, and therefore the maximum upper surface velocity also decreases.

Another effect of small h_0 is that, the characteristic air pressure attenuation time τ will decrease rapidly as h_0 increases (see Eq.(11) and Fig.5). So T_i decreases to zero, and the flow becomes smooth. As shown in Fig.5, τ decreases to zero at $h_0 \approx 50mm$, so that the characteristic threshold should be between h_c and $\lambda + h_c$, in qualitative agreement with our experiment. And we conclude that the reason for the disappearance of ticking behavior in Wu [10] and Veje's [11] open-top hourglass experiment is that their granular packing height is small.

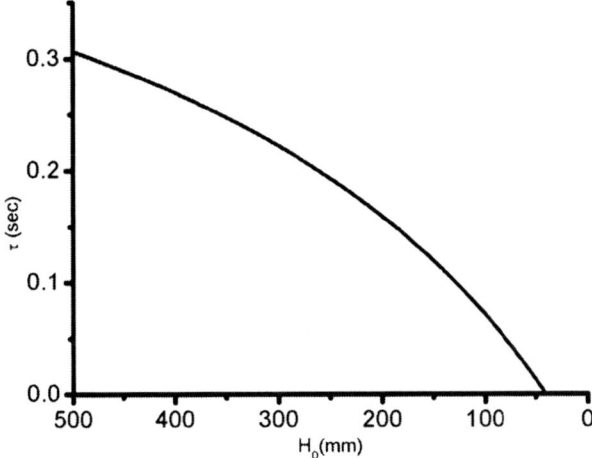

Fig. 5. The variation of τ with h_0, calculated from Eq.(11).

5 Conclusion

Our experimental and theoretical studies on the particle discharge process from an open-top capillary pipe reveal some exceptions for the common belief that the oscillatory motion results solely from arching of beads at the orifice and that the outflux is not sensitive to the filling height. When the filling height is above a threshold, the outflux fluctuates greatly with time and the bulk dense granular flow above the capillary shows stop-and-go motion. When the filling height falls to the threshold, led by a transitional stage, the outflux and the bulk movement become stable.

We measure the dropping velocity variation of the upper surface. And with a heuristic theory, we show that the granular compaction and interstitial air pressure effect are responsible for the behavior. The prediction using our theory agrees well with the experimental observation. We also conclude that the transitional filling height threshold is between the typical capillary part height h_c and the characteristic height of granular static pressure $\lambda + h_c$.

This work is supported by the National Natural Science Foundation of China (Grant No. 10274074, 10532060, 10404025), the National Basic Research Program of China (Grant No. 2006CB705500), and the Australian Research Council through a Discovery Project Grant.

References

1. Yan X, Shi Q, Hou M et al., Phys. Rev. Lett. 2003 **91(1)**:014302.
2. Hu M B, Wu Q S, Jiang R, Chin. Phys. Lett., 2003, **20(7)**:1091.
3. Chen W, Hou M, Lu K, Phys. Rev. E, 2001, **64**: 061305.
4. Hou M, Chen W, Zhang T, Lu K, Phys. Rev. Lett. 2003,**91(20)**:204301.
5. Beverloo W A, J. Chem. Eng. Sci., 1961, **15**:260.
6. Khelil A, Roth J, Eur. J. Mech. B /Fluids, 1994, **13**:57.
7. Talbot T, Viot P, Phys. Rev. Lett., 2002, **89(6)**:064301.
8. Reydellet G, Rioual F, Clement E, Europhys. Lett., 2000, **51**:27.
9. Goldhirsch I, Zanetti G, Phys. Rev. Lett., 1993, **70(11)**:1619.
10. Wu X L, Maloy K J, Phys. Rev. Lett., 1993, **71(9)**:1363.
11. Veje C T, Dimon P, Phys. Rev. E, 1997, **56(4)**: 4376.
12. Moriyama O, Kuroiwa N and Matsushita M, Phys. Rev. Lett., 1998, **80**:2833.
13. Raafat T, Hulin J P, Herrmann H J, Phys. Rev. E, 1996, **53(5)**:4345.
14. Aider J L, Sommier N, Raafat T, Hulin J P, Phys. Rev. E, 1999, **59(1)**:778.
15. To K, Lai P Y, Pak H K, Phys. Rev. Lett., 2001, **86(1)**: 71.
16. de Gennes P G, Rev. Mod. Phys. 1999, **71**: S374.
17. Bao D S, Zhang X S, Xu G L, Pan Z Q, Tang X W, Lu K Q, Phys. Rev. E, 2003, **67**:062301.
18. Duran J, Mazozi T, Luding S, Clement E, Rejchenbach J, Phys. Rev. E, 1996, **53(2)**: 1923.
19. Guyon E, Oger L, Plona T J, J. Phys. D, 1987, **20**: 1637.

Part II

Transport in Biological Systems

From Intracellular Traffic to a Novel Class of Driven Lattice Gas Models

Hauke Hinsch[1], Roger Kouyos[2], and Erwin Frey[1]

[1] Arnold Sommerfeld Center and CeNS, Department for Physics,
  Ludwig-Maximilians-Universität München, Theresienstrasse 37, D-80333
  München, Germany
[2] Theoretical Biology, Eidgenössische Technische Hochschule Zürich,
  Universitätsstrasse 16, CH-8092 Zürich, Switzerland

Summary. Motor proteins are key players in intracellular transport processes and biological motion. Theoretical modeling of these systems has been achieved by the use of step processes on one-dimensional lattices. After a comprehensive introduction to the total asymmetric exclusion process and some analytical tools, we will give a review on different lines of research attracted to the aspects of this systems. We will focus on the generic properties of a coupling between the exclusion process and Langmuir bulk kinetics that induce topological changes in the phase diagram and multi-phase coexistence.

1 Introduction

The identification of motion as a manifestation of biological life dates back to the earliest records of science itself. The Greek physician Erasistratos of Ceos studied biological motion on the length scale of muscles already in the 3rd century BC. He imagined muscles to function in the way of a piston contracting and relaxing from pneumatic origin. It was not until the invention of the microscope in the 17th century by van Leeuwenhoek that this theory could be devalidated with Swammerdams observation that muscles contract at constant volume.

Concerning biological motion on a microscopic scale, scientists favored concepts of "living forces" for many centuries until this was finally ruled out by the observations of the Scottish botanist Robert Brown in 1828 who found all kind of matter to undergo erratic motion in suspensions. A satisfactory explanation was provided by Einstein in 1905 by the interaction with thermally fluctuating molecules in the surroundings. However, the molecular details remained unknown in the fog of low microscope resolution. Modern experimental techniques [1] have lately revealed the causes of sub-cellular motion and transport.

Today we know that every use of our muscles is the collective effort of a class of proteins called myosin that "walk" on actin filaments. Generally spoken, we refer to all proteins that convert the chemical energy of ATP (adenosine-triphospate) in a hydrolysis reaction into mechanical work as molecular motors. These motors are highly specialized in their tasks and occur in a large variety: ribosomes move along mRNA strands while translating the codons into proteins, dynein is responsible for cilia motion and axonal transport, and kinesins play a key role in cytoskeletal traffic and spindle formation (for an overview see [2] and references therein).

While the exact details of the molecular structure and function of motor proteins [3] remain a topic of ongoing research, on a different level attention was drawn to phenomena that arise out of the collective interaction of many motors. Early research along this line was motivated by mRNA translation that is managed by ribosomes. Ribosomes are bound to the mRNA strand with one subunit and step forward codon by codon. The codon information is translated into corresponding amino acids that are taken up from the cytoplasm and assembled into proteins. To increase the protein synthesis many ribosomes can be bound to the same mRNA strand simultaneously. This fact might induce collective properties as was first realized by MacDonald [4] who set up a theoretical model for the translation of highly expressed mRNA. The importance of effects caused by the concerted action of many motors can be deduced from a very simple example that has yet drastic consequences: the slow down of ribosomes due to steric hindrance caused by another ribosome in front – comparable to an intracellular traffic jam that might significantly slow down protein synthesis.

A theoretical approach to collective phenomena in intracellular traffic will try to simplify the processes of molecular motion down to a single step rate rather than focus on the chemical or mechanical details on the molecular level of motor steps. Then it becomes possible to model and analyze the behavior of several motors with the tools of many-body and statistical physics. We will start this review with a short introduction on this single step model in Sec. 2 before we introduce the total asymmetric exclusion process (TASEP) as a theoretical model for intracellular transport. Sec. 3 describes the stationary states and density distributions and their phase diagram as a function of boundary conditions. After a review on several recent extensions in Sec. 4, we will focus on the competition of TASEP and bulk dynamics in Sec. 5. Before concluding, Sec. 6 contains further recent developments.

2 Model and Methods

In the quest for a theoretical model for the motion of molecular motors the first and simplest choice may be the use of a Poisson process. The "Poisson stepper" is assumed to be an extensionless object advancing stochastically in discrete steps along a one-dimensional periodic lattice. The process is unidirectional as the position of the stepper can be described by $x(t) = a\, n(t)$ with the discrete step size a and the random variable $n(t)$ being a sequence of growing integers. Step events occur stochastically with a rate r constant in both space and time. Consequently, the average time between two steps is then given by the dwell time $\tau = 1/r$ and the probability to find the "Poisson stepper" at a position n after time t by the Poisson distribution [5].

After we have defined a model for the translocation of a single motor, we proceed with our original task which aims at the understanding of collective properties of many motors. Of course, more elaborate models have been established [6, 7] that account for several rate limiting steps – examples are the ATP supply or the availability of amino acids for ribosomal mRNA translation. However, the very basic "Poisson stepper" is chosen for reasons of simplicity and in order to prevent unnecessary molecular details from masking collective effects. Still, the validity and limitations of this simplification have to be kept in mind.

Being supplied with the dynamics of a single motor, a stage for the concerted action of many can now be set up. This was first done in a pioneering work by MacDonald [4] and is now widely know as the total asymmetric exclusion process (TASEP). It consists of a one-dimensional lattice (Fig. 1) with N sites labeled by $i = 1, \cdots, N$ and with a spacing of $a = L/N$, where L is the total length of the lattice. For convenience, L is often set to 1 and the lattice spacing then referred to as $\varepsilon = 1/N$.

Particles have an extension of the size of the lattice spacing and are subjected to hard core exclusion due to steric hindrance. Therefore the occupation number n_i of site i can only take the values 0 or 1. Particles on the lattice attempt jumps to their right neighboring site with a rate r, which will be set to unity in the following. Hereby, a reference time scale is set. The effective frequency of jumps can be much smaller than r when attempted jumps are rejected due to an already occupied target site. The attempted jump rate to the left is zero, since we deal with a total asymmetric exclusion process, in contrast to the asymmetric exclusion process or the symmetric exclusion process, where the jump rate to the left is non-zero or even equal to the jump rate to the right.

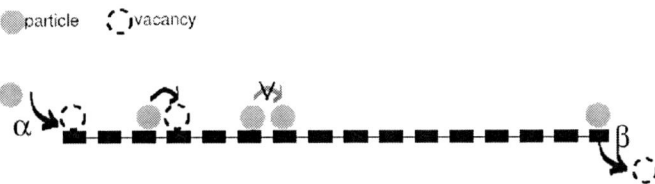

Fig. 1. Schematic model of TASEP (particles are injected with rate α, move exclusively to the right subject to hard-core exclusion, and are removed with rate β)

Unless one uses periodic boundary conditions, specific dynamic rules have to be defined at the boundaries, which play a crucial role in the solution of the process. Among different other conditions (reflective, open with a blockage) the most common type are open boundaries, which we will use as well: at the left boundary ($i = 1$) particles attempt to attach with a rate α, while they detach at the right boundary ($i = N$) with rate β. This is equivalent to two additional sites $i = 0$ and $i = N + 1$ at the boundaries, which are connected to the system by the bulk dynamics described above, and are constantly set to the density α and $1 - \beta$ respectively.

In spite of its simplicity, TASEP shows a wide range of interesting properties. Since it was propelled into the scope of statistical physicists, it has become a paradigm for non-equilibrium physics. In contrast to equilibrium systems it lacks detailed balance but evolves into a steady state where a non-vanishing current is maintained between boundaries. Upon varying these boundary conditions, TASEP was found to exhibit phase transitions which – following general theorems [8] – are not even allowed for one-dimensional equilibrium systems in the absence of long-range interactions. However, the analysis of non-equilibrium systems is considerably complicated by the lack of universal concepts like the Boltzmann-Gibbs ensemble theory. Feasible methods exist nevertheless and will be explained in the next section.

3 Density and Current in Stationary States

In analyzing exclusion processes research can focus on a multitude of different properties. The probably most obvious to address is the density and current distribution in the stationary state. This is motivated by two reasons. On the one hand, one intuitively attributes a strong importance to density information with respect to the biological background as e.g. the ribosome density is connected to the rate of protein synthesis. On the other hand, promising experimental techniques can measure motor densities and may allow for validation of theoretical models. Of course, quite extensive research has also been attracted to a multitude of different properties like correlation functions [9, 10], relaxation properties [11] or super-diffusive spreading of fluctuations [12] which will not be the topic of this review. We will focus on analytical methods (supported by numerical simulations) that are designed to investigate spatial density distributions in the stationary state of the system. To this end we will introduce some basic tools that have proven useful in the exploration of TASEP properties. These are based on mean-field approximations and reproduce many results that can also be derived exactly. We are well aware that this approach neglects correlations as included in the exact solutions that have been achieved for the TASEP density profile by applying either recursion relations [13] or a quantum Hamilton formalism with Bethe ansatz [14].

3.1 Quantum Mechanics and Statistic Properties

As an introduction we will outline some general statistical properties. At any given moment, the system can be found in a certain configuration μ made up of the occupation numbers at each lattice site. The next occurring stochastic event (i.e. the jump of one particle to a neighboring site) will therefore change the system to another configuration μ'. The transition probability $p_{\mu \to \mu'}$ is independent of the way the system had reached the initial configuration. Since there is no memory of the system's history, but any transition probability solely depends on the preceding state, TASEP is a Markov process. In order to describe the system's evolution, we can then use a master equation for the probability to find the system in a certain state.

$$\frac{dP(\mu)}{dt} = \sum_{\mu' \neq \mu} \left[w_{\mu' \to \mu} P_{\mu'}(t) - w_{\mu \to \mu'} P_\mu(t) \right] , \tag{1}$$

where the $w_{\mu \to \mu'}$ are the transition rates from one configuration μ to another μ'.

How can we now translate this general property into a description of TASEP? To this end we will use a convenient notation, which applies methods from the quantum mechanics toolbox in order to formulate the master equation in terms of operators. It was introduced as "quantum Hamiltonian formalism" and allows for exact solutions [14]. We introduce operators $\hat{n}_i(t)$, which act as occupation number operators, measuring the presence ($n_i = 1$) or absence ($n_i = 0$) of a particle at site i. This results in the Heisenberg equation (for an introduction see e.g. [14])

$$\frac{d}{dt} \hat{n}_i(t) = \hat{n}_{i-1}(t)[1 - \hat{n}_i(t)] - \hat{n}_i(t)[1 - \hat{n}_{i+1}(t)] , \tag{2}$$

where the first term on the right hand side constitutes the jump of a particle from the left neighboring site to site i (and thus a particle gain) and the second term a jump from site i to the adjacent lattice site on the right (a particle loss). Note the intrinsic exclusion constraint in both terms that prevents jump events if the destination site is occupied, i.e. the expression in brackets equals zero. If one expresses these gains and losses in current, it becomes convenient to use the current operator

$$\hat{j}_i(t) = \hat{n}_i(t)[1 - \hat{n}_{i+1}(t)] . \tag{3}$$

This allows to rewrite (2) as a discretized form of a continuity equation with the discrete divergence $\nabla \hat{j}_i(t) = \hat{j}_i(t) - \hat{j}_{i-1}(t)$:

$$\frac{d}{dt} \hat{n}_i + \nabla \hat{j}_i(t) = 0 . \tag{4}$$

Similar equations for the boundaries are readily derived in the same way. Since we are interested in the average density on a certain lattice site, we need to compute the time (or ensemble) average of the operators. Equation (2) gives an equation of motion for the operator and allows to solve for the time evolution of $\langle \hat{n}_i(t) \rangle$. In executing the ensemble average of (2) two-point correlation functions like $\langle \hat{n}_{i-1}(t)(1 - \hat{n}_i(t)) \rangle$ appear on the right hand side. These correlation functions again are connected to higher order correlations via their equations of motion. The resulting infinite series of correlation functions can be solved exactly for special cases only. Generally, one is required to use mean-field approaches.

3.2 Mean-Field Solution

The rather blurry term "mean-field theory" is based on the concept of using time or space averages (e.g. by neglecting temporal or spatial fluctuations) and has found a wide range of applications in statistical physics (see [15]). In this chapter, we will explain its implementation for the TASEP and show a possible solution. To point out the use of mean-field theory in TASEP, we look again at the average of the operator $\hat{n}_i(t)$. We are interested in the stationary state and therefore averaging signifies either a time or an ensemble average, since the system is ergodic. Then the average returns the density at the considered site i as $\varrho_i = \langle \hat{n}_i \rangle$. Performing the average over (2), leaves us with the difficulty of the infinite series of correlation functions mentioned earlier. The mean field approximation consists now in neglecting any correlations by setting e.g. $\langle \hat{n}_i \hat{n}_j \rangle = \langle \hat{n}_i \rangle \langle \hat{n}_j \rangle$ (see [16]). In our case, we obtain for the current

$$\langle \hat{n}_i(t)(1 - \hat{n}_{i+1}(t)) \rangle = \langle \hat{n}_i(t) \rangle (1 - \langle \hat{n}_{i+1}(t) \rangle) . \tag{5}$$

This allows then, to rewrite (2) in the stationary state $(d\varrho_i(t)/dt = 0)$ as

$$0 = \varrho_{i-1}(1 - \varrho_i) - \varrho_i(1 - \varrho_{i+1}) . \tag{6}$$

Obviously, (6) could easily be solved numerically, since it forms a system of N difference equations. However, it is possible to reduce the set of equations to arrive at an explicit solution by making two assumptions. First, note that the stationary state condition $d\varrho_i(t)/dt = 0$ implies a conservation of current throughout the bulk, as can be seen from (6). Ergo, we just have to solve one equation out of the set, to determine the stationary, homogeneous current (and density). For that purpose, we use the continuum approximation, which turns the spatial lattice variable quasi-continuous. This is achieved for a large number N of lattice sites on a lattice of normalized length $L = 1$. The distribution of sites then approaches a continuum as $\varepsilon = L/N \ll 1$ and $x = i/N$ is rescaled to the interval $0 \leq x \leq 1$. Thereby an expansion in powers of ε is allowed.

$$\varrho(x \pm \varepsilon) = \varrho(x) \pm \varepsilon \partial_x \varrho(x) + \frac{1}{2} \varepsilon^2 \partial_x^2 \varrho(x) + O(\varepsilon^3) \tag{7}$$

Using this expansion in (6) and the corresponding equations for the boundaries, results in the following first-order differential equation if we neglect all terms with higher orders in ε:

$$(2\varrho - 1)\partial_x\varrho = 0 . \qquad (8)$$

The corresponding boundary conditions are $\varrho(0) = \alpha$ and $\varrho(1) = 1 - \beta$. Because we have a first order differential equation that needs to satisfy two boundary conditions, we are evidently concerned with an over-determined boundary value problem. There are three solutions to (8): $\varrho_{\text{bulk}}(x) = 1/2$ does not satisfy either boundary condition (except for the special case $\alpha = \beta = 1/2$), while $\varrho(x) = C$ can satisfy either the left or the right boundary condition, resulting in $\varrho_\alpha(x) = \alpha$ and $\varrho_\beta(x) = 1 - \beta$, respectively.

3.3 Phase Diagram and Domain Wall Theory

To obtain a general solution satisfying the boundary conditions, both solutions need to be matched. Consider the case $\alpha, \beta < 1/2$. To meet the boundary condition at both sides, the global density function $\varrho(x)$ has to be ϱ_α (ϱ_β) in an environment close to the left (right) boundary. Since both ϱ_α and ϱ_β are uniform, the two solutions do not intersect. At this point, we have to go beyond mean-field theory and assume that at any given time both solutions are valid in non-overlapping areas of the system. Where those areas border, they are connected by a sharp domain wall (DW) at position x_w (see Fig. 2)

$$\varrho(x) = \begin{cases} \varrho_\alpha & \text{for} \quad 0 \le x \le x_w , \\ \varrho_\beta & \text{for} \quad x_w \le x \le 1 . \end{cases} \qquad (9)$$

From the dynamics of this domain wall [17] we can deduce important information about the system. The key point of domain wall theory is the identification of particle currents as the cause of domain wall motion. If for example the current $j_\alpha = \varrho_\alpha(1 - \varrho_\alpha)$ of the left density solution ϱ_α is higher than the

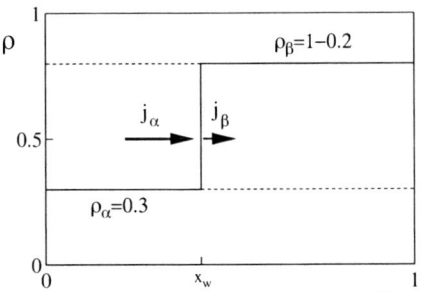 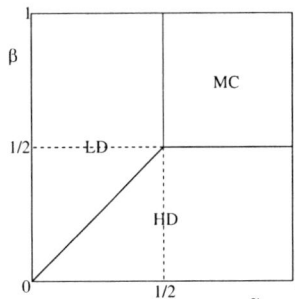

Fig. 2. (**left**) Schematic density distribution for $\alpha = .3, \beta = .2$: in this situation the particle current j_α exceeds the current j_β, thus carrying more particles to the domain wall which is then shifted to the left. (**right**) Phase diagram of TASEP in α, β phase space shows a low density (LD), high density (HD) and maximal current (MC) phase

current in the right part of the system (corresponding to $\alpha > \beta$), particles are transported faster to the DW from the left end then they can head on to the right. Thus, the domain wall is shifted to the left. The system is filled up, until finally the whole bulk density has taken the value of ϱ_β except for a small boundary layer [3] at the system's left boundary. The opposite happens, if $j_\beta > j_\alpha$, which is the case for $\beta < \alpha$. Hence we can empirically state that the boundary with the smaller rate acts as a bottleneck and imposes its density distribution on the system. To quantify this behavior, one can use a traveling wave solution [17] of the form $\varrho(x - Vt)$ to obtain the domain wall velocity as

$$V = \frac{j_\beta - j_\alpha}{\varrho_\beta - \varrho_\alpha} = \beta - \alpha . \tag{10}$$

To arrive at a phase diagram in α, β-space we will analyze the bulk density. A positive DW velocity is obtained for $\alpha < \beta$ and pushes the DW to the right side of the system resulting in a bulk density smaller $1/2$ called low-density phase (LD). On the contrary, $\beta < \alpha$ leads to a high-density phase (HD), as illustrated in the phase diagram (Fig. 2). LD and HD phase are connected by a first-order transition. Note, that this phase diagram is equally obtained by the above mentioned exact methods.

The transition line $\alpha = \beta < 1/2$ in the phase diagram requires special treatment. The velocity of the domain wall yields zero here, but Monte Carlo simulations show that the DW actually will make random steps to either side. This behavior is caused by the stochasticity of the input and output events that are described as rate processes. Hence, the DW makes random steps to either side with equal probability, being nothing else than the famous random walk in a domain with reflecting boundaries. An average over a sufficiently long time will therefore result in a homogeneous probability density over space. The stationary density profile will just be a linear slope connecting the two boundary conditions:

$$\varrho(x) = \alpha + (1 - \beta - \alpha)x . \tag{11}$$

Note that mean-field theory neglects fluctuations and hence fails to predict this density distribution, but gives the result of a stationary domain wall. However, coming back to the picture of the domain wall as a random walker, the linear density distribution can readily be made plausible. Interpreting the random walk as free diffusion of the domain wall and keeping in mind that the density constraints $\varrho(0) = \alpha$ and $\varrho(1) = 1 - \beta$ hold at all times, the linear density profile is easily derived as solution of the one dimensional diffusion equation with boundary conditions.

Finally, consider an increase of e.g. α to values that exceed $1/2$, while $\beta > 1/2$ (crossing the boundary from the upper left to the upper right quadrant

---

[3] The boundary layer is necessary in order to fulfill both boundary conditions. Its extend is finite for small systems, but will vanish in the thermodynamic limit $N \to \infty$.

in the phase diagram). In this situation particles are removed sufficiently fast from the system and the supply of particles on the left side acts as a bottleneck limiting density and current. For $\alpha < 1/2$ any increase in α results in an increased current (think of a highway, where an additional car will result in a higher overall traffic). But above a certain value $\alpha_C = 1/2$ any increase in particle input will not increase the current further [4] but will cause the current to diminish again (as an additional car will further slow down traffic during rush hour). As a result the bulk will keep its current maximum at $\varrho = 1/2$ and a boundary layer will form at the left side of the system to match the boundary condition. This is of course nothing else than the bulk solution ϱ_{bulk} with two boundary layers and was baptized maximal current phase (MC). It is reached via second-order transitions for values $\alpha > 1/2$ and $\beta > 1/2$. A more rigorous treatment of this behavior can be gained by computing the collective velocity of the particles [17].

4 Biologically Motivated Generalizations of TASEP

The adoption of lattice gas models for biological systems was followed by a variety of efforts to fit TASEP to different realistic environments. For that purpose some of the simplifying assumptions TASEP is based on had to be questioned. For example, experimental observations have shown that ribosomes typically cover an area on the mRNA that exceeds the lattice spacing by a multiple. To account for this situation the particles in TASEP have to extend over several lattice sites. However it was found that this does not change the phase diagram quantitatively [18]. Another direction of research went back to the original field of MacDonald to elucidate the importance of initiation and prolongation of ribosomes [19] or formation of mRNA loops to facilitate the back transport of ribosomes from the termination site [20].

While particle interactions in TASEP is limited to hard-core potentials, even a small increase in the interaction radius – as it could be caused by charged molecules – leads to qualitative changes in the phase diagram [21]. It was shown that lattice gases with short-range repulsive interaction exhibit a density-current relation with two local maxima in contrast to simple TASEP that leads to one maximum. This behavior results in a qualitative change of the phase diagram that is enriched by four more regions one being a minimal-current phase.

Further work was dedicated to the scenario of the interaction of different species of molecular motors that move in opposite directions. This can either happen on the same filament when two different particles are able to surpass each other with a jump rate that differs from the jump rate of either particle

---

[4] As $j = \varrho(1 - \varrho) = \varrho - \varrho^2$ has a maximum at $\varrho = 1/2$. This density-current relation is a convenient tool to characterize traffic processes and its plot is often referred to as fundamental diagram.

to a free site [22] or on two adjacent one-dimensional filaments [23]. In both cases, spontaneous symmetry breaking was observed.

Intracellular transport along cytoskeletal filaments has also served as a source of inspiration for driven lattice gas models. While in the TASEP model motors can only bind and unbind on the left and the right boundary respectively, cytoskeletal motors are known to detach from the track to the cytoplasm [2] where they perform Brownian motion and subsequently reattach to the track. The interplay between diffusion in the cytoplasm and directed motion along the filament was studied [24] both in open and closed compartments, focussing on anomalous drift and diffusion behavior, and on maximal current and traffic jams as a function of the motor density.

In [25] it has been realized that the on-off kinetics may not only give rise to quantitative changes in the transport efficiency but also to a novel class of driven lattice gas models. It was shown that the interplay between bulk on-off kinetics and driven transport results in a stationary phase exhibiting phase separation. This was achieved by an appropriate scaling of the on-off rates that ensures that particles travel a finite fraction on the lattice even in the limit of large systems. Then, particles spend enough time to "feel" the their mutual interaction and, eventually, produce collective effects. In the following section, we will review the results of these studies [25, 26].

5 Phase Coexistence

The essential features of cytoskeletal transport are the possibility of bulk attachment and detachment and a finite residence time on the lattice. The latter can be understood as a effect of thermal fluctuations that may overcome the binding energy of the motors that is only of the order of several $k_B T$. Hence, attachment and detachment is a stochastic process whose dynamic rules have to be defined. Parmeggiani et $al.$ [25] chose to use Langmuir kinetics (LK) known as adsorption-desorption kinetics of particles on a one- or two-dimensional lattice coupled to a bulk reservoir [27]. Particles can adsorb at empty sites and desorb from occupied sites and microscopic reversibility demands that the kinetic rates obey detailed balance leading to an evolution towards an equilibrium steady state describable by standard concepts of equilibrium statistical mechanics. In this sense the choice of LK is especially tempting as we are now faced with the competition of two representatives of both equilibrium and non-equilibrium systems. The system – in the following referred to as TASEP/LK – is defined as follows: the well-known TASEP is extended with the possibility of particles to attach to the filament with rate ω_A and to detach from an occupied lattice site to the reservoir with rate ω_D. According to the type of ensemble (canonical, grand canonical) the reservoir is either finite or infinite. Here, the reservoir is assumed to be infinite and homogeneous throughout space and time. The density on a lattice reached in the equilibrium state of LK is only dependent on the ratio $K = \omega_D / \omega_A$

and is completely uncorrelated in both space and time for neglection of any particle interaction except hard-core. This is justified by the assumption that the diffusion in the cytoplasm is fast enough to flatten any deviations of the homogeneous reservoir density. The resulting density profile on the lattice is homogeneous and given as Langmuir isotherm $\varrho_L = K/(K+1)$.

If we now consider the combination of TASEP and LK into the model displayed in Fig. 3, attention has to be paid to the different statistical nature of both processes. TASEP evolves into a non-equilibrium state carrying a finite current. Since particles are conserved in the bulk, the system is very sensitive to the boundary conditions, whereas LK as a equilibrium process is expected to be robust to any boundary effects especially for large systems. Combining both processes would thus lead to a trivial domination of LK as the bulk rates ω_A and ω_D that apply to a large number of bulk sites become predominant over the rates α and β that only act on the two boundary sites. To observe any interesting behavior (i.e. real interplay) between the two dynamics, one needs competition. A prerequisite for the two processes to compete are comparable jump rates. To ensure that the rates are of the same order independently of the system size, an appropriate scaling is needed. To this end a N-independent global detachment rate Ω_D is introduced, while the local rate per site scales as

$$\omega_D = \frac{\Omega_D}{N} . \tag{12}$$

For the attachment one proceeds similarly. What does this scaling signify physically? For an explanation it is instructive, to have a look at the time scales involved: a particle on the lattice will perform a certain move on an average time scale, which is the inverse of that moves rate. Therefore, a particle spends an average time $\tau \approx 1/\omega_D$ on the lattice before it detaches. Bearing in mind that the TASEP jump rate is set to unity, a particles will jump to its adjacent site after a typical time of one unit time step. Therefore the particle will travel a number $N_T = 1/\omega_D$ of sites before leaving the lattice. Compared to the lattice length this corresponds to a fraction of $n_T = N_T/N = 1/(N\omega_D)$. In order to keep this fraction finite in the thermodynamic limit, ω_D needs to scale as defined in (12). Only if the fraction n_T is finite, a given particle can

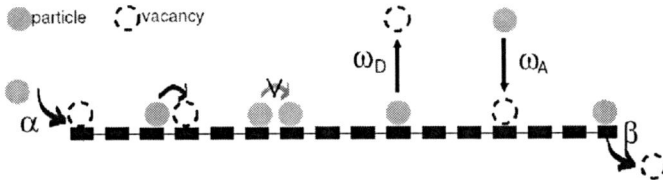

Fig. 3. Schematic model of TASEP/LK: the TASEP is extended by possible particle attachment and detachment in the bulk with rate ω_A, ω_D

experience interaction with other particles and give rise to collective phenomena [26].

5.1 Mean Field Solution of TASEP/LK

To obtain density and current distributions of the TASEP/LK, we use again
a mean-field approach and proceed along the lines of Sec. 3.2.

First of all, we need to account for the Langmuir kinetics. This is done
by adding the following terms to the Heisenberg equation (2) of the simple
TASEP to obtain

$$\frac{d}{dt}\hat{n}_i(t) = \hat{n}_{i-1}(t)(1-\hat{n}_i(t))-\hat{n}_i(t)(1-\hat{n}_{i+1}(t))-w_D\hat{n}_i(t)+w_A(1-\hat{n}_i(t)) . \quad (13)$$

where the first added term captures the detachment events and the latter the
attachment. Neglecting correlations as done before gives for the stationary
state

$$0 = \varrho_{i-1}(1 - \varrho_i) - \varrho_i(1 - \varrho_{i+1}) - w_D\varrho_i + w_A(1 - \varrho_i) . \quad (14)$$

The boundary conditions are not altered by LK. Using again the power series
expansion (7) and keeping in mind the scaling of the on and off rates w as in
(12), we obtain the following ODE:

$$\frac{\varepsilon}{2}\partial_x^2\varrho + \partial_x\varrho(2\varrho - 1) - w_D\varrho + w_A(1 - \varrho) = 0 . \quad (15)$$

The ratio $K = w_A/w_D$ between the attachment and detachment rates will
prove an important parameter in the analysis of this differential equation.
Since the case $K \neq 1$ is considerably complicated in its mathematical analysis
we refer the reader to reference [26] and restrain our discussion to the case
$K = 1$. In the thermodynamic limit ($\varepsilon \to 0$) and with $w_D = w_A = \Omega$ the
ODE(15) simplifies to first order:

$$(\partial_x\varrho - \Omega)(2\varrho - 1) = 0 . \quad (16)$$

Obviously, there are two solutions to this general ODE problem. The homogeneous density $\varrho_L = 1/2$ given by the Langmuir isotherm and the linear slope
$\varrho(x) = \Omega x + C$. The value of C is determined by the boundary conditions and
leads to the two solutions $\varrho_\alpha(x) = \alpha + \Omega x$ and $\varrho_\beta(x) = 1 - \beta - \Omega + \Omega x$. The
complete density profile $\varrho(x)$ is the combination of one or several of the three
densities above. Depending on how they are matched, we distinguish several
phases as explained in the following.

5.2 Phase Diagram and Density Distributions

The only area in the phase diagram that does not change compared to simple
TASEP is the upper right quadrant. This is not surprising since the maximal

current phase is a bulk controlled regime anyway. Therefore the additional bulk dynamics with the Langmuir isotherm at $\varrho_L = 1/2$ do not result in any changes in the density distribution. In this case, non-equilibrium and equilibrium dynamics do not compete but cooperate.

As in TASEP different solutions can be matched in various ways, the simplest being the connection by a domain wall between the left solution ϱ_α and the right solution ϱ_β. Depending on the current distribution, two possibilities have to be distinguished. As both solutions are non-homogeneous, the corresponding currents j_α and j_β will be strictly monotonic (Fig. 4 (left)). If the currents equal each other inside the system at a position x_w, the DW is localized at this position, as a displacement to either side would result in a current inequality that drives the DW back to x_w (see Fig. 4 (left)). Ergo, the TASEP/LK exhibits multi-phase existence of low and high density regions (LD-HD phase) in the stationary state on all time-scales, opposed to TASEP where this behavior is only observed for short observation times. Recently, this DW localization has been observed experimentally [28].

If the matching of left and right current is not possible inside the system, the known LD and HD phases are found. This is the case for one boundary condition being considerably larger than the other and is evidently depending

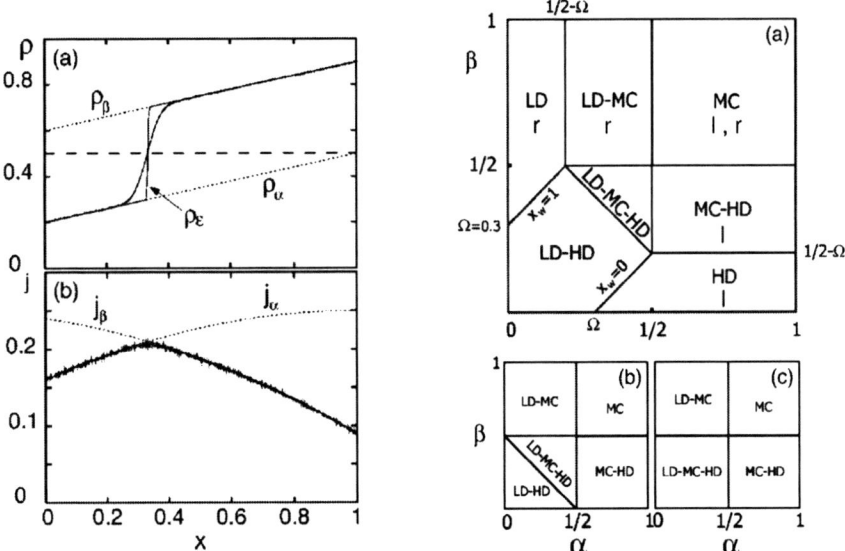

Fig. 4. (*left*) The DW connects the two densities ρ_α and ρ_β (*both dashed*) and is localized at the point where the correspondent currents j_α and j_β match. Note the finite extend of the DW (localization length) that is only produced for Monte-Carlo simulations (*solid wiggly line*) and is not captured by mean-field results (*solid line*). (*right*) Topological changes in the phase diagram of TASEP/LK for (a) $\Omega = 0.3$, (b) $\Omega = 0.5$, (c) $\Omega = 1$, from [26]

quantitatively on the slope of the density solutions. This slope is determined by the ratio of the TASEP step rate and the bulk interchange rate Ω. For large Ω any density imposed by the boundary relaxes fast against the Langmuir isotherm of $\varrho_L = 1/2$ resulting a steep slope of the density profile.

This fact allows for the existence of two other phases with multi-regime coexistence. We could imagine a scenario in which the boundary imposed density solutions decay fast enough towards the isotherm to enable a three-regime coexistence of low density, maximal current and high density (LD-MC-HD phase). Furthermore, a combination of a MC phase with a boundary layer on one side and a LD or HD region of finite extend at the other boundary can be imagined. Not all these phases will be realized for every value of Ω. Instead, the phase topology of two-dimensional cuts through the α, β, Ω-phase space changes. An example is shown in Fig. 4 (right).

5.3 Domain Wall Theory

After we have derived an analytical solution for the density profile based on the mean-field differential equation (15), we now have a closer look on the domain wall and its stochastic properties.

As mentioned before, the density profile exhibits a discontinuity at x_w that is actually a finite size continuous transition between the high and low density for systems of finite size. Only upon increasing the system size $N \to \infty$ a sharp transition between the left and the right solution occurs. However, this is only due to the fact that the lattice spacing decreases to $\varepsilon \to 0^+$. So compared to the lattice length L, usually normalized to 1, the domain wall is discontinuous, while on the length scale of lattice sites it will still have a finite extend. This extend is an intrinsic statistical feature and is usually referred to as localization length.

The domain wall in its random walk behavior can be either subjected to equal rates (unbiased) as in TASEP or the rates of movement to the left/right can be different (biased). In general, the rates will not only be different, but also depend on the space variable. To begin with, we will show a way how to derive these rates by taking into account fluctuations of particle number [17, 29]. Consider a situation where all events that can change the particle number (α, β, Ω) have a typical time scale that is considerably larger than that of jump processes on the lattice. In this case, the time between any entry and exit events is so long, that the system has enough time to "rearrange" (to reach a temporary steady state) in between . Then it is possible to identify jump rates $\omega_l(x)$ $(\omega_r(x))$ for DW movement to the left (right) with the overall rate for entry and exit of any particle at any site. Specifically, if a particle enters the system, the DW is shifted to the left by a distance of $\approx \varepsilon / [\varrho_\beta(x_w) - \varrho_\alpha(x_w)]$. Therefore the rate for the DW to move one lattice site to the left is $\omega_l = \omega_{entry} / [\varrho_\beta(x_w) - \varrho_\alpha(x_w)]$.

If the density distribution $\varrho(x)$ is known analytically, it is possible to calculate ω_{entry}, the overall rate of particle entrance, as the sum over all possible entrance events

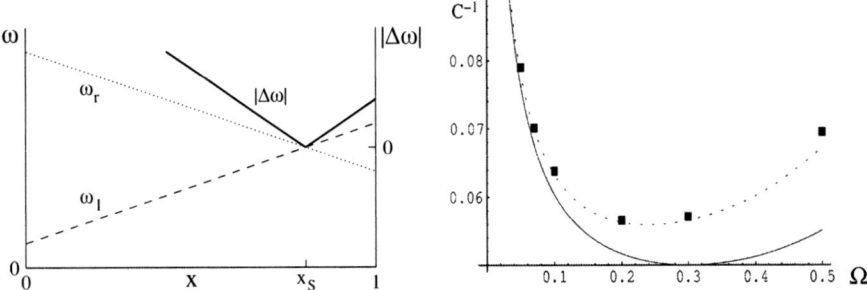

Fig. 5. (left) The DW will be localized at the point where the jump rate to the left ω_l and right ω_r intersect. This can be interpreted as an potential $|\Delta\omega|$. **(right)** Variance of DW probability distribution over Ω for $\alpha = \beta = 0.01$ and $\alpha = \beta = 0.1$ in a system of $N = 500$, predictions according to [30] (*solid*) and based on particle number fluctuations (*dashed*) compared to MC simulations (*dots*)

$$\omega_{\text{entry}} = \alpha(1 - \alpha) + \int_0^1 dx \, \omega_A(1 - \varrho(x)) \, . \tag{17}$$

The first term captures entrance events from the left boundary reservoir and the integral accounts for the Langmuir kinetics. The first multiplicand in both terms is the attempted rate of a jump, whereas the difference in brackets states the probability of the destination site to be vacant. Along the same lines, the exit rate is computed. As we know the analytical density distribution as $\varrho(x) = \alpha + \Omega x + \Delta\Theta(x - x_w)$ with the Heavyside function Θ and the DW height [5] $\Delta = 1 - \alpha - \beta - \Omega$, we can execute the integrals to obtain

$$\omega_{\text{entry}} = \alpha(1 - \alpha) + \Omega(\beta + \frac{\Omega}{2}) + x\Omega\Delta \, . \tag{18}$$

Knowing these rates, one can complete the description of the domain wall as a random walker by calculating the position dependent jump rates to the left and right $\omega_l(x)$ and $\omega_r(x)$. The two rates constitute an effective potential displayed in Fig. 5. The DW will always be driven to that point x_S in the system where $\omega_{\text{exit}}(x_S) = \omega_{\text{entry}}(x_S)$. This position is the center of the DW probability density in a stochastic picture and the stationary DW position in a mean-field picture. It is in agreement with mean field results and yields

$$x_S = \frac{\Omega - \alpha + \beta}{2\Omega} \, . \tag{19}$$

The other quantity of interest which - in contrast to the DW position - cannot be determined by mean-field calculations is the localization length. In order to compute this quantity, fluctuations have to be taken into account as we will

---

[5] The term -1Ω accounts for the diminished height that is caused by the Langmuir kinetics on the whole length 1 of the system.

show in the following. If we use the notation $p(x)$ for the probability that the domain wall is at position x, then the condition for a stationary DW reads in the continuum limit

$$\omega_r(x)p(x) = \omega_l(x+\varepsilon)p(x+\varepsilon) . \tag{20}$$

Introducing now $y(x) = \omega_l(x)p(x)$ and approximate $y'(x) = |y(x+\varepsilon)-y(x)|/\varepsilon$ we obtain:

$$y'(x) + Ny(x)\left(1 - \frac{\omega_r(x)}{\omega_l(x)}\right) = 0 . \tag{21}$$

The solution then is given by

$$p(x) = \frac{\tilde{p}(x)}{Z} = \frac{1}{Z\omega_l(x)} \exp[-N \int_{x_0}^{x} dx'(1 - \frac{\omega_r(x)}{\omega_l(x)})] . \tag{22}$$

where Z accounts for normalization. In general, Z is not available explicitly, but it has been shown [30] that the unnormalized probability function can be approximated by a Gaussian

$$\tilde{p}(x) \propto e^{-C(x-x_S)^2} , \tag{23}$$

where C is given by the second order derivative of the exponent in (22) as

$$C = \frac{1}{2}\frac{d^2}{dx^2}\left[N \int_{x_0}^{x} dx'\left(1 - \frac{\omega_r(x)}{\omega_l(x)}\right)\right] = \frac{N(\omega_l - \omega_r)'(x_S)}{2\omega_l(x_S)} . \tag{24}$$

Hence, the variance $\sigma = \sqrt{1/(2C)}$ of the domain wall can be easily be obtained provided that the jump rates $\omega_l(x)$ and $\omega_r(x)$ are available. Evans et al. [30] have assumed those rates to be $\omega_{l,r}(x) = j_{\alpha,\beta}/(\varrho_\beta - \varrho_\alpha)$. Kouyos has shown [31] that using the rates (17) derived from the fluctuations of particle number, one arrives at more accurate results compared to Monte Carlo simulations (see Fig. 5). In this case C evaluates to

$$C = \frac{2N\Omega\Delta}{\alpha(1-\alpha) + \beta(1-\beta) + \Omega} . \tag{25}$$

As the width of the DW is given by $\sigma = 1/\sqrt{2C}$, the localization of the DW scales with $N^{-1/2}$.

6 Conclusions and Outlook

Much in the same way as MacDonald's pioneering paper [4] on mRNA translation, recent work on kinesin motors walking along microtubules has spurred progress in nonequilibrium transport phenomena. The ubiquitous exchange of material between the cytoplasm and the molecular track, which originally was thought to only lead to quantitative modifications of the dynamics [24], has recently been identified [25] as the source for qualitatively new phenomena such as phase separation. This introduced a completely new class of nonequilibrium transport models.

These lattice gas models are characterized by a scaling of the on-off rates with system size which enables competition of driven motion and equilibrium Langmuir kinetics. Hereby, a finite residence time on the lattice ensures co-operative effects to establish multi-phase coexistence, localized shocks and an enriched phase behavior compared to prior TASEP results.

There are now various routes along which one could proceed. The first one is to add more realistic features of the molecular motors, such as the fact that they are dimers [32] or that there is more than one chemo-mechanical state [28]. Such investigations are crucial for a quantitative understanding of intracellular traffic in various ways. One might ask how robust the features of minimal models are with respect to the addition of more molecular details. In the case of dimers the answer is far from obvious since the non-equilibrium dynamics of dimer adsorption shows rich dynamic behavior with anomalously slow relaxation towards the equilibrium state [27]. How this combines with the driven transport along the molecular track was recently analyzed thoroughly [32]. While correlation effects due to the extended nature of dimers invalidate a simple mean-field picture it was found that an extended mean-field scheme can be developed which quantitatively describes the stationary phases. Surprisingly, the topology of the phase diagram and the nature of the phases is similar to the minimal model with monomers. The physical origin of this robustness can be traced back to the form of the current-density relation which exhibits only a single maximum.

The second line of research generalizing the minimal model [25] asks for the effect of interactions, more than one molecular traffic lane, "road blocks" such as microtubule associated proteins and various other kinds of "disorder", bi-directional traffic, coupling of driven and diffusive transport and the like on the stationary density profiles and the dynamics. In almost every instance it is found that this leads to an even richer behavior with new phenomena emerging.

For TASEP it is known that isolated defects (slow sites) depending on their strength may either give rise to local density perturbations for low particle densities or yield macroscopic effects for densities close to the carrying capacity [33]. The interplay between coupling to the motor reservoir in the cytoplasm and the fluctuations in the capacity limit due to disorder along the track gives rise to a number of interesting collective effects [34].

Coupling two lanes by allowing particle exchange at a constant rate along the molecular track also results in novel phenomena. Similar to equilibrium phase transitions described by field theories with two coupled order parameters higher order critical points may emerge [35].

Coupling diffusive and driven transport, the origin of new phenomena is due to a competition of different processes of comparable time scales. The qualitative failure of mean-field theory in some of these systems [36] comes as quite a surprise, since mean-field has proven to predict phase diagrams for large systems with an astonishing accuracy in the lattice gas models mentioned above.

References

1. A.D. Mehta, M. Rief, J.A. Spudich, D.A. Smith, R.M. Simmons: Science **283**, 1689 (1999)
2. J. Howard: *Mechanics of Motor Proteins and the Cytoskeleton* (Sinauer, Sunderland, 2001)
3. M. Schliwa and G. Woehlke: Nature **422**, 759 (2003)
4. C.T. MacDonald, J.H. Gibbs, A.C.Pipkin: Biopolymers **6**, 1 (1968)
5. N.G. van Kampen: *Stochastic Processes in Physics and Chemistry* (North Holland, Amsterdam, 1981)
6. A. Parmeggiani, C.F. Schmidt: in *Function and Regulation of Cellular Systems* (Birkhäuser, 2004)
7. F. Jülicher, A. Ajdari, J. Prost: Rev. Mod. Phys. **69**, 1269 (1997)
8. N.D. Mermin, H. Wagner: Phys. Rev. Lett. **17**, 1133 (1966)
9. B. Derrida, M.R. Evans: J. Physique I **3**, 311 (1993)
10. G. Schütz: Phys. Rev. E **47**, 4265 (1993)
11. M. Dudzinski, G.M. Schütz: J. Phys. A **33**, 8351 (2000)
12. H. van Beijeren, R. Kutner, H. Spohn: Phys. Rev. Lett. **54**, 2026 (1985)
13. B. Derrida, E. Domany, D. Mukamel: J. Stat. Phys. **69**, 667 (1992)
14. G. Schütz: in *Phase Transitions and Critical Phenomena* vol 19 edited by C. Domb, J. Lebowitz (Academic, London 2001)
15. N. Goldenfeld: *Lectures on phase transitions and the renormalization group* (Perseus, Reading, 1992)
16. G.D. Mahan: *Many Particle Physics* (Kluwer, 2000)
17. A.B. Kolomeisky, G.M. Schütz, E.B. Kolomeisky, J.P. Straley: J. Phys. A **31**, 6911 (1998)
18. L.B. Shaw, P.K.P. Zia, K.H. Lee: Phys. Rev. E **68**, 021910 (2003)
19. R. Heinrich, T.A. Rapoport: J. Theor. Bio. **86**, 279 (1980)
20. T. Chou: Biophys. J **86**, 755 (2003)
21. P. Popkov, G. M. Schütz: Europhys. Lett. **48**, 257 (1999)
22. M.R. Evans, D.P. Foster, C Godreche, D Mukamel: Phys. Rev. Lett. **74**, 208 (1995)
23. V. Popkov, I. Peschel: Phys. Rev. E **64**, 026126 (2001)
24. R. Lipowsky, S. Klumpp, T.M. Nieuwenhuizen: Phys. Rev. Lett. **87**, 108101 (2001)
25. A. Parmeggiani, T. Franosch, E. Frey: Phys. Rev. Lett. **90**, 86601 (2003)
26. A. Parmeggiani, T. Franosch, E. Frey: Phys. Rev. E **70**, 46101 (2004)
27. E. Frey, A. Vilfan: Chem. Phys. **284**, 287 (2002)
28. K. Nishinari, Y. Okada, A. Schadschneider, D. Chowdhury: Phys. Rev. Lett **95**, 118101 (2005)
29. L. Santen, C. Appert: J. Stat. Phys **106**, 187 (2002)
30. M.R. Evans, R. Juhász, L. Santen: Phys. Rev. E **68**, 026117 (2003)
31. R. Kouyos, Diploma Thesis, ETH Zürich (2004)
32. P. Pierobon, T. Franosch, E. Frey: Phys. Rev. E (in press); cond-mat/0603385
33. G. Tripathy and M. Barma: Phys. Rev. E **58**, 1911 (1998)
34. P. Pierobon, M. Mobilia, R. Kouyos, E. Frey: Phys. Rev. E (in press); cond-mat/0604356
35. T. Reichenbach, T. Franosch, E. Frey: Phys. Rev. Lett. **97**, 050603 (2006)
36. H. Hinsch, E. Frey: Phys. Rev. Lett. **97**, 095701 (2006)

Traffic Phenomena in Biology: From Molecular Motors to Organisms

Debashish Chowdhury[1], Andreas Schadschneider[2], and Katsuhiro Nishinari[3]

[1] Department of Physics, Indian Institute of Technology, Kanpur 208016, India.
[2] Institut für Theoretische Physik, Universität zu Köln D-50937 Köln, Germany
[3] Department of Aeronautics and Astronautics, Faculty of Engineering, University of Tokyo, Hongo, Bunkyo-ku, Tokyo 113-8656, Japan.

Summary. Traffic-like collective movements are observed at almost all levels of biological systems. Molecular motor proteins like, for example, kinesin and dynein, which are the vehicles of almost all intra-cellular transport in eukaryotic cells, sometimes encounter traffic jam that manifests as a disease of the organism. Similarly, traffic jam of collagenase MMP-1, which moves on the collagen fibrils of the extracellular matrix of vertebrates, has also been observed in recent experiments. Traffic-like movements of social insects like ants and termites on trails are, perhaps, more familiar in our everyday life. Experimental, theoretical and computational investigations in the last few years have led to a deeper understanding of the generic or common physical principles involved in these phenomena. In particular, some of the methods of non-equilibrium statistical mechanics, pioneered almost a hundred years ago by Einstein, Langevin and others, turned out to be powerful theoretical tools for quantitative analysis of models of these traffic-like collective phenomena as these systems are intrinsically far from equilibrium. In this review we critically examine the current status of our understanding, expose the limitations of the existing methods, mention open challenging questions and speculate on the possible future directions of research in this interdisciplinary area where physics meets not only chemistry and biology but also (nano-)technology.

1 Introduction

Motility is the hallmark of life. What distinguishes a *traffic-like* movement from all other forms of movements is that motile elements move on *"tracks"* or *"trails"*. However, in sharp contrast to vehicular traffic, the tracks and trails, which are the biological analogs of roads, can have nontrivial dependence on time during the typical travel time of the motile elements. What makes biological traffic even more unusual is that in many cases the motile elements themselves not only create the tracks but also modify their lengths as well as shape and, in some extreme cases, even leave behind a trail of destruction by wiping out the track as they move forward.

We are mainly interested in the *general principles* and common trends seen in the mathematical modeling of collective traffic-like movements at different levels of biological organization [1]. Although the choice of the physical examples and modelling strategies are biased by our own works and experiences, we put these in a broader perspective by relating these with works of other research groups. We begin at the lowest level, starting with intracellular biomolecular motor traffic on filamentary rails. Then we present brief summaries of recent works on the traffic of molecular motors along the collagen fibrils in the extra-cellular matrix and those on transport of micron-size cargo by uni-cellular micro-organisms. We end our review by discussing the collective traffic-like terrestrial movements of social insects, particularly, ants, on their trails.

2 Theoretical Approaches

In recent years many individual-based models of biological traffic have been formulated in discretized space. While in some models the dynamics of the system has been formulated in terms of differential equations assuming continuous time, in many others the dynamics has been implemented in terms of "update rules" in discrete time steps in the spirit of cellular automata and lattice gas models [2–4]. Most of these models are extensions of a class of particle-hopping models which were earlier successfully used in the context of vehicular traffic [5]. The *asymmetric simple exclusion process (ASEP)* [6] is one of the simplest particle-hopping models. In the ASEP particles can hop (with some probability or rate) from one lattice site to a neighbouring one, but only if the target site is not already occupied by another particle. "Simple Exclusion" thus refers to the absence of multiply occupied sites. Generically, it is assumed that the motion is "asymmetric" such that the particles have a preferred direction of motion.

For such driven diffusive systems the boundary conditions turn out to be crucial. If periodic boundary conditions are imposed, i.e., the sites 1 and L are made nearest-neighbours of each other, all the sites are treated on the same footing. If the boundaries are open, then a particle can enter from a reservoir and occupy the leftmost site ($j = 1$), with probability α, if this site is empty. In this system a particle that occupies the rightmost site ($j = L$) can exit with probability β. The ASEP has been studied extensively in recent years and is now well understood [6, 7]. Both parallel and random-sequential updating schemes have been studied extensively in the literature.

The average number of motile elements that arrive at (or depart from) a fixed detector site on the track per unit time interval is called the *flux*. One of the most important transport properties is the relation between the flux and the density of the motile elements; a graphical representation of this relation is usually referred to as the *fundamental diagram*. If the motile elements interact mutually only via their steric repulsion their average speed

v would decrease with increasing density because of the hindrance caused by each on the following elements. On the other hand, for a given density c, the flux J is given by $J = cv(c)$, where $v(c)$ is the corresponding average speed. At sufficiently low density, the motile elements are well separated from each other and, consequently, $v(c)$ is practically independent of c. Therefore, J is approximately proportional to c if c is very small. However, at higher densities the increase of J with c becomes slower. At high densities, the sharp decrease of v with c leads to a decrease, rather than increase, of J with increasing c. Naturally, the fundamental diagram of such a system is expected to exhibit a maximum at an intermediate value of the density.

3 Intra-Cellular Traffic of Cytoskeletal Molecular Motors

Intracellular transport is carried by molecular motors which are proteins that can directly convert the chemical energy into mechanical energy required for their movement along filaments constituting what is known as the cytoskeleton [8, 9]. Three superfamilies of these motors are kinesin, dynein and myosin; majority of these motors are two-headed. Most of the kinesins and dyneins are like "porters" in the sense that these move over long distances carrying cargo along the filamentary tracks without getting completely detached; such motors are called *processive*. On the other hand, the conventional myosins and a few unconventional ones are nonprocessive; they are like "rowers".

These cytoskeleton-based molecular motors play crucially important biological functions, e.g., in axonal transport in neurons. The mechano-chemistry of single cytoskeletal motors and the mechanism of their motility have been investigated both experimentally and theoretically for quite some time [10–12].

Often a single microtubule (MT) is used simultaneously by many motors and, in such circumstances, the inter-motor interactions cannot be ignored. In this article we shall focus mostly on the effects of mutual interactions of these motors on their collective spatio-temporal organisation and the biomedical implications of such organisations. Fundamental understanding of these collective physical phenomena may also expose the causes of motor-related diseases (e.g., Alzheimer's disease) [13–16] thereby helping, possibly, also in their control and cure. The bio-molecular motors have opened up a new frontier of applied research — "bio-nanotechnology". A clear understanding of the mechanisms of these natural nano-machines will give us clue as to the possible design principles that can be utilized to synthesize artificial nanomachines.

Derenyi and collaborators [17, 18] developed one-dimensional models of interacting Brownian motors. They modelled each motor as a *rigid rod* and formulated the dynamics through Langevin equations for each such rod assuming the validity of the overdamped limit; the mutual interactions of the rods were incorporated through the mutual exclusion.

The model considered by Aghababaie et al. [19] is not based on TASEP, but its dynamics is a combination of Brownian ratchet and update rules in discrete time steps. In this model the filamentary track is discretized and the motors are represented by *field-driven* particles in the spirit of the particle-hopping models. The hopping probabilities of the particles are obtained from the instantaneous form of the local time-dependent potential. No site can accommodate more than one particle at a time. Each time step consists of either an attempt of a particle to hop to a neighbouring site or an attempt that can result in switching of the potential from flat to sawtooth form or vice-versa. Both forward and backward movement of the particles are possible. However, neither attachment of new particles nor complete detachment of existing particles were allowed.

The fundamental diagram of the model [19], computed imposing periodic boundary conditions, is very similar to those of TASEP. This observation indicates that further simplification of the model proposed in ref. [19] is possible to develope a minimal model for interacting molecular motors. Indeed, the detailed Brownian ratchet mechanism, which leads to a noisy forward-directed movement of the *field-driven* particles in the model of Aghababaie et al. [19], is replaced in some of the more recent theoretical models [20–28] by a TASEP-like probabilistic forward hopping of *self-driven* particles [29]. In these simplified versions, none of the particles is allowed to hop backward and the forward hopping probability is assumed to capture most of the effects of biochemical cycle of the enzymatic activity of the motor. The explicit dynamics of the model is essentially an extension of that of the asymmetric simple exclusion processes (ASEP) [6, 30] that includes, in addition, Langmuir-like kinetics of adsorption and desorption of the motors.

In the model of Parmeggiani et al. [24], the molecular motors are represented by particles whereas the sites for the binding of the motors with the

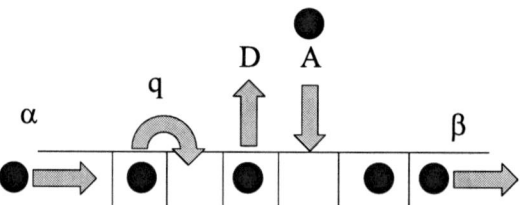

Fig. 1. A schematic description of the TASEP-like model introduced in ref. [24] for molecular motor traffic. Just as in TASEP, the motors are allowed to hop forward, with probability q. In addition, the motors can also get "attached" to an empty lattice site, with probability A, and "detached" from an occupied site, with probability D from any site except the end points; the rate of attachment at the entry point on the left is α while that at the exit point on the right is β.

cytoskeletal tracks (e.g., microtubules) are represented by a one-dimensional discrete lattice. Just as in TASEP, the motors are allowed to hop forward, with probability q, provided the site in front is empty. However, unlike TASEP, the particles can also get "attached" to an empty lattice site, with probability A, and "detached" from an occupied site, with probability D (see fig.1) from any site except the end points. The state of the system was updated in a random-sequential manner. Carrying out Monte-Carlo simulations of the model, applying open boundary conditions, Parmeggiani et al. [24] demonstrated a novel phase where low and high density regimes, separated from each other by domain walls, coexist [26, 27]. Using a mean-field theory (MFT), they interpreted this spatial organization as traffic jam of molecular motors.

A cylindrical geometry of the model system was considered by Lipowsky, Klumpp and collaborators [20–22] to mimic the microtubule tracks in typical tubular neurons. The microtubule filament was assumed to form the axis of the cylinder whereas the free space surrounding the axis was assumed to consist of N_{ch} channels each of which was discretized in the spirit of lattice gas models. They studied concentration profiles and the current of free motors as well as those bound to the filament by imposing a few different types of boundary conditions. This model enables one to incorporate the effects of exchange of populations between two groups, namely, motors bound to the axial filament and motors which move diffusively in the cylinder. They have also compared the results of these investigations with the corresponding results obtained in a different geometry where the filaments spread out radially from a central point.

A novel feature of the model of Klein et al. [31] (see Fig. 2) is that the lattice site at the tip of a filament is removed with a probability W per unit time provided it is occupied by a motor; the motor remains attached to the newly exposed tip of the filament with probability p (or remains bound with the removed site with probability $1 - p$). Thus, p may be taken as a measure of the processivity of the motors. This model clearly demonstrated a dynamic accumulation of the motors at the tip of the filament arising from the processivity.

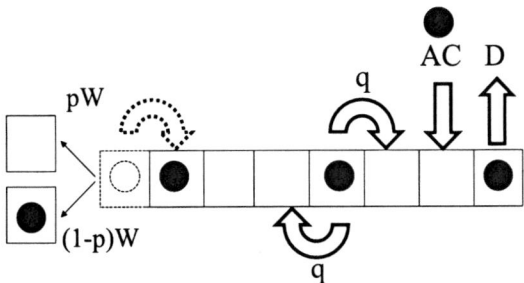

Fig. 2. A schematic description of the model suggested by Klein et al. [31] for motor induced depolymerization of cytoskeletal filaments.

3.1 Traffic of Interacting Single-Headed Motors KIF1A

The models of intracellular traffic described so far are essentially extensions of the asymmetric simple exclusion processes (ASEP) [6, 30] that includes Langmuir-like kinetics of adsorption and desorption of the motors. In reality, a motor protein is an enzyme whose mechanical movement is loosely coupled with its biochemical cycle. In a recent work [32], we have considered specifically the *single-headed* kinesin motor, KIF1A [33–37]; the movement of a single KIF1A motor was modelled earlier with a Brownian ratchet mechanism [38, 39]. In contrast to the earlier models [21, 24, 26, 28] of molecular motor traffic, which take into account only the mutual interactions of the motors, our model explicitly incorporates also the Brownian ratchet mechanism of individual KIF1A motors, including its biochemical cycle that involves *adenosine triphosphate(ATP) hydrolysis*.

The ASEP-like models successfully explain the occurrence of shocks. But since most of the bio-chemistry is captured in these models through a single effective hopping rate, it is difficult to make direct quantitative comparison with experimental data which depend on such chemical processes. In contrast, the model we proposed in ref. [32] incorporates the essential steps in the biochemical processes of KIF1A as well as their mutual interactions and involves parameters that have one-to-one correspondence with experimentally controllable quantities. Thus, in contrast to the earlier ASEP-like models, each of the self-driven particles, which represent the individual motors KIF1A, can be in two possible internal states labelled by the indices 1 and 2. In other words, each of the lattice sites can be in one of three possible allowed states (Fig. 3): empty (denoted by 0), occupied by a kinesin in state 1, or occupied by a kinesin in state 2.

Good estimates for the parameters of the model could be extracted by analyzing the empirical data [32]. Assuming that each time step of updating corresponds to 1 ms of real time, we performed simulations up to 1 minute. In the limit of low density of the motors we have computed, for example, the mean speed of the kinesins, the diffusion constant and mean duration of the

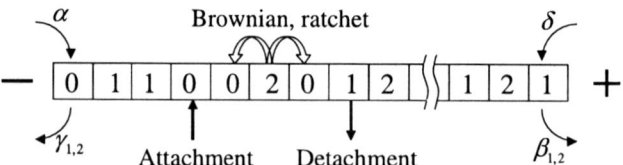

Fig. 3. A 3-state model for molecular motors moving along a MT. 0 denotes an empty site, 1 is K or KT and 2 is KD. Transition from 1 to 2, which corresponds to hydrolysis, occurs within a cell whereas movement to the forward or backward cell occurs only when motor is in state 2. At the minus and plus ends the probabilities are different from those in the bulk.

movement of a kinesin on a microtubule from simulations of our model; these are in excellent *quantitative* agreement with the corresponding empirical data from single molecule experiments.

Using this model we have also calculated the flux of the motors in the mean field approximation imposing periodic boundary conditions. Although the system with periodic boundary conditions is fictitious, the results provide good estimates of the density and flux in the corresponding system with open boundary conditions.

In contrast to the phase diagrams in the $\alpha - \beta$-plane reported by earlier investigators [21, 25, 26], we have drawn the phase diagram of our model in the $\omega_a - \omega_h$ plane by carrying out extensive computer simulations for realistic parameter values of the model with open boundary conditions. The phase diagram shows the strong influence of hydrolysis on the spatial distribution of the motors along the MT. In particular, the position of the immobile shock depends on the concentration of the motors as well as that of ATP; the shock moves towards the minus end of the MT with the increase of the concentration of kinesin or ATP or both. The formation of the shock has been established by our direct experimental evidence; our findings on the domain wall are in qualitative agreement with the corresponding experimental observations [32].

This work has been discussed in more detailed in our separate article [40] in this proceedings.

4 Intra-Cellular Traffic of Nucleotide-Based Motors

Helicases and polymerases are the two classes of nucleotide-based motors that have been the main focus of experimental investigations. In this section, we discuss only the motion of the ribosome along the m-RNA track. Historically, this problem is one of the first where TASP-like model was successfully applied to a biological system.

In a living cell *ribosomes* translate the genetic information 'stored' in the *messenger-RNA (mRNA)* into a program for the synthesis of a protein. mRNA is a long (linear) molecule made up of a sequence of triplets of nucleotides; each triplet is called a *codon* (see Fig. 4). The genetic information is encoded in the sequence of codons. A ribosome, that first gets attached to the mRNA chain, "reads" the codons as it translocates along the mRNA chain, recruits the corresponding amino acids and assembles these amino acids in the sequence thereby synthesizing the protein for which the "construction plan" was stored in the mRNA. Once the synthesis is completed, the ribosome gets detached from the mRNA. Thus, the process of "translation" of genetic information stored in mRNA consists of three steps: (i) *initiation*: attachment of a ribosome at the "start" end of the mRNA, (ii) *elongation*: of the polypeptide (protein) as the ribosome moves along the mRNA, and (iii) *termination*: ribosome gets detached from the mRNA when it reaches the "stop" codon.

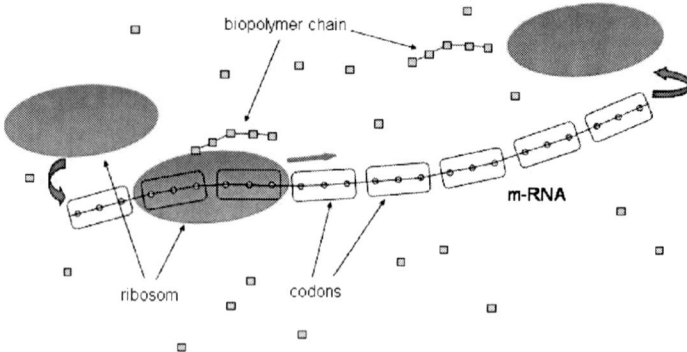

Fig. 4. The process of biopolymerization: Ribosomes attach to mRNA and read the construction plan for a biopolymer which is stored in the genetic code formed by the sequence of codons. None of the codons can be read by more than one ribosome simultaneously.

In order to model the traffic of ribosomes on a m-RNA track, let us denote each of the successive codons by the successive sites of a one-dimensional lattice where the first and the last sites correspond to the start and stop codons. The ribosomes are much bigger (20-30 times) than the codons. Therefore, neighbouring ribosomes attached to the same mRNA can not read the same information or overtake each other. In other words, any given site on the lattice may be covered by a single ribosome or none. Let us represent each ribosome by a rigid rod of length ℓ_r. If the rod representing the ribosome has its left edge attached to the i-th site of the lattice, it is allowed to move to the right by one lattice spacing, i.e., its left edge moves to the site $i + 1$ provided the site $i + \ell_r$ is empty. In the special case $\ell_r = 1$ this model reduced to the TASEP. Although the model was originally proposed in the late sixties [41], significant progress in its analytical treatment for the general case of arbitrary ℓ_r could be made only three decades later; even the effects of quenched disorder has also been considered in the recent literature [42–46].

As mentioned above, a ribosom is much bigger than a base triplet. However, modifying the ASEP by taking into account particles that occupy more than one lattice site does not change the structure of phase diagram [41].

5 Extracellular Transport: Collagen-Based Motors

The extracellular matrix (ECM) [47] of vertebrates is rich in collagen. Monomers of collagen form a triple-helical structure which self-assemble into a tightly packed periodic organization of fibrils. Cells residing in tissues can secret matrix metalloproteases (MMPs), a special type of enzymes that are capable of degrading macromolecular constituents of the ECM. The most notable among these enzymes is MMP-1 that is known to degrade collagen. The collagen fibril contains cleavage sites which are spaced at regular intervals of 300 nm. The collagenase MMP-1 cleaves all the three α chains of the collagen monomer at a single site.

Breakdown of the ECM forms an essential step in several biological processes like, for example, embryonic development, tissue remodelling, etc. [47]. Malfunctioning of MMP-1 has been associated with wide range of diseases [48]. Therefore, an understanding of the MMP-1 traffic on collagen fibrils can provide deeper insight into the mechanism of its operation which, in turn, may give some clue as to the strategies of control and cure of diseases caused by the inappropriate functions of these enzymes.

Saffarian et al. [49] used a technique of two-photon excitation fluorescence correlation spectroscopy to measure the correlation function corresponding to the MMP-1 moving along the collagen fibrils. The measured correlation function strongly indicated that the motion of the MMP-1 was not purely diffusive, but a combination of diffusion and drift. In other words, the "digestion" of a collagen fibril occurs when a MMP-1 executes a biased diffusion processively (i.e., without detachment) along the fibril. They also demonstrated that inactivation of the enzyme eliminates the bias but the diffusion remains practically unaffected. They claimed that the energy required for the active motor-like transport of the MMP-1 comes from the proteolysis (i.e., degradation) of the collagen fibrils.

There is a close relation between the traffic of MMP-1 on collagens and the "burnt-bridge model" introduced by Mai et al. [50]. In the burnt bridge model (see Fig. 5), a "particle" performs a random walk on a *semi-infinite one-dimensional* lattice that extends from the origin to $+\infty$. Each site of the lattice is connected to the two nearest neighbour sites by links; a fraction c of these links are called "bridges" and these are prone to be burnt by the random walker. A bridge is burnt, with probability p, if the random walker either crosses it *from left to right* or *attempts to cross if from right to left* [50, 51]. In either case, if the bridge is actually burnt, the walker stays on the right of the burnt bridge and cannot cross it any more. The hindrance against leftward motion, that is created by the burnt bridges, is responsible for the overall rightward drift of the random walker. Mai et al. [50] studied the dependence of the average drift velocity v on the parameters p and c by computer simulation. They also derived approximate analytical forms of these dependences in the two limits $p \ll 1$ and $p \simeq 1$ using a continuum approximation.

Saffarian et al. [49] also carried out computer simulations of a two-dimensional model of the MMP-1 dynamics on collagen fibrils which is essentially a two-dimensional generalization of the burnt-bridge model. By comparing the results of their simulations with their experimental observations, Saffarian et al. they concluded that the observed biased diffusion of the MMP-1 on collagen fibrils can be described quite well by a Brownian ratchet mechanism [38, 39].

Fig. 5. Schematic representation of the one-dimensional burnt bridge model of MMP-1 dynamics proposed in ref. [50].

6 Cellular Traffic

A *Mycoplasma mobile* (MB) bacterium is an uni-cellular organism. Each of the pear-shaped cells of this bacterium is about 700 nm long and has a diameter of about 250 nm at the widest section. Each bacterium can move fast on glass or plastic surfaces using a *gliding* mechanism. In a recent experiment [52] narrow linear channels were constructed on lithographic substrates. The channels were typically 500 nm wide and 800 nm deep. Note that each channel was approximately twice as wide as the width of a single MB cell. The channels were so deep that none of the individual MB cells was able to climb up the tall walls of the channels and continued moving along the bottom edge of the walls of the channels. In the absence of direct contact interaction with other bacteria, each individual MB cell was observed to glide, without changing direction, at an average speed of a few microns per second.

When two MB cells made a contact approaching each other from opposite directions within the same channel, one of the two cells gave way and moved to the adjacent lane. However, in a majority of the cases, two cells approaching each other from the opposite directions simply passed by as if nothing had happened; this is because of the fact that the width of the channel is roughly twice that of the individual MB cell. Moreover, when two cells moving in the same direction within a channel collided with each other, the faster cell moved to the adjacent lane after the collision.

Hiratsuka et al. [52] attached micron-sized beads on the MB cells using biochemical technique and demonstrated that the average speed of each MB cell remained practically unaffected by the load it was carrying. In contrast to the nonliving motile elements discussed in all the preceding sections, the cells are the functional units of life. Therefore, the MB cells have the potential for use in applied research and technology as "micro-transporters". More recently, the unicellular biflagellated algae *Chlamydomonas reinhardtii* (CR), which are known to be phototactic *swimmers*, have been shown to be even better candidates as "micro-transporter" as these can attain average speeds that is about two orders of magnitude higher than what was possible with MB cells [53]. However, to our knowledge, the effects of mutual interactions of the CR cells on their average speed at higher densities has not been investigated.

7 Traffic in Social Insect Colonies: Ants and Termites

From now onwards, we shall study traffic of multi-cellular organisms, particularly, ants which are social insects. The ability of the social insect colonies to function without a leader has attracted the attention of experts from various disciplines [54–61]. Insights gained from the modeling of the colonies of such insects are finding important applications in computer science (useful optimization and control algorithms) [62], communication engineering [63], artificial "swarm intelligence" [64] and micro-robotics [65] as well as in task partitioning, decentralized manufacturing [66–71] and management [72]. the collective terrestrial movements of ants have close similarities with the other traffic-like phenomena considered here. When observed from a sufficiently long distance the movement of ants on trails resemble the vehicular traffic observed from a low flying aircraft.

Ants communicate with each other by dropping a chemical (generically called *pheromone*) on the substrate as they move forward [73–75]. Although we cannot smell it, the trail pheromone sticks to the substrate long enough for the other following sniffing ants to pick up its smell and follow the trail. [73]. Both the continuum model developed by Rauch et al. [76] and the CA model introduced by Watmough and Edelstein-Keshet [77] were intended to address the question of formation of the ant-trail networks by foraging ants. Couzin and Franks [78] developed an individual based model that not only addressed the question of self-organized lane formation but also elucidated the variation of the flux of the ants.

In the recent years, we have developed discrete models [79–82] that are not intended to address the question of the emergence of the ant-trail [83], but focus on the traffic of ants on a trail which has already been formed. We have developed models of both unidirectional and bidirectional ant-traffic by generalizing the totally asymmetric simple exclusion process (TASEP) [6, 84, 85] with parallel dynamics by taking into account the effect of the pheromone.

In our model of uni-directional ant-traffic the ants move according to a rule which is essentially an extension of the TASEP dynamics. In addition, a second field is introduced which models the presence or absence of pheromones (see Fig. 6). The hopping probability of the ants is now modified by the presence of pheromones. It is larger if a pheromone is present at the destination site. Furthermore, the dynamics of the pheromones has to be specified. They are created by ants and free pheromones evaporate with probability f per unit time. Assuming periodic boundary conditions, the state of the system is updated at each time step in two stages. In stage I ants are allowed to move while in stage II the pheromones are allowed to evaporate. In each stage the *stochastic* dynamical rules are applied in parallel to all ants and pheromones, respectively.

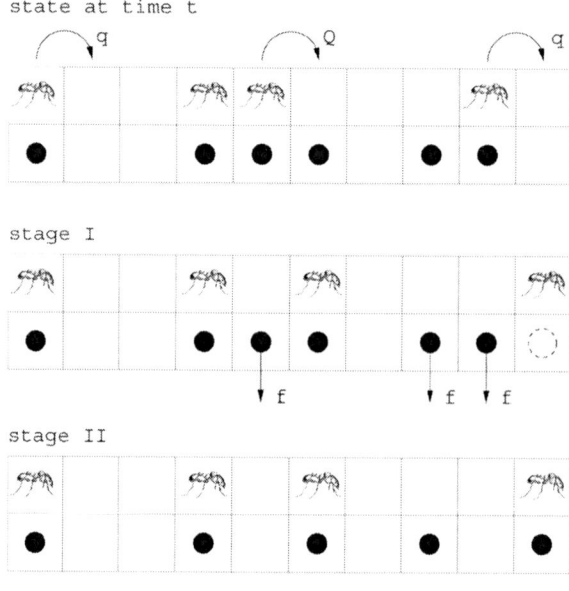

state at time t

Fig. 6. Schematic representation of typical configurations of the uni-directional ant-traffic model. The symbols • indicate the presence of pheromone. This figure also illustrates the update procedure. Top: Configuration at time t, i.e. *before stage I* of the update. The non-vanishing probabilities of forward movement of the ants are also shown explicitly. Middle: Configuration *after* one possible realisation of *stage I*. Two ants have moved compared to the top part of the figure. The open circle with dashed boundary indicates the location where pheromone will be dropped by the corresponding ant at *stage II* of the update scheme. Also indicated are the existing pheromones that may evaporate in *stage II* of the updating, together with the average rate of evaporation. Bottom: Configuration *after* one possible realization of *stage II*. Two drops of pheromones have evaporated and pheromones have been dropped/reinforced at the current locations of the ants.

One interesting phenomenon observed in the simulations is coarsening. At intermediate time usually several non-compact clusters are formed. However, the velocity of a cluster depends on the distance to the next cluster ahead. Obviously, the probability that the pheromone created by the last ant of the previous cluster survives decreases with increasing distance. Therefore clusters with a small headway move faster than those with a large headway. This induces a coarsening process such that after long times only one non-compact cluster survives.

A similar coarsening phenomenon has been observed also in the bus-route model [86, 87]. In fact, the close relation between our model of uni-directional traffic on ant-trails and the bus-route model has been pointed out earlier [88].

In the bus route model, each bus stop can accommodate at most one bus at a time; the passengers arrive at the bus stops randomly at an average rate λ and each bus, which normally moves from one stop to the next at an average rate Q, slows down to q, to pick up waiting passengers [86, 87].

In vehicular traffic, usually, the average speed of the vehicles decreases *monotonically* with increasing density because the inter-vehicle interactions tend to hinder each other's movements. In contrast, in our models of ant-traffic the average speed of the ants varies *non-monotonically* with their density over a wide range of small values of f because of the coupling of their dynamics with that of the pheromone. This uncommon variation of the average speed gives rise to the unusual dependence of the flux on the density of the ants in our models of ant-traffic. Furthermore, the flux does not exhibit the particle-hole symmetry which is a characteristic of the TASEP. Details of the models and results on both uni-directional and bi-directional traffic of ants on trails are given in the article of John et al. in this proceedings [89]. The experimental data reported in the pioneering experimental work of Burd and collaborators [90] were too scattered to test our theoretical predictions. However, more accurate recent data [91, 92] establish both the non-monotonic variation of the average speed with density as well as the formation of cluster by the ants.

8 Summary and Conclusion

In this article we have reviewed our current understanding of traffic-like collective phenomena in living systems, starting from the smallest level of intra-cellular bio-molecular motor transport and ending at the level of the traffic of social insects like, for example, ants. We have restricted our attention to those theoretical works where, in the spirit of particle-hopping models of vehicular traffic, the language of cellular automata or extensions of TASEP has been used. The success of this modelling strategy has opened up a new horizon in traffic science and, we hope, we have provided a glimpse of an exciting frontier of interdisciplinary research.

Acknowledgements

It is our great pleasure to thank Yasushi Okada, Alexander John and Ambarish Kunwar for enjoyable collaborations on the topics discussed in this article.

References

1. D. Chowdhury, A. Schadschneider and K. Nishinari, Phys. of Life Rev. **2**, 318 (2005).
2. S. Wolfram, *Theory and Applications of Cellular Automata* (World Sci., 1986); *A New Kind of Science* (Wolfram Research Inc., 2002)
3. B. Chopard and M. Droz, *Cellular Automata Modeling of Physical Systems* (Cambridge University Press, 1998).
4. J. Marro and R. Dickman, *Nonequilibrium Phase Transitions in Lattice Models* (Cambridge University Press, 1999).
5. D. Chowdhury, L. Santen, and A. Schadschneider, Phys. Rep. **329**, 199 (2000).
6. G.M. Schütz: *Exactly Solvable Models for Many-Body Systems*, in C. Domb and J.L. Lebowitz (eds.), *Phase Transitions and Critical Phenomena*, Vol. 19 (Academic Press, 2001).
7. M.R. Evans and R.A. Blythe, Physica **A313**, 110 (2002).
8. J. Howard, *Mechanics of Motor Proteins and the Cytoskeleton*, (Sinauer Associates, 2001) .
9. M. Schliwa (ed.), *Molecular Motors*, (Wiley-VCH, 2002).
10. G. Oster and H. Wang, in ref. [9].
11. M.E. Fisher and A.B. Kolomeisky, Proc. Natl. Acad. Sci. **98**, 7748 (2001).
12. R.D. Astumian, Appl. Phys. A **75**, 193 (2002).
13. M. Aridor and L.A. Hannan, Traffic **1**, 836 (2000); **3**, 781 (2002).
14. N. Hirokawa and R. Takemura, Trends in Biochem. Sci. **28**, 558 (2003)
15. E. Mandelkow and E.M. Mandelkow, Trends in Cell Biol. **12**, 585 (2002).
16. L.S. Goldstein, Proc. Natl. Acad. Sci. **98**, 6999 (2001); Neuron **40**, 415-425 (2003). **28**, 558 (2003); Curr. Op. Neurobiol. **14**, 564-573 (2004).
17. I. Derenyi and T. Vicsek, Phys. Rev. Lett. **75**, 374 (1995).
18. I. Derenyi and A. Ajdari, Phys. Rev. E **54**, R5 (1996).
19. Y. Aghababaie, G.I. Menon and M. Plischke, Phys. Rev. E **59**, 2578 (1999).
20. R. Lipowksy, S. Klumpp, and Th. M. Nieuwenhuizen, Phys. Rev. Lett. 87, 108101 (2001).
21. R. Lipowksy and S. Klumpp, Physica A 352, 53 (2005).
22. M.J.I. Müller, S, Klumpp and R. Lipowsky, J. Phys. Cond. Matt. **17**, S3839 (2005) and references therein.
23. S. Klumpp and R. Lipowsky, this proceedings.
24. A. Parmeggiani, T. Franosch, and E. Frey, Phys. Rev. Lett. **90**, 086601 (2003); Phys. Rev. E **70**, 046101 (2004).
25. E. Frey, A. Parmeggiani and T. Franosch, Genome Informatics **15(1)**, 46 (2004) and references therein.
26. M.R. Evans, R. Juhasz, and L. Santen, Phys. Rev. E **68**, 026117 (2003).
27. R. Juhasz and L. Santen, J. Phys. A **37**, 3933 (2004).
28. V. Popkov, A. Rakos, R.D. Williams, A.B. Kolomeisky, and G.M. Schütz, Phys. Rev. E **67**, 066117 (2003).
29. F. Schweitzer: *Brownian Agents and Active Particles*, Springer Series in Synergetics (Springer 2003).
30. B. Schmittmann and R.P.K. Zia, in C. Domb and J.L. Lebowitz (eds.), *Phase Transitions and Critical Phenomena*, Vol. 17 (Academic Press, 1995).
31. G.A. Klein, K. Kruse, G. Cuniberti and F. Jülicher, Phys Rev. Lett. **94**, 108102 (2005).

32. K. Nishinari, Y. Okada, A. Schadschneider and D. Chowdhury, Phys. Rev. Lett. **95**, 118101 (2005).
33. Y. Okada and N. Hirokawa, Science **283**, 1152 (1999).
34. Y. Okada and N. Hirokawa, Proc. Natl. Acad.Sci. USA **97**, 640 (2000).
35. Y. Okada, H. Higuchi, and N. Hirokawa, Nature, **424**, 574 (2003).
36. R. Nitta, M. Kikkawa, Y. Okada, and N. Hirokawa, Science **305**, 678 (2003).
37. Y. Okada, K. Nishinari, D. Chowdhury, A. Schadschneider, and N. Hirokawa (to be published).
38. F. Jülicher, A. Ajdari, and J. Prost, Rev. Mod. Phys. **69**, 1269 (1997).
39. P. Reimann, Phys. Rep. **361**, 57-265 (2002).
40. K. Nishinari, Y. Kanayama, Y. Okada, P. Greulich, A. Schadschneider and D. Chowdhury, in this proceedings.
41. C. MacDonald, J. Gibbs, and A. Pipkin, Biopolymers **6**, 1 (1968); C. MacDonald and J. Gibbs, Biopolymers **7**, 707 (1969)
42. L.B. Shaw, R.K.P. Zia and K.H. Lee, Phys. Rev. E **68**, 021910 (2003).
43. L.B. Shaw, J. P. Sethna and K.H. Lee, Phys. Rev. E **70**, 021901 (2004).
44. L.B. Shaw, A.B. Kolomeisky and K.H. Lee, J. Phys. A **37**, 2105 (2004).
45. G. Lakatos and T. Chou, J. Phys. A **36**, 2027 (2003).
46. T. Chou and G. Lakatos, Phys. Rev. Lett. **93**, 198101 (2004).
47. H. Nagase and J. F. Woessner, J. Biol. Chem. **274**, 21491 (1999).
48. M. Whittaker and A. Ayscough, Celltransmisions **17**, 3 (2001).
49. S. Saffarian, I. E. Collier, B.L. Marmer, E.L. Elson and G. Goldberg, Science **306**, 108 (2004).
50. J. Mai, I.M. Sokolov and A. Blumen, Phys. Rev. E **64**, 011102 (2001).
51. T. Antal and P.L. Krapivsky, cond-mat/0504652.
52. Y. Hiratsuka, M. Miyata and T. Q. P. Uyeda, Biochem. Biophys. Res. Commun. **331**, 318 (2005).
53. D. B. Weibel, P. Garstecki, D. Ryan, W. R. DiLuzio, M. Mayer, J. E. Seto and G. M. Whitesides, Proc. Nat. Acad. Sci. USA, **102**, 11963 (2005).
54. E. Bonabeau, G. Theraulaz, J.L. Deneubourg, S. Aron and S. Camazine, Trends in Ecol. Evol. **12**, 188 (1997)
55. C. Anderson, G. Theraulaz and J.L. Deneubourg, Insect. Sociaux **49**, 99 (2002)
56. Z. Huang and J.H. Fewell, Trends in Ecol. Evol. **17**, 403 (2002).
57. E. Bonabeau, Ecosystems **1**, 437 (1998).
58. G. Theraulaz, J. Gautrais, S. Camazine and J.L. Deneubourg, Phil. Trans. Roy. Soc. Lond. A **361**, 1263 (2003).
59. J. Gautrais, G. Theraulaz, J.L. Deneubourg and C. Anderson, J. Theor. Biol. **215**, 363 (2002).
60. L. Edelstein-Keshet, J. Math. Biol. **32**, 303 (1994).
61. G. Theraulaz, E. Bonabeau, S.C. Nicolis, R.V. Sole, V. Fourcassie, S. Blanco, R. Fournier, J.L. Joly, P. Fernandez, A. Grimal, P. Dalle and J.L. Deneubourg, Proc. Natl.Acad. Sci. **99**, 9645 (2002).
62. M. Dorigo, G. di Caro and L.M. Gambardella, Artificial Life **5(3)**, 137 (1999); Special issue of Future Generation Computer Systems dedicated to ant-algorithms (2000).
63. E. Bonabeau, M. Dorigo and G. Theraulaz, Nature **400**, 39 (2000).
64. E. Bonabeau, M. Dorigo and G. Theraulaz, *Swarm Intelligence: From Natural to Artificial Intelligence* (Oxford University Press, 1999).
65. M.J.B. Krieger, J.B. Billeter and L. Keller, Nature **406**, 992 (2000).

66. F.L.W. Ratnieks and C. Anderson, Insectes Sociaux **46**, 95 (1999).

67. C. Anderson and F.L.W. Ratnieks, Am. Nat. **154**, 521 (1999).

68. F.L.W. Ratnieks and C. Anderson, Am. Nat. **154**, 536 (1999).

69. C. Anderson and F.L.W. Ratnieks, Insectes Sociaux **47**, 198 (2000).

70. C. Anderson and D.W. McShea, Biol. Rev. **76**, 211 (2001).

71. C. Anderson and F.L.W. Ratnieks, in: *Complexity and complex systems in industry*, eds. I.P. McCarthy and T. Rakotobe-Joel, (University of Warwick, U.K.), 92 (2000).

72. E. Bonabeau and C. Meyer, Harvard Business Review (May), 107 (2001).

73. E.O. Wilson, *The Insect Societies* (Belknap, Cambridge, USA, 1971); B. Hölldobler and E.O. Wilson, *The Ants* (Belknap, Cambridge, USA, 1990)

74. S. Camazine, J.L. Deneubourg, N. R. Franks, J. Sneyd, G. Theraulaz, E. Bonabeau: *Self-organization in Biological Systems* (Princeton University Press, 2001).

75. A.S. Mikhailov and V. Calenbuhr, *From Cells to Societies: Models of Complex Coherent Action* (Springer, 2002).

76. E.M. Rauch, M. M. Millonas and D.R. Chialvo, Phys. Lett. A **207**, 185 (1995).

77. J. Watmough and L. Edelstein-Keshet, J. Theor. Biol. **176**, 357 (1995).

78. I.D. Couzin and N.R. Franks, Proc. Roy Soc. London B **270**, 139 (2003).

79. D. Chowdhury, V. Guttal, K. Nishinari, A. Schadschneider, J. Phys. A:Math. Gen. **35**, L573 (2002)

80. K. Nishinari, D. Chowdhury, A. Schadschneider, Phys. Rev. E **67**, 036120 (2003)

81. A. Kunwar, D. Chowdhury, A. Schadschneider and K. Nishinari, accepted in J. Stat. Mech.

82. A. John, A. Schadschneider, D. Chowdhury and K. Nishinari, J. Theor. Biol. **231**, 279 (2004).

83. D. Helbing, F. Schweitzer, J. Keltsch, P. Molnar: Phys. Rev. **E56**, 2527 (1997)

84. B. Derrida, Phys. Rep. **301**, 65 (1998)

85. B. Derrida and M.R. Evans, in: *Nonequilibrium Statistical Mechanics in One Dimension*, ed. V. Privman (Cambridge University Press, 1997)

86. O.J. O'Loan, M.R. Evans, M.E. Cates, Europhys. Lett. **42**, 137 (1998); Phys. Rev. E**58**, 1404 (1998).

87. D. Chowdhury, R.C. Desai, Eur. Phys. J. B**15**, 375 (2000).

88. A. Kunwar, A. John, K. Nishinari, A. Schadschneider and D. Chowdhury, J. Phys. Soc. Jap. **73**, 2979 (2004).

89. A. John, A. Kunwar, A. Namazi, A. Schadschneider, D. Chowdhury, and K. Nishinari, in this proceedings.

90. M. Burd, D. Archer, N. Aranwela and D.J. Stradling, Am. Nat. **159**, 283 (2002).

91. M. Burd et al. (2005) unpublished.

92. A. John et al. (2005) unpublished.

Cooperative Behaviour of Semiflexible Polymers and Filaments

Jan Kierfeld, Pavel Kraikivski, Torsten Kühne, and Reinhard Lipowsky

Max Planck Institute of Colloids and Interfaces, Science Park Golm, 14424 Potsdam, Germany

Summary. Semiflexible polymers and filaments play an important role in biological and chemical physics. The cooperative behaviour of interacting filaments and the internal bending modes of a single filament give rise to various equilibrium phase transitions, such as bundling and adsorption, which are reviewed in this article. In motility assays, filaments are adsorbed and driven by motor proteins, which are anchored to a planar two-dimensional substrate. We present a simulation model for the active filament dynamics in this non-equilibrium system.

1 Introduction

Stiff, filamentous polymers play an important role in biological and chemical physics. Such polymers have a considerable bending rigidity, which gives rise to persistence lengths comparable to or larger than their contour lengths. These semiflexible polymers exhibit a variety of cooperative phenomena, which we want to discuss in this article. These transitions result from the competition of several energies in the system, i.e., the bending energy, the thermal energy, interaction energies, and external driving forces. In biological systems, driving forces can arise from the activity of molecular motors which perform directed walks on cytoskeletal filaments.

First we will discuss the equilibrium phase transition that leads to the formation of filament bundles in the presence of attractive interactions, which can arise from crosslinking proteins or unspecific interactions [3]. In eukaryotic cells, the most important building blocks of the cytoskeleton are microtubules and filamentous actin (F-actin). Actin filaments have a persistence length $L_p \simeq 30\mu m$ [1], microtubules are much stiffer with a persistence length $L_p \sim mm$ [2]. In the cortex of the cell, actin filaments form a dense meshwork which is responsible for many of the viscoelastic properties of the cell. Another important morphology that is found in the cell are filament bundles [4], which, e.g., support cell protrusions and serve as stress fibres. Both meshworks and bundles are hold together by different actin-binding crosslinking proteins [4, 5].

Actin bundling crosslinkers possess two adhesive end domains which bind to filaments by weak bonds; crosslinker mediated interactions therefore allow a reversible formation of actin bundles, which can be regulated by the concentration of crosslinkers in solution. Solution of actin filaments and crosslinking proteins have been studied *in vitro* in a number of recent experiments [6–8]. In these studies it has been observed that bundle formation in F-actin solutions containing crosslinking molecules requires a threshold crosslinker concentration above which F-actin bundles become stable against the thermal fluctuations of filaments and a phase containing networks of bundles separates.

Another important equilibrium phase transition of polymers is their adsorption to an attractive planar surface. For semiflexible polymers or filaments, the adsorption transition is similar to the binding of two filaments but represents a distinct universality class [9]. Various single molecule methods have been applied to adsorbed semiflexible polymers because both visualization and manipulation are easier for adsorbed polymers with a large diameter, such as DNA [10, 11]. These polymers are generically semiflexible because stronger entropic or enthalpic interactions along their backbone increase the bending rigidity. The thermally activated dynamics of single filaments adsorbed on structured substrates has been discussed in Refs. [12]. Here, we will focus on the adsorption behaviour of filaments in motility assays. In such an assay, cytoskeletal filaments are adsorbed and driven over a two-dimensional, planar substrate by motor proteins whose tails are anchored to the substrate [13]. In order to obtain adsorption, a critical density of motor proteins is needed in close analogy to the critical crosslinker concentration for the formation of a filament bundle.

Motility or gliding assays are a standard biochemistry assay to characterize motor proteins, which is based on measuring the active dynamics of adsorbed filaments. In biological cells, small forces generated by motor proteins organize and rearrange cytoskeletal filaments and give rise to active, non-equilibrium filament dynamics, which plays an important role for cell division, motility, and force generation [17]. Whereas conventional "passive" polymer dynamics is governed by thermal fluctuations [18], active filament dynamics is characterized by a constant supply of mechanical energy by motor proteins, which hydrolyze adenine triphosphate (ATP). Motility assays are model systems, which allow to study active filament dynamics in a controlled manner. By analyzing the transport velocities of *single* filaments gliding over the substrate, information can be obtained about basic properties of molecular motors such as their maximal velocity. We introduce a simulation model for motility assays, which refines previous models [14–16] and contains semiflexible filaments, motor heads, and polymeric motor tails as separate degrees of freedom.

This article is organized as follows. In section 2 the formation of filament bundles via crosslinker-mediated attractive interactions is discussed. The adsorption of a filament onto an adhesive surface is considered in section 3. In particular, we discuss the filament adsorption on a planar two-dimensional substrate covered with anchored motor proteins, which represents the geometry used in motility assays. In section 4, we introduce a model for the active filament dynamics in motility assays and present recent simulation results.

2 Filament Bundles

We consider N filaments with bending rigidity κ in a solution containing crosslinking molecules with two adhesive end groups. The persistence length of such a filament is $L_p = 2\kappa/T$, where T is the temperature in energy units. This system exhibits a critical crosslinker concentration, $X_1 = X_{1,c}$, which separates two different concentration regimes. For $X_1 < X_{1,c}$, the filaments are unbound and uniformly distributed within the compartment. For $X_1 > X_{1,c}$, the filaments form either a single bundle, which represents the true ground state of the system as in Fig. 1(a) and (c) , or several sub-bundles, which represent metastable, kinetically trapped states as in Fig. 1(b). Furthermore, as we decrease the crosslinker concentration from a value above $X_{1,c}$ towards a value below $X_{1,c}$, the bundles undergo a discontinuous unbinding transition at $X_1 = X_{1,c}$. The existence of a single, discontinuous unbundling transition can be established by analytic methods for $N = 2$ filaments [9] and by Monte Carlo (MC) simulations for larger bundles containing up to $N = 20$ filaments.

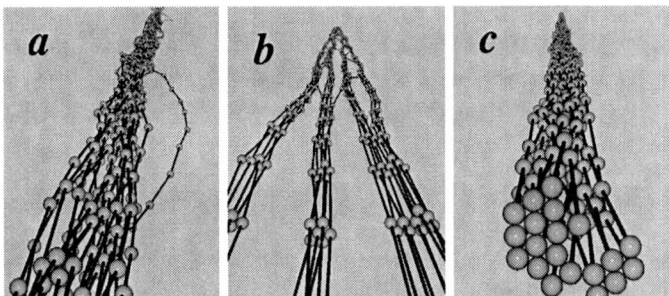

Fig. 1.   Monte Carlo snapshots of bundles with $N = 20$ filaments. (a) Close to the unbinding transition in the bundled phase. (b) Deep in the bound phase, the bundle tends to segregate due to slow kinetics and filament entanglement. (c) The equilibrium shape of the bundle is roughly cylindrical.

2.1 Model

The filaments are oriented along one axis, say the x-axis, and can be parametrized by two-dimensional displacements $\mathbf{z}_i(x)$ $(i = 1, ..., N)$ perpendicular to the x-axis, with $0 < x < L$, where L is the projected length of the polymer. This parametrization is appropriate provided the longitudinal correlation length is small compared to L_p. We discretize the filament into segments of length $a_\parallel$, i.e., $x_k = ka_\parallel$ and $\mathbf{z}_{i,k} = \mathbf{z}_i(x_k)$. The presence or absence of a crosslinker molecule at segment k of filament i is described by the occupation number $n_{i,k} = n_i(x_k) = 0, 1$. The filament-crosslinker system is described by the Hamiltonian

$$\mathcal{H} = \sum_i \left[\mathcal{H}_{b,i}\{\mathbf{z}_i\} + \mathcal{H}_1\{n_i\} \right] + \sum_{i,j} \mathcal{H}_2\{\mathbf{z}_i - \mathbf{z}_j, n_i, n_j\} , \qquad (1)$$

where the first contribution $\mathcal{H}_{b,i} = \int_0^L dx \frac{1}{2}\kappa \left(\partial_x^2 \mathbf{z}_i \right)^2$ contains the bending energies of the filaments. The term $\mathcal{H}_1$ describes the intrafilament interactions of linkers. We consider a lattice gas of linkers with hard-core repulsion adsorbing on a filament with $\mathcal{H}_1 = \sum_k a_\parallel W n_{i,k}$ where $W < 0$ is the adhesive energy (per length) of one linker end group. The third contribution $\mathcal{H}_2$ describes the pairwise interactions between filaments i and j and is given by

$$\mathcal{H}_2 = \sum_k a_\parallel \left[V_r(\Delta \mathbf{z}_{ij,k}) + \frac{1}{2}(n_{i,k} + n_{j,k} - 2n_{i,k}n_{j,k})V_a(\Delta \mathbf{z}_{ij,k}) \right] \qquad (2)$$

where $\Delta \mathbf{z}_{ij,k} \equiv \mathbf{z}_{i,k} - \mathbf{z}_{j,k}$. The first term is the hard-core repulsion of filaments that is independent of the linker occupation with a potential $V_r(\mathbf{z}) = \infty$ for $|\mathbf{z}| < \ell_r$ and $V_r(\mathbf{z}) = 0$ otherwise where ℓ_r is of the order of the filament diameter. The second term is the linker-mediated attraction and is non-zero if either segment k of filament i or segment k of filament j carries a linker. Then the other filament is attracted by a linker-mediated potential $V_a(\mathbf{z})$. The latter filament gains the additional energy $|W|$ if $|\Delta \mathbf{z}_{ij,k}| \leq \ell_a$, where the potential range ℓ_a is of the order of the linker size. This attraction is modelled by the potential well

$$V_a(\mathbf{z}) = W \quad \text{for } 0 < |\mathbf{z}| - \ell_r < \ell_a , \quad V_a(\mathbf{z}) = 0 \quad \text{otherwise.} \qquad (3)$$

We can perform the partial trace over the crosslinker degrees of freedom $n_{i,k}$ in the grand-canonical ensemble to obtain an effective interaction between filaments. Each crosslinker has two adhesive ends. The first adhesive end adsorbs on a filament and establishes the standard Langmuir-type adsorption equilibrium with a linker concentration per site $X_1 \equiv \langle n_{i,k} \rangle_1 = Kc_x/(1 + Kc_x)$ where the average is taken with the Hamiltonian $\mathcal{H}_1$. X_1 is thus determined by the concentration c_x of linkers in solution, where K is the equilibrium constant of the association reaction of the crosslinker with the filament. Tracing over weakly bound linkers with $|W| \ll T/a_\parallel$, we end up with effective pairwise linker-mediated filament interactions, i.e., $\bar{\mathcal{H}}_2 \approx \frac{1}{2}\sum_k a_\parallel [V_r(\Delta \mathbf{z}_{ij,k}) + \bar{V}_a(\Delta \mathbf{z}_{ij,k})]$, which have the same functional form as the bare interactions; the short-range attractive part $\bar{V}_a$ is of the form (3) with a strength $\bar{W} \approx 2X_1 W$ proportional to the linker concentration on the filament. Pairwise filament interactions with potentials of the form (3) are generic and do not only arise from crosslinkers but also from van-der-Waals, electrostatic, or depletion forces.

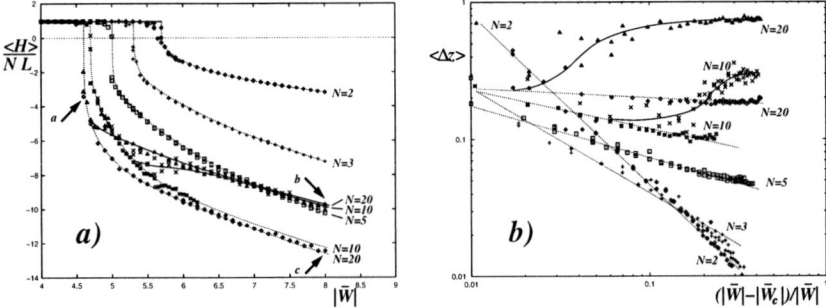

Fig. 2. MC data for $N = 2, 3, 5, 10, 20$ identical filaments (with persistence length $L_p = 200$, contour length $L = 100$, potential range $\ell_a = 0.001$, and hard core radius $\ell_r = 0.1$; all lengths are in units of Δx, lines are guides to the eye). For $N = 10, 20$ two branches of data are shown corresponding to two different initial conditions; in the lower branch we prepared a compact cylindrical configuration, in the upper branch (thick lines) we arranged filaments initially in a plane. (a) Mean energy $\langle \mathcal{H} \rangle / NL$ per filament (in units of T) as a function of the effective potential strength $|\bar{W}|$ (in units of $T/\Delta x$). Arrows correspond to the snapshots in Fig. 1. (b) Logarithmic plot of the mean filament separation $\langle \Delta z \rangle \equiv \langle |\Delta \mathbf{z}_{ij}| - \ell_r \rangle$ (in units of Δx) as a function of the reduced potential strength $(|\bar{W}| - |\bar{W}_c|)/|\bar{W}|$.

2.2 Discontinuous Unbundling Transition

We have studied bundle formation by MC simulations for up to $N = 20$ identical filaments ($\kappa_i = \kappa$) using the effective Hamiltonian $\mathcal{H} = \sum_i \mathcal{H}_{b,i} + \sum_{i,j} \bar{\mathcal{H}}_2$. The MC simulations can be used to determine the locus and order of the unbinding transitions, at which the mean energy $\langle \mathcal{H} \rangle$ exhibits a discontinuity, see Fig. 2a. To gain further insight into bundle morphologies, we also measure the mean segment separation $\langle |\Delta \mathbf{z}_{ij}| - \ell_r \rangle$, see Fig. 2b, which is directly proportional to the mean bundle thickness that can be determined by optical microscopy in experiments.

Our MC simulations confirm that, for bundles containing up to $N = 20$ filaments, there is a single, *discontinuous* unbinding transition, see Fig. 2a. In the presence of a hard-core repulsion, the critical potential strength $\bar{W}_c$ saturates to a N-independent limiting value for large N. As can be seen in Fig. 1a typical bundle morphologies close to the transition are governed by pair contacts of filaments. The bundle thickness, as given by the mean segment separation $\langle |\Delta \mathbf{z}_{ij}| - \ell_r \rangle$, stays *finite* up to the transition, see MC data in Fig. 2b. For increasing N, an increasing number of higher moments $\langle (|\Delta \mathbf{z}_{ij}| - \ell_r)^m \rangle$ remains finite at the transition [all moments $m < 2(N-1)(3N-4)/3$ remain finite] showing that the critical thickness fluctuations of large bundles become small.

Deep in the bundled phase, i.e., for large $|\bar{W}|$, our MC simulations show that bundles do not always reach their equilibrium shape. Small sub-bundles containing typically $N \sim 5$ filaments form easily, start to entangle, and further equilibration is kinetically trapped suggesting that the bundle is in a "glass" phase. Fig. 1b shows the segregation into sub-bundles in a typical configuration and Fig. 2a shows the corresponding rise in the mean bundle energy per filament which approaches the $N = 5$ result. In Fig. 2b the pronounced rise of the mean separation for $N > 5$ with increasing potential strength and with increasing N is due to the segregation. This behaviour is reminiscent of the experimentally observed F-actin structures consisting of networks of small bundles [7]. Only when starting from a sufficiently compact initial state, bundles relax towards the equilibrium form in the MC simulation, which is a roughly cylindrical bundle with a hexagonal filament arrangement as shown in Fig. 1c. In contrast to the segregated form, the bundle thickness and the mean energy per filament of the equilibrium form decrease with increasing N, as can be seen in Fig. 2.

The critical potential strength $\bar{W}_c$ corresponds to a critical crosslinker concentration $X_{1,c}$. For weakly bound linkers $|W| \ll T/a_{\parallel}$, we have a simple linear relation $\bar{W} \approx 2X_1 W$ such that $X_{1,c} \approx \bar{W}_c/2W$. The corresponding relation for strongly bound linkers is more complicated.

Our simulations use periodic boundary conditions and treat very long and essentially parallel filaments. In order to include translational and rotational entropy we can map the ensemble of semiflexible filaments considered here onto an ensemble of rigid rods of finite length L and diameter $a_{\perp}$ at a certain concentration c. The effective pairwise attraction (per length) J is given by the bundling free energy of the filaments with $J \sim \bar{W}_c - \bar{W} > 0$ for $|\bar{W}| > |\bar{W}_c|$. Using the results of Refs. [19], we find that the hard rod system separates into a high-density nematic phase and a low-density nematic or isotropic phase above a critical attraction, which is in qualitative agreement with the experimental results in Refs. [6–8].

3 Filament Adsorption

The adsorption transition of a single filament onto a planar substrate is qualitatively similar to the bundle formation for $N = 2$ filaments in 1+1 dimensions, where the one-dimensional perpendicular distance $z(x)$ from the surface is analogous to a one-dimensional separation between filaments. The adsorption transition can be solved analytically [9], which reveals that unbinding and desorption represent two distinct universality classes with different critical exponents.

Here we want to consider the adsorption of a filament with persistence length $L_p = 2\kappa/T$ on a planar two-dimensional substrate where molecular motors are adsorbed with an areal density σ. Each motor can bind to a filament within a capture radius w and a binding energy $W_m < 0$. In contrast to the case of the annealed crosslinker ensemble considered previously, the motors represent a *quenched* ensemble of adsorption points. In the following, we consider the typical experimental situation of a rather uniform coverage with motor proteins and also neglect effects from filament fluctuations parallel to the surface. Then the array of motors gives rise to an average adsorption potential $\bar{V}_{ad}(z)$ of the same functional form as the potential (3) with a potential strength $\bar{W}_{ad} = W_m\sigma w$, the hard substrate at $\ell_r = 0$ and ℓ_a of the order of the capture radius w. On length scales comparable or smaller than L_p, the semiflexible polymer is only weakly bent by thermal fluctuations and its configurations are governed by the effective Hamiltonian

$$\mathcal{H}_{ad} = \int_0^L dx \left[\frac{\kappa}{2}(\partial_x^2 z)^2 + \bar{V}_{ad}(z(x)) \right] . \tag{4}$$

We consider the limit of long filaments $L\bar{W}_{ad} \gg T$, which can exhibit a desorption transition. Using the model (4), this desorption transition has been studied by transfer matrix techniques in Refs. [9]. The critical potential strength for desorption is $\bar{W}_{ad,c} = -cT/w^{2/3}L_p^{1/3}$ corresponding to a critical motor density $\sigma_c = T/W_m w^{5/3}L_p^{1/3}$, where $c \approx \sqrt{3\pi}/2 \approx 1.5$. For motor densities above this critical density, filaments adsorb onto the substrate with anchored motors against the thermal fluctuations of filaments. The critical motor density for adsorption is decreasing with increasing filament rigidity κ. The transfer matrix treatment shows that the free energy difference between adsorbed and unbound state vanishes as $|\Delta f| \approx |\bar{W}_{ad,c}||w|/\ln|w|^{-1}$ where $w \equiv (\bar{W}_{ad} - \bar{W}_{ad,c})/\bar{W}_{ad,c}$. Therefore, the correlation length $\xi_\| = T/|\Delta f| \propto |w|^{-\nu}$ diverges with an exponent $\nu = 1 + \log$. The weak bending approximation is valid as long as gradients are small, i.e., $\langle(\partial_x z)^2\rangle \sim \xi_\|/L_p \lesssim 1$, which is fulfilled for $|\bar{W}_{ad} - \bar{W}_{ad,c}| \gtrsim T/L_p$, which typically applies to stiff filaments such as microtubules adsorbed by kinesins.

4 Motility Assays for Motor Proteins

We consider a motility assay, where filaments are connected to the substrate by anchored motors of sufficient density $\sigma > \sigma_c$. In the presence of ATP, the motor heads start to perform a directed walk on the filaments, which induces active dynamics of adsorbed filaments.

4.1 Model

Our microscopic model for motility assays describes filaments, motor heads, and polymeric motor tails as separate degrees of freedom [22]. One end of the motor tail is anchored to the substrate and the motor head on the other end can bind to a filament in the correct orientation due to the tail flexibility. Once bound the motor head moves along the filament thereby stretching the polymeric tail, which gives rise to a loading force acting both on the motor head and the attached filament. This force feeds back onto the motion of the

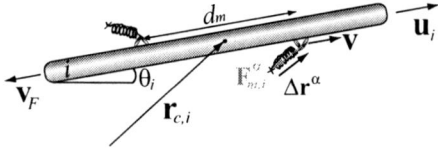

Fig. 3. Schematic top view of a filament i in the motility assay with two motors attached. The configuration of filament i is specified by the position $\mathbf{r}_{c,i}$ of its center of mass an its orientation angle θ_i; $\mathbf{u}_i = (\cos\theta_i, \sin\theta_i)$ is the orientational unit vector of the filament. $\mathbf{F}^\alpha_{m,i}$ is the force arising from the stretched polymeric tail of motor α, which has an end-to-end vector $\Delta\mathbf{r}^\alpha$. The polymeric tail is stretched by the motor head moving with velocity v, see eq. 6. d_m denotes the distance between attached motors.

bound motor head, which moves with a load-dependent motor velocity [20, 21]. Filaments follow an overdamped dynamics with external forces arising from the stretched motor tails and the repulsive filament-filament interaction.

To proceed, let us consider N rigid filaments of length L on a planar two-dimensional substrate [23]. The configuration of filament i $(i = 1, ..., N)$ is then specified by the position of its center of mass $\mathbf{r}_{c,i}$ and its orientation angle θ_i, see Fig. 3. The overdamped translational and rotational dynamics of each filament i is described by stochastic Langevin-type equations of motion [22]

$$\boldsymbol{\Gamma} \cdot \partial_t \mathbf{r}_{c,i} = \sum_{\alpha=1}^{N_i} \mathbf{F}^\alpha_{m,i} + \sum_{j=1}^{N} \mathbf{F}_{r,ij} + \boldsymbol{\zeta}_i$$

$$\Gamma_\theta \partial_t \theta_i = \sum_{\alpha=1}^{N_i} M^\alpha_{m,i} + \sum_{j=1}^{N} M_{r,ij} + \zeta_{\theta,i} \ , \tag{5}$$

where N_i is the number of motor heads attached to filament i and indexed by α. $\mathbf{u}_i = (\cos\theta_i, \sin\theta_i)$ is the orientational unit vector of filament i. $\boldsymbol{\Gamma}$ is the matrix of translational friction coefficients, $\boldsymbol{\Gamma} = \Gamma_\| \mathbf{u}_i \otimes \mathbf{u}_i + \Gamma_\perp (\mathbf{I} - \mathbf{u}_i \otimes \mathbf{u}_i)$ [18], where $\mathbf{I}$ is the unit matrix, and Γ_θ is the rotational friction coefficient. $\Gamma_\|$, $\Gamma_\perp$ and Γ_θ are the friction coefficients of the passive filament dynamics. $\boldsymbol{\zeta}_i(t)$ and $\zeta_{\theta,i}(t)$ are the translational and the angular components of the Gaussian distributed thermal random forces. $\mathbf{F}^\alpha_{m,i}$ is the force arising from the stretched tail of motor α. The end-to-end vector of the polymeric tail is $\Delta\mathbf{r}^\alpha \equiv \mathbf{r}^\alpha_i - \mathbf{r}^\alpha_0$, where the motor tail is anchored at $\mathbf{r}^\alpha_0$ and the head position is $\mathbf{r}^\alpha_i$. We model the polymeric tail as freely jointed chain such that $\mathbf{F}^\alpha_{m,i}$ is pointing in the direction $-\Delta\mathbf{r}^\alpha$ and its absolute value is obtained by inverting the force-extension relation of a freely jointed chain [24]. There is also a corresponding torque due to the motor activity, $M^\alpha_{m,i} = |(\mathbf{r}^\alpha_i - \mathbf{r}_{c,i}) \times \mathbf{F}^\alpha_{m,i}|$. The interaction forces $\mathbf{F}_{r,ij}$ and torques $M_{r,ij}$ are due to the purely repulsive interactions between filaments i and j corresponding to a hard-rod interaction for filaments of diameter D.

The dynamics of motor heads is described by a deterministic equation of motion, which has the form

$$\partial_t x_i^\alpha = v(\mathbf{F}_{m,i}^\alpha) \,, \tag{6}$$

where $|x_i^\alpha| \le L/2$ defines the position of the motor α on the rod i, i.e., $\mathbf{r}_i^\alpha = \mathbf{r}_{c,i} + x_i^\alpha \mathbf{u}_i$, i.e., the filament polarity is such that the motor head moves in the direction $\mathbf{u}_i$. The motor velocity v is a function of the loading force $\mathbf{F}_{m,i}^\alpha$ which builds up due to stretching of the motor tail. We use a force-velocity relation with a maximum value v_{max} for forces $\mathbf{F}_{m,i}^\alpha \cdot \mathbf{u}_i \ge 0$ pulling the motor forward, a linear decrease for forces $\mathbf{F}_{m,i}^\alpha \cdot \mathbf{u}_i < 0$ pulling the motor backwards, and $v = 0$ for $\mathbf{F}_{m,i}^\alpha \cdot \mathbf{u}_i < -F_{st}$, where F_{st} is the stall force [20, 21]. We assume that the motor binds to the filament when the distance between the position of the fixed end of the motor tail at $\mathbf{r}_0^\alpha$ and the filament is smaller than the capture radius w. Apart from the stall force F_{st} the motor is also characterized by its detachment force F_d, above which the unbinding rate of the motor head becomes large. For simplicity we assume in our model that the motor head detaches whenever the force $F_{m,i}^\alpha$ exceeds a threshold value F_d. We consider the case of processive motors with a high duty ratio close to unity, i.e., motors detach from a filament only if they reach the filament end or if the force F becomes larger than the detachment force F_d.

4.2 Simulation

Using the above model we performed simulations of gliding assays for a random distribution of motors with a surface density σ and periodic boundary conditions. At each time step Δt we update the motor head position x_i^α and filament position by using the discrete version of the equations of motion (5) and (6). The parameter values that we choose for the simulations are comparable with experimental data on assays for conventional kinesin. The simulation results presented in this article have been obtained for assays with quadratic geometry and size $25\mu m^2$ with rigid filaments of length $L = 1\mu m$ and diameter $D = L/40$. We simulate at room temperature $T = 4.28 \times 10^{-3} pN\,\mu m$. Friction coefficients are $\Gamma_\perp = 2\Gamma_\| = 4\pi\eta L/\ln(L/D)$ and $\Gamma_\theta = \Gamma_\| L^2/6$, where η is the viscosity of the surrounding liquid. We use a value $\eta = 0.5 pN\,s/\mu m^2$ much higher than the viscosity of water, $\eta_{water} \sim 10^{-3} pN\,s/\mu m^2$, which allows to take larger time steps and decreases the simulation time. We checked that this does not affect results. We use a maximum motor speed of $v_{max} = 1\mu m s^{-1}$ and a stall force of $F_{st} = 5pN$. The capture radius for motor proteins is $w = 10^{-2}\mu m$ and the length of the fully stretched motor tail $L_m = 5 \times 10^{-2}\mu m$.

The motion of a *single* filament with contour length L is characterized by stochastic switching between rotational and translational diffusion if no motors are attached, directed translation in rotationally diffusing directions if one motor is attached, and directed translation in one direction if two or more motors are attached. The relative frequency of these types of motion depends on the mean number of motors attached to the filament or the mean distance $\langle d_m \rangle$ between bound motors and, thus, on the surface motor concentration

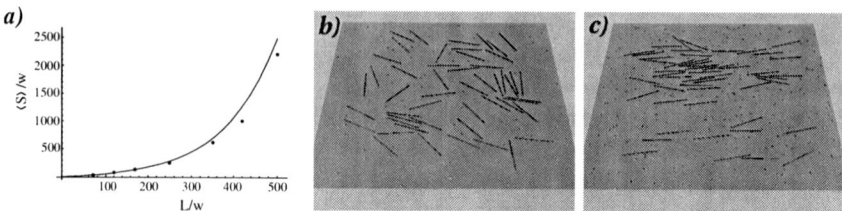

Fig. 4. (a) Simulation results for average distance $\langle S \rangle$ traveled by a filament between successive rotations as a function of the filament length L for high motor concentration. The solid line is the analytical result (7) as derived in Ref. [14]. (b) and (c): Snapshots of a gliding assay of rodlike filaments with filament density $\rho = 2/L^2$ on a motor coated substrate with randomly distributed motors and periodic boundary conditions. For detachment forces $F_d = F_{st}$, we find (b) an isotropic phase at low motor surface density $\sigma w L = 0.03$ and (c) active nematic ordering at high motor surface density $\sigma w L = 0.09$.

σ [14]. In the limit of high motor concentration a filament has two or more bound motors on average and $\langle d_m \rangle \sim 1/\sigma w$. The single filament performs a persistent walk with a persistence length [14]

$$\xi_p = \frac{\langle S \rangle}{\langle \Delta\theta^2 \rangle} = \frac{1}{\langle \Delta\theta^2 \rangle} \frac{L + 2\langle d_m \rangle}{L + 3\langle d_m \rangle} \frac{\langle d_m \rangle^2}{L} \left(e^{L/\langle d_m \rangle} - 1 - \frac{L}{\langle d_m \rangle} \right) \qquad (7)$$

where $\langle S \rangle$ is the mean distance traveled by a filament between successive rotations and $\langle \Delta\theta^2 \rangle^{1/2} = 3\sigma/\sigma L^2$ the mean angle at rotations. The theoretical result (7) is confirmed by our simulation as shown in Fig. 4a. The mean filament velocity $v_F = \langle |\dot{\mathbf{r}}_{c,i}| \rangle$ can be obtained by simultaneously equating the filament friction force with the total motor driving force and the filament velocity with the motor velocity in the steady state, which gives $v_F = v_{max}(1 + \Gamma_\| v_{max} \langle d_m \rangle / L F_{st})^{-1}$. This relation is confirmed by our simulations.

Our results for the simulation of *many* filaments with hard-core interactions indicate that the motility assay exhibits *active nematic ordering* if the motor density σ is increased as can be seen in the two simulation snapshots Figs. 4b and c.

References

1. A. Ott, M. Magnasco, A. Simon, and A. Libchaber, Phys. Rev. E **48**, R1642 (1993); J. Käs, H. Strey, and E. Sackmann, Nature **368**, 226 (1994).
2. F. Gittes, B. Mickey, J. Nettleton, and J. Howard, J. Cell Biol. **120**, 923 (1993).
3. J. Kierfeld, T. Kühne and R. Lipowsky, Phys. Rev. Lett. **95**, 038102 (2005).
4. J.R. Bartles, Curr. Opin. Cell Biol. **12**, 72 (2000).
5. K.R. Ayscough, Curr. Opin. Cell Biol. **10**, 102 (1998); S.J. Winder, *ibid.* **15**, 14 (2003).
6. M. Tempel, G. Isenberg, and E. Sackmann, Phys. Rev. E **54**, 1802 (1996).
7. O. Pelletier, E. Pokidyshevam L.S. Hirst, N. Bouxsein, Y. Li, and C.R. Safinya, Phys. Rev. Lett. **91**, 148102 (2003).
8. M.L. Gardel, J.H. Shin, F.C. MacKintosh, L. Mahadevan, P. Matsudaira, and D.A. Weitz, Science **304**, 1301 (2004).
9. J. Kierfeld and R. Lipowsky, Europhys. Lett. **62**, 285 (2003); J. Phys. A: Math. Gen. **38**, L155 (2005).
10. S.S. Sheiko and M. Möller, Chem. Rev. **101**, 4099 (2001).
11. N. Severin, J. Barner, A. A. Kalachev and J.P. Rabe, Nano Lett. **4**, (2004) 577.
12. P. Kraikivski, R. Lipowsky, and J. Kierfeld, Europhys. Lett. **66**, 763 (2004); Eur. Phys. J. E **16**, 319 (2005); Europhys. Lett. **71**, 138 (2005).
13. J. Scholey, *Motility assays for motor proteins*, Meth. Cell Biology **39**, (Academic Press, New York, 1993).
14. T. Duke, T.E. Holy, and S. Leibler, Phys. Rev. Lett. **74**, 330 (1994).
15. M.R. Faretta and B. Basetti, Europhys. Lett. **41**, 689 (1998).
16. F. Gibbons, J.-F. Chauwin, M. Despósito, and J.V. José, Biophys. J. **80**, 2515 (2001).
17. J. Howard, *Mechanics of Motor Proteins and the Cytoskeleton* (Sinauer Associates, Inc., Sunderland, 2001).
18. M.Doi and S.F.Edwards, *The Theory of Polymer Dynamics* (Clarendon, Oxford, 1986).
19. M. Warner and P.J. Flory, J. Chem. Phys. **73**, 6327 (1980); A.R. Khokhlov and A.N. Semenov, J. Stat. Phys. **38**, 161 (1985).
20. C. M. Coppin, D. W. Pierce, L. Hsu, R. D. Vale, Proc. Natl. Acad. Sci. USA **94**, (1997) 8539.
21. S. M. Block, C. L. Asbury, J. W. Shaevitz, M. J. Lang, Proc. Natl. Acad. Sci. USA **100**, (2003) 2351.
22. P. Kraikivski, R. Lipowsky, and J. Kierfeld, Phys. Rev. Lett. **96**, 258103 (2006).
23. The model can be extended to deformable filaments by modeling each filament as a set of N_s segments connected by elastic springs and hinges, see P. Kraikivski, Ph.D. thesis, Universität Potsdam, 2005.
24. J. Kierfeld, O. Niamploy, V. Sa-yakanit, and R. Lipowsky, Eur. Phys. J. E **14**, 17 (2004).

Traffic of Molecular Motors

Stefan Klumpp, Melanie J. I. Müller, and Reinhard Lipowsky

Max-Planck-Institut für Kolloid- und Grenzflächenforschung,
Wissenschaftspark Golm, 14424 Potsdam, Germany

Summary. Molecular motors perform active movements along cytoskeletal filaments and drive the traffic of organelles and other cargo particles in cells. In contrast
to the macroscopic traffic of cars, however, the traffic of molecular motors is characterized by a finite walking distance (or run length) after which a motor unbinds
from the filament along which it moves. Unbound motors perform Brownian motion
in the surrounding aqueous solution until they rebind to a filament. We use variants
of driven lattice gas models to describe the interplay of their active movements, the
unbound diffusion, and the binding/unbinding dynamics. If the motor concentration
is large, motor-motor interactions become important and lead to a variety of cooperative traffic phenomena such as traffic jams on the filaments, boundary-induced
phase transitions, and spontaneous symmetry breaking in systems with two species
of motors. If the filament is surrounded by a large reservoir of motors, the jam length,
i.e., the extension of the traffic jams, is of the order of the walking distance. Much
longer jams can be found in confined geometries such as tube-like compartments.

1 Introduction

The traffic of vesicles, organelles, protein complexes, messenger RNA, and
even viruses within the cells of living beings is driven by the molecular motors of the cytoskeleton which move along cytoskeletal filaments in a directed
fashion [1–3]. There are three large classes of cytoskeletal motors, kinesins and
dyneins which move along microtubules, and myosins which move along actin
filaments. These motors use the free energy released from the hydrolysis of
adenosinetriphosphate (ATP), which represents their chemical fuel, for active
movement and to perform mechanical work. They move in discrete steps in
such a way that one molecule of ATP is used per step. Typical step sizes are
~ 10 nm, typical motor velocities are in the range of μm/sec.

Since the interior of cells is quite crowded and motors are strongly attracted by the filaments, which leads to relatively large motor densities along
the filaments, it is interesting to study the collective traffic phenomena which
arise from motor–motor interactions, in particular the formation of traffic

jams due to the mutual exclusion of motors from filament sites. To study these cooperative phenomena theoretically we have introduced new variants of driven lattice gas models [4] which have been studied extensively during the last years both by our group [4–11] and by several other groups [12–17] and which will be described below. These models are related to lattice gas models for driven diffusive systems and exclusion processes as studied in the context of non-equilibrium phase transitions [18–21] and highway traffic [22, 23]. Since molecular motors can be studied in a systematic way using biomimetic systems which consist of a small number of components (such as motors, filaments, and ATP), they can also serve as model systems for the experimental investigation of driven diffusive systems.

Although the traffic of cargo particles pulled by molecular motors within cells is remarkably similar to the macroscopic traffic on streets or rails, there is an important difference which is a direct consequence of the nanoscale size of molecular motors: The motor–filament binding energy can be overcome by thermal fluctuations which are ubiquitous on this scale, and molecular motors therefore have a finite walking distance or run length after which they unbind from the filament along which they move. This walking distance is typically of the order of 1 μm for a single motor molecule.[1] Likewise, unbound motors which diffuse freely in the surrounding aqueous solution, can bind to a filament and start active movement. In contrast to highway traffic, where additional cars enter only at on-ramps, i.e. at specific locations, binding of molecular motors occurs along the full length of the filaments. In addition to stepping along a one-dimensional track and mutual exclusion, lattice models for the traffic of molecular motors must therefore also describe the dynamics of motor–filament binding and unbinding as well as the diffusive movement of the unbound motors.[2]

In contrast to the transport properties of single motor molecules which have been studied extensively during the last 15 years [1, 2], the traffic phenomena in many-motor systems have only recently attracted the interest of experimentalists and are still largely unexplored from the experimental point of view. The quantity of main interest has so far been the profile of the bound motor density along a filament. Density profiles with a traffic jam-like accumulation of motors at the end of filaments have been observed in vivo for a kinesin-like motor which was overexpressed in fungal hyphae [28, 29]. Recently, motor traffic jams have also been observed in biomimetic in vitro systems using both conventional kinesin (kinesin 1) [30] and the monomeric kinesin KIF1A (kinesin 3) [17].

---

[1] In order to transport a cargo actively over larger distances as, e.g., in the axon of a nerve cell, several motors work together in a cooperative fashion. We have recently shown that 7–8 motors are sufficient for processive transport over distances in the centimeter range as necessary in axons [24].

[2] These processes have not been taken into account in earlier studies of exclusion effects in many-motor systems which were based on ratchet models [25–27].

In the following, we will give a short overview over the lattice models for molecular motors and discuss the motor traffic in various systems which differ mainly in the compartment geometry and the arrangement of filaments. In section 4, we address the length of motor jams on filaments and argue that in the presence of a large motor reservoir this jam length is typically of the order of the walking distance. Longer jams are found in confined geometries as discussed in section 5. In the last section of the paper, we briefly review our results for systems with two motor species.

2 Lattice Models for Molecular Motor Traffic

To describe the interplay of the movements of bound and unbound motors, we have introduced a class of lattice models which incorporate the active movement of bound motors, the passive diffusion of unbound motors, and the motor–filament binding and unbinding dynamics [4]. These models can also account for motor–motor interactions such as their mutual exclusion from binding sites of the filament.

We describe the motor movements as random walks on a (in general, three-dimensional) cubic lattice as shown in Fig. 1(a). Certain lines on this lattice represent the filaments. The lattice constant is taken to be the motor step size ℓ which for many motors is equal to the filament periodicity. When a motor is localized at a filament site, it performs a biased random walk. Per unit time τ, it makes forward and backward steps with probabilities α and β, respectively. With probability γ, the motor makes no step and remains at the same site. The latter parameter is needed to account for the fact that if the lattice constant is given by the motor step size, unbound diffusion over this scale is much faster than an active step of a bound motor. Finally, the motor hops to each of the adjacent non-filament sites with probability $\epsilon/6$ and unbinds from the filament. The total unbinding probability per unit time is $\epsilon_0 = n_{ad}\epsilon/6$ with the number n_{ad} of adjacent non-filament sites which is given by $n_{ad} = 4$ and $n_{ad} = 3$ for filaments in solution and filaments immobilized to a surface, respectively.

At non-filament sites, the motor performs a symmetric random walk and hops to all neighboring sites with probability $1/6$ per time τ. This choice of the hopping rate for unbound motor movements implies that the time scale τ is given by the diffusion coefficient D_{ub} of unbound motors via $\tau \equiv \ell^2/D_{ub}$. If it reaches a filament site, it binds to the filament with the sticking probability π_{ad}.

The behavior at the filament end has to be specified separately. We consider two possibilities: (i) active unbinding of motors at the filament end where motors at the last filament site make a forward step with probability α as at the other filament sites, but the latter step brings them to the forward non-filament neighbor site, so that their total unbinding probability is $\epsilon_0 + \alpha$, and (ii) thermal unbinding where a motor at the last filament site does not make

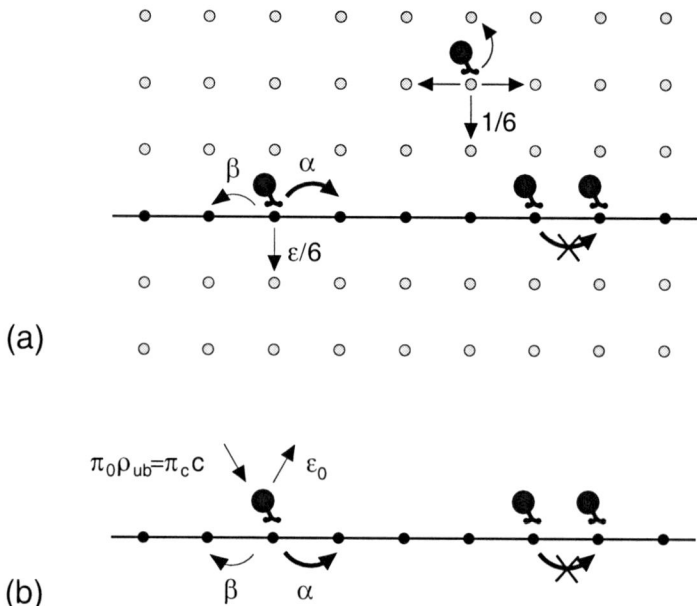

Fig. 1. Lattice model for molecular motor traffic: (a) Molecular motors step in a biased fashion along a filament (black line). With probability $\epsilon/6$, a motor unbinds from the filament by stepping to an adjacent non-filament site. Unbound motors perform symmetric random walks and, when reaching a filament site, rebind to it with probability π_{ad}. Mutual exclusion implies that motors cannot step to lattice sites that are already occupied by another motor. (b) In some situations, the unbound motor density ρ_{ub} (or the corresponding concentration c) can be considered as constant. In that case, bound motors move along the filament as in (a) and unbind from it with probability $\epsilon_0 = n_{\mathrm{ad}}\epsilon/6$. Binding of a motor to an empty filament site occurs with probability $\pi_0\rho_{\mathrm{ub}} = n_{\mathrm{ad}}\pi_{\mathrm{ad}}\rho_{\mathrm{ub}}/6$ which can also be expressed as $\pi_c c$ as discussed in section 4.

a forward step, but has an increased waiting probability $\gamma' = \gamma + \alpha$. In that case, unbinding occurs with probability ϵ_0.

The hopping rates can be chosen in such a way that one incorporates the measured transport properties of single motors such as velocity, diffusion coefficient, and average walking distance before unbinding from the filament [4, 10].

Finally, motor-motor interactions can easily be incorporated into these models. In the following we mainly consider the mutual exclusion of motors from lattice sites. Exclusion is most important at filament sites (since the motors are strongly attracted to these sites), but in principle also applies to unbound motors. In the last section of this article, we consider cooperative binding of motors to the filament. In that case, the binding and unbinding probabilities depend on the occupation of the nearest neighbor filament sites.

When considering many-motor systems, one is often interested in the motor densities and currents profiles rather than the single-motor trajectories. The quantities of main interest are then the bound and unbound motor densities ρ_b and ρ_{ub}. If gradients of the unbound motor density along the direction parallel to a filament can be neglected – either because unbound diffusion is very fast or because the space available for unbound diffusion is large, so that motors remain unbound for a long time before rebinding to the filament – the unbound density can be treated as constant. In that case, one obtains a one-dimensional model for the filament which is coupled to a reservoir of unbound motors with constant motor density as studied in Refs. [7, 12, 13, 15, 17]. Per unit time τ, a motor on the filament unbinds with probability ϵ_0 and binding of a motor from the reservoir to an empty lattice site occurs with probability $\pi_0 \rho_{ub} \equiv n_{ad} \pi_{ad} \rho_{ub}/6$, as shown in Fig. 1(b). This situation will be discussed in section 4.

3 Motor Traffic in Tube-Like Compartments

The motor traffic through tube-like compartments in which one or several filaments are aligned parallel to the cylinder axis represents a simple system which mimics the transport in axons. We have studied tube-like systems with various kinds of boundary condition: closed systems [4, 9, 11], periodic boundary conditions [6], open boundaries coupled to motor reservoirs [6], and half-open systems [11].

The simplest case is given by periodic boundary conditions which can be solved exactly [6]. In this case, the stationary probability distribution is given by a product measure; the bound and unbound motor densities are constant and satisfy the radial equilibrium condition

$$\pi_{ad}\rho_{ub}(1-\rho_b) = \epsilon\rho_b(1-\rho_{ub}) \approx \epsilon\rho_b, \tag{1}$$

where the last approximation usually holds under experimentally accessible conditions, where the unbound density is small, but the bound motor density can be of the order of one motor per binding site. The bound motor current is given by $J = v_b\rho_b(1-\rho_b)$ with the bound motor velocity v_b. As a function of the bound motor density or of the total number N of motors within the tube, it exhibits a maximum and decreases for high motor densities due to motor jamming as shown in Fig. 2(a).

If the tube is coupled to motor reservoirs at its orifices, the motor traffic exhibits boundary-induced phase transitions related to those of the one-dimensional asymmetric simple exclusion process (ASEP) [6]. As for the ASEP, a low-density, high-density and maximal-current phase are present, and correspond to situations where the bottleneck which limits the transport is given by the left boundary, the right boundary or the interior of the tube, respectively. In all three phases, the motor densities are approximately constant and satisfy radial equilibrium sufficiently far from the boundaries. The

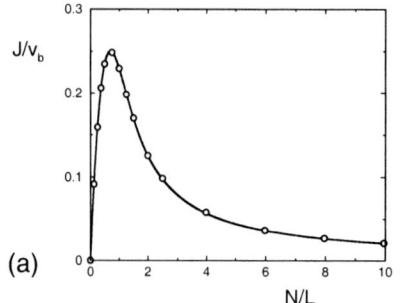
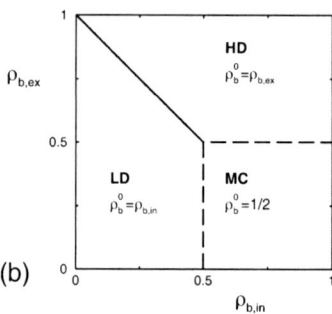

Fig. 2. (a) Bound motor current J for a filament in a tube with periodic boundary conditions as a function of the number N/L of motor particles per length within the tube. (b) Phase diagram for an open tube coupled to motor reservoirs with densities $\rho_{b,in}$ and $\rho_{b,ex}$ at the left and right end of the tube.

location of the transition lines within the phase diagram is quite sensitive to the precise choice of the boundary conditions and can be shifted by tuning the model parameters. A particularly simple case is obtained if we impose radial equilibrium at the boundaries. In this case, the phase diagram, which is shown in Fig. 2(b), is independent of the motor transport properties and the geometric parameters of the tube, and the phase diagram corresponds exactly to the ASEP phase diagram.

4 Traffic Jams on Filaments in Contact with a Large Motor Reservoir

From an experimental point of view, the simplest system, for which one can study molecular motor traffic, is given by one or several immobilized filaments which are in contact with a solution with a certain motor concentration. For typical in vitro systems, unbound motor diffusion is very fast and the space available for unbound diffusion is large, so that we can describe the unbound motors by a constant density ρ_{ub}. In the following, we will use dimensionful quantities with units typically used by the experimentalist, and therefore characterize the unbound motors by the concentration c, which is typically in the nano-molar range, rather than by the local volume fraction ρ_{ub}. In these units, the rate for the binding of an unbound motor to an empty filament site is given by $\pi_c c$ where π_c is the second-order binding rate. It is related to the binding rate in density units via $\pi_c c = \pi_0 \rho_{ub}$ with $\pi_0 \equiv n_{ad} \pi_{ad}/6$ and is most conveniently expressed in terms of the dissociation constant $K_d \equiv \epsilon_0/\pi_c$ which has the dimension of concentration and is typically of the order of ~ 100 nM.

   If the filament is long compared to the motor walking distance, the bound motor density is constant except for the regions close to the filament end and given by the equilibrium of the binding/unbinding dynamics, $\rho_b^{(0)} = c/(K_d + c)$ as shown in Fig. 3(a) and (b) for thermal and active unbinding at the filament end, respectively, using parameters for conventional kinesin. Likewise, the current along this part of the filament is given by $J_0 = v_b \rho_b^{(0)}(1 - \rho_b^{(0)})$. If the motors unbind thermally at the filament end, a (rather short) traffic jam forms at the filament end, where the motors accumulate. Note that no jam is obtained if the motors unbind actively as shown in Fig. 3(b).

The length of the jam region can be defined as $L_* \equiv L - x_*$ with the filament length L and the position x_* of the jam end, where the density starts to deviate from the equilibrium value $\rho_b^{(0)}$. An estimate of L_* can be obtained from the balance of currents

$$J_0 - J_{\text{end}} = \epsilon_0 \int_{L-L_*}^{L} dx \left[\rho_b - \frac{c}{K_d}(1 - \rho_b) \right], \tag{2}$$

where J_{end} is the forward current at the last filament site. Eq. (2) leads to the jam length

$$L_* = \Delta x_b \frac{J_0 - J_{\text{end}}}{v_b} \left[\bar{\rho}_{b,\text{jam}} - \frac{c}{K_d}(1 - \bar{\rho}_{b,\text{jam}}) \right]^{-1}, \tag{3}$$

where $\bar{\rho}_{b,\text{jam}}$ is the average bound density in the jam region and Δx_b is the walking distance of the motors as given by $\Delta x_b \equiv v_b/\epsilon_0$.

For thermal unbinding of motors at the filament end, $J_{\text{end}} = 0$. If we estimate the density within the jam by the maximal value, $\bar{\rho}_{b,\text{jam}} \simeq 1$, Eq. (3) leads to

$$L_* \simeq \Delta x_b \frac{J_0}{v_b} \leq \Delta x_b/4 \tag{4}$$

for the jam length in agreement with simulations which show that $L_* \simeq \Delta x_b J/v_b$ with a prefactor close to one.

This estimate shows that, for filaments in contact with a solution with constant motor concentration, the jam length L_* is of the order of the walking distance Δx_b. Longer jam lengths can arise (i) if the unbinding rate ϵ decreases with increasing density or (ii) if a gradient in the concentration of unbound motors is build up [4, 9] which increases binding to the filament in the jam region and thus also increases the last term of Eq. (3).

5 Geometry-Enhanced Traffic Jams in Closed Compartments

If filaments are embedded into closed compartments, the motor current along these filaments leads to the build-up of density gradients within these compartments [4, 9, 11]. These gradients are particularly pronounced in tube-like compartments where all filaments are aligned in parallel and with the same

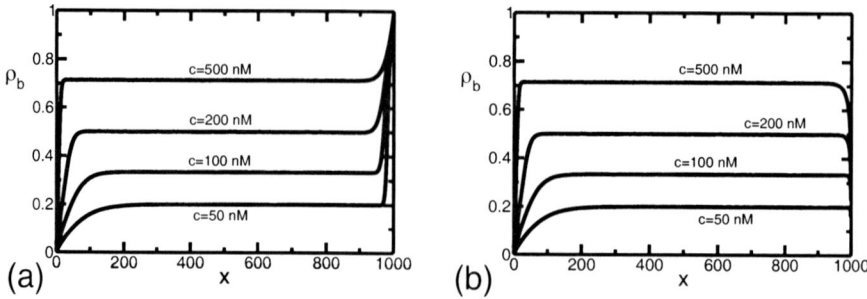

Fig. 3. Profiles of the bound motor density ρ_b on a filament in contact with a solution with constant unbound motor concentration c as a function of the coordinate x parallel to the filament for (a) thermal and (b) active unbinding at the filament end. Note that a traffic jam only occurs for thermal unbinding at the filament end and that this jam is rather short, of the order of the walking distance. The parameters are as appropriate for kinesin, $\epsilon_0 = 1/s$, $K_d = 200\text{nM}$, $v_b = 1\mu m/s$, and for a microtubule of length $8\mu m$.

orientation along the tube axis. For low motor densities, the motors are essentially localized at that end of the tube towards which their active movements are directed. If exclusion can be neglected, the bound and unbound motor densities decrease exponentially if one moves away from this tube end. The length scale ξ of the exponential decrease is given by $\xi = D_{ub}\phi\epsilon/(v_b\pi_{ad})$, the ratio of the distance unbound motors diffuse before rebinding and the walking distance of bound motors. In this expression, v_b and D_{ub} are the bound velocity and the unbound diffusion coefficient of the motors, respectively, and ϕ is the cross-section of the tube. A constant unbound motor density is a good approximation if $\xi \gg L$, i.e., for large unbound diffusion coefficients D_{ub} and for large tube radii.

If the overall motor density is increased in these systems, the region in which the motors are localized develops into an extended crowded domain, see Fig. 4. The length L_* of this domain defines the jam length for these systems. In contrast to the systems discussed in the previous section, the jam length can be larger than the walking distance and increases with increasing overall motor concentration until the crowded domain spreads over the full tube length. In this crowded domain, the density profiles can approximately be described by local radial equilibrium. For the case of a half-open tube, which is very similar to the closed tube, but more easily accessible to analytical methods, the latter approximation shows that the jam length scales essentially as $L_* \sim 1/v_b$ [11] rather than $L_* \sim \Delta x_b \sim v_b$ as for a filament in contact with a constant unbound motor density. The jam length is given by

$$L_* = \frac{\phi D_{ub}\pi_{ad}}{v_b\epsilon}G(\epsilon/\pi_{ad}, \rho_{b,in}), \tag{5}$$

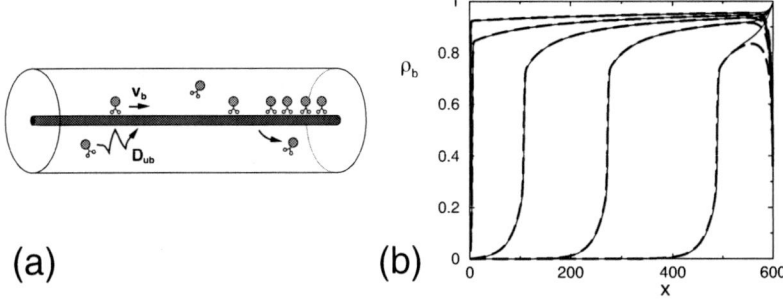

ρ_b

(a)

(b)

Fig. 4. (a) Motor traffic within a closed tube. The current of bound motors which move along a filament with velocity v_b is balanced by a diffusive current of unbound motors which diffuse back with the diffusion coefficient D_{ub}. (b) Corresponding profiles of the bound motor density ρ_b as a function of the coordinate x along the filament. A traffic jam domain at the right end of the tube builds up both for thermal and active unbinding of motors from the 'last' filament site (solid and dashed lines, respectively). With increasing overall motor concentration, the crowded domain spreads to the left.

where G is a function of the ratio of the unbinding and binding probabilities and of the bound motor density $\rho_{b,in}$ in the reservoir to which the tube is coupled at its open end [11]. If the boundary density is sufficiently close to one, G behaves as $G \approx -\ln(1 - \rho_{b,in})$ and the jam length diverges logarithmically with $1 - \rho_{b,in}$. For the closed tube, G is determined by an integral constraint which fixes the total number of motors within the tube.

In addition, the traffic jam is present for both thermal and active unbinding at the filament end [9] as shown in Fig. 4(b). This means that the crowded domains are due to a combination of the motor behavior at the last filament site and the motor accumulation in the region close to the filament end. The latter accumulation is strongly geometry-dependent.

We have also studied centered or aster-like filament systems [9]. In this case, the accumulation of motors in the center of an aster is much weaker than in tube-like systems and, in fact, determined by a power law rather than by an exponential. As for filaments in contact with a reservoir with constant unbound motor density, traffic jams are obtained only for thermal unbinding at the filament end, but not for active unbinding. In addition, when the overall motor density is increased, the traffic jams remain short in this case. The main effect of an increase in the overall motor density is a flattening of the density profile.

6 Symmetry Breaking and Traffic Lanes in Systems with Two Motor Species

Each molecular motor moves either towards the plus- or towards the minus-end of the corresponding filament, but different types of motors move into opposite directions along the same filament. In this situation, cooperative binding of the motors to the filament – in such a fashion that a motor is more likely to bind and less likely to unbind next to a bound motor moving in the same direction, while it is less likely to bind and more likely to unbind next to a motor with opposite directionality – leads to spontaneous symmetry breaking [7]: If the motor–motor interactions, which we characterize by a single interaction parameter q, are stronger than a certain critical interaction strength q_c, one motor species occupies the filament, while the other one is largely excluded from it. This symmetry breaking has been found both for tube-like compartments with periodic boundary conditions and for systems with a constant unbound motor density. In the latter case, symmetry breaking occurs, independent of the choice of the boundary conditions provided that the system size or the filament length is large compared to the motors' walking distance. Note that, in contrast to the previously reported example for symmetry breaking in a driven diffusive system, the 'bridge model' [31], the symmetry breaking here is not boundary-induced.

Symmetry breaking has two interesting consequences. First, it implies that for $q > q_c$ there is a discontinuous phase transition if the relative concentrations of the two motor species are varied. This transition is accompanied by hysteresis, which is again not boundary-induced, in contrast to the hysteresis which was reported recently for another driven diffusive system [32]. Second, if several filaments are aligned in parallel and with the same orientation, this symmetry breaking leads to the spontaneous formation of traffic lanes for motor traffic with opposite directionality [7].

7 Concluding Remarks

The traffic phenomena in systems with many molecular motors can be described by stochastic lattice gas models which are similar to asymmetric exclusion processes, but have the additional property that the motors can unbind from the filamentous track and diffuse in the surrounding fluid. These systems exhibit a variety of cooperative phenomena and, in addition to their importance for our understanding of the traffic within cells and for prospective applications in nanotechnology, provide promising model systems for the experimental study of driven diffusive systems.

References

1. J. Howard: *Mechanics of Motor Proteins and the Cytoskeleton* (Sinauer Associates, Sunderland 2001)
2. M. Schliwa, editor: *Molecular motors* (Wiley-VCH, Weinheim 2003)
3. M. Schliwa, G. Woehlke: Nature **422**, 759 (2003)
4. R. Lipowsky, S. Klumpp, T. M. Nieuwenhuizen: Phys. Rev. Lett. **87**, 108101 (2001)
5. T. M. Nieuwenhuizen, S. Klumpp, R. Lipowsky: Europhys. Lett. **58**, 468 (2002)
6. S. Klumpp, R. Lipowsky: J. Stat. Phys. **113**, 233 (2003)
7. S. Klumpp, R. Lipowsky: Europhys. Lett. **66**, 90 (2004)
8. T. M. Nieuwenhuizen, S. Klumpp, R. Lipowsky: Phys. Rev. E **69**, 061911 (2004)
9. S. Klumpp, T. M. Nieuwenhuizen, R. Lipowsky: Biophys. J. **88**, 3118 (2005)
10. R. Lipowsky, S. Klumpp: Physica A **352**, 53 (2005)
11. M. J. I. Müller, S. Klumpp, R. Lipowsky: J. Phys.: Condens. Matter 17, S3839 (2005)
12. A. Parmeggiani, T. Franosch, E. Frey: Phys. Rev. Lett. **90**, 086601 (2003)
13. M. R. Evans, R. Juhász, L. Santen: Phys. Rev. E **68**, 026117 (2003)
14. V. Popkov, A. Rákos, R. D. Willmann, A. B. Kolomeisky, G. M. Schütz: Phys. Rev. E **67**, 066117 (2003)
15. G. Klein, K. Kruse, G. Cuniberti, F. Jülicher: Phys. Rev. Lett. **94**, 108102 (2005)
16. C. M. Arizmendi, H. G. E. Hentschel, F. Family: Physica A **356**, 6 (2005)
17. K. Nishinari, Y. Okada, A. Schadschneider, D. Chowdhury: Phys. Rev. Lett. **95**, 118101 (2005)
18. S. Katz, J. L. Lebowitz, H. Spohn: J. Stat. Phys. **34**, 497 (1984)
19. J. Krug: Phys. Rev. Lett. **67**, 1882 (1991)
20. A. B. Kolomeisky, G. M. Schütz, E. B. Kolomeisky, J. P. Straley: J. Phys. A: Math. Gen. **31**, 6911 (1998)
21. Y. Kafri, E. Levine, D. Mukamel, G. M. Schütz, J. Török: Phys. Rev. Lett. **89**, 035702 (2002)
22. K. Nagel, M. Schreckenberg: J. Phys. I France **2**, 2221 (1992)
23. D. Chowdhury, L. Santen, A. Schadschneider: Phys. Rep. **329**, 199 (2000)
24. S. Klumpp, R. Lipowsky: Proc. Natl. Acad. Sci. USA 102, 17284 (2005)
25. I. Derényi, T. Vicsek: Phys. Rev. Lett. **75**, 374 (1995)
26. I. Derényi, A. Ajdari: Phys. Rev. E **54**, R5 (1996)
27. Y. Aghababaie, G. I. Menon, M. Plischke: Phys. Rev. E **59**, 2578 (1999)
28. S. Konzack: Funktion des Kinesin Motorproteins KipA bei der Organisation des Mikrotubuli-Cytoskeletts und beim polaren Wachstum von Aspergillus nidulans. Ph.D. thesis, Universität Marburg (2004)
29. S. Konzack, P. E. Rischitor, C. Enke, R. Fischer: Mol. Biol. Cell **16**, 497 (2005)
30. C. Leduc, O. Campàs et al.: Proc. Natl. Acad. Sci. USA **101**, 17096 (2004)
31. M. R. Evans, D. P. Foster, C. Godrèche, D. Mukamel: Phys. Rev. Lett. **74**, 208 (1995)
32. A. Rákos, M. Paessens, G. M. Schütz: Phys. Rev. Lett. **91**, 238302 (2003)

Stochastic Modelling and Experiments on Intra-Cellular Transport of Single-Headed Molecular Motors

Katsuhiro Nishinari[1], Yuko Kanayama[1], Yasushi Okada[2], Philip Greulich[3], Andreas Schadschneider[3], and Debashish Chowdhury[4]

[1] Department of Aeronautics and Astronautics, Faculty of Engineering, University of Tokyo, Hongo, Bunkyo-ku, Tokyo 113-8656, Japan.
[2] Department of Cell Biology and Anatomy, Graduate School of Medicine University of Tokyo, Hongo, Bunkyo-ku, Tokyo 113-0033, Japan.
[3] Institut für Theoretische Physik, Universität zu Köln D-50937 Köln, Germany
[4] Department of Physics, Indian Institute of Technology, Kanpur 208016, India.

Summary. We develop a theoretical model of intra-cellular transport by mutually interacting molecular motors KIF1A that takes into account the hydrolysis of ATP explicitly. A remarkable feature of this model is that all its parameters can be directly determined from experimental data. Our results are in excellent quantitative agreement with the empirical data, and we also provide experimental evidence for the existence of domain walls in our in-vitro experiment.

1 Introduction

Active transportation of mitochondria and vesicles is made possible by motor proteins, like kinesin and dynein, which move on filamentary tracks called microtubules (MT) [1]. It is quite important to study the collective behaviour of motors because their malfunction may be related to some diseases, e.g., Alzheimer's disease [2]. Therefore a fundamental understanding of the collective transportation is expected to help in the control and cure of such diseases.

The molecular motors are regarded as self-driven particles [3], like vehicles moving on a road, which move on MT faster than diffusion by using the fuel of adenosine triphosphate (ATP) hydrolysis. Some of the most recent theoretical models of interacting molecular motors [4–7] utilize this similarity between molecular motor traffic on MT and vehicular traffic on highways [8]. In those models the dynamics is essentially an extension of that of the asymmetric simple exclusion processes (ASEP) [9] and additionally includes Langmuir-like kinetics of adsorption and desorption of the motors.

In reality, a motor protein is an enzyme whose mechanical movement is regulated by its biochemical cycle. Thus recently we have proposed a Brownian ratchet model that includes the biochemical cycle of ATP hydrolysis [10], which is not explicitly taken into account in the earlier models. In the model we specifically consider the *single-headed* kinesin motor, KIF1A [11] and try to make detailed *quantitative* predictions which can be tested experimentally. We have shown that, in the low-density regime, where inter-motor interactions are rare, predictions of the proposed model are in excellent quantitative agreement with the corresponding results obtained in laboratory experiments on single KIF1A motors. Moreover, the spatio-temporal organization of the motors predicted by the same model in the high-density limit is also in qualitative agreement with the corresponding experimental observations [10]. In this paper we show the detailed estimations of parameters used in this model by using experimental results [11–13], and make extensive simulations in the biologically-admissible range of these parameters.

2 A Model of Two Mechanical States of KIF1A

A single protofilament of MT is modelled by a one-dimensional lattice of L sites each of which corresponds to one KIF1A-binding site on the MT; the lattice spacing is equivalent to 8 nm which is the separation between the successive binding sites on a MT [1].

Four biochemical states are involved in each elementary cycle of the molecular motor (Fig. 1), that is, bare kinesin (K), kinesin bound with ATP (KT), kinesin bound with the products of hydrolysis, i.e., adenosine diphosphate (ADP) and phosphate (KDP), and, finally, kinesin bound with ADP (KD) after releasing phosphate. Both K and KT bind firmly to the MT. KDP has a very short lifetime and the release of phosphate, i.e., transition from KDP to KD, triggers the detachment of kinesin from MT; KD is also bound to the MT, but can execute Brownian motion along the track. Finally, KD releases ADP at the next binding site on the MT utilizing a Brownian ratchet mechanism, and thereby returns to the state K. Thus, from the mechanical point of view we can distinguish two states of KIF1A during the cycle (Fig. 1), i.e., the rigorously bound state ('state 1') and the Brownian moving state on MT ('state 2'). Thus it is natural to consider that each kinesin is represented by a particle with two possible mechanical states labelled by the indices 1 and 2 which capture the *rigorously* bound and Brownian moving states of KIF1A, respectively.

Moreover, attachment of a motor to the MT occurs stochastically whenever a binding site on the latter is empty. Detachment of a motor happens stochastically only at the transition from the state 1 to state 2, which is confirmed by a recent experiment [14]. Thus, each of the lattice sites can be in one of three possible allowed states: empty (denoted by 0), occupied by a kinesin in state 1, or occupied by a kinesin in state 2.

For the dynamical evolution of the system, one of the L sites is picked up randomly and updated according to the rules given below together with the corresponding probabilities:

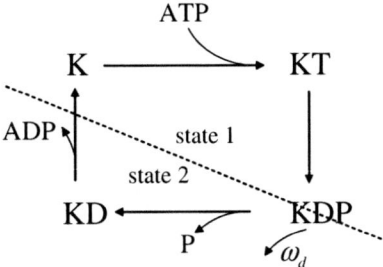

Fig. 1. The biochemical cycle of a single KIF1A motor. Two mechanical states can be distinguished, as shown by the broken line. State 1 is a rigorously bound state and in state 2 Brownian motion is possible.

$$\text{Attachment}: \quad 0 \to 1 \ \text{with} \ w_a dt \tag{1}$$

$$\text{Detachment}: \quad 1 \to 0 \ \text{with} \ w_d dt \tag{2}$$

$$\text{Hydrolysis}: \quad 1 \to 2 \ \text{with} \ w_h dt \tag{3}$$

$$\text{Ratchet}: \quad \begin{cases} 2 \to 1 \ \text{with} \ w_s dt \\ 20 \to 01 \ \text{with} \ w_f dt \end{cases} \tag{4}$$

$$\text{Brownian motion}: \quad \begin{cases} 20 \to 02 \ \text{with} \ w_b dt \\ 02 \to 20 \ \text{with} \ w_b dt \end{cases} \tag{5}$$

The ends of the MT protofilament are known to have a structural conformation different from that in its middle region, and the probabilities of detachment and attachment at the two ends of the MT may be different from those at any bulk site. We choose α and δ, instead of w_a, as the probabilities of attachment at the left and right ends, respectively. Similarly, we take γ_1 and β_1, instead of w_d, as probabilities of detachments at the two ends. Finally, γ_2 and β_2, instead of w_b, are the probabilities of exit of the motors through the two ends by random Brownian movements. Note that the rate constants w_f, w_s and w_b are strictly related with the corresponding physical processes in the Brownian ratchet mechanism of a single KIF1A motor [10].

Let us denote the probabilities of finding a KIF1A molecule in the states 1 and 2 at the lattice site i at time t by the symbols r_i and h_i, respectively. In mean-field approximation the master equations for the dynamics of motors in the bulk of the system are given by

$$\frac{dr_i}{dt} = w_a(1 - r_i - h_i) - w_h r_i - w_d r_i + w_s h_i + w_f h_{i-1}(1 - r_i - h_i), \tag{6}$$

$$\frac{dh_i}{dt} = -w_s h_i + w_h r_i - w_f h_i(1 - r_{i+1} - h_{i+1})$$
$$- w_b h_i(2 - r_{i+1} - h_{i+1} - r_{i-1} - h_{i-1}) + w_b(h_{i-1} + h_{i+1})(1 - r_i - h_i) \tag{7}$$

The corresponding equations for the boundaries, which depend on the rate constants α, δ, γ_i and β_i for entry and exit, are similar.

3 Estimation of Parameters

From experimental data [11–13], good estimates for the parameters of the suggested model can be obtained. We will assume that one timestep corresponds to 1 ms. From the experimental observations of Brownian ratchet motion we have $\omega_f/\omega_s \simeq 3/8$, and the rate of releasing ADP of a single motor is given by $\omega_s + \omega_f \simeq 0.2$ ms^{-1} This gives the individual estimates $\omega_s \simeq 0.145$ ms^{-1} and $\omega_f \simeq 0.055$ ms^{-1}. The detachment rate is given by $\omega_d \simeq 0.0001$ ms^{-1}, which is found to be independent of the kinesin population. On the other hand, the attachment rate depends on the concentration C (in M) of the kinesin motors. The equilibrium constant for attachment and detachment is $\omega_d/(\omega_a/C) = 10$nM, we have $\omega_a = 10^7$ C/M·s. In typical eucaryotic cells *in-vivo* the kinesin concentration C can vary between 10 and 1000 nM. Therefore, the allowed range of ω_a is 0.0001 ms$^{-1} \leq \omega_a \leq 0.01$ ms^{-1}. Moreover the experimental data on the Michaelis-Menten type kinetics of hydrolysis [1] suggest that

$$\frac{1}{V} = \frac{1}{V_{\text{max}}}(1 + \frac{K_m}{T})\tag{8}$$

where V is the reaction rate for the hydrolysis, and $1/V_{\text{max}}$ is the shortest reaction time for the hydrolysis which is observed as 9ms. K_m is the rate constant given by 0.1 mM, and T (in mM) is the concentration of ATP. Since the duration of the state 2 is estimated as 5ms, the reation $1/\omega_h + 5 = 1/V$ holds. Thus we have

$$\omega_h^{-1} \simeq \left[4 + 9\frac{0.1}{T}\right]\text{ms}\tag{9}$$

so that the allowed biologically relevant range of ω_h is $0 \leq \omega_h \leq 0.25$ ms^{-1}. Finally the rate ω_b^{-1} must be such that the Brownian diffusion coefficient D is of the order of 72000 nm^2/s; using the relation $\omega_b \sim D/(8\text{nm})^2$, we get $\omega_b \simeq 1.125$ ms^{-1}. Note that in [12] the Brownian diffusion coefficient is of the order of 40000 nm^2/s, but this must be rescaled to 40000 × 9(total time of a hydrolysis) / 5(time in the state 2 in a hydrolysis) = 72000 in our simulation because only the state 2 can move on the MT.

The predictions of the model for the mean speed of the kinesins etc. are in excellent agreement with single-molecul experiments [10].

4 Simulations and Domain Wall Formation

In contrast to the phase diagrams in the $\alpha - \beta$-plane reported in earlier investigations [4, 5, 7], we have determined the phase diagram of our model in the $\omega_a - \omega_h$ plane by carrying out extensive computer simulations for realistic parameter values of the model with open boundary conditions (Fig. 2). The phase diagram shows the strong influence of hydrolysis on the spatial distribution of the motors along the MT. For very low ω_h no kinesins can exist in state 2; the kinesins, all of which are in state 1, are distributed rather

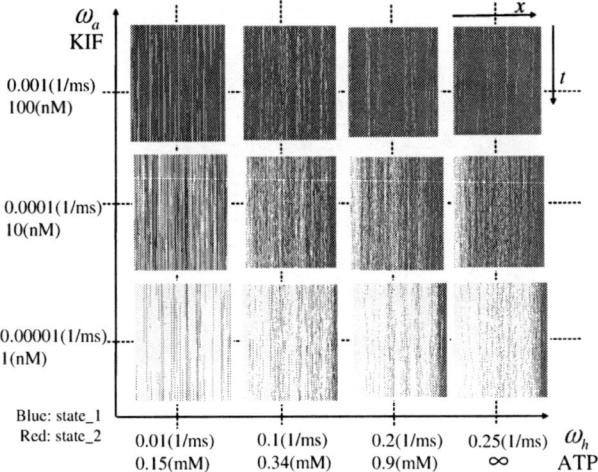

Fig. 2. Phase diagram of the model in the $\omega_h - \omega_a$ plane, with the corresponding values for ATP and KIF1A concentrations given in brackets. These quantities are controllable in experiment. The boundary rates are $\alpha = \omega_a, \beta_{1,2} = \omega_d, \gamma_{1,2} = \delta = 0$. We see the formation of an immobile shock, whose position depends on both ATP and KIF1A concentrations.

homogeneously over the entire system. In this case the only dynamics present is due to the Langmuir kinetics. Even a small, but finite, rate ω_h is sufficient to change this scenario. In this case both the density profiles of kinesins in state 1 and 2 exhibit a shock. As for the ASEP-like models with Langmuir kinetics [4, 5], these shocks are localized. Moreover we have found that the position of the immobile shock depends on the concentration of the motors as well as that of ATP; the shock moves towards the minus end of the MT with increasing kinesin or ATP concentration (Fig. 2).

5 Experiments on Domain Wall Formation on MT

Finally, we present direct experimental evidence that supports the formation of a shock. Imaging of kinesin motor movement was carried out as described previously [11]. Microtubules labeled with a green fluorescent dye Bodipy (Molecular Probes) were immobilized on the top surface of the flow cell. Recombinant KIF1A protein labeled with a red fluorescent dye AlexaFluor (Molecular Probes) was then introduced to the flow cell at 100 pM concentration along with 2 mM ATP. Our standard motility buffer (imidazole 50 mM, Mg-acetate 5 mM, EGTA 5 mM, K-acetate 50 mM, Triton X-100 1%) was supplemented with paclitaxcel 10 uM, ATP-regeneration system and oxygen scavenger system. Green and red fluorescent images of microtubules and KIF1A kinesins were separately obtained with our custom-made fluorescent microscope and image-intensified CCD camera (Roper).

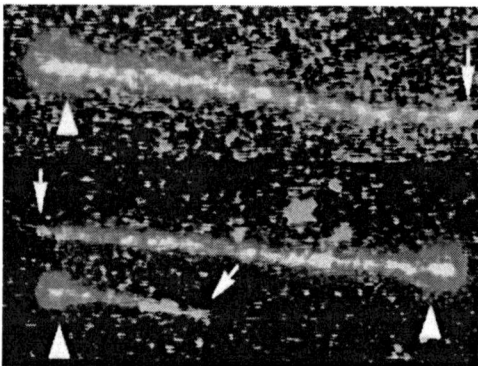

Fig. 3. Formation of comet-like accumulation of kinesin at the end of a MT. Fluorescently labeled KIF1A (red) was introduced to MT (green). Arrows are the minus end and triangles are the plus end of MT. As predicted theoretically, a domain wall is formed on the MT at high concentrations of KIF1A.

The "comet-like structure", shown in Fig. 3, is the collective pattern formed by the red fluorescent labelled kinesins where a domain wall separates the low-density region from the high-density region. As predicted by the model, the position of the domain wall depends on both ATP and KIF1A concentrations. Currently a more quantitative comparison is performed. The results will be presented elsewhere.

References

1. M. Schliwa (ed.), *Molecular Motors*, (Wiley-VCH, 2002).
2. N. Hirokawa and R. Takemura, Trends in Biochem. Sci. **28**, 558 (2003).
3. D. Helbing, Rev. Mod. Phys. **73**, 1067 (2001).
4. A. Parmeggiani, T. Franosch, and E. Frey, Phys. Rev. Lett. **90**, 086601 (2003).
5. M.R. Evans, R. Juhasz, and L. Santen, Phys. Rev. E **68**, 026117 (2003).
6. V. Popkov, A. Rakos, R.D. Williams, A.B. Kolomeisky, and G.M. Schütz, Phys. Rev. E **67**, 066117 (2003).
7. R. Lipowsky, S. Klumpp, and T. M. Nieuwenhuizen, Phys. Rev. Lett. 87, 108101 (2001).
8. D. Chowdhury, L. Santen, and A. Schadschneider, Phys. Rep. **329**, 199 (2000).
9. G.M. Schütz, in C. Domb and J.L. Lebowitz (eds.), *Phase Transitions and Crititcal Phenomena*, Vol. 19 (Academic Press, 2001).
10. K. Nishinari, Y. Okada, A. Schadschneider and D. Chowdhury, Phys. Rev. Lett. **95**, 118101 (2005).
11. Y. Okada and N. Hirokawa, Science **283**, 1152 (1999).
12. Y. Okada and N. Hirokawa, Proc. Natl. Acad.Sci. USA **97**, 640 (2000).
13. Y. Okada, H. Higuchi, and N. Hirokawa, Nature, **424**, 574 (2003).
14. R. Nitta, M. Kikkawa, Y. Okada, and N. Hirokawa, Science **305**, 678 (2004).

Traffic on Bidirectional Ant Trails: Coarsening Behaviour and Fundamental Diagrams

Alexander John[1], Ambarish Kunwar[2], Alireza Namazi[1], Andreas Schadschneider[1], Debashish Chowdhury[2], and Katushiro Nishinari[3]

[1] Institut für Theoretische Physik, Universität zu Köln, Germany
[2] Department of Physics, Indian Institute of Technology, Kanpur, India
[3] Department of Aeronautics and Astronautics, Faculty of Engineering, University of Tokyo, Japan

Summary. We investigate traffic on preexisting ant trails using minimal cellular automaton models. We focus on generic properties of the models like the coarsening of particles and the fundamental diagrams. Crucial differences between the bi- and the unidirectional model are also discussed. However, based on the coarsening behaviour both models belong to the same universality class. Furthermore it will be shown how coarsening in both models can be understood in terms of different kinds of dynamically induced disorder.

1 Introduction

The occurrence of different kinds of spatio-temporal patterns has been investigated in various traffic systems like vehicular traffic [1, 2], pedestrians dynamics [3] or biological transport [4, 5]. In order to reproduce and understand the empirically observed phenomena various model approaches have been proposed. Recently cellular automata models have become popular which allow to capture the most important features of a system in an intuitive way [1, 6].

Apart from specific applications, the models are also interesting from a more theoretical point of view. Surprisingly models introduced to describe traffic on existing uni- and bidirectional ant trails were found to exhibit anomalous features in their fundamental diagrams [7–9]. These are related to the spatio-temporal organization of the ants, as a result of different kinds of coarsening processes [7, 10]. We will show that these processes can be understood as effects of effective disorder. Disorder has already been investigated in various driven systems [11–16], but in our case it is not assigned statically either to a certain particle or a lattice site. Instead the disorder is formed dynamically through the collective motion of the particles themselves.

Starting from a random initial state, the uni- and the bidirectional models will exhibit clustering of particles. Measuring density-density correlation functions will allow us to describe these processes quantitatively in time. The stationary state is characterised by the fundamental diagrams, i.e. the density-dependence of the flux and the average velocity. Each of the diagrams exhibits unusual features on its own in both models. We will show that this is caused by the same mechanisms which have already governed the coarsening process.

A more detailed discussion and comparison with empirically observed phenomena as well as applications to pedestrian dynamics can be found in [9, 17, 18].

2 The Unidirectional Model

The unidirectional model [7] is a generalization of the totally asymmetric simple exclusion process (TASEP) [16, 19] considering ants as particles. In the TASEP particles move forward with rate q to the next site ahead only if this site is empty (hard-core exclusion). The ant trail model takes into account the increase of the walking speed due to the effects of *chemotaxis*, the relevant form of communication between the ants in this context [20]. In addition to hard-core exclusion, another coupling between neighbouring particles is introduced. A moving particle will create a mark (a 'pheromone' in the language of chemotaxis) at the site it leaves. If a second particle tries to hop to that site, its hopping rate is increased to $Q > q$. The pheromone marks have a finite lifetime and 'free' pheromones (i.e. pheromones at sites without a particle) evaporate with rate f (see Fig. 1).

For simplicity, periodic boundary conditions and random sequential dynamics are used as the investigated effects seem to be quite stable against special choice of dynamics [7, 10] or boundary conditions [21].

2.1 Coarsening Behaviour

Starting from an initial state where particles are distributed randomly, the formation of clusters (see Fig. 2) is observed during the time evolution of the

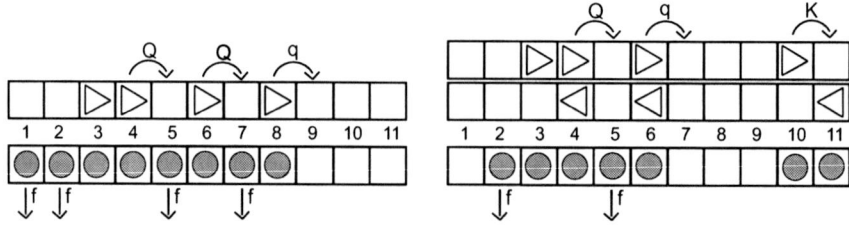

Fig. 1. Illustration of the definitions for the uni- and bidirectional model: ants moving to the right ▷, ants moving to the left ◁, pheromone marks •. Java applets can be found at *www.thp.uni-koeln.de/ant-traffic*

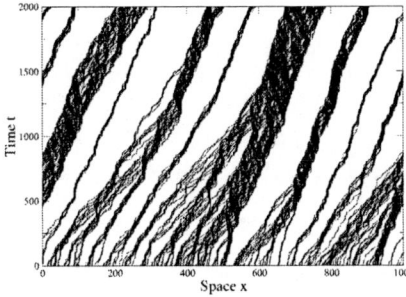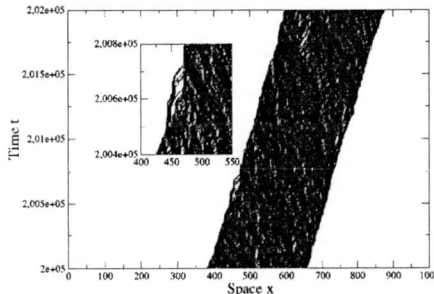

Fig. 2. Space-time plots for the unidirectional model ($Q = 0.9$, $q = 0.2$, $f = 0.002$, $\rho = 0.2$): On the left the formation of small moving clusters can be seen. The plot on the right shows the stationary state.

system. One also observes an increase of the velocity of particles when catching up with preceding ones. The probability that a particle finds a pheromone mark decreases with increasing distance to the preceding particle. Therefore the velocity of a cluster does not depend on its size etc., but only on the distance to the cluster ahead. Thus clusters 'accelerate' with decreasing distance. At later times this process ends up in one single but moving cluster which comprises all particles. This cluster is denoted as 'loose cluster' [8] since it is not compact. Typically it is an alternating sequence of empty and occupied sites, i.e. the cluster is much larger than the number of particles it consists of. In a finite system this loose cluster can resolve due to fluctuations and reform at a different position.

A quantitative characterization of the coarsening process can be obtained from equal-time density-density correlations. Following a method already used in [22] one defines a suitably normalized correlation function

$$C(r,t) = \frac{1}{\rho(1-\rho)}\left(\frac{1}{L}\sum_{i=1}^{L}\langle n(i,t)(n(i+r,t)\rangle - \rho^2\right) \tag{1}$$

which is determined numerically by averaging over different initial conditions. Here $n(i,t) = 0, 1$ is the occupation number of site i at time t and $\rho = N/L$ the particle density determined by the number of particles N and the system length (number of sites) L.

Since no other high density areas outside of the cluster exist, $C(r,t)$ reaches its minimum $C_{\min} = -\frac{\rho}{1-\rho}$ for r larger than the cluster size l. But as $C(r,t)$ is symmetric with respect to $r = L/2$, this method is limited to cluster sizes $l < L/2$.

Starting from a random initial state, at early times only short-ranged correlations exist and $C(r,t) = 0$ for large separations r (Fig. 3). With increasing time, the range of correlations is also increasing. $C(r,t)$ is positive for short distances due to the existence of small clusters. For large distances, $C(r,t)$ becomes negative. The location $R(t)$ of the first zero-crossing of $C(r,t)$ (see

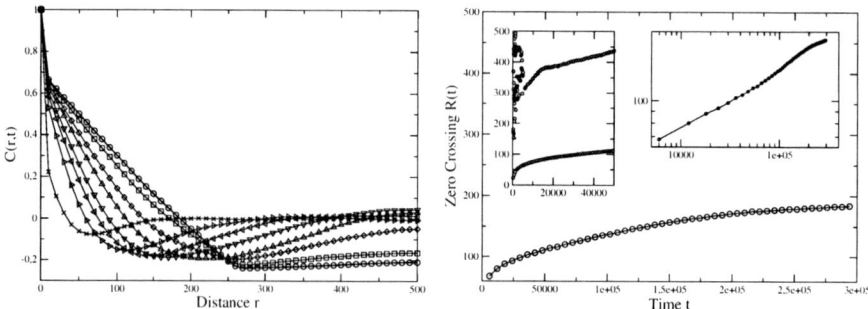

Fig. 3. Density-density correlations $C(r,t)$ of the unidirectional model ($Q = 0.9$, $q = 0.2$, $f = 0.002$, $\rho = 0.2$) at different times $t = 300 \cdot 10^3 (\circ)$, $200 \cdot 10^3 (\square)$, $100 \cdot 10^3 (\diamond)$, $50 \cdot 10^3 (\triangle)$, $20 \cdot 10^3 (\triangledown)$, $10 \cdot 10^3 (\triangleleft)$, $5 \cdot 10^3 (\triangleright)$, $2 \cdot 10^3 (\times)$. The time evolution of $C(r,t)$ can be seen on the left. The right plot shows the location $R(t)$ of the first zero crossing. The insets show the location of the higher zero crossings and a double-logarithmic plot of $R(t)$.

Fig. 3) can be considered as a measure for the average length of clusters at time t.

As $R(t)$ describes the average distance of uncorrelated particles obviously clustering leads to an increase of the range of correlations in time (Fig. 3). Uncorrelated particles can only be found at growing distances $r > R(t)$.

2.2 Fundamental Diagrams

An important quantity for the characterization of the stationary state is the fundamental diagram, i.e. the relation between flux and density, or equivalently, between average velocity and density.

The most surprising feature is the non-monotonic dependence of the velocity on density for small evaporation rates f (see Fig. 4, left) [7, 8]. For larger f the velocity shows a strictly monotonic decrease known e.g. from vehicular traffic [1]. It has its origin in the hindrance effect that each additional particle has on the others. In contrast, for small f the velocity increases sharply with density until it reaches the curve for $f = 0$. This is due to the velocity enhancement through the presence of the pheromones. It only occurs at small values of the evaporation rate since otherwise the lifetime of the trace is not long enough to be felt at small densities (large particle separations). At higher densities the pheromone marks become less important and the dynamics is dominated by the exclusion principle. The velocity is almost constant at small to intermediate densities. The regime of constant velocity resembles to observations made for models with particle-wise disorder [12–14]. In the present case this disorder is not quenched, but can change dynamically due to the absence or presence of pheromone marks. At sufficiently high densities, particle-wise disorder vanishes, leading to the usual fundamental diagram known from the TASEP.

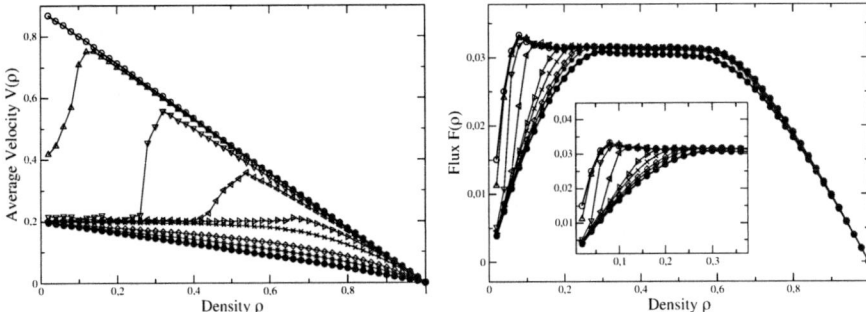

Fig. 4. Velocity-density relation for the uni- (left) and flux-density relation for bidirectional model (right) for $Q = 0.9$, $q = 0.2$, $K = 0.1$ and evaporation probabilities $f = 0(\circ)$, $0.0002(\triangle)$, $0.0008(\triangledown)$, $0.002(\triangleleft)$, $0.008(\triangleright)$, $0.02 (\times)$, $0.08(\diamond)$, $0.2(*)$, $1(\bullet)$.

3 The Bidirectional Model

The bidirectional model for traffic on ant trails [10] can be considered as two *coupled* unidirectional models with particles moving in opposite directions. It consists of two lattices with L sites and N_σ particles ($\sigma = \rightarrow, \leftarrow$) for each direction. The dynamics in each of the lattices is identical to that of the unidirectional model. Only in the case where the corresponding target site of the other lattice is also occupied, the motion occurs with a different probability K (see Fig. 1). This allows to take into account the effects of interactions between oppositely moving particles. By construction the bidirectional model reduces to the unidirectional one if either $N_\rightarrow = 0$ or $N_\leftarrow = 0$. Here we will focus on the symmetric case $N_\rightarrow = N_\leftarrow$, which already shows the generic features.

The choice of K determines the nature of the coupling between the two directions. For $q < Q < K$ the holes (i.e. empty sites) on the lattice in opposite direction lead to slowing down, whereas for $K < q < Q$ this is done by particles. The latter case takes into account the slowing down of counterflowing ants due to the exchange of information [23].

Instead of considering a model with two pheromone trails [17], one for each direction, we consider here a variant where the oppositely moving particles create a common pheromone trail. Here a pheromone mark evaporates with rate f only if both cells are empty (Fig. 1).

3.1 Coarsening Behaviour

Analogous to the unidirectional model, particles are distributed randomly in the initial state. During time-evolution clusters are formed (see Fig. 5) which, in contrast to the unidirectional model, are localized. Due to exchange of particles, some clusters grow whereas others shrink. At later times, one large cluster survives which is localized at a position that only fluctuates slightly. Outside this large cluster smaller ones with shorter lifetimes also exist.

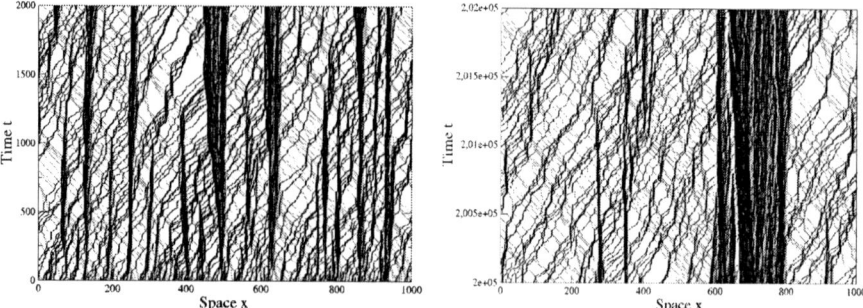

Fig. 5. Space-time plots for the bidirectional model ($Q = 0.9$, $q = 0.2$, $K = 0.1$, $f = 0.002$, $\rho = 0.2$): The left plot shows the formation of small localized clusters out of the random distribution of particles in the initial state. At late times most small clusters have disappeared and one large localized cluster has survived.

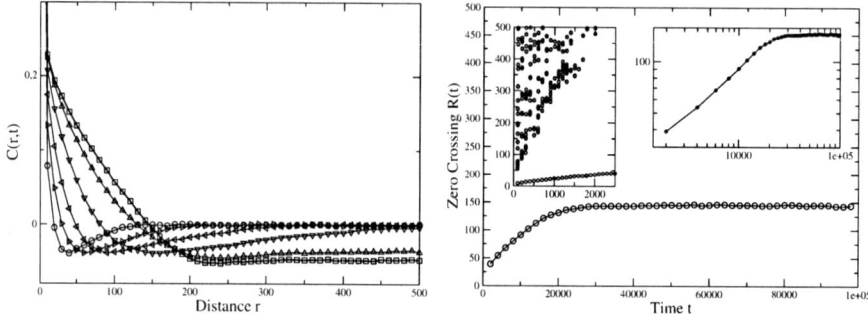

Fig. 6. Density-density correlation function of the bidirectional model ($Q = 0.9$, $q = 0.2$, $K = 0.1$, $f = 0.002$, $\rho = 0.2$) at different times $t = 100 \cdot 10^3(\square)$, $20 \cdot 10^3(\triangle)$, $10 \cdot 10^3(\triangledown)$, $5 \cdot 10^3(\triangleleft)$, $3 \cdot 10^3(\triangleright)$, $2 \cdot 10^3$ ($\circ$). On the left, the time evolution of the density-density correlations is depicted. The right figure shows first zero-crossings vs. time. The insets show later crossings and a double-logarithmic plot of $R(t)$.

For capturing the dynamics of high-density regimes (clusters), second-class particles [11] have already been employed successfully. The dynamics of these particles can be defined in such a way, that they will occupy the boundary areas between the low- and high-density regimes [24]. Here we will again follow the method already used in [22]. The results (see Fig. 6) are very similar to those observed in the unidirectional case discussed in Sec. 2.1.

3.2 Fundamental Diagrams

The generic property of this model can be seen in the behaviour of the flux. Here, at intermediate densities, the flux becomes almost independent of the density and f leading to a characteristic plateau (see Fig. 4, right). This plateau region coincides with the regime of phase separation, namely the localized large cluster. With increasing density, the cluster grows larger. Therefore effectively the extension of the defect formed by particles moving in the opposite direction becomes larger which also explains the small slope of the plateau region. This does not occur in the well-studied models of constant defect length [15], where the flux attains a constant value over intermediate densities. But the value of this constant flux decreases with increasing defect length. This effect can also be seen in our model, since an increasing density also leads to an increase of the defect length.

The defect strength is determined by the difference between the effective unhindered hopping rate q_{eff} and the counterflow exchange rate K. Since q_{eff} decreases with increasing evaporation rate f, this difference is smaller for large f. Therefore the defect strength is smaller and the regime of constant flux starts at higher densities, like in systems with lattice-wise defects [11, 15].

4 Discussion and Summary

The investigation of the spatio-temporal organization of particles in models inspired by uni- and bidirectional ant trails has revealed interesting coarsening processes. Density-density correlation functions were used in order to identify the formation of high-density areas, namely clusters of particles during the early stages of the time evolution.

Coarsening in both models exhibits common features. The growth of the largest cluster in time is described by a power-law, similar to what has been observed in [22, 25]. The exponents for the uni- and bidirectional model appear to be the same, $z = \frac{1}{3}$ at early and $z = \frac{1}{2}$ at intermediate times. Based on this both models belong to the same universality class.

Also the number of clusters depends on $R(t)$. This is indicated by taking into account the higher-order zero-crossings of the density-density correlation function. Using the first zero-crossing one gets the minimum average distance of uncorrelated clusters. This implies the existence of some kind of periodic structure. So in general, zero-crossings can be expected at $R_n = nR < \frac{L}{2}$, with $n \in [1, \frac{L}{2R}]$ with n being interpreted as the number of clusters at a given instance of time.

In the unidirectional model at the end only one "loose" cluster emerges comprising all particles in the system. A main feature in that state is the regime of constant average velocity which is known from particle hopping models with particle-wise disorder [14]. With increasing density, the effective disorder dissolves which then leads to the observed non-monotonicity. For densities close to $\rho = 1$, the stationary state becomes very similar to that of the TASEP.

The main feature of the stationary state in the bidirectional model is the occurrence of a plateau in the fundamental diagram, similar to systems with lattice-wise disorder [15]. However, in the present model the disorder is not static, but created dynamically by the particles moving in the opposite direction.

Acknowledgements

One of the authors (AJ) acknowledges support from the German Academic Exchange Service (DAAD) through a joint Indo-German research project. He thanks R. Gadagkar and D. Chowdhury for kindly providing hospitality during the preparation of this manuscript.

References

1. D. Chowdhury, L. Santen, A. Schadschneider, Phys. Rep. **329**, 199 (2000).
2. D. Helbing, Rev. Mod. Phys. **73**, 1067 (2001).
3. M. Schreckenberg and S.D. Sharma (Eds.), *Pedestrian and Evacuation Dynamics* (Springer, 2001).
4. D. Chowdhury, K. Nishinari, A. Schadschneider, Phase Trans. **77**, 601 (2004).
5. S. Camazine, J.L. Deneubourg, N.R. Franks, J. Sneyd, G. Theraulaz, E. Bonabeau, *Self-organization in Biological Systems* (Princeton University Press, 2001).
6. K. Nagel and M. Schreckenberg, J. Phys. I France **2**, 2221 (1992).
7. D. Chowdhury, V. Guttal, K. Nishinari, and A. Schadschneider, J. Phys. A: Math. Gen. **35**, L573-L577 (2002).
8. K. Nishinari, D. Chowdhury, A. Schadschneider, Phys. Rev. E **67**, 036120 (2003).
9. A. John, A. Schadschneider, D. Chowdhury, and K. Nishinari, J. Theor. Biol. **231**, 279 (2004).
10. A. Schadschneider, D. Chowdhury, A. John, and K. Nishinari: in *Traffic and Granular Flow '03*, S.P. Hoogendoorn, S. Luding, P.H.L. Bovy, M. Schreckenberg, and D.E. Wolf (Eds.) (Springer, 2003).
11. S.A. Janowsky and J.L. Lebowitz, Phys. Rev. A **45**, 618 (1992).
12. M.R. Evans, Europhys. Lett. **36**, 13 (1996); J. Phys. A **30**, 5669 (1997)
13. J. Krug and P.A. Ferrari, J. Phys. A **29**, L465 (1996).
14. J. Krug, Braz. J. Phys. **30**, 97 (2000).
15. G. Tripathy and M. Barma, Phys. Rev. Lett. **78**, 3039 (1997).
16. G.M. Schütz, *Phase Transitions and Critical Phenomena, Vol. 19*, eds. C. Domb and J.L. Lebowitz, (Academic Press, 2000).
17. A. John, A. Kunwar, A. Namazi, D. Chowdhury, K. Nishinari, and A. Schadschneider, in *Pedestrian and Evacuation Dynamics '05* (Springer, 2006).
18. A. Schadschneider, A. Kirchner, and K. Nishinari, Applied Bionics and Biomechanics **1**, 11 (2003)
19. B. Derrida and M.R. Evans: in *Nonequilibrium Statistical Mechanics in One Dimension*, ed. V. Privman (Cambridge University Press, 1997).
20. B. Hölldobler and E.O. Wilson, *The Ants* (Belknap, Cambridge, USA, 1990).
21. A. Kunwar, A. John, K. Nishinari, A. Schadschneider, and D. Chowdhury, J. Phys. Soc. Jpn. **73**, 2979 (2004).
22. D. Chowdhury and R.C. Desai, Eur. Phys. J. B **15**, 375 (2000).
23. M. Burd and N. Aranwela, Insectes Sociaux **50**, 3-8 (2003).
24. A. John, A. Schadschneider, D. Chowdhury, and K. Nishinari, in preparation
25. O.J. O'Loan, M.R. Evans, and M.E. Cates, Phys. Rev. E **58**, 1404 (1998); Europhys. Lett. **42**, 137 (1998).

Phase Diagram of Group Formation in 2-D Optimal Velocity Model

Yuki Sugiyama[1], Akihiro Nakayama[2], and Eiji Yamada[1]

[1] Graduate School of Information Science, Nagoya University, Nagoya 464-8601, Japan

[2] Department of Physics and Earth Sciences, University of Ryukyus, Okinawa 903-0213, Japan

Summary. We present a mathematical model for self-driven particles (2-D Optimal Velocity Model), which can reproduce a big variety of patterns in group formation and phase transition physics, by changing only two control-parameters and the number of particles. We show the formation of several typical patterns and their phase diagrams. The model can be applied to the study of group formation of collective bio-motions such as schools of fish.

1 Introduction

Basic idea and physical interests: The original 1-dimensional Optimal Velocity (OV) model is a mathematical model for traffic flow [1, 2], which succeeds in reproducing the fundamental properties of freeway traffic flow. At the same time, it is one of the simplest dynamical models describing a collective motion of self-driven particles as well. It is introduced as

$$\frac{d^2}{dt^2} x_n(t) = a \left\{ V(\Delta x_n(t)) - \frac{d}{dt} x_n(t) \right\}. \tag{1}$$

x_n is the position of the nth particle, and $\Delta x_n = x_{n+1} - x_n$ is the headway distance. a is a sensitivity constant (inverse of relaxation time). $V(\Delta x_n)$, the so-called OV-function, determines the optimal velocity depending on the distance, which has a form such as $V(\Delta x) = \alpha\{\tanh \beta(\Delta x - b) + c\}$. The model represents a simple problem: if the particles try to adjust their velocity to the optimal velocity, what happens? The answer is that under some conditions the homogeneous movement can not be maintained, but instead a flow of collective patterns appears and preserves its shape.

When we extend the OV model to higher dimensions, the model can be applied to several kinds of collective motions of interacting particles, such as granular flow in liquid, pedestrians, evacuation dynamics and collective bio-motions or group formation of organisms, etc.

General properties of the 2-D OV model: In the 2-dimensional model, the negative sign of attractive force in one dimension is not simply extended to repulsive forces. The two characteristics in the OV model independently appear in 2-D space: the following behavior induced by the attractive force, and the exclusion behavior induced by the repulsive force. A wide variety of models is naturally given only by extending the dimensionality. We can introduce a new parameter (denoted by c in the following), which controls the distance dominated by the attractive or repulsive forces. Thus, the model has one additional control parameter, besides the sensitivity a and the number of particles N (or the average particle-density).

2 Modeling Collective Bio-Motion in 2-D OV Model

Mathematical formulation of the 2-D model: The equation of motion for a particle with the index i is given by

$$\frac{d^2}{dt^2}\mathbf{x}_i(t) = a\left\{\sum_j \mathbf{V}(\Delta\mathbf{x}_{ij}(t)) - \frac{d}{dt}\mathbf{x}_i(t)\right\}, \qquad (2)$$

where bold letters represent two-dimensional vectors. $\mathbf{x}_i = (x_i, y_i)$ and $\mathbf{x}_j = (x_j, y_j)$ are the positions of ith and jth particles, respectively, and $\Delta\mathbf{x}_{ij} = \mathbf{x}_j - \mathbf{x}_i$. $\mathbf{V}(\Delta\mathbf{x}_{ij})$ expresses the interaction between two particles with the following form:

$$\mathbf{V}(\Delta\mathbf{x}_{ij}) = f(r_{ij})(1 + \cos\varphi)\,\mathbf{n}_{ij}, \qquad (3)$$
$$f(r_{ij}) = \alpha\{\tanh\beta(r_{ij} - b) + c\}, \qquad (4)$$

where $r_{ij} = |\mathbf{x}_j - \mathbf{x}_i|$, $\cos\varphi = (x_j - x_i)/r_{ij}$ and $\mathbf{n}_{ij} = (\mathbf{x}_j - \mathbf{x}_i)/r_{ij}$. The strength of the interaction depends on the distance r_{ij} between the ith and jth particles, and on the angle φ between the directions of $\mathbf{x}_j - \mathbf{x}_i$ and the current velocity $\frac{d}{dt}\mathbf{x}_i$. Due to the term $(1 + \cos\varphi)$, a particle receives stimulus more sensitively from the proceeding particles than the following ones. Eq. (4) has the same form as the OV function for the 1-D model for studying all cases in a unified way.

Repulsive/attractive interaction: If we add a constant term $a\mathbf{V}_0$ to the r.h.s. in Eq. (2) and set $c = -1$ (meaning that $f < 0$), the interaction is of repulsive type. $\mathbf{V}_0$ is a constant vector which expresses a "desired velocity". A particle moves with the desired velocity, if it is isolated. The interaction between pedestrians or particles moving in a liquid through a pipe, is described by this case[3]. In general, $\mathbf{V}_0$ is an external flow in the background such as a "tide". In the case $c = 1.0$, the force is attractive for all r_{ij}.

---

[3] The linear stability of this type of model has been analyzed. The phase diagram and the flow-pattern in each phase are shown in [7].

In the case $1 > c > -1$, both repulsive and attractive interactions coexist. Actually, the parameter c controls the border of the regions with repulsive and attractive interactions. For example, for $\alpha = 1/4, \beta = 4, b = 1$ (the values used in our simulations): In the case $c = -0.5$, the force between two particles is repulsive for $r_{ij} < 1.14$ and attractive for $r_{ij} > 1.14$. In the case $c = 0$, the border is 1. In such cases, the model is suitable for collective bio-motions.

Variation in types of scope (interaction): First, the ith particle receives stimuli from neighbor particles within a circle $r_{ij} < R$. The interaction is cut off by setting $\mathbf{V}(\Delta \mathbf{x}_{ij}) = 0$ outside the circle. Under this condition, we study two types of scope (interaction). **Type-1**: a particle receives forces from all particles within its circle. **Type-2**: a particle receives forces from the nearest neighbor particle in each sector around the particle. (We divide the range within its circle into six sectors for simplicity.) Though we concentrate our simulations on Type-1 in this paper, the parameter c brings a big variety in pattern formation.

3 Patterns of Group Formation

In general the homogeneous flow is not stable in 2-dimensional models even for a small number of particles (low density) when an external flow does not exist in the background. In the case with external flow the stability of the homogeneous flow depends on the parameter c. In the simulations in this paper, no external flow exists. We note that if we set the homogeneous flow in the initial condition at a density lower than $1/R$, it is preserved as in Fig. 1(a). It is trivial owing to the cut-off of interaction and the corresponding phase is not interesting. The simulations are performed in a 2-dimensional space with periodic boundary.

Type-1, c=1 (attractive force only): For high density $(> 1/R)$, particles move almost randomly (Fig. 1(b)). Small "platoons" are observed as a group formation. The members of the groups are successively changing.

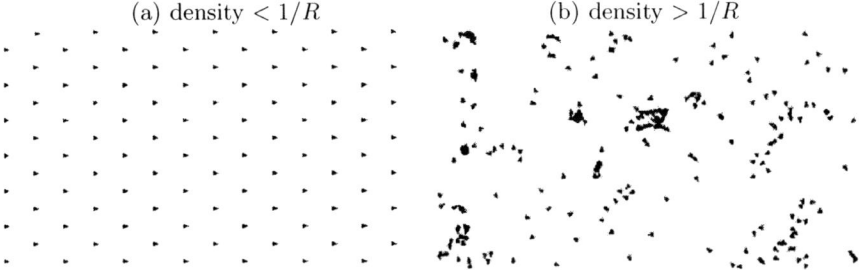

(a) density $< 1/R$ (b) density $> 1/R$

Fig. 1. The patterns of flow in Type-1, $c = 1$. Each triangle represents a particle and its direction of movement.

Type-1, c=-1 (repulsive force only): In this case there are three stable patterns depending on N as shown in the phase diagram in Fig. 2. The region of large $N(> 1/R)$ is divided into three phases. For the region $N < 350 \sim 480$, particles are moving randomly with no group formation (Fig. 2(a)). For the region $360 \sim 480 < N < 550$, particles can not move and stay at each position being uniformly distributed just like a crystal solid (Fig. 2(b)). Moreover, each particle is always changing its direction. For much higher density, $N > 550$, two or three particles form a "bound state" (Fig. 2(c)), and then slowly change their positions. After enough relaxation time, particles form a specific pattern like the "shirakawa dune" seen in a Japanese garden (Fig. 2(d)).

In contrast to the attractive force (the case $c = 1$), the repulsive force (the case $c = -1$) is important for the emergence of a "crystal" phase. While, the attractive force forms a local pattern, such as a "platoon".

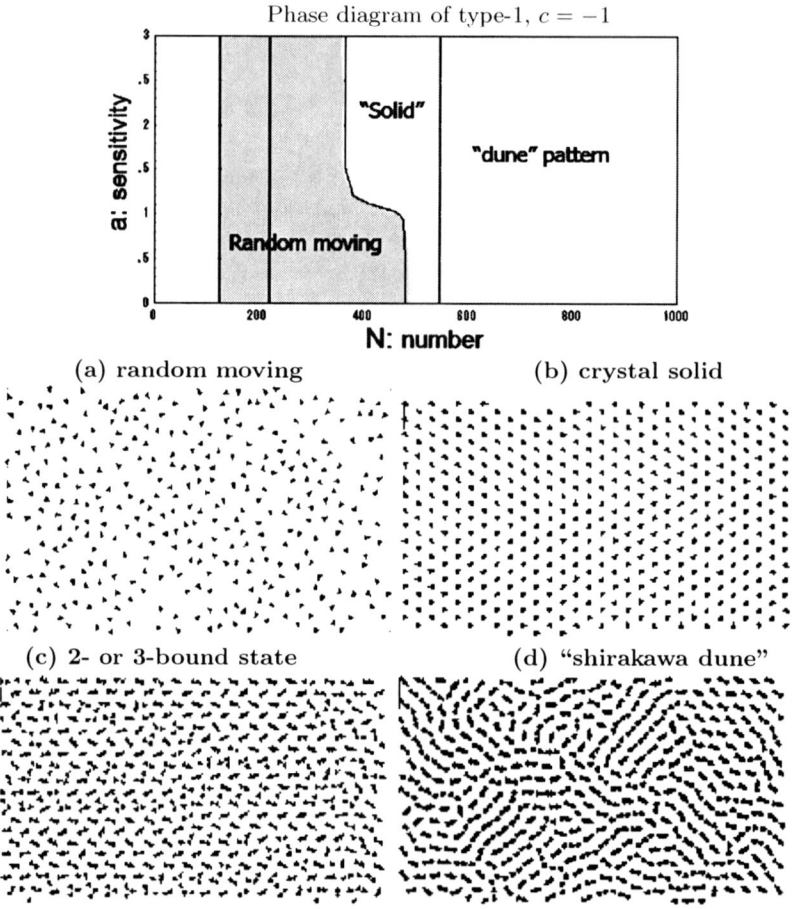

Phase diagram of type-1, $c = -1$

a: sensitivity

"Solid"

"dune" pattern

Random moving

N: number

(a) random moving   (b) crystal solid

(c) 2- or 3-bound state   (d) "shirakawa dune"

Fig. 2. Phase diagram and patterns emerging in type-1, $c = -1$.

Type-1, c=0 (repulsive for $r_{ij} < R/2$; attractive for $R/2 < r_{ij} < R$):
In this case, two stable local patterns of group formation appear. The phases
are characterized by these objects. In contrast to the case $c = -1$, the parameter space (N, a) is divided into three phases depending on a (sensitivity), not
on N (number of particles) as shown in Fig. 3.

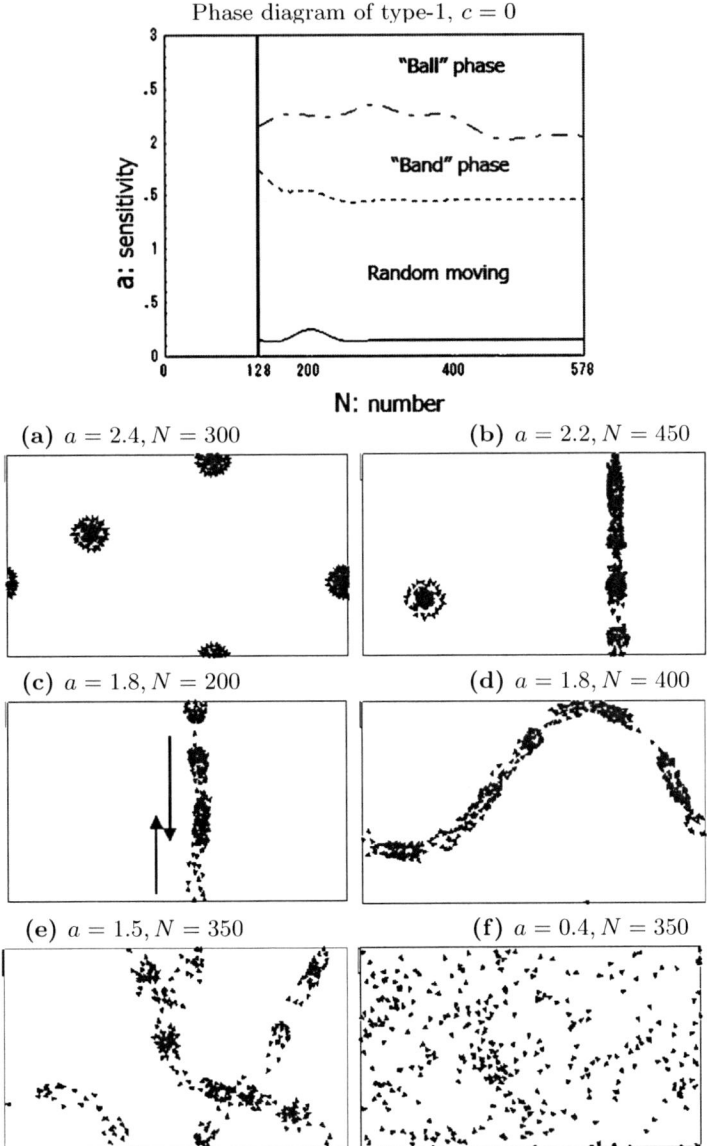

(a) $a = 2.4, N = 300$ (b) $a = 2.2, N = 450$

(c) $a = 1.8, N = 200$ (d) $a = 1.8, N = 400$

(e) $a = 1.5, N = 350$ (f) $a = 0.4, N = 350$

Fig. 3. Phase diagram and patterns emerging in type-1, $c = -1$.

For high sensitivity ($a > 2.3$), the specific local pattern of group formation is emerged. We call it "ball". Particles are actively moving within a ball, which is formed by the non-equilibrium balance of attractive and repulsive interactions. A ball can move slowly without decaying. It has a limit capacity of the number of particles in it. So, as the number of particles N becomes larger, the number of balls increases (Fig. 3(a)).

As the sensitivity becomes lower ($1.5 < a < 2.3$), balls gradually decay, and connect with each other. Thus, particles form a "band". At large N near critical a between the ball and band phases, ball and band coexist (Fig. 3(b)). Inside a band, particles move together as groups in opposite directions (Fig. 3(c)). If N becomes larger in this phase, the length of a band becomes longer (Fig. 3(d)).

The smaller the sensitivity becomes ($a < 1.5$), the weaker the tightness of forming a band. The particles in a band become loosely bounded and a band diffuses and decays (Fig. 3(e)). Finally, at much lower sensitivity no band or ball pattern can be observed and all particles move randomly with no structure (Fig. 3(f)).

4 Summary and Discussion

The repulsive force is important for the emergence of crystal structures, while the attractive force forms a local pattern but of tiny size. We expect that a local pattern with finite size appears in a hybrid case such as $c = 0$. We succeeded in forming "ball" and "band" patterns in Type-1 model. We remark that the another expected patterns like schools of fish are successfully observed in the Type-2 model (nearest neighbor interaction). The patterns of group formation emerging in Types 1 and 2 are quite different.

References

1. M. Bando, K. Hasebe, A. Nakayama, A. Shibata, Y. Sugiyama: Japan J. of Ind. and Appl. Math, **11**, 203 (1994); Phys. Rev. E**51**, 1035 (1995); with K. Nakanishi: J. Phys. I France **5**, 1389 (1995).
2. Y. Sugiyama: In *Traffic and Granular Flow*, D. E. Wolf, M. Schreckenberg A. Bachem (eds.), (World Scientific, 1996) pp. 137-149.
3. H.-S. Niwa: J. Theor. Biol. **171** 123 (1994); **188** 47 (1996)
4. T. Vicsek, A. Czirók, D. Helbing: In *Traffic and Granular Flow '99*, D. Helbing, H.J. Herrmann, M. Schreckenberg, D. E. Wolf (eds.), (Springer, 2000) pp. 147-159 and references therein.
5. Y. Sugiyama, A. Nakayama, K. Hasebe: In *Pedestrian and Evacuation Dynamics* M. Schreckenberg, S.D. Sharma (eds.) (Springer, 2001) pp. 155-160
6. A. Nakayama, Y. Sugiyama: In *Modeling of Complex Systems*, P.L. Garrido, J. Marro (eds), (American Institute of Physics), (2003) pp. 107-110
7. A. Nakayama, K. Hasebe, Y. Sugiyama: Phys. Rev. E **71**, 036121 (2005)

On the Harmonic–Mean Property of Model Dispersive Systems Emerging Under Mononuclear, Mixed and Polynuclear Path Conditions

Adam Gadomski[1], Natalia Kruszewska[1], Marcel Ausloos[2], and
Jakub Tadych[1]

[1] University of Technology and Agriculture, Bydgoszcz PL85-796, Poland
[2] SUPRATECS, University of Liège, Liège B-4000, Belgium

Summary. The goal of our study is to make use of the (fractally-defined) harmonic-mean criterion (HMC) as an indicator of proper/improper matter nucleation–transportation and matter-(non)densification tasks realized over certain thermodynamic-kinetic pathways in d–dimensional environments. We investigate three dynamic processes: self-avoiding random walk (SAW), cluster-cluster aggregation (CCA) and diffusion-limited aggregation (DLA). They are all considered as dispersive systems characteristic of excluded-volume effect (EVE). From our mean-field investigation it turns out that the HMC shows that SAW and CCA belong to the same kinetic (or, dispersive chemical kinetics) class, whereas DLA does not since it is realized over a mixed (non-homogeneous) thermodynamic-kinetic pathway. Our findings clearly reveal that the dimension two appears to be kinetically optimal for SAW and CCA but cast again some serious doubts on whether the so-called DLA $2D$ "paradigm" is here a well-posed problem.

1 Introduction

In this work, we are going to show that certain model dispersive systems, manifesting aggregation-desaggregation effects, underlie the same mean characteristics, whereas some other do not.

The mean characteristics we have in mind are generally related with the two-point harmonic-mean (HM), $\nu_k^{(d=2)}$, defined as

$$\frac{1}{\nu_k^{(d=2)}} = \frac{1}{2}\left(\frac{1}{\nu_k^{(d=1)}} + \frac{1}{\nu_k^{(d=3)}}\right), \tag{1}$$

where $\nu_k^{(d=2)}$, appears to play the role of the HM and is at the same time calculated as an average (logarithmically defined) speed [1] of a k-process embedded in $d = 2$. It is related by eq. (1) with its corresponding values $v_k^{(d=1)}$ and $v_k^{(d=3)}$, also calculated as the average speeds of the same type but taken in neighboring Euclidean sub- and super-spaces $d = 1$ and $d = 3$, respectively.

For a method of calculation of the average speeds, $v_k^{(d=j)}$, $j = 1, 2, 3$, see [1, 2]. For a definition of the HM, applied to model nanoparticle formation, see [3]. A k-process realized in the d-dimensional space, here $d = 1, 2, 3$, is said to obey the harmonic-mean criterion (HMC) iff its average speeds, inserted into eq. (1), yield the average speed in $d = 2$ exactly as HM of the two other remaining speeds, taken for $d = 1$ and $d = 3$, respectively. It ultimately means that such a (harmonic) mean quantity at $d = 2$ becomes exactly an average, evaluated by a statistical-mechanical method as the ensemble-average in the same geometric space, cf. [1–3]. By offering such a definition we automatically infer that the mean harmonicity fulfilled in the domain of the speed of the process implies a sufficient kinetic optimality of it, i.e. that the process in question goes smoothly in the Euclidean space in which it is embedded, here in $d = 2$.

Such a convergence/non-convergence of two types of the statistical measures mentioned above would implicitly resemble a type of dimension-influenced ergodic hypothesis. Its fulfillment/non-fulfillment in $d = 2$, herein specifically formulated for the model dispersive systems under consideration, demands that the HM of purely algebraic nature can/cannot equal the (fractally-defined) average. The average is taken over a certain coupled configurational-temporal space of each embedding k-process. See, especially eq. (7) and eq. (8) of sec. 2.3, and additionally [1, 2] for some argumentation. The k-processes that we analyzed throughout the present paper are [4]:

- a (topologically) linear system/polymer, i.e. the self-avoiding walk (SAW);
- a branched system/polymer, i.e. the diffusion-limited aggregation (DLA);
- a network-like/polymeric system, i.e. the cluster-cluster aggregation (CCA).

The SAW is defined as a chain of monomers that undergo an attraction-repulsion Lennard-Jones type interaction scheme measured along the chain (see, Fig. 1) [5].

The DLA is defined by a trial random walker that randomly samples an available space until it meets some accretion center at which it remains ultimately captured (see Fig. 2) [7].

The CCA is defined by pairwise cluster-cluster interactions via the cluster's surface, where the inter-cluster space is usually recovered upon an adequate raise of the temperature as a cooperation compaction-relaxing effect between late stage growing and mechanical relaxation modes (Fig. 3) [1].

A common physics-involving feature of all above listed systems can be termed excluded-volume effect (EVE): None of them can self-overlap, and some remaining interspace is always left during its evolution in the available isotropic space. In turn, a common theoretical framework of all of them is that their

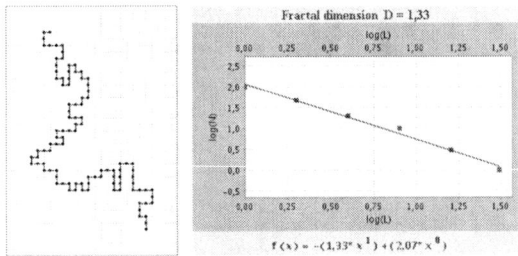

Fig. 1. Some arbitrary small-scale square-lattice realization of SAW for n=100 steps (left), and its fractal dimension depicted at the top of a log-log plot (right) [6]

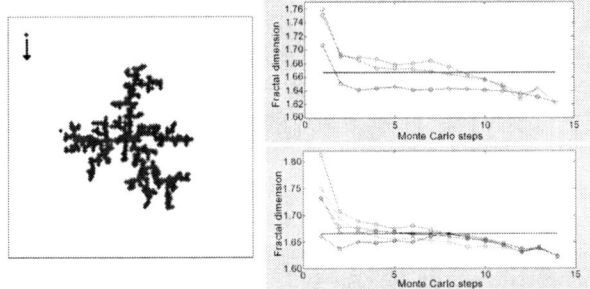

Fig. 2. Small-scale computer realization of $2D$ DLA model for the number of random walkers $N_{RW} = 9$ for thousand particles incorporated by the cluster (left) and its type of lattice, and type of accretion-seed dependent (from top to bottom) fractal characteristics. The characteristics quite fairly accommodate to the basic horizontal line which is the Meakin's estimate of DLA fractal dimension $D_{DLA} \approx \frac{5}{3}$ (right) [8]

dynamics undergo a Smoluchowski-type dynamics [5] while their scaling properties are quite satisfactorily described by the mean-field approach, for example, the SAW enjoys the Flory-Fisher (F-F) mean-field approach [9], with a well known d–dimensional scaling formula

$$\overline{\lambda_k}(t) \propto t^{\nu_k^{(d)}}, \tag{2}$$

where $\overline{\lambda_k}$ is an ensemble-averaged linear characteristic [2, 5, 9], t the time (or quite equivalently: the degree of polymerization) and $\nu_k^{(d)}$ a scaling exponent, here $\nu_{SAW}^{(d)} = 3/(d+2)$. In short, the F-F approach begins with a certain Gibbs-Boltzmann type construction of the statistical sum, Z, of the radially distributed segments (monomers) of a polymeric chain immersed in a solution. The segments are considered as a continuous cloud of homogeneous density, so that each segment "feels" the same density around it. The potential interaction of segments is a pairwise (repulsive) interaction - it enters then the statistical

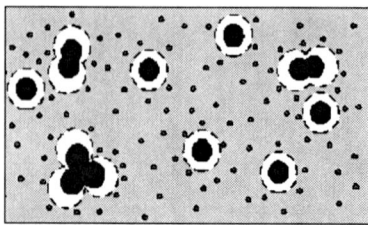

Fig. 3. Schematics of some assemblies of molecular clusters: Cluster-Cluster Aggregation (CCA) occurs. The light realms around each cluster can be considered as depletion zones, characteristic of colloid and protein aggregations, see [1, 2] and refs. therein. The small dark dots are monomers

sum, which is included in the free energy of the chain, $F = -\beta^{-1}\ln Z$ (for β, see eq. (3) below). After minimalizing F over the chain length one notices that a scaling relation of the type of eq. (2) solves the problem with the scaling exponent $\nu_{SAW}^{(d)}$ stated as above, cf. [9] (chapter 3).

Some important motivation to the present study states that the HMC is practically used in colloid (e.g., aerosol) science, here a silicon nanoparticle formation [3]. The silicon nanoparticle formation is a type of coagulation process, which is also a complicated nucleation-transportation problem. Growth of the coagulate is modeled by means of a difference scheme, which is by aerosol scientists called a standard "two-point" method. According to this method, the growth rate of the number of particles in a bin is a result of flowing in some particles from a neighboring bin as well as of flowing out a certain number of particles from the bin of interest. A kind of simple discrete master equation can be written down. It includes a quantity which is named the effective particle transfer rate between the bins. This rate, exactly as in our HMC is designed as a two-point HM, being composed of two other growth rates (precisely, as in our case), one of them coming from the free molecular motion of an aerosol particle, and the other from dealing with the aerosol as a continuum [3]. This way, an analogy between using the HM-s in our and aerosol system is completed, especially when we would compare it with CCA, which is - similarly to aerosol particle formation - a surface-involving process, see Appendix A.2 in [3] and eqs. (10) and (11) therein.

The paper is organized as follows. In Section 2, we give an overview of the common theoretical framework, finally emphasizing the fulfillment of HMC for CCA. In Section 3, we present SAW and CCA as dynamic processes that, even though being realized over two different pathways, will be shown to obey the HMC. In Section 4, we demonstrate that all known mean-field estimates of DLA speed do not obey the HMC, and attribute this shortage to the fact that DLA is realized over a mixed (non-homogeneous) kinetic pathway, or shows up a nonlinear EVE. Moreover, we draw some special attention to the fact that the proposed HMC should not be confused with the well-known harmonic-measure criterion for DLA which gives a well-established multifractal spectrum for the growth probability, properly yielding the single fractal dimension of a DLA cluster at the highest growing mode. (As a promising fact one would report here is that the numerical estimate of $\nu_{DLA}^{(d)} = 6/5d$ by Meakin [10, 11] obeys the HMC, which essentially means that the numerically obtained $\nu_{DLA}^{(d)}$ exactly conforms to the eq. (1), cf. eq. (8) in subsection 2.3.) The last section (Section 5) includes the main results of the paper and their discussion.

2 Common Theoretical Framework

In this study, we deal with three essentially different nucleation–transportation and matter-compaction involving problems. In spite of their clearly different dynamical behaviors there is at least one common effect which can be assigned to all of them: The EVE, well described by the F-F as well as by Fokker-Planck and Smoluchowski (F-P&S) dynamic frameworks [5, 9]. In case of SAW it is due to the repulsive part of the interactions between monomers from which the chain is made of. As concerns CCA, it can be revealed by calculating the total volume of the system for both low and high temperature matter aggregations, and then by seeing that there is a matter expansion exclusively in the case of the high temperature CCA [1]. While looking at DLA, in turn, one can immediately see that EVE is generically present in the process because the branches of the DLA microstructure do not overlap each other - they rather behave as separated SAW-s, nucleated at one nucleation seed, which always appears to be a cluster-surface (interface) process in a d-dimensional space [12].

2.1 Fokker-Planck and Smoluchowski Type Dynamics

For revealing the presence of the EVE as a dynamic phenomenon a standard way assumes the F-P&S type equation to be fulfilled. For most of dispersive systems it is usually based on the form of the matter current (see also Appendix for details)

$$J(x,t) = -D(x)\left(\frac{\partial f}{\partial x} + \beta\frac{dU(x)}{dx}f\right), \tag{3}$$

where $\beta = 1/k_BT$, k_B is the Boltzmann's constant (in case of dispersive soft-matter systems $dU(x)\cdot\beta \sim 1$ typically holds), T is the temperature, t - time, D - diffusion coefficient (it usually depends also upon the parameter d - the space dimension; in general: $D \equiv D(x,t;d)$ [1, 5]), $U \equiv U(x)$ - potential, $f \equiv f(x,t)$ - concentration of the constituting entities characteristic of SAW, CCA and DLA, respectively. Here, the entities constituting SAW are monomers whereas for CCA one has molecular, e.g. protein clusters and for DLA there can be atoms, molecules or macromolecules sometimes [4]. As concerns the variable x, it is to be specified according to each of the three analyzed processes separately.

For SAW the "drifted" diffusion is realized along a spacial-coordinate axis, so that x stands for the current position of the monomer, for example, that of either hydrophilic or hydrophobic type [5, 6, 13].

For CCA model, actually based on [1], the diffusion is always realized along an axis of cluster sizes, thus, x is here typically the volume of a single cluster. For DLA there are not so many conclusive analytic studies pointing to the F-P&S dynamics. One of the exceptions found in literature would be that of Fokker-Planck dynamics for the needles emerging during a model (1+1)-DLA. In this approach, each needle undergoes a one-dimensional random walk, starting from a horizontal line, and obeying the standard DLA rules, resulting in favoring taller needles at the cost of their non-tall neighbors. As a consequence of the model formulated in such way a most relevant stochastic variable, x, is defined by means of an excess length of the two neighboring teeth of $(1 + 1)$–DLA microstructure [12].

In view of the above, the three processes under consideration might have another interesting dynamic feature in common: They are two-state Kramers-type processes with a weak surmountable barrier given by $U(x)$ accounting for EVE. It means that they are mesoscopic systems underlying basic rules of non-equilibrium thermodynamics [14].

2.2 Mean Field Approach

In order to explain the mean-field (MF) approach let us rest again on the well known F-F procedure [9]. First, the F-F as each MF approach neglects the fluctuations of the monomer concentration of a SAW. Second, it takes into account two main contributions to a SAW free energy: an entropic, due to elasticity-influenced conformational changes of the chain, as well as the enthalpic coming from repulsion between non-neighboring monomers, which inevitably leads to EVE. Third, the interactions "seen" by such a procedure are always binary interactions - it results in f^2 (Van der Waals) contribution in the free energy [5], cf. Appendix. The three above stated assumptions, after minimizing the free energy of the SAW with respect to its size, and after letting a similarity relation to be a solution of the resulting equation, lead to a straightforward derivation of the SAW exponent [5, 9]

$$\nu_{SAW}^{(d)} = \frac{3}{d+2},\tag{4}$$

where d is a dimension of the Euclidean space. *Mutatis mutandis*, we can provide the CCA as well as the DLA exponents. For CCA we get [1]

$$\nu_{CCA}^{(d)} = \frac{1}{d+1},\tag{5}$$

whereas for DLA one typically[1] obtains [11, 17]

$$\nu^{(d)}_{\text{DLA}} = \frac{d+1}{d^2+1}.$$ (6)

2.3 HMC of the Fractally-Defined Speed in Dimension d

To compare somehow the speeds of the three processes analyzed, assumed that their descriptions are all based on the same MF Van-der-Waals (dispersive-force) type approximation, see the preceding subsection and the Appendix, let us define the speed of the process in a fractal-like manner

$$\nu^{(d)}_k = \frac{\ln\overline{\lambda}_k(t)}{\ln t}; \qquad \frac{t}{t^{(k)}_0} \gg 1,$$ (7)

where k indicates SAW, CCA and DLA respectively, $\overline{\lambda}_k$ stands for a characteristic length of each of k-processes, $t^{(k)}_0 > 0$ - the initial instant of each of k-processes, t - as above. Note that the late time condition $\frac{t}{t^{(k)}_0} \gg 1$ holds.

Notice that all $\overline{\lambda}_k$-s are fully derivable from the F-P&S dynamic characteristics of each system of interest [1, 5, 18]. Moreover, note that formally the F-P&S dynamics define a stochastic process which is a nonequilibrium (drifted) process - its stochasticity implies that t is an "active" (kinetic) variable. The F-F type MF approach, in turn, being based on the minimization of the free energy, implies that some N (e.g., the polymerization degree of the SAW) becomes a crucial but thermodynamic variable. By postulating such a definition of the speed (cf., eq. (7)) we somehow claim that N be equivalent to (or, at least, proportional to) t which should be true in a late stage of the aggregation-compaction process when the system presumably arrives at one of its quasi-equilibrium states.

While analyzing CCA [1] we have found that the so defined speed of CCA obeys a two-point HMC, namely

$$\frac{2}{\nu^{(2)}_{\text{CCA}}} = \frac{1}{\nu^{(1)}_{\text{CCA}}} + \frac{1}{\nu^{(3)}_{\text{CCA}}}.$$ (8)

From eq. (8) it is seen that the HMC unquestionably points to a special relevance of the Euclidean dimension d, here with emphasizing the role played by $d = 2$ for kinetically optimal path of any process obeying eq. (8). In the next section of the paper we are wondering whether and why it is true (or

---

[1] There exists at least one more estimate for $\nu^{(d)}_{\text{DLA}} = \frac{6+5d}{8+5d^2}$ obtained in [15] which, in turn, uses some other refined MF approach. Note that $\left[\nu^{(d)}_{\text{DLA}}\right]^{-1}$ is nonlinear in d, which is the case of eq. (6) too, cf. [16], in which the number of branches of a DLA microstructure increases with d arriving, however, at a saturation effect in $d = 2$ but not for $d > 2$

not) for SAW and DLA, which is equivalent to examining the necessary and sufficient conditions of existence of (8). (For convincing the reader that the CCA lies in the same class of pairwise-interaction driven processes, such as that of SAW and DLA, we encourage her/him to consult the Appendix.)

We find the HMC important. Firstly, because the HM inherently involves a logistic competition (Malthus-type) effect between (squared) geometric, G, matter-aggregation resources and its arithmetic, A, linear counterpart. For accepting it simply realize that based upon eq. (8) one may rewrite it as follows: $\nu_{CCA}^{(2)} = [\nu_{CCA}^{G}]^2/\nu_{CCA}^{A}$, where ν_{CCA}^{G} and ν_{CCA}^{A} denote, respectively, the geometric and arithmetic means composed of $\nu_{CCA}^{(1)}$ and $\nu_{CCA}^{(3)}$. Such a presence of nonlinear (geometric) matter-aggregation resources and of their linear (arithmetic) counterparts involved in $\nu_{CCA}^{(2)}$ as a ratio of them both, is reminiscent of some competition effect of mean matter-aggregation speeds, ν_{CCA}^{G} and ν_{CCA}^{A}, reflected as chemical reaction rates of second and first order (possibly of the broken order in between, Section 5, eq. (11)), respectively. Secondly, it indicates the dimension two ($d = 2$) as a relevant dimension - this is why throughout the whole paper we have allowed ourselves to illustrate our work by some small-scale $2D$ numerical simulations [6, 8] - in which some processes may go kinetically optimally but other ones, as we will see apparently of DLA-type, may not be optimal.

3 SAW & CCA as Processes Realized over Mono- and Polynuclear Paths

The SAW, see Fig. 1, is clearly realized over a single kinetic pathway which we call a mononuclear path. It is because the SAW object formed may serve as a singular nucleation seed *per se*. Moreover, two or more SAW-s may form a molecular cluster, for example a dimer ($N_{SAW} = 2$). Such a formation procedure, resting upon a creation of $N_{SAW} = 2$ and $N_{SAW} > 2$ clusters, we wish to call a polynuclear path - clearly, the high-temperature polynuclear path in which any matter-compaction effect is typically relaxed [1, 6].

As is mentioned in subsection 2.3 the CCA obeys the HMC. It is easy to check by inserting (4) into (8), that also SAW with its characteristic exponent, which is now according to the definition given by eq. (7) the speed of SAW process, obeys the HMC too, just in the way shown by eq. (8). We attribute this fact to the observation that both SAW and CCA go over kinetic pathways of non-mixed low-energy-barrier states: the SAW goes by addition of monomers only, and CCA goes purely by some linkage of clusters (see, Fig. 3). The addition as well as the linkage can typically be either first-order or second-order chemical reactions, see discussion in the preceding section. Therefore their speeds, assumed that each process is realized in a homogeneous (structure-less) d-dimensional space, fulfill the same (HMC) criterion.

4 DLA as a Process Realized over Some Mixed Path

The HMC is hardly fulfilled in case of DLA described in terms of MF. One can easily prove by inserting (6) into (8) that here the HMC fails. It is even the case of the refined estimate of ν_{DLA} (see, footnote 1). Such a behavior may be attributed to the fact that DLA can be thought of as a dynamic process in which quite many SAW-s are nucleated on a single nucleation center (see, Figs. 1 and 2). This can be anticipated as a certain intermediate stage since this way one can only create a branched molecular cluster but neither a (topologically) linear chain nor some network-like assemblage of clusters. This difference strongly suggests that DLA is realized over an intermediate, or better said, over some mixed kinetic pathway, because we can observe therein some common signatures of both SAW and CCA. Also, the EVE must be of different type than the corresponding EVE - s characteristic of SAW and CCA. Thus, this entices us to state that the F-F type MF approach which one applied to get $\nu_{\mathrm{DLA}}^{(d)}$ is insufficient and probably must be completed by taking into account fluctuations, i.e. going visibly beyond the MF approach. Some confirmation arises from large-scale computer simulations by Meakin and coworkers [10]. They estimated [11]

$$\nu_{\mathrm{DLA}}^{(d)} = \frac{6}{5d},\tag{9}$$

which surprisingly obeys the HMC stated by eq. (8). Note that $[\nu_{\mathrm{DLA}}^{(d)}]^{-1}$ is a linear function of d, so is also the case of eq. (4) and eq. (5) but it is certainly not true for the reciprocal of $\nu_{\mathrm{DLA}}^{(d)}$ taken from eq. (6).

At this place, let us clearly state that the HMC proposed in the present study should not be confused with a harmonic-measure criterion so often mentioned for DLA, especially realized in $d = 2$ [18]. This criterion refers to a theoretical description of $2D$ DLA in terms of the theory of analytic functions. It extracts the growth probability distribution in such a way that inclusion of a Brownian particle at a tip of the branch of the DLA cluster is much more probable than having it landed on any fjord between two neighboring branches. Thus, it has no obvious relation to the d-dimensional HMC that we propose to use for showing here a kinetic non-optimality of $2D$ DLA, whereas the harmonic-measure criterion mostly points to the self-similarity of a DLA cluster [4, 18], which we take for granted (Fig. 2).

5 Results and Discussion

Let us point out that the basic common features of three model dispersions considered under F-F type scaling approximation are the following:

1. Excluded-volume effect (EVE)
   - for SAW and DLA it leads to a long-time superdiffusive behavior:
     $\overline{\lambda_k}(t) \propto t^{\nu_k^{(d)}}$, where $1 > \nu_k^{(d)} \geq \frac{1}{2}$ $(d = 1, 2, 3)$;

- for CCA the total volume of the system is an unconserved quantity [1]; this way, the EVE is manifested in cluster-cluster aggregations; but the analogous scaling behavior is subdiffusive: $\overline{\lambda}_k(t) \propto t^{\nu_k^{(d)}}$, where $0 < \nu_k^{(d)} \leq \frac{1}{2}$ ($d = 1, 2, 3$) since cluster-cluster aggregation takes more time than typically a non-cluster-cluster processes, such as SAW or DLA. It has been summarized in a picturesque way in Fig. 4.

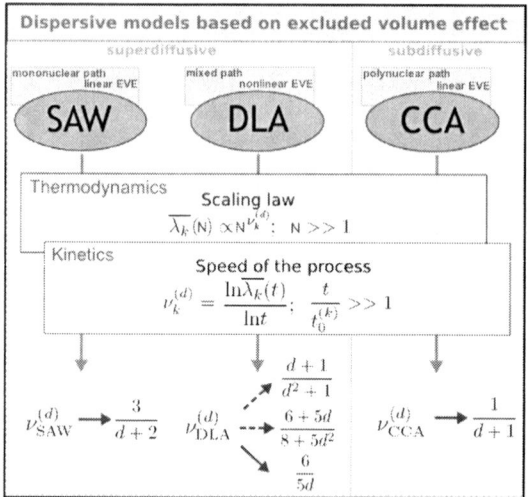

Fig. 4. Summarizing scheme for the three embedding processes, manifesting some EVE-caused dynamic disorder, with possibilities of entering the chaotic matter regime, especially when $d \to \infty$ (a need for applying then an M-point HMC might appear, where $M > 2$) [2]. It proclaims, that kinetically optimal k-processes emerge when $\left[\nu_k^{(d)}\right]^{-1}$ is linear in d, cf. eq. (10). The dashed arrows indicate kinetic non-optimality seen in terms of the MF approach

2. The main discrimination procedure that emphasizes whether the HMC becomes eventually effective, or appears to be ineffective[2], may rest upon a fundamental observation, namely that
   - SAW and CCA are realized over uniquely defined, mononuclear (one-chain viz cluster of monomers) or polynuclear (many-cluster based) kinetic pathways [1, 5, 13];
   - DLA is always realized over some mixed pathway: many SAW-s are clustered on one nucleation seed, what makes a difference when formally applying the HMC [4, 7, 10, 12]. It seems that a nonlinear branching, typically loopless topology would make some difficulties when one

---

[2] Thus, the HMC can also be thought of to be a measure of some (in)effectiveness of the k-process embedded in the subsequent Euclidean space

tried to kinetically optimize the process leading to creation of such a structure, especially in the two-dimensional embedding space.

- Notice formally, that a critical dimension for SAW, d_c, reads $d_c = 4$ (no effective pairwise interaction) [9], whereas for CCA $d_c = 1$ (surface-independent case), cf. [1]. But for DLA d_c takes on a fractional value - when evaluating it based on eq. (6), i.e. comparing it to one-half (standard-diffusion exponent), it reads $d_c \cong 2.41$, which also makes a formal difference between integer-influenced d_c of SAW and CCA and its non-integer counterpart for DLA [4, 11].

Moreover, from an algebraic point of view, we have observed that for the three processes under consideration it is sufficient if the (numerical) estimate of the fractally-defined inverse speed of the process $(1/\nu_k^{(d)})$ is a linear function of dimension d, such as

$$D_f^{(d)} := \frac{1}{\nu_k^{(d)}} = A \cdot d + B, \tag{10}$$

where A, B are constants. If this is not the case, like in Tokayama-Kawasaki or Hentschel [15, 17] estimates, the HMC clearly favoring $d = 2$ realization space, fails. Thus, it is an unambiguously stated criterion likely favoring the EVE which is a common feature of the three processes considered, but arises naturally as interaction vs elasticity enthalpic-entropic effect in case of SAW and CCA, but rather purely diffusionally in case of DLA, i.e. when the enthalpic part is missing.

In particular, for DLA it would imply that any approach attempting for getting a final estimate of either a d–dependent fractal dimension[3] $(D_f^{(d)})$ or some d–dependent fractally-defined speed of the DLA process, should be a non mean-field approximation, enabling the influence of thermal (β being x-dependent ?) and/or athermal (density changes) fluctuations in the final outcome [19, 20]. It must lead to a more realistic description of the process, addressing an interaction of the DLA cluster with its surroundings, careful inspection of the sticking rules, that would make a DLA structure more mechanically compact or tenuous, as well as use of its multiparticle variants [8]. Moreover, the system has to be thermodynamically checked at least for presence of non-equilibrium steady states as well as for its consistency with the theorem of minimum entropy production [14]. It can be done by certain renormalization-group considerations [21]. Another option could be to resort to some numerical approaches that view DLA process in terms of some deterministic particle trajectories, and that the cluster is build on their realizations [22].

Bear in mind that, although DLA is widely recognized as a Laplacian growth model the two above mentioned approaches [21, 22] do not need the Laplace

---

[3] Note, however, that for CCA $D_f^{(d)} = \frac{1}{\nu_{CCA}^{(d)}} = d + 1$, cf. eq (7), is not of fractal-dimension form. It is rather a superdimension, or according to ref. [9], a measure of the random close-packing [1, 2] (A standard version of the approach [1] is, however, presented for non-fractal objects.)

equation to be solved for obtaining the speed of the DLA in the way shown by our paper. As is known, the $2D$ Laplacian growth is not well-formulated mathematically in $d = 2$ [23] and the problem needs logarithmic corrections [16]. It seems that in $d = 2$ the stationarity of the diffusion field is not fully guaranteed, especially under the presence of an absorbing boundary (sink) somehow violating its (external) harmonicity. As a consequence, at least certain matter fluctuations near the boundary can be suspected to occur - therefore some non mean-field DLA of turbulent type [24] looks more confident, and a suitable modification of $\nu_{\mathrm{DLA}}^{(d)}$ could emerge from such a proposal. An open question remains whether it will confirm the estimate by Meakin, eq. (9). If so, it will then formally give us the main message coming from the HMC, namely that in $d = 2$ only those processes may go thermodynamic-kinetically optimally for which the reciprocals of their speeds are linear functions of the space dimension d, eq. (10), cf. [16]. Therefore, a quite accurate numerical realization of DLA by Meakin, and even his followers [18, 19], point to $D_f^{(d)} = 5d/6$ to be a reliable value [4, 10].

When looking at the problem in terms of the HMC, the above results suggest an isotropic realization in space of $d = 2$ for SAW [25] and CCA [1, 2] but somehow discourages the HMC for DLA-type practical experiments on a plane, i.e. as in the case of thin-films realizations of DLA patterns [4, 26].

Moreover, looking at our three processes, manifesting EVE-caused dynamic disorder, cf. Fig. 4, one would interpret some of our findings in terms of dispersive (fractal-like) chemical reaction kinetics, especially when the characteristic length $\overline{\lambda_k}$ would be of order of the kinetic mean free path of the system which is often the case met in condensed phases [27] or in aerosols [3]. In such a Loschmidt-type limit [28], one would presume that a product of the (dispersive) chemical reaction rate coefficient, $\kappa \equiv \kappa(t)$, and $\overline{\lambda_k}$ will likely arrive at

$$\overline{\lambda_k} \times \kappa \sim \delta_k(d), \qquad (11)$$

where a t–independent constant $\delta_k(d) \propto \nu_k^{(d)}$, where again the argumentation about (non)linearity of $\delta_k(d)^{-1}$ can be used to distinguish between the SAW/CCA and DLA different 'kinetic universality' classes, cf. eq. (10). The above, but confined to CCA only, could provide an alternative view of the time-dependent kinetics of certain more specific, e.g. nucleated-polymerization processes such as model prion growth [27, 29].

To sum up, in this paper it has been shown that, within the mean-field approximation, in order to fulfill the d-dimensional HMC ($d = 1, 2, 3$), physically meaning that the system evolves toward optimal kinetic conditions in $d = 2$, a linear EVE (see, Fig. 4), to be quantitatively characterized by the dimensionless potential $\beta \times U(x)$ [5, 13] from the Smoluchowski-type equation (3), has to be shown up by the system - it is just the case of SAW and CCA but, unfortunately, $2D$ DLA likely suffers a nonlinear EVE, i.e. it possesses a totally branched internal meandric and fjords-involving microstructure at all length scales (Fig. 2), hardly penetrable by, say, a testing particle that has no additional (third) dimension at its disposal to eventually escape from it, or to successfully percolate through it, once entering one of its fjords in either direction [4, 7, 16, 18, 19, 24]. Therefore, eq. (10) appears to be a central HMC-oriented result of our study [2, 10, 17, 18, 23]. Thus, the Euclidean space $d = 2$ does not seem optimal from a thermodynamic-kinetic point of view for DLA and is attributed to non-fulfillment (violation of eq. (10)) of the dimension-dependent matter-aggregational ergodic hypothesis (see, sec. 1) - a case that discourages application-oriented activities, oppositely to $2D$ SAW [25] or CCA [1] colloid type applications that can be found elsewhere, e.g. in membrane science [26].

Acknowledgements

N.K. and A.G. thank Dr. Th. Pöschel for his kind hospitality at HUB of Berlin. Part of M.A. work is supported through an Action de Recherches Concertée Program of the University of Liège, i.e. ARC 02/07-293. (Un)folding-and-aggregation aspects of model SAW and CCA based polymers are subject to a KBN grant 2P03B 03225(2003-2006) regulations. A.G. acknowledges a year 2006 support by the ESF Programme Stochastic Dynamics: Fundamentals and Applications (STOCHDYN).

Appendix

For the class of CCA processes [1] the local continuity equation

$$\frac{\partial}{\partial t}f(v,t) + \frac{\partial}{\partial v}J(v,t) = 0, \tag{12}$$

where x is now v - the volume of a molecular cluster ($v^{(d-1)/d}$ stands for its 'reactive' surface), and $f(v,t)$ is the distribution function of the clusters at time t, is the conservation law which, after inserting $J(v,t)$ in eq. (12), leads to the F-P&S dynamic framework.

It can be found useful to transform the F-P&S equation into its possibly simple functional representation. For doing so, let us express the matter flux, eq. (3), in the following form [30]

$$J(v,t) = -(B(v)\frac{\delta F(f)}{\delta f(v,t)} + \beta D(v)\frac{\partial}{\partial v}\frac{\delta F(f)}{\delta f(v,t)}). \tag{13}$$

Here $\delta F(f)/\delta f(v,t)$ stands for the functional derivative, and the free-energy functional $F(f)$ looks as follows [31]

$$F(f) = (1/2)f(v,t)\int C(v-v')f(v',t)dvdv', \tag{14}$$

where above (cf. (13)) $B(v) = b(v)dU(v)/dv$ has been used ($b(v)$ - the mobility, linearly proportional to $D(v)$). If one takes the kernel, C, $C(v-v') = \delta(v-v')$, one obtains

$$F(f) = (1/2)[f(v,t)]^2, \tag{15}$$

which because of the power 2 in (15), unambiguously suggests binary interactions between clusters, as is, for example, assumed in Van der Waals (real)
gases between the gas molecules in the framework of a mean-field description [28]. For other details of the functional-based approach to CCA, see [31].

References

1. A. Gadomski et al.: Chemical Physics **310**, 153 (2005)
2. A. Gadomski, M. Ausloos: 'Agglomeration/Aggregation and Chaotic Behaviour'.
   In: *The Logistic Map and the Route to Chaos: From the Beginning to Modern Applications*, ed. by M. Ausloos, M. Dirickx (Springer Complexity Program Series:
   Understanding Complex Systems, Berlin 2006) pp. 275-294
3. S.S. Talukdar, M.T. Swihart: J. Aerosol Sci. **35**, 889 (2004)
4. H. J. Herrmann: Phys. Rep. **136**, 153 (1986)
5. M. Doi , S.F. Edwards: *The Theory of Polymer Dynamics* (Oxford University
   Press, Oxford, 1986)
6. N. Bąkowska: M.Sc. Thesis, University of Technology and Agriculture, Bydgoszcz
   (2005)
7. T.A. Witten, L.M. Sander: Phys. Rev. Lett. **47**, 1400 (1981)
8. J. Tadych: M.Sc. Thesis, University of Technology and Agriculture, Bydgoszcz
   (2005)
9. R. Zallen: *The Physics of Amorphous Solids* (John Wiley and Sons, New York
   1983)
10. P. Meakin: Phys. Rev. A **27**, 1495 (1983)
11. S. Tolman, P. Meakin: Phys. Rev. A **40**, 428 (1989)
12. M.-O. Bernard et al.: Phys. Rev. E **64**, 041401 (2001)
13. G. Favrin et al.: Biophysical J. **85**, 1457 (2003)
14. J.M.G. Vilar, J.M. Rubí: Proc. Natl. Acad. Sci. USA **98**, 11081 (2001)
15. H.G.E. Hentschel, J.M. Deutsch: Phys. Rev. A **29**, 1609 (1984)
16. S. Schwarzer at al.: Phys. Rev. E **53**, 1795 (1996)
17. M. Tokayama, K. Kawasaki: Phys. Lett. A **100**, 337 (1983)
18. T.C. Halsey: Physics Today **53**, 36 (2000)
19. N. Vandewalle, M. Ausloos: Phys. Rev. E **51**, 597 (1995)
20. R. Lambiotte, M. Ausloos: Phys. Rev. E, **73**, 011105 (2006)
21. T. Nagatani: J. Phys. A: Math. Gen. **20**, 6603 (1987)
22. Y. Taguchi: J. Phys. A: Math. Gen. **21**, 4235 (1988)
23. J.M. Deutch, P. Meakin: J. Chem. Phys. **74**, 2093 (1983)
24. R.C. Ball, E. Somfai: Phys. Rev. Lett. **89**, 135503 (2002)
25. S.C. Bae et al.: Current Opinion Solid State Mater. Sci. **5**, 327 (2001)
26. A. Danch, A. Gadomski: J. Molec. Liquids **86**, 249 (2000)
27. A. Plonka: *Dispersive Kinetics* (Kluwer, Dordrecht 2001)
28. A. Bader, L. Parker: Physics Today **54**, 45 (2001)
29. T. Pöschel et al.: Biophysical J. **85**, 3460 (2003)
30. G. Giacomin et al.: Nonlinearity **13**, 2143 (2000)
31. A. Gadomski et al.: Materials Science, (in press)

Part III

Pedestrians

Pedestrian Free Speed Behavior in Crossing Flows

Winnie Daamen and Serge P. Hoogendoorn

Delft University of Technology, Stevinweg 1, 2628 CN Delft, The Netherlands

Summary. Insights into pedestrian behavior, and tools to predict this behavior, are essential in the planning and design of public pedestrian facilities, such as transfer stations, shopping malls, and airports. Both macroscopic features of pedestrian flows and microscopic walking behavior underlying these features are important.

This paper discusses findings from an experiment with crossing pedestrian flows, and in particular the free speed distribution of the participating pedestrians. Available free speed estimation methods developed for car traffic appear to be not suited for pedestrian traffic. This paper presents a dedicated adaptation of a method used for car traffic, with satisfactory results in pedestrian crossing flows.

1 Introduction

Free speed or desired speed is the speed a pedestrian walks with when he or she is not hindered by other pedestrians. The free speed differs among pedestrians, among types of walking infrastructure, and among external conditions. This is due to the characteristics of pedestrians (age, gender, physical abilities), characteristics of walking infrastructure (grade, length, width, type of facility), and weather and other external conditions. Since the exact relation between these characteristics is not known, free speeds are usually described as a stochastic variable with a distribution.

Free speed and its distribution play an important role in many traffic flow models. To illustrate: the free speed distribution is an input for gas-kinetic models [3, 4], while many microscopic simulation models draw free speeds of individual pedestrians from free speed distributions [5, 6].

Insights into free speeds are also important from the viewpoint of design of facilities and public transport timetables. Walking times between origins and destinations in a facility can be derived, giving insight into the efficiency of a facility with respect to minimizing walking efforts.

The aim of this paper is to derive a free speed distribution for crossing pedestrian flows. The data on which the distributions are estimated come from large-scale laboratory walking experiments (see [2] and Fig. 1). Since an appropriate approach to derive free speed distributions for (crossing) pedestrian traffic is not available, we apply one developed for car traffic [7, 8]. The results are rather unsatisfying, so we adapt the criterion to determine whether a pedestrian is constrained or not. Estimations using this new criterion appear to be much more promising.

Fig. 1. Overview of crossing flows experiment

2 Free Speed Estimation

Estimating free speed distributions is not as straightforward as it looks like. Pedestrians are either walking at their free speed or following another pedestrian. This suggests that only those pedestrians walking freely need to be considered in deriving the free speed distribution. However, pedestrians having a relatively high free speed have a higher probability of being constrained than pedestrians with a relatively low free speed. This method will lead to underestimation of the free speeds.

Since existent free speed estimation methods all have their own drawbacks [1], Hoogendoorn [7] recently developed a new estimation approach for car traffic referred to as the modified Kaplan-Meier approach [9]. This approach is based on the concept of censored observations [10] using a non-parametric method to estimate the parameters of the free speed distribution. Applying this method for pedestrian traffic with crossing flows leads to the cumulative distribution functions for free speed shown in Fig. 2. This figure shows the measured speeds of all pedestrians ($F(v)$), the speeds of the unconstrained pedestrians ($F^0(v^0)$) and the estimated free speed distribution ($F_{\text{mod.Kaplan}-\text{Meier}}$).

The results appear to be unsatisfactory. First, the median free speed is 1.35 m/s, which is much lower than the free speeds measured for other pedestrian experiments. Second, all distributions are nearly similar, which is neither what we expect nor what our other experiments show. Reason for this might be the use of a headway criterion to determine the probability of pedestrians being constrained. This headway criterion only accounts for pedestrians walking in front of the considered pedestrian. This leads in crossing flows to an underestimation of constrained pedestrians, since significant hindrance is caused by pedestrians from aside. In the ensuing we will develop a new criterion for this constrainedness.

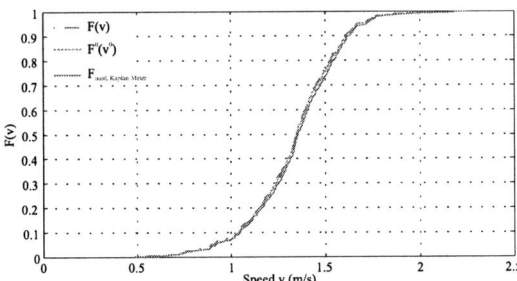

Fig. 2. Estimated free speed distributions for crossing flows according to [7].

3 New Criterion for Constrainedness of a Pedestrian

The probability θ_p that a pedestrian p is constrained is directly related with the presence of other pedestrians q_i. Not only the distance between pedestrians is important to determine the hindrance, also the time aspect is: someone getting very close over a few seconds will give less hinder than someone currently at the same close distance. In car traffic, two notions are known in this respect, namely time-to-collision and post encroachment time [11]. Here, we look at the distance between two pedestrians and how this distance varies over time, assuming that both pedestrians maintain their current walking speed and angle of movement.

Figure 3a shows a hypothetic situation, with four pedestrians present in the observation area (arrows indicate their current walking speed). The aim is to determine θ_p, depending on pedestrians q_1, q_2 and q_3. Since pedestrians are anisotropic (they will mainly react to pedestrians in front of them), only pedestrians q_1 and q_2 will be considered in the approach. Extrapolating current speeds, the distance between the centers of the pedestrians is calculated over time (see Fig. 3b).

In effect, Fig. 3b shows two criteria for the constrainedness of a pedestrian, namely distance between pedestrians and the moment that a specific distance occurs. The fuzzy approach is very suited to describe θ_p as a probability of a pedestrian being constrained, varying between 0 (free flowing) and 1 (constrained) [8]. This approach is also able to handle a combination of criteria, as is the case here: a pedestrian is more constrained when the distance to another pedestrian is smaller ("proximity") and the moment this occurs is closer ("urgency"). In Fig. 3c the probability to be constrained due to a specific distance between pedestrian p and pedestrians q_1 and q_2 respectively is plot over time. The specific relations for the membership functions for proximity θ^P and urgency θ^U are shown in Fig. 4.

For a given distance d between two pedestrians p and q_i the membership $\theta^P_{p,q_i}(d)$ is determined as well as the membership $\theta^U_{p,q_i}(h)$ for the moment h on which this distance d occurs. The probability that pedestrian p is constrained due to a specific pedestrian q_i depends on both $\theta^P_{p,q_i}(d)$ and $\theta^U_{p,q_i}(h)$:

$$\theta_{p,q_i}(d, h) = \theta^P_{p,q_i}(d)\theta^U_{p,q_i}(h) \tag{1}$$

The function $\theta^P_{p,q_i}(d)$ can be interpreted as the probability that two pedestrians having an intermediate distance d are constrained, while the function $\theta^U_{p,q_i}(h)$ can be interpreted as the probability that a pedestrian experiencing this distance at a specific time moment h is constrained. The joint probability

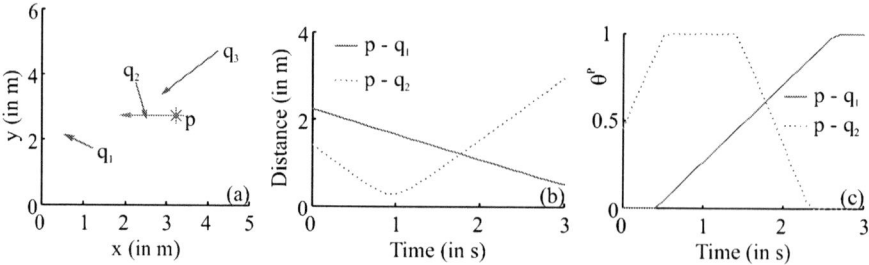

Fig. 3. Conflict area of pedestrian p (a), distance between pedestrian p and pedestrians q_1 and q_2 (b) and probability of pedestrian p being constrained by pedestrians q_1 and q_2 over time (c).

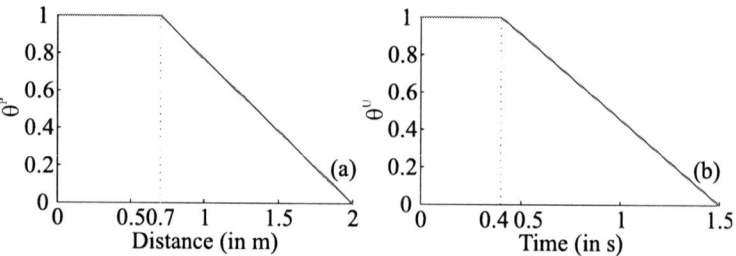

Fig. 4. Membership functions for proximity (a) and urgency (b).

can be determined from these functions under the assumption that both are independent.

For each pair of pedestrians (in the example $p - q_1$ and $p - q_2$) the maximum $\theta_{p,q}$ needs to be calculated. It might be argued that this is the minimum distance between the pedestrians. However, if two pedestrians have nearly the same angle of movement, this point would be at the end of the area ($\theta^P =$ max; $\theta^U = 0$), whereas at the current moment, pedestrian p may be already hindered ($0 < \theta^P <$max; $\theta^U > 0$). Therefore, θ_{p,q_i} is determined over the complete predicted time period and the maximum is assigned to this pair of pedestrians:

$$\theta_{p,q_i} = \max \left(\theta^P_{p,q_i}(d) \theta^U_{p,q_i}(h) \right) \tag{2}$$

To determine θ_p we need to know which pedestrian q_i is most constraining pedestrian p and assign the corresponding θ_{p,q_i} to pedestrian p. To do this, we take the maximum of θ_{p,q_i} for each pedestrian q_i on the area:

$$\theta_p = \max_{q_i} \left(\theta_{p,q_i} \right) \tag{3}$$

4 Free Speed Distribution in Crossing Flows

Applying the criterion described in the previous section leads to the estimation results shown in Fig. 5. The estimated free speed distribution has significantly moved to the right, resulting in a median free speed of 1.6 m/s (in other experiments we found median free speeds of 1.55 m/s). Also, the difference between the speed distribution of the unconstrained pedestrians and the distribution for all pedestrians has grown, since fewer pedestrians are considered unconstrained (using the same new criterion).

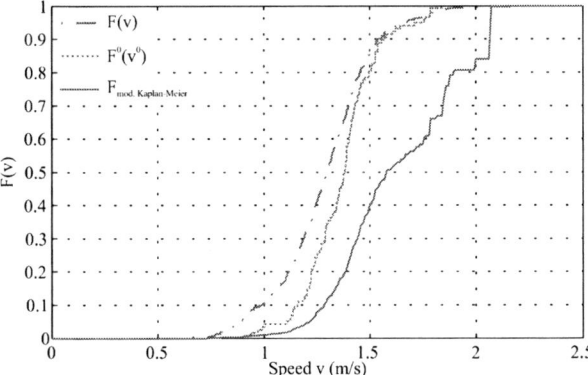

Fig. 5. Estimated free speed distributions applying the new criterion.

5 Conclusions

We have shown how the approach to estimate free speed distributions for car traffic can be improved to make it applicable for crossing pedestrian flows. The criterion determining the probability of a pedestrian being constrained has been based on a fuzzy approach, using the distance between pedestrians and the time moment this distance occurs as parameters of the membership functions. Application of this method on data from laboratory experiments shows much better results than the original method (median free speed of 1.6m/s in pedestrian crossing flows compared to 1.55m/s in other experiments).

Despite the promising results, additional research has to be performed. One of the points is to investigate the choice of parameters of the membership functions. Applying this approach on data from other experiments, we can compare the free speed distributions and adapt the parameters so that the distributions are similar. After all, the same pedestrians participated in the experiments. Also, the sensitivity of the approach for the choice of parameters needs to be investigated. In addition, the form of the membership functions might be varied.

Acknowledgements

This research has been funded by the Social Science Research Council (MaGW) of the Netherlands Organisation for Scientific Research (NWO).

References

1. H. Botma: The Free Speed Distribution of Drivers: Estimation Approaches. In: *Five years Crossroads of theory and practice.* ed. by P. Bovy (Delft University Press, Delft 1999) pp. 1-22.
2. W. Daamen, S.P. Hoogendoorn: Transp. Res. Rec. 1828, (2003) pp. 20-30.
3. D. Helbing: Complex Systems 6 (1992) pp. 391-415.
4. S.P. Hoogendoorn, P.H.L. Bovy: Transp. Res. Rec. 1710, (2000) pp. 28-36.
5. J. Barcelo, E. Codina, J. Casas, J.L. Ferrer, D. García: Journal of Intelligent and Robotic Systems, Theory and Applications 41 (2-3), (2005) pp. 173-203.
6. S.P. Hoogendoorn, P.H.L. Bovy: Transp. Res. Part B 38 (2004) pp. 169-190.
7. S.P. Hoogendoorn: Transp. Res. Part B 39 (2005) pp. 709-727.
8. S.P. Hoogendoorn: Vehicle-Type and Lane-Specific Free Speed Distributions on Motorways; A Novel Estimation Approach Using Censored Observations, Proceedings of the Annual Meeting of the Transportation Research Board. CD-ROM. Transportation Research Board, National Research Council, Washington, D.C. (2005).
9. E. Kaplan, P. Meier. Jrl. Am. Stat. Assoc. 53, (1958) pp. 457-481.
10. W. Nelson: *Applied Life Time Analysis* (Wiley, New York 1982).
11. A.R.A. Van der Horst: *A Time-Based Analysis of Road User Behaviour in Normal and Critical Encounters.* PhD thesis, Delft University of Technology, The Netherlands (1990).

The Fundamental Diagram of Pedestrian Movement Revisited – Empirical Results and Modelling

Armin Seyfried[1], Bernhard Steffen[1], Wolfram Klingsch[2], Thomas Lippert[1], and Maik Boltes[1]

[1] Central Institute for Applied Mathematics, Forschungszentrum Jülich GmbH, 52425 Jülich, Germany
[2] Institute for Building Material Technology and Fire Safety Science, University of Wuppertal, Pauluskirchstrasse 7, 42285 Wuppertal, Germany

Summary. The simplest system for the investigation of the fundamental diagram for pedestrians is the single-file movement. We present experimental results for this system and discuss the observed linear relation between the velocity and the inverse of the density. For the modelling we treat pedestrians as self-driven objects moving in a continuous space. On the basis of a modified social force model we analyze qualitatively the influence of various approaches for the interactions of pedestrians on the resulting velocity-density relation. The one-dimensional model allows focusing on the role of the required length and remote force. We found that the reproduction of the typical form of the fundamental diagram is possible if one considers the increase of the required length of a person with increasing current velocity. Furthermore we demonstrate the influence of a remote force on the velocity-density relation.

1 Introduction

Empirical studies of pedestrian streams can be traced back to the year 1937 [1]. To this day a central problem is the relation between density and flow or velocity. This dependency is termed the fundamental diagram and has been the subject of many investigations from the very beginning, see references in [2,3]. One simple system is the uni-directional movement of pedestrians in a plane without bottlenecks. In this context the fundamental diagram of Weidmann [2] is frequently cited. It is part of a review work and the author summarized 25 different investigations for the determination of the fundamental diagram. Apart from the fact, that with growing density the velocity decreases, the relation shows a non-trivial form. Weidmann notes that different authors choose different approaches to fit their data, indicating that the dependency is not completely analyzed. A multitude of possible effects can be considered which may influence the dependency. For instance we re-

fer to passing maneuvers, internal friction, self-organization phenomena like
marching in steps [4] or ordering phenomena like the 'zipper' effect [5]. A
reduction of the degrees of freedom helps to restrict possible effects and allows an improved insight to the problem. Thus we choose a one-dimensional
system for this investigation. Furthermore, the fundamental diagram is used
for the evaluation of microscopic models for pedestrian movement [6–9]. The
models can be classified in two categories: the cellular automata models [10–
14] and models in continuous space [15–18]. We focus on models continuous
in space, which differ substantially with respect to the 'interaction' between
the pedestrians and thus to the update algorithms as well. The social force
model for example assumes amongst others a repulsive force with remote action between the pedestrians [15, 19–22]. Other models treat pedestrians by
implementing a minimum inter-person distance, which can be interpreted as
the radius of a hard body [17, 18]. To concentrate on the influence of the
required space and the remote action on the velocity-density relation we introduce in a one-dimensional model different approaches for the interaction
between the pedestrians. This contribution summarizes parts of two articles.
The reader may consult [3, 23] for more detailed discussions and additional
results.

2 Experiment

2.1 Description

Our aim is the measurement of the relation between density and velocity
for the single-file movement of pedestrians. To facilitate this with a limited
amount of test persons also for high densities and without boundary effects,
we choose a experimental set-up similar to the set-up in [24].

The corridor, see Figure 1, is build up with chairs and ropes. The width of the
passageway in the measurement section is $0.8\,m$. Thus passing is prevented
and the single-file movement is enforced. The circular guiding of the passageway gives periodic boundary conditions. The length of the measured section
is $l_m = 2\,m$ and the whole corridor $l_p = 17.3\,m$. The experiment is located in
the auditorium 'Rotunde' at the Central Institute for Applied Mathematics
(ZAM) of the Research Centre Jülich. The group of test persons is composed of
students of Technomathematics and staff of ZAM. To enable measurements at
different densities we execute six cycles with $N = 1, 15, 20, 25, 30, 34$ numbers
of test persons in the passageway.

The measurement of the flow characteristics is based on video recordings with
a DV camera (PAL format, 25 fps) of the measured section. These recordings
were analyzed frame-wise, see Figure 1. For every person i we collect the
entrance time (of the ear) in the measured section t_i^{in} and the exit time t_i^{out}.
These two times allow the calculation of the individual velocities $v_i^{man} = l_m/(t_i^{out} - t_i^{in})$ and the the density ρ_i^{man}. We regard a crossing of an individual

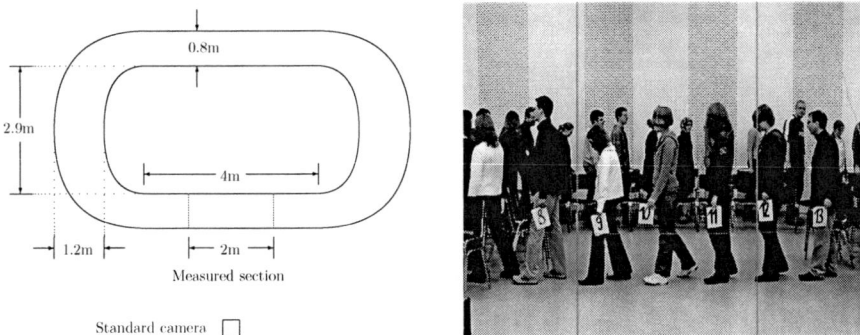

Fig. 1. Left: Experimental set-up for the measurement of the velocity-density relation for the single-file movement. Right: One frame of the cycle with $N = 30$. The two vertical lines mark the measured section.

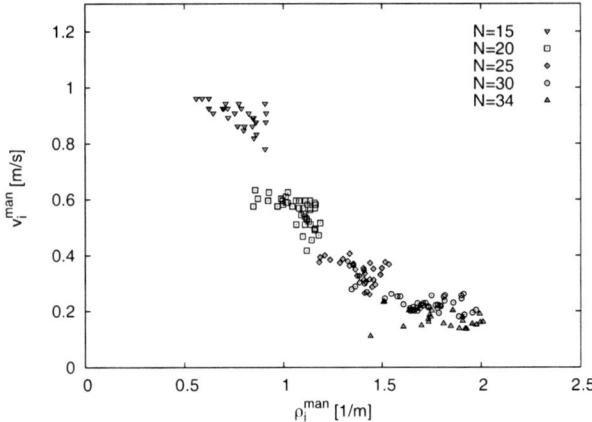

Fig. 2. Dependency between individual velocity and density for single-file movement

pedestrian with velocity v_i^{man} as one statistical event, which is associated to the density ρ_i^{man}. While ρ_i^{man} is the mean value of the density during the time-slice $[t_i^{in}, t_i^{out}]$. For a more detailed discussion and an example for the time-development of ρ we refer to [3].

Figure 2 shows the distribution of the events $(v_i^{man}, \rho_i^{man})$ of the cycles with $N = 15, 20, 25, 30$ and 34. We exclude the data where the influence of the starting phase and opening of the passageway are apparent.

2.2 Empirical Results

For the single-file movement the distance to the next pedestrian in front can be regarded as the required length d of a pedestrian to move with velocity v.

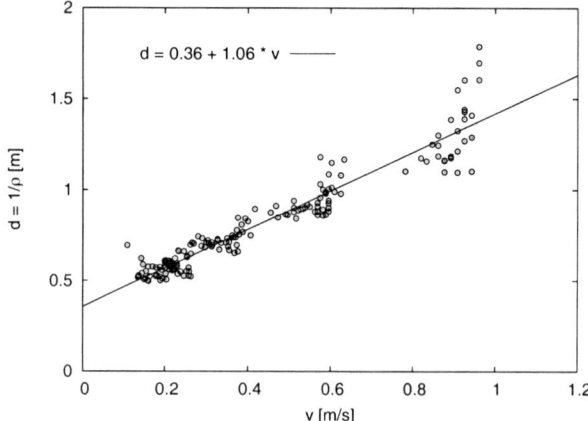

Fig. 3. Dependency between required length and velocity according to the data from the cycles with $N = 15, 20, 25, 30$ and 34. A linear relationship gives the best fit to the data.

Considering that in a one-dimensional system the harmonic average of this quantity is the inverse of the density, $d = 1/\rho$, one can investigate the relation between required length and velocity by means of the velocity-density relation for the single-file movement.

Fig. 3 shows the dependency between required length and velocity. We tested several approaches for the function $d = d(v)$ and found that a linear relationship with $d = 0.36 + 1.06\,v$ gives the best fit to the data. According to [2] the step length is a linear function of the velocity[3] only for $v \gtrsim 0.5\,m/s$. Thus it is surprising, that the linearity for entire distance holds even and persists for velocities smaller than $0.5\,m/s$. Possible explanations will be discussed later. The comparison of the relation between velocity and density for the single-file movement with the movement in a plane according to Weidmann shows a surprising conformity, see [3]. This conformance indicates that two-dimensional specific properties, like internal friction and other lateral interferences, have no strong influence on the fundamental diagram at least at the density domains considered. Instead, the visual analysis of the video recordings suggests that the following 'microscopic' properties of pedestrian movement determine the relation between velocity and density. At intermediate densities and velocities the step length is reduced with increasing density. The distance to the pedestrian in front is related to the step length as well as to the safety margin to avoid contact with the pedestrian in front. Both, step length and safety margin are connected with the velocity. At high densities and small velocities we observed that small groups pass into marching in lock-step. Furthermore

---

[3] Lower average velocities arise from a combination of a lower step frequency and a shorter step length.

the utilization of the available place is optimized. This is achieved by some persons setting their feet far right and left of the line of movement, giving some overlap in the space occupied with the pedestrian in front. While at intermediate densities and relative high velocities the pedestrians are concentrated on their movement, this concentration is reduced at smaller velocities and leads to a delayed reaction on the movement of the pedestrian in front. The marching in lock-steps and the optimized utilization of the available space, which compensate the slower step frequency, are possible explanations that the linearity between the required length and the velocity holds even for velocities $v < 0.5\,m/s$.

3 Modelling

3.1 Modification of the Social Force Model

The social force model was introduced in [15]. It models the one-dimensional movement of a pedestrian i at position $x_i(t)$ with velocity $v_i(t)$ and mass m_i by the equation of motions

$$\frac{dx_i}{dt} = v_i \qquad m_i \frac{dv_i}{dt} = F_i = \sum_{j \neq i} F_{ij}(x_j, x_i, v_i). \tag{1}$$

The summation over j accounts for the interaction with other pedestrians. We assume that friction at the boundaries and random fluctuations can be neglected and thus the forces are reducible to a driving and a repulsive term $F_i = F_i^{drv} + F_i^{rep}$. According to the formulation in [20] we choose

$$F_i^{drv} = m_i \frac{v_i^0 - v_i}{\tau_i} \quad \text{and} \quad F_i^{rep} = \sum_{j \neq i} -\nabla A_i \left(\|x_j - x_i\| - d_i \right)^{-B_i}. \tag{2}$$

Here v_i^0 is the intended speed and τ_i controls the acceleration. The hard core d_i reflects the size of pedestrian i acting with a remote force on other pedestrians. Without other constraints a repulsive force which is symmetric in space can lead to velocities which are in opposite direction to the intended speed. Furthermore, it is possible that the velocity of a pedestrian can exceed the intended speed through the impact of the forces of other pedestrians. In a two-dimensional system this effect can be avoided through the introduction of additional forces like a lateral friction, together with an appropriate choice of the interaction parameters. In a one-dimensional system, where lateral interferences are excluded, a loophole is the direct limitation of the velocities to a certain interval [15, 19]. Another important aspect is the dependency between the space requirement d_i and the current velocity v_i. In [6, 7] it was observed that in cellular automata models it makes a big difference whether a pedestrian occupies all cells passed in one time-step or not. Further, other authors suggested that the space requirement or step length is correlated with the speed [15, 17, 25]. In Section 2 we quantify empirically the relation between the required length d for one pedestrian to move with velocity v. Summing up, for the modelling of regular motions of pedestrians we modify the reduced one-dimensional social force model in order to meet the following properties: the force is always pointing in the direction of the intended velocity v_i^0; the movement of a pedestrian is only influenced by effects which are directly positioned in front; the required length d of a pedestrian to move with velocity v is $d = a + bv$. For detailed discussion we refer to [23]. To investigate the influence of the remote action both a force which treats pedestrians as simple hard bodies and a force according to Eq. (2), where a remote action is present, will be introduced. For simplicity we set $v_i^0 \geq 0$, $x_{i+1} > x_i$ and the mass of a pedestrian to $m_i = 1$.

Hard bodies without remote action

$$
F_i(t) = \begin{cases} \frac{v_i^0 - v_i(t)}{\tau_i} & : \quad x_{i+1}(t) - x_i(t) > d_i(t) \\ -\delta(t)v_i(t) & : \quad x_{i+1}(t) - x_i(t) \leq d_i(t) \end{cases} \quad \text{with} \quad d_i(t) = a_i + b_i v_i(t)
$$

$$(3)$$

The force which acts on pedestrian i depends only on the position, its velocity, and the position of the pedestrian $i + 1$ in front. As long as the distance between the pedestrians is larger than the required length d_i, the movement of a pedestrian is only influenced by the driving term. If the required length at a given current velocity is larger than the distance the pedestrian stops (i.e. the velocity becomes zero). This ensures that the velocity of a pedestrian is restricted to the interval $v_i = [0, v_i^0]$ and that the movement is only influenced by the pedestrian in front. The definition of d_i is such that the required length increases with growing velocity.

Hard bodies with remote action

$$
F_i(t) = \begin{cases} G_i(t) & : \quad v_i(t) > 0 \\ \max(0, G_i(t)) & : \quad v_i(t) \leq 0 \end{cases}
$$

$$(4)$$

with

$$
G_i(t) = \frac{v_i^0 - v_i(t)}{\tau_i} - e_i \left(\frac{1}{x_{i+1}(t) - x_i(t) - d_i(t)} \right)^{f_i}
$$

and

$$
d_i(t) = a_i + b_i v_i(t).
$$

Again the force is only influenced by actions in front of the pedestrian. By means of the required length d_i, the range of the interaction is a function of the velocity v_i. Two additional parameters, e_i and f_i, have to be introduced to fix the range and the strength of the force. Due to the remote action one has to change the condition for setting the velocity to zero. The above definition assures that pedestrian i stops if the force would lead to a negative

velocity. With the proper choice of e_i and f_i and sufficiently small time steps this condition gets active mainly during the relaxation phase. Without remote action this becomes important. The pedestrian can proceed when the influence of the driving term is large enough to get positive velocities.
This different formulation of the forces requires different update algorithms, see the discussion in [23].

3.2 Model Results

To enable a comparison with the empirical fundamental diagram from Section 2 we choose a system with periodic boundary conditions and a length of $L = 17.3\,m$. The values for the intended speed v_i^0 are distributed according to a normal-distribution with a mean value of $\mu = 1.24\,m/s$ and $\sigma = 0.05\,m/s$. In a one-dimensional system the influence of the pedestrian with the smallest intended speed masks jamming effects which are not determined by individual properties. Thus we choose a σ which is smaller than the empirical value and verified with $\sigma = 0.05, 0.1, 0.2\,m/s$, that a greater variation has no influence to the mean velocities at larger densities. For the parameters τ, a, b, e and f we choose identical values for all pedestrians (for a discussion see [23]). According to [22], $\tau = 0.61\,s$ is a reliable value.
Figure 4 shows the relation between the mean values of walking speed and density for hard bodies without remote action, according to the interaction introduced in Eq. (3). To demonstrate the influence of a required length dependent on velocity we choose different values for the parameter b. With $b = 0$

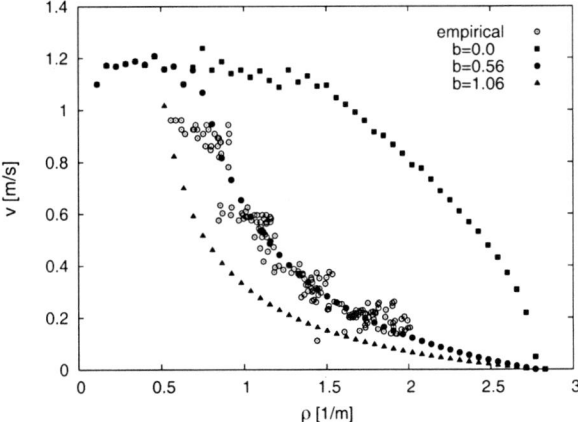

Fig. 4. Velocity-density relation for hard bodies without remote action according to Eq. (3) in comparison with the empirical data. The filled squares result from simple hard bodies with $a = 0.36\,m$ and $b = 0$. The introduction of a required length leads to a good agreement with the empirical data for $b = 0.56\,s$.

one gets simple hard bodies and the required length is independent of the velocity. In this case one gets a negative curvature of the function $v = v(\rho)$. The velocity-dependence controls the curvature and $b = 0.56\,s$ results in a good agreement with the empirical data. With $b = 1.06\,s$ we found a difference between the velocity-density relation predicted by the model and the empirical fundamental diagram. The reason for this discrepancy is that the interaction and the equation of motion do not describe the individual movement of pedestrians correctly. To illustrate the influence of the remote force, we fix the parameter $a = 0.36\,m$, $b = 0.56\,s$ and set the values which determine the remote force to $e = 0.07\,N$ and $f = 2$.

The fundamental diagram for the interaction with remote action according to Eq. (4) is presented in Fig. 5. The influence is small if one considers the velocity-dependence of the required length. But with $b = 0$ one gets a qualitatively different fundamental diagram. The increase of the velocity can be expected due to the effective reduction of the required length. The gap at $\rho \approx 1.2\,m^{-1}$ is surprising. It is generated through the development of distinct density waves, see [23], as are well known from highways. From experimental view we have so far no hints to the development of strong density waves for pedestrians, see Section 2. The width of the gap can be changed by variation of the parameter f which controls the range of the remote force. Near the gap the occurrence of the density waves depends on the distribution of the individual velocities, too.

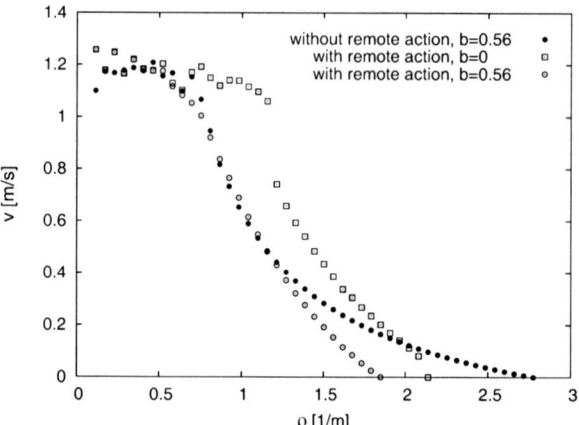

Fig. 5. Velocity-density relation for hard bodies with remote action according to Eq. (4) in comparison with hard bodies without a remote action (filled circles). Again we choose $a = 0.36\,m$. The parameter $e = 0.07\,N$ and $f = 2$ determine the remote force. With $b = 0$ one gets a qualitatively different fundamental diagram and a gap for the resulting velocities.

4 Summary and Outlook

In the investigation presented we determine the empirical relation between velocity and density for the single-file movement of pedestrians. The comparison of this fundamental diagram with the literature data for the movement in a plane shows a surprising agreement, see [3]. The conformance indicates that the internal friction and other lateral interferences, which are excluded in the single-file movement, have no influence on the relation at the density domains considered. The visual analysis of the video recording gives hints to possible effects, like the self-organization through marching in step, the optimized utilization of the available space at low velocities and the velocity dependence of step-length and safety margin. The data shows a linear relation between the velocity and the inverse of the density, which can be regarded as the required length of one pedestrian to move.

For the modelling we investigate the influence of the required length and remote action on the resulting velocity-density relation. We have introduced a modified one-dimensional social force model, which takes into account that the required length for moving with a certain velocity depends on the current velocity. The model-parameter can be adjusted to yield a good agreement with the empirical fundamental diagram. The remote action has a sizeable influence on the resulting velocity-density relation only if the required length is independent of the velocity. In this case one observes distinct density waves, which lead to a velocity gap in the fundamental diagram. For the model parameter b which correlates the required length with the current velocity, we have found that without remote action the value $b = 0.56\,s$ results in a velocity-density relation which is in good agreement with the empirical fundamental diagram. However, from the same empirical fundamental diagram one determines $b = 1.06\,s$. We conclude that a model which reproduces the right macroscopic dependency between density and velocity does not necessarily describe correctly the microscopic situation, and the space requirement of a person at average speed is much less than the average space requirement. This discrepancy may be explained by the "short-sightedness" of the model. Actually, pedestrians adapt their speed not only to the person immediately in front, but to the situation further ahead, too. This gives a much smoother movement than the model predicts. The investigation presented provides a basis for a careful extension of the modified social force model and an upgrade to two dimensions including further interactions.

References

1. W. M. Predtetschenski and A. I. Milinski, Personenströme in Gebäuden - Berechnungsmethoden für die Projektierung, Verlagsgesellschaft Rudolf Müller, Köln-Braunsfeld (1971)
2. U. Weidmann, Transporttechnik der Fußgänger, Schriftenreihe des IVT Nr. 90, zweite ergänzte Auflage, ETH Zürich (1993)
3. A. Seyfried, B. Steffen, W. Klingsch and M. Boltes, J. Stat. Mech. P10002 (2005)
4. P. D. Navin and R. J. Wheeler, Traf. Engin. 39, 31-36 (1969)
5. S. P. Hoogendoorn and W. Daamen, Transp. Sci. 39/2, 0147-0159 (2005)
6. T. Meyer-König, H. Klüpfel and M. Schreckenberg, Assessment and analysis of evacuation processes on passenger ships by microscopic simulation, in M. Schreckenberg and S. D. Sharma (eds.), Pedestrian and Evacuation Dynamics, Springer, Berlin (2002)
7. A. Kirchner, H. Klüpfel, K. Nishinari, A. Schadschneider and M. Schreckenberg, J. Stat. Mech. P10011 (2004)
8. S. P. Hoogendoorn, P. H. L. Bovy and W. Daamen, Microscopic pedestrian wayfinding and dynamics modelling, in M. Schreckenberg and S. D. Sharma (eds.), Pedestrian and Evacuation Dynamics, Springer, Berlin (2002)
9. www.rimea.de
10. M. Muramatsu, T. Irie and T. Nagatani, Physica A 267, 487-498 (1999)
11. V. J. Blue and J. L. Adler, J. Transp. Res. Board 1678, 135-141 (2000)
12. C. Burstedde, K. Klauck, A. Schadschneider and J. Zittartz, Physica A 295, 507-525 (2001)
13. K. Takimoto and T. Nagatani, Physica A 320, 611-621 (2003)
14. A. Keßel, H. Klüpfel, J. Wahle and M. Schreckenberg, Microscopic simulation of pedestrian crowd motion, in M. Schreckenberg and S. D. Sharma (eds.), Pedestrian and Evacuation Dynamics, Springer, Berlin (2002)
15. D. Helbing and P. Molnár, Phys. Rev. E 51, 4282-4286 (1995)
16. S. P. Hoogendoorn and P. H. L. Bovy, Transp. Res. Rec. 1710, 28-36 (2000)
17. P. Thompson and E. Marchant, Fire Safety Journal 24, 131 (1995)
18. V. Schneider and R. Könnecke, Simulating evacuation processes with ASERI, in M. Schreckenberg and S. D. Sharma (eds.), Pedestrian and Evacuation Dynamics, Springer, Berlin (2002)
19. P. Molnár, Modellierung und Simulation der Dynamik von Fußgängerströmen, Shaker, Aachen (1996)
20. D. Helbing, I. Farkas and T. Vicsek, Phys. Rev. Lett. 84, 1240-1243 (2000)
21. D. Helbing, I. Farkas, and T. Vicsek, Nature 407, 487-490 (2000)
22. T. Werner and D. Helbing, The social force pedestrian model applied to real life scenarios, in E. R. Galea (ed.), Pedestrian and Evacuation Dynamics, CMS Press, London (2003)
23. A. Seyfried, B. Steffen and T. Lippert, Physica A 368, 232-238 (2006)
24. B. D. Hankin and R. A. Wright, Operational Research Quarterly 9, 81-88 (1958)
25. J. L. Pauls, Suggestions on evacuation models and research questions, Conference Proceedings of the 3rd Int. Symp. on Human Behaviour in Fire (2004).

Flow-Density Relations for Pedestrian Traffic

Winnie Daamen and Serge P. Hoogendoorn

Delft University of Technology, Stevinweg 1, 2628 CN Delft, The Netherlands

Summary. This paper discusses the validity of first-order traffic flow theory to describe two-dimensional pedestrian flow operations in case of an oversaturated bottleneck upstream of which a large high-density region has formed. Pedestrians passing the same cross-section inside of the congested region appear to encounter different flow conditions. In the lateral center, high densities and low speeds are observed. However, on the boundary of the congested region, pedestrians may walk in nearly free flow conditions. Visualising pedestrian flow data in the flow-density plane results in a large scatter of points having similar flows (bottleneck capacity), but different densities. Observations on congestion of pedestrian traffic over the total width of the cross-section are found to belong to a set of different fundamental diagrams instead of a single one. This has consequences for the estimation of the fundamental diagram describing pedestrian traffic.

1 Introduction

Insight into pedestrian behaviour is essential in the planning and design process of public pedestrian facilities, such as transfer stations, airports, inner cities and shopping malls. Managing pedestrian flows through these facilities requires knowledge of pedestrian flow characteristics as well as of the walking behaviour underlying these characteristics. Designers of pedestrian facilities often use flow characteristics to determine levels-of-service on specific parts of the facility. Given the fact that, on average, pedestrians behave the same under similar average conditions, a statistical relation exists between speed, flow and density - the fundamental diagram. Many researchers have reported their empirical findings on this particular aspect, including the flow-density relation for various types of infrastructure, flow composition, etc.

First-order pedestrian traffic flow theory combines the use of the fundamental diagram with the conservation of pedestrians, which holds equally for car traffic. This theory has among other things been applied in the pedestrian simulation tool SimPed [1]. Other continuum theories describing pedestrian traffic have been derived by Helbing [4], Hughes [8], and Hoogendoorn and Bovy [6].

Based on data collected in laboratory experiments, we derive fundamental diagrams using cumulative curves for two cross-sections. Next, we discuss the congested state observations in the flow-density diagram. To find an explanation for seemingly confusing results, we focus on the lateral positions where pedestrians pass a specific cross-section and look at the corresponding speeds. We show that the physical form of the congestion influences the observations of the flow on the total cross-section. This indicates an essential difference in the use of the fundamental diagram in pedestrian traffic compared to car traffic.

2 Experimental microscopic Pedestrian Data

The trajectory data used in this paper have been collected during controlled walking behaviour experiments performed in 2002 in a large hallway of the Faculty of Civil Engineering and Geosciences. Ten experiments have been conducted, where the participating pedestrians had different ages and genders. For more information on these experiments, see [2].

This paper only considers the narrow bottleneck experiment, in which the observed area had a length of 10 m and a width of 4 m. The experiment lasted about 15 min. The narrow bottleneck experiment is characterised by the presence of a bottleneck having a length of 5 m and a width of 1 m. The bottleneck width is such that pedestrians inside of the bottleneck are not able to pass each other. As pedestrian demand increases, it will exceed the capacity of the bottleneck at some moment. From that time onward, congestion appears just upstream of the bottleneck moving further up-stream over time (see Fig. 1; bottleneck on the left, pedestrians walk from right to left). After decreasing the demand, congestion resolved in due time. Individual pedestrian data has been extracted from digital video footage, allowing pedestrian trajectories to be determined with high accuracy [7].

Fig. 1. Overview of the narrow bottleneck experiment.

3 Derivation of the Fundamental Diagram

In this section, fundamental diagrams are derived from cumulative flow plots. A cumulative plot of pedestrians is a function $N(x,t)$ that represents the counted number of pedestrians passing a cross-section x from an arbitrary starting moment. The flow q measured at a cross-section x during a time period from t_1 to t_2 equals:

$$q(x, t_1, t_2) = \frac{N(x, t_2) - N(x, t_1)}{t_2 - t_1} \tag{1}$$

At each time instant t when a pedestrian passes cross-section x, the speed u of this pedestrian is measured as well. The density k at spot x and instant t is then derived from the fundamental relation between speed, flow, and density:

$$k(x, t) = \frac{q(x, t)}{u(x, t)} \tag{2}$$

These data have been aggregated for a fixed number of pedestrians (here $N = 30$) passing cross-section x. Usually, an aggregation is performed on a fixed period of time, but this may result in a very low number of observations when flows are low. Rather than fixed interval lengths, we consider fixed sample sizes. That is, starting at some time instant t_n, the number of observed pedestrians is accumulated until a fixed number N of observations are collected. Then, the average of the relevant traffic flow variable can be determined. From a statistical point of view, this method offers important merits [5]. Among other things, the averages have comparable statistical accuracy, independent of the occurring flow and flow composition.

Fundamental diagrams have been constructed, based on the data of the narrow bottleneck experiment. The cross-sections on which speed and flow are derived are situated both inside ($x = 4$ m) and in front ($x = 7$ m) of the bottleneck (see Fig. 2d). Figure 2 also shows observations in the three phase-spaces (speed-density phase-space in Fig. 2a; speed-flow phase-space in Fig. 2b; flow-density phase-space in Fig. 2c).

Although congestion occurs, pedestrians continue walking with speeds higher than 0.4 m/s. All three phase-spaces indicate high variance during congestion, whereas we would have expected a smaller range of observations. Specifically the large scatter at different densities for similar flows (equal to the bottleneck capacity) is remarkable (see dotted ellipse in Fig. 2c). One might hypothesize that walkers adapt their following and speed choice behaviour if confronted with high density and low speed conditions during a longer period of time. However, it appears that speeds, flows, and densities do not change during the period that congestion occurs.

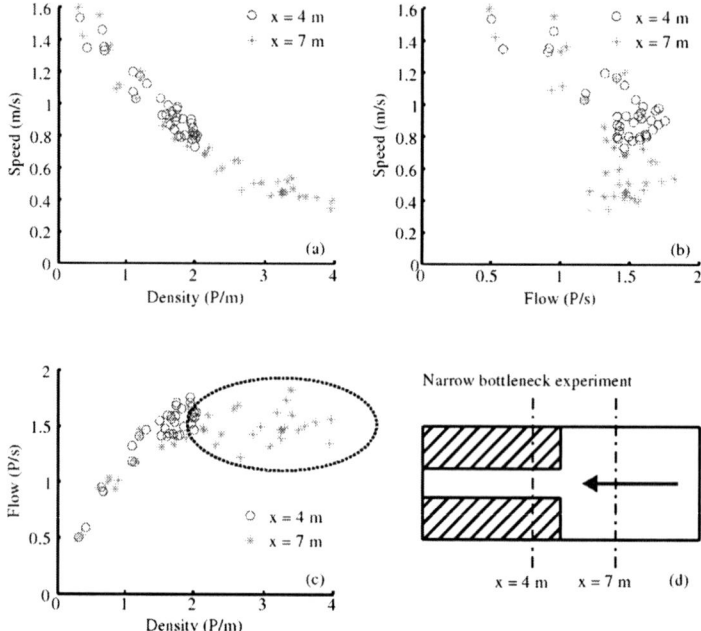

Fig. 2. Observations in the three phase-spaces of the narrow bottleneck experiment for two cross-sections: (a) Speed - density relation, (b) Speed - flow relation, (c) Flow - density relation, and (d) Overview of the experiment

4 First-Order Pedestrian Traffic Flow Theory

Let us reconsider Fig. 2, and focus on the large variance of the congested measurements in the flow-density plane. This variance cannot be contributed to random noise, representing changes in pedestrian behaviour during congestion, since the variation between the points is too large. In the following, an explanation for this phenomenon is presented.

First, let us consider a related situation in car traffic. Figure 3a shows the flow-density relation common for car traffic [3]. The situation is a lane drop of a two-lane road, being similar to the narrow bottleneck experiment for pedestrians. Two fundamental diagrams are applied: one for the two-lane part of the road upstream of the bottleneck (solid line) and another one (grey stripe-dotted line) inside the bottleneck, behind the lane drop. When flows are small, the free flow part of the fundamental diagram is observed (solid ellipse in Fig. 3a). When the flow becomes higher than the bottleneck capacity, congestion occurs upstream of the bottleneck. Observations are found on the congestion branch of the two lane fundamental diagram for the solid cross-section (dotted ellipse on the right in Fig. 3a) as well as on the capacity part of the single lane

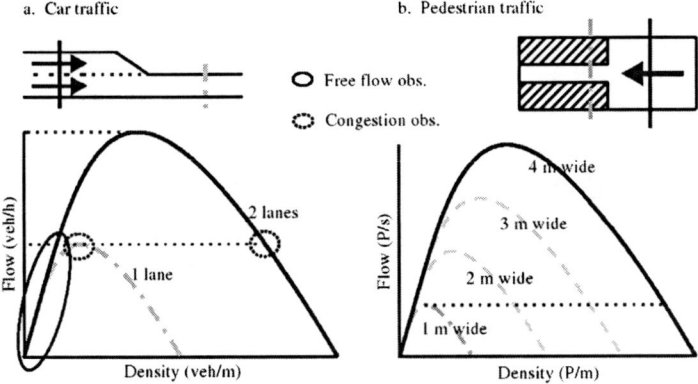

Fig. 3. Traffic flow theory for car traffic (a) and pedestrian traffic (b).

fundamental diagram for the grey stripe-dotted cross-section (dotted ellipse on the left in Fig. 3a).

Also in the narrow bottleneck experiment, flows and densities have been determined at two cross-sections (see Fig. 3b). The solid fundamental diagram is valid for the total width upstream of the bottleneck ($w = 4$ m), while the grey stripe-dotted fundamental diagram applies inside of the bottleneck ($w = 1$ m). However, when pedestrians do not occupy the complete width of the area upstream of the bottleneck, another fundamental diagram applies. Therefore, for each width between 1 m (bottleneck width) and 4 m (complete width of the area upstream of the bottleneck) a different fundamental diagram seems to apply (grey striped lines in Fig. 3b).

The width used by pedestrians upstream of the bottleneck is thus very important to be able to derive a proper fundamental diagram. When pedestrians use the complete width upstream of the bottleneck homogeneously, the fundamental diagram for a walkway of 4 m wide may be applied. This is similar to car traffic flow theory. However, the question is whether pedestrians do occupy the complete width of the area upstream of the bottleneck, and if they do, whether the distribution over the width is homogeneous.

Figure 1 shows that pedestrians form a funnel-shaped group while waiting to enter the bottleneck. Only part of the cross-section is thus occupied. However, Fig. 1 is only a snap shot, so Fig. 4 shows the average spatial form that waiting pedestrians adopt during the total congestion period. Figure 4 only shows the area upstream of the bottleneck, which is located on the left side of the figure between the lateral positions $y = 1.5$ m and $y = 2.5$ m. According to the figure, nearly all pedestrians pass the area just upstream of the bottleneck (between $x = 5$ m and $x = 5.5$ m). The further upstream of the bottleneck, the larger width is occupied by pedestrians. However, the outsides of the funnel are only used by about 10% of the pedestrians (in terms of density), while most pedestrians use the center of the cross-section. The scale values

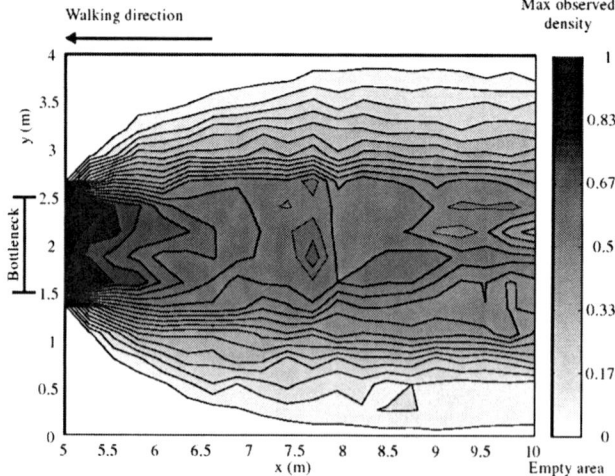

Fig. 4. Average density of pedestrians over 2-D space during congestion upstream of the bottleneck.

for the density are indicated as percentages of the maximum observed density on the right hand side of Fig. 4, where 1 indicates that all pedestrians have passed this location, and 0 indicates that none of the pedestrians have passed.

Figure 5 shows another macroscopic characteristic of pedestrian flows, namely speed. Flow and speed are directly observed on the cross-section, whereas density is derived from these characteristics. The speeds are aggregated (again over small groups) in relation to the (discrete) lateral positions of pedestrians passing cross-section $x = 7$ m, situated upstream of the bottleneck. During congestion, speeds observed at a cross-section vary significantly, depending

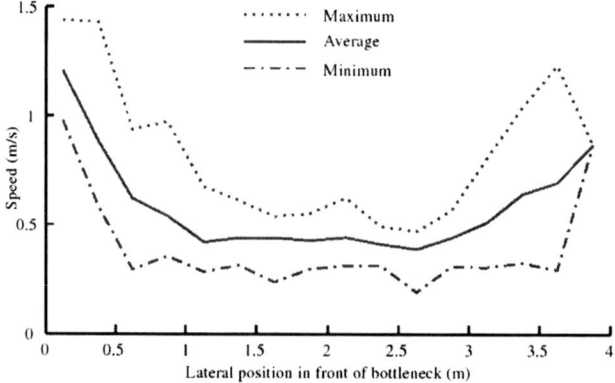

Fig. 5. Speeds as function of the lateral position in a cross-section upstream of the bottleneck during congestion.

on the pedestrian's lateral position. At the outsides of the funnel, pedestrians encounter (nearly) free flow conditions thus being able to maintain a higher speed. In the center of the flow pedestrians walk in congestion, with corresponding low speeds.

The previous paragraphs have indicated that it is not meaningful to de-rive a single flow-density relation for the complete width of an area upstream of the bottleneck. In fact, a number of points may be distinguished, forming together a single observation in the flow-density phase-space in Fig. 2. This is made clear in Fig. 6.

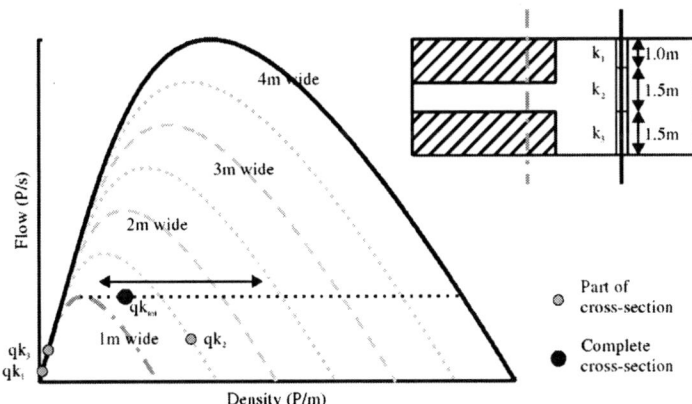

Fig. 6. Composition of a measurement point in the flow-density diagram for the complete cross-section.

As an example, we will distinguish three equilibrium regions, having similar speed and flow. The two outer regions (with densities k_1 and k_3) are more or less free flow, whereas congestion occurs in the lateral center (with density k_2). The three observations are indicated in the flow-density diagram by grey dots. However, the observation in the flow-density plane in Fig. 2 is based on a combination of these three equilibrium points. The result is that the aggregate observation (indicated by the black dot in Fig. 6) may be located anywhere on the horizontal dotted line and does not be-long to a specific fundamental diagram. The congestion branch of the fundamental diagram therefore cannot be estimated using aggregate observations for the complete width of a cross-section upstream of the bottleneck.

5 Conclusions

We have discussed results of dedicated experiments conducted to gain more insights into walking behaviour. This paper focuses on the applicability of fundamental diagrams to describe pedestrian flow operations in congestion.

One of the experiments involved a narrow bottleneck. Pedestrian demand was so large that congestion occurred. Based on observations at two cross-sections (one inside of the bottleneck and one upstream of the bottleneck), fundamental diagrams are shown describing pedestrian traffic.

It turns out that during congestion, pedestrians form a funnel-shaped group upstream of the bottleneck. Pedestrians meet different conditions at a specific cross-section (at the same time), depending on their lateral position. In the lateral center, high densities and low speeds are observed. However, on the boundary of the congested region, pedestrians may walk in nearly free flow conditions and literally walk around the congestion. An observation in the flow-density diagram thus consists of several observations, each describing a smaller lateral range of the cross-section, in which similar, homogeneous conditions occur.

We can therefore state that the congestion branch of the fundamental diagram cannot be estimated using aggregate observations for the complete width of a cross-section upstream of the bottleneck. Instead, homogeneous parts should be distinguished in a cross-section. The width of each homogeneous part also identifies the fundamental diagram a specific (aggregate) observation belongs to.

Acknowledgements

This research has been funded by the Social Science Research Council (MaGW) of the Netherlands Organization for Scientific Research (NWO).

References

1. Daamen, W.: *SimPed: A Pedestrian Simulation Tool for Large Pedestrian Areas.* In Conference Proceedings EuroSIW. CDROM. Simulation Interoperability Standards Organization, Orlando, Florida (2002)
2. Daamen, W.: *Modelling Passenger Flows in Public Transport Facilities.* PhD Thesis, Delft University of Technology. Delft University Press, Delft (2004)
3. Daganzo, C.F.: *Fundamentals of Transportation and Traffic Operations.* Pergamon, New York (1997)
4. Helbing, D.: *A Fluid Dynamic Model for the Movement of Pedestrians.* Complex Systems **6**, 391-415 (1992)
5. Hoogendoorn, S.P.: *Multiclass Continuum Modelling of Multilane Traffic Flow.* PhD Thesis, Delft University of Technology. Delft University Press, Delft (1999)
6. Hoogendoorn, S.P., and P.H.L. Bovy: *Gas-Kinetic Modeling and Simulation of Pedestrian Flows.* In Transportation Research Record: Journal of the Transportation Research Board, No. **1710**, TRB, National Research Council, Washington, D.C., pp. 28-36 (2000)
7. Hoogendoorn, S.P., and W. Daamen: *Extracting Microscopic Pedestrian Characteristics from Video Data.* In Proceedings of the Annual Meeting of the Transportation Research Board. CD-ROM. Transportation Research Board, National Research Council, Washington D.C. (2002)
8. Hughes, R.L.: *A Continuum Theory for the Flow of Pedestrians.* Transportation Research Part B **36**, 507-535 (2002)

Avoiding Inefficient Oscillations in Intersecting Vehicle and Pedestrian Flows by a Speed Limit

Rui Jiang[1,2] and Dirk Helbing[1]

[1] Institute for Economics and Traffic, Dresden University of Technology, Andreas-Schubert-Str. 23, D-01062 Dresden, Germany
[2] University of Science and Technology of China, Hefei 230026, P.R.China

Summary. This paper studies the coupled vehicle-pedestrian delay problem, taking into account the interactions between vehicles and pedestrians. In a previous paper, we found that for a large pedestrian arrival probability, coupled inefficient oscillations of pedestrian and vehicle flows emerge when pedestrians cross the street with a small time gap to approaching cars (aggressive pedestrians), while both pedestrians and vehicles benefit, when they keep some overcritical time gap (careful pedestrians). In this paper, we take into account the fact that the crossing time of a pedestrian group increases with its size. Our simulations show that when the crossing time of the pedestrian group is considered, the situation of careful pedestrians changes qualitatively. While vehicle and pedestrian flows for a low vehicle flow rate and small pedestrian arrival probability are efficient, oscillations occur when the vehicle flow rate and/or the pedestrian arrival probability increase. We propose a variable speed limit implemented by a LED display to avoid the inefficiency of vehicle and pedestrian flows for careful pedestrians.

1 Introduction

In a recent paper [1] (see also [2] for an analytical investigation), we have investigated the problem of interacting vehicle and pedestrian flows, a problem that has not been thoroughly studied in the past. In a way, the problem can be viewed as two dynamically coupled queues, which cannot be served at the same time, as pedestrians must cross the street at times when no vehicle passes and vice versa. It is found that for a large pedestrian arrival probability, coupled oscillations of pedestrian and vehicle flows emerge when the pedestrians cross the street with a small time gap to approaching cars (aggressive pedestrians), while both pedestrians and vehicles benefit, when they keep some overcritical time gap, i.e. if their safety coefficient is high (careful pedestrians). This may be interpreted as a slower-is-faster effect [3].

In our previous paper, however, we have assumed that the crossing time of the pedestrian group is the same as that for a single pedestrian. This is not so realistic. It would probably be more realistic to assume that the crossing time of a pedestrian group increases with the number of the pedestrians in the group. Apart from this improvement, we will study the case of relatively high vehicle flow rates, which has been neglected before.

It is found that, when the crossing time of the pedestrian group is considered, a different dynamics is found for careful pedestrians. While there is no inefficient vehicle and pedestrian flow for small vehicle and pedestrian arrival rates, inefficient oscillations occur when the vehicle flow rate and/or pedestrian arrival probability increase. In order to get rid of this inefficiency, we propose to use a speed limit.

Our contribution is organized as follows. In the next section, the model and results of our previous paper are briefly reviewed. In section 3, we present new results, taking into account the size-dependent crossing time of a pedestrian group. In section 4, we present the implementation of the speed limit, which can remove the inefficient oscillations. Conclusion are given in section 5.

2 Model and Previous Results

Our model to simulate interacting vehicle and pedestrian flows is as follows: First, within one incremental time step dt of 0.1s, we assume the arrival of one pedestrian along the roadside at a given crossing point O with probability p. An arriving pedestrian checks the traffic situation (Fig. 1(a)). We suppose that, when the safety criterion $d > d_0 + \sigma t_r v_n$ is satisfied, the pedestrian will cross the road. Here, σ is a safety coefficient, d_0 is the minimum safety distance of pedestrians, t_r the time needed for a pedestrian to traverse the street, d the distance to point O from the nearest upstream vehicle and v_n its velocity. We assume that when no other pedestrian is on the road, the safety coefficient chosen by a pedestrian is $\sigma = \sigma_0$. When other pedestrians are crossing the road, he or she is encouraged to cross as well and, therefore, chooses a smaller safety coefficiency $\sigma = \sigma_1$.

The computer simulation is carried out as follows: First, we scan the positions of the vehicles and find the nearest vehicle n upstream of point O. If a pedestrian is on the street, the net distance to the next object is specified

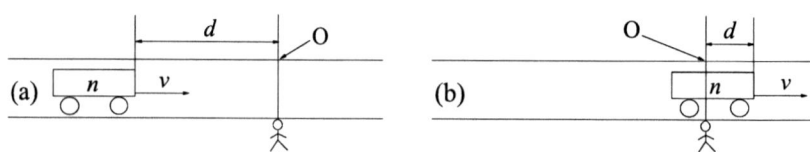

Fig. 1. Sketch of the vehicle-pedestrian delay system. In (b), the vehicle blocks point O, thus, the pedestrian cannot cross. This is consistent with our criterion, because d is negative and the condition $d > d_0 + \sigma t_r v_n$ is not met.

as $\Delta x = x_O - x_n(t)$, and the velocity of the object ahead into the driving direction is $v = 0$. Otherwise the distance and velocity are given by the next vehicle ahead, i.e. $\Delta x = x_{n+1} - x_n - l_{n+1}$ and $v = v_{n+1}$ (here l is the vehicle length). This enters the equation of vehicle motion according to the intelligent driver model (IDM) [4].

In each simulation time step, if a new pedestrian occurs at point O, he or she crosses the road if the safety criterion $d > d_0 + \sigma t_r v_n$ is satisfied. Otherwise, he or she will wait until the next time step. In a previous paper [1], we have assumed that if there is more than one pedestrian waiting at the roadside, the pedestrian group will need the time t_r to cross the road, independently of the number of pedestrians in the group.

In our simulations, open boundary conditions and parameter values as in Ref.[1] are used. The simulations show that, for aggressive pedestrians, the average delay time essentially remains constant when the arrival probability p is sufficiently small. When p increases, it begins to increase with growing values of p. Then, after reaching a maximum, it goes down with a further increase of the arrival rate p (Fig. 2). However, for careful pedestrians, the average delay is always small (Fig. 3).

A detailed investigation shows that the increase in the average delay of aggressive pedestrians originates from an alternating vehicle and pedestrian flow occuring at large pedestrian arrival probabilities (see Fig. 2(a) in [1]). Nevertheless, when p is very large, the stopped vehicle queue may occupy the whole length of the road section upstream of the crossing point. This results in a limitation and an eventual decrease of the average delay, as many pedestrians will cross the road together after a maximum waiting time. Careful pedestrians cannot stop vehicles. As a result, no oscillations occur, which is more efficient for pedestrians and vehicles. The simulations show that there exists a critical value $\sigma_1 \approx 0.96$ of the safety coefficient σ at which a transition between the two different system behaviors takes place.

3 Inefficient Oscillations Caused by Careful Pedestrians

In this section, the crossing time is set to $t_r = t_{r,0} + \beta(N_w - 1)$. Here t_r is the number of time steps needed to cross the road for a pedestrian group of size N_w, and $t_{r,0} = 20$ is the number of time steps needed to cross the road for a single pedestrian. In our simulations, the β parameter is set to 0.5. If we set $\beta = 0$, the improved model reduces to the original one.

Let us first consider the case of aggressive pedestrians. Our simulations show that, when the increased crossing time of pedestrian groups is taken into account, only slight changes occur (see filled triangles in Fig. 2).

Let us consider next the case of careful pedestrians. Fig. 3 shows the simulation results. One can see that for low vehicle flow rates the average delay does not change for small arrival probabilities p. However, when the arrival probability p is very high, oscillations occur (Fig. 4). This is because when p is very large,

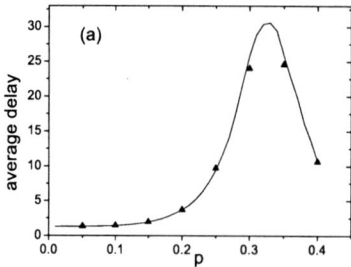 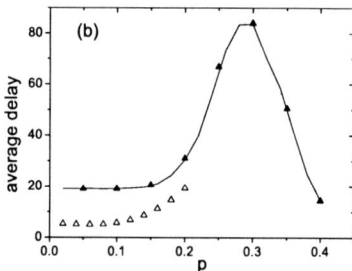

Fig. 2. Average time delay time to aggressive pedestrians ($\sigma_1 = 0.6$) as a function of their arrival probability p. In (a), the vehicle flow is 370 veh/h, while in (b) the vehicle flow is 810 veh/h. The solid lines represent the results of the original model, the filled triangles the results of our improved model without speed limit, and the empty triangles our results for the improved model with speed limit.

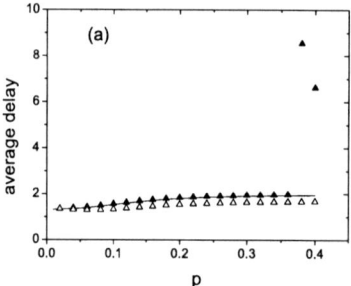 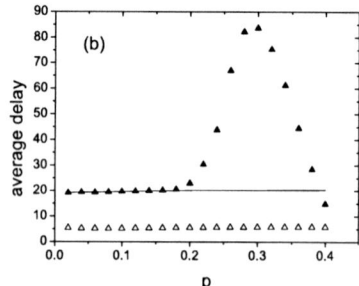

Fig. 3. Average time delay to careful pedestrians ($\sigma_1 = 1.0$) as a function of their arrival probability p for a vehicle flow rate of (a) 370 veh/h and (b) 810 veh/h.

a large group of waiting pedestrian can form in a short time. This group needs quite some time to cross the road and therefore, it can stop vehicles.

With an increase in the vehicle flow rate, the transition from efficient flows to inefficient oscillations occurs at smaller values of p (Fig. 3(b)). This is because with an increase in the vehicle flow rate, time gaps large enough to allow a crossing of pedestrians become less frequent. As a result, even for the same pedestrian arrival probability p, the group of waiting pedestrians usually becomes larger in our improved model when the vehicle flow rate is high. Therefore, oscillations occur easier.

4 Implementation of a Speed Limit

From the above section, we know that the reason for inefficient oscillations is the formation of large pedestrian group. Therefore, in order to avoid the inefficiency of alternating vehicle and pedestrian flows in the case of careful pedestrians, it is necessary to reduce the size of waiting pedestrian groups. This can be achieved by a variable speed limit implemented as follows.

A LED display is placed upstream of the crossing point. It enforces a speed reduction with a cycle period τ. While the speed limit is turned off for a fraction α of the time period, it slows down vehicles for the remaining fraction $(1 - \alpha)$ of the cycle time τ.

In our simulations, the LED display is placed 50 m upstream of the crossing point. Its period is set to $\tau = 30$ s. α is set to the value 0.67, i.e., the speed limit is activated for 10 s in a period. When the speed limit is turned on, vehicles within 50 m upstream of the LED will slow down: their maximum velocity decreases to the speed limit $v_{max,s}$ displayed on the LED display. In our simulations, $v_{max,s}$ is set to 5 m/s.

In Fig. 3(b), we show the simulation results for a speed limit. One can see that not only the oscillations are removed at large values of p, but also the average delay decreases dramatically compared with the original model. In Fig. 4(b), we show a spacetime plot of the vehicle trajectories. Compared with Fig. 4(a), one can see that a speed limit increases the frequency of large time gaps. Before reaching the time t_0, the speed limit has already generated a large time gap, which is available at the crossing point at time t_1. In this way, no large pedestrian group builds up. Consequently, vehicles are not anymore stopped by pedestrians, and inefficient oscillations are avoided.

In Fig. 3(a), we show the simulation results for a speed limit and a low flow rate. One can see that the oscillations disappear for large values of the pedestrian arrival probability p. Compared with the original model, the average delay decreases only slightly. This is because large time gaps occur frequently in this model even without a speed limit.

Finally, we point out that, when the vehicle flow is high, a speed limit can decrease the average delay also for small arrival probabilities p of aggressive pedestrians, although the oscillations cannot be removed (see Fig. 2(b)).

5 Conclusions

In this contribution, we have extended a previous model for the interactions of pedestrians and vehicles by considering the size-dependent crossing time of pedestrian groups. Our simulations show that this changes the system dynamics qualitatively for careful pedestrians. While vehicle and pedestrian flows are efficient for low vehicle flow rates and small arrival probabilities p, inefficient alternating flows occur when the vehicle flow rate and/or the pedestrian arrival probability p increases. However, for aggressive pedestrians, there is no qualitative change.

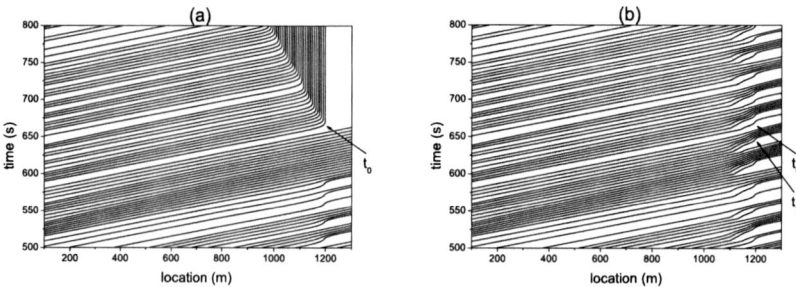

Fig. 4. Spacetime plots of vehicle trajectories for $p = 0.3$, $\sigma_1 = 1.0$ and a vehicle flow rate of 680 veh/h. The simulation results are (a) for the improved model without speed limit (b) for the improved model with a speed limit. In (a), one can see that, before time t_0, the time gaps are so small that no pedestrian can cross at point O for a long time period. Therefore, a lot of pedestrians are waiting to cross when a large time gap becomes available at time t_0. This large pedestrian group needs long to cross the road, so that vehicles will form long queues.

In order to avoid the inefficiency of oscillatory vehicle and pedestrian flows in the case of careful pedestrians, we have proposed to implement a speed limit. It is shown that in this way the inefficiency of vehicle and pedestrian flows is removed for careful pedestrians. Furthermore, it is shown that, even when the vehicle flow is high and the pedestrian arrival probability p is small, a speed limit can decrease the average delay both for careful pedestrians and aggressive pedestrians. This is reached by generating larger time gaps between successive vehicles.

Acknowledgements

The authors thank for partial financial support by the Chinese National Natural Science Foundation (Grant No. 10404025,10272101, 10532060), the National Basic Research Program of China (2006CB705500), the Alexander von Humboldt Foundation, and the German Research Foundation (DFG project He 2789/7-1).

References

1. R. Jiang, D. Helbing, P.K. Shukla and Q.S. Wu, Physica A **368**, 568 (2006).
2. D. Helbing, R. Jiang and M. Treiber, Phys. Rev. E**72**, 046130 (2005).
3. D. Helbing, I.J. Farkas and T. Vicsek, Nature **407**, 487 (2000).
4. M. Treiber, A. Hennecke and D. Helbing, Phys. Rev. E **62**, 1805 (2000).

Microscopic Calibration and Validation of Pedestrian Models: Cross-Comparison of Models Using Experimental Data

Serge P. Hoogendoorn and Winnie Daamen

Transport & Planning Department, Delft University of Technology, 2628 CN Delft, The Netherlands

Summary. This contribution proposes a new approach to estimate model parameters of microscopic pedestrian models using individual pedestrian trajectory data. To this end, a generic approach is proposed that enables parameter identification for microscopic models in general and in particular for walker models. The application results provide new insight into the behavior of individual pedestrians, inter-pedestrian differences, as well as the resulting pedestrian flow characteristics. By comparing different models of increasing complexity, it is investigated which of the model amendments are significant from a statistical point of view and which are not. It is shown that besides anisotropy, finite reaction times play an important role in correctly describing microscopic walking behavior. The implications of these findings in the microscopic description of pedestrians flows are considered by studying the predicted flow operations at a narrow bottleneck. It turns out that the finite reaction times have a significant effect on the pedestrian flow operations.

1 Introduction

In general, calibration of microscopic pedestrian models is performed by comparing aggregate model outcomes (flows, speeds, densities, etc.), predicted macroscopic relations (e.g. speed density curves), or emerging spatio-temporal patterns (dynamic lane formation, formation of diagonal strips in crossing flows) with macroscopic empirical data (if available) or expert opinion (does the model act as expected). In doing so, it has been shown that a number of pedestrian flow models are able to predict macroscopic flow conditions with reasonable accuracy; see for example [1–3].

There are many reasons why a macroscopic approach to microscopic model calibration will not always yield the desired results. For example, inter-pedestrian differences expressed by the variability in model parameters cannot be determined using macroscopic data. Furthermore, it is unclear if microscopic models are able to describe individual walking behavior accurately, or if they mainly provide a reasonable 'average' macroscopic prediction. This being the case, there is no way to assess whether the behavioral assumptions underlying a microscopic model are valid or not. If not, it is doubtful whether the microscopic model is sufficiently generic and able to predict pedestrian behavior in other situations than the model was calibrated for.

This main contribution of this paper is the development of an estimation approach for calibrating microscopic models using pedestrian trajectory data. These data can and have been collected using a variety of systems, such as video and infrared [4, 5]. A continuous time microscopic walker model is generalized to include different aspects of walking that are deemed important. It is emphasized that this is *not a benchmarking study* in the sense that all microscopic pedestrian models are cross-compared. We also do not consider route choice behavior.

We have focused on the microscopic walker model NOMAD [1], which is similar to the well-known social-forces model [2]. An important contribution of the work presented in the paper pertains to new insights into important processes in walking behavior, which is determined by cross-comparing the effects of different model extensions. That is, we determine which of the model parameters are important in predicting microscopic pedestrian behavior, as well as considering their statistical properties. In doing so, we will study two potentially important aspects of walking behavior, namely *anisotropy* and *finite reaction times*. Lastly, since we establish model parameters for individual pedestrians, we will show inter-pedestrian differences in walking behavior and their consequences for pedestrian flow operations.

2 Considered Walker Models

In this contribution, we focus on a simplified version of the model of [1], which is similar to the original social-forces model [2]. This section presents the basic model and the amendments that are considered for further analysis. It is noted that many extensions to both the NOMAD model and the social-forces model have been proposed. For the purpose of this paper, it is however not necessary to consider these.

2.1 Basic Model

The basic model predicts the two-dimensional acceleration vector $\mathbf{a}_p(t)$ as

$$\mathbf{a}_p(t) = \frac{\mathbf{v}_p^0 - \mathbf{v}_p(t)}{T_p} - A_p \sum_{q \in Q_p} \mathbf{u}_{pq}(t) e^{-\frac{d_{pq}(t)}{R_p}} \tag{1}$$

where $\mathbf{v}_p$ denotes the velocity of pedestrian p; Q_p denotes the set of pedestrians q that influence pedestrian p. Finally, we have

$$d_{pq}(t) = \|\mathbf{r}_q(t) - \mathbf{r}_p(t)\| \quad \text{and} \quad \mathbf{u}_{pq}(t) = \frac{\mathbf{r}_q(t) - \mathbf{r}_p(t)}{d_{pq}(t)} \tag{2}$$

The basic model has four *pedestrian specific parameters*, namely the free speed $V_p^0 = \|\mathbf{v}_p^0\|$, the acceleration time T_p, the interaction constant A_p and the interaction distance R_p that are to be estimated from data. Note that the desired walking direction $\mathbf{e}_p^0 = \mathbf{v}_p^0/V_p^0$ is determined by pedestrian route choice and is assumed known.

2.2 Instantaneous Model Including Anisotropy

Anisotropy implies that pedestrians will only – or at least mainly – react to pedestrians in front of them. The basic model has been amended to include anisotropy as follows [1]:

$$\mathbf{a}_p(t) = \frac{\mathbf{v}_p^0 - \mathbf{v}_p(t)}{T_p} - A_p \sum_{q \in Q_p} \mathbf{u}_{pq}(t) e^{-\frac{d_{pq}^*(t)}{R_p}} \mathbf{1}_{\mathbf{u}_{pq}(t) \cdot \mathbf{v}_p(t) > 0} \tag{3}$$

with

$$d_{pq}^*(t) = \frac{\mathbf{u}_{pq}(t) \cdot \mathbf{v}_p(t)}{\|\mathbf{v}_p(t)\|} + \eta_p \frac{\mathbf{u}_{pq}(t) \cdot \mathbf{w}_p(t)}{\|\mathbf{v}_p(t)\|} \tag{4}$$

where $\mathbf{w}_p(t)$ is the vector perpendicular to the velocity vector $\mathbf{v}_p(t)$, with the same length as $\mathbf{v}_p(t)$; $\eta_p > 1$ is a pedestrian specific factor that describes differences in pedestrian reaction to stimuli directly in front and stimuli from the sides of the pedestrians, which is to be estimated from the available microscopic data. The indicator function $\mathbf{1}_{\mathbf{u}_{pq}(t) \cdot \mathbf{v}_p(t) > 0}$ is one if pedestrian q is in front $(\mathbf{u}_{pq}(t) \cdot \mathbf{v}_p(t) > 0)$ if p and zero otherwise $(\mathbf{u}_{pq}(t) \cdot \mathbf{v}_p(t) < 0)$.

2.3 Model Including Finite Reaction Time

To determine if the reaction time can be neglected or not, the final model considered is a retarded (or delayed) model:

$$\mathbf{a}_p(t + \tau_p) = \frac{\mathbf{v}_p^0 - \mathbf{v}_p(t)}{T_p} - A_p \sum_{q \in Q_p} \mathbf{u}_{pq}(t) e^{-\frac{d_{pq}^*(t)}{R_p}} \mathbf{1}_{\mathbf{u}_{pq}(t) \cdot \mathbf{v}_p(t) > 0} \tag{5}$$

where $\tau_p > 0$ is the pedestrian-specific reaction time (or rather, perception-response time) to be estimated from the microscopic data. Pedestrians are thus assumed to have a delayed response to the observations they make at time instant t. We expect that the reaction times will be between 0.1 s and 0.8 s.

2.4 Approach to Model Estimation

The parameters to be estimated are the free speed V_p^0, the acceleration time T_p, the interaction factor A_p, the interaction distance R_p, and the reaction time τ_p. The latter is determined by considering all plausible reaction time values – i.e. between 0.1 s and 0.8 s – and afterwards determining which value yields the best performance (in terms of log-likelihood). For the anisotropy factor η_p, different values have been considered and cross-compared to test which yields good model performance, after which one fixed value was chosen *for all pedestrians*; this was done to keep the number of parameters small. The available observations are pedestrian trajectories (the location $\mathbf{r}_p(t_k)$ as a function of time instant t_k, for $k = 1, \ldots, n$) of all pedestrians p. From these data, all relevant quantities can be derived either directly or by applying finite differences. For the data considered in the remainder, observations are present each 0.1 s (i.e. $t_k = t_0 + 0.1k$).

2.5 Maximum Likelihood Estimation

Most continuous time microscopic walker models, including those considered in this contribution can be expressed in the following form:

$$\mathbf{a}_p(t + \tau_p) = \mathbf{f}_p(\mathbf{v}_p(t), \mathbf{r}_q(t) - \mathbf{r}_p(t), \ldots | \theta_p) + \varepsilon_p(t) \tag{6}$$

where θ_p denotes the set of unknown parameters, including the reaction time. The error vector $\varepsilon_p(t)$ is introduced to reflect errors in the modeling, similar to the error term used in multivariate linear regression. Note that the error vectors $\varepsilon_p(t)$ are generally serially correlated (i.e. $\varepsilon_p(t_k)$ and $\varepsilon_p(t_{k-1})$ have a large positive correlation). For now, we assume that the error term is normally distributed with mean zero and standard deviation σ_p (pedestrian specific). Since we can determine all relevant variables (positions, distances, speeds, relative speeds) directly from available experimental data, we can use Eq. (6) to determine a prediction for the retarded acceleration directly from the data. The prediction is clearly dependent on the model parameters to be estimated and can be compared with the observed acceleration $\mathbf{a}_p^{obs}(t + \tau_p)$. According to the model, the difference between the prediction and the observation follows the normal distribution with mean 0 and standard deviation σ_p. The likelihood L_k of a single prediction step, say from time t_k to time t_{k+1}, is related directly to the probability density $g(\varepsilon)$ of the normal distribution. More specifically:

$$L_k(\theta_p, \sigma_p^2) = \frac{1}{\sigma_p\sqrt{2\pi}} \exp\left(-\frac{\left(\mathbf{a}_p^{obs}(t + \tau_p) - \mathbf{a}_p(t + \tau_p)\right)^2}{2\sigma_p^2}\right) \quad \text{s.t.} \tag{7}$$

$$\mathbf{a}_p(t + \tau_p) = \mathbf{f}_p(\mathbf{v}_p^{obs}(t), \mathbf{r}_q^{obs}(t) - \mathbf{r}_p^{obs}(t), \ldots | \theta_p) \tag{8}$$

Considering an entire sample of subsequent acceleration observations and neglecting correlations between subsequent samples (serial correlation), the likelihood of the observation given the model parameters becomes:

$$L = L(\theta_p, \sigma_p^2) = \prod_{k=1}^{n} \frac{1}{\sigma_p\sqrt{2\pi}} \exp\left(-\frac{\left(\mathbf{a}_p^{obs}(t + \tau_p) - \mathbf{a}_p(t + \tau_p)\right)^2}{2\sigma_p^2}\right) \tag{9}$$

where n denotes the number of time-instants t_k for which an observation is available. The log-likelihood equals:

$$\tilde{L}(\theta_p, \sigma_p^2) = -\frac{n}{2} \ln\left(2\pi\sigma_p^2\right) - \frac{1}{2\sigma_p^2} \sum_{k=1}^{n} \left(a_p^{obs}(t + \tau_p) - a_p(t + \tau_p)\right)^2 \quad (10)$$

Maximum-Likelihood (ML) estimation involves finding the parameters that maximize the (log-) likelihood. A necessary condition for the optimum allows determination of the standard deviation:

$$\frac{\partial \tilde{L}(\theta_p, \sigma_p^2)}{\partial \sigma_p^2} = 0 \Rightarrow \hat{\sigma}_p^2 = \frac{1}{n} \sum_{k=1}^{n} \left(a_p^{obs}(t + \tau_p) - a_p(t + \tau_p)\right)^2 \quad (11)$$

From Eq. (11) we see that the ML estimate for the variance of the error term is given by the MSE of the predictions and the observations. For the remaining parameters, the ML estimates can be determined by numerical optimization, i.e.

$$\hat{\theta}_p = \arg \max \tilde{L}(\theta_p, \hat{\sigma}_p^2) \quad (12)$$

with

$$\tilde{L}(\theta_p, \hat{\sigma}_p^2) = -\frac{n}{2} \ln\left(\frac{2\pi}{n} \sum_{k=1}^{n} \left(a_p^{obs}(t + \tau_p) - a_p(t + \tau_p)\right)^2\right) - \frac{n}{2} \quad (13)$$

This expression shows that maximization of the log-likelihood is equivalent to minimization of the mean squared error (MSE).

2.6 Covariance Estimates

To approximate the covariance matrix of the estimated parameters, we can use the so-called *Cramér-Rao lower bound* [6], stating that:

$$var\left(\theta_p\right) \geq -E\left(\nabla^2 \tilde{L}\right) \quad (14)$$

Since ML is asymptotically efficient, we can show that the asymptotic variance of the parameters is given by the right-hand side of Eq. (14); see [6]. In the remainder, this approximation is used to determine an estimate for the covariance of the estimated parameters.

The Cramér-Rao bound provides important insights into the statistical properties of the models by providing estimates for the model standard error and the statistical correlation between the parameter estimates. The standard-errors can be used to determine whether the model parameters are not equal to zero in a statistical sense. The correlation matrix provides additional insight into the statistical properties of the estimates, for instance by explaining large standard errors cause by large correlation between estimates.

We will use the so-called *likelihood-ratio test* to test whether the one model is better than the other. We emphasize that the log-likelihood test accounts for the number of parameters via the degrees of freedom, thereby enabling the comparison of simple and complex models.

2.7 Inter-Pedestrian Parameter Correlation

Besides the correlation in the parameter estimates determined via the Cramér-Rao bound, inter-pedestrians differences can be determined. In the ensuing, this is achieved by computing the mean, variance and inter-personal correlation of the individual parameter estimates for the different pedestrians. These statistics provide insight into the behavioral differences between pedestrians, and the inter-pedestrian correlation between the parameter estimates.

3 Experimental Data Used for Model Calibration

Microscopic pedestrian data is gradually becoming more available. Different observation techniques have been developed recently that enable collecting the time-space behavior of individual pedestrians. Examples are video [4] and infrared sensors [5]. The trajectory data used in the ensuing of this contribution have been collected from walking experiments performed in 2002. These walking experiments were conducted in a large hallway of the Faculty of Civil Engineering and Geosciences. The group of pedestrians participating consisted of people of different ages and genders. Using a digital camera mounted at the ceiling of the hallway and dedicated software to process the digital footage into pedestrian trajectories, all trajectories of the pedestrians participating in the experiments were determined. For the remainder of this contribution, data from the narrow bottleneck experiment are used. The narrow bottleneck experiment was characterized by a high pedestrian demand trying to pass through a narrow bottleneck of 1 m width. Since the demand was larger than the bottleneck capacity, the bottleneck became oversaturated resulting in congestion. For a detailed description of the experiment, the data and their characteristics, we refer to [4].

4 Estimation Results

To get a general impression of the model performance as well as a better understanding of the important behavioral processes underlying walking behavior, this section presents the results of cross-comparing the model predictions with a *zero-acceleration reference model* (i.e. assuming constant velocities for all pedestrians). The parameters of the models have been estimated from the experimental data discussed in the preceding section. The performance of the models is cross-compared based on the overall relative increase in the log-likelihood compared to the null-log likelihood (stemming from the zero-acceleration model), and the percentage of models that passed the likelihood ratio test.

Tab. 1 shows an overview of the estimation results. The table shows that the average differences between model performances are considerable. Especially

Model type	% improvement L̇	% passing LR-test
Basic model	6.4%	71%
Anisotropic model	7.7%	76%
Retarded anisotropic model	19.7%	83%

Table 1. Overview of estimation results.

the difference between the instantaneous models and the retarded models is relatively large (improvement of the log-likelihood of 6.4% and 7.7% for the respective instantaneous models, compared to 19.7% in case of the retarded anisotropic model).

4.1 Basic Model Results

From Tab. 1, we can conclude that the basic model passes the likelihood ratio test (LR test) in 71% of all cases. For the remaining 29%, the basic model *did show a higher log-likelihood than the null-model* (which assumes no acceleration), but the improvement was not large enough to pass the LR test. Tab. 2 shows an overview of the statistics of the parameter estimates. The estimates are plausible in terms of their magnitude. The table shows that the inter-pedestrian differences are most prominently reflected by the differences in the acceleration times T_p and the interaction distance R_p, as can be concluded from the CoV (Coefficient of Correlation) values of 0.32 and 0.50 respectively. Furthermore, it turns out that the inter-pedestrian correlations between the parameter estimates are generally small, except for the positive correlation between free speed and interaction distance (0.49).

Parameter	Mean	Stand. dev	CoV	Correlation amongst parameters			
				V_p^0	T_p	A_p	R_p
V_p^0 (m/s)	1.34	0.21	0.16	1	-0.20	0.02	0.49
T_p (s)	1.09	0.35	0.32		1	0.10	-0.36
A_p (m/s^2)	11.96	0.23	0.02			1	-0.16
R_p (m)	0.16	0.08	0.50				1

Table 2. Statistics of parameter estimates for the basic model.

4.2 Anisotropic Model

As concluded before, the anisotropic model with instantaneous reaction outperforms the basic model, although the improvement is rather limited. The LR test shows that the model improves significantly with respect to the naïve zero-acceleration model in case of 76% of all pedestrians considered.

Tab. 3 shows an overview of the parameter estimates for all individual estimates which passed the likelihood ratio test, as well as the correlation between the individual estimates. The results show that especially the acceleration time T_p has a relatively large standard deviation (judging from the CoV values), implying that the inter-pedestrian differences in acceleration times are large. This holds equally for the interaction distance R_p.

The correlation between the parameters reveals a considerable relation between the free speed and the interaction distance (positive correlation of 0.62), implying that on average, pedestrians having a large free speed V_p^0 have a large acceleration distance R_p. An interpretation of this (statistical) result might be that pedestrian with a high free speed have the tendency to better anticipate on pedestrians further away from them. However, the explanatory performance of the model is still limited, and care should be taken in interpreting the estimation results.

Other high correlations are found between the acceleration time T_p and the interaction factor A_p (negative correlation of 0.54), and between the interaction factor A_p and the interaction distance R_p (positive correlation of 0.46).

Parameter	Mean	Stand. dev	CoV	Correlation amongst parameters			
				V_p^0	T_p	A_p	R_p
V_p^0 (m/s)	1.32	0.22	0.17	1	-0.23	0.28	0.62
T_p (s)	0.96	0.24	0.25		1	-0.54	-0.32
A_p (m/s^2)	11.46	0.56	0.05			1	0.46
R_p (m)	0.33	0.09	0.27				1

Table 3. Statistics of parameter estimates for anisotropic model.

If we compare the estimates of the basic model with the estimates of the anisotropic model, we see that the estimates are similar, except for the interaction distance R_p. In the anistropic model, the interaction distance R_p is on average twice as large as in the non-anistropic model. Regarding the inter-pedestrian parameter differences, it turns out that the variability in the acceleration time T_p reduces substantially.

4.3 Retarded Anisotropic Model

The statistical analysis clearly reveals the importance of the finite reaction time in walking behavior modeling, as shown from the improvements of the model performances as indicated by Tab. 1: besides the fact that 83% of the considered cases passes the LR test, we can also see that the log-likelihood improvement over the naïve zero-acceleration model of 19.7% is much higher than for the non-retarded models (6.4% and 7.7% respectively).

Tab. 4 provides an overview of the average parameter values, their standard deviation and the inter-pedestrian correlation between the parameter estimates. It shows that in particular the standard deviations - and thus the inter-pedestrian differences - of the acceleration times T_p and interaction distances A_p are relatively large. Also note the medium inter-pedestrian differences in the reaction time τ_p (mean of 0.28 s and standard deviation of 0.07 s).

Parameter	Mean	Stand. dev	CoV	Correlation amongst parameters				
				V_p^0	T_p	A_p	R_p	τ_p
V_p^0 (m/s)	1.34	0.23	0.17	1	0.23	0.39	0.57	-0.02
T_p (s)	0.74	0.23	0.31		1	-0.23	-0.06	0.44
A_p (m/s^2)	11.33	0.64	0.06			1	0.36	-0.46
R_p (m)	0.35	0.11	0.31				1	-0.17
τ_p (s)	0.28	0.07	0.25					1

Table 4. Statistics of parameter estimates for the retarded anisotropic model.

As for the instantaneous models, from Tab. 4 we observe that the free speed V_p^0 and the interaction distance R_p are positively correlated (0.57). It also turns out that the reaction time τ_p and the interaction factor A_p are negatively correlated, implying that the reaction time and the interaction factor are to a certain extent mutually exclusive.

To gain more insight into the distributions of the parameter estimates, Fig. 1 shows histograms of the estimates, from which inter-pedestrian differences can be observed clearly. Also the shape of the parameter distributions becomes apparent. Note that all parameter distributions appear to be skewed rather than symmetric.

Fig. 2 shows the distribution of the reaction time estimates. Again the distribution appears to be skewed. The median reaction time equals 0.3 s; few pedestrians have a reaction time which is larger than the median of 0.3 s.

Let us finally note that the parameter estimates determined by applying the approach to data from other experiments are consistent (wide bottleneck, one-directional flow, bi-directional flow, etc.).

5 Consequences for Pedestrian Flow Modelling

The estimation results presented in the previous section show some interesting issues that may have important implications for microscopic pedestrian flow modeling. In particular, these findings relate to:

1. Inter-pedestrian differences as expressed by the variability in the parameter estimates.
2. Correlation between parameter estimates.
3. Importance of delays in the correct description of microscopic behavior.

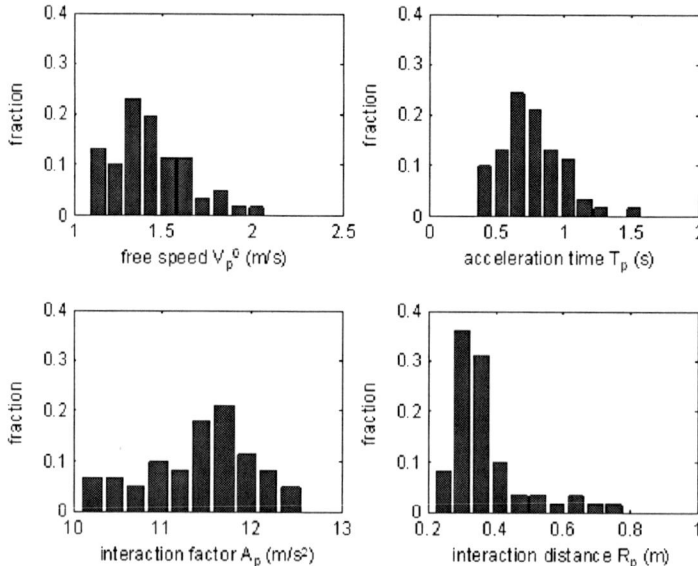

Fig. 1. Parameter distributions for 82 pedestrians using an anisotropy factor of 8 (retarded model).

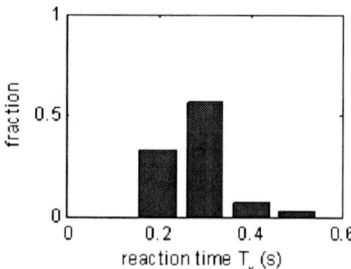

Fig. 2. Distribution of the reaction time estimates.

Regarding the first point, the current models can in general be amended easily when sufficient insight has been gained into the distribution of the parameters for the pedestrian population to be simulated. The distribution that needs to be used will be dependent on amongst other things gender distribution, trip purpose, and external conditions (such as weather). The variability in pedestrian behavior will cause macroscopic properties of the pedestrian flow to become stochastic. This holds in particular for the bottleneck capacity and the jam-densities (and consequently the queue lengths). The extent in which this occurs in beyond the scope of this paper. The second issue is of particular interest when generating pedestrians in a microscopic model. When

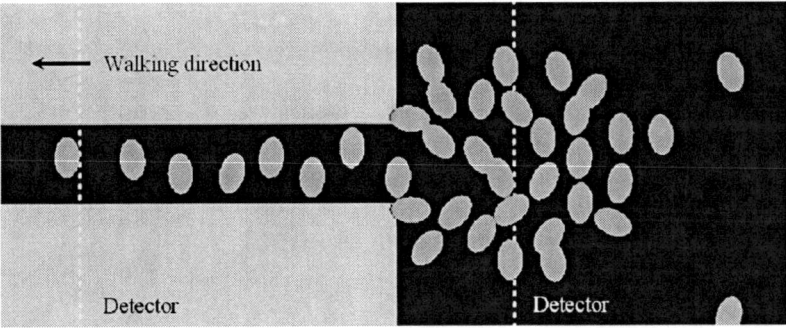

Fig. 3. Snapshot of situation at the narrow bottleneck as determined by the adapted NOMAD pedestrian simulation model. The figure shows the two detectors that have been used to compute the queue discharge rate. In the example shown, congestion is detected at detector 2. As a result, the flow observed at detector 1 is labeled as an observation of the queue discharge rate.

the behavioral parameters are generated, it is important to take into account the correlations during the parameter generation process.

With respect to the third point, considerable changes in the properties of the pedestrian flow dynamics are expected. Amongst the phenomena that are likely to occur are instabilities in the flow, congestion probabilities, reduction in the queue discharge rate once congestion sets in, etc.

To test whether this hypothesis is indeed correct, we have modified our model to include finite reaction times, and applied the modified simulation model on a simple test-case example (a narrow bottleneck). The bottleneck has a width of 1.0 m. From experimental observations, we expect a bottleneck capacity of 1.6 P/s (see for [4] details). The pedestrian demand equals 1.8 P/s. The parameters used in the simulation correspond to the estimates determined in the previous sections of this contribution. Fig. 3 shows a snapshot of a simulation for the considered scenario. The figure shows the locations of the detectors that have been used to collect the synthetic data on which our analyses are based.

When comparing the instantaneous and the retarded models, several things become apparent. For one, the flow becomes more unstable, more erratic. The considered flow operations are near oversaturation (volume to capacity almost equal to 1). From our simulation results, it turns out that before congestion occurs, the bottleneck capacity is higher than once congestion has set in (capacity drop). This phenomenon is quite common in car traffic, but apparantly it is also present in pedestrian flow operations. The reasons for this capacity drop is the fact that when congestion has set in, pedestrians are dispersed over part of the width of the walking area, i.e. they are not standing right in front of the bottleneck. When arriving at the bottleneck, they need to turn as well as to accelerate. This process is likely to be affected by the reaction time of the pedestrians: when the response to prevailing traffic conditions is retarded, both turning and acceleration will be delayed and thus capacity of the bottleneck is likely to decrease.

We also found that the predicted probability of breakdown occurring is larger when using retarded models than when using the instantaneous model. More specifically, the instantaneous model predicts congestion occurring in 54% of all simulations. For the retarded model with a reaction time of 0.3 s, congestion occurs more frequently, namely in 68% of all situations.

6 Conclusions and Recommendations

This contribution puts forward a generic approach for calibration of microscopic models. Using trajectory data, the approach enables estimation of the pedestrian-specific parameters of different walker models. The approach provides insight into the statistical properties of the estimates, as well as the performance of the models to which the calibration approach is applied.

In applying the approach to data collected during walking experiments, inferences could be made regarding the behavioral processes that are to be included in the modeling to ensure a realistic description of walking. It turns out that besides anisotropy, *finite reaction times* play an important role in correctly describing microscopic walking behavior. Furthermore, *inter-pedestrian differences* in walking behavior have been shown. Based on these findings, a small scale study was performed regarding the changes in the dynamic properties of the pedestrian flow model. It was observed that including a reaction time has a significant effect on the pedestrian flow operations, in particular with respect to the congestion probabilities.

Future research will entail estimating parameters of more advanced models. Furthermore, we will further study the macroscopic properties of microscopically calibrated pedestrian models. Further research should reveal how the observed finite reaction time and the inter-pedestrian differences affect the dynamic flow properties.

References

1. S. Hoogendoorn and P. Bovy: Optimal Control Applications and Methods **24**, 153-172 (2003)
2. D. Helbing, I. Farkas, and T. Vicsek: Nature **407**, 487-490 (2000)
3. H. Klüpfel, M. Schreckenberg and T. Meyer-König: in *Traffic and Granular Flow '03*, pages 357-372, Eds.: S. Hoogendoorn, S. Luding, P. Bovy, M. Schreckenberg, and D. Wolf (Springer 2005)
4. S. Hoogendoorn and W. Daamen: Transportation Science **39**(2), 147-159 (2005)
5. J. Kerridge, R. Kukla, A. Willis, A. Armitage, D. Binnie, and L. Lei: in *Traffic and Granular Flow '03*, pages 383-392, Eds.: S. Hoogendoorn, S. Luding, P. Bovy, M. Schreckenberg, and D. Wolf (Springer 2005)
6. G. Casella and R. Berger: *Statistical Interference* (Duxbury Press, Belmont, California, 1990)

The Simulation of Crowds at Very Large Events

Hubert Klüpfel

TraffGo HT GmbH, Falkstraße 73–77, 47057 Duisburg, Germany
e-mail: `kluepfel@traffgo-ht.com`

Summary. In this article, we show two examples for the application of pedestrian flow simulation and analysis: the World Youth Day 2005 (WYD) in Cologne and the (non-emergency) egress from a football stadium. Various circumstances are specific for religious events. The persons might perform rituals and therefore the patterns of movement or gathering are governed by rules that go beyond simple necessity or comfort. Furthermore, the persons are usually very much attracted by the (idealistic) aim of their pilgrimage. The final service at the WYD in Cologne, celebrated by the Pope, was the major event during the WYD. The paper is divided into three parts: The first section is concerned with the World Youth Day and the second with the egress from a football stadium. The final section summarizes the results, provides recommendations and concludes with the most important implications for the field of crowd dynamics simulation.

1 Description of the Model

1.1 Model Characteristics

The model is extensively described in [1–3]. It is similar to the model used in [4, 5] to simulate competitive egress behavior apart from the friction and the dynamic floor field. The model characteristics can be summarized as follows:

1. The geometry is respresented as a regular grid of quadratic cells where walls are represented as non-accessible cells. The cell size is 40cm.
2. Persons move on these cells. Their velocity may vary between 2 and 5 cells per time step. The length of a time step is 1 second.
3. Diagonal movement is possible and the diagonal distance is correctly accounted for (by a factor of $\sqrt{2}$).
4. The update is shuffled sequential [6], which is equivalent to an iterative conflict resolution [5].
5. There are as many static potentials (called floor fields in [4, 5, 7]) as there are exits. Each potential measures the distance to its exit.

6. There is no dynamic floor field.
7. The transition probabilities are given by $e^{\Delta p}$, where Δp is the difference between the potential of the current and the destination cell. The exit has potential 0.

1.2 Parameter Settings

The following table 1 contains the parameter values for the standard population used in the examples in sections 2 and 3. It is important to note that the reaction time distribution was deliberately chosen to be very low in order to get a worst case scenario. It is well known from empirical observations [8] that immediate detection of and reaction to an alarm leads to the highest rates of congestion.

Parameter	Minimum	Maximum	Mean	Std. Dev.	Unit
Free Walking Speed	2	5	3	1	m/s
Dawdling Probability	0	0.3	0.15	0.05	-
Reaction Time	0	10	5	2	s

Table 1. Parameters of the standard population.

2 World Youth Day

The World Youth Day took place in August 2005 in Cologne, Germany. The final event was a service with Pope Benedict XVI. It was held on a large ground (around 92 ha) with a stage in the centre. The geometry is shown in fig. 1. Altogether around 700 to 800 thousand pilgrims were expected. Apart from roads and public transportation systems, footpaths played an eminent role in the mobility concept.

Concerning pedestrian motion, two cases must be distinguished: the normal case of getting to and back from the area and the emergency case, when part of the site or the complete area has to be evacuated. In order to estimate the performance of the roads and footpaths in case of an emergency evacuation, several simulations were performed. One example is shown below. Since the size of the area is about 2 km in East-West and about 1.6 km in North-South direction, a complete evacuation is in most cases neither sensible nor feasible. The scenarios considered were accidents with trucks or a fire. In these cases, the strategy was to evacuate the field directly affected and one neighbouring field (the rectangular areas defined by the roads and footpaths – cf. Fig. 1). The scenario shown in fig. 2 is a fire on or near the central stage and a partial evacuation of the so called pilgrim field near the location of the incident. After

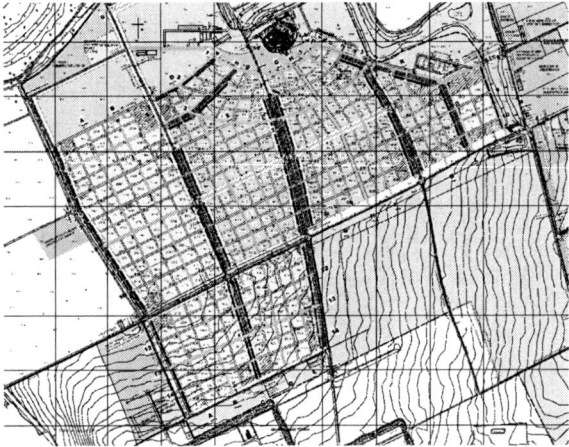

Fig. 1. World Youth Day Premises: The "Marienfeld" near Kerpen and Cologne.

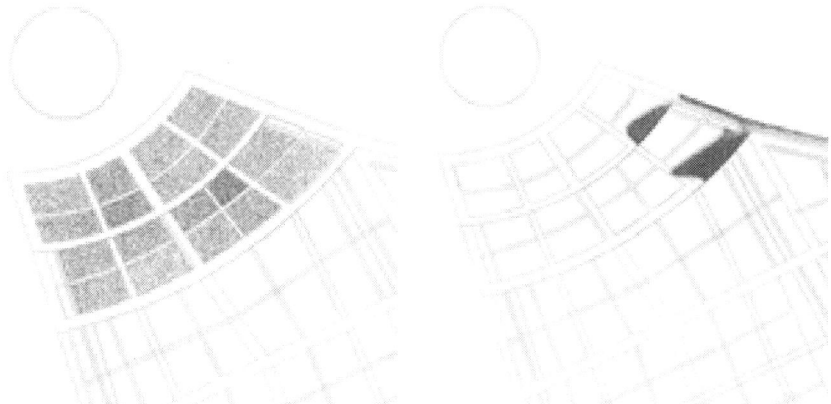

Fig. 2. World Youth Day Simulation: Initial Distribution of 36,000 persons (left) for the scenario "fire near the stage". The area is 32,000 sqm, i.e. 32 ha and the density ca. 1.1 persons/sqm. The right picture shows the situation after 30 minutes.

30 minutes the major part of the afflicted area can be evacuated. However, it takes another one and a half hours to get the persons completely off the area in this simulation. Since the surrounding pilgrim fields are also filled with people, there is a strong need to contain the threat within a few fields (the small rectangles defined by the paths and roads – cf. fig. 1). Otherwise the complete area would have to be evacuated which took more than 4 hours in the simulation. There was, however, open space several times larger than the "Marienfeld" around the area which is a necessary condition for a safe evacuation in the first place.

3 Egress from a Football Stadium

Concerning quantitative verification, movement patterns provide a valuable tool to investigate the reliability of simulation results. In the following, video footage is compared to simulations, especially concerning overall egress time (non-emergency). The video footage was taken at an international match between Germany and Scotland in Dortmund (Westfalenstadion). The results described here are an extension of [10], where simulation results for the stands in the four corners of the stadium were investigated by simulations. This reference also contains further information on the model and its application. Furthermore, [11] contains an in depth description and comparison of different modeling approaches for pedestrian dynamics and especially evacuation simulation.

 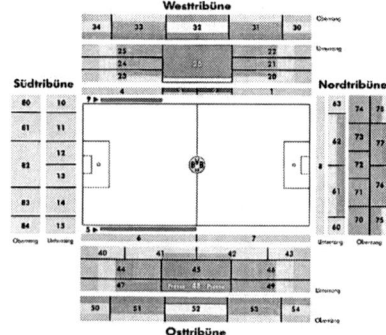

Fig. 3. The Westfalenstadion Dortmund: Outside view and general arrangement plan (Borusia Dortmund KGaA, www.borussia-dortmund.de).

In fig. 4 the first six minutes of the video footage and the first three minutes of the simulation are shown. The reason for the different time spans is that the real persons react slower. However, due to their effectivenes and group formation which is not represented in the simulation, the motion is more synchronized than in the simulation. Therefore, the snapshots were chosen such that the situations are comparable even though the times might be different. For the second half of the egress shown in fig. 5 this difference vanishes and after 13 minutes, the situation is very much alike for reality and simulation. It is remarkable that after less than 15 minutes, the normal egress is nearly complete. One important pattern that can be identified is the sequence of egress from the rows. The lower rows are emptied first. This pattern is represented nicely by the simulation.

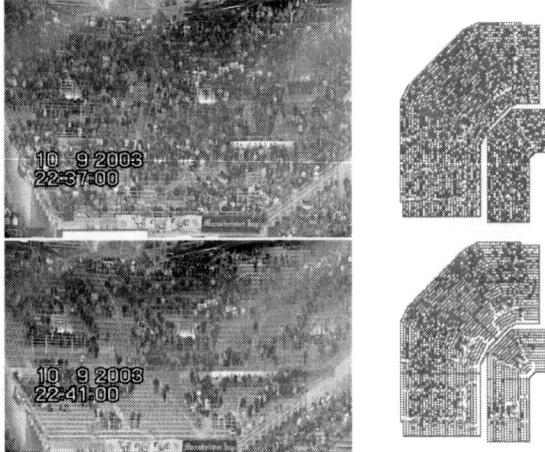

Fig. 4. The Westfalenstadion Dortmund: Comparison of the results for the video analysis (left column) and the simulation (right column) at the beginning of the egress. The video snapshots are taken at (from top to bottom) $t = 2$ and $t = 6$ minutes for the videos and $t = 20$ seconds and $t = 3$ minutes for the simulation.

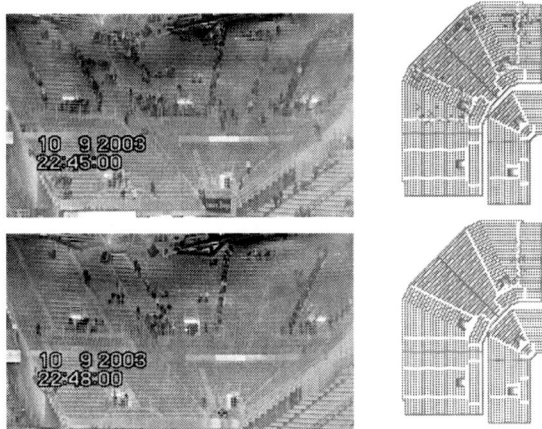

Fig. 5. The Westfalenstadion Dortmund: Comparison of the results for the video analysis (left column) and the simulation (right column). The video (left column) and simulation snapshots (right column) are taken at (from top to bottom) $t = 10$ minutes and $t = 13$ minutes.

4 Conclusions

We have presented egress simulations (and in one case video analysis) of two different events with large numbers of pedestrians. An important aspect in the egress from football stadiums are V-like shapes that are formed because the egress from the lower seating rows is faster. For events with several hundred thousand participants like the World Youth Day 2005 in and near Cologne an evacuation of the complete area is usually not advisable, as can be seen from the simulations presented and from the fact that the overall evacuation time was estimated to be at least 4 hours.

Acknowledgements. I would like to thank the Kerpen fire brigade, especially Mr. Graß and Mr. Cordier for many discussions and in depth information on the emergency precautions and safety concept. Tim Meyer-König is the developer and implementer of the simulation tool PedGo that was used for the simulations. The video footage at the Westfalenstadion was done by TraffGo GmbH in cooperation with Michael Schreckenberg's group at the University Duisburg-Essen.

References

1. H. Klüpfel, T. Meyer-König, J. Wahle, and M. Schreckenberg. Microscopic simulation of evacuation processes on passenger ships. In S. Bandini and T. Worsch, editors, *ACRI 2000*, pages 63–71, London, 2000. Springer.
2. H. Klüpfel. *A Cellular Automaton Model for Crowd Movement and Egress Simulation*. PhD thesis, University Duisburg–Essen, 2003. available from http://deposit.ddb.de/cgi-bin/dokserv?idn=96883180x.
3. T. Meyer-König, H. Klüpfel, and M. Schreckenberg. A microscopic model for simulating mustering and evacuation processes onboard passenger ships. In *Proceedings of the International Emergency Management Society Conference*, 2001. CD-Rom.
4. A. Kirchner, H. Klüpfel, K. Nishinari, A. Schadschneider, and M. Schreckenberg. Simulation of competitive egress behaviour. *Physica A*, 324:689–697, 2002.
5. A. Kirchner, H. Klüpfel, K. Nishinari, A. Schadschneider, and M. Schreckenberg. Discretization effects and influence of walking speed in cellular automata models for pedestrian dynamics. *J. Stat. Mech.: Theory and Experiment*, P10011, 2004.
6. M. Wölki, A. Schadschneider, and M. Schreckenberg. Asymmetric exclusion processes with shuffled dynamics. *J. Phys. A* 39, 33–44, 2006.
7. C. Burstedde, K. Klauck, A. Schadschneider, and J. Zittartz. Simulation of pedestrian dynamics using a 2-dimensional cellular automaton. *Physica A* 295, 507–525, 2001.
8. D.A. Purser and M. Bensilium. Quantification of behaviour for engineering design standards and escape time calculations. *Safety Science*, 38(2), 2001.
9. S. Hoogendoorn, S. Luding, P.H.L. Bovy, M. Schreckenberg, and D.E. Wolf, editors. *Traffic and Granular Flow '03*, Berlin, 2005. Springer.
10. H. Klüpfel and T. Meyer-König. Simulation of the evacuation of a football stadium. In Hoogendoorn et al. [9], pages 423–430.
11. H. Klüpfel and T. Meyer-König. Models for crowd movement and egress simulation. In Hoogendoorn et al. [9], pages 357–372.

Transport-Equilibrium Schemes for Pedestrian Flows with Nonclassical Shocks

Christophe Chalons

Université Paris 7 - Denis Diderot & Laboratoire J.-L Lions, U.M.R. 7598,
Université Pierre et Marie Curie, Boîte courrier 187, 75252 Paris Cedex 05, France.
E-mail: chalons@math.jussieu.fr

Summary. This paper deals with the numerical approximation of the solutions of a macroscopic model for the description of the flow of pedestrians. Solutions of the associated Riemann problem are known to be possibly *nonclassical* in the sense that the underlying discontinuities may well violate Oleinik inequalities, which makes their numerical approximation very sensitive. This study proposes to apply the Transport-Equilibrium strategy proposed in [2] for computing nonclassical solutions of scalar conservation laws to this framework. Numerical evidences are proposed.

1 Introduction

In this paper, we are interested in the numerical approximation of weak solutions of a scalar conservation law arising in the description of the flow of pedestrians. The model under consideration has been introduced recently by Colombo and Rosini in [8]. It is based on the well-known Lighthill-Whitam [14] and Richards [15] model and reads

$$\begin{cases} \partial_t \rho + \partial_x q(\rho) = 0, & \rho(x,t) \in \mathbb{R}, \ (x,t) \in \mathbb{R} \times \mathbb{R}^{+*}, \\ \rho(x,0) = \rho_0(x), & x \in \mathbb{R}, \end{cases} \tag{1}$$

where $\rho \geq 0$ is the pedestrian density and $q : \mathbb{R}^+ \to \mathbb{R}^+$ is the flow function. The form of equation (1) is an immediate consequence of two basic assumptions, namely the conservation of the total number of pedestrians and a given speed law v which depends on density $\rho \in [0, R]$ only (R denotes the maximal density). Recall that $q = \rho v$. However, this model was first dedicated to car flows and so is not able to reproduce important features of pedestrian flows, at least when considering typical concave increasing-decreasing flow functions. For instance let us mention the *overcompression phenomenon* in a crowd or the *fall of pedestrians* in the outflow through a door of a crowd in panic. In order to overcome this difficulty, Colombo and Rosini [8] first proposed to modify the typical shape of the flow function q by introducing another characteristic density $R^* > R$ for the maximal density in exceptional situations of panic. The flow function now looks like a concave-convex and increasing-decreasing function on $[0, R]$ and a convex-concave and increasing-decreasing function on $[R, R^\star]$, see Fig. 1 below. In particular, discontinuities satisfying the usual Rankine-Hugoniot conditions but violating the standard admissibility entropic conditions such as Oleinik inequalities (see (6) below) are present in the model. Then, the same authors defined a unique Riemann solver using such nonclassical shocks. The main motivation in considering nonclassical solutions is to allow panic states ($\rho \in [R, R^\star]$) to appear in a initially calm situation ($\rho \in [0, R]$), because of a sharp increase in the density for instance. Note that the *maximum principle* in classical solutions prevents such panic situations from arising. We refer the reader to [12] for a general theory of classical and nonclassical solutions.

From a numerical point of view, the numerical approximation of nonclassical solutions is known to be very challenging and still constitutes (at least generally speaking) an open problem nowadays (see for instance [4–6, 9, 10, 13], but also [1] and the references within). Very recently, a new efficient numerical strategy has been proposed in [2] for computing nonclassical solutions of scalar conservation laws. Roughly speaking, the corresponding finite volume scheme is based on two steps, namely an *Equilibrium step* which aims to put at stationary equilibrium nonclassical discontinuities when present, and a *Transport step* for propagating these discontinuities. n this paper, we thus propose to adapt the Transport-Equilibrium scheme developed in [2] to the present setting. We refer the reader to [3] for details about the slight (but important) difference between both algorithms. Importantly, we will see that the resulting scheme still provides numerical solutions in full agreement with exact ones.

2 Governing Equation and Closure Relation

The model we consider for the description of the flow of pedestrians has been recently introduced and studied by Colombo and Rosini [8]. It is based on the well-known Lighthill-Whitam [14] and Richards [15] model and reads

$$\begin{cases} \partial_t \rho + \partial_x q(\rho) = 0, & q(\rho) = \rho v(\rho), \quad (x,t) \in \mathbb{R} \times \mathbb{R}^{+*}, \\ \rho(x,0) = \rho_0(x), & x \in \mathbb{R}, \end{cases} \tag{2}$$

where ρ is the pedestrian density, q is the flux function and v the speed of pedestrians. For simplicity, initial data ρ_0 is assumed to be made of two constant states ρ_l and ρ_r, separated by a discontinuity located at point $x = 0$:

$$\rho_0(x) = \begin{cases} \rho_l & \text{if } x < 0, \\ \rho_r & \text{if } x > 0. \end{cases} \tag{3}$$

Equation (2) expresses the conservation of the number of pedestrians in the space domain, while the speed v is assumed to depend only on the density ρ

by means of the so-called *fundamental relation*. In the context of car flows, it is pretty classical to consider that the function $\rho \in [0, R] \to q(\rho)$ is concave, with $q(0) = q(R) = 0$, R being the maximal density, and reaches its maximum value at a critical density $R_M \in [0, R]$: $q(R_M) = \max_{\rho \in [0,R]} q(\rho)$. Here and in order to take into account some important features of human flows, like the *overcompression phenomenon* or the *fall of pedestrians* due to panic for instance, we follow [8] and take a flux function q whose form is given on Figure 1.

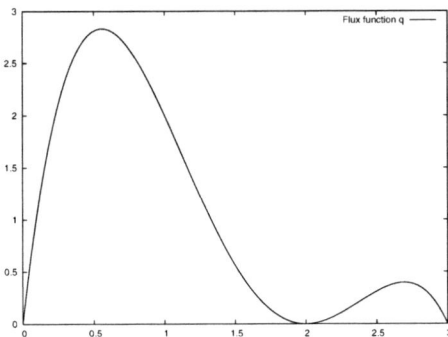

Fig. 1. Closure relations: $\rho \to q(\rho)$

Two remarkable values R and $R^\star$ ($R < R^\star$) are now considered for the density ρ: the first one represents a natural bound of ρ in situations with little or not panic ($\rho \in [0, R]$), and the second one is a maximal value of ρ in situations of great panic ($\rho \in [R, R^\star]$). The flow function q now admits in each of these regions a maximum value: $R_M \in [0, R]$ with $q(R_M) = \max_{\rho \in [0,R]} q(\rho)$ and $R_M^\star \in [R, R^\star]$ with $q(R_M^\star) = \max_{\rho \in [R,R^\star]} q(\rho)$ as well as an inflection point: $R_I \in [0, R]$ with $q''(R_I) = 0$ and $R_I^\star \in [R, R^\star]$ with $q''(R_I^\star) = 0$, while $q(0) = q(R) = q(R^\star) = 0$. Choosing (without loss of generality in the forthcoming developments)

$$q(\rho) = -\rho(\rho - R)^2(\rho - R^\star), \quad R = 2, \quad R^\star = 3, \qquad (4)$$

the following values are easily found:

$$R_M \simeq 0.5570, \quad R_M^\star \simeq 2.6930, \quad R_I \simeq 1.1208, \quad R_I^\star \simeq 2.3792. \qquad (5)$$

3 The Riemann Solver

Let us now turn to the definition of a Riemann solution for eqs. (2-4). First of all, let us recall that there exists a unique *classical* solution for (2-4), that is a weak solution selected by the validity of Oleinik inequalities across discontinuities separating ρ_- and ρ_+:

$$\frac{q(\rho) - q(\rho_-)}{\rho - \rho_-} \geq \frac{q(\rho_+) - q(\rho_-)}{\rho_+ - \rho_-}, \qquad \text{for all } \rho \text{ between } \rho_- \text{ and } \rho_+. \qquad (6)$$

Moreover, this solution obeys the following maximum principle property

$$\rho(x, t) \in [\min(\rho_l, \rho_r), \max(\rho_l, \rho_r)], \quad \text{for all} \quad x \in \mathbb{R} \quad \text{and} \quad t \geq 0. \qquad (7)$$

See for instance [12] for details. In this context, imagine that ρ_l and ρ_r belong to the calm region $[0, R]$. Then, due to the maximum principle (7), no panic state may be found in the *classical* solution. From a practical point of view, such a panic state is nevertheless expected to appear in certain situations when ρ_r is very close to R. That is the reason for which *nonclassical* solutions violating both the maximum principle property and Oleinik inequalities across discontinuities are introduced in the present framework. With this in mind and following [8] and [12], we define two functions $\psi : [0, R^\star] \to [R, R^\star]$ and $\Phi : [0, R] \to [0, R]$, related to the graph of function q in the (ρ, q)-plane, as follows:

- $\psi(\rho)$ is such that the line L_ρ that passes through the points with coordinates $(\rho, q(\rho))$ and $(\psi(\rho), q(\psi(\rho)))$ is tangent to the graph of function q at point $(\psi(\rho), q(\psi(\rho)))$
- $\Phi(\rho)$ is such that this line intersects the curve $q = q(\rho)$ at a further point with coordinates $(\Phi(\rho), q(\Phi(\rho)))$.

An illustration is given in Figure 2 (left) where both the function q and the line $L_{\rho=0.5}$ are plotted. The right part of Figure 2 is concerned with the graph of both functions ψ and Φ. In particular, note that Φ is not defined in the whole domain $[0, R]$ simply because the additional intersection point between L_ρ and q is sometimes realized below 0.

Then, introducing two real thresholds s and Δs such that

$$s \in]0, R_M[\quad \text{and} \quad \Delta s \in]0, R - s[, \qquad (8)$$

Colombo and Rosini [8] defined a unique Riemann solution for (2-4), which coincides with the *classical* solution except in the next three situations:

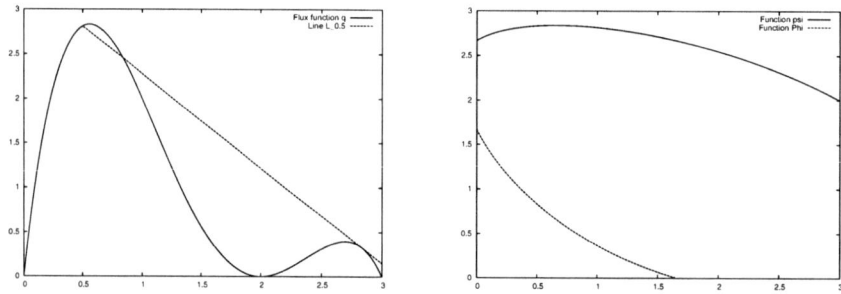

Fig. 2. Function q and line $L_{0.5}$ (left); functions ψ and Φ (right)

(*i*) (ρ_l, ρ_r) belongs to A with

$$A = \{(\rho_l, \rho_r) \in [0, R^\star]^2 \ / \ s \leq \rho_l \leq R, \ \Phi(\rho_l) < \rho_r \leq R, \ (\rho_r - \rho_l) > \Delta s\}, \quad (9)$$

(*ii*) (ρ_l, ρ_r) belongs to B with

$$B = \{(\rho_l, \rho_r) \in [0, R^\star]^2 \ / \ \rho_r > R, \ \rho_r > \rho_l, \ \rho_r \leq \psi(\rho_l)\}, \quad (10)$$

(*iii*) (ρ_l, ρ_r) belongs to C with

$$C = \{(\rho_l, \rho_r) \in [0, R^\star]^2 \ / \ \rho_r > R, \ \rho_r > \rho_l, \ \rho_r > \psi(\rho_l)\}. \quad (11)$$

The last two situations aim at defining the solution when the right state ρ_r belongs to the panic area, *i.e.* $\rho_r > R$. More precisely, if ρ_l and ρ_r are such that (ρ_l, ρ_r) belongs to B the Riemann solution contains a nonclassical shock connecting ρ_l to $\psi(\rho_l)$ followed by the classical Riemann solution associated with initial states $\psi(\rho_l)$ and ρ_r. And if ρ_l and ρ_r are such that (ρ_l, ρ_r) belongs to C, the Riemann solution is a nonclassical shock connecting ρ_l to ρ_r. In this paper, we will focus (without restriction) on the first situation $(\rho_l, \rho_r) \in A$ which explains that if the left state ρ_l is sufficiently large ($\rho_l \geq s$) and faces a right state ρ_r which is pretty far ($\rho_r - \rho_l > \Delta s$) from ρ_l and already close to the panic region ($\Phi(\rho_l) < \rho_r \leq R$), then a panic state is created due to the sharp increase in the density so that the maximum principle property is violated, and the corresponding solution is made of a nonclassical shock connecting ρ_l to $\psi(\rho_l)$ followed by the classical Riemann solution associated with initial states $\psi(\rho_l)$ and ρ_r.

Before addressing the numerical approximation of these Riemann solutions, it is worth noticing that function ψ plays the part of a *kinetic function* that manages the transitions between calm and panic. Finally, we refer the reader to [8] for additional properties of interest satisfied (or not) by the Riemann solver proposed in this section.

4 Numerical Approximation

Aim of this section is the description of the so-called Transport-Equilibrium schemes recently proposed by the author for approximating nonclassical solutions of conservation laws. The work proposed in [2] is concerned with scalar conservation laws, while the case of systems will be treated in a subsequent paper. Applying these schemes to the model of pedestrian flows under consideration is the main objective of this section. We begin by introducing some notations.

Let Δt and Δx be the time and the space steps. Introducing the interfaces $x_{j+1/2} = j\Delta x$ for $j \in \mathbb{Z}$ and the intermediate times $t^n = n\Delta t$ for $n \in \mathbb{N}$, we classicaly seek at each time t^n an approximation ρ_j^n of the solution $x \to \rho(x, t^n)$ on each interval $C_j = [x_{j-1/2}; x_{j+1/2}), \ j \in \mathbb{Z}$. In this context, we assume as given a two-point (without loss of generality) numerical flux function $(u,v) \rightarrow g(u,v)$ consistent with the flux function q, and we set $\lambda = \Delta t / \Delta x$.

In order to motivate the need of a particular treatment when numerically dealing with nonclassical solutions, let us have a look on what happens when considering the following classical conservative scheme:

$$\rho_j^{n+1} = \rho_j^n - \lambda(g_{j+1/2} - g_{j-1/2}), \quad j \in \mathbb{Z}, \tag{12}$$

with $g_{j+1/2} = g(\rho_j^n, \rho_{j+1}^n)$ for all $j \in \mathbb{Z}$. Figure 3(left) shows the solution generated by a standard relaxation numerical flux g (see (19) below) when initial states ρ_l and ρ_r are such that $(\rho_l, \rho_r) \in A$. In other words, the exact solution contains a transition between calm and panic regions that should be observed via a *nonclassical* shock connecting ρ_l to $\psi(\rho_l)$. On the contrary, we note that the numerical solution is *classical* and then remains entirely calm. This means that the update formula (12) is not able to create by itself the panic state $\psi(\rho_l)$. The same failure would be observed with $(\rho_l, \rho_r) \in B$. Then, *one must enforce the appearance of the corresponding discontinuity between ρ_l and $\psi(\rho_l)$ when it is relevant, that is when $(\rho_l, \rho_r) \in A$ and when $(\rho_l, \rho_r) \in B$.* Actually, it turns out that the conservative scheme (12) is not either able to properly capture the exact *nonclassical* solution when $(\rho_l, \rho_r) \in C$, that is in the case when the Riemann initial data (3) should be simply propagated at speed $\sigma(\rho_l, \rho_r)$ given by Rankine-Hugoniot jump relation:

$$\sigma(\rho_l, \rho_r) = \frac{q(\rho_r) - q(\rho_l)}{\rho_r - \rho_l}. \tag{13}$$

Instead, a classical solution is observed on Figure 3 (right). *Then, one must enforce the initial data to be simply propagated (at the right speed !) when it is relevant, that is when $(\rho_l, \rho_r) \in C$.*

These observations led us to replace (12) with the following nonconservative update formula:

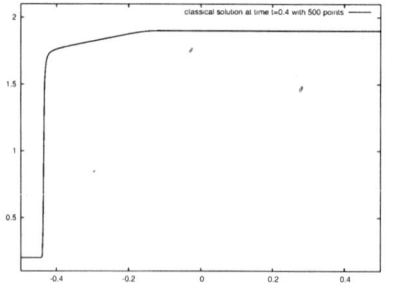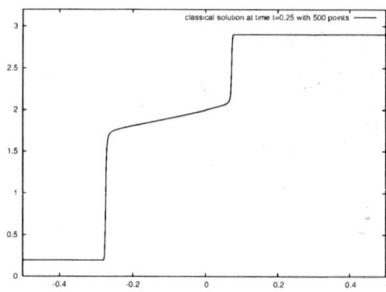

Fig. 3. Solution generated by standard relaxation method when $(\rho_l, \rho_r) \in A$ (left) and $(\rho_l, \rho_r) \in C$ (right).

$$\rho_j^{n+1-} = \rho_j^n - \lambda(g_{j+1/2}^L - g_{j-1/2}^R), \quad j \in \mathbb{Z}, \tag{14}$$

where the numerical fluxes $g_{j+1/2}^L$ and $g_{j+1/2}^R$ have to be suitably defined. The very idea is to modify the numerical flux $g_{j+1/2}$ by means of two fluxes $g_{j+1/2}^L$ and $g_{j+1/2}^R$ each time that a nonclassical shock appears in the solution of the Riemann problem (2)-(3)-(4) associated with $\rho_l = \rho_j^n$ and $\rho_r = \rho_{j+1}^n$. Otherwise, $g_{j+1/2}$ will be unchanged. According to whether the nonclassical shock connects states ρ_l and $\psi(\rho_l)$ or not, that is depending on if $(\rho_j^n, \rho_{j+1}^n) \in A \cup B$ or $(\rho_j^n, \rho_{j+1}^n) \in C$, we will use different formulas. More precisely, we set for all $j \in \mathbb{Z}$:

$$g_{j+1/2}^L = \begin{cases} g(\rho_j^n, \rho_j^n) & \text{if } (\rho_j^n, \rho_{j+1}^n) \in A \cup B \cup C, \\ g(\rho_j^n, \rho_{j+1}^n) & \text{otherwise,} \end{cases} \tag{15}$$

and

$$g_{j+1/2}^R = \begin{cases} g(\psi(\rho_j^n), \rho_{j+1}^n) & \text{if}(\rho_j^n, \rho_{j+1}^n) \in A \cup B, \\ g(\rho_{j+1}^n, \rho_{j+1}^n) & \text{if}(\rho_j^n, \rho_{j+1}^n) \in C, \\ g(\rho_j^n, \rho_{j+1}^n) & \text{otherwise.} \end{cases} \tag{16}$$

The aim of $g_{j+1/2}^L$ is to keep at the next time step the same value ρ_j^n in the cell C_j since ρ_j^n always coincides with the left state of the nonclassical shock in the Riemann solution associated with $\rho_l = \rho_j^n$ and $\rho_r = \rho_{j+1}^n$. The aim of $g_{j+1/2}^R$ is double. First, to keep the same value ρ_{j+1}^n in the cell C_{j+1} when $(\rho_j^n, \rho_{j+1}^n) \in C$, since ρ_{j+1}^n coincides in this case with the right state of the nonclassical shock in the Riemann solution associated with $\rho_l = \rho_j^n$ and $\rho_r = \rho_{j+1}^n$. And then, to force the value $\psi(\rho_j^n)$ to appear in the cell C_{j+1} when $(\rho_j^n, \rho_{j+1}^n) \in A \cup B$.

With these definitions, we easily check for instance that discontinuities separating two states ρ_- and ρ_+ such that $(\rho_-, \rho_+) \in C$ are kept at stationary equilibrium by formulas (14-16). See also [2] or [3] for more details. More generally, we are thus bound to introduce a dynamic step in order to make moving the values that we previously forced to be present in specific cells.

Recall that the speed of propagation $\sigma(\rho_-, \rho_+)$ of a discontinuity between ρ_- and ρ_+ is given by Rankine-Hugoniot conditions (13). We then decide to define at each interface $x_{j+1/2}$ a speed of propagation $\sigma_{j+1/2}$:

$$\sigma_{j+1/2} = \begin{cases} \sigma(\rho_j^{n+1-}, \rho_{j+1}^{n+1-}) & \text{if } (\rho_j^n, \rho_{j+1}^n) \in A \cup B \cup C, \\ 0 & \text{otherwise,} \end{cases} \tag{17}$$

and solve locally (at each discontinuity $x_{j+1/2}$) a transport equation with speed $\sigma_{j+1/2}$. In order to get a new approximation ρ_j^{n+1} at time $t^{n+1} = t^n + \Delta t$, we propose to pick up randomly on interval $[x_{j-1/2}, x_{j+1/2}[$ a value in the juxtaposition of the solutions of these transport equations at time Δt that we choose sufficiently small to avoid wave interactions. In particular, such a sampling strategy prevents the emergence of spuriousintermediate values with respect to those obtained at time t^{n+1-}. See again [2] or [3] for more details. Given a well distributed random sequence (a_n) within interval $(0, 1)$, it amounts to set:

$$\rho_j^{n+1} = \begin{cases} \rho_{j-1}^{n+1-} & \text{if } a_{n+1} \in [0, \lambda\sigma_{j-1/2}^+[, \\ \rho_j^{n+1-} & \text{if } a_{n+1} \in [\lambda\sigma_{j-1/2}^+, 1 + \lambda\sigma_{j+1/2}^-[, \\ \rho_{j+1}^{n+1-} & \text{if } a_{n+1} \in [1 + \lambda\sigma_{j+1/2}^-, 1[, \end{cases} \tag{18}$$

with $\sigma_{j+1/2}^+ = \max(\sigma_{j+1/2}, 0)$ and $\sigma_{j+1/2}^- = \min(\sigma_{j+1/2}, 0)$ for all $j \in \mathbb{Z}$. The description of our numerical strategy is now completed.

5 Numerical Experiments

This section proves the good design of the transport-equilibrium scheme we have proposed. To that purpose and without restriction, we consider a relaxation scheme as a basic numerical flux g, that is

$$g(u, v) = \frac{1}{2}(q(u)+q(v)) + \frac{a(u, v)}{2}(u-v) \quad \text{with} \quad a(u, v) = \max_{[\min(u,v),\max(u,v)]} |q'|, \tag{19}$$

(see [11] for instance) and we use the following standard CFL condition for computing time step Δt at each time iteration:

$$\Delta t = \frac{1}{2} \cdot \frac{\Delta x}{\max_j |a(\rho_j^n, \rho_{j+1}^n)|}.$$

Following a proposal by Collela [7], we consider the van der Corput random sequence (a_n) defined by

$$a_n = \sum_{k=0}^{m} i_k 2^{-(k+1)},$$

where $n = \sum_{k=0}^{m} i_k 2^k$, $i_k = 0, 1$, denotes the binary expansion of the integers $n = 1, 2, \ldots$. Closure relations for the numerical simulations are as follows. First of all, the flux function q is chosen as in (4) (see also (5) and Fig. 1 for the graph of the function q). Then, thresholds s and Δs are chosen to be

$$\Delta s = \Phi(0) = \frac{5}{3}, \quad s = \frac{1}{2}(R - \Delta s) = \frac{1}{6},$$

so that condition (8) holds true. As last, we mention that the computations are performed on two grids, containing respectively 100 ($\Delta x = 0.01$) and 500 ($\Delta x = 0.002$) points per unit interval. Let us now consider two typical behaviors of the Riemann solution given in Section 3. We refer the reader to [3] for additional numerical tests.

In Test 1, we choose $\rho_l = 0.2$ and $\rho_r = 1.9$ so that it is easily checked that $(\rho_l, \rho_r) \in [0, R]^2$ and $(\rho_l, \rho_r) \in A$. In such a situation, panic arises and the solution is composed of a nonclassical discontinuity between $\rho_l = 0.2$ and $\psi(\rho_l) \simeq 2.7744$, followed by a classical part made of a rarefaction wave and a classical shock attached to the rarefaction. We observe on Fig. 4 (left) that our algorithm properly captures this nonclassical solution. Note also that the nonclassical discontinuity from ρ_l to $\psi(\rho_l)$ is sharp: there is no point in the profile. For the sake of comparison, Fig. 4 (right) shows again that the usual relaxation scheme defined by update formula (12) generates a (classical) solution which lies entirely in interval $[0, R]$ and so is far from the expected one. What proves both the need of modifying classical conservative approaches and the validity of our strategy.

In Test 2, we take $\rho_l = 0.2$ and $\rho_r = 2.9$ so that we have now $(\rho_l, \rho_r) \in C$. By Section 3, the solution is a single nonclassical shock connecting ρ_l to ρ_r. Fig. 5 (left) shows that our algorithm again sharply captures this nonclassical discontinuity. Actually, note that it and Glimm's random choice scheme are identical for this test case since the equilibrium step is clearly transparent. The solution obtained with the standard relaxation scheme is plotted again on Fig. 5 (right).

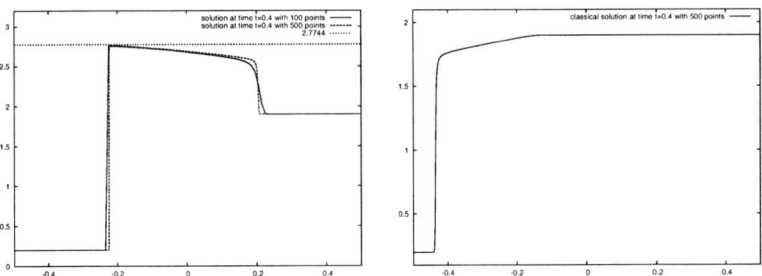

Fig. 4. Test 2 - nonclassical solution (left) and classical solution (right)

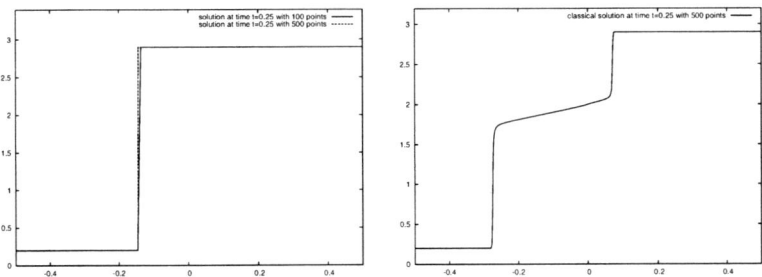

Fig. 5. Test 5 - nonclassical solution (left) and classical solution (right)

6 Conclusion

An efficient numerical strategy has been presented for computing nonclassical solutions of a particular scalar conservation law for the simulation of human flows. Our approach turns out to be nonconservative, but measures in [3] have shown that the loss of mass is extremely low, while numerical solutions fully agree with exact ones.

References

1. Chalons C., *Bilans d'entropie discrets dans l'approximation numérique des chocs non classiques. Application aux équations de Navier-Stokes multi-pression 2D et à quelques systèmes visco-capillaires*, PhD Thesis, Ecole Polytechnique, (2002).
2. Chalons C., *Transport-Equilibrium Schemes for Computing Nonclassical Shocks. I. Scalar Conservation Laws*, Preprint of the Laboratoire Jacques-Louis Lions (2005).
3. Chalons C., *Numerical approximation of a macroscopic model of pedestrian flows*, Preprint of the Laboratoire Jacques-Louis Lions (2005).
4. Chalons C. and LeFloch P.G., *A fully discrete scheme for diffusive-dispersive conservation laws*, Numerisch Math. **89**, 493-509 (2001).
5. Chalons C. and LeFloch P.G., *High-order entropy conservative schemes and kinetic relations for van der Waals fluids*, J. Comput. Phys. **167**, 1-23 (2001).
6. Chalons C. and LeFloch P.G., *Computing undercompressive waves with the random choice scheme. Nonclassical shock waves*, Interfaces and Free Boundaries **5**, 129-158 (2003).
7. Collela P., *Glimm's method for gas dynamics*, SIAM J. Sci. Stat. Comput. **3**, 76-110 (1982).
8. Colombo R.M. and Rosini M.D., *Pedestrian Flows and Nonclassical Shocks*, Mathematical Methods in the Applied Sciences **28**(13), 1553-1567 (2005).
9. Hayes B.T. and LeFloch P.G., *Nonclassical shocks and kinetic relations : Scalar conservation laws*, Arch. Rational Mech. Anal. **139**, 1-56 (1997).
10. Hayes B.T. and LeFloch P.G., *Nonclassical shocks and kinetic relations : Finite difference schemes*, SIAM J. Numer. Anal. **35**, 2169-2194 (1998).
11. Lattanzio C. and Serre D., *Convergence of a relaxation scheme for hyperbolic systems of conservation laws*, Numer. Math. **88**, 121-134 (2001).
12. LeFloch P.G., *Hyperbolic Systems of Conservation Laws: The theory of classical and nonclassical shock waves*, E.T.H. Lecture Notes Series, Birkhäuser (2002).
13. LeFloch P.G. and Rohde C., *High-order schemes, entropy inequalities, and nonclassical shocks*, SIAM J. Numer. Anal. **37**, 2023-2060 (2000).
14. Lighthill M.J. and Whitham G.B., *On kinematic waves. II. A theory of traffic flow on long crowded roads*, Proc. Roy. Soc. London, Ser. A. **229**, 317-345 (1955).
15. Richards P.I., *Shock waves on the highway*, Operations Res. **4**, 42-51 (1956).

Part IV

Networks and Urban Traffic

Decision-Making and Transport Costs in Complex Networks

Sean Gourley and Neil F. Johnson

Physics Department, Clarendon Laboratory, Parks Road, Oxford, OX1 3PU, U.K.

Summary. We analyse the effects of agents' decisions on the creation of, and reaction to, congestion on a centralised network with ring-and-hub topology. The system dynamics are driven by an interplay between the creation of, and the transition between, particular stable states which arise as the network is varied. Our results show that the existence of congestion in a network is a dynamic process which is as much dependent on the agents' decisions as it is on the structure of the network itself.

1 Introduction

Traffic is an interesting example of an interacting multi-particle system (i.e. cars) on a non-trivial topological network (i.e. roads). Many traffic studies have taken the view that cars follow automata-like rules. This is probably a good approximation for dealing with traffic which is already on a particular road – however it does not address the arguably more fundamental question of *why* those cars, or rather their drivers, chose that route in the first place. Indeed with in-car access to real-time traffic monitoring already available and likely to become more prevalent in the future, an understanding how motorists' individual decisions affect the traffic patterns which emerge on road networks, is of great practical importance [1]. Here we consider such a question, in a common real-world scenario in which there is a choice between choosing a route through a potentially crowded central region, or instead choosing the safe but long option of an outside road.

The quickest route across the network is easy to determine when you are the only agent on the network. However this is rarely the case in real-world problems, where you have multiple agents all trying to mimimise the time/cost of traversing the network. When this happens you see congestion at the major shortcut points. If the affected agents then react in a similar way in order to avoid the bottle-neck, an even bigger congestion point can then arise at a different location on the network. Congestion arises then not solely as a result of the network topology, but rather as a result of the dynamic interplay between the structure of the network and the decisions of the agents who are using it.

Ashton *et al* presented an exactly solvable model of a ring-and-hub network [5] which extended the model of Ref. [6] to include congestion costs on a central hub or hubs. In this paper we introduce a model for describing the effects of agents' decisions on the creation of congestion within such a ring-and-hub topology. Hence this work generalized the model introduced by Ashton *et al*, by introducing *decision-making agents* onto the network. Agents use their own strategies to make inductive decisions about the future behaviour of the system in order to find the cheapest pathway across it. Our model lends itself to real life situations such as communication across social/business networks, flow of data across the internet, air traffic, or any situation where competing agents have to navigate a network where congestion is a factor. Indeed, the study of the functional properties of networks is gaining increased attention across a range of disciplines [1–4]. We use this hub and spoke model not only as a tool to solve specific traffic issues, but also as a platform upon which we can understand the general principles of this class of problems.

2 System Set-Up

The simulation consists of N agents and a central hub of capacity L. Each of the N agents are connected to their nearest neighbours by an undirected link of unit length. These links form a peripheral pathway around the outside of the network. The agents also have the possibility of being connected to another point on the network through the central hub. If this pathway exists it is known as the hub pathway and the number of these in the network is defined to be λ. Through these sets of connections the agents form a combined ring-and-hub topology, i.e. a hub and spoke network, which can be seen in Fig. 1.

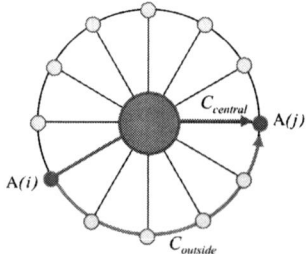

Fig. 1. Our model network with the nodes connected to nearest neighbours around the outside and the central hub located in the middle. Agent $A(i)$ is randomly connected to another point on the network $A(j)$ and the costs of the transport are shown as $C_{\text{central}}, C_{\text{outside}}$.

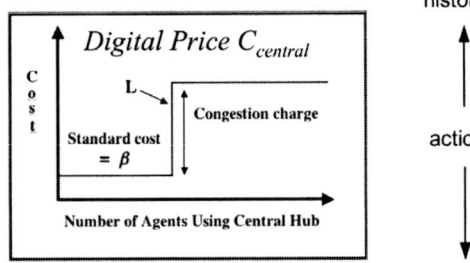

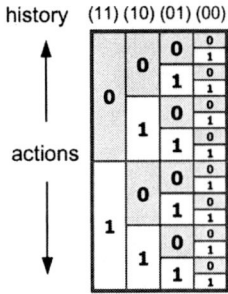

Fig. 2. (a) Digital pricing structure for the central hub with capacity L. A standard cost of $\beta = 1$ is applied before the capacity reached. After capacity is reached the cost of using the hub increases by fixed amount cc. **(b)** The $m = 2$ set of strategies allocated to agents in this simulation. Each column represents a different realisation of the history string, and each row then contains a full set of actions for each history string. Action 1 corresponds to using the central hub, and action 0 is to take the peripheral path.

Each agent $A(i)$ must transport himself (e.g. a car containing himself, or a message) from one location on the hub and spoke network to a randomly selected final destination $A(j)$ at another point on the network. If the agent $A(i)$ is connected to the central hub they have the option of using this resource with an associated cost $C_{central}$, or they can use the peripheral pathway constructed from connections between nearest neighbours at a cost of C_{out}. The goal of each agent is to transport the object/data to its final destination with the minimum cost incurred to the individual agent.

There are costs associated with each decision and these are given below by (1), where the cost of using the central hub is a variable cost that is dependent both on the actions of the agents within the group and the capacity of the hub. The central hub has a finite capacity given by L, if this capacity is reached then the hub is congested and a congestion charge cc (time/money) is imposed on all traffic through the hub. There are several ways to implement the congestion charge depending on the system being modelled, but for the purposes of this paper we will choose a digital cost structure (as shown in Fig. 2a), where each connection to the hub is $\frac{1}{2}$ a unit length and the congestion charge only applies when $> L$ agents use the central hub.

In contrast to the variable pricing structure of the central hub, the cost of using the peripheral pathway is determined only by the number of nodes traversed and as such is a fixed cost. There is a cost of $\beta = 1$ associated with traveling between two neighbouring nodes on the network. The transport costs across the network are then given by

$$C_{\text{central}} = \begin{cases} 2\left(\frac{\beta}{2}\right) & \text{if } N_{\text{central}} \leq L \\ cc + \beta & \text{if } N_{\text{central}} > L \end{cases}$$

$$C_{\text{out}} = n\beta \tag{1}$$

Where n is the number of nodes traversed, cc is the congestion charge, which can vary between 0 and $N/2$, and β is the standard or unit cost. The connections between nodes on the network are undirected links of unit length; hence data can travel either way round the network to its destination. The maximum distance an object can travel is then $N/2$. The connections to the central hub are directed and unique to the agent, hence only agents directly connected to the hub are able to use it. If an agent is not connected to the central hub, they are forced to use a peripheral pathway.

In order to make their decisions as to whether to use the central hub, each agent is randomly assigned $s = 2$ strategies from a pool of binary strategies. For $m = 2$ these strategies take the form (1011), where each digit in the strategy sequence corresponds to an action associated with the history string of the same position $(11, 10, 01, 00)$ i.e. for history string (11) we have action 0. Here action action 1 denotes a decision to use the central hub, whilst 0 corresponds to not using the central hub. The strategy table (shown in Fig. 2b) is self-similar in nature, and for every node visited by the global history string, another column of the table is accessed to reveal differences in strategies.

At each time-step in the game every agent with a connection to the central hub must make a decision whether or not to use the central hub, the decision can be summarized as "through the middle, or around the outside?" Their decision is dictated by the relative success of the two strategies that they hold. The agents make their decision based on the action associated with their highest scoring strategy, and if the two strategies are tied then the agent will flip a coin to decide. If the agent chooses action 1 and $C_{\text{central}} < C_{\text{out}}$ then the agent has made the correct decision and the success of their strategy will be reinforced with an increase in it's virtual points score of $+1$ points, else if $C_{\text{out}} > C_{\text{central}}$ then the strategy will be penalised by -1 points. The reverse applies if the agents high scoring strategy predicts action 0. At time $t = t + 1$, using the newly updated strategies, the agent again makes a decision about the cheapest pathway to use, and the above process is repeated.

3 Results

3.1 Variable Network Structure

At the start of the simulation there are no connections to the central hub and each agent is only connected to their nearest neighbours. At each time-step a randomly chosen agent is connected to the central hub such that $\lambda = \lambda + 1$. With this new network in place, the strategies and destinations are then reassigned amongst the agents and the simulation is repeated with the

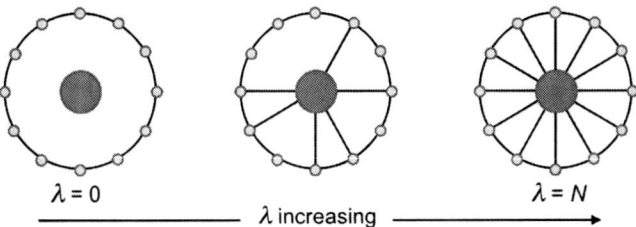

$\lambda = 0$ $\lambda = N$

λ increasing

Fig. 3. Network schematic, initially $\lambda = 0$ and only nearest neighbour connections exist. As λ increases connections are added from origin to destination until the network is fully connected.

agents competing to minimise their costs over the new network. This process is repeated until all agents have a connection to the central hub. A schematic of this process is shown in Fig. 3 as λ increases from 0 to $\lambda = N$. We have run this simulation with $N = 101$ agents, a memory length of $m = 2$ and $s = 2$ strategies assigned per agent. The central hub has a capacity given by L and a congestion charge determined by the digital price structure shown above in Fig. 2a. The simulation is run for 10,000 time-steps which constitutes one run, with each value of λ representing the average of 1000 runs.

The global cost per agent of transportation across the network is defined as $g(\lambda)$, and is displayed in Fig. 4a for various values of of cc in a network with $L = 40$. The transport cost, is initially the same for all values of cc and starts at $g(0) \sim 25$. This value is then the cost of transporting data/objects across the network with no connections to the central hub, and the general expression is given by

$$g(0) = \frac{1}{2} \left(\frac{N}{2} \right) \beta .$$
(2)

As λ increases, g decreases linearly for all values of cc up to a critical point at $\lambda \sim 50$. The curves for the various values of cc then diverge and follow their own pathways. Using this information we can divide the plots into two groups; high penalty $cc > 25$ and low penalty $cc < 25$. For the high penalty group $g(\lambda)$ increases rapidly after the critical point until the emergence of a stable state at $\lambda \sim 65$ where g stays relatively constant before increasing further as λ tends to N. For the low penalty group an increase in $g(\lambda)$ after the critical point is also observed, however after this initial increase the cost of transportation across the network again begins to fall as connections are added.

In Fig. 4b we see the plot of the probability of the central hub being crowded, $\gamma(\lambda)$. For values of $\lambda < 50$, $\gamma = 0$ as the central hub is never overcrowded. Above $\lambda = 50$ we observe, for some of the systems, the emergence of a second stable state at $\gamma = 0.5$. For the low penalty systems little or no time is

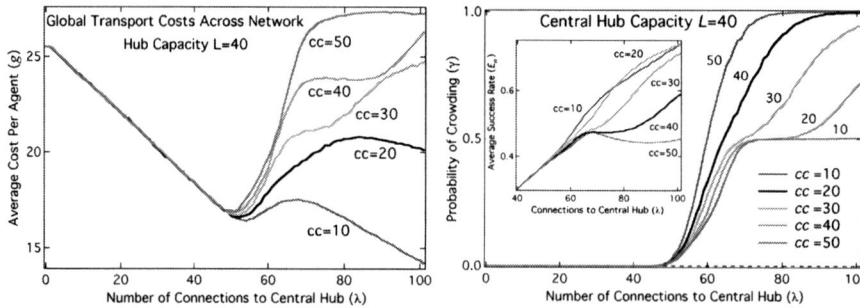

Fig. 4. (a) This graph displays the results from the simulation showing the global transport cost across the network $g(\lambda)$ as a function of the number of agents connected to the central hub. **(b)** This graph shows the probability of the central hub being crowded (γ). Inset shows the agents success rate in predicting the cheapest pathway across the network.

spent in the $\gamma = 0.5$ state. Whereas for high penalty systems as cc increases, the amount of time spent in this state also increases, and when $cc = 50$ the system does not leave the $\gamma = 0.5$ state. The inset of Fig. 4b shows the average success rate of the agents in predicting the correct transport pathway across the network. Following the critical point, agents in systems with $cc < 25$ initially enjoy a continued increase in average success rates. However for systems with $cc > 25$ the increase in average success rate only occurs for higher values of λ.

3.2 Variable Capacity Hubs

The simulation was also run for networks with varying central hub capacities. The size of the central hub was increased from $L = 10$ to $L = 70$ in increments of $L = 10$ and the simulation was repeated for each realisation of the network. The global transport costs per agent is shown for these systems in Fig. 5a with a congestion charge of $cc = 30$. The main features to note include the initial decrease in $g(\lambda)$ that is present for each curve irrespective of the capacity of L. This decrease continues as links are added up to a critical number of connections $\lambda_{\text{critical}}$. Above this value $g(\lambda)$ starts to increase. By increasing the size of L we observe a delay in the onset of the critical point and corresponding non-linear portion of the curves. For values of $L > 75$ (not shown here) we do not see this critical point and observe a continuous linear decrease in $g(\lambda)$ as λ is increased up to $\lambda = N$.

For low capacity hubs with $L < 40$ the global transport cost saturates at $g(\lambda) = \frac{N}{4}$. Once in this saturated state, adding extra links only serves to marginally decrease transport costs. It is interesting to note that despite the differences in where the critical point occurs, the same basic shape of the

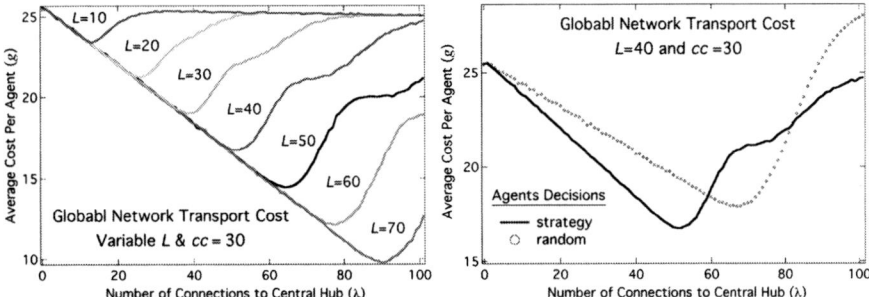

Fig. 5. (a) This graph shows the global transport cost across the network g(λ) for systems with varying central hub capacities. Where $cc = 30$ and the hub capacity varies from $L = 10$ to $L = 70$. **(b)** This graph shows the comparison of global transport costs between two systems with two types of agents. The first (*solid line*) has agents that use global information and strategies to make their decisions. The second (*dotted line*) comprises of agents that make random decisions to determine their best route.

curves is preserved (for the portions that can be seen), offset dependent on the size of L. Though the existence of a plateau is only observed for $30 < L < 50$, as below this threshold the transition to the saturated state occurs too rapidly, and for capacities higher than this, the peripheral nodes are fully connected to the central hub before the plateau emerges.

3.3 Random Agents

The results in Sects. 3.1 and 3.2 were obtained for systems where each agent had access to both strategies and global information. In this section we ran simulations where the agents did not have access to global information or strategies. The agents were thus forced to make random decisions about which path to take around the network. In Fig. 5b we compare the global transport costs of these strategy based systems with the costs of comparative random systems. The set-up for both systems is otherwise the same with $cc = 30$ and $N = 101$ in both networks.

In each system the initial decrease in $g(\lambda)$ is observed up to a critical point $\lambda_{\text{critical}}$. However, for the system with random decision making agents, the slope of $g(\lambda)$ is not as great as for the system where agents have access to both strategies and global information. Due to the differences in slope, $\lambda_{\text{critical}}$ is larger for the random agent system occurring at $\lambda_{\text{critical}} \sim 70$, compared to a value of $\lambda_{\text{critical}} \sim 50$ for the strategy based system. It is also interesting to note that for the random system there is no intermediate states present after the onset of the critical point. This is contrasted with the strategy based systems, where a stable state emerges for $\lambda > \lambda_{\text{critcal}}$ in the high penalty systems. In terms of efficiency, the strategy based system outperforms the random agent system for all values of λ except in the region $60 < \lambda < 80$.

3.4 Fixed Strategies and Destinations

In this version of the game the agents' destinations and strategies were allocated at the start and these remained constant throughout the simulation. Connections were then added from the peripheral nodes to the central hub one at a time. For every new connection added to the network the simulation was repeated, with the same strategies and destinations, and $g(\lambda)$ determined. This data is shown in Fig. 6 as the solid line, whilst each of the markers in the background represent a single realisation of the simulation where the agents' destinations and strategies are randomly allocated as described in Sect 3.1. This set of results in Fig. 6 clearly reveals the dynamic interplay between stable states, where adding connections to the hub does not greatly influence the global transport cost, and critical points, where adding links can have a dramatic effect on the global cost.

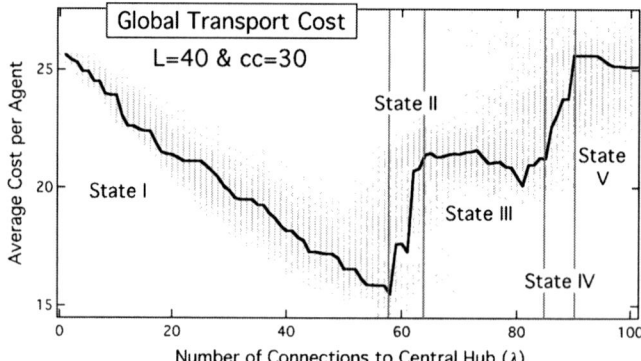

Fig. 6. Network with $L = 40$, $cc = 30$, $m = 2$, and $N = 101$. Graph shows how $g(\lambda)$ varies as connections are added to the hub, for fixed allocation of strategies and destinations amongst the agents (*solid line*). The background data show the range of values for various strategy/destination distributions.

State I is a stable state and is defined by the near linear decrease in $g(\lambda)$. Closer inspection of Fig. 6 reveals a step-wise decrease in cost for State I, where the addition of some connections to the hub does not reduce $g(\lambda)$ across the network. This step-wise decrease arises because some links are allocated to agents who hold strategies that do not allow them to take advantage of the cheaper central route across the network. The critical point $\lambda_{\text{critical}}$ (discussed in more detail in Sect. 4), is clearly shown here at $\lambda = 57$. The result illustrates the sharpness of the transition, where for $\lambda < 57$ the central hub is undersubscribed, but adding one more link results in a crowding of the central hub, and the system immediately jumps into State II. The system only resides in State II for a brief time before the more stable State III is reached.

State III is a stable state, and the global transport cost remains relatively constant as λ increases more agents gain access to the central hub. At $\lambda = 83$ the addition of a single connection results in another sharp transition that moves the system into State IV, where the addition of links to the central hub results in a rapid increase in g. This increase in global transport cost across is halted upon entering State V, where the addition of new connections has little effect on g. We also observe the symmetry of the transitions between States IV-V and States II-III. In the next section we will discuss the underlying phenomena driving the effects we observe in Figs. 4-6, looking particularly at the emergence of the unique stable states and the sharp transitions between them.

4 Analysis

We can divide the agents into two groups based on the distance that they have to travel and the size of the congestion charge, where N_{short} denotes the average number of 'short trip' agents and N_{long} the average number of 'long trip' agents. N_{long} agents have to travel a distance that is greater than $cc + 1$. Hence it is always advantageous for the long trip agents to use the central hub irrespective of the other agents actions. The short trip agents N_{short} have a distance to travel which is less than the size of the congestion charge. Hence for short trip agents, using the central hub will only be cheaper than the peripheral pathway provided that the central hub is *not* congested. However if the central hub is congested, then it will prove cheaper to use the peripheral pathway. The correct decision for short trip agents is then dependent on the collective actions of the group. Since the agents' final destinations are distributed randomly we get for $cc < N/2$:

$$N_{short} = \left(\frac{\text{cost of using crowded hub}}{\text{cost of maximum path across network}} \right) N_{total} + \beta \qquad (3)$$

$$= \left(\frac{cc}{\frac{N_{total}}{2}} \right) N_{total} + 1$$

$$= 2cc + 1, \qquad (4)$$

$$N_{long} = N_{total} - N_{short}$$

$$= N_{total} - 2cc - 1. \qquad (5)$$

If $cc < N/2$ then there are no long trip agents and $N_{short} = N_{total}$. Because the size of the congestion charge remains constant, the sizes of these two groups stays fixed for the duration of the game. There are then two separate history stings μ associated with the game μ_{short} and μ_{long} one for each group of agents.

The agents in each group can then effectively be treated as a cohesive unit, whose actions are jointly determined by their respective values of μ and the initial strategy distribution.

4.1 Un-Crowded Central Hub: State I

The game starts with $\lambda = 0$ and the hub is under-subscribed with no congestion, this state will be called State I. The history string for both groups of agents is then $\mu_{\text{long}} = \mu_{\text{short}} = (0000...)$, where 0 denotes the global uncrowded result. This history string only visits one node on the De Bruijn graph (00) (Fig. 7a) and as such results in a compression of the strategy space from the original pool of 16, to just two. The two strategies are then $(0|1)$ and $(0|0)$, where the first term denotes the history and the second the agent's action. Because the history string consists of consecutive 0's the virtual point scores for the two strategies diverge linearly over time as shown in Fig. 7b. This gives rise to the state level diagram shown in Fig. 7c with two states corresponding to the two different strategies $(0|1)$ and $(0|0)$. Because each agent is randomly assigned $s = 2$ strategies, and plays the higher scoring of the two, the populations of the levels is $3N/4$ and $N/4$ respectively. Where the average number of agents taking action 1 (population of top level in Fig. 7c) is defined as $N(1)$ and $N(0)$ is then the average number of agents taking action 0 (population of bottom level). Because the long trip and short trip agents behave in the same fashion in this state;

$$\frac{N(1)_{\text{long}}}{N_{\text{long}}} = \frac{N(1)_{\text{short}}}{N_{\text{short}}} = \frac{3}{4} \tag{6}$$

Thus for every link that is added to the central hub, the usage of the central hub $(N(1))$ increases on average by 3/4 agents, giving us for State I;

$$g(\lambda) = (\text{average peripheral cost}) - \lambda(N_{\text{central}}/N_{\text{total}})C_{\text{central}}$$
$$= \frac{N}{4} - \frac{3\lambda}{4}\beta \tag{7}$$

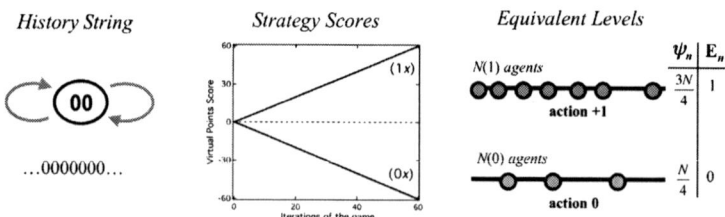

History String   *Strategy Scores*   *Equivalent Levels*

Fig. 7. (a) The path across the de Bruijn graph for State 1. **(b)** The corresponding virtual points scores for the two classes of strategies used by agents to make their decisions, **(c)** Equivalent levels, along with associated populations and success rates.

Where the slope given by equation (7) is in good agreement with that seen in Figs. 4a and 5a. The slope of the graph for the random agents in Fig. 5b is not as steep because for this system the second term in (7) is reduced to $(\lambda\beta)/2$. This process of transport cost reduction continues as long as the system remains in the un-crowded state with $N(1) < L$. We can then define a critical point to be the point where the system has a 50% chance of $N(1) > L$. Thus the critical point where the system moves out of State I and into the crowded regime in State II occurs when $N(1) = L$, or when

$$\lambda_{\text{critical}} = \frac{4}{3}L \ . \tag{8}$$

Using (8) for systems with hub capacities of $L = (10, 20, 30, 40, 50, 60, 70)$ gives us critical points of $(13.3, 26.6, 39.9, 53.3, 66.6, 79.98, 93.3)$ respectively. These values are in excellent agreement with the minimum values of $g(\lambda)$ seen in Fig. 5a.

The change in state corresponds to a change in history string for short trip agents $\mu_{\text{short}} = (0001...)$. However the history string for the long trip agents remains fixed at $\mu_{\text{long}} = (0000...)$. The long trip agents maintain this history string for the duration of the game irrespective of what state the system is in. Upon leaving State I, the two groups of agents then cease to behave as one unit and we need to consider their actions independently. The number of agents using the central hub is given by

$$N(1) = N(1)_{\text{long}} + N(1)_{\text{short}} \ . \tag{9}$$

Because μ_{long} remains fixed for the game, the average number of long trip agents using the central hub is simply dependent on the size of the congestion charge and the number of connections to the central hub:

$$N(1)_{\text{long}} = \frac{3}{4}N_{\text{long}}\lambda$$
$$= \frac{3}{4}(N_{\text{total}} - 2cc - 1)\lambda \ . \tag{10}$$

In contrast to the long trip agents, $N(1)_{\text{short}}$ is more difficult to calculate, since μ_{short} varies depending on which state the system resides in. In order to determine the short trip agents' contribution to the central hub congestion, we must then analyse the system dynamics for States II and above.

4.2 Stable States and Noise

In State II (see Fig. 8a) the system visits two extra nodes (01) and (10). Visiting the nodes has two effects, the first is that for part of the time the central hub is crowded and a congestion charge is applied to all agents using it. This crowding is responsible for the increase in $g(\lambda)$ which occurs after the critical point. The increase in $g(\lambda)$ is a gradual one since the probability of residing in the State II is goverend by a binomial distribution. The second effect of the nodes is to increase the number of strategies available to the agents, which in turn changes the number of levels/bands in the state level diagram. Here the total number of levels in increased to six in State II (see Fig. 8b) from the original two in State I.

The two nodes can thus be thought of as providing extra degrees of freedom for the system and as such they resolve the $(1xx)$ strategy into four new strategies (111, 110, 101, 100). This extra resolution means that for node (10) a percentage of the agents that were taking action 1 before the transition are now taking action 0, and as such $N(1)$ must initially be less that L in State II. The same strategy splitting effect occurs for the $(0xx)$ group of strategies, which if $s = 1$ would cancel the above effects and the system would move out of this state. However because $s = 2$ the distribution of strategies is top heavy for the strategies with $(00 \mid 1)$. As a result of this, the splitting of strategies acts to create a buffer by reducing $N(1)$. This means that as more links are added to the central hub the new state will remain stable until the 2nd critical point is reached.

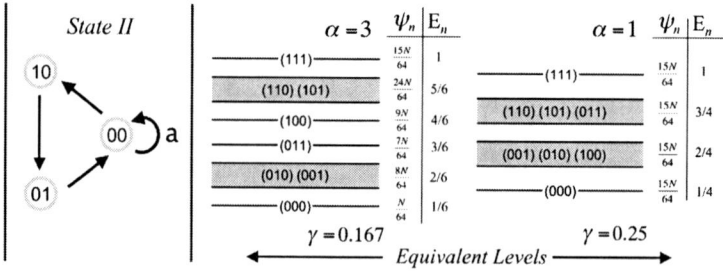

Fig. 8. (a) de Bruijn graph for State II, system visits three nodes and α is the number of times the system returns to (00) node. (b) State level diagram for State II, the wide bands represent groups of strategies with the same mean success rate.

In State II, the system traverses the de Bruijn graph in a cyclic fashion, returning to the (00) node each time (Fig. 8). However after taking this pathway the original ordering of the strategies is different. The system then resets itself by returning to the (00) node α times. As L increases the value of α decreases, from $\alpha = 3$ initially to $\alpha = 1$ immediately before leaving State II. This change in α reduces the number of unique levels in the state diagram, and alters the make-up of the strategies within them (Fig. 8b).

The wider bands in the state level diagram represent sets of strategies which have the same average success rate over time (E_n), but vary about E_n during the cycle around the graph. The strategies will vary about this mean, but on two out of the three nodes they will have equal virtual point scores. On these nodes agents that then hold two strategies from within the same band are forced to flip a coin in order to decide which strategy to play. If the agents' two tied strategies make the same predictions for the particular node, then it does not increase the disorder in the system. However if the two strategies make different predictions, flipping the coin does influence their action and this agent falls into a new group of agents which we shall call the undecided group, denoted here by $N(\frac{1}{2})$. There are thus three elements in the system, $N(1)$ the number of agents choosing action 1 with certainty, $N(0)$ the number of agents choosing action 0 with certainty and $N(\frac{1}{2})$ the number of agents whose decision is determined by chance. Because the long trip agents effectively only have two non-equal strategies to play, the $N(\frac{1}{2})$ agents come exclusively from the short trip population. The relative sizes of these three groups of agents determine which state the system will reside in and hence the probability that the central hub is crowded.

A global '1' result on the (10) node will shift the system out of State II and into State III. In order to determine the probability of this occurring we need to consider the size of the $N(1)$ and $N(\frac{1}{2})$ components for this state. We will look specifically at the $\alpha = 1$ distribution since this is the last cycle the system visits before moving into State III. Analysis of ψ_n and the band levels in Fig. 8b gives us,

$$N\left(\frac{1}{2}\right) = \frac{8\lambda}{64}N_{\text{short}} \quad \text{for } \alpha = 1$$

and using Eqns. (9,10) in conjunction with Fig. 8b, we have;

$$N(1) = \left(\frac{34}{64}N_{\text{short}} + \frac{3}{4}N_{\text{long}}\right)\lambda$$

If $N(\frac{1}{2})$ is greater than the difference between the resource level and $N(1)$, then the system will be randomly 'kicked out' of the stable cycle in State II and will briefly reside in State III before returning. This process is shown in Fig. 9a. If we define the buffer as,

$$\Delta = L - N(1) - N\left(\frac{1}{2}\right).$$

We can then use binomial probability distribution to determine the likelihood that the system will be 'randomly' perturbed into a new state. Doing this gives us

$$P_{\text{perturb}} = \sum_{k=\Delta}^{N(\frac{1}{2})} P\left(k \mid N\left(\frac{1}{2}\right)\right) = \sum_{k=\Delta}^{N(\frac{1}{2})} \frac{N\left(\frac{1}{2}\right)!}{k!(N\left(\frac{1}{2}\right) - k)!}\left(\frac{1}{2}\right)^k\left(1 - \frac{1}{2}\right)^{N(\frac{1}{2})-k}.$$

The effect of the $N(\frac{1}{2})$ agents is to act as noise which can randomly perturb the system into a range of intermediate states between States II and III, these can be seen clearly in Fig. 9b for $0.25 < \gamma < 0.5$.

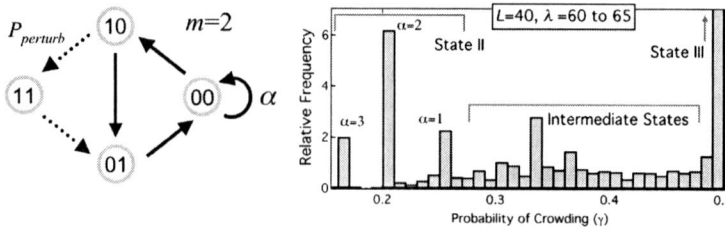

Fig. 9. (a) de Bruijn graph for the intermediate state between States II and III. Probability of visiting the (11) node given by P_{perturb}. (b) The histogram reveals the existence of intermediate states of γ for $60 < \lambda < 65$.

4.3 High γ States

The movement into State III corresponds to a re-ordering of the strategy bands, which compress from the multiple bands in State II to just one band in State III. This new state is known as the Eulerian trail (Fig. 10a), since it visits each node exactly twice during it's cycle. The compression of the strategies into one band serves two purposes. Firstly, since all the strategies have the same mean value of E_n, the number of agents having to flip coins to make their decisions increases, so $N(\frac{1}{2})$ increases. The second effect is another example of the buffering process, which reduces $N(1)$ as all the strategies compress into one state. The two processes make it difficult to perturb the system into a new cycle and the Eulerian trail is thus an attractor in this system. This stability is evidenced by the fact (see Fig. 4b) that the sole addition of central hub connections for short trip agents cannot drive the system out of this state. Because there is only one strategy band, every new short trip agent connected to the hub has a 0.5 probability of them picking action 1. The addition of short trip agents serves only to reinforce the Eulerian trial and in order to move into State IV, the system needs long trip agents to be connected to the central hub. This effect can be seen in Fig. 3b, where it is only when cc is reduced such that N_{long} increases, do we get significant movement out of State III.

The movement into states above State III is then increasingly driven by long trip agents and their associated action bias. As λ increases the system moves through the high γ states IV and V. As the system progresses through these states the cycles around the de Bruijn graph are left shifted (see Fig. 10) as the agents actions lead to a predominately crowded central hub. The global history cycles for States IV and V shown in Fig. 10 exhibit similar behaviour to those of States I and II. Where the system resets itself α' times on the (11) node, and each cycle around the de Bruijn graph has a unique state-level diagram associated with it. Because of the inherent symmetry of the de Bruijn graph the earlier analysis of States I and II proves useful for our understanding of these high γ states.

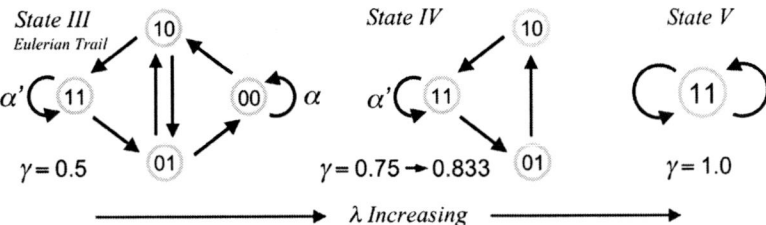

Fig. 10. Here we see the high γ states, (a) Eulerian trail, (b) State IV, (c) State V. These cycles around the de Brujin graph for States IV and V are mirror images of the cycles in states I and II.

The de Brujin cycles for States IV and V correspond to the mirror images of States II and I respectively. The difference here is that the strategy levels for these states are inverted, with the top ranking strategy bands now at the bottom, and vice versa. The population ψ_n of each of the n levels remains the same. Whilst this reordering of strategies does not affect the long trip agents' decisions, it does act to reduce the likelihood of short trip agents using the central hub. Due to the inverted nature of the state-level diagrams, the average number of short trip agents using the central hub in State IV is then equal to the number of agents electing to take the peripheral path in State II:

$$N(1)_{\text{short(IV)}} = N(0)_{\text{short(II)}} = \frac{22}{64}\lambda. \tag{11}$$

Likewise for State V and State I, giving us;

$$N(1)_{\text{short(V)}} = N(0)_{\text{short(I)}} = \frac{\lambda}{4}. \tag{12}$$

Long trip agents continue to provide an addition of 3/4 agents per link as given by (10), and in essence drive the transitions between through these higher states. Hence systems with more long trip agents will move out of these states at lower values of λ, and this is observed in Fig. 4b. We can substitute (11) and (12) for each of these states into (9) to determine the total number of agents using the central hub. From this equation for $N(1)$ it is then possible to determine the likelihood of the central hub being crowded. Each new state that the system visits has a unique cost g, and γ associated with it. We can derive $\gamma(\lambda)$ for each realisation of the network using the expressions for $N(1)$ that we have derived for each state. Using this result in conjunction with a careful analysis of the state-level diagrams the transport costs can be determined for each state. This gives us a general expression for the global transport cost:

$$g(\lambda) = \sum_{n=1}^{i} g(\lambda)_n P(\lambda)_n. \tag{13}$$

Where $g(\lambda)_n$ is the cost associated with State n, and $P(\lambda)_n$ is the probability of residing in this state. Elsewhere we will present a detailed analysis of these states and the transitions between them.

5 Summary

This paper has shown that the agents in the network can be divided into two groups, where the size of each group is determined by the maximum path length across the network and cc. The addition of new links to the network brings about two main processes that serve to drive the system into new states. The first of these processes is the addition of $N(1)$ agents who use the central hub with *certainty*. The second process is a more unpredictable one with the addition of 'random' $N(\frac{1}{2})$ agents to the system. If $N(1)+N(\frac{1}{2}) > L$, there is a finite probability that the $N(\frac{1}{2})$ agents will randomly perturb the system into new states, but these states are not permanent. However for any node on the graph if $N(1) > L$, the change is permanent and a new state is formed that has a unique and 'predictable' cycle associated with it. This new state and the associated cycle around the de Bruijn graph has the effect, when compared with the previous state, of changing the history strings for the N_{short} agents, which in turn re-orders the agents strategies. This re-ordering of strategies reduces the number of short trip agents using the central hub, and in effect acts as a buffer to produce stable states in the system.

Acknowledgements

We acknowledge the financial support of the EU through the MMCOMNET project on Complex Systems.

References

1. D. Helbing, M. Schönhof, H.-U. Stark, and J. A. Holyst, Adv. in Complex Systems **8**, 87 (2005)
2. R. Albert and A.L. Barabasi, Phys. Rev. Lett. **85**, 5234 (2000).
3. D.J. Watts and S.H. Strogatz, Nature **393**, 440 (1998)
4. D. S. Callaway, M. E. J. Newman, S. H. Strogatz, and D. J. Watts, Phys. Rev. Lett. **85**, 5468 (2000)
5. D. Ashton, T. Jarrett and N. Johnson, Phys. Rev. Lett. **94**, 0508701 (2005).
6. S.N. Dorogovtsev and J.F.F. Mendes, Europhys. Lett. **50**, 1 (2000).

Specifica of Fundamental Diagram in Urban Traffic

Kai-Uwe Thiessenhusen and Peter Wagner

Deutsches Zentrum für Luft- und Raumfahrt, Institut für Verkehrsforschung,
Rutherfordstr. 2, 12489 Berlin, Germany

Summary. The relation between the fundamental quantities flow, density, and speed is well studied empirically in "undistorted" traffic, i.e. traffic on highways or major non-urban roads. The situation on urban roads is more complicated since urban traffic is influenced by a number of effects. Especially, the situation at intersections (traffic lights, merging traffic) is of striking importance.

While the flow between two intersections can be considered as roughly constant, the density and the speed are quantities that vary rapidly along the road. By analyzing data from stationary sensors (measuring flow and local speed) as well as Floating Car Data (measuring travel time, i.e. averaged speed) we found that the speed on many urban roads does not depend on the corresponding flow (respectively the density) alone. For a given flow, the speeds at different times of the day may differ. This effect can be observed for local speeds as well as for speeds averaged along edges or over an area. Very often the maximum in the flow at morning peak precedes the speed minimum. The time difference between these two events is up to one hour varying from road to road. Similar observations have been found in averages over an urban area.

A possible explanation for the observed relation between flow and speed is – besides the influence of traffic signalling – that a more heterogeneous traffic (with a higher fraction of vehicles merging into different directions) leads to slower speeds.

1 Introduction

Traffic is described by fundamental relations between the characteristic quantities, traffic flow q, density k, and speed v (see, e.g., [1, 2]). In the following, we concentrate on flow-speed relations because these quantities are measured directly in the data available here. The density can in principle be derived from flow and speed. However, simply using q/v as an estimate can yield strange results for small speeds. The fundamental relations are studied well for "undistorted" systems, e.g. highway traffic. Fig. 1 shows as a characteristic example for freeway traffic the flow speed relation from data obtained by a stationary sensor on a highway.

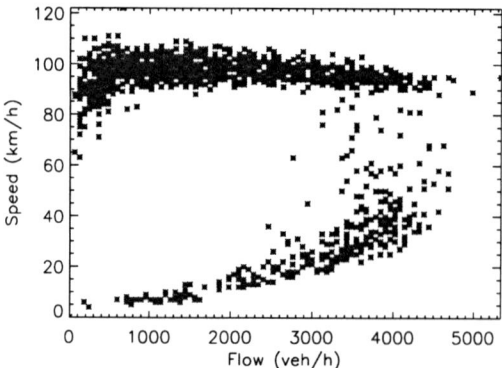

Fig. 1. Example for the flow-speed relation on a highway. The data are taken from a stationary detector on a highway close to Munich for one day.

The flow-speed plot consists of two branches. The upper branch with high speeds represents the free-flow-state, the lower branch represents the congested regime. In the free flow state the speed is roughly the same and does not depend (or depends only weakly) on flow when the flow is low. For large flows, the speed becomes smaller with growing flow. In the congested regime, flow and speed are reduced until the extreme case of total traffic breakdown. Except for special situations, e.g. locally reduced road capacity, highway data typically have a similar structure as described here.

Urban traffic, however, can be strikingly different. Caused by traffic signals, merging traffic at intersections, parked vehicles and the like, the speed of a single vehicle, and with it, the speed profile $v(x)$ along a road can undergo rapid transitions. A typical speed profile is displayed in Fig. 2.

The most important intersections in Fig. 2 are at position 0.3 km, 0.75 km, and 1.45 km, respectively. The averaged speed is reduced before these intersections. In contrast, the averaged flow can be expected to change only slightly between two adjacent nodes.

2 The Data

This paper studies two different classes of data. The first one is from 280 stationary detectors maintained by the Verkehrsmanagementzentrale (VMZ) Berlin. The detectors are distributed in the main road (non-highway) network in the centre of Berlin. Commuter roads leading out off the city centre are covered as well as connections within the inner city. Data from the highway network are not taken into account. Each sensors averages flow and speed over sample intervals of 5 minutes.

The other source is floating car data (FCD) from taxis of the Berlin company Cityfunk [3, 4]. The vehicles are equipped with a GPS device which transmits

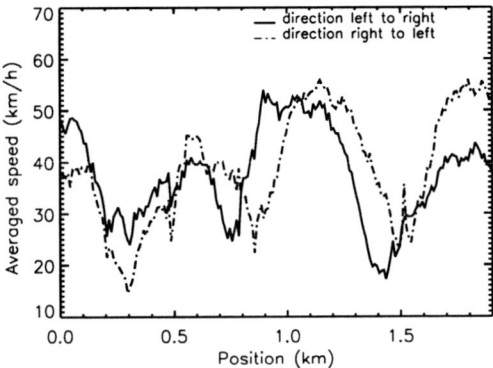

Fig. 2. Speed profiles along a road section in Berlin. The speed is plotted as function of position for vehicles moving into positive (solid) and negative x-direction (dashed). The data are obtained from floating car data. The speed values are averaged over 39 working days and the time from 6:00 to 18:00.

the positions to the headquarter in a fixed time interval, typically 30s. The positions are matched on a digital map. From this, travel times and averaged speeds can be obtained. Different from the stationary detectors that measure flow and speed simultaneously, PVD do not give direct information about traffic flow.

We considered data from a three month period, 13 weeks, in summer 2003. Unless stated otherwise, the plots refer to 39-day averages of all Tuesdays, Wednesdays, and Thursdays in the analyzed time period.

3 Observations

Fig. 3 shows the daily profile of speed and flow averaged over a 10 km × 10 km area in the centre part of Berlin and 39 working days (Tuesday, Wednesday, and Thursday). The flow values are obtained from 120 stationary sensors; the speed values from the sensors as well as from FCD.

The speed profiles obtained by both methods are qualitatively very similar. The speed minima are in the afternoon and morning rush hours. However, both data sets differ by an obvious offset of approximately 15 km/h. Typically, stationary detectors are located away from larger intersections. (Their positions are selected in order to avoid problems to detect vehicles waiting at the traffic signal or from averaging over merging and non-merging vehicles.) As a consequence of such a choice of positions the averaged speed from the sensors must be larger than the averaged speed in the whole network. In contrast, FCD contain real travelling times and are therefore a better indicator for the speed average.

The flow profile shows the typical peaks in the morning and the afternoon rush hours with a dominating afternoon peak. The absolute flow values depend on

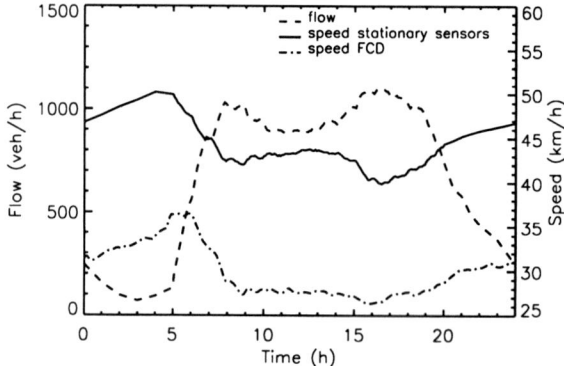

Fig. 3. Shown is the speed from stationary sensors (solid line) and floating car data (dashed dotted line) as function of time averaged over 39 working days from the inner part of Berlin. The flow averaged over 39 days and 120 stationary sensors is plotted with dashed lines.

the selection of detectors for averaging. The qualitative structure of flow and speed profiles however is also recovered when averaging over different detector ensembles.

There is another, less eye-catching, difference between highway and urban traffic visible in Fig. 3. The speed on highways is in the free flow regime roughly independent from flow for low and moderate flow values. In contrast either FCD as well as stationary sensors observe a different situation at low flow times (i.e., at night) in urban traffic. There is a continuous increase in speed until a maximum at about 4 in the morning, roughly at time of flow minimum. At this time, the speed values obtained by both methods are about 5 km/h larger than in the late evening.

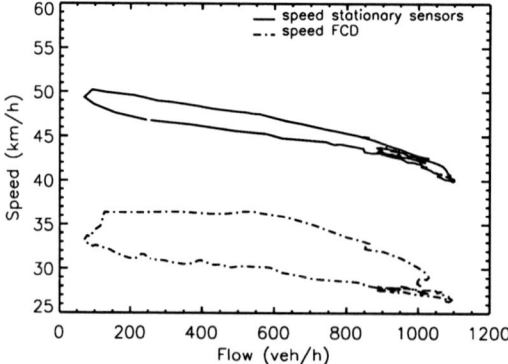

Fig. 4. Speed vs. flow from the same data as in Fig. 3. To obtain the data for FCD, the travel speeds have been plotted against the flows measured by the stationary detectors.

In Fig. 4 right, the averaged speed is plotted as function of flow. Different points on the curves are from different times of the day. Obviously, there are two different branches; the same flows are correlated with different speeds at different times of the day. With other words, one and the same speed can be brought into accordance with totally different flows at different times of the day.

Midnight is the gap in the left part of the curves in the plot. The loop is clockwise, upper branches consists of data from the night and morning hours. Typically, speeds at morning are higher than at other times later the day. A similar loop (in some sense hysteretic) structure appears in the floating car data.

The situation at one single sensor strongly depends on the type of road it is located on. Nevertheless, similar effects as described above can also be observed for many single detectors. One example is given in Figs. 5 and 6 for a commuter road leading into the city. Therefore, the morning flow peak (in the plot averaged over 39 working days) is the dominating one. Correspondingly, the global speed minimum is in the morning hours.

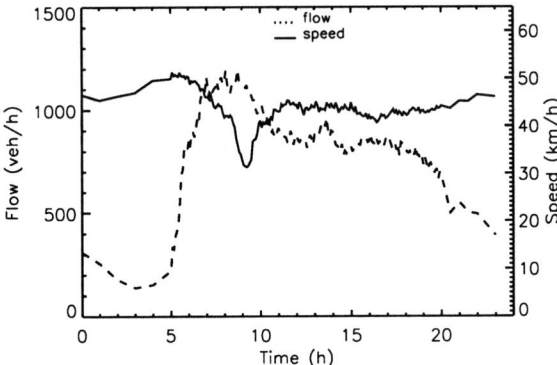

Fig. 5. 39 day average of daily profiles of flow (dashed line) and speed (solid line) obtained from a stationary sensor on a commuter road.

Fig. 6 (solid line) shows the speed as function of flow from the same data as in Fig. 5. Also, a loop structure as in Fig. 4 appears. At morning hours, the speed typically is higher than at times with the same flow later the day. Here, the speed minimum takes place in the late morning. The afternoon data are concentrated in a cluster (with flow values of about 900 veh/h and speed of about 45 km/h).

The dotted line contains data from one single day. The basic shape and the loop structure can be observed in the same way as in the 39 day average. In this example, a beginning of congestion can be observed. It illustrates that the loop like structure has nothing to do with the congestions branch as displayed in Fig. 1.

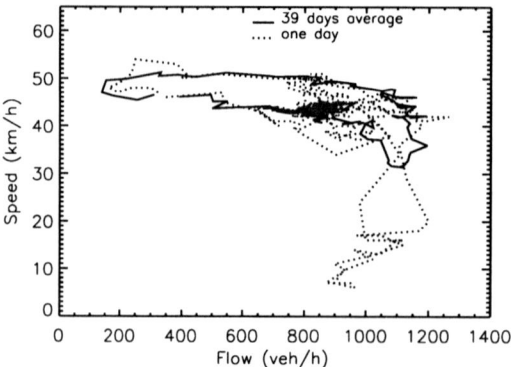

Fig. 6. Speed vs. flow at the same sensor as in Fig. 5. The 39-day average (solid line) and data from one single day (dotted) are plotted.

In this example (Fig. 5), the morning flow maximum takes place before the morning speed minimum. This turns out to be a rather frequent effect in urban traffic. To make this observation more quantitative, the distribution of flow maxima and speed minima has been analyzed. For each five minutes interval of the data, and for data from 150 sensors (all sensors in the data set with a maximum flow above 1000 veh/h) in the regarded time interval of 39 days, the number of flow maxima has been counted and plotted as function over the time of the day. For the speeds, the minimum speed is treated accordingly. The result is displayed in Figure 7.

Nearly all flow maxima (solid line) take place either between 6:00 and 10:00 in the morning or between 14:00 and 18:00 in the afternoon peak time, re-

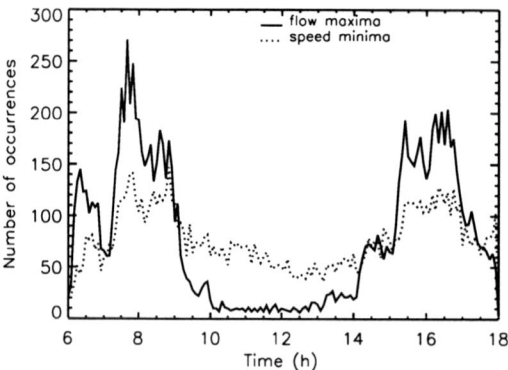

Fig. 7. Frequency of the number of occurrences of flow maximum (solid line) and speed minimum (dashed line) at a certain time as function of time. Data from 150 sensors and 39 working days have been taken into account.

spectively. The speed minima (dashed line) are more widespread. This is an indication that more sources than the local flow alone influence the speed.

Fig. 8 shows the distribution of temporal differences between flow maximum and speed minimum. The number of occurrences of a certain temporal difference as function of this difference is plotted either for the morning flow maxima (solid line) and afternoon maxima (dashed). The maxima for both curves are not exactly at 0, but at −5 minutes. Except for this, the afternoon curve is roughly symmetric with respect to zero. This is not the case for morning data. The dominance of occurrences at negative x-values indicate the flow maximum often tends to precede the speed minimum.

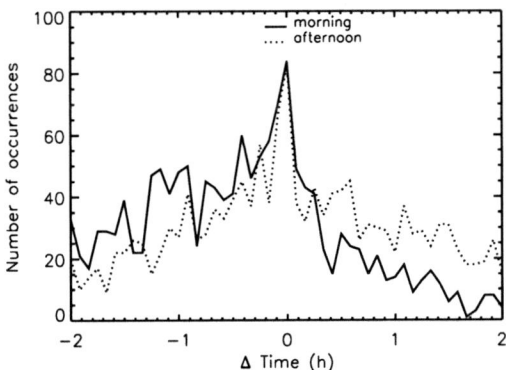

Fig. 8. Frequency of the occurrence of temporal difference between flow maximum and speed minimum as function of the temporal difference. Negative x-values indicate that the flow maximum precedes the speed minimum. The solid line is for data from the morning (flow maximum between 6:00 and 10:00), the dashed line for afternoon data (flow maximum between 14:00 and 18:00). Data from 150 loops and 39 days are taken into account.

4 Discussion

Urban traffic is affected by many influences. Interestingly, this is reflected in the fundamental relation between the characteristics quantities flow and speed (or density, respectively), which display a large amount of different flavours of the same underlying theme.

One feature which can be observed at many urban roads is that one and the same flow can lead to different speed at different times of the day even in the non-congested regime. This is the case either for single measurements at one fixed position as well as for local and/or temporal averages. Among the different features observable in this context, the most prominent is that the flow maximum in the morning rush hour tends to precede the speed minimum. In some situations it is easily imaginable and well-known that an increase in flow precedes a speed reduction. E.g., this can be the case when congestion is originating or if there is a larger number of vehicles waiting at a traffic signal and the speed is measured downstream. However, these and similar situations lead to an offset in the order of a few minutes only. Also, they can be expected to take place in similar ways in the morning and afternoon peak hours. It is possible that such effects are responsible for the observed shift in the order five minutes between flow and speed (Fig. 8). However this effect cannot explain larger offsets in the morning data (and the lack of such features in the afternoon).

Also, hysteretic effects described previously [5, 6] which are mainly caused by drivers' behaviour take place on different time scales.

So far, we do not have a conclusive answer to the question where the observed effect results from. But we have hypotheses: the first one assumes, that traffic in the early morning rush hour possibly is more homogeneous. A large fraction of commuter vehicles follow the same routes over relatively long distances. This could lead two a smaller amount of vehicles crossing the lanes in order to merge.

The second hypothesis is that the number of vehicles stopping outside parking zones seems to grow only later the day in the morning rush hour. Such vehicles block a lane for a certain time, leading to a decreased speed. It seems intuitively clear that both of these effects are principally able to reduce the road capacity, and, therefore, the speed. However, it can not been concluded from the available data in which quantity these effects are really important or if other reasons must be taken into account to explain the observations.

It has been demonstrated by an empirical analysis that urban traffic differs from freeway traffic. Although there are several candidates for the cause of this difference, it remains to explicitly demonstrate the culprit for these changes. To better understand these effects are of uttermost importance for both traffic flow simulations as well as for any endeavour to predict traffic states. Microsimulation models designed to simulation traffic flow probably have to be extended. However, a much better data base is needed to obtain a sufficient input for such simulations.

References

1. M.J. Lighthill, J.B. Whitham: Proc. Roy. Soc. London **A 229**, 281-345 (1955)
2. C.F. Daganzo, Fundamentals of Transportation and Traffic Operations. Pergamon, New York (1997)
3. R.P. Schäfer, K.U. Thiessenhusen, P. Wagner: A traffic information system by means of real-time floating-car data. ITS World Congress 2002, Chicago
4. R.P. Schäfer, K.U. Thiessenhusen, E. Brockfeld, P. Wagner: Analysis of travel times and routes on urban roads by means of floating-car data. European Transport Conference PTRC, September 4-9 2002, Cambridge, UK
5. J. Treiterer, J.A. Myers: The Hysteresis Phenomena in Traffic Flow, Proceedings of the 6th International Symposium on Transportation and Traffic Flow Theory, pp 13-38 (1974)
6. H.M. Zhang: Transp. Res. **B 33**, 1-23 (1999).

A Fluidodynamic Model for Traffic in a Road Network

Mauro Garavello[1] and Benedetto Piccoli[2]

[1] Dipartimento di Matematica e Applicazioni, Università di Milano Bicocca, Via
   R. Cozzi 53 - Edificio U5, 20125 Milano (Italy). (mauro.garavello@unimib.it).
[2] Istituto per le Applicazioni del Calcolo "M. Picone", Viale del Policlinico 137,
   00161 - Roma, Italy (piccoli@iac.rm.cnr.it).

Summary. This paper is concerned with a fluidodynamic model for traffic flow.
More precisely, we consider a single conservation law, deduced from the conservation
of the number of cars, defined on a road network that is a collection of roads with
junctions. The evolution problem is underdetermined at junctions, hence we choose
to have some fixed rules for the distribution of traffic plus an optimization criteria
for the flux. We prove existence of solutions to the Cauchy problem and we show
that the Lipschitz continuous dependence by initial data does not hold. Our method
is based on wave front tracking approach, see [4].

1 Introduction

This paper deals with a fluidodynamic model of heavy traffic on a road network. More precisely, we consider the conservation law formulation proposed
by Lighthill and Whitham [15] and Richards [16]. This nonlinear framework
is based simply on the conservation of cars and is described by the equation:

$$\varrho_t + f(\varrho)_x = 0, \tag{1}$$

where $\varrho = \varrho(t,x) \in [0, \varrho_{max}]$, $(t,x) \in \mathbf{R}_+ \times \mathbf{R}$, is the *density* of cars, $v =
v(t,x)$ is the *velocity* and $f(\varrho) = v\,\varrho$ is the *flux*. This model is appropriate to
reveal shock formation as it is natural for conservation laws, whose solutions
may develop discontinuities in finite time even for smooth initial data (see
[4]). In most cases one assumes that v is a function of ϱ only and that the
corresponding flux is a concave function.
We deal with a network of roads, as in [7, 11, 12, 14]. This means that we have
a finite number of roads modeled by intervals $[a_i, b_i]$ (with one of the two
endpoints possibly infinite) that meet at some junctions. For endpoints that
do not touch a junction (and are not infinite), we assume to have a given
boundary data and solve the corresponding boundary problem, as in [1, 2].
The key role is played by junctions at which the system is underdetermined

even after prescribing the conservation of cars, that can be written as the Rankine-Hugoniot relation:

$$\sum_{i=1}^{n} f(\varrho_i(t, b_i)) = \sum_{j=n+1}^{n+m} f(\varrho_j(t, a_j)), \tag{2}$$

where ϱ_i, $i = 1, \ldots, n$, are the car densities on incoming roads, while ϱ_j, $j = n + 1, \ldots, n + m$, are the car densities on outgoing roads. In [14], the Riemann problem, that is the problem with constant initial data on each road, is solved maximizing a concave function of the fluxes and it is proved existence of weak solutions for Cauchy problems with suitable initial data of bounded variation. In this paper we assume that:

(A) there are some prescribed preferences of drivers, i.e. the traffic from incoming roads is distributed on outgoing roads according to fixed coefficients;

(B) respecting (A), drivers choose so as to maximize fluxes.

To deal with rule (A), we fix a traffic distribution matrix

$$A = \{\alpha_{ji}\}_{j=n+1,\ldots n+m,\ i=1,\ldots,n} \in \mathbf{R}^{m \times n},$$

such that

$$0 < \alpha_{ji} < 1, \qquad \sum_{j=n+1}^{n+m} \alpha_{ji} = 1, \tag{3}$$

for each $i = 1, \ldots, n$ and $j = n + 1, \ldots, n + m$, where α_{ji} is the percentage of drivers arriving from the i-th incoming road that take the j-th outgoing road. Notice that with only the rule (A) Riemann problems are still underdetermined. This choice represents a situation in which drivers have a final destination, hence distribute on outgoing roads according to a fixed law, but maximize the flux whenever possible. We are able to solve uniquely Riemann problems, under suitable conditions on the matrix A, and then to construct solutions to Cauchy problems for networks with simple junctions, i.e. junctions with two incoming roads and two outgoing ones. Recall that the solution depends on the way the Riemann problem is solved. In some other papers, see [10, 11], other Riemann solvers at junctions are studied. In particular in [11], the authors solve the Riemann problem at junction substituting the rule (B) with a maximization of a quadratic functional, that permits to eliminate some technical assumptions of this paper. Moreover D'Apice, Manzo and Piccoli in [10] proposed to solve the Riemann problem by inverting the order of the rules (A) and (B). The solution constructed in this way depends in a Lipschitz continuous way from the initial datum.

The main technique to construct a solution is using the wave-front tracking algorithm and controlling the total variation of the flux. We refer the reader to [4] for the general theory of conservation laws and for a discussion of wave front tracking algorithms.

The main difficulty in solving systems of conservation laws is the control of the total variation, see [4]. It is easy to see that for a single conservation law the total variation is decreasing, however in our case it may increase due to interaction of waves with junctions.

There is a natural lack of symmetry for *big waves* (i.e. waves crossing the value σ) and *bad data* at junctions, since the role of entering roads is different from that of exiting ones. Similarly, for scalar conservation laws with discontinuous coefficients, one has to use a definition of strength for discontinuities of the coefficient, seen as waves, that is not symmetric but depends on the sign of the jump in the solution. This is enough to control the total variation in that case, on the contrary our problem is more delicate. In fact, the variation can still increase due to interactions of waves with junctions. The bounded quantity is the total variation of the flux. We prove this fact for junctions with only two incoming roads and two outgoing ones. Unfortunately the total variation of the flux is not equivalent to the total variation of ϱ, since $f'(\sigma) = 0$, and so it is not sufficient to prove existence of solutions. Therefore some compactness argument is used together with a bound of big waves near junctions.

Our techniques are quite flexible, so we can deal with time dependent coefficients for the rule (A). In particular, we can model traffic lights and also in this case the control of total variation is extremely delicate. An arbitrarily small change in the coefficients can produce waves whose strength is bounded away from zero. Still it is possible to consider periodic coefficients, a case of particular interest for applications. We can also deal with roads with different fluxes: this can be treated in the same way with the necessary notational modifications.

There is an interesting ongoing discussion on hydrodynamic models for heavy traffic flow. In particular some models using systems of two conservation laws have been proposed, see [3, 8, 13]. We do no treat this aspect.

The paper is organized as follows. In Sec. 2 we give the definition of weak entropic solution and, following rules (A) and (B), we introduce an admissibility condition at junctions. In Sec. 3 we state the result about the existence and uniqueness of admissible solutions for the Riemann Problem in a junction, then using this we describe the construction of the approximants for the Cauchy Problem (see Sec. 4). In Sec. 5 we give the bound on the total variation of the flux and existence of admissible solutions for the Cauchy Problem with suitable initial data. In Sec. 6 we show with a counterexample that the Lipschitz continuous dependence with respect to initial data does not hold.

2 Basic Definitions

We consider a network of roads, that is modeled by a finite collection of intervals $I_i = [a_i, b_i] \subset \mathbf{R}$, $i = 1, \ldots, N$, $a_i < b_i$, possibly with either $a_i = -\infty$ or $b_i = +\infty$. On each road consider the model proposed by Lighthill-Whitham-Richards

$$\varrho_t + f(\varrho)_x = 0. \tag{4}$$

where ϱ denotes the density of cars in roads. Hence the datum is given by a finite collection of functions ϱ_i defined on $[0, +\infty[\times I_i$.

On each road I_i we want ϱ_i to be a weak entropic solution, that is for every function $\varphi : [0, +\infty[\times I_i \to \mathbf{R}$ smooth with compact support on $]0, +\infty[\times]a_i, b_i[$

$$\int_0^{+\infty} \int_{a_i}^{b_i} \left(\varrho_i \frac{\partial \varphi}{\partial t} + f(\varrho_i) \frac{\partial \varphi}{\partial x} \right) dx dt = 0, \tag{5}$$

and for every $k \in \mathbf{R}$ and every $\tilde{\varphi} : [0, +\infty[\times I_i \to \mathbf{R}$ smooth, positive with compact support on $]0, +\infty[\times]a_i, b_i[$

$$\int_0^{+\infty} \int_{a_i}^{b_i} \left(|\varrho_i - k| \frac{\partial \tilde{\varphi}}{\partial t} + \mathrm{sgn}\, (\varrho_i - k)(f(\varrho_i) - f(k)) \frac{\partial \tilde{\varphi}}{\partial x} \right) dx dt \geq 0. \tag{6}$$

It is well known that, for equation (1) on $\mathbf{R}$ and for every initial data in L^∞, there exists a unique weak entropic solution depending in a continuous way from the initial data in L^1_{loc}.

We assume that the roads are connected by some junctions. Each junction J is given by a finite number of incoming roads and a finite number of outgoing roads, thus we identify J with $((i_1, \ldots, i_n), (j_1, \ldots, j_m))$ where the first n–tuple indicates the set of incoming roads and the second m–tuple indicates the set of outgoing roads. We assume that each road can be incoming road at most for one junction and outgoing at most for one junction.

Hence the complete model is given by a couple $(\mathcal{I}, \mathcal{J})$, where $\mathcal{I} = \{I_i : i = 1, \ldots, N\}$ is the collection of roads and $\mathcal{J}$ is the collection of junctions.

Fix a junction J with incoming roads, say $I_1, \ldots, I_n$, and outgoing roads, say $I_{n+1}, \ldots, I_{n+m}$. A weak solution at J is a collection of functions $\varrho_l : [0, +\infty[\times I_l \to \mathbf{R}, l = 1, \ldots, n + m$, such that

$$\sum_{l=0}^{n+m} \left(\int_0^{+\infty} \int_{a_l}^{b_l} \left(\varrho_l \frac{\partial \varphi_l}{\partial t} + f(\varrho_l) \frac{\partial \varphi_l}{\partial x} \right) dx dt \right) = 0, \tag{7}$$

for every $\varphi_l, l = 1, \ldots, n+m$ smooth having compact support in $]0, +\infty[\times]a_l, b_l]$ for $l = 1, \ldots, n$ (incoming roads) and in $]0, +\infty[\times [a_l, b_l[$ for $l = n+1, \ldots, n+m$ (outgoing roads), that are also *smooth across the junction*, i.e.

$$\varphi_i(\cdot, b_i) = \varphi_j(\cdot, a_j), \quad \frac{\partial \varphi_i}{\partial x}(\cdot, b_i) = \frac{\partial \varphi_j}{\partial x}(\cdot, a_j), \quad i = 1, ..., n, \ j = n+1, ..., n+m.$$

Definition 1. *Let $\varrho = (\varrho_1, \ldots, \varrho_{n+m})$ be such that $\varrho_i(t, \cdot)$ is of bounded variation for every $t \geq 0$. Then ϱ is an admissible weak solution to (1) related to the matrix A, satisfying (3), at the junction J if and only if the following properties hold:*

1. ϱ is a weak solution at the junction J;
2. $f(\varrho_j(\cdot, a_j+)) = \sum_{i=1}^{n} \alpha_{ji} f(\varrho_i(\cdot, b_i-))$, for each $j = n+1, ..., n+m$;
3. $\sum_{i=1}^{n} f(\varrho_i(\cdot, b_i-))$ is maximum subject to 2.

For every road $I_i = [a_i, b_i]$, if $a_i > -\infty$ and I_i is not the outgoing road of any junction, or $b_i < +\infty$ and I_i is not the incoming road of any junction, then a boundary data $\psi_i : [0, +\infty[\to \mathbf{R}$ is given. In this case we ask ϱ_i to satisfy $\varrho_i(t, a_i) = \psi_i(t)$ (or $\varrho_i(t, b_i) = \psi_i(t)$) in the sense of $[1, 2]$. The treatment of boundary data can be done in the same way as in $[1, 2]$, thus we treat the case without boundary data. All the stated results hold also for the case with boundary data with obvious modifications.

Our aim is to solve the Cauchy problem on $[0, +\infty[$ for a given initial and boundary data as in next definition.

Definition 2. *Given $\bar\varrho_i : I_i \to \mathbf{R}$, $i = 1, \ldots, N$, L^∞ functions, a collection of functions $\varrho = (\varrho_1, \ldots, \varrho_N)$, with $\varrho_i : [0, +\infty[\times I_i \to \mathbf{R}$ continuous as functions from $[0, +\infty[$ into L^1_{loc}, is an admissible solution if ϱ_i is a weak entropic solution to (1) on I_i, $\varrho_i(0, x) = \bar\varrho_i(x)$ a.e., at each junction ϱ is a weak solution and is an admissible weak solution in case of bounded variation.*

On the flux f we make the following assumption

$(\mathcal{F})$ $f : [0, 1] \to \mathbf{R}$ is smooth, strictly concave (i.e. $f'' \le -c < 0$ for some $c > 0$), $f(0) = f(1) = 0$. Therefore there exists a unique $\sigma \in]0, 1[$ such that $f'(\sigma) = 0$ (that is σ is a strict maximum).

3 The Riemann Problem

For a scalar conservation law a Riemann problem is a Cauchy problem for an initial data of Heaviside type, that is piecewise constant with only one discontinuity. One looks for centered solutions, i.e. $\varrho(t, x) = \phi(\frac{x}{t})$, which are the building blocks to construct solutions to the Cauchy problem via wave front tracking algorithm. These solutions are formed by continuous waves called rarefactions and by traveling discontinuities called shocks. The speed of waves are related to the values of f', see $[4]$.

Analogously, we call Riemann problem for the road network the Cauchy problem corresponding to an initial data that is piecewise constant on each road. The solutions on each road I_i can be constructed in the same way as for the scalar conservation law, hence it suffices to describe the solution at junctions. Because of finite propagation speed, it is enough to study the Riemann Problem for a single junction.

Consider a junction J in which there are n roads with incoming traffic and m roads with outgoing traffic, and a traffic distribution matrix A. For simplicity we indicate by

$$(t, x) \in \mathbf{R}_+ \times I_i \mapsto \varrho_i(t, x) \in [0, 1], \quad i = 1, ..., n, \tag{8}$$

the densities of the cars on the roads with incoming traffic and

$$(t, x) \in \mathbf{R}_+ \times I_j \mapsto \varrho_j(t, x) \in [0, 1], \quad j = n + 1, ..., n + m \tag{9}$$

those on the roads with outgoing traffic, see Figure 1.

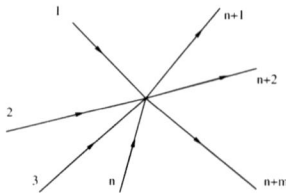

Fig. 1. A junction.

We need some more notation:

Definition 3. *Let $\tau : [0, 1] \to [0, 1]$ be the map such that:*

1. $f(\tau(\varrho)) = f(\varrho)$ for every $\varrho \in [0, 1]$;
2. $\tau(\varrho) \neq \varrho$ for every $\varrho \in [0, 1] \setminus \{\sigma\}$.

Clearly, τ is well defined and satisfies

$$0 \leq \varrho \leq \sigma \Longleftrightarrow \sigma \leq \tau(\varrho) \leq 1, \qquad \sigma \leq \varrho \leq 1 \Longleftrightarrow 0 \leq \tau(\varrho) \leq \sigma.$$

To state the main result of this section we need some assumption on the matrix A satisfied under generic conditions. Let $\{e_1, \ldots, e_n\}$ be the canonical basis of $\mathbf{R}^n$ and for every subset $V \subset \mathbf{R}^n$ indicate by $V^\perp$ its orthogonal. Define for every $i = 1, \ldots, n$, $H_i = \{e_i\}^\perp$, i.e. the coordinate hyperplane orthogonal to e_i and for every $j = n + 1, \ldots, n + m$ let $\alpha_j = (\alpha_{j1}, \ldots, \alpha_{jn}) \in \mathbf{R}^n$ and define $H_j = \{\alpha_j\}^\perp$. Let $\mathcal{K}$ be the set of indices $k = (k_1, ..., k_\ell)$, $1 \leq \ell \leq n - 1$, such that $0 \leq k_1 < k_2 < \cdots < k_\ell \leq n + m$ and for every $k \in \mathcal{K}$ set

$$H_k = \bigcap_{h=1}^{\ell} H_{k_h}.$$

Letting $\mathbf{1} = (1, \ldots, 1) \in \mathbf{R}^n$, we assume

(C) for every $k \in \mathcal{K}$, $\mathbf{1} \notin H_k^\perp$.

The following theorem describes the solution to Riemann problems at junction. For a proof see [7].

Theorem 1. *Consider a junction J, assume that the flux $f : [0, 1] \to \mathbf{R}$ satisfies $(\mathcal{F})$ and the matrix A satisfies condition (C). For every $\varrho_{1,0}, ..., \varrho_{n+m,0} \in [0, 1]$, there exists a unique admissible centered weak solution, in the sense of Definition 1, $\varrho = (\varrho_1, ..., \varrho_{n+m})$ to (1) at the junction J such that*

$$\varrho_1(0, \cdot) \equiv \varrho_{1,0}, \; ......, \; \varrho_{n+m}(0, \cdot) \equiv \varrho_{n+m,0}.$$

Moreover, there exists a unique $(n+m)-$tuple $(\hat{\varrho}_1, ..., \hat{\varrho}_{n+m}) \in [0, 1]^{n+m}$ such that

$$\hat{\varrho}_i \in \begin{cases} \{\varrho_{i,0}\} \cup]\tau(\varrho_{i,0}), 1], & if \; 0 \leq \varrho_{i,0} \leq \sigma, \\ [\sigma, 1], & if \; \sigma \leq \varrho_{i,0} \leq 1, \end{cases} \qquad i = 1, ..., n, \qquad (10)$$

and

$$\hat{\varrho}_j \in \begin{cases} [0, \sigma], & if \; 0 \leq \varrho_{j,0} \leq \sigma, \\ \{\varrho_{j,0}\} \cup [0, \tau(\varrho_{j,0})[, & if \; \sigma \leq \varrho_{j,0} \leq 1, \end{cases} \qquad j = n+1, ..., n+m. \qquad (11)$$

Fixed $i \in \{1, ..., n\}$, if $\varrho_{i,0} \leq \hat{\varrho}_i$, we have

$$\varrho_i(t, x) = \begin{cases} \varrho_{i,0}, & if \; x < \frac{f(\hat{\varrho}_i) - f(\varrho_{i,0})}{\hat{\varrho}_i - \varrho_{i,0}} t + b_i, \; t \geq 0, \\ \hat{\varrho}_i, & if \; x > \frac{f(\hat{\varrho}_i) - f(\varrho_{i,0})}{\hat{\varrho}_i - \varrho_{i,0}} t + b_i, \; t \geq 0, \end{cases} \qquad (12)$$

and, if $\hat{\varrho}_i < \varrho_{i,0}$,

$$\varrho_i(t, x) = \begin{cases} \varrho_{i,0}, & if \; x \leq f'(\varrho_{i,0})t + b_i, \; t \geq 0, \\ (f')^{-1}((x - b_i)/t), & if \; f'(\varrho_{i,0})t + b_i \leq x \leq f'(\hat{\varrho}_i)t + b_i, \; t \geq 0, \\ \hat{\varrho}_i, & if \; x > f'(\hat{\varrho}_i)t + b_i, \; t \geq 0. \end{cases} \qquad (13)$$

Fixed $j \in \{n+1, ..., n+m\}$, if $\varrho_{j,0} \leq \hat{\varrho}_j$, we have

$$\varrho_j(t, x) = \begin{cases} \hat{\varrho}_j, & if \; x \leq f'(\hat{\varrho}_j)t + a_j, \; t \geq 0, \\ (f')^{-1}((x - a_j)/t), & if \; f'(\hat{\varrho}_j)t + a_j \leq x \leq f'(\varrho_{j,0})t + a_j, \; t \geq 0, \\ \varrho_{j,0}, & if \; x > f'(\varrho_{j,0})t + a_j, \; t \geq 0, \end{cases} \qquad (14)$$

and, if $\hat{\varrho}_j < \varrho_{j,0}$,

$$\varrho_j(t, x) = \begin{cases} \hat{\varrho}_j, & if \; x < \frac{f(\varrho_{j,0}) - f(\hat{\varrho}_j)}{\varrho_{j,0} - \hat{\varrho}_j} t + a_j, \; t \geq 0, \\ \varrho_{j,0}, & if \; x > \frac{f(\varrho_{j,0}) - f(\hat{\varrho}_j)}{\varrho_{j,0} - \hat{\varrho}_j} t + a_j, \; t \geq 0. \end{cases} \qquad (15)$$

4 The Wave-Front Tracking Algorithm

Once the solution to a Riemann problem is provided, we are able to construct piecewise constant approximations via wave-front tracking algorithm. The construction is very similar to that for scalar conservation law, see [4], hence we briefly describe it.

Let $\bar{\varrho} = (\varrho_1, \ldots, \varrho_N)$ be a piecewise constant map defined on the road network. We want to construct a weak solution of (1) with initial condition $\varrho(0, \cdot) \equiv \bar{\varrho}$. We begin by solving the Riemann Problems on each road in correspondence of the jumps of $\bar{\varrho}$ and the Riemann Problems at junctions determined by the values of $\bar{\varrho}$ (see Theorem 1). We split each rarefaction wave into a rarefaction fan formed by rarefaction shocks, that are discontinuities traveling with the Rankine-Hugoniot speed. We always split rarefaction waves inserting the value σ (if it is in the range of the rarefaction). Moreover, we let any rarefaction shock with endpoint σ have velocity zero.

When a wave interacts with another one we simply solve the new Riemann Problem. Instead, when a wave reaches a junction, we solve the Riemann Problem at the junction. The number of waves may increase only for interactions of waves at junctions. Since the speeds of waves are bounded, there are finitely many waves on the network at each time $t \geq 0$. We call the obtained function *an approximate wave front tracking solution*. Given a general initial data, we approximate it by a sequence of piecewise constant functions and construct the corresponding approximate solutions. If they converge in L^1_{loc}, then the limit is a weak entropic solution on each road, see [4] for a proof.

5 Existence of Solutions

Assume that every junction has exactly two incoming roads and two outgoing ones. This hypothesis is crucial, because the presence of more complicate junctions provokes additional increases of the total variation of the flux. The case where junctions have at most two incoming roads and at most two outgoing roads can be treated in the same way. So, for each junction J, the matrix A, defined in the introduction, takes the form

$$A = \begin{pmatrix} \alpha & \beta \\ 1 - \alpha & 1 - \beta \end{pmatrix}, \tag{16}$$

where $\alpha, \beta \in]0, 1[$ and $\alpha \neq \beta$, so that (C) is satisfied.

From now on we fix an approximate wave front tracking solution ϱ, defined on the road network. The following lemma gives an estimate of the total variation of the flux of an approximate wave-front tracking solution. The proof is contained in [7] and it is based on analyzing what happens in term of the total variation of the flux when a wave interacts with a junction.

Lemma 1. *Consider a road network* $(\mathcal{I}, \mathcal{J})$. *For some* $K > 0$, *we have*

$$Tot.\,Var.\,(f(\varrho(t+, \cdot))) \leq e^{Kt}\,Tot.\,Var.\,(f(\varrho(0+, \cdot)))$$

for each $t \geq 0$.

The previous estimate permits to prove the following theorem.

Theorem 2. *Fix a road network* $(\mathcal{I}, \mathcal{J})$. *Given* $C > 0$ *and* $T > 0$, *there exists an admissible solution defined on* $[0, T]$ *for every initial data* $\bar{\varrho} \in cl\{\varrho : TV(\varrho) \leq C\}$, *where cl indicates the closure in* L^1_{loc}.

Proof. (Sketch) Fix a sequence of initial data $\bar{\varrho}_\nu$ piecewise constant such that $TV(\bar{\varrho}_\nu) \leq C$ for every $\nu \geq 0$ and $\bar{\varrho}_\nu \to \bar{\varrho}$ in L^1_{loc} as $\nu \to +\infty$. For each $\bar{\varrho}_\nu$ we consider an approximate wave-front tracking solution ϱ_ν such that $\varrho_\nu(0, x) = \bar{\varrho}_\nu(x)$ and rarefactions are split in rarefaction shocks of size $\frac{1}{\nu}$.

For every road I_i, we have to consider the zone $D^i_1(\varrho_\nu)$ where the solution is influenced only by the initial datum and the zone $D^i_2(\varrho_\nu)$, where the solution is influenced by junctions. On D^i_1, by classical arguments, we have that $\varrho_\nu \to \varrho$ in L^1_{loc}, with ϱ admissible solution to the Cauchy problem.

On D^i_2, we have, up to a subsequence, $\varrho_\nu \rightharpoonup^* \varrho$ weak* on L^1 and, $f(\varrho_\nu) \to \bar{f}$ in L^1 for some $\bar{f}$. It is possible to prove, see [7], that there are at most two big waves on D^i_2 for every time, hence, splitting the domain D^i_2 in a finite number of pieces where we can invert the function f, we obtain $\varrho_\nu \to f^{-1}(\bar{f})$ in L^1. Together with $\varrho_\nu \rightharpoonup^* \varrho$ weak* on L^1, we conclude that $\varrho_\nu \to \varrho$ strongly in L^1.

6 Lipschitz Continuous Dependence

In this section we present a counterexample to the Lipschitz continuous dependence by initial data with respect to the L^1-norm. The continuous dependence by initial data with respect the L^1-norm remains an open problem. The counterexample is constructed using shifts of waves as in the spirit of [5], to which we refer the reader for general theory. The result is contained in the following proposition.

Proposition 1. *Let* $C > 0$, J *be a junction and let* $(\varrho_{1,0}, \ldots, \varrho_{4,0})$ *be an equilibrium configuration as in Lemma 4. There exist two piecewise constant initial data satisfying the equilibrium configuration at* J *such that the* L^1*-distance between the corresponding two solutions increases by the multiplication factor* C.

This proposition is based on the following lemmata, which describe how a shift propagates through a junction. For a proof, see [7].

Lemma 2. *Let us consider in a road two waves, with speeds* λ_1 *and* λ_2 *respectively, that interact together at a certain time* $\bar{t}$ *producing a wave with speed* λ_3. *If the first wave is shifted by* ξ_1 *and the second wave by* ξ_2, *then the shift of the resulting wave is given by*

$$\xi_3 = \frac{\lambda_3 - \lambda_2}{\lambda_1 - \lambda_2}\xi_1 + \frac{\lambda_1 - \lambda_3}{\lambda_1 - \lambda_2}\xi_2. \tag{17}$$

Moreover we have that

$$\Delta\varrho_3\,\xi_3 = \Delta\varrho_1\,\xi_1 + \Delta\varrho_2\,\xi_2, \tag{18}$$

where $\Delta\varrho_i$ *are the signed strengths of the corresponding waves.*

Lemma 3. *Consider a junction J with incoming roads I_1 and I_2 and outgoing roads I_3 and I_4. If a wave on a road I_i ($i \in \{1, \ldots, 4\}$) interacts with J without producing waves in the same road I_i and if ξ_i is the shift of the wave in I_i, then the shift ξ_j produced in a different road I_j ($j \in \{1, \ldots, 4\} \setminus \{i\}$) satisfies:*

$$\xi_j \left(\varrho_j^+ - \varrho_j^- \right) = \frac{\Delta\gamma_j}{\Delta\gamma_i} \xi_i \left(\varrho_i^+ - \varrho_i^- \right), \tag{19}$$

where $\Delta\gamma_l$ ($l \in \{i, j\}$) represents the variation of the flux in the road I_l and ϱ_l^-, ϱ_l^+ ($l \in \{i, j\}$) are the states at J in the road I_l respectively before and after the interaction.

Lemma 4. *There exists an initial datum given by $(\varrho_{1,0}, \varrho_{2,0}, \varrho_{3,0}, \varrho_{4,0})$, that is an equilibrium configuration at J, a wave $(\bar{\varrho}_2, \varrho_{2,0})$ on road I_2, waves $(\varrho_{3,0}, \varrho_3^*)$ with shift $\xi_{3,0}$ and $(\varrho_3^*, \bar{\varrho}_3)$ on road I_3 such that the followings happen in chronological order:*

1. *the initial distance in L^1 is $\xi_{3,0} |\varrho_{3,0} - \varrho_3^*|$;*
2. *the wave $(\varrho_{3,0}, \varrho_3^*)$ in I_3 with shift $\xi_{3,0}$ interacts with J;*
3. *waves are produced only in I_2 and I_4;*
4. *the wave on road I_2 interacts with $(\bar{\varrho}_2, \varrho_{2,0})$ producing a new wave;*
5. *the new wave from road I_2 interacts with J;*
6. *waves are produced only in I_3 and I_4;*
7. *in I_4 the L^1-distance after the interactions, is equal to*

$$2\frac{1-\beta}{\beta} |\xi_{3,0} \left(\varrho_3^* - \varrho_{3,0} \right)|,$$

and the L^1-distance on road I_3 is equal to $\xi_{3,0} |\varrho_{3,0} - \varrho_3^*|$.

References

1. D. Amadori: NoDEA **4** (1997), pp. 1–42.
2. F. Ancona, A. Marson, Nonlinear Analysis, **35** (1999), pp. 687–710.
3. A. Aw, M. Rascle: SIAM J. App. Math., **60** (2000), pp. 916–938.
4. A. Bressan: *Hyperbolic Systems of Conservation Laws — The One-dimensional Cauchy Problem* (Oxford University Press 2000)
5. A. Bressan, G. Crasta, B. Piccoli: Amer. Math. Soc. Memoir, **146** (2000)
6. A. Bressan, A. Marson: Comm. Part. Diff. Equat. **20** (1995), p. 1491.
7. G. M. Coclite, M. Garavello, B. Piccoli: SIAM J. Math. Anal. **36** (2005), pp. 1862–1886.
8. R. M. Colombo: SIAM J. Appl. Math. **63** (2002), pp. 708–721.
9. C. Dafermos: *Hyperbolic Conservation Laws in Continuum Physics* (Springer, 1999)
10. C. D'Apice, R. Manzo, B. Piccoli: (2005)
11. M. Garavello, B. Piccoli: Commun. Math. Sci. **3** (2005), pp. 261–283.
12. M. Garavello, B. Piccoli: Commun. Partial Diff. Eq. **31** (2006), pp. 243–275.
13. J. M. Greenberg: SIAM J. Appl. Math. **62** (2001), pp. 729–745.
14. H. Holden, N. H. Risebro: SIAM J. Math. Anal. **26** (1995), pp. 999–1017.
15. M. J. Lighthill, G. B. Whitham: Proc. Roy. Soc. London A **229** (1955), 317.
16. P. I. Richards: Oper. Res. **4** (1956), pp. 42–51.

Approach to Critical Link Analysis of Robustness for Dynamical Road Networks

Victor L. Knoop, Serge P. Hoogendoorn, and Henk J. van Zuylen

Delft University of Technology, Faculty of Civil Engineering and Geosciences –
Transport & Planning, Stevinweg 1, 2628 CN, Delft, The Netherlands
v.l.knoop@citg.tudelft.nl, s.hoogendoorn@citg.tudelft.nl,
h.j.vanzuylen@citg.tudelft.nl

Summary. This article presents a study of the performance decrease caused by blocking one link. We predict which types of links are important for the well functioning of the network as a whole. Traffic is simulated on a regional network in which the links are blocked sequentially. Each of these outcomes is compared to the outcome of the fully operational network. We argue that traffic dynamics including spillback are important in the reduction of performance. We assessed the performance decrease for two situations, one in which people stick to their everyday routes and one in which they will adapt their routes to the new situation. A blocking of motorways turns out to influence the traffic flow most. In the situation that people stick to their everyday routes, urban roads close to the motorway are equally important: spillback effects causes delays also for through traffic on the motorway.

1 Introduction

The reliability and the robustness of traffic networks are important performance indicators, both from the perspective of the traveler (travelers prefer reliable and robust networks, reference) and network operators [1, 2]. Methods to gain insight into impacts on network performance of large accidents, terrorist attacks, flooding, etc., are hence very valuable.

The main research question considered in this contribution is on which type of links the network performance relies. In other words, if there is a incident on a certain link, the service level of the network will reduce. Which are the type of links that reduce the number of arrived vehicles the most when blocked?

We used the number of arrived vehicles as performance measure. Rather than the average or total time spent in the network, where one should correct for the number of vehicles still queuing, this is a more absolute measure for a fixed-time simulation [3].

This is already done in static models for a simple network [4, 5]. There, risk-adverse route choice is studied in a fully analyzable static network of a few links. It analyzes a game between road users and failing links. If links fail too often, users will include the risk in their route choice. This has been further developed by introducing extra costs for a link with high densities [6, 7]; this is still a small test network. We applied these concepts on a real size network with time dependent demand and tried to find typical characteristics of vulnerable parts of a existing road network; general conclusions about reliable network types can be found in [8]. We simulated the flow of a time dependent demand on the whole network and thus taking network dynamics and spillback effects into account.

The symbols used in the article are listed in table 1.

Symbol	Meaning
b	blocked link
L	set of all links in the network
G	the network
G_b	the network with link b blocked
π	path
$\pi^*(G)$	equilibrium paths in network G.
$A(\pi, G)$	number of travelers arrived at their destination at the end of the simulation given route choice π and network G
$A(\pi^*(G), G_b)$	number of travelers arrived at their destination at the end of the simulation if link b is blocked and the route choice is based on a complete network
$A(\pi^*(G_b), G_b)$	number of travelers arrived at their destination at the end of the simulation if link b is blocked and the route choice is based on the actual traffic situation

Table 1. List of used symbols

2 Mathematical Formulation of the Problem

The problem can be formulated mathematically as a two-level optimization problem. There is a fixed demand of travelers for each origin-destination pair od. At one hand, the travelers aim to change to the route with the shortest travel time. The equilibrium state is such that traveler i had no possibility to change unilaterally change his route and reduce the travel time he perceives at the moment of choice [9]. This optimum is called a Wardrop optimum [10, 11]. We will refer to this optimum using $\pi^*_{od}(G)$, where G is the network on which the route choice is based. The number of travelers that arrive on their destination within the fixed time frame, using path π given network G is referred to as $A(\pi, G)$.

An effect counteracting the optimization of the travelers is the (partial) closing of one of the links. The network disturbed by the closing (partial) closing of link b is called G_b.

A link is a unreliable element of the network if the number of arrived travelers is much reduced if there is an incident on that link. At this level, the aim is to *minimize* the number of arrived travelers A. So, the aim is to find the link b such that the least travelers arrive in the fixed time frame.

Of course, the optimal route choice is dependent on the network and therefore on the blocked link. Thus, we define a Wardrop optimum of the route choice in case of a network with link b blocked:

$$\pi_{od}^*(G_b) \tag{1}$$

This can be considered as a two-player game: one group of players is the collective of travelers, the destroyer of the network is the other player. In game theory, one speaks of leaders and followers meaning the one who moves first and the one responding on that move, respectively. Here also, we could speak of leaders and followers. Then, two cases can be considered: one in which the travelers lead and the destroyer follows and the other one in which the destroyer leads and the travelers follow.

2.1 Scenario 1: Travelers Lead, Destroyer Follows

In the scenario that the travelers lead, the first move is to be made by the travelers. They find one route choice optimized for the case in which the network is intact. In this scenario, they will stick to the routes $\pi^*(G)$. This scenario describes the situation where neither information is given, nor an extraordinary traffic situation will make the travelers change their route.

Now, the performances of the destroyed network can be calculated. For each blocked link b, we could calculate

$$A(\pi^*(G), G_b). \tag{2}$$

The most critical link b^* can be found by minimizing the number of arrivals

$$b^* = \underset{b}{\mathrm{argmin}}\left(A\left(\pi^*\left(G\right), G_b\right)\right). \tag{3}$$

2.2 Scenario 2: Destroyer Leads, Travelers Follow

This scenario describes a situation where the people are fully informed about the traffic conditions and delays. They are not informed about the incident, though. That is to say, the vehicles calculate their travel time based on speeds. As long as there are no vehicles queuing yet, they will not foresee any problems in their route choice.

Mathematically, this is represented by letting the first move to the destroyer and letting the travelers respond to this first move. The final function to evaluate is

$$A(\pi^*(G_b), G_b). \tag{4}$$

As the travelers will change their actions, i.e. their routes, on the action of the destroyer, this game is a Stackelberg game. The most critical link b^* in this scenario can be found by optimizing

$$b^* = \operatorname*{argmin}_b \left(A \left(\pi^* \left(G_b \right), G_b \right) \right). \tag{5}$$

3 Model

The simulation model we used had to be able to simulate spill back-effects well. A substantial part of the discovered delay is namely caused by the spill back effects of a traffic jam. Furthermore, we had to use a dynamic model with different time periods to simulate the morning peak with an acceptable accuracy.

Therefore, we used the macro traffic simulation model DSMART, developed at the TU Delft. It differentiates vehicles with different destinations. The model does not distinguish between user classes, vehicle types nor between road lanes. A traffic jam for vehicles wanting to make a turn will also block the main stream of the traffic. This corresponds to a driving style where drivers wanting to take the exit do not keep the outer lane in a traffic jam.

The stochastic route choice model is done by a probit model. For each link, the travel time is calculated based on the link speeds. At the base travel time is calculated by dividing the length by the speed, an error is added. The expected travel time on which the route choice is based, is randomly drawn from a normal distribution of travel times with the base travel time as mean and with a standard deviation of 10%.

For the case without route choice, the calculation is based on 20 samples from this route choice. Each period, 20 fastest routes from each node to each destination will be determined from this stochastic process. Those routes are considered representative for all travelers. For half of the travelers, the new route choice is based upon these routes; the other half will stick to the routes of the previous period.

In the scenario of adapted route choice, the used model is comparable. Due to computational limitations – now, the route choice has to be recalculated for every simulation – we reduced the number of samples for this scenario from 20 to 8.

The network consists of both motorways and urban links. The urban road network is not very detailed, but the main urban roads are modeled, as well as single origin/destination links representing a quarter/district, see also Fig. 2.

4 Simulations

The network we simulated are the roads around Rotterdam, a Dutch city with 600.000 inhabitants (see Fig. 1). Fig. 2 shows the model of this network. As this figure shows, the motorway network is completely modeled. The underlying road network is just partly modeled.

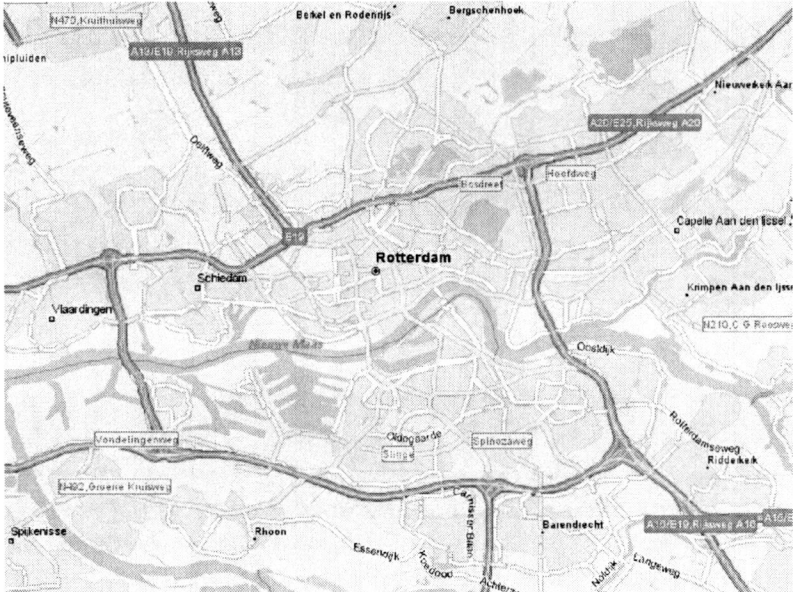

Fig. 1. Map of the Rotterdam area.

The simulated time is the morning peak, from 6.30 to 9.30 am. At the start of the simulation, the network is empty. The network consists of 468 transport links, 44 origin links and 44 destination links; the connections are provided by 239 nodes.

As all macroscopic models, DSMART supposes the speed of vehicles constant during a time step. The chosen time step is 15 seconds; for longer times, one cannot assume the vehicle speeds to be constant. Furthermore, serious problems arise if vehicles can travel more then a link length within one time step [12]. On the other hand, further reducing the time step to values of 10, 5 or even 1 second would increase the total calculation time. The simulation time for a scenario with a fixed route choice (in mathematical terms, calculate $A(\pi^*(G), G_b)$ or $A(\pi^*(G), G_b)$ with $\pi^*(G)$ precalculated) is a around five minutes on a Pentium 4, 2.8 GHz with 512 MB RAM.

Of course, if the route choice has to be optimized too, calculate $A(\pi^*(G_b), G_b)$ or $A(\pi^*(G_b), G_b)$ takes longer. One of these calculations route choice optimization will take around twenty minutes of calculation time.

Fig. 2. The model of the network.

These simulation times may seem fair, but remember that calculation of both $A(\pi^*(G), G_b)$ and $A(\pi^*(G_b), G_b)$ are needed for all $b \in L$, so 468 times. All together, it is over a week of calculation time.

5 Results

The types of links that cause the most delay when blocked are different for the both cases studied. In Fig. 3 the number of arrivals for both cases are shown. An asterisk in the figure is the network performance in the case of a blocked link – each asterisk is a different position of the blockage. For each position of the blockage, there are two scenarios possible. In one, the travelers stick to their fixed route and in the other one, they will adjust their route choice. The results, the number of travelers that reached their destination, of both scenarios are plotted on the axes.

The points are located under the diagonal (dashed black line) of the graph. That means that in almost all cases, i.e. for almost all of the blocked links, the number of arrivals is higher if people are allowed to take an other route. Some statistics about these results are that in the case of no rerouting, in average $1.6 \cdot 10^6$ travelers arrive, with a standard deviation of $5 \cdot 10^5$ travelers, which equals 30%. In the case of flexible routes, these numbers are $1.9 \cdot 10^6$, $3 \cdot 10^5$ and 16% respectively.

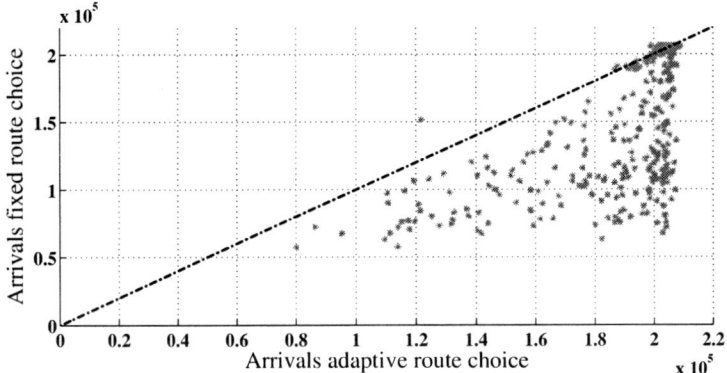

Fig. 3. Comparison of arrivals

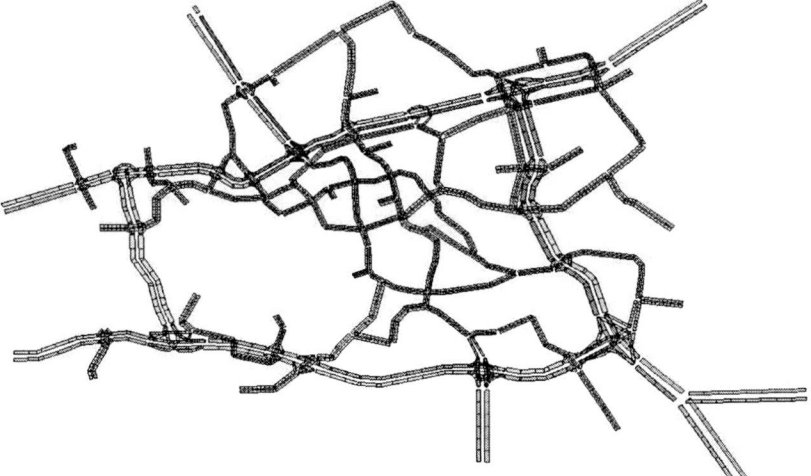

Fig. 4. Loss of arrivals at destination if link is blocked – case without rerouting.

In Fig. 4 and Fig. 5 it is indicated how valuable links are. The color is a measure of the travelers not arriving if that particular link is blocked. If a blockage of a certain link is of no influence on the number of arrived travelers, the link is green. If a link is red, it means that the least people arrive if that link is blocked. The scale for the color is the same in both figures. Comparing both figures, one can remark that Fig. 4 is much redder than Fig. 5, meaning that more people arrive. In Fig. 6 the difference of these two is plotted. In case the adaptivity of the route choice does improve the performance of a network (disturbed by the blocking of link b) much, the link (b) is colored green. The redder link b is, the less advantageous the rerouting was for the performance of the network in case of a blockage of link b.

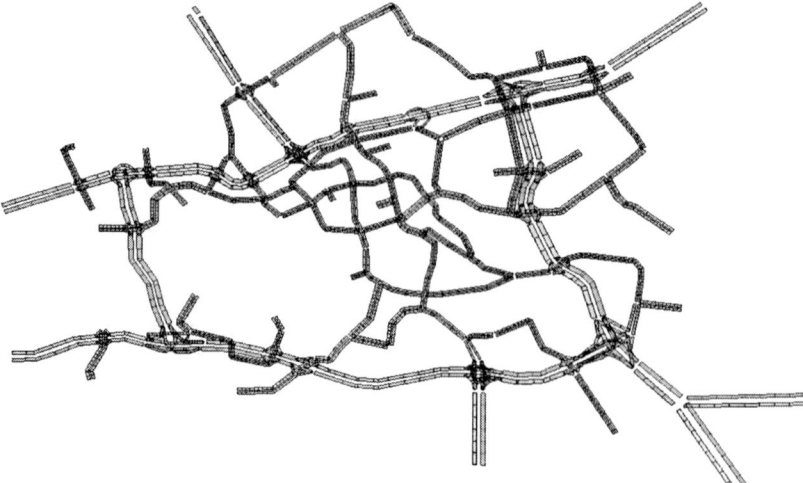

Fig. 5. Loss of arrivals if link is blocked – case with rerouting.

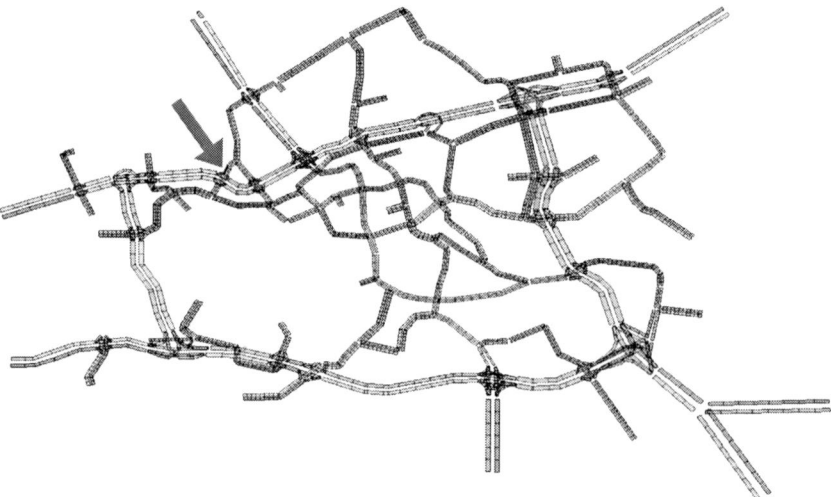

Fig. 6. Difference in arrivals between adaptive route choice and fixed routes.

Then, three categories of links can be identified. Firstly, the links that do not harm the flow in either case. These can be found in the upper left corner of the graph. In both scenarios, fixed routes and rerouting, the number of arrivals are high. These are unused links.

The second category that can be found are the links that reduce the number of arrived travelers is both cases, both with rerouting and without rerouting. These are represented by asterisks in the lower left corner When analyzing these points, these links turn out to be motorway links, mostly close to a destination area.

The third area is the most interesting, this is the lower right corner of the graph. At the links represented by the asterisks there, rerouting can improve the network performance a lot. The number of arrived travelers is namely low in the case with fixed route choice and high with a flexible route choice. These links are non-motorway links close to a destination and close to a motorway too.

This can be explained by the fact that there are many alternatives in an urban road network. But, as people take their usual routes, they got stuck. Spillback effects then cause the flow on the motorway to be blocked too.

These spillback effect cause the motorway to be completely blocked. Therefore, the delay for the though traffic could be bigger than in the case of motorway link itself being blocked.

Fig. 6 shows also which link is the one that is above the diagonal in Fig. 3. In one case the network will perform significantly better if the routes are *not* adapted. The link for which this is the case is indicated with a blue arrow in Fig. 6. This might be a consequence of the location of this link and the busy interchange downstream for the alternative route. If the routes are adapted, the traffic will flow by an alternative route. The best alternative will lead the traffic to an already congested interchange. There, this extra flow can harm the flow of through traffic. Here, the difference between system optimum and user optimum might play a role. It could be better for all if some will not perform their best.

6 Conclusions

We conclude that the motorway links are vulnerable for the performance of a road network, even if the route choice can be adjusted to the faced traffic situation. Furthermore, we conclude that severe congestion can be avoided if travelers are rerouted if there is a blockage of a urban link close close to their destination.

Spillback effects turn out to have a big impact on the occurred congestion. Some urban roads are critical links because of congestion spillback effects to the motorway.

The links for which congestion can be avoided by rerouting are links for which there are parallel alternatives. We therefore conclude that in a network with lots of parallel routes could be regarded as robust with regards to incidents.

The calculations are done in a single lane macroscopic simulation. In regions with good discipline, drivers might queue for an exit at the outer lane, so letting through traffic the possibility to pass. Therefore, a multi lane simulation model with an appropriate modeling of lane change behavior at ramps [13] is required.

For each level of detail, though, calculation time will increase. A way to reduce the total time of computation is to find vulnerable links in an other way than calculating the network performance with all links sequentially blocked. A possible way to search for these links is with an genetic algorithm. This will be a direction for future work.

Acknowledgements

This research is sponsored by the Delft Transport Research Center and the NWO/Connekt research programme AMICI.

References

1. E.A.I. Bogers, F. Viti, S.P. Hoogendoorn: Joint modeling of ATIS, habit and learning impacts on route choice by laboratoy simulator experiments, TRB 2004, prepared paper for Transportations Research Records 2005
2. Uncovering the contribution of travel time reliability to dynamic route choiceusing real-time loop data, Henry X. Liu et al., Transportation Research Part A: Policy and Practice (vol. 38 issue 6), 435-453, 2004
3. A.Hegyi: Model Predictive Control for Integrating Traffic Control Measures. PhD Thesis, TRAIL series, Delft (2004)
4. Michael G.H. Bell: A game theory approach to measure the performance reliability of transport networks, Transportation Research B **34**, 533-545, 2000
5. Michael G.H. Bell, Chris Cassir: Risk-adverse user equilibrium traffic assignment: an application of game theory, Transportation Research B **36**, 671-681 (2002)
6. Pamela M. Murray-Tuite and Hani S. Mahmassani: A Methodology for the Determination of Vulnerable Links in a Transportation Network, prepared paper for TRR; TRB 2004
7. Pamela M. Murray-Tuite and Hani S. Mahmassani: Identification of Vulnerable Transportation Infrastructure and Household Decision Making Under Emergency Evacuation Conditions, Research Report SWUTC/05/167528-1 (2005)
8. Réka Albert, Hawoong Jeong and Albert-László Barabási: Error and attack tolerance of complex networks, Nature **406**, 378-382 (2000)
9. Huey-Kuo Chen: *Dynamic Travel Choice Models* (Springer, Heidelberg, 1999)
10. J. G.Wardrop: Some theoretical aspects of road traffic research. In Proceedings of the Institute of Civil Engineers, Part II, volume 1, pages 325-378 (1952)
11. A. Haurie and P. Marcotte: On the relationship between Nash–Cournot and Wardrop equilibria, Networks **15**, 295–308 (1985)
12. S.P. Hoogendoorn, G. Hegeman, and T. Dijker: Traffic Flow Theory and Simulation. Course reader. TU Delft (2006)
13. D. Ngoduy, S.P. Hoogendoorn, H.J. Van Zuylen: A new continuum model for freeway with on- an off-ramp explains different traffic congested states (accepted for TRB 2006)

Traffic Flow in Bogotá

Luis Olmos and José Daniel Muñoz

Universidad Nacional de Colombia, Bogotá D.C.

Summary. We introduce cellular automaton models for both cars alone [1] and mixed traffic (cars and buses) on motorways in Bogotá. Our model includes three elements: hysteresis between acceleration and braking gaps, a delay time in the acceleration, and instantaneous braking. In addition, we include a lane changing rule and the disordered behavior of Bogotan bus drivers. The parameters of our model were obtained from direct measurements on a car and a bus in this city. We use this model to simulate the flux-density fundamental diagram for a single-lane road with car traffic and a two-lane road with mixed traffic, and compare the results with experimental data. Our simulations are in very good agreement with experimental measurements, and reproduce both the shape and the value of the maximal flux. Moreover, they show that the causes of the measured high fluxes are the short gaps that the Bogotan drivers are used to maintain to the car ahead (the agressive driving that is typical for this city).

1 Introduction

In the 1990's, the urban tranport system in Bogotá was characterized by a severe congestion and a quite poor road network condition. There was a high occurrence of accidents and the mean travel time was about 1.5 hours between home and work. Since 1998, the city administration has tried to solve this problem by introducing transportation strategies such as a mass transportation system (TransMilenio), almost 250 kilometers of bike paths, pedestrian bridges everywhere, restrictions on the use of private cars at rush hours (Pico y placa) and a great number of campaigns in favor of a better civic culture. However, the number of deaths in automobile accidents is still high (about 700 per year). In addition, the imprudent and aggressive driving of the urban bus drivers in Bogotá is well known. They stop everywhere, although very nice bus stops have been built. Actually, in this city it is not unusual to see competition between buses on the same route or passengers which are picked up on the left lane of the road. For this reason, and keeping in mind that one half of the accidents are due to buses, it is interesting to study the effect of buses in Bogotá's traffic flow.

Since the construction of the STCA model [2, 3], cellular automata (CA) have been applied with success to traffic simulations. These models are able to reproduce the macroscopic properties of highway traffic from the microscopic behavior of each car. Some recent works have kept in mind the driving particularities of each place to simulate a more realistic traffic flow [1, 4–7]. Moreover, some of them have extended these models to multilane highways, with both symmetric or asymmetric lane changing rules, depending on the road density that can reproduce the density inversion between right and left lanes near maximum flow.

In [1] we proposed a cellular automaton model for the traffic flow in Bogotá, with parameters that were taken from measurements inside a car running in Bogotá and with results that were in agreement with experimental flow-density diagrams. Our model includes three elements. The first one is the set of gaps the driver uses to decide to brake (brake gap gap_{brake}) or accelerate (acceleration gap gap_{accel}). They are, in general, different (hysteresis) and both depend on the speed. The second element is the time it takes the car to reach the next discrete speed value (retarded acceleration, t_{up}). The last one is an instantaneous brake reaction that we have observed when the car ahead brakes. Hereby we extend this work to simulate two-lane highways with mixed traffic (cars and buses). As for cars, the driving parameters for buses are obtained from real measurements, plus simple symmetric changing-lane rules. The goal is to study the effect of the ratio between cars and buses on the flow-density fundamental diagram.

The paper proceeds as follows. In Sec. 2 we show a detailed description of our model, including tables with parameters for both cars and buses and lane-changing rules. Sec. 3 shows the effect of changing from a one-lane to a two-lane simulation with cars alone on the flow-density diagram and compares them against the experimental results of our previous work. Sec. 4 shows the simulation results for a two-lane road with mixed traffic at several mixtures of cars and buses. Finally, Sec. 5 contains the main conclusions and discussions of this work.

2 Model Description

In our model the highway is represented by an array of two lanes of length L with periodic boundary conditions. Each site of the array is a cell of length 2.75 m. Vehicles can only have integer speed values, $v = 0, 1, ..., v_{\text{max}}$. With a speed unit $v_{\text{unity}} = 10$ km/h, our model takes $v_{\text{max}} = 7$ for cars and $v_{\text{max}} = 5$ for buses. This corresponds to time steps of $t_{\text{step}} = 0.9$ s, which is near the typical driver's reaction time. A car occupies two consecutives cells: the car length (4.5 m) plus the distance between cars in a jam (1 m). A bus occupies three cells: the bus length (7 m) plus a distance between buses in a jam of

1.25 m. Thus, the maximal number of vehicles in one lane of the highway is given by $N = \frac{L}{\%\text{cars} \cdot 2 + \%\text{buses} \cdot 3}$.

At time t the n-th vehicle is completely defined by its type (car or bus), its position $x_n(t)$, its velocity $v_n(t)$ and its brake-light status, $b_n(t)$, which is 1 or 0 and indicates whether the driver braked or not at the previous timestep $t-1$ [8]. The effective gap is defined as $gap = \Delta x(t) + \Delta v(t)$, where $\Delta x(t) = x_{n+1}(t) - x_n(t) - 1$ is the number of empty cells to the vehicle ahead and $\Delta v(t) = v_{n+1}(t) - v_n(t)$ is the relative speed of the car ahead. The car position is the last cell occupied by the vehicle.

As already mentioned, our model includes three elements: the hysteresis between braking and acceleration gaps, the time to accelerate and the instantaneous braking. The three parameters gap_{brake}, gap_{accel} and t_{up} are functions of speed and represent the drivers' driving. For each vehicle type, these parameters were experimentally found [1] and are summarized in table 1.

Cars					Buses			
Speed	gap_{brake}	gap_{accel}	t_{up}		Speed	gap_{brake}	gap_{accel}	t_{up}
0	0	3	1		0	0	4	1
1	3	4	1		1	4	6	1
2	3	5	1		2	6	7	2
3	4	5	1		3	8	9	2
4	5	6	2		4	8	10	3
5	6	7	2		5	9	12	3
6	6	8	2					
7	7	9	2					

Table 1. Driving parameters for Bogotá.

The lane-changing rule considers a symmetric incentive criterion. That is, one considers to change lanes only when the speed of the car ahead (v_{ahead}) is lower that my own speed (v). In this case, one computes the distance to the first vehicle ahead ($\Delta x_{\text{front}}(t)$) and the first vehicle behind me ($\Delta x_{\text{back}}(t)$) on the other lane. Now, if the gap on my own lane is less than the gap to the car ahead on the other lane ($x_{\text{back}}(t) + v_{\text{back}}(t) \leq x(t)$) and my position is less than the future position of the vehicle behind me on the target lane ($x(t) < x_{\text{back}}(t) + v_{\text{back}}(t)$), then I change to the same position at the target lane. The last condition is a security criterion to prevent accidents.

Summarizing, all vehicles execute in parallel the following set of rules:

- Compute its $gap(t)$.
- *lane–changing rule*: in case $v_{\text{ahead}}(t) < v(t)$ (*incentive critera*) and also $\Delta x(t) < \Delta x_{\text{front}}(t)$ and $x_{\text{back}}(t) + v_{\text{back}}(t) < x(t)$ (*security criterion*), then change lane (let $x(t) = x(t)$ on the other lane) and compute again its $gap(t)$.
- Determine its parameters gap_{accel}, gap_{brake} and t_{up} from table 1.

  – *normal braking*: if $gap(t) < gap_{brake}$, decelerate to the maximal speed
     $v(t+1)$ such that $gap'_{brake} < gap(t) < gap'_{accel}$, where gap'_{brake} and
     gap'_{accel} are the parameters at speed $v(t+1)$. In addition, let $delay = 0$
     and turn on brake-lights ($b_n(t+1) = 1$).
  – If $gap \geq gap_{accel}$, then
     · *instantaneous braking*: if $gap(t) \leq gap_{accel} + 2$ and the brake lights
       of the vehicle ahead are on ($b_{n+1}(t) = 1$), let $v(t+1) = v(t) - 1$
       (brake), turn on brake-lights ($b_n(t+1) = 1$) and let $delay = 0$.
     · *accelerate*: else, turn off brake-lights ($b_n(t+1) = 0$) and
       · If $delay = t_{up}$, let $v(t+1) = v(t) + 1$ (accelerate) and let
         $delay = 0$
       · Else, let $delay = delay + 1$ and preserve $v(t+1) = v(t)$.
  – Otherwise, let $delay = 0$, turn off the brake-lights ($b_n(t+1) = 0$) and
     preserve $v(t+1) = v(t)$.
• Finally, move v cells ahead,

$$x(t+1) = x(t) + v(t+1). \tag{1}$$

The counter *delay* defines whether t_{up} has been completed. The variable
$b_{n+1}(t)$ defines the brake light status of the vehicle ahead. The *instantaneous
braking* rule represents the braking reaction we have observed when the vehicle ahead also brakes. This reaction is observed for all distances but is just
included in the gaps when $gap \leq gap_{accel}$. Thus, we have included it as an
additional rule only if $gap_{accel} \leq gap(t) \leq gap_{accel} + 2$ through a brake-light
in each vehicle.

3 Effect of the Lane-Changing Rule

For all simulations in this paper, $L = 3000$ cells and the system starts with an
initial configuration of N vehicles, with random distributions of speeds and
positions and all brake-lights turned off. In order to prevent traffic accidents
at start, we test for each vehicle if $v_{ini} \leq \Delta x$. In such case, $v_{ini} = \Delta x$ as an
additional first step.
First, we want to investigate whether the lane-changing rule modifies the
fundamental diagram. Fig. 1 shows the fundamental diagram from our two-
lane model with cars alone and compares with the results from our previous
single-lane model. It also includes measurements on Bogotá's highways [1] and
results from a STCA model plus our retarded acceleration. As expected, we
do not observe any difference between the single- and two-lane models, i.e.
the rule is well implemented.
Nevertheless, this figure deserves some discussion. It shows that all models,
STCA + *retarded acceleration* included, are in good agreement with experimental data. As we have shown before [1], the value of maximal flow we have
obtained, $q_{max} = 1.320(4)$ cars/timestep (88 cars/minute) is much larger than

those measured on german highways (66 cars/minute) [9]. This suggests that the small gaps to the car ahead (the *aggressive driving* that is so characteristic for Bogotá) make the traffic flux more efficient. The price may be a high rate of automobile accidents, but this is area of future studies.

In addition, we found in [1] that for small single-lane systems, and due to the periodic-boundary conditions, there are some configurations that do not relax, remaining forever at a mean velocity that is larger than the average velocity for the relaxed system, and generate a spurious peak in the flow-density diagram. These configurations disappear when the second lane is added to the model.

4 Mixed Traffic Flow

Now we include buses and cars on the highway traffic. Fig. 2 shows the fundamental diagram for several mixtures of cars and buses. For buses alone we obtain values of maximal flow, $q_{max} = 1.173(3)$ (78 buses/minute), and maximal-flow density, $\rho(q_{max}) = 0.24(2)$, that are smaller than those for cars alone. This is clear, just because buses are slower and longer than cars.

When the percentage of buses lies between 100% and 10%, the shape of the fundamental diagram does not show any important changes. In this range the buses restrict the normal car flow and the system behaves like a road with buses only. It is interesting to see that in the free-flow regime the average speed is $v_{aver} = 5$ cells/timestep (50 km/h), i.e. the system relaxes to configurations where one or many clusters of cars are lead by one or more buses. When the bus percentage is lower than 10% we observe some changes in the diagram shape. In this case the fundamental diagram transforms from a bus-traffic behavior to a car-traffic one. For the free-flow region the average velocity increases from 50 km/h to 70 km/h for bus percentages around 5%. Finally,

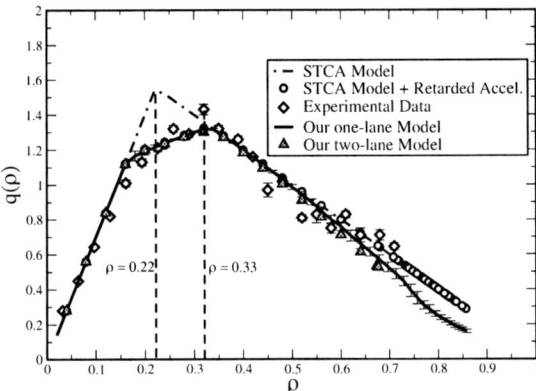

Fig. 1. Comparison of the two-lane simulation with the single-lane results [1].

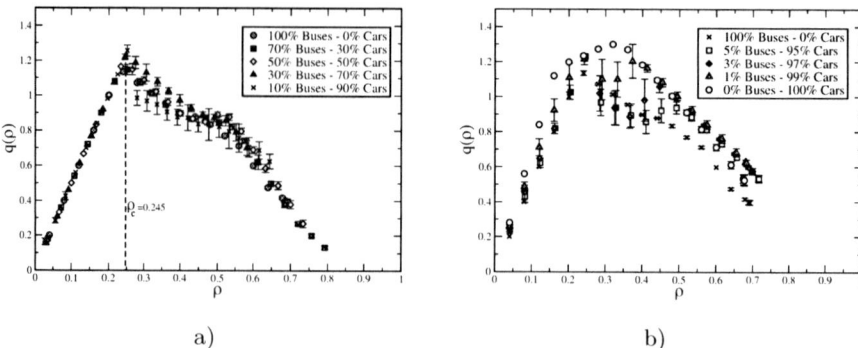

Fig. 2. Fundamental diagram for a two-lane highway with mixed traffic: a) bus percentages between 100% and 10%, b) bus percentages between 5% and 0% .

equilibrium times also change when mixed traffic is introduced. For both buses and cars alone we found equilibrium times that are around five times shorter than for mixtures. This is a good point to take into account in future phase-transition studies of this kind of systems.

5 Conclusions

The simulations of a two-lane highway with cars alone are in good agreement with the real measurements. In this case, the value of critical density is $\rho_c = 0.32(4)$ and the maximal flow is $q(\rho_c) = 1.320(4)$ cars/timestep (88 cars/minute). These values are larger than those measured in other cities due to the small gaps that Bogotan drivers use to maintain with the car ahead, even at large speeds. The inclusion of buses on the highway has a great effect on the fundamental diagram. When the percentage of buses is higher than 10% the critical density decreases to $\rho_c = 0.24(2)$ and the maximal flow also decreases to $q(\rho_c) = 1.17(2)$ buses/timestep (78 buses/minute), which are the values for a highway with buses alone. In other words, the presence of buses on the lane is catastrophic on the car flow, and this can explain the constant congested traffic in Bogotá, where cars and buses share the same lane.

Having a cellular automaton for mixed traffic opens a wide spectrum of future applications. For example, questions like how much the flux will improve if buses stop just in bus stations, or if bus lanes were implemented, can be directly addressed with this model. These are problems of great interest in Bogotá, where pedagogical campaigns have been one of the main ingredients of success to address the traffic problem, and they will be topic of future work. Finally, our model is easy to implement, and its parameters can be measured without expensive equipment. These features make this model a good candidate to investigate vehicular flow in developing countries, where traffic is much more complex and where each city has its own peculiarities and problems.

References

1. L. E. Olmos and J. D. Muñoz: Int. J. Mod. Phys. C **15**, 1397, (2004).
2. K. Nagel and M. Schreckenberg: J. Phys. I (France) **I2**, 2221 (1992).
3. K. Nagel: Phys. Rev. E **53**, 4655 (1996).
4. *Traffic and Granular Flow*, edited by D.E. Wolf, M. Schreckenberg and A. Bachem (World Scientific, Singapore, 1995).
5. P. Wagner: "Traffic simulations using cellular automata: Comparison with reality". In [4], pp.199-203.
6. A. Schadschneider and M. Schreckenberg: J. Phys. A **26**, L679 (1997).
7. R. Chrobok, S.F. Hafstein, A. Pottmeier: *OLSIM:A New Generation of Traffic Information Systems*. University of Duisburg-Essen. `www.autobahn.nrw.de`.
8. W. Knospe, L. Santen, A. Schadschneider, and M. Schreckenberg: J. Phys. A **33**, L477 (2000).
9. K. Nagel, D.E. Wolf, P. Wagner, and P. Simon: Phys. Rev. E **58**, 1425 (1998).

Simulating Pedestrian-Vehicle Interaction in an Urban Network Using Cellular Automata and Multi-Agent Models

Abhimanyu Godara[1], Sylvain Lassarre[2], and Arnaud Banos[3]

[1] Dept. of Civil Eng., Institute of Technology, Banaras Hindu University, 221005 Varanasi, India (abhimanyu_godara@rediffmail.com)
[2] GARIG, Institut National De Recherche Sur Les Transports Et Leur Securite, Champs Sur Marne, France (lassarre@inrets.fr)
[3] SET Laboratory, University of Pau, France (arnaud.banos@univ-pau.fr)

Summary. Agent-based and cellular automata models have been widely used in an efficient and effective way for studying granular traffic, but rarely considering the combined effect and interactions of pedestrians and vehicles in urban networks. So from this point of view an attempt has been made to develop a virtual urban environment which considers both vehicular and pedestrian traffic and the interactions arising from their behavior. This paper presents details of the model we have developed. For vehicular traffic a cellular automata model, combining and appropriately modifying (e.g. to account for the pedestrian movement) BML, NaSch and ChSch models, is considered. Pedestrian traffic is simulated using simple behavioral rules combined with an agent-based approach. Different constraints affecting the mobility of the whole system are considered, which can be seen and even changed by the user in the simulated environment. The model belongs to the microscopic category where pedestrians/vehicles behave in their environment by making a sequence of decisions. The interactions among vehicles and pedestrians are also incorporated which signifies various effects, ranging from accident risk of pedestrians to the generation of traffic jams. NetLogo which is a multi-agent based modeling language is used as the programming platform for the simulation.

1 Introduction

Traffic flow is a subject of interdisciplinary [1] interest at the present time, both for the very real problems of congestion on busy highways [2], as well as an example of a complex system whose behaviour has yet to be fully understood. The investigation of such systems started as early as 1934 by Greenshields with a study on traffic capacity [3]. Since its inception, the study of traffic flow was based on stochastic processes [4]. In 1955 Lighthill and

Whitham described the existence of density waves as well as shock waves [5] in a continuous model of traffic flow.

Pedestrian simulation has only recently received more attention in the context of crowd evacuation management, and panic situation analysis [6-11]. The ability of predicting movements of pedestrians is valuable in many contexts. Apart from panic situations, capturing the behaviour of pedestrians is of growing importance in architecture, urban planning, land use, marketing and traffic operations.

The development of Intelligent Transportation Systems (ITS) has triggered important research activities in the context of behavioral dynamics. Several new models (driving and travel behavior models), new simulators (traffic simulators, driving simulators) and new integrated systems to manage various elements of ITS, have been proposed in the past decade [12-19]. With regard to pedestrians, the focus of ITS has mainly been on safety issues and so modeling pedestrian movements in detail has rarely been considered.

A certain number of work recently tried to explore this field, in particular from the point of view of complex systems, but rarely are those tackling the problem of the dynamics of pedestrian displacements in interaction with road traffic, which is however the key dimension of the urban living environment.

Pedestrian simulation is a new area of safety and health research employing contemporary technology in a form traditionally used in areas of vehicular transportation, skill acquisition and defense [14]. This paper discusses the design considerations of developing such a simulator, which provides scope for multi-modal research in the fields of safety, health and transportation.

In order to fill this gap, this paper presents some glimpse on how to contribute, through creation of a multi-agent simulator supplied with ethopsychologic observations in real situations. Such a virtual laboratory would indeed allow, on a quasi experimental basis, to find solutions of various problems related to this area.

The objective of the research is the development of a computer simulation model for pedestrian and vehicular movement in architectural and urban space as an animation. The characteristics of the model is the ability to visualize the movement of each pedestrian in a plan as an animation. Based on response to various safety and health-related scenarios, the participant makes decisions regarding the effect of the virtual built environment on his safety, health and comfort. The findings can be reintroduced into field conditions allowing improvements in public health and safety. So architects and designers can easily find and understand the problems in their design projects.

Cellular automata (CA) are alternatives to differential equations in an attempt to model transportation systems. CA's are dynamical systems in which space and time are discrete. A cellular automaton consists of a regular grid of cells, each of which can be in one of a finite number of k possible states, updated synchronously in discrete time steps according to a local, identical interaction rule. On the other hand, developed in the context of artificial intelligence, agent-based simulation has been widely used in the context of traffic simulation. It provides a great deal of flexibility, as the behavior of each element in the system can be modeled independently, and complex interactions can be captured. In our context, each pedestrian/vehicle is an "agent". The behavior of each agent can be modeled as a sequence of specific choices, such as the destination, the itinerary, an overall direction, or where to put the next step. Discrete choice models in general, and random utility models in particular, are disaggregate behavioral models designed to forecast the behavior of individuals in choice situations [15]. So by analyzing all the above reasons we have adopted these two approaches, namely cellular automata and agent-based simulation, in our research.

2 Modeling Vehicular and Pedestrian Traffic Movements

2.1 CA Model for Cars

Following the prescription of the NaSch model, we allow the speed V of each vehicle to take one of the $V_{max} + 1$ integer values $V = 0, 1, 2, \ldots, V_{max}$. For urban systems we do not want to have V_{max} more than 72 km/h. So we are taking maximum speed as 3 (22.5 m/s as each cell is 7.5 m in length as in the NaSch model). Suppose V_n is the speed of the nth vehicle at time t while moving in any direction (different from NaSch/ChSch/BML model where vehicles move either towards east or towards north and number of cars is fixed on a given road). Also we want each car to slow down at the intersection (considering the distance P_n to the closest signalized intersection, or S_n, the distance to the closest unsignalized intersection) and decide regarding the turning movements (to get homogenity). The above assumption is true considering the fact that drivers become more cautious and reduce their speed at intersections to avoid any kind of collisions with other vehicles. At each discrete time step $t \rightarrow t + 1$, the arrangement of N vehicles is updated in parallel according to the following *driving rules:*

Step 1: Acceleration.
If $V_n < V_{max}$, the speed of the n-th vehicle is increased by one, i.e., $V_n \rightarrow V_n + 1$.

Step 2: Deceleration due to 1) other vehicles, 2) intersections (non-signalised and signalised), 3) pedestrians on the street (D = the minimum gap between the car under consideration and the pedestrian in front of it on the road (if any) in the radius vision-cars where vision-cars is the vision for cars. It is directly related to the distance up to which a driver can see in the urban network while driving).

Step 3: Randomization.
If $V_n > 0$, the speed of the car under consideration is decreased randomly by unity (i.e, $V_n \rightarrow V_n - 1$) with probability p ($0 \leq p \leq 1$); p, the random deceleration probability, is identical for all the vehicles and does not change during updating.

Step 4: Movement.

Each vehicle moves forward with the given speed, i.e. $X_n \rightarrow X_n + V_n$, where X_n denotes the position of the n-th vehicle at any time t.

The major changes are made in step 2 which reflects the interaction among vehicles and pedestrians. Step 4 shows that there are no more north-bound or east-bound vehicles, the speed of each car is updated simultaneously without any specific classification.

2.2 Model for Pedestrians

In our model randomly located pedestrians converge at the same point, i.e. the destination, by path-finding. The path-finding of a pedestrian is influenced by two new concepts which we have introduced: 1) an attraction field and 2) a punctuated equilibrium effect in our pedestrian model. We will discuss them one by one before presenting the rules for walking.

Attraction Field

Attraction represents how marked-crosswalks influence the behavior of pedestrians, i.e. if a marked-crosswalk is at such a distance that it can attract the attention of pedestrians while crossing so as to avoid midblock crossings. In simple words we can say that a crosswalk will generate a sort of magnetic field in its nearby area (sidewalk, because pedestrians will not be influenced by this field once he/she starts crossing) and it will attract agents (pedestrians) in this area. So the pedestrians will move on the sidewalk until they find a marked-crosswalk, and will then decide to cross. Otherwise, if the attraction field is not strong enough or if the pedestrian is out of the range of the field (i.e. the distance to the crosswalk is too large), to attract the attention then pedestrians will choose midblock-crossings. This is actually what happens in the real world also where the pedestrian is influenced by a crosswalk if it is not too far from him.

Punctuated Equilibrium Effect

This effect comes into the picture at the time when a pedestrian arrives at a cell where he/she has to decide about the crossing. We do not want our model to be quite deterministic where each pedestrian has only one option to choose. Now based on this concept the pedestrian will analyze the situation and make a decision about crossing/no-crossing (Yield/No-Yield). Yield means that a pedestrian will stop at the cell and will analyze the situation globally/locally and will decide to cross once he/she feels that it is safe to cross otherwise (in No-Yield) he will start crossing without taking care of the surrounding conditions.

We tried to keep our pedestrian model as simple as possible by using only a minimal set of rules but at the same time these rules are able to predict pedestrian behavior quite accurately. At each discrete time step $t \rightarrow t+1$, the arrangement of N pedestrians is updated in parallel according to the following "rules". The rules are divided in two categories: 1) pedestrian moving on sidewalk and 2) pedestrian entering a crossing.

1. **Pedestrian moving on sidewalk:**
    Step 1: Path-finding
    In this step a pedestrian will choose the direction in which he has to move (depending on a vision radius).
    Case I: When the pedestrian is in the attraction field (sidewalk $\rightarrow$ crosswalk)
    Case II: When pedestrian is not in the attraction field ($\rightarrow$ sidewalk or mid-block crossing)
    Step 2: Checking (street crossing)

2. **Pedestrian entering a crossing:**
    Step 1: Decision-making
    pt = proportion of pedestrians having yielded up to time t (global)
    If $p > \epsilon$ (noise) then,
    (If $pt > .33$, pedestrian will Yield
    If $pt = .33$, the pedestrian will continue to do whatever he was doing earlier, i.e. Yield or No-Yield.
    If $pt < .33$, the pedestrian will not Yield (No-Yield).)
    else the pedestrian will choose randomly from the option Yield/No-Yield.
    Step 2: Yield/No-Yield
    If a pedestrian has yielded, then he will stop and check for any moving cars on the street in his/her radius of vision and if he detects any moving cars, he will stop and allow the cars to pass.
    Else, s/he will just cross without taking care of cars, i.e. blindly.
    Step 3: Movement
    The pedestrian will move forward with his/her walking speed.

3 Results

In this section we will be discussing some fundamental results of our model. Most of the results are quite interesting to observe in the sense that they give a picture that is generally found in the urban complex systems, arising from a lot of interactions among the agents which belong to different categories.

3.1 Effect of Yield/No-Yield Pedestrians on Accidents

The proportion pt of yielding pedestrians can be varied using the parameter $\epsilon = En$ (noise in the system).
As it can be seen from the above graphs as the noise in the system increases the system becomes more or less unstable (more randomization takes place) and proportion of intelligent (yielding) pedestrians also decreases which gives rise to more number of accidents. One can expect these results in the real world too where pedestrians are not bothered about the cars, and thus there chances/risk of getting involved in an accident also increases.

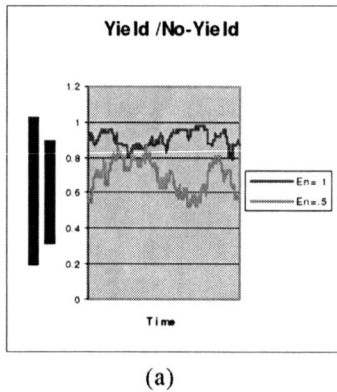
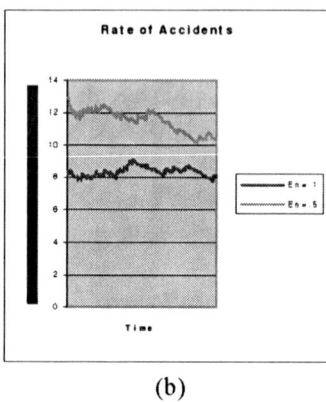

(a)   (b)

Fig. 1. (a) Proportion of yielding pedestrians with time for varying noise En in the system (b) Rate of accidents for varying noise En in the system.

3.2 Effect of Maximum Speed of Cars on Accidents

Here we will study the effect of varying speed on accident rate by keeping other parameters constant.
One can expect these results because as the speed of car increases the pedestrian/cars become more vulnerable to be involved in an accident.

3.3 Effect of Vision-Car/Pedestrians

In the earlier examples we have kept the car vision and pedestrian vision constant, i.e. 2 cells. Now we will show how changing vision can affect other conditions and interactions.
Changing the vision has direct impact on the distance to which a pedestrian/car can see. So if vision is larger (as explained earlier), the strength of the attraction field increases. Hence chances of pedestrians to choose the

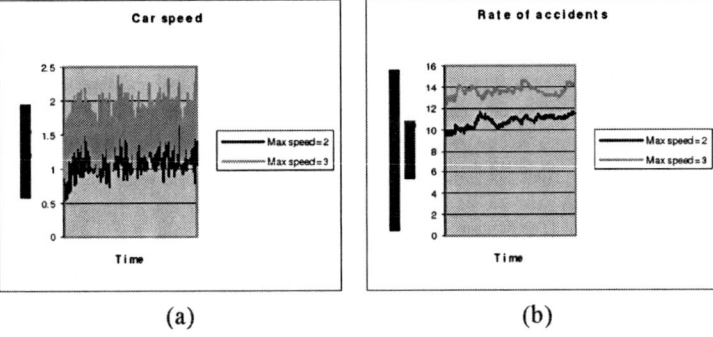

Fig. 2. (a) Average speed variation with change in maximum speed; (b) Rate of accidents for different maximum speeds of cars. The other parameters are the same for both cases.

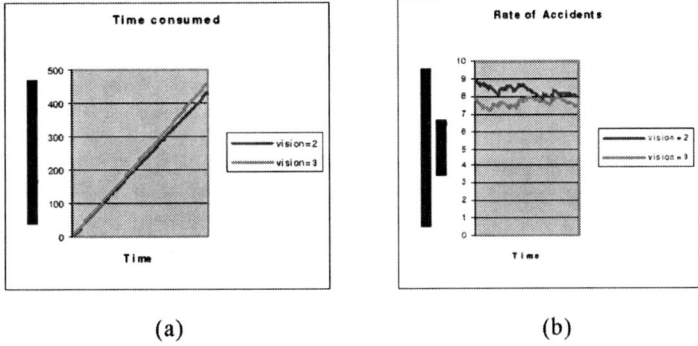

Fig. 3. Effect of changing vision on (a) time spent by pedestrians on sidewalk (b) rate of accidents.

crosswalk also increase, which will make pedestrians spend more time on the sidewalk rather than making midblock-crossings which are now suppressed by the attraction field.

4 Conclusions

Pedestrian and vehicular simulation is a new area of safety and health research employing contemporary technology in a form traditionally used in areas of vehicular transportation, skill acquisition and defense. This paper discusses the design considerations of developing such a simulator, which provides scope for multi-modal research in the fields of safety, health and transportation. The model is being developed to create simulated environments whereby the participant of the study responds to simulated environments as though he or she is actually in the field. Under these conditions, the simulated environment can be manipulated to further research in many aspects of pedestrian facility design with low-risk to participants in the study. Based on the response to various safety and health-related scenarios, the participant makes decisions regarding the effect of the virtual built environment on his safety, health and comfort. The findings can be reintroduced into field conditions allowing improvements in public health and safety. Other current needs for research in pedestrian environments can be examined so that the simulator can be designed with enough flexibility to support various research needs. This paper discusses the considerations in the design of the simulator, which accommodate a variety of current pedestrian research needs related to improving both the safety of the pedestrian and vehicular environments designed for transportation corridors, as well as defining the nature of walking facility design required by the health industry for preventive and curative use.

References

1. N. Gartner, H. Mahmassani, C. H. Messer, R. Cunard and A. Rathy, *Traffic Flow Theory: A State-of-Art-Report,* published by Transportation Research Board Committee on Traffic Flow Theory and Characteristic (1987).
2. D. Chowdhury, L. Santen and A. Schadschneider, Phys. Rep. 329, 199 (2000).
3. B.D. Greenshields, Highways Research Record 14, 448 (1934).
4. W.F. Adams, *Road traffic considered as a Random Series,* J. Inst. Civil. Engin., London, 1936.
5. M.J. Lighthill and G.B. Whitham, Proc. Royal Soc. Lond. **A229**, 317 (1955)
6. D. Helbing, L. Farkas, T. Vicsek, Nature 407(28), 487 (2000).
7. D. Helbing, *Verkehrsdynamik. Neue Physikalische Modellierungskonzepte,* Springer, Berlin, 1997 (in German).
8. D. Helbing, Rev. Mod. Phys. **73**, 1067 (2001).
9. D. Helbing, P. Molnar, Phys. Rev. E 51, 4282 (1995).
10. G. Antonini, M. Bierlaire, M. Weber: *Simulation of Pedestrian Behaviour using a Discrete Choice Model Calibrated on Actual Motion Data,* (2004).
11. J.R. Naderi, B. Raman, *Design Considerations in Simulating Pedestrian Environments.* Report submitted to Texas Transportation Institute. (2001).
12. F. Feurty: *Simulating the Collision Avoidance Behavior of Pedestrians,* Master's degree thesis, Dept. of Electronics Engg., University of Tokyo (2000).
13. C. Burstedde, K. Klauck, A. Schadschneider, J. Zittartz: Physica A 295, 507 (2001).
14. M. E. Ben-Akiva, Bergman, A. J. Daly, Ramaswamy R.: *Modeling inter-urban route choice behaviour,* in J. Volmuller and R. Hamerslag (Eds.) Proceedings from the ninth international symposium on transportation and traffic theory, pp. 299-330, VNU Science Press, Utrecht, Netherlands (1984).
15. M. E. Ben-Akiva, S. R. Lerman: *Discrete Choice Analysis: Theory and Application to Travel Demand,* MIT Press, Cambridge, Ma, (1985).
16. J. Wu, M. Brackstone, M. McDonald: Transp. Res. Part C 11, 463 (2003).
17. A. B. Downey, G. Gay: *How to Think Like a Computer Scientist,* Logo version, (2003).
18. D. E. Wolf: *Cellular automata for traffic simulations,* Physica A 263, 438 (1999).
19. F. Weifeng, Y. Lizhong, F. Weicheng: Physica A 321, 633 - 640 (2003).

Multi-Phase Signal Setting and Capacity of Intersections

Chang Yulin, Zhang Peng, Mao Lin, and Gong Zhen

School of Automobile and Traffic Engineering, Jiangsu University, Zhenjiang, Jiangsu, P.R. China

Summary. The multi-phase signal control method is one of the important measures to enhance intersection capacities and alleviate urban traffic problems. This paper employs the stopping-line-method to study the capacity model of cross multi-phase signalized intersections and analyses the relevant change of the intersection capacity and cycle length. It studies the intersection capacity in detail under two normal situations, one with one straight lane and one left-turn lane, and the other with two straight lanes and one left-turn lane. Finally, a practical intersection is chosen and its phase design is improved by the method proposed in the paper.

1 Introduction

With the sustaining and rapid development of our national economy, urban automobile possession, traffic volumes, and traffic demand are increasing drastically. Traffic congestion of different degrees appears universally in many metropolis. Because an intersection, as the joint of road networks, joins traffic flows from different directions, besides, due to the factors such as red light time loss and mixed driving of automobile and the non-mobile, intersection capacity is far lower than roads capacity. As a result, the intersection becomes the bottleneck of retaining the excessive traffic flows from the roads, and the sector of high accident occurrence in urban road networks. In order to assure the traffic security of intersections and make full use of the intersection capacity, it is an important measure to operate scientific management and control at intersections.

At present the intersection traffic control in the cities of our country mostly applies signal control methods, either the two-phase signal control method or the multi-phase signal control method. Multi-phase signal control method is a unified name of the control method for more than two signal phases. It separates traffic flow in time, decreases the traffic conflict spots at intersections, and improve traffic orders and security when vehicles and pedestrians pass intersections. In [1] the multi-phase signal control method has been compared with the two-phase signal control method under the aspects of traffic conflict, capacity, service levels etc. They conclude that the former is an effective way to improve intersection security and service levels.

2 Capacity of Multi-Phase Signalized Intersection

Looking into a cross-signalized intersection, Approach i is composed of Approach 1, Approach 2, Approach 3, and Approach 4 from four different directions. It is assumed that every approach has special left-turn and straight signals. For a signalized intersection, the opposite approaches often apply the same signal phase, that is, the signal phase of Approach 1 and Approach 3 is the same, and the signal phase of Approach 2 and Approach 4 is the same. To simplify calculations, Approach 1 and Approach 2 are taken as representatives. If the total signal cycle length is denoted by L, the green time of Approach i from direction j is recorded as l_{ij} ($i = 1, 2$; $j = 1$ is straight, $j = 2$ is left turn), the yellow time is recorded as c_{ij}, and when the vehicle flow of Approach i from direction j is passing the intersection stopping line, if the time spent by the first vehicle is recorded as t_{cij} and the time spent by the following vehicle is recorded as t_{fij}. Then the following formula can be given:

$$\sum_{i,j=1}^{2} (l_{ij} + c_{ij}) = L.\tag{1}$$

Let $c = \sum_{i,j=1}^{2} c_{ij}$, then $\sum_{i,j=1}^{2} l_{ij} + c = L$. The yellow interval c_{ij} is between 2 and 4 s. The parameter c is the sum of yellow times in one cycle. It is a constant related to the intersection geometry, which mainly depends on the intersection size and the lane setting method.

If the number of the lanes of Approach i from direction j is denoted by m_{ij}, then the number of vehicles of the approach from the direction in one signal cycle can be calculated by the following formula:

$$s_{ij} = m_{ij} \left(\frac{l_{ij} - t_{cij}}{t_{fij}} + 1 \right) = m_{ij} \left(\frac{l_{ij}}{t_{fij}} - \frac{t_{cij} - t_{fij}}{t_{fij}} \right).\tag{2}$$

If one signal cycle length is L, then the number of vehicles that can pass the intersection in unit time, i.e. the capacity of the intersection, is

$$N = \frac{S}{L} = \frac{1}{L} \sum_{i,j=1}^{4} s_{ij} = \frac{2}{L} \sum_{i,j=1}^{2} s_{ij} = 2 \sum_{i,j=1}^{2} \frac{m_{ij}}{L} \left(\frac{l_{ij}}{t_{fij}} - \frac{t_{cij} - t_{fij}}{t_{fij}} \right).\tag{3}$$

3 Distribution of Headways at Intersections

When the signal light changes from red to green, the first vehicle in the queue needs some reaction time and acceleration time. Therefore, the headway of the first vehicle is the longest and the headway of the following vehicle decreases one by one. But the headway of the vehicle behind the fourth is basically the same (Fig. 1).

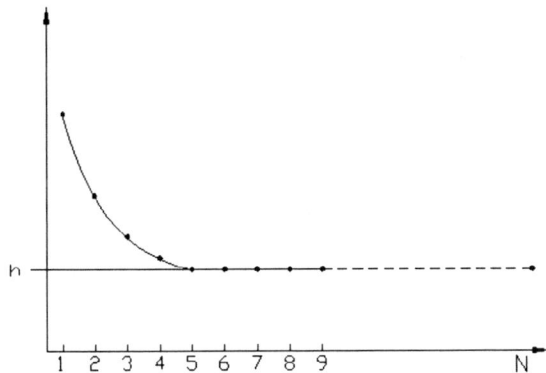

Fig. 1. The following car's gap at different positions

Based on the measured result, the average headway of the first vehicle is about 5 s, that of the vehicle behind the fourth is about 2 s, and the excessive headway of the front vehicles is about 6 s. To simplify calculations, the excessive headway of the front vehicle is recorded as the headway of the first vehicle. Therefore, when the queuing vehicles are passing the intersection after the signal changes from red to green, the headway of the first vehicle is about 8 s and the headway of the following vehicle is about 2 s. In addition, there is little difference between the headway of the straight, left-turn and right-turn vehicles. Therefore, t_{cij} can be recorded as 8 s and t_{fij} can be recorded as 2 s in practical calculations.

4 Two Types of Simplified Situations

4.1 One Straight Lane and One Left-Turn Lane at Every Approach

In one type of simplification it is assumed that grade and shape of the two crossing roads of the intersection are basically the same: There is only one straight lane and one left-turn lane at every approach. We assume that the following car headways t_{fi1} $(i = 1, 2, 3, 4)$ of the straight vehicle flow (denoted by t_{f1}) and the following car headways t_{fi2} $(i = 1, 2, 3, 4)$ of the left-turn vehicle flow (denoted by t_{f2}) at every approach of the intersection are the same. Introducing the parameter $\alpha = \sum_{i,j=1}^{2} \frac{t_{cij}-t_{fij}}{t_{fij}}$ which is determined by the geometrical features of the intersection and the vehicle performance, the intersection capacity (see eq. (3)) becomes

$$N = \frac{2}{L} \left(\sum_{j=1}^{2} \frac{1}{t_{fj}} \sum_{i=1}^{2} l_{ij} - \alpha \right). \tag{4}$$

If the following car headway of the straight vehicle is approximated by the following car's headway of the left-turn vehicle, that is $t_{f1} = t_{f2} = t_f$, and according to eq. (1), then eq. (4) is simplified as following:

$$N = \frac{2}{L} \left(\frac{1}{t_f} \sum_{i,j=1}^{2} l_{ij} - \alpha \right) = \frac{2}{L} \left(\frac{L-c}{t_f} - \alpha \right) = 2 \left(\frac{1}{t_f} - \frac{1}{L} \left(\frac{c}{t_f} + \alpha \right) \right) \tag{5}$$

Eq. (5) indicates that the capacity of the signalized intersection is related to the signal cycle length and the reciprocal of Cycle Length L, i.e. the capacity is larger for longer cycle lengths L. But when the signal cycle length reaches a certain value, further increase of the signal cycle length does not affect the capacity much. The capacity N is limited to $2/t_f$ when $L \to \infty$. If the yellow interval c_{ij} is taken as 2 s, then c is 8 s. If $t_{c1} - 8$ s, $l_{c2} = 7$ s, $t_{f1} = 2$ s, and $t_{f2} = 2$ s in eq. (5), then it is simplified to

$$N = \left(1 - \frac{30}{L} \right) (pcu/s). \tag{6}$$

In the above discussion of the intersection capacity, the case of right-turn vehicles is not taken into account. If a special right-turn lane is designed at an intersection, the intersection capacity will only have to add the capacity of the special right-turn lane, because the right-turn signal phase can be designed connecting with the other phases and does not take up of the entire signal period. If the right-turn lane and the straight lane share the same approach, the total capacity of the intersection will decrease a little, due to the interact of straight vehicles and right-turn vehicles.

4.2 Two Straight Lanes and One Left-Turn Lane at Every Approach

For many common intersections in urban roads, the proportion of the straight vehicle flow is greater than the left-turn and the right-turn vehicle flow. Thereby, the number of straight lanes is more than right-turn and left-turn lanes at many intersections. One of the usual situations is that the number of straight lanes is 2, the number of right-turn lanes and left-turn lanes is both 1. The capacity of the intersection on this condition is discussed as following.

If $m_{i1} = 2$, $m_{i2} = 1$, $t_{fij} = 2$ s, $t_{cij} = 8$ s in eq. (3), and the yellow interval c_{ij} is taken as 2 s on the restrictive condition of eq. (1), then

$$N = 1 + \frac{l_{11} + l_{21} - 42}{L} \ (pcu/s).$$ (7)

The change of the capacity of the above two intersections is analyzed as following. Taking the difference of eqs. (7) and (6), then dividing it by eq. (6), the growth rate of capacity is given by

$$\frac{\Delta N}{N} = \frac{l_{11} + l_{21} - 12}{2(L - 30)}.$$ (8)

If the number of straight lanes increases from 1 to 2 when the number of straight vehicles is more than the number of left-turn vehicles and the straight phase is longer than the left-turn phase, the capacity of the intersection can increase by about 50%.

5 Model in the Case of Vehicle Arrival Rate

The factor of vehicle delay needs to be considered during designing the signal phase of an intersection in practice. Generally, as the signal period is longer, the capacity is larger. But meanwhile the caused vehicle delay is longer. As the signal period is shorter, the vehicle delay is shorter, but the capacity of the intersection is smaller. Therefore, in order to make vehicle delay the shortest, during designing the signal phase, we only have to discuss the case that the intersection capacity is larger than vehicle arrival rate.

The situation of non-saturated traffic flow, that is, the case that the arrival traffic flow is smaller than the intersection capacity, is discussed as following. Because the same signal phase is often applied in the opposite approaches, when traffic flow from a certain direction is large, the main traffic demand needs to be met during designing the signal phases. The average arrival rate of Approach i from direction j is supposed to be q_{ij}. To make the arrival traffic flow can pass the intersection in the signal period, the design of signal phase length must satisfy the following condition:

$$m_{ij} \left(\frac{l_{ij}}{t_{fij}} - \frac{t_{cij} - t_{fij}}{t_{fij}} \right) > q_{ij} L, \qquad (i, j = 1, 2).$$ (9)

After dividing eq. (9) by m_{ij}, adding the divided results based on i and j, and finally dividing the total result by the cycle length L, one arrives at

$$N = 2 \sum_{i,j=1}^{2} \frac{1}{L} \left(\frac{l_{ij}}{t_{fij}} - \frac{t_{cij} - t_{fij}}{t_{fij}} \right) > 2 \sum_{i,j=1}^{2} \frac{q_{ij}}{m_{ij}}.$$ (10)

Eq. (10) indicates that the capacity of the intersection is larger than the vehicle arrival rate.

In the case of a non-saturated flow rate, the design of signal phases must make the remaining time after vehicles of all the approaches from all the directions have passed the intersection as balanced as possible. That is, the following values should be made as same as possible:

$$\left(\frac{l_{ij}}{t_{fij}} - \frac{t_{cij} - t_{fij}}{t_{fij}} \right) - \frac{q_{ij}}{m_{ij}} L, \qquad (i, j = 1, 2). \tag{11}$$

Mathematically this can be formulated as

$$\min \sum_{i,j=1}^{2} \left(\left(\frac{l_{ij}}{t_{fij}} - \frac{t_{cij} - t_{fij}}{t_{fij}} \right) - \frac{q_{ij}}{m_{ij}} L \right)^2 \tag{12}$$

$$s.t. \sum_{i,j=1}^{2} l_{ij} = L - c. \tag{13}$$

Using Lagrangian multiplication factors, one obtains

$$l_{ij} = \frac{t_{fij}^2}{\sum_{i,j=1}^{2} t_{fij}^2} \left(L - c - \sum_{i,j=1}^{2} \left(t_{cij} - t_{fij} + \frac{q_{ij} t_{fij}}{m_{ij}} L \right) \right)$$

$$+ t_{cij} - t_{fij} + \frac{q_{ij} t_{fij}}{m_{ij}} L. \tag{14}$$

Eq. (14) is just the formula to design phases of all the directions. It can be simplified a bit in practice. The difference between the following car headway of the left-turn and straight traffic flow at intersections is not large after the traffic flow is stable. Therefore, Eq. (14) can be simplified as

$$l_{ij} = \frac{1}{4}(L - c) + \left((t_{cij} - t_{fij}) - \frac{1}{4} \sum_{i,j=1}^{2} (t_{cij} - t_{fij}) \right)$$

$$+ t_f L \left(\frac{q_{ij}}{m_{ij}} - \frac{1}{4} \sum_{i,j=1}^{2} \frac{q_{ij}}{m_{ij}} \right). \tag{15}$$

This can even be further simplified. In case that the difference between the interval when the first vehicle and the following vehicle of the straight and left-turn traffic flow at approaches are passing the stopping-line is not large, the second part of the right side of the equation can be eliminated:

$$l_{ij} = \frac{1}{4}(L - c) + t_f L \left(\frac{q_{ij}}{m_{ij}} - \frac{1}{4} \sum_{i,j=1}^{2} \frac{q_{ij}}{m_{ij}} \right). \tag{16}$$

Eq. (16) indicates for the cross intersection designed with a special left-turn signal, the signal phase length l_{ij} is determined by parameters such as the cycle length L, yellow interval c, following car headway t_f, and the main traffic volume q_{ij}/m_{ij} and the cycle length L is determined by eq. (10). If one assumes all the values of t_{fij} are the same and denoted by t_f, according to eqs. (1), (4), (9), then the cycle length can be determined as the following condition:

$$\frac{1}{L}\left(\frac{L-c}{t_f}-\alpha\right) > \sum_{i,j=1}^{2}\frac{q_{ij}}{m_{ij}}.\tag{17}$$

In practice, Eq. (17) is changed into an equation which can be applied to calculate the cycle length L.

6 Conclusion

(1) The capacity of a signalized intersection increases with signal cycle length. If the cycle length is too short, the capacity will decrease too much. If the cycle length is too long, the capacity will increase slowly and delay will increase. (2) If the right-turn traffic flow is not taken into account, the intersection capacity will increase by about 50% when approaches are broadened from one straight lane and one left-turn lane to two straight lanes and one left-turn lane. Therefore, broadening approaches is one of the effective ways to improve the intersection capacity. (3) In the design of phases at intersections, the cycle length and the phase length are related to the number of approaches, following car headway, yellow interval, starting time loss etc.

References

1. Zhou Tongmei, Du Yaxun. *Study and Application of multi-phase signal control.* Journal of Chinese People's Public Security University (Natural Science Edition) **4**, 47-51 (2003)

Stability of Flows on Networks

Alexander P. Buslaev, Alexander G. Tatashev, and Marina V. Yashina

Moscow State Automobile and Road Technical University, 64, Leningradsky pr., Moscow, Russia

Summary. The problems of traffic flow forecasting on complex traffic networks are still almost not explored. However these problems are very actual for scientists as well as for traffic engineers. In this paper we consider problems of stability of particle (car) flows on networks. The definitions of critical, stable and unstable flow states on networks are obtained as properties of solutions of nonlinear differential equations on graphs. For networks with different geometry the necessary and sufficient conditions of flow stability on networks are found. The perspective problems of exploration of qualitative properties of flows on networks are formulated.

1 Introduction

The rapid growth of the motorization level in the world provoked interest of scientists and engineers in problems of stability of traffic flows [1-5]. However, more difficult problems, such as forecasting of stable states of flows on complex traffic networks, are still almost not explored [6-8].

In this paper we consider problems of stability of particle (car) flows on networks. A network is an oriented graph with edges corresponding to road sections and vertices corresponding to road junctions. A state of flow on networks is defined by the vector-function of densities $\rho(t) = \{\rho_i(t)\}$. The time-dependence of each coordinate $\rho(t)$ is simulated by a system of differential equations and expresses the following physical principle: *the rate of density change on an edge is proportional to the difference of the intensities (flux) of cumulative input and output flows from the edge.*

The flow intensity $q_i(t)$ on edge i is a function of the density $\rho_i(t)$, which is defined by the fundamental diagram $q_i(t) = \lambda_i \rho_i(t)(\rho_i^* - \rho_i(t))$, where ρ_i^* is the maximal density on edge i. Then $q_i^* = \lambda_i \left(\rho_i^*\right)^2 / 4$ is the maximal intensity *(the highway capability of the edge)*. At last if l_i is the length of edge i, then $C_i^* = l_i \rho_i^*$ is the *edge capacity* and $C^* = \sum_i C_i^*$ the *network capacity*.

The flow regime $\bar{\rho} = \bar{\rho}(t)$ *in the time* t_0 *is called critical,* if an i-edge exists with $\rho_i(t_0) = \rho_i^*$. According to the physical principle of the model the flow regime will be critical also for $t \geq t_0$, because $q_i(t) \equiv 0$, $t \geq t_0$.

The flow regime is called T-critical, if for $t \geq T$ the flow regime is critical,
otherwise *the flow regime is called T-uncritical. So the flow state $\bar{\rho}(t)$ is called*
T-critical point or T-uncritical point accordingly. Let $T_*(\bar{\rho}(0))$ denote min$\{\tau \geq$
0/ the flow regime is critical for $t \geq \tau\}$.

The point $\bar{\rho}(0)$ is called stable T-uncritical point, if all points belonging to
some neighborhood of $\bar{\rho}(0)$ and considered as initial states will be *T-uncritical.*
Otherwise this T-uncritical point is called unstable T-uncritical point.

Let $\bar{\rho}(0)$ *be a stationary uncritical point of flow,* i.e. the uncritical flow state is
not changing during time. Then $\bar{\rho}(0)$ is an ∞-uncritical point as $T_*(\bar{\rho}(0)) = \infty$.
An *uncritical flow state $\bar{\rho}(0)$ is locally stable,* if for small changes of the flow
state the flow returns to the state $\bar{\rho}(0)$ when $t \to \infty$. Clearly $\bar{\rho}(0)$ is a stationary point.

We consider *closed and open networks.* Let the *flow mass C be the total quan-*
tity of particles on the network. If the network is closed then C is constant. If
the network is open then particles can arrive to and depart from the network,
thus $C = C(t)$. It is clear that the open network can be considered as part of
a more complex closed network.

2 Closed Networks

2.1 Flow on Ring Consisting of Two Identical Sections

Let us consider a unidirectional flow on a ring consisting of two identical
sections (see Fig. 1). Let $l = l_1 = l_2$ be the lengths of sections, $\rho^* = \rho_1^* = \rho_2^*$
be the maximal densities. Then $C_1^* = C_2^* = \rho^* l$ are the capacities of sections
and $C^* = 2\rho^* l$ is the network capacity. In the interval of admissible values
$\rho \in [0, \rho^*]$ the dependence of the intensity on density is $\lambda \rho(\rho^* - \rho)$. In this
case the exact expression for the flow density dependence can be found.

The flow densities on the sections satisfy a system of two differential equations
and a normalization condition expressing the constancy of the flow mass.
Whether the flow regime is critical or not, $T_*(\bar{\rho}(0)) = \infty$, will depend on the
ratio between the ring capacity C^* and the flow mass C. It means that:

If the flow mass is less than half of the ring capacity, $C < C^*/2$, then the flow

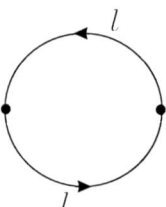

Fig. 1. Ring consisting of two identical sections

*regime is stable ∞-uncritical. In this case flow state converges to stationary
regime $\rho_1 = \rho_2 = C/2$ when t grows and $\left| \rho_i(t) - \frac{C}{2} \right| \searrow 0$, $i = 1, 2$.
If $C > C^*/2$, $\rho_1(0) \neq \rho_2(0)$, then the flow regime is critical for any initial admissible conditions and* $T_*(\rho_0) \leq \frac{l}{2\lambda\rho^*\left(\frac{2C}{C^*}-1\right)} \ln \frac{(\rho^* - \frac{C\rho^*}{C^*})}{(\rho_0 - \frac{C\rho^*}{C^*})}$, *where* $\rho_0 = \rho_1(0)$.

Indeed, since in the case of identical sections $\rho_1(0) + \rho_2(0) = C/l$, then either
$\rho_1(0) = \rho_2(0) = C/2l$, or on one of the sections the initial density is greater
then $C/2l$. In the first case the flow regime is unstable uncritical regime (the
stationary point). In the second case the flow state is critical. Let us find the
time required to reach the critical regime. Assume that $\rho_0 = \rho_1(0) > C/2l >
\rho_2(0)$. Until the critical regime is not reached the time-dependence of the flow
density on the first section is given by $\rho(t) = \frac{C}{2l} + (\rho_0 - \frac{C}{2l})e^{\frac{2\lambda}{l}\left(\frac{C}{l}-\rho^*\right)t}$. For the
time $T_*(\rho_0)$ to reach the critical regime we have $\frac{C}{2l} + (\rho_0 - \frac{C}{2l})e^{\frac{2\lambda}{l}\left(\frac{C}{l}-\rho^*\right)T_*} =
\rho^*$, i.e. $e^{\frac{2\lambda}{l}\left(\frac{C}{l}-\rho^*\right)T_*} = \frac{\rho^* - \frac{C}{2l}}{\rho_0 - \frac{C}{2l}}$ and $\frac{2\lambda}{l}\left(\frac{C}{l} - \rho^*\right)T_* = \ln \frac{\rho^* - \frac{C}{2l}}{\rho_0 - \frac{C}{2l}}$. Thus we get

$$T_*(\rho_0) = \frac{l}{2\lambda\rho^*\left(\frac{2C}{C^*}-1\right)} \ln \frac{(\rho^* - \frac{C\rho^*}{C^*})}{(\rho_0 - \frac{C\rho^*}{C^*})}.$$

If $C = C^*/2$, *then the flow state is stationary and unstable at any initial
condition* $l(\rho_1(0) + \rho_2(0)) = C$.

2.2 Ring Consisting of Two Non-Identical Sections

Let us consider the flow on a ring consisting of two non-identical sections
$(i = 1, 2)$ with lengths l_i, maximal densities ρ_i^* and intensities $q_i = \lambda_i\rho_i(\rho_i^* - \rho_i)$
(see Fig. 2).
The system of differential equations for the flow is then

$$\begin{aligned}
l_1\frac{d\rho_1}{dt} &= -\lambda_1\rho_1(\rho_1^* - \rho_1) + \lambda_2\rho_2(\rho_2^* - \rho_2) \\
l_2\frac{d\rho_2}{dt} &= \lambda_1\rho_1(\rho_1^* - \rho_1) - \lambda_2\rho_2(\rho_2^* - \rho_2).
\end{aligned} \tag{1}$$

We have the condition of flow mass constancy

$$l_1\rho_1 + l_2\rho_2 = C, \tag{2}$$

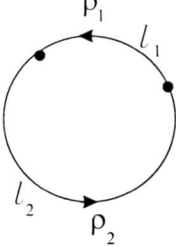

Fig. 2. A ring consisting of two non-identical sections.

and the set of admissible densities $\mathcal{D}$:

$$0 \le \rho_1 \le \rho_1^*, 0 \le \rho_2 \le \rho_2^*. \tag{3}$$

The stationary points are the solutions of the equation

$$-\lambda_1 \rho_1 (\rho_1^* - \rho_1) + \lambda_2 \rho_2 (\rho_2^* - \rho_2) = 0. \tag{4}$$

We put $x = \rho_1 - \rho_1^*/2$, $y = \rho_2 - \rho_2^*/2$. Hence in (4) we obtain $\lambda_1 (x^2 - (\rho_1^*)^2/4) - \lambda_2 (y^2 - (\rho_2^*)^2/4) = 0$ or

$$-\lambda_1 x^2 + \lambda_2 y^2 = -\lambda_1 (\rho_1^*)^2/4 + \lambda_2 (\rho_2^*)^2/4 = -q_1^* + q_2^*. \tag{5}$$

Assume that $q_2^* > q_1^*$, i.e. the highway capacity of the second section is greater than the capacity of the first section. Then (5) is a hyperbola with focuses on the y−axis, i.e. in initial coordinates on $\mathcal{D}$ (Fig. 3).
The direction field of velocities $(\dot{\rho}_1, \dot{\rho}_2)$ is parallel to the line $l_1 \rho_1 + l_2 \rho_2 = C$, and its direction is defined by the sign $\dot{\rho}_1$ in Fig. 3.
In the point $(\rho_1^*, 0)$ the normal to the line (4) is $(\lambda_1 \rho_1^*, \lambda_2 \rho_2^*)$ and in the point $(0, \rho_2^*)$ is symmetrical and is equal to $(-\lambda_1 \rho_1^*, -\lambda_2 \rho_2^*)$. *Assume that*

$$\frac{\lambda_2 \rho_2^*}{\lambda_1 \rho_1^*} > \frac{l_2}{l_1}, \tag{6}$$

i.e. the tangent of the angle of slope of the normal to the hyperbola in the point $(\rho_1^*, 0)$ is greater than the tangent of the angle of slope of the normal to (2). Then when C changes the line (2) can not meet one of hyperbola branches (4) twice. Thus, we get the qualitative sketch shown in Fig. 4.
Condition (6) is equivalent to

$$v_2^* = \frac{\lambda_2 \rho_2^*}{l_2} > \frac{\lambda_1 \rho_1^*}{l_1} = v_1^*. \tag{7}$$

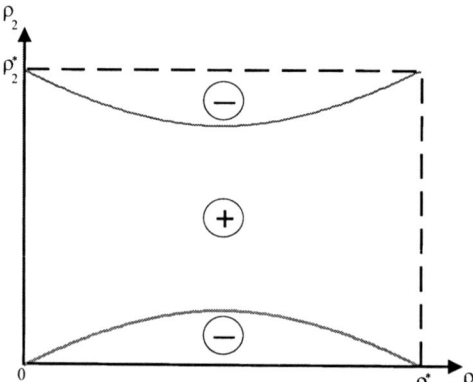

Fig. 3. Signs of $\dot{\rho}$

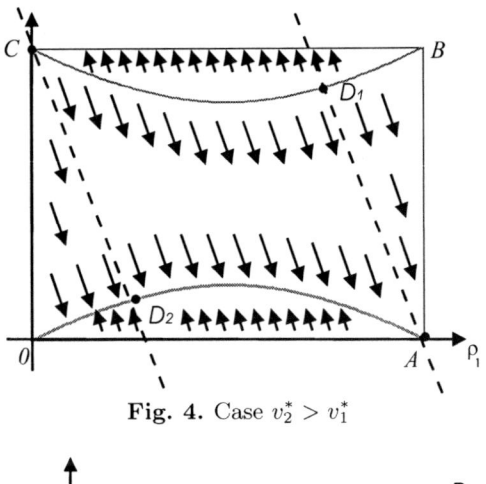

Fig. 4. Case $v_2^* > v_1^*$

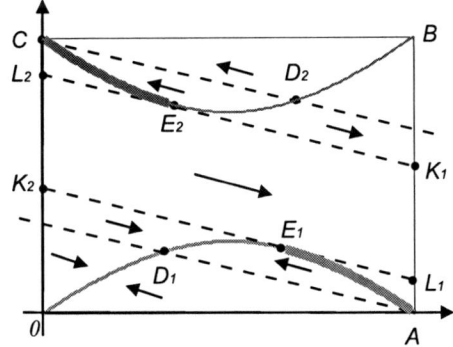

Fig. 5. Case $v_2^* < v_1^*$

Condition (7) means that *the rate of density change on the second section in the neighborhood of the boundaries* $[0, \rho_2^*]$ *is greater than the corresponding rate on the first section.* The opposite case, $(v_2^* < v_1^*)$, is shown in Fig. 5.

The distinctive feature of the second case is *the existence of stable and unstable fragments of stationary points on each hyperbola branch. Otherwise, in the first case, when* $v_2^* > v_1^*$ *(Fig. 4), the upper branch is unstable and the lower branch is stable.*

Let us assume that $q_1^* = q_2^*$, i.e. $\lambda_1(\rho_1^*)^2 = \lambda_2(\rho_2^*)^2$. The hyperbola (4) degenerates to the pair of lines $\lambda_1 \left(\rho_1 - \rho_1^*/2\right)^2 = \lambda_2 \left(\rho_2 - \rho_2^*/2\right)^2$. It is equal to

$$\sqrt{\lambda_1} \left|\rho_1 - \frac{\rho_1^*}{2}\right| = \sqrt{\lambda_2} \left|\rho_2 - \frac{\rho_2^*}{2}\right|. \tag{8}$$

These lines meet the opposite vertices of the rectangle $\mathcal{D}$ (Fig. 6).

Depending on the velocity ratio v_1^* *and* v_2^*, *(7), we have either the upper branch* CDB, *or the right stationary branch* ADB *unstable, and the corresponding adjunct is stable.*

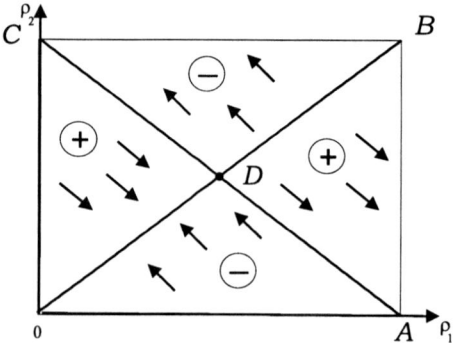

Fig. 6. Degeneration, $q_1^* = q_2^*$

3 Open Unidirectional Edge

Let us consider an edge with length l, which receives a flow of constant intensity q (Fig. 7). Due to the fact that as a rule an edge in the network has an

Fig. 7. Elementary open section

exit, the intensity of the output flow is fixed. Thus,

$$l\frac{d\rho}{dt} = \Theta\left(\rho_{\max} - \rho\right) - \min(q_{\max}, f(\rho)), \tag{9}$$

where $\Theta\left(\rho\right) = \{1; \rho > 0; 0; \rho \leq 0\}$, $f(\rho) = \lambda\rho(\rho^* - \rho)$, $0 \leq \rho \leq \rho^*$.
It is clear, that if $q > \min(q_{\max}, q^*)$, than the flow is critical and

$$T_*(\rho(0)) \leq \frac{l(\rho^* - \rho)}{q - \min(q_{\max}, q^*)}, \tag{10}$$

where $q^* = \frac{\lambda(\rho^*)^2}{4}$ is highway capability of the edge.
Then, let us assume that $q \leq \min(q_{\max}, q^*)$. At the beginning $q^* \leq q_{\max}$. Then
$\min(q_{\max}, f(\rho)) = f(\rho)$, and equation (9) reads $l\frac{d\rho}{dt} = q - f(\rho)$.
It is obvious that the point $(\rho_1(q), q)$ (Fig. 8) is a stable stationary point, the
point $(\rho_2(q), q)$ is an unstable stationary point and $\rho(0) \in (\rho_2(q), \rho^*)$ is the
set of critical points. So

$$l\int_{\rho(0)}^{\rho^*} \frac{d\rho}{q - f(\rho)} = \int_0^{T_*(\rho(0))} dt = T_*\left(\rho(0)\right). \tag{11}$$

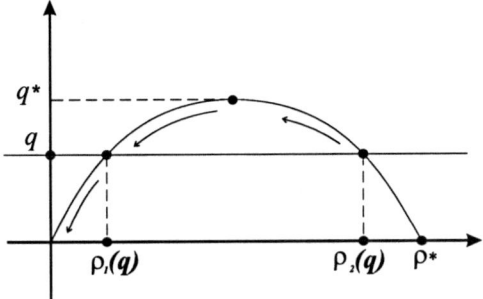

Fig. 8. $q \le q^* \le q_{max}$

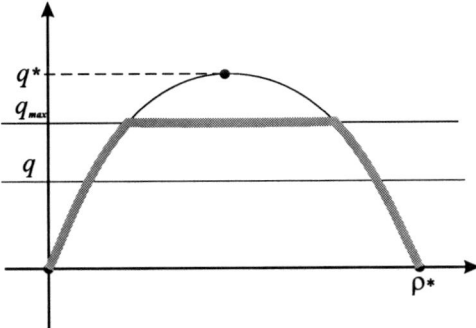

Fig. 9. $q \le q_{max} \le q^*$

Let us consider the remaining case $q_{max} < q^*$, i.e. $q < q_{max} < q^*$. The main qualitative characteristics of the flow's behaviour are similar to that of the previous case, with the only difference that instead of $f(\rho)$ (Fig. 8), $\min (f(\rho), q_{max})$ is used (Fig. 9).

4 Problems

4.1 Ring of n Sections

Let us consider unidirectional movement on a ring consisting of n sections ($n > 2$,) and l is the section length (Fig. 10).
The following statements are true:

- If $C > nC^*/2$, C^* is *capacity of a section, then the flow regime becomes critical in finite time for any initial conditions,* except for the case of stationary points.
- If $C < C^*$, *i.e. flow mass is less than a section capacity, then the movement is uncritical and converges to the stationary point.*
- However, if $C^* < C < nC^*/2$, *then qualitative characteristics of the flow depend on its initial conditions* $\bar{\rho}(0) = (\rho_1(0), \cdots, \rho_n(0))$, $l \sum_{i=1}^{n} \rho_i(0) = C.$

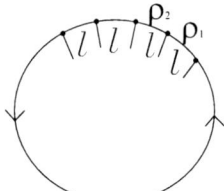

Fig. 10. Ring of n sections

For example, when $n = 2m$ and $\rho_{2i-1}(0) = \rho_1$, $\rho_{2i}(0) = \rho_2$, $i = 1, \cdots, m$ and the initial conditions are periodic, then the vector $\bar{\rho}(t)$ will be periodic due to its uniqueness at any point of time t.

Thus, *the flow on a ring with $2m$ sections will be equivalent to the flow on m rings of 2 sections.* Therefore at $l(\rho_1 + \rho_2) < C^*$ the movement is ∞-uncritical, $T_* = \infty$, but at $l(\rho_1 + \rho_2) > C^*$, except for stationary conditions, the movement is critical.

The following problems are to be explored: How to describe a flow in the common case n and at any initial conditions? What are sufficient conditions for $\bar{\rho}(0)$ at which the flow converts to the critical regime in a finite period of time?

4.2 Ring of 2 Sections with Control

Another generalization of a ring model of 2 sections is unidirectional movement on a ring of 2 sections with "traffic lights" (Fig. 11). In this model an alternation of 2 phases is considered. During the first phase, the movement of particles from one section to another is allowed, and during the second phase it is prohibited. The flow can be controlled by choosing the length of phases to prevent the transition to the critical regime.

What other stationary states can be generated and in what ways by using above described controls?

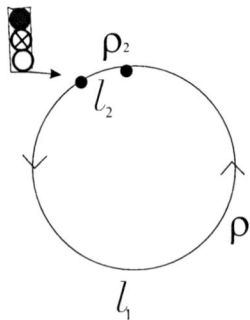

Fig. 11. Circular movement with control

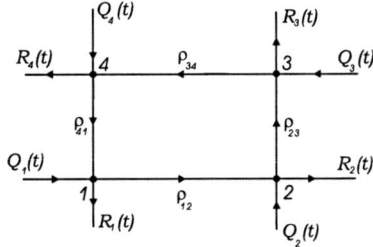

Fig. 12. Crossroads without mixing

4.3 Crossroads

Let us consider the model in Fig. 12. Assume that the edges and main diagrams of the graph are identical. The input flows have intensities $Q_i(t)$, $i = 1, 2, 3, 4$. In nodes 1-4 the Boolean control functions $U_i(t)$ are introduced ($i = 1, 2, 3, 4$) with $U_i(t) = 0$, if horizontal movement is allowed, and $U_i(t) = 1$ if vertical movement is allowed (Fig. 12).

The problem is in the description of control methods, ensuring uncritical regimes and fixed intensities of output flows $R_i(t)$, $i = 1, \ldots, 4$.

References

1. A.P. Buslaev, A.V. Novikov, V.M. Prikhodko, A.G. Tatashev, M.V. Yashina: Stochastical and Imitation Approch to Traffic Movement. (Mir, Moscow 2003)
2. V.N. Lukanin, A.P. Buslaev, A.V. Novikov, M.V. Yashina: Traffic Flows Modelling and the Evaluation of Energy-Ecological Parameters. Part I. Int. J. of Vehicle Design (2001)
3. V.N. Lukanin , A.P. Buslaev, A.V. Novikov, M.V. Yashina: Traffic Flows Modelling and the Evaluation of Energy-Ecological Parameters. Part II. Int. J. of Vehicle Design (2001)
4. I. Lubashevski, R. Mahnke, P. Wagner, S. Kalenkov: Phys. Rev. E **66**, 016117 (2002).
5. I. Lubashevsky, P. Wagner, R. Mahnke: Eur. Phys. J. B **32**, 243–247 (2003)
6. Yu.V. Pokorny, E.N. Povorotova, O.M. Penkin: On spectre of some vector boundary problems. Problems of the qualitative theory differential equations. ed. by V.M. Matrosov (Nauka, Novosibirsk 1988)
7. S. Nicaise: Some results on spectral theory over networks, applied to nerve impulse transmission. Lecture Notes in Math. **1171**, pp. 532–541 (Springer 1985)
8. A.P. Buslaev, A.G. Tatashev, M.V. Yashina: On properties of a class of systems of non-linear differential equations on graphs. Vladikavkaz Math. J. **4** (2004)
9. A.P. Buslaev, V.M. Prikhodko, A.G. Tatashev, M.V. Yashina: Deterministic-stochastic flow model. (2005) ArXiv.org/0504139
10. A.P. Buslaev, A.G. Tatashev, M.V. Yashina: Traffic flow stochastic model 2*2 with discrete set of states and continuous time. (2004) ArXiv.org/0405471

Laboratory Experiments with Nagel-Schreckenberg Algorithm

Thorsten Chmura[1,2], Thomas Pitz[1,2], and Michael Schreckenberg[3]

[1] Laboratory for Experimental Economics, University of Bonn, 53113 Bonn, Germany
[2] Shanghai Jiao Tong University, Antai School of Managment, Shanghai 200052, People's Republic of China
[3] Department of Traffic and Transportation, University of Duisburg-Essen

Summary. A new software environment (NETSIM) is presented that can be used for interactive experimental studies concerning the route-choice behaviour of human actors in different scenarios. It is also possible to create scenarios in which human actors interact with software agents. Since the treatments in laboratory experiments are well controlled, the behaviour of subjects in situations of economically relevant decision-making can be analysed more thoroughly. We describe an experimental setup in which the Nagel Schreckenberg Algorithm for vehicle dynamics is used.

1 Introduction

The quality of traffic systems is a decisive factor in the wealth and economic growth of modern societies. In order to satisfy the need for mobility especially in areas with a high volume of traffic, current traffic problems have to be identified and solved without reducing the mobility and location quality of the economic area. A multitude of insights and methods of solution have already been presented by different scientific disciplines. Of particular interest is the development of intelligent traffic information systems [1,2,12]. However, such concepts for transport policy can only be successful if they are accepted by the traffic participants. It is not yet evident whether more information for the traffic participants positively influences the traffic flow [3]. Traffic participants who receive too much information tend to build simple heuristics [10]. Thus, over-reactions can emerge which cause additional fluctuations [3,4,19]. In order to gain a better understanding of such reactions, it is necessary to study the learning behaviour of traffic participants in more detail. The insights achieved in such a study should be used to analyse the effectiveness of technical solutions in advance and thus to avoid cost-intensive field experiments.

The understanding of the individual behaviour of traffic participants is essential for the development and optimisation of intelligent transport and traffic information systems. While these systems have in part achieved a high technical standard, the reactions of traffic participants in complex traffic networks are so far largely unexplored.

Experimental studies [7,8,16,17] concerning the route-choice behaviour of traffic participants in simple scenarios are already available. The behaviour model developed in these studies formed the basis of multi-agent systems which simulate the route-choice behaviour of traffic participants. It was possible to show in simple scenarios that the theory of intensifying learning known from technical literature [9] is in a slightly modified form suitable for the prediction of human behaviour.

Due to the initial successes achieved in this field, these investigations concerning route-choice behaviour are to be carried out in more complex and realistic scenarios. The compilation of data for the analysis of behavioural phenomena is not easy because the reactions of traffic participants to traffic news or even to new technologies can hardly be estimated and vary strongly between individuals. In order to cope with this problem, researchers are mainly concentrating on two methods at the moment. On the one hand, one attempts to determine the needs and wishes of car drivers with the help of extensive surveys. On the other hand, there is the concept to let driving simulators take route-choice decisions. One investigates the backgrounds of the route-choice taken with the help of virtual simulations. In the following, the methods and achieved results shall be shortly summarised and compared with approaches of Experimental Economic Research.

A simple and cost-saving method is to deduce the reactions of car drivers to a specific situation with the aid of *questionnaires*. Studies can be carried out with several hundreds of test subjects (e.g. [14]), which is a condition for representative results. Nevertheless, the evaluation of questionnaires is very time-consuming and in part error-prone. In addition to that, the participants are not necessarily confronted with a realistic picture of the situation. For instance, decisions are taken under a high pressure of time during the drive. This can hardly be simulated in a questionnaire situation.

Critics of the survey method developed *route-choice simulators* such as IGOR [5] and VLADIMIR [6] which physically represent a scenario and thus offer the test subjects a more realistic picture (e.g. [13,5,6]). The development of such a simulator is time-consuming and expensive. One advantage is that the results are directly electronically recorded, can be processed easily and are thus less prone to errors. However, its disadvantage is the lack of interaction with other traffic participants. This is completely neglected so that learning processes in a changing environment cannot be analysed.

Experimental Economics is an empirical discipline whose data basis is generated in fundamentally replicable experimental sessions. In these sessions, test subjects are confronted with an economically relevant situation, e.g. a negotiation situation, a simulated market or a modelled traffic situation. In order to

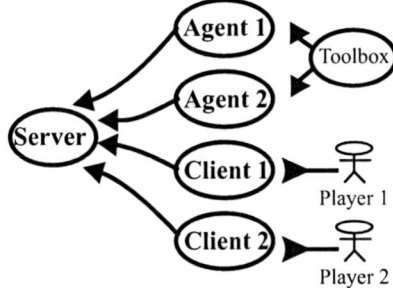

Fig. 1. Architecture of NETSIM, with 2 agents and 2 human players.

ensure appropriate incentive structures, the test subjects are paid in relation to their performance. Success does not only depend on one's own decisions but also decisively on the interaction with other participants.

Of particular importance is the analysis of learning behaviour in recurring similar decision-making situations. In the case of traffic sciences, one could for instance consider the daily drive to work during which the person in question can choose between two alternative routes. Several books about experiments concerning route-choice behaviour are already available [4,11]. Additionally, route-choice experiments in simple scenarios in which the test subjects were paid in a performance-related way were analysed in the context of a BMBF research project [16,17,18].

In order to analyse more complex and realistic scenarios, we have developed NETSIM as a software environment that can be used for interactive experimental studies of the route-choice scenario with (and without) the behaviour of human actors. The simulated traffic flow is calculated by a server. It is possible to create scenarios in which human actors interact with software agents. Since the treatments in laboratory experiments are well controlled, the behaviour of subjects in situations of economically relevant decision-making can be analysed more thoroughly. Figure 1 shows an example with two agents and two human players.

2 Simulation Model

This paper gives a report of an experimental setup with a microsimulation transition model for the cars on the roads. The traffic flow was simulated by a cellular automaton as described in the following subsections dealing with different models. The roads are represented by an array of length L. The velocity v of each car is an integer with $0 \leq v \leq v_{\max}$. The gap of a car is equal to the number of empty cells in front of it. In each period, each car moves according to the following rules [15]:

Definition (Nagel Schreckenberg Algorithm):

- **Acceleration:** If $v < v_{max}$ and the gap is less than $v + 1$, the velocity increases by 1 $[v \to v + 1]$.
- **Slow down:** If the gap g of a car is less than $v + 1$ $(v + 1 < g)$ the car slows down to $v' = g$. $[v \to g]$.
- **Random velocity change:** With probability p, the velocity v (if $v > 0$) decreases by 1 $[v \to v - 1]$.
- Each car moves v cells forward.

The Nagel-Schreckenberg model is used to describe emergent effects like traffic jams in traffic networks.

3 Experimental Setup

In this paper, we focus on the route-choice in a generic two route scenario, which has already been investigated previously (e.g. [11,16,17]).

Subjects are informed that in each of k trips they have to make a choice between a main road M and a side road S for travelling from A to B. If a car reaches B, the actor who sent this car receives the value of its travel time. After the subject has chosen a road, the car moves on that road from A to B. For the calculation of the travel time of the cars, we used the simulation algorithm described in section 2. The simulations produce emergent effects like traffic jams. The subject was given different information about the past and current traffic scenarios. Therefore, the reactions to this information could be tested in a well controlled environment.

We report an experimental setup with 18 subjects in each session. These subjects are told that S is longer than M. In the experiment, the length of road M was 60 and the length of road S was 90 cells. After a subject has got a car, he or she has a limited time of 20 sec. to come to a decision and send his or her car either on road M or road S. If the limited time passes without the subject's taking a decision, his or her car will automatically be sent to a by-pass with a substantially higher travel time.

Four experiments with 6 sessions were played. The travel time in the experiments was calculated by the Nagel Schreckenberg algorithm. The subjects receive the following information about each of their cars on their computer screens:

- Number of the car
- Chosen Road
- Travel Time
- Payoff (depends on the travel time)
- Only in experiment II: Average travel time of the last 5 cars (distributed over all players) which passed L

Technical Background:

In each session, one "simulated day" was played. A "simulated day" is divided in 5 periods which differ in the volume of traffic:

  I. Low volume of traffic L
  II. High volume of traffic H
  III. Low volume of traffic L
  IV. High volume of traffic H
  V. Low volume of traffic L.

One day with 24 hours is listed in the following table (Fig. 2). In this model, each of the 18 players got 17 cars per day.

Hour	1	2	3	4	5	6	7	8	9	10	11	12	13	14	15	16	17	18	19	20	21	22	23	24	Σ
Traffic	L	L	L	L	L	L	H	H	H	H	H	H	L	L	L	L	L	H	H	H	H	H	L	L	
cars/player	1		1		1		1	1	1	1	1	1		1		1	1	1	1	1	1		1		17

Fig. 2. Traffic volume per simulated day.

Algorithm: Allocation of the Cars

To ensure that each of the players got exactly 17 cars, the set of 18 players was divided in 2 groups G and H with 9 players each. Each of the players is chosen randomly and allocated a new car which has to be sent by the player to one of the two roads.

In Fig. 3 (Fig. 4), 20 (32) periods of a part with high (low) traffic are shown. In this part, each of the 18 players was activated exactly one time. One cell represents one period in the transition algorithm. In one period, each car on the two roads moves exactly one time according to the transition algorithm. In the programme, the length of one period was 4 sec.

The decision time of each player is shown by 5 horizontally connected cells with a grey margin. Therefore, in the experiments the length of the decision time of each player was 20 sec.

In cells marked with letter G (H), players from group G (or H respectively) are activated to take their decision. From period 1 (8) to period 13 (20), each player of group G (H) was randomly activated exactly one time.

If a player overruns the decision time, his car is automatically sent to a bypass with substantially higher travel time. If the player chooses a road during his decision time, his car is sent to a pool. If there are more than 5 cars in the pool, his car is also sent to the bypass. All the cars in the pool are sent fifo to the chosen road. The cars on the road move according to the transition algorithm to the end of the road. If a car reaches its destination, the player receives information about its travel time.

4 Experimental Results

Figure 5 shows a typical session with the Nagel Schreckenberg algorithm. The x-axis represents the periods when a car is entering the road. The length

Period	1	2	3	4	5	6	7	8	9	10	11	12	13	14	15	16	17	18	19	20
	G	G	G	G	G			H	H	H	H	H								
		G	G	G	G	G			H	H	H	H	H							
			G	G	G	G	G			H	H	H	H	H						
				G	G	G	G	G			H	H	H	H	H					
					G	G	G	G	G			H	H	H	H	H				
						G	G	G	G	G			H	H	H	H	H			
							G	G	G	G	G			H	H	H	H	H		
								G	G	G	G	G			H	H	H	H	H	
									G	G	G	G	G			H	H	H	H	H

Fig. 3. Section with high traffic

Period	1	2	3	4	5	6	7	8	9	10	11	12	13	14	15	16	17	18	19	20	21	22	23	24	25	26	27	28	29	30	31	32
	G	G	G	G								H	H	H	H	H																
		G	G	G	G	G							H	H	H	H	H															
			G	G	G	G	G							H	H	H	H	H														
				G	G	G	G	G							H	H	H	H	H													
					G	G	G	G	G							H	H	H	H	H												
						G	G	G	G	G							H	H	H	H	H											
							G	G	G	G	G							H	H	H	H	H									
								G	G	G	G	G								H	H	H	H	H							
									G	G	G	G	G									H	H	H	H	H	H					

Fig. 4. Section with low traffic

of the vertical bars below (above) is the travel time of a car on the longer side (shorter main) road. The bars above the period axis refer to cars on the shorter main road and the bars below refer to cars on the longer side road.

One can see in the table in Fig. 6 that in each session the players prefer to choose the shorter main road. In Fig. 5, one can see that in the first periods the travel time on the shorter main road is lower because there are only a few cars on the road. Due to an overreaction of the players, the average travel time on the shorter main road is higher than that on the side road.

In each experiment except for one, the travel time on M is higher than on S but in experiments with traffic information, the travel time on the main

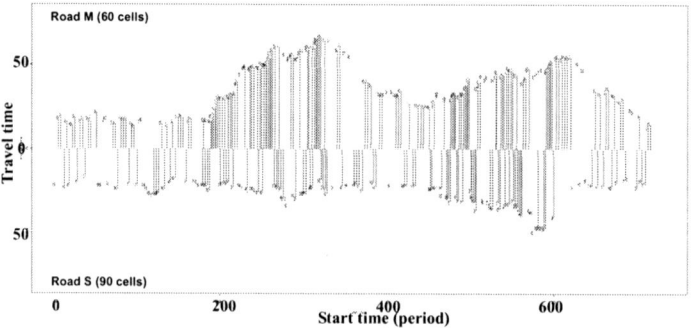

Fig. 5. Typical session in experiment I.

	Mean Travel Time on M	Mean Travel Time on S	Mean Travel Time on M and S	Travel Time M -Travel Time S	Number of Players on M	Number of Players on S	Number of Road Changes	Intervals of Road Entry time on M
tr I1	61.29	24.15	47.68	37.14	196	110	7.56	3.88
tr I2	42.65	28.18	36.51	14.47	177	129	9.00	4.13
tr I3	37.77	26.49	32.94	11.28	175	131	8.94	4.22
tr I4	45.09	23.18	36.64	21.91	188	118	7.83	3.99
tr I5	54.10	22.81	40.98	31.29	179	127	7.83	4.02
tr I6	37.50	34.87	36.18	2.63	152	154	8.61	4.28
Ex I	46.40	26.61	38.49	19.79	177.83	128.17	8.33	4.09
tr II1	38.75	25.27	32.83	13.48	172	134	8.67	4.21
tr II2	33.59	23.94	29.65	9.65	182	124	7.61	4.37
tr II3	34.79	23.82	30.81	10.97	196	110	7.89	4.36
tr II4	30.56	26.39	28.59	4.17	162	144	9.06	4.44
tr II5	41.76	29.83	36.78	11.93	179	127	8.11	4.13
tr II6	28.28	32.12	30.19	-3.84	155	151	8.27	4.47
Ex II	34.62	26.90	31.475	7.73	174.33	131.67	8.33	4.33

Fig. 6. Experiment I and II with Nagel Schreckenberg Algorithm.

road is even significantly higher. The null-hypothesis could be rejected by a Wilcoxon-Mann-Whitney-Test on the significance level of 5% (one-sided). One can still find this relation when one compares the mean travel time on both roads. On the side roads, we cannot verify a significantly differing travel time. The first assumption was that in experiments with information, players chose the main road less often. But as you can see, the players preferred the main road also in experiments with information. In fact, we could not verify a significant difference between the frequencies of the roads chosen.

However, we notice a small but significant difference in the mean intervals of the road entry times of the cars on M. The higher mean interval of road entry times on M in Experiment II causes the distribution of the cars on the main road to be more effective.

We find the contrary relation on the side road. Nevertheless, the road entry time is not low enough to significantly increase the travel time on S. One can also exemplify the relation between the mean travel time on M and the mean interval of the road entry time by the following scatter diagram (Fig. 7).

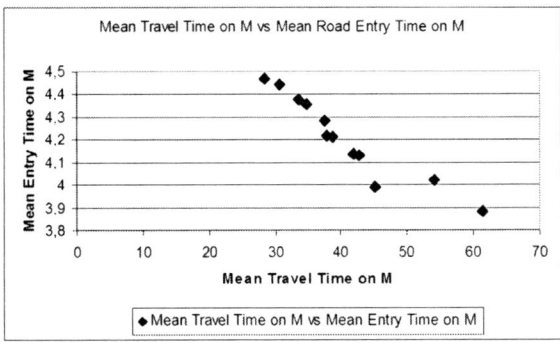

Fig. 7. Scatter diagram: Travel time vs. road entry.

5 Conclusion

We have described an experimental setup that can be used for interactive studies of the route-choice behaviour of human actors in different scenarios. We have shown that with this software, it is also possible to create scenarios in which human actors interact with software agents. Since the treatments in laboratory experiments are well controlled, the behaviour of subjects in situations of economically relevant decision-making can be analysed more thoroughly. In the future, we will run the laboratory experiments with the two transition models in which human participants will play against agents.

References

1. Adler, J.L. and V.J. Blue: Transp. Res. **C 6**, 157-172 (1998).
2. Barfield, W. and T.A. Dingus: *Human Factors in Intelligent Transportation Systems.* Lawrence Erlbaum Associates Inc., Mahwah, New Jersey (1998).
3. Ben-Akiva, M., A. de Palma, and I. Kaysi: Transp. Res. **A 25**, 251-266 (1991).
4. Bonsall, P. W. and T. Parry: In: Proc. of 18th Summer Annual Meeting of PTRC, Seminar H, pp. 113-124, PTRC, London, (1990).
5. Bonsall, P.: Transportation **19**, 1-23 (1992).
6. Bonsall, P.W., P.E. Firmin, M.E. Anderson, I.A. Palmer, P.J. Balmforth: Proceedings 4th International Conference on Survey Methods in Transport, Steeple Aston, Oxford, 170-191 (1996)
7. Chmura, T., T. Pitz, M. Schreckenberg: in *Traffic and Granular Flow '03*, S.P. Hoogendoorn, S. Luding, P.H.L. Bovy, M. Schreckenberg, and D.E. Wolf (Eds.), (Springer, Heidelberg, 2005).
8. Chmura, T., T. Pitz, M. Schreckenberg: in *Traffic and Granular Flow '03*, S.P. Hoogendoorn, S. Luding, P.H.L. Bovy, M. Schreckenberg, and D.E. Wolf (Eds.), (Springer, Heidelberg, 2005).
9. Erev, I. and A. E. Roth: American Economic Review **88**(4), 848 (1998).
10. Gigerenzer, G., P.M. Todd and ABC Research Group (eds.): *Simple heuristics that make us smart*, New York: Oxford University Press (1999).
11. ITSA, "What is ITS?", ITS America On-line Document (1998)
12. Iida, Y., T. Akiyama, and T. Uchida. Transp. Res. **B 26**, 17-32 (1992).
13. Koutsopoulos, H.N., A. Polydoropoulou, M. Ben-Akiva: Transp. Res. **C 3**, 143-159 (1995).
14. LISB: Leit- und Informationssystem Berlin. Schlußbericht, Berlin (1991).
15. Nagel, K., and M. Schreckenberg: J. Physique **I 2**, 2221-2229 (1992).
16. Selten, R., M. Schreckenberg, T. Chmura, T. Pitz, J. Wahle: In: *Traffic and Granular Flow '01*, M. Fukui, Y. Sugiyama, D. E. Wolf (Eds.), (Springer, Heidelberg), 325-331 (2002)
17. Selten, R., M. Schreckenberg, T. Chmura, T. Pitz, S. Kube: Experiments and Simulations on Day-to-Day Route Choice Behaviour. CESIFO Working Paper No. 900, Munich (2003).
18. Selten, R., M. Schreckenberg, T. Chmura, T. Pitz: in: R. Selten, M. Schreckenberg (Eds.), *Traffic and Human Behaviour*, 1-21. (Springer, Heidelberg, 2004).
19. Wahle, J., A.L. Bazzan, F. Klügl, M. Schreckenberg: Physica **A 287**, 669 (2000).

Part V

Traffic Flow: Theory

Phase Transitions in Stochastic Models of Flow

Martin R. Evans

SUPA, School of Physics, The University of Edinburgh,
Mayfield Road, Edinburgh EH9 3JZ, Scotland

Summary. In this talk I will review some very simple models of nonequilibrium systems known as the 'Asymmetric Exclusion Process' and the 'Zero-Range Process'. These involve particles hopping stochastically on a lattice and thus form stochastic models of flow. Systems driven out of equilibrium can often exhibit behaviour not seen in systems in thermal equilibrium - for example phase transitions in one-dimensional systems. I shall show how examples of such transitions may be interpreted as jamming transitions in the context of traffic flow. More generally I shall discuss other instances of the condensation transition which is the phenomenon of a finite fraction of the driven conserved quantity condensing into a small spatial region. Criteria for the occurrence of condensation may be formulated and the detailed properties of the condensate such as its fluctuations have recently been elucidated.

1 Introduction

In this talk I shall discuss some very simple models of stochastic flow of of particles, which can be thought of as representing granular flow or vehicular traffic or even as biophysical entities such as molecular motors. These models have been studied from a theoretical viewpoint with the aim of elucidating the properties of nonequilibrium systems, which I shall discuss further below. However it is encouraging to note that the models have more recently been applied in a number of contexts reported in this conference (see for example the talk of D. Chowdhury [1] at this meeting).

First let me review what is meant by a nonequilibrium system by contrasting with the idealisation of an equilibrium system in which: i) a system has reached a state in which its properties are stationary in time ii) the system is in equilibrium with its environment with respect to exchange of energy or particles or volume. These conditions are rarely met and a system will be nonequilibrium by virtue of: i) not yet having stationary properties and relaxing towards thermal equilibrium ii) being stationary but being held away from thermal equilibrium. In the latter case the system is driven by its environment rather than being in equilibrium with it. The steady state of the system will usually not be described by Gibbs-Boltzmann statistical weights, rather it will be a nonequilibrium steady state.

A well studied class of systems with nonequilibrium steady states are those with a conserved quantity driven through the system. The presence of a current within the steady state ensures that detailed balance is not satisfied. Such systems are known as driven diffusive systems [2, 3].

In order to illustrate how phase separation can trivially occur even in one-dimensional nonequilibrium systems let us consider a very simple example which serves to illustrate the class of models to be discussed in the rest of this paper. The significance is that separation is precluded in one-dimensional equilibrium systems under quite general conditions.

In this example studied in [4] particles hop forward stochastically on a one-dimensional lattice with periodic boundary conditions. An exclusion interaction implies that no two particles can occupy the same site. The hopping rates between neughbouring sites is unity except for one 'defect bond' where the hop rate is $r < 1$—see Fig. 1. This bond could be thought of as an obstruction or bottleneck in a traffic flow scenario. For low enough global density of particles one finds a traffic jam or high density region of particle behind the defect and a low density region in front of the defect. Thus very simply we see that the defect can induce phase separation into two macroscopic regions of different density in this driven system. Also note that further around the lattice the high density and low density regions must meet and at this point there will be shock which is a sharp change in density over a microscopic distance. A related system would be where there is one defect particle which hops more slowly than the rest. Again, for low enough global density of particles one expects a traffic jam behind the slow particle, hence phase separation. As we shall see in the next section this mobile-defect system has a steady state that one can solve exactly.

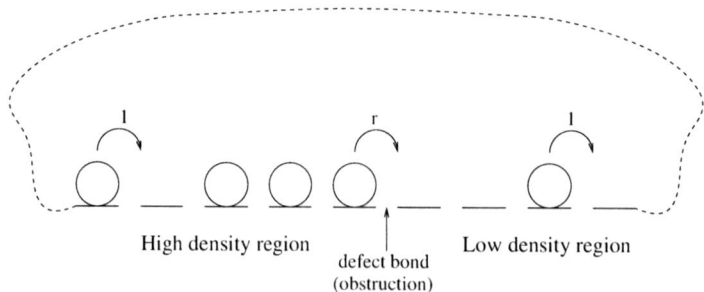

High density region   Low density region

defect bond
(obstruction)

Fig. 1. A simple driven system illustrating phase separation into high density and low density region. Periodic boundary conditions by the dashed region are indicated and one would find a 'shock' where the high density region meets the low density region further around the lattice in the dashed region.

1.1 Types of Stochastic Dynamics

It is useful to make a brief digression to discuss the different types of stochastic dynamics and how one implements these in a stochastic simulation [5].
Let us consider a model where a particle may hop forward stochastically if the site in front is empty

$$\text{in} \quad \Delta T \qquad \text{with probability} \quad p\Delta T \qquad (1)$$

A system of such hopping particles with at most one particle per site is known as an Exclusion Process. The continuous time limit is $\Delta T \rightarrow 0$ then p becomes the hopping *rate* of a particle. In this limit at most one hopping event occurs in each time ΔT. Note that the rate can be greater than one. The implementation of continuous time dynamics can be done through a random sequential algorithm. Here a particle is picked at random and then it hops forward (if an empty site is available) with relative probability p_{rel}. Let us illustrate what is meant by relative probability by considering a more complicated situation where there is an additional process which may occur as well as hopping forward with rate p. To be specific let us consider annihilation of the particle with rate a. Then to simulate both the processes one would choose the ratio of the relative probabilities for the two events to be $p_{rel}/a_{rel} = p/a$ and the sum of the two relative probabilities to be unity $p_{rel} + a_{rel} = 1$. The latter condition ensures that an event occurs when the particle is randomly selected, thus economising on random numbers. The time between updates corresponds (on average) to $\Delta T = \frac{p_{rel}}{p} \frac{1}{M}$ where M is the total number of particles.
An alternative implementation of continuous time dynamics is to use a random number to generate the time to the next event. This is particlularly convenient in more complicated systems when certain rates which effectively control the system become very small. Such an algorithm is variously known as 'continuous time Monte Carlo' [5], BKL algorithm [6], Gillespie algorithm [7] or even 'kinetic Monte Carlo'.
On the other hand in many simulations of granular material and traffic flows discrete time is favoured. In this case many events can happen in the same update e.g. many particles can hop forward simultaneously. Setting the time between updates as $\Delta T = 1$ then implies that $p \leq 1$. The case $p = 1$ corresponds to a deterministic limit where particles hop forward if an empty site is available. In this limit features not supported in the stochastic case may appear. For example in an exclusion process under parallel dynamics in the deterministic limit and for low enough density, a configuration of particles where each particle has an empty site ahead will be an absorbing state.

2 The Zero-Range Process

The Zero-Range Process (ZRP) was introduced some years ago by Spitzer [8] and recent interest has been reviewed in [9, 10]. In this section we define the model and present results for the steady state (more detail can be found in [9]). In later sections I will review more recent results obtained in collaboration with Satya Majumdar and Royce Zia.

To begin with we consider a one-dimensional lattice of N sites with sites labelled $i = 1 \ldots N$ and periodic boundary conditions (site $N+1=$ site 1). Each site can hold an integer number of indistinguishable particles. The configuration of the system is specified by the occupation numbers m_i of each site i. The total number of particles is denoted by M and is conserved under the dynamics. The dynamics of the system is given by the rates at which a particle leaves a site i (one can think of it as the topmost particle—see Figure 2a) and moves to the left nearest neighbour site $i-1$. The hopping rates $u(m)$ are a function of m the number of particles at the site of departure. Some particular cases are: if $u(m) = m$ then the dynamics of each particle is independent of the others; if $u(m) = $ const   for $m > 0$ then the rate at which a particle leaves a site is unaffected by the number of particles at the site (as long as it is greater than zero).

As illustrated in Figure 2 there exists an exact mapping from a ZRP to an asymmetric exclusion process, which as discussed above is a driven system where there is at most one particle per site. The idea is to consider the particles of the ZRP as the holes (empty sites) of the exclusion process. Then the sites of the ZRP become the moving particles of the exclusion process. Note that in the corresponding exclusion process we have M particles hopping on a lattice of $M + N$ sites. A hopping rate in the ZRP, $u(m)$, which is dependent on m corresponds to a hopping rate in the exclusion process which depends on the length of the gap to the particle in front. So the particles can feel each other's presence and one can have a long-range interaction.

The important attribute of the ZRP is that it has a *factorised steady state*. By this it is meant that the steady state probability $P(\{m_i\})$ of finding the

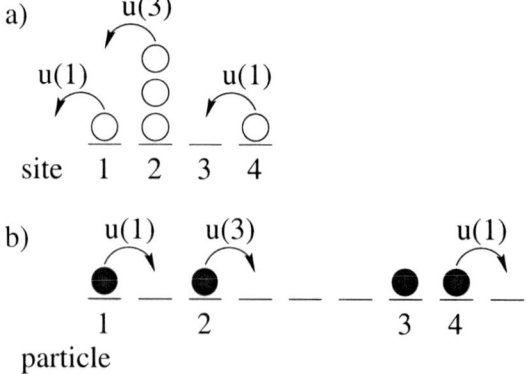

Fig. 2. Mapping between the zero-range process and the asymmetric exclusion process

system in configuration $\{m_1, m_2 \ldots m_N\}$ is given by a product of factors $f(m_i)$ along with a constraint which fixes $\sum_i m_i = M$

$$P(\{m_i\}) = \frac{1}{Z(M,N)} \prod_{i=1}^{N} f(m_i)\delta(\sum_i m_i - M) . \tag{2}$$

Here the normalisation $Z(M, N)$ is introduced so that the sum of the probabilities for all configurations, with the correct number of particles M, is one:

$$Z(M, N) = \sum_{m_1, m_2 \ldots m_N} \delta(\sum_i m_i - M) \prod_{i=1}^{N} f(m_i) \tag{3}$$

The normalisation may usefully be considered as the analogue of a canonical partition function of a thermodynamic system [11]. One may conveniently compute (3) numerically by using the recursion

$$Z(M, N) = \sum_{m=0}^{M} f(m)Z(M - m, N - 1) . \tag{4}$$

It is important to realise that due to the constraint of fixed particle number the single-site weight $f(m)$ is not the same as the single-site mass probability distribution $p(m)$ which would be calculated as

$$p(m) = \frac{f(m)Z(M - m, N - 1)}{Z(M, N)} . \tag{5}$$

In other words, although the steady state factorises, the constraint of fixed particle number still induces correlations between sites.

In the basic model described above, $f(m)$ is given by

$$f(m) = \prod_{n=1}^{m} \frac{1}{u(n)} \quad \text{for} \quad m \geq 1$$
$$= 1 \quad \text{for} \quad m = 0 \tag{6}$$

To prove (2,6) one simply considers the stationarity condition on the probability of a configuration (probability current out of the configuration due to hops is equal to probability current into the configuration due to hops):

$$\sum_i \theta(m_i)u(m_i)P(m_1 \ldots m_i \ldots m_N) =$$

$$\sum_i \theta(m_i)u(m_{i+1}+1)P(m_1 \ldots m_i-1, m_{i+1}+1 \ldots m_N) . \tag{7}$$

The step function $\theta(m_i)$ highlights that it is the sites with $m > 1$ that allow exit from the configuration (lhs of (7)) but also allow entry to the configuration (rhs of (7)). It is straightforward to show that (7) is satsified when the steady state is of the form (2) [9].

We can easily generalise to consider an heterogeneous system by which we mean that the hopping rates are site dependent: the hopping rate out of site i when it contains m_i particles is $u_i(m_i)$. It is easy to check that the steady state still factorises and the single-site weights are simply modified to

$$f_i(m) = \prod_{n=1}^{m} \frac{1}{u_i(n)} \quad \text{for} \quad m \geq 1 . \tag{8}$$

We now return to the motivation for studying the ZRP. Firstly there exist some exact mappings of particular nonequilibrium systems onto the ZRP. Examples include the repton model of polymer dynamics under periodic boundary conditions [12]; the drop-push model for the dynamics of a fluid moving through backbends in a porous medium [13]; clustering in shaken granular gases (which furnishes a pleasing experimental example of condensation phenomemon to be discussed in section 3); the exchange of monomers between protein filaments [14]; cluster-cluster aggregation in surface growth [19].

More generally, however, one may think of the sites of the ZRP as representing domains of some driven system—this is most natural within the exclusion process interpretation of the ZRP (Figure 1). The domains may have some internal structure, for example further degrees of freedom, but this is all integrated out, and one is left with an effective dynamics of exchange of length between domains. An early example of this use of the ZRP was in the context of the Bus Route Model [17]. By using the ZRP as the effective description, a general criterion for phase separation in one-dimensional driven systems has been developed [15, 16]. Within this description phase separation is manifested by the emergence of one large domain and this corresponds to the phenomenon of condensation in the ZRP which we now discuss.

3 Condensation Transitions

A class of phase transitions which may occur in models such as the ZRP is that involving spatial condensation, whereby a finite fraction of the constituent particles condenses onto the same site [17–19]. Of particular interest is the fact that transitions may occur in one-dimensional systems and that for a factorised steady state one may analyse the transition exactly.

To analyse the condensation transitions which may occur it is simplest to use the grand canonical distribution where $p_i(n)$ is approximated by

$$p_i \propto z^n f_i(n) \tag{9}$$

and the fugacity z is fixed by the equation for the average density $\rho = M/N$

$$\rho = \frac{1}{N} \sum_i \langle m_i \rangle \quad \text{where} \quad \langle m_i \rangle = \sum_n m p_i(n) \tag{10}$$

In the thermodynamic limit

$$N \to \infty \quad \text{with} \quad M = \rho N \,, \tag{11}$$

where the density ρ is held fixed, one expects the grand canonical distribution to be exact if one can solve for z. Thus the condensation mechanism reduces to the question of whether one can satisfy (10). Although there are some subtle differences between the heterogeneous and homogeneous cases to be described below, the basic mechanism is as follows. First note that each $\langle m_i \rangle$ is a monotonically increasing function of z. Thus as ρ increases the required value of z increases. However there is a maximum value that z can take so that $\langle m_i \rangle$ converges. If at the maximum value of z (10) takes a finite value $\rho_c = \rho(z_{\max})$, then for $\rho > \rho_c$ (10) cannot be satisfied and we have condensation. We expect the excess number of particles $N(\rho - \rho_c)$ to condense onto a single site.

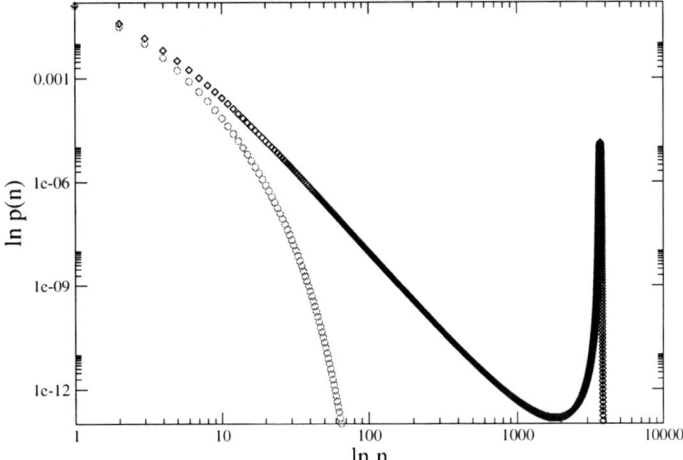

Fig. 3. ln-ln plot of the single-site distribution $p(n)$ vs. particle number n. The data are obtained by iterating the recursion relation (4) for $Z_{L,N}$ for $L = 1000$ and $b = 5$. The circles represent $\rho = 1/4$ where the system is in the fluid phase; the diamonds represent $\rho = 4$ where the system is in the condensed phase.

3.1 Heterogeneous Case

To give an idea of how a condensation transition may occur we consider the case $u_i(m) = u_i$ for $m > 0$ *i.e.* the hopping rate does not depend on the number of particles at a site. In this case f_i is given by

$$f_i(m) = \left(\frac{1}{u_i} \right)^m \,. \tag{12}$$

The mapping to an ideal Bose gas is evident: the M particles of the ZRP are viewed as Bosons which may reside in N states with energies E_i determined by the site hopping rates: $\exp(-\beta E_i) = 1/u_i$. Thus the ground state corresponds to the site with the lowest hopping rate. We can sum a geometric series to calculate $\langle m_i \rangle = z/(u_i - z)$ then taking the large N limit allows the sum over i to be written as an integral

$$\rho = \int_{u_{\min}}^{\infty} du P(u) \frac{z}{u - z} \tag{13}$$

where $P(u)$ is the probability distribution of site hopping rates with $u_{\min}$ the lowest possible site hopping rate. The maximum allowed value of z is then $u_{\min}$. Interpreting $P(u)$ as a density of states, equation (13) corresponds to the condition that in the grand canonical ensemble of an ideal Bose gas the number of Bosons per state is ρ. The theory of Bose condensation tells us that when certain conditions on the density of low energy states pertain we can have a condensation transition. Then (10) can no longer be satisfied and we have a condensation of particles into the ground state, which is here the site with the slowest hopping rate.

A very simple example is to have just one 'slow site' i.e. $u_1 = p < 1$ while the other $N - 1$ sites have hopping rates $u_i = 1$ when $i > 1$. Using the mapping to an exclusion process (see Fig. 2), this corresponds to one slow particle. One can show [20, 21] that for a high density of particles in the ZRP (low density of particles in the corresponding asymmetric exclusion process) we have a condensate since site 1 contains a finite fraction of the particles. In the low density phase the particles are evenly spread between all sites.

A simple interpretation of heterogeneous condensation in a traffic flow context is to think of the slow particle of the exclusion process as a slow vehicle such as a tractor or agricultural vehicle. Then, in a city situation where the density of vehicles is high the fact that the vehicle is slow does not limit its speed rather it is the high density of traffic. However, with a lower overall density of traffic e.g. on a country road the slow vehicle will limit the flow of the traffic and a traffic jam will be generated behind the slow vehicle and empty road ahead. This corresponds to the condensed phase.

3.2 Homogeneous Case

We now consider the homogeneous ZRP where the hopping rates $u(m)$ are site independent. Then $\langle m_i \rangle$ is independent of i and (10) reads

$$\rho = \langle m \rangle \propto \sum_m m z^m f(m) \tag{14}$$

Let us take for example $u(m) = \beta(1+\gamma/m)$ so that the hopping rate decreases to an asymptotic value β as the mass at the site increases. In this case one can show from (6) that $f(m)$ behaves for large m as

$$f(m) \sim \beta^{-m} m^{-\gamma} \tag{15}$$

and the maximum allowed value of z is $z_{\max} = \beta$.

Then to have condensation we require that (14) is finite for $z = \beta$ which implies that $\sum_m m^{1-\gamma}$ is convergent and therefore $\gamma > 2$. Thus the condition for condensation is simply that $u(m)$ decays to β more slowly than $\beta(1 + 2/m)$. Note that in the homogeneous system the particle excess $N(\rho - \rho_c)$ condenses onto a spontaneously selected site. Therefore there is a spontaneoous symmetry breaking associated with the transition. The condensation mechanism is also different from the heterogeneous (or Bose-Einstein) case in that one cannot access the condensed phase by taking $z \nearrow z_{\max}$ in a system size dependent way.

A simple interpretation of homogeneous condensation in a traffic flow context is to think of the particles of the exclusion process as buses and the sites as bus stops [17]. Then, the movement of a bus between bus stops is slower when the distance to the next bus ahead is large simply because there will be more passengers waiting at the bus stop. this corresponds to $u(n)$ decreasing with n. Condensation then corresponds to the buses clustering together into a large jam with stops empty of buses (and full of waiting passengers) lying ahead.

4 General Mass Transport Model

So far we have considered the ZRP and taken advantage of the fact that it has a factorised steady state. Actually there are some other models known to have factorised steady states for example the Asymmetric Random Average Process (ARAP) [22–24] which has been used a simple model for force propagation in granular media and traffic flow. In that case sites contain a quantity of mass that is a continuous rather than discrete variable and the dynamics is a discrete time updating where

$$m_j(T + 1) = (1 - r_j(T))m_j(T) + r_{j-1}(T)m_{j-1}(T) \tag{16}$$

Here $r_j(T)$ is a random number between 0 and 1. Thus at an update a random fraction $r_j(T)$ of the mass at site j is transferred to the neighbouring site ahead. The steady state factorises if the distribution of r has the form of a Beta distribution $\sim r^p(1 - r)^q$.

A natural question is under what conditions does a model obtain a factorised steady state? To answer this we will consider a general mass transport model where the mass may be continuous and also time is discrete (the cases of discrete mass and continuous time may be considered as special limits) [25]. Thus we consider a one-dimensional lattice of N sites with periodic boundary conditions: associated with each site is a mass m_i which is most generally a continuous variable. The total mass is given by $M = \sum_{i=1}^{N} m_i$. The dynamics is as follows: in each time-step, at each site i, mass μ_i drawn from a distribution

$\phi_i(\mu_i|m_i)$ 'chips off' the mass m_i, and moves to site $i - 1$. Thus the master equation reads

$$P_{T+1}(\{m\}) =$$

$$\prod_{i=1}^{N} \int_0^\infty \mathrm{d}m_i' \int_0^{m_i'} \mathrm{d}\mu_i \ \phi_i(\mu_i|m_i') \prod_{j=1}^{N} \delta(m_j - m_j' + \mu_j - \mu_{j+1}) \, P_T(\{m'\}) \quad (17)$$

We show in [25] that a necessary and sufficient condition for the steady state to factorise is

$$\phi_i(\mu|m) = \frac{v(\mu) \, w_i(m - \mu)}{[v * w_i](m)} \,, \quad (18)$$

in which case the single-site steady-state weights are given by

$$f_i(m) = [v * w_i](m) \equiv \int_0^m \mathrm{d}\mu \, v(\mu) \, w_i(m - \mu) \,. \quad (19)$$

Here v and w_i are arbitrary functions but v must be the same for each site. Note that $\phi_i(\mu|m)$ factorises into a function of the mass which moves, $v(\mu)$, and a function $w_i(\sigma)$ of the mass which stays, $\sigma = m - \mu$.

Let us stress that this simple condition (18) determines a very general class of mass transport models with factorised steady states. This class encompasses both continuous and discrete mass, as well as parallel and random sequential dynamics. This approach provides a unified perspective of all previously known models and includes the ZRP and the ARAP as special cases [26]. Although (18) is appealingly simple it is not an explicit test in that given a particular $\phi(\mu|m)$ it may not be obvious if it is of the form (18). Happily a simple explicit test for factorisation for any given $\phi(\mu|m)$ can easily be obtained from (18) as explained in [26].

5 The Nature of the Condensate

In section 3.2 for homogeneous condensation we noted that the grand canonical distribution can only describe the system in the fluid (non-condensed) phase. In order to analyse the condensate (corresponding to the 'bump' in $p(n)$ in Figure 3) one needs to calculate $Z(M, N)$ within the canonical ensemble rather than grand canonical i.e. to evaluate (3). A full analysis has recently been accomplished [27, 28] by writing (3) as a contour integral and evaluating the integral in the large N limit. Furthermore for some special cases an exact closed form has been obtained for all system sizes [27, 28]. Some properties of $p(n)$ in the canonical ensemble have also been calculated in [29].

Here we will present a simple picture where the shape and fluctuations of the condensate bump can be predicted from the properties of sums of random variables.

Consider a system with continuous mass for which

$$Z(M,N) = \int_0^\infty dm_1 \int_0^\infty dm_2 \ldots \int_0^\infty dm_N f(m_1)f(m_2)\ldots f(m_L)\delta(\sum_i m_i - M)$$

(20)

Without loss of generality we can let

$$\int_0^\infty dm f(m) = 1$$

(21)

which allows us to interpret $f(m)$ as the probability distribution for the jumps of a random walker taking only positive steps. Then $Z(M,N)$ is simply the probability that a walker taking positive jumps with distribution $f(m)$ reaches M after N jumps.

Now let us assume

$$\int_0^\infty dm\, mf(m) = \mu_1 < \infty \quad \text{i.e.} \quad f(m) \sim m^{-\gamma} \quad (\gamma > 2)$$

(22)

Then if $N\mu_1 > M$ the jumps are typically all $O(1)$ whereas if $N\mu_1 < M$ then (20) will be dominated by trajectories of the walker where one jump is $O(M - N\mu_1)$. This single extensive jump corresponds to the condensate Further one can calculate the condensate fluctuations which are given by the fluctuation of the large jump. Denoting the large jump by m_{cond} one deduces that if $\gamma > 3$, $\overline{\Delta m_{\text{cond}}} \sim N^{1/2}$ whereas if $3 > \gamma > 2$ $\overline{\Delta m_{\text{cond}}} \sim N^{1/(\gamma-1)}$. Therefore for $\gamma > 3$ we have gaussian fluctuations and a normal condensate whereas for $3 > \gamma > 2$ we have anomalous fluctuations.

A schematic phase diagram in the ρ–γ plane showing the fluid phase, anomalous condensate and normal condensate is illustrated in Fig. 4.

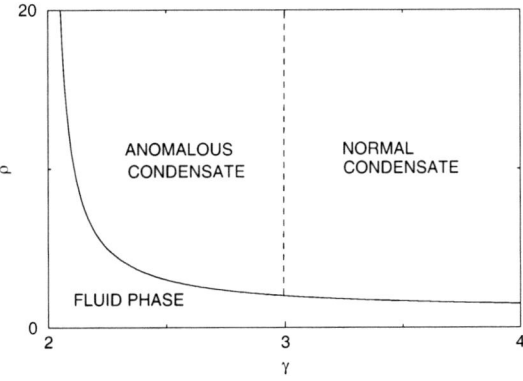

Fig. 4. Schematic phase diagram in the ρ–γ plane. The full line represents the critical density $\rho_c(\gamma)$.

6 Further Developments in Theory of Condensation

As a conclusion let us mention some recent developments in the study of the ZRP. Firstly, one can generalise the ZRP to contain two or more species of particle whose hop rates are function of the number of particles of each species at the departure site. Under certain conditions on these hop rates a factorised steady state is obtained [30, 31]. The multispecies system allows new condensation mechanisms whereby one species can induce condensation in the other [32].

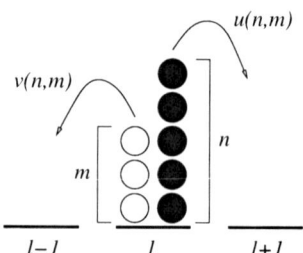

Fig. 5. A two species ZRP with hop rates $v(m, n)$ to the left for one species and $u(m, n)$ to the right for the other.

Secondly, the dynamics of condensation has been of considerable interest. Starting from some random initial condition a coarsening process ensues from which a single condensate ultimately emerges. A variety of approaches have been used to study this process both for heterogeneous condensation [21], [33] and homogeneous condensation [34–36]. Also in the case of homogeneous condensation the 'flip time' for the condensate to dissolve and reform on another site has been calculated in [29].

Finally let us mention that since the ZRP is used to model systems of 'mesoscopic size' it would be of interest to have a detailed and general analysis of the finite size effects associated with the condensation mechanisms.

Acknowledgements

It is a pleasure to thank the following colleagues with whom some of the work described in this talk has been carried out Mike Cates, Tom Hanney, Yariv Kafri, Erel Levine, Satya Majumdar, PK Mohanty, David Mukamel, Owen O'Loan and Royce Zia.

References

1. D. Chowdhury, A. Schadschneider, and K. Nishinari, these proceedings
2. S. Katz, J.L. Lebowitz and H. Spohn, *Phys. Rev. B* **28**, 1655 (1983); *J. Stat. Phys* **34**, 497 (1984)
3. B. Schmittmann and R.K.P. Zia, *Statistical Mechanics of Driven Diffusive Systems* vol. 17 of Domb and Lebowitz series, Academic Press, U.K. (1995)
4. S.A. Janowsky and J.L. Lebowitz, *Physical Review A* **45**, 618 (1992)
5. M.E.J. Newman and G.T. Barkema, *Monte Carlo Methods in Statistical Physics*, Oxford University Press, U.K., (1999)
6. A.B. Bortz, M.H. Kalos and J.L. Lebowitz, *J. Comp. Phys* **17**, 10 (1975)
7. D.T. Gillespie, J. Phys. Chem. **81**, 2340 (1977)
8. F. Spitzer, *Advances in Math.* **5** 246 (1970)
9. M.R. Evans and T. Hanney, *J. Phys. A: Math. Gen.* **38**, R195-R239 (2005)
10. M.R. Evans, *Pramana* **64** 859 (2005)
11. M.R. Evans and R.A. Blythe, *Physica A* **313** 110 (2002); R. A. Blythe, *PhD Thesis* University of Edinburgh, U.K. (2001)
12. J.M.J van Leeuwen and A. Kooiman, *Physica* **A 184**, 79 (1992)
13. M. Barma and R. Ramaswamy in *Non-linearity and Breakdown in Soft Condensed Matter*, edited by K.K. Bardhan, B. K. Chakrabarti and A. Hansen (Springer, Berlin 1993) p.309
14. D. Biron and E. Moses, Biophys. J. **86**, 3284 (2004)
15. Y. Kafri, E. Levine, D. Mukamel, G.M. Schütz and J. Török, *Phys. Rev. Lett.* **89**, 035702 (2002)
16. M.R. Evans, E. Levine, P.K. Mohanty and D. Mukamel, *Euro. Phys. J. B* **41**, 223 (2004)
17. O.J. O'Loan, M.R. Evans and M.E. Cates, *Phys. Rev. E.* **58**, 1404 (1998)
18. P. Bialas, Z. Burda and D. Johnston, Nucl. Phys. B **493**, 505 (1997)
19. S.N. Majumdar, S. Krishnamurthy, M. Barma, *Phys. Rev. Lett.* **81**, 3691 (1998); *Phys. Rev. E* **61**, 6337 (2000)
20. M. R. Evans, *Europhys. Lett.* **36**, 13 (1996)
21. J. Krug and P.A. Ferrari, *J. Phys. A: Math. Gen.* **29**, L465 (1996)
22. J. Krug and J. Garcia, *J. Stat. Phys.* **99**, 31 (2000)
23. R. Rajesh and S. N. Majumdar, *J. Stat. Phys.* **99**, 943 (2000); *Phys. Rev. E* **64**, 036103 (2001).
24. F. Zielen and A. Schadschneider, *J. Stat. Phys* **106**, 173 (2002); *J. Phys. A: Math. Gen* **36**, 3709 (2003)
25. M.R. Evans, S.N. Majumdar and R.K.P. Zia, *J. Phys. A: Math. Gen.* **37**, L275 (2004)
26. R.K.P. Zia, M.R. Evans and S.N. Majumdar, *J. Stat. Mech. : Theor. Exp.* (2004) L10001
27. S.N. Majumdar, M.R. Evans and R.K.P. Zia, *Phys. Rev. Lett* **94**, 180601 (2005)
28. M.R. Evans, S.N. Majumdar and R.K.P. Zia, J. Stat. Phys. **123**, 357 (2006)
29. C. Godreche and J.M. Luck, *J. Phys. A* **38**, 7215-7237 (2005)
30. M. R. Evans and T. Hanney, *J. Phys. A: Math. Gen.* **36**, L441 (2003)
31. S. Grosskinsky and H. Spohn, *Bull. Braz. Math. Soc.* **34**, 489-507 (2003)
32. T. Hanney and M. R. Evans, *Phys. Rev. E* **69**, 016107 (2004)
33. K. Jain and M. Barma, *Phys. Rev. Lett.* **91**, 135701 (2003)
34. E.K.O. Hellén and J. Krug, *Phys. Rev. E* **66**, 011304 (2002)
35. S. Grosskinsky, G.M. Schütz, H. Spohn, *J. Stat. Phys.* **113**, 389 (2003)
36. C. Godrèche, *J. Phys. A: Math. Gen.* **36**, 6313-6328 (2003)

Metastability of Traffic Flow
in Zero-Range Model

Jevgenijs Kaupužs[1], Reinhard Mahnke[2], and Rosemary J. Harris[3]

[1] Institute of Mathematics and Computer Science, University of Latvia, LV–1459 Riga, Latvia; kaupuzs@latnet.lv
[2] Institut für Physik, Universität Rostock, D–18051 Rostock, Germany; reinhard.mahnke@uni-rostock.de
[3] Institut für Festkörperforschung, Forschungszentrum Jülich, D–52425 Jülich, Germany; r.harris@fz-juelich.de

1 Introduction

The development of traffic jams in vehicular flow is an everyday example of the occurence of phase separation in low–dimensional driven systems, a topic which has attracted much recent interest [1–4]. In [5] the existence of phase separation is related to the size-dependence of domain currents and a quantitative criterion is obtained by considering the zero-range process (ZRP) as a generic model for domain dynamics. We use zero-range picture to study the phase separation in traffic flow in the spirit of the probabilistic (master equation) description of transportation [6]. Significantly, we find [7] that prior to condensation studied in previous works [8,9] the system can exist in a homogeneous metastable state and we provide estimates of critical cluster size and mean nucleation time. Finally, we calculate the fundamental flux-density diagram which includes a metastable branch. Metastability and hysteresis effects have been observed in real traffic [10,11]. For previous work focusing on the description of jam formation as a nucleation process, see [12,13].

2 The Model

We consider a model of traffic flow, where cars are moving along a circular road. We divide the whole road of total length L into cells of size ℓ. Each cell can be either empty or occupied by a car. In distinction to most cellular automaton models, we consider the development of our system in continuous time. The first car in each cluster (uninterrupted string of n occupied cells) is allowed to move forward by one cell with transition rate w_n. This model can be directly mapped to the zero-range process (ZRP) (see also [1]). Each vacancy (empty cell) in the original model is related to a box in the zero-range model. The number of boxes is fixed, and each box can contain an arbitrary number of particles (cars), which is equal to the size of the cluster located to the left (if cars are moving to the right) of the corresponding vacancy in the original model. If this vacancy has another vacancy to the left, then it means that the box is empty. Since the boundary conditions are periodic in the original model, they remain periodic also in the zero-range model. In this representation, one particle moves from a given box to the right with transition rate w_n, which depends only on the number of particles n in this box. Our choice

$$w_n = w_\infty \left(1 + b/n^\sigma\right) \quad \text{for} \quad n \geq 2 , \tag{1}$$

is motivated by the slow-to-start effect — the longer a car has been stationary the larger the probability of a delay when starting (cf. [14–17]). The transition rate w_1 is given separately as a constant describing the freely moving cars, whereas w_n with $n \geq 2$ represents the escaping from a jam of size n.

3 Master Equation

In the grand canonical ensemble where the total number of particles is allowed to fluctuate, the stationary distribution over the cluster–size configurations is the product of independent distributions for individual boxes [18, 19]. Assuming this product ansatz also in the dynamics, one arrives at the mean-field master equations [4]

$$\partial_t P(n,t) = \langle w \rangle P(n-1,t) + w_{n+1} P(n+1,t) - [\langle w \rangle + w_n] P(n,t) \quad (n \geq 1),$$
$$\partial_t P(0,t) = w_1 P(1,t) - \langle w \rangle P(0,t) , \tag{2}$$

where $\langle w \rangle (t) = \sum_{k=1}^{\infty} w_k P(k,t)$ is the mean inflow rate in a box. The above mentioned factorisation is an exact property of the stationary state of the grand canonical ensemble or, alternatively, of an infinitely large system [18]. Hence, in these cases, the master equations (2) give the exact stationary state while providing a mean–field approximation to the dynamics of reaching it. The stationary solution $P(n)$ corresponding to $\partial P(n,t)/\partial t = 0$ can be found recursively, starting from $n = 0$. It yields the known result [4, 18, 19]

$$P(n) = P(0) \langle w \rangle^n \prod_{m=1}^{n} \frac{1}{w_m} \tag{3}$$

for $n > 0$, where $P(0)$ is found from the normalisation condition.
Denoting by M the number of boxes, which corresponds to the number of vacancies in the original model, the mean number of cars on the road is given by $\langle N \rangle = M \langle n \rangle$, where $\langle n \rangle = \sum_{n=1}^{\infty} n P(n)$ is the average number of particles in a box. Note that in the grand canonical ensemble the total number of cars as well as the length of the road L fluctuate. For the mean value, measured

in units of ℓ, we have $\langle L \rangle = M + \langle N \rangle$, and the average density of cars is $c = \langle N \rangle / \langle L \rangle = \langle n \rangle / (1 + \langle n \rangle)$. Hence, we have the following relation

$$\frac{c}{1-c} = \left[\sum_{n=1}^{\infty} n \langle w \rangle^n \prod_{m=1}^{n} \frac{1}{w_m} \right] \times \left[1 + \sum_{n=1}^{\infty} \langle w \rangle^n \prod_{m=1}^{n} \frac{1}{w_m} \right]^{-1} \tag{4}$$

from which the stationary mean inflow rate $\langle w \rangle$ can be calculated at a given average density c.

If $\sigma > 1$ in (1), as well as for $b \leq 2$ at $\sigma = 1$, Eq. (4) has a solution for any density $0 < c < 1$. This implies that the homogeneous phase is stable in the whole range of densities, i. e., there is no phase transition in a strict sense. If $\sigma < 1$, as well as for $b > 2$ at $\sigma = 1$, $\langle w \rangle / w_\infty$ reaches 1 at a critical density $0 < c_{cr} < 1$, and there is no physical solution of (4) for $c > c_{cr}$. This means that the homogeneous phase cannot accommodate a larger density of particles and condensation takes place at $c > c_{cr}$. This behaviour underlies the known criterion for phase separation in one–dimensional driven systems [5].

4 Metastability

Suppose that at the initial time moment $t = 0$ the system is in a homogeneous state with overall density slightly larger than c_{cr}. Here we study the development of such a state in the mean–field approximation provided by (2). With this initial condition, the mean inflow rate in a box $\langle w \rangle$ is slightly larger than that at $c = c_{cr}$, i. e., $\langle w \rangle = w_\infty + \varepsilon$ holds with small and positive ε. Hence, only large clusters with $w_n < w_\infty + \varepsilon$ have a stable tendency to grow, whereas any smaller cluster typically (except a rare case) fluctuates until it finally dissolves. In other words, the initially homogeneous system with no large clusters can stay in this metastable supersaturated state for a long time until a large stable cluster appears due to a rare fluctuation.

Neglecting the fluctuations, the time development of the size n of a cluster is described by the deterministic equation $dn/dt = \langle w \rangle - w_n$. According to this equation, the undercritical clusters with $n < n_{cr}$ tend to dissolve, whereas the overcritical ones with $n > n_{cr}$ tend to grow, where the critical cluster size n_{cr} is given by the condition $\langle w \rangle = w_{n_{cr}}$. Using (1) yields

$$n_{cr} \simeq \left(\frac{b}{\langle w \rangle / w_\infty - 1} \right)^{1/\sigma}. \tag{5}$$

Assuming that the distribution of relatively small clusters contributing to $\langle n \rangle$ is quasi–stationary, i. e., that the detailed balance (equality of the terms in (2) describing opposite stochastic events) for these clusters is almost reached before any cluster with $n > n_{cr}$ has appeared, we have

$$\frac{c}{1-c} \simeq \left[\sum_{n=1}^{n_{cr}} n \langle w \rangle^n \prod_{m=1}^{n} \frac{1}{w_m} \right] \times \left[1 + \sum_{n=1}^{n_{cr}} \langle w \rangle^n \prod_{m=1}^{n} \frac{1}{w_m} \right]^{-1} \tag{6}$$

[instead of (4)] relating the current $\langle w \rangle$ in (5) to the density c. The critical cluster size n_{cr} is found numerically by solving (5) and (6) consistently.

Based on the mean-field dynamics we have evaluated the mean (nucleation) time, which the system spends in the metastable state, as the first passage time to reach the overcritical cluster size $n_{cr}+1$ in one of the boxes. Assuming an exponential form for the first passage time distribution $\mathcal{P}(t) \sim \exp\left(-w_{esc}t\right)$ in one box, where w_{esc} is the escaping rate from the region $n \in [0, n_{cr}]$, we arrive at a very simple expression

$$\langle T \rangle_M \simeq \langle T \rangle_1 / M \tag{7}$$

relating the mean first passage time (or nucleation time) to reach the over-critical cluster size $n_{cr}+1$ in a system of M boxes ($\langle T \rangle_M$) with that of one box. The latter can be calculated easily by the known formula [20]

$$\langle T \rangle_1 = \sum_{n=0}^{n_{cr}} \left[\langle w \rangle \tilde{P}(n) \right]^{-1} \sum_{m=0}^{n} \tilde{P}(m) , \tag{8}$$

where $\tilde{P}(0) = 1$ and $\tilde{P}(n) = \prod_{k=1}^{n}(\langle w \rangle / w_k)$ with $n > 1$ represent the un-normalised stationary probability distribution. Since the exponential decay of the first passage time distribution is a long–time behaviour, the estimate (7) is valid only for large enough mean nucleation times $\langle T \rangle_M$.

Further on we have assumed $w_\infty = 1/\tau_\infty = 1$ by choosing the time constant τ_∞ as a time unit. We have tested our mean-field predictions by comparing them with Monte Carlo (MC) simulations for a system of $M = 10^5$ boxes. The simulations, starting from random uniform initial condition, show clear evidence for the existence of a metastable state prior to condensation — see Fig. 1 presenting three MC runs. We have evaluated the distribution of cluster sizes (for small clusters) averaged over the metastable state of one such run and have found a very good agreement with Eq. (3) with $\langle w \rangle = w_{n_{cr}}$, thus supporting the assumption of quasi–stationarity. Fig. 2 shows a comparison between the simulated and the predicted by mean-field equations (5) to (7) values of n_{cr} and $\langle T \rangle_M$ depending on the density c. Note that the mean-field values diverge at $c \to c_{cr}$. We find that our mean–field theory fairly accurately reproduces the critical cluster size but systematically underestimates the nucleation time. Nevertheless, it provides a good qualitative description of the metastable state and its dependence on density thus representing an important first step towards more refined theories.

5 Fundamental Diagram

The relation between density c and flux j of cars is known as the fundamental diagram of traffic flow. The average stationary flux can be calculated as $j = \sum_{n=1}^{\infty} Q(n)\, w_n$, where $Q(n)$ is the probability that there is a car in a given cell

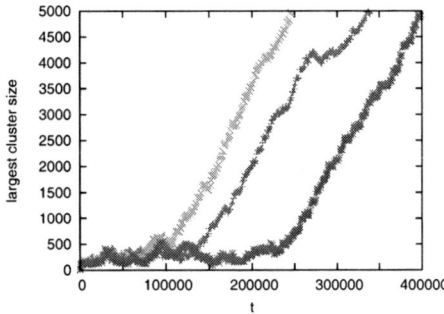

Fig. 1. Largest cluster size versus time for $\sigma = 0.5$, $b = 1$, $w_1 = 5$, $c = 0.66$, $M = 10^5$ ($c_{cr} \simeq 0.56$). Results from three independent MC runs are shown (cf. [7]).

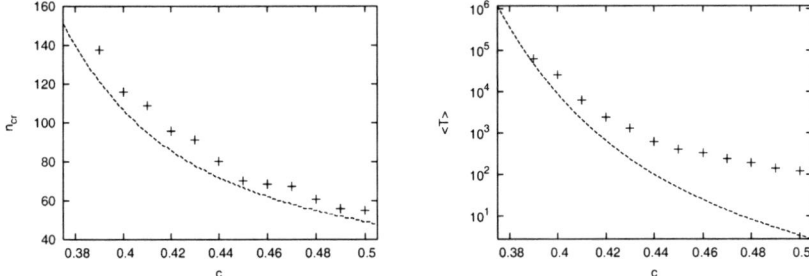

Fig. 2. Critical cluster size (left) and the mean nucleation time (right) versus density for $\sigma = 0.5$, $b = 3$, $w_1 = 5$, $M = 10^5$ ($c_{cr} \simeq 0.27$). Crosses show simulation data (averaged over 10 Monte Carlo histories), dashed lines are predictions of the mean–field equations (cf. [7]).

(in the original model) which can move forwards with the rate w_n. Relating the latter to the stationary distribution $P(n)$ and to the fraction of cells occupied by the cars which are allowed to move, we obtain

$$j = (1 - c) \langle w \rangle . \tag{9}$$

This relation can be used also to calculate the flux in the metastable state, which is quasi-stationary. In this case $\langle w \rangle$ is calculated using (6).

The resulting fundamental diagrams for two different sets of control parameters, i. e., $\sigma = 1$, $b = 6$, $w_1 = 10$ and $\sigma = 0.5$, $b = 3$, $w_1 = 5$ are shown in Fig. 3. As we see, the shape of the fundamental diagram, as well as the critical density and location of the metastable branch depend remarkably on the values of these parameters. We therefore believe that by suitable variation of parameters our simple model can reproduce some important features of real traffic flow. There is much scope for further investigation, both analytical and computational.

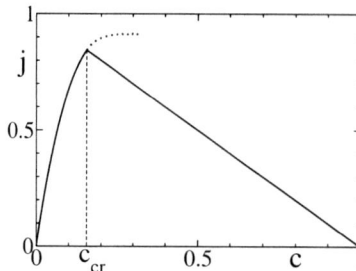 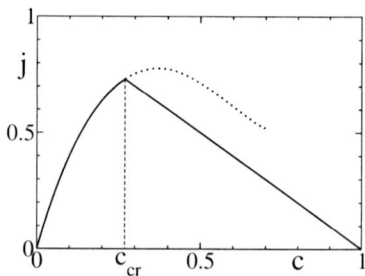

Fig. 3. The fundamental (flux–density) diagram for two different sets of control parameters: $\sigma = 1$, $b = 6$, $w_1 = 10$ (left); $\sigma = 0.5$, $b = 3$, $w_1 = 5$ (right). The branches of metastable homogeneous state are shown by dotted lines, the critical densities c_{cr} are indicated by vertical dashed lines.

Acknowledgements

One of us (J. K.) acknowledges support by the Academic Exchange Program via Rostock University.

References

1. M. R. Evans, these proceedings
2. D. Mukamel, Phase transitions in nonequilibrium systems, in *Soft and Fragile Matter: Nonequilibrium Dynamics, Metastability and Flow*, edited by M. E. Cates and M. R. Evans, Institute of Physics Publishing, Bristol, 2000
3. G. M. Schütz, J. Phys. A: Math. Gen. **36**, R339 (2003)
4. M. R. Evans, T. Hanney, J. Phys. A: Math. Gen. **38**, R195 (2005)
5. Y. Kafri, E. Levine, D. Mukamel, G. M. Schütz, J. Török, Phys. Rev. Lett. **89**, 035702 (2002)
6. R. Mahnke, J. Kaupužs, I. Lubashevsky, Phys. Rep. **408**, 1–130 (2005)
7. J. Kaupužs, R. Mahnke, R. J. Harris, Phys. Rev. E **72**, 056125 (2005)
8. S. Grosskinsky, G. M. Schütz, H. Spohn, J. Stat. Phys. **113**, 389 (2003)
9. C. Godrèche, J. Phys. A **36**, 6313 (2003)
10. D. Chowdhury, L. Santen, A. Schadschneider, Phys. Rep. **329**, 199 (2000)
11. D. Helbing, Rev. Mod. Phys. **73**, 1067 (2001)
12. R. Mahnke, J. Kaupužs, Phys. Rev. E **59**, 117 (1999)
13. R. Mahnke, J. Kaupužs, V. Frishfelds, Atmos. Res. **65**, 261 (2003)
14. O' Loan, M. R. Evans, M. E. Cates, Phys. Rev. E **58**, 1404 (1998)
15. M. Takayasu, H. Takayasu, Fractals **1**, 860 (1993)
16. S. C. Benjamin, N. F. Johnson, P. M. Hui, J. Phys. A **29**, 3119 (1996)
17. R. Barlovic, L. Santen, A. Schadschneider, M. Schreckenberg, Eur. Phys. J. B **5**, 793 (1998)
18. F. Spitzer, Adv. Math. **5**, 246 (1970)
19. M. R. Evans, Braz. J. Phys. **30**, 42 (2000)
20. C. W. Gardiner, *Handbook of Stochastic Methods for Physics, Chemistry, and the Natural Sciences*, Springer, Berlin, 1983, 1994

Extension of Cluster Dynamics to Cellular Automata with Shuffle Update

David A. Smith and R. Eddie Wilson

Bristol Centre for Applied Nonlinear Mathematics, Department of Engineering Mathematics, University of Bristol, Queen's Building, University Walk, Bristol BS8 1TR, United Kingdom

Summary. The random shuffle update method for the asymmetric exclusion process (ASEP) is introduced and the cluster dynamics technique is extended in order to analyse its dynamics. A sequence of approximate models is introduced, the first element of which corresponds to the classical parallel update rule whose two-cluster dynamics is reviewed. It is then shown how the argument may be extended inductively to solve for the two-cluster probabilities for each element of the sequence of approximate models. A formal limit is then taken, and macroscopic velocities and flow rates are derived.

1 Introduction

This paper is concerned with cellular automata of Nagel-Schreckenberg type [1], with the maximum velocity parameter v_{max} set equal to one. This type of model is sometimes referred to as the Asymmetric Exclusion Process (ASEP) [2, 3]. In this well-known set-up, space is discretised into a one-dimensional array of cells each of which is either empty or occupied by exactly one agent, and each agent moves according to a pair of very simple microscopic rules:

1. If the cell immediately downstream is occupied, remain stationary. (R1)
2. If the cell downstream is unoccupied, move forward into it with probability p, $0 < p \leq 1$. (R2)

The only remaining subtlety (and the subject of this paper) concerns the precise order in which rules (R1,2) are applied.

We consider the dynamics of rules (R1,2) under the *shuffle update* scheme, which has received very little attention in the literature to date [4–6]. At each time step in this scheme, a random order is generated which contains each agent exactly once. Rules (R1,2) are then applied to each individual agent in turn, according to this order, and the system is updated incrementally as each agent takes its turn. After all agents have had their turns, a new random order is generated and the next time step begins.

The shuffle update is similar to the random sequential scheme [7] in that the occupancy of cells is updated incrementally as each agent applies its rules and consequently, the shuffle update does not require conflict resolution to preserve single-occupancy (even in multi-dimensional extensions). However, the shuffle update enjoys the modelling advantage that individual agents never receive large numbers of consecutive turns, hence the possibility of unphysical velocities is eliminated.

In this paper we give an outline of how the *two-cluster* analysis of Schrecken-berg *et al* [8], which analyses (R1,2) under the parallel update scheme, can be extended to the more complicated case of the shuffle update. The argument here is more involved than [8] because under the shuffle update, it is possible for large blocks of contiguous agents to move forward in a single time step, if their turns are served in upstream order.

Due to this increase in complexity, we use a sequence of approximations to the full model. We define the *truncated process of order n* to mean that rules (R1,2) are applied under the shuffle update scheme, with the proviso that the opportunity to move is offered only to agents who are in the first n positions of a contiguous block at the beginning of the time step. Our procedure is thus to explain briefly how two-cluster dynamics works for $n = 1$ and then explain how to extend it inductively to any truncated process of order n. Finally we let $n \to \infty$.

2 Two-Cluster Analysis for $n = 1$

This method is described fully in [8]. We suppose that the occupancy of neighbouring cell pairs is independent and we seek to compute the so-called *two-cluster* probabilities P_2 for all possible combinations of occupancy of two adjacent cells, i.e., the probabilities of two adjacent cells having states $(1,0)$, $(0,1)$, $(1,1)$ and $(0,0)$, where 0 and 1 denote empty and occupied respectively. It can be shown that all such two-cluster probabilities can be calculated from $y := P_2(1,0)$ and hence the goal is to seek this quantity.

The strategy is to list all possible configurations S at time $t^* - 1$ which can give rise to $(1,0)$ in a monitored two-cluster at time t^*. If we can calculate the transition probability $W(S)$ for each configuration, in addition to the probability $P(S)$ of the configuration itself (which is usually expressed in terms of y), and if we assume the process has reached statistical stationarity, then we may employ conditional probability to write $y = \sum_S P(S)W(S)$, which for the truncation of order n we re-write and express in the form $f_n(y; c, p) = 0$ where c is the mean density and p is the parameter of rule (R2).

The truncated model with $n = 1$ is identical to the parallel update rule of Nagel and Schreckenberg [1], and for this standard case the two-cluster calculations are derived in detail in [8]. Note further that the two-cluster method has been shown to be exact in this case, in the sense that the spatial independence assumption for neighbouring two-clusters is exact.

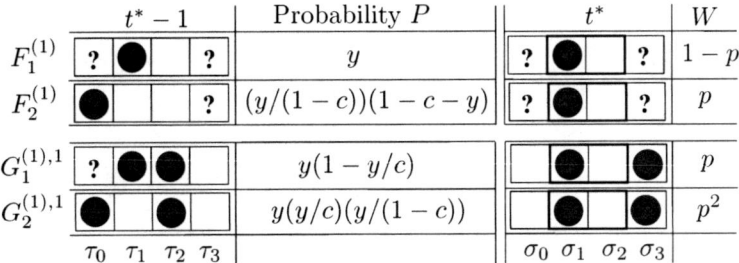

	$t^* - 1$			Probability P		t^*		W
$F_1^{(1)}$	? ●	?		y	? ●	?		$1-p$
$F_2^{(1)}$	●	?		$(y/(1-c))(1-c-y)$	? ●	?		p
$G_1^{(1),1}$	? ● ●			$y(1-y/c)$	●	●		p
$G_2^{(1),1}$	●	●		$y(y/c)(y/(1-c))$	●	●		p^2
	τ_0 τ_1 τ_2 τ_3				σ_0 σ_1 σ_2 σ_3			

Fig. 1. The list of all possible transitions to a $(\sigma_1, \sigma_2) = (1, 0)$ two-cluster (highlighted in bold) at time step t^*, for the truncated process of order $n = 1$. The list is identical to that for the parallel update rule. Probabilities for the window states at time step $t^* - 1$ are denoted by P; transition probabilities are denoted by W. Cells marked by ? can be either occupied or empty, with no effect on the P or W calculation: The state probability contribution is just a factor of one, and the movement or lack of movement of an agent in this cell cannot affect the monitored (σ_1, σ_2) two-cluster. The families of left hand column states labelled by $F_i^{(n)}$, $G_i^{(n),m}$ are the building blocks of the inductive process that follows later.

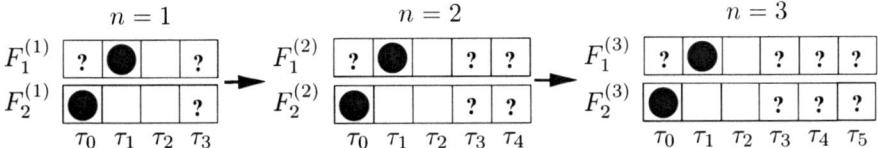

$n = 1$		$n = 2$		$n = 3$	

Fig. 2. The propagation of $F_i^{(n)}$ states as n increases. These states correspond to the left-hand column in Fig. 1 for $n = 1$. The added right hand cell in each window takes the value ? meaning that it can be either occupied or empty but we need not consider which, since it has no effect on the ability or probability to produce $(\sigma_1, \sigma_2) = (1, 0)$.

For the case $n = 1$, Fig. 1 gives a listing of the relevant configurations and their probabilities P and transition probabilities W. Rather than construct $f_1(y; c, p)$ and analyse its zeroes, we instead show now how inductive arguments may be used to extend Fig. 1 to truncated processes of arbitrarily large order n.

3 Inductive Construction for Truncated Processes

We now generalise to look at the truncated process for arbitrary order n. As n is increased, we need to consider bigger families of states, because there are more ways of obtaining $(\sigma_1, \sigma_2) = (1, 0)$. The states also have wider windows, because as n increases, (σ_1, σ_2) can be affected by more sites further downstream. The key is to build the families of cell windows inductively from those with lower n. Our choice of labels for the states was chosen with this process in mind, and we treat the $F_i^{(n)}$ and $G_i^{(n),m}$ states separately as they extend in quite different ways, summarised in Figs. 2–4.

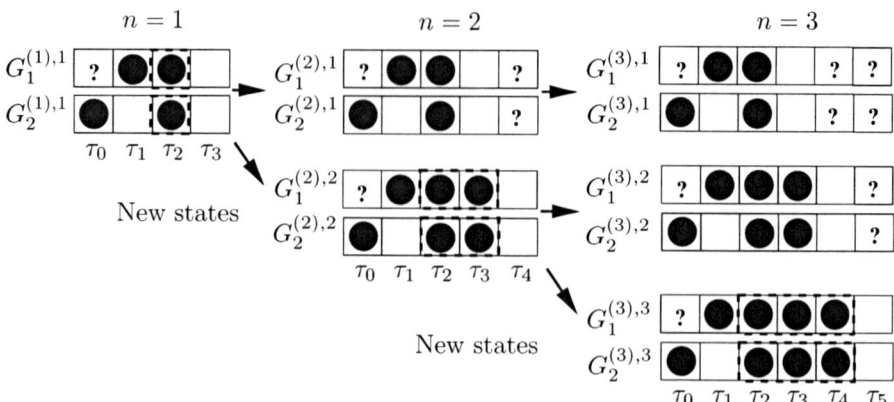

Fig. 3. Propagation of the $G_i^{(n),m}$ states as n increases. Existing states breed new ones as well as propagating in the same manner as the $F_i^{(n)}$ states (shown in Fig. 2). The characteristic feature of states which breed is that all cells with the dashed outline should be filled. For any given n, $F_i^{(n)}$ and $G_i^{(n),m}$ encompass all states capable of producing $(\sigma_1, \sigma_2) = (1, 0)$ at the next time step. The breeding of $G_i^{(n),m}$ means that the number of states increases by two each time n is increased by one.

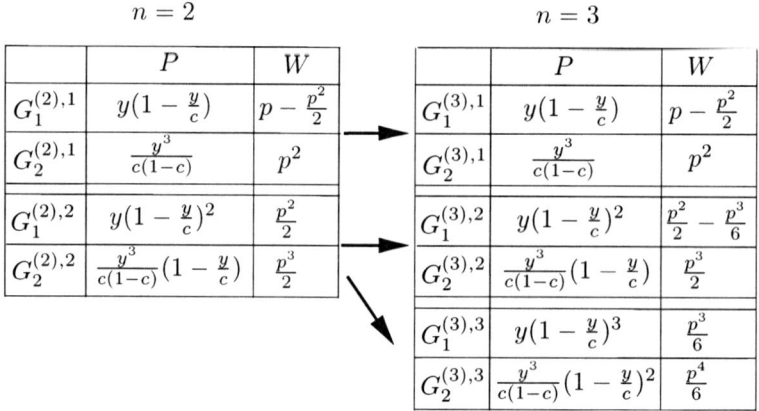

Fig. 4. The state and transition probabilities for the $G_i^{(n),m}$ for increasing n. We see the state probabilities unaltered along the propagating (horizontal) arrows, and gaining a factor of the conditional probability $1 - y/c$ on the breeding (diagonal) arrow. Transition probabilities gain a factor of p/n on a diagonal arrow, while $W(G_1^{(n),n-1}) = W(G_1^{(n-1),n-1})(1 - p/n)$. All others remain unchanged on the horizontal arrows.

By considering Fig. 2, we observe that the $F_i^{(n)}$ states and their probabilities do not change in any substantive way, so that $P(F_i^{(n)}) = P(F_i^{(1)})$, $W(F_i^{(n)}) = W(F_i^{(1)})$, for all n.

Now we look at Fig. 3 and consider the $G_i^{(n),m}$ states. Let us look at the $n = 2$ case and see how the P and W values relate to those for $n = 1$. Note $P(G_i^{(2),1}) = P(G_i^{(1),1})$, since a **?** contributes a factor of one to the probability, and $P(G_i^{(2),2}) = (1 - y/c)P(G_i^{(1),1})$, gaining a factor of the conditional probability $1 - y/c$ from the extra agent appearing in the blocks at τ_3. For the transition probabilities we find $W(G_2^{(2),1}) = W(G_2^{(1),1})$. Moreover $W(G_i^{(2),2}) = (p/2)W(G_i^{(1),1})$, because one more agent is required to move, gaining a factor of p, whilst the $1/2$ comes from the probability that the agents will update in an order which will allow the second agent to move. Finally $W(G_1^{(2),1}) = (1 - p/2)W(G_1^{(1),1})$ because it has become necessary to specify that the second agent in the block does not move, although it now can. These $n = 2$ probabilities are shown in Fig. 4. By looking at both Figs. 3 and 4, we see that the $(n = 1) \mapsto (n = 2)$ transition involved the *breeding* of states. This process generalises inductively to give the breeding behaviour at higher n values. Non-breeding $G_i^{(n),m}$ states propagate unaltered in the manner of the $F_i^{(n)}$ states.

4 General Solution for the Two-Cluster Probability

By employing conditional probability we may write

$$y_n = \sum_i P(F_i^{(1)})W(F_i^{(1)}) + \sum_{i,m} P(G_i^{(n),m})W(G_i^{(n),m}) , \tag{1}$$

where y_n denotes $P_2(1,0)$ for the truncated process of order n and $F_i^{(n)}$, $G_i^{(n),m}$ are the families of states described in the previous section. This formula, on substitution of the relevant quantities, may be rearranged in the form $f_n(y; c, p) = 0$, where

$$f_n(y_n; c, p) = p - \frac{py}{1 - c} + \sum_{i=1}^n \left(\frac{-y}{c} \right)^i \sum_{j=i}^n \frac{p^j}{j!} \binom{j-1}{i-1} \left(1 - \frac{py}{1 - c} \right). \tag{2}$$

By letting $n \to \infty$ we solve $f(y; c, p) = 0$ for the two-cluster probability of the full process, where

$$f(y; c, p) = -(1 - p) + \frac{1}{c - y} \left(1 - \frac{py}{1 - c} \right) \left(c - ye^{p(1 - y/c)} \right). \tag{3}$$

In general, this equation appears to have a unique solution for y, but requires numerical solution.

5 Steady State Velocities and Flow Rates

We now find the mean velocity and flow rate in terms of the two-cluster probability y. We proceed by using y to find the probability distribution of the length of the block to which an agent chosen at random belongs. This quantity may be constructed from conditional two-cluster probabilities in the form

$$\mathcal{P}_l = \left(\frac{y}{c}\right)^2 \left(1 - \frac{y}{c}\right)^{l-1} , \tag{4}$$

where l is the length of the block in question. We now use the fact that the agent is equally likely to be in any position within the block, and we sum the probability that it moves from the kth position over all $k = 1, 2, \ldots, l$ positions. By using (4), we thus obtain the mean velocity (equivalent to the probability that an agent chosen at random moves) in the form

$$\hat{v} = \sum_{l=1}^{\infty} l\mathcal{P}_l \frac{1}{l} \sum_{k=i}^{l} P(\text{block serves at least } k \text{ agents})$$

$$= \frac{y}{c - y} \left(\exp\left(\frac{p}{c}(c - y)\right) - 1\right) . \tag{5}$$

We can therefore write down flow rate

$$q = \frac{cy}{c - y} \left(\exp\left(\frac{p}{c}(c - y)\right) - 1\right) , \tag{6}$$

as the product of the system density and mean velocity.

6 Conclusion

By extending the two-cluster analysis of the well-known parallel update model [1, 8], we have been able to derive expressions for steady state distributions and flow rates of the more complicated shuffle update case. The results found here agree with the recent paper of Wölki et al [5] who employed a car-oriented mean field (COMF) method, as opposed to the site-oriented (SOMF) method that we use. However, it remains to be shown whether exactness, i.e. the spatial independence of neighbouring clusters, holds for the full model and the truncated processes of order $n > 1$.

References

1. K. Nagel, M. Schreckenberg: Journal de Physique I **2**, 2221 (1992)
2. N. Rajewsky, L. Santen, A. Schadschneider, M. Schreckenberg: J. Stat. Phys. **92**, 151 (1998)
3. D. Chowdhury, L. Santen, A. Schadschneider: Phys. Rep. **329**, 199 (2000)
4. G. Lunt: Cellular Automaton Models for Flows Without Momentum. Masters Thesis, University of Bristol (2001)
5. M. Wölki, A. Schadschneider, M. Schreckenberg: J. Phys. A: Math. Gen. **39**, 33 (2006)
6. D. Smith, R. E. Wilson: preprint
7. B. Derrida, E. Domany, D. Mukamel: J. Stat. Phys. **69**, 667 (1992)
8. M. Schreckenberg, A. Schadschneider, K. Nagel, N. Ito: Phys. Rev. E **51**, 2939 (1995)

Asymmetric Exclusion Processes with Non-Factorizing Steady States

Marko Wölki[1], Andreas Schadschneider[2], and Michael Schreckenberg[1]

[1] Universität Duisburg-Essen, Theoretische Physik, 47057 Duisburg, Germany
[2] Universität zu Köln, Institut für Theoretische Physik, 50937 Köln, Germany

Summary. The asymmetric simple exclusion process (ASEP) with periodic boundary conditions is investigated for shuffled dynamics. In this type of update, in each discrete timestep all particles are updated in a random sequence. It is shown that in contrast to all other updates studied previously, the ASEP with shuffled update does not have a product measure steady state apart from some simple limits. Approximative formulas for the steady-state distribution and fundamental diagram are derived that are in very good agreement with simulation data.

1 Introduction

The asymmetric simple exclusion process (ASEP) can be considered as the simplest discrete traffic model [1]. It captures essential features like unidirectional motion and hard-core exclusion. In the case of parallel dynamics it corresponds to the $v_{\max} = 1$ limit of the Nagel-Schreckenberg model [2]. For the ASEP a lot of analytical results exist, both for open and periodic boundary conditions; for a review see [3]. The model describes a particle system on a discrete one-dimensional lattice. Particles are allowed to hop one site to their right, supposed that it is empty. The usual dynamics is in continuous time: the random-sequential update. Moreover, discrete-time update schemes have been studied: backward- and forward-ordered sequential updates, sublattice-parallel and fully-parallel dynamics (for an overview, see [4]). We analyze another update scheme, the 'shuffled update' that has originally been introduced in a two-dimensional cellular automaton describing pedestrian dynamics [5, 6]. In this type of update in each timestep the update sequence is determined by a random permutation of the particle numbers. Note that the shuffled update is different from random-sequential dynamics which is generically used for the ASEP. Whereas the latter describes stochastic processes in continuous time which can be realized by updating a randomly chosen particle in each timestep, the shuffled dynamics combines elements of discrete updates and dynamics in continuous time. It is discrete in the sense that there is a well-defined timestep during which each particle is updated exactly once. On the other hand, the order of updating the particles is not fixed. E.g. it may happen that a specific particle is updated last during a timestep and first during the next one! Despite this important difference, the shuffled and random-sequential updates share certain similarities. In contrast to the ordered updates with fixed order, they do not have a deterministic limit, even for hopping probabilities $p = 1$. However, in the random-sequential case the dynamics depends on p only in a trivial way, since by rescaling time always $p = 1$ can be chosen. This is not possible in the shuffled case. We therefore expect a non-trivial p-dependence of the results, as in the other discrete-time updates.

2 ASEP with Shuffled Dynamics

We consider a one-dimensional lattice with L sites and periodic boundary conditions. Each site may either be occupied by one of the N particles, labelled $i = 1, 2, ..., N$, or it may be empty. Therefore the particles are distinguishable. In each discrete timestep a random permutation $\pi(1, ..., N)$ of the particle labels equals the update sequence. If the right neighboring cell is empty, the relevant particle moves one site to the right with probability p; if it is occupied, the particle stays in its cell.

Figure 1 shows a part of a large system consisting of six cells and four particles (numbered $1, 2, 3, 4$ without loss of generality), at time t (left) and $t+1$ (right). The drawn update sequence is $..., 3, ..., 4, ..., 1, ..., 2, ...,$ where the ellipsis indicate that other particle numbers belonging to different clusters[3] can be chosen inbetween (for the cluster depicted, only the relative positions of the numbers of its particles in the sequence are of interest). Particle 3, chosen first, can not move since the cell in front is occupied by 2, and similar for particle 4. Considering the case $p = 1$, particle 1 then moves deterministically to the right. Then 2 also moves, because it was drawn after 1. Although particle 4 is drawn after 3, it can not move, since both were drawn before 1 and 2.

Fig. 1. Shuffled update of a cluster consisting of four particles numbered from right to left. The drawn sequence is $3, 4, 1, 2$ and $p = 1$.

---

[3] A cluster is defined as the sequence of occupied cells between two consecutive holes (unoccupied cells).

3 Analytic Description of the Steady State

Due to the translational invariance of the system with periodic boundary conditions a configuration can be denoted by $(n_1, n_2, \ldots, n_N)$ where n_i is the number of empty cells in front of particle i. It is assumed that the probability for such a configuration in the thermodynamic limit $(N, L \to \infty$, with $N/L =$ const.$= \rho)$ factorizes into N single-particle probabilities P_n, i.e. $\mathcal{P}(n_1, n_2, \ldots, n_N) = \prod_{i=1}^{N} P_{n_i}$. This is usually not exact and constitutes the so-called car-oriented mean-field (COMF) theory, successfully applied to traffic flow models [7, 8] previously. For a slightly different approach, see [9]. The factorization assumption includes that the probability $Q(k)$ for an arbitrary particle to have a string of exactly $k - 1$ particles in front is approximated by

$$Q(k) = P_0^{k-1}(1 - P_0). \tag{1}$$

The factor $(1 - P_0)$ arises from the fact that the rightmost particle of the string has per definition at least one hole in front. Since every random sequence can be drawn with the same probability $(1/N!)$ in the beginning of a certain timestep, one can calculate the probability with which a particle has moved at the end of the timestep in a given configuration. If the particle has at least one hole in front (with probability $Q(1) = 1 - P_0$) it moves with probability p. If it has a string of $k-1$ particles in front (with probability $Q(k)$, $k = 2, 3 \ldots$), the situation is slightly more sophisticated. The first k particles have to be chosen in the order from the right to the left (with probability $1/k!$) and each of these particles has to move (with probability p^k). Thus the considered particle moves with probability $p^k/k!$. The average velocity $\langle v \rangle$ is obtained by summing over all possible events and using (1):

$$\langle v \rangle = \sum_{k=1}^{\infty} \frac{p^k}{k!} Q(k) = \begin{cases} p, & \text{for } P_0 = 0, \\ \frac{1-P_0}{P_0}(\exp(pP_0) - 1), & \text{for } P_0 > 0. \end{cases} \tag{2}$$

The average velocity $\langle v \rangle$ is related to the flow J via the particle density ρ and yields the so-called fundamental diagram

$$J(\rho) = \rho\langle v \rangle. \tag{3}$$

To express J or $\langle v \rangle$ through ρ we need to know P_0. The probabilities P_n can be calculated straightforwardly and are given by [10, 11]

$$P_0 = \frac{p(2\rho - 1) - ((1 + p)\rho - 1)\langle v \rangle}{p\rho(1 - \langle v \rangle)}, \tag{4}$$

$$P_n = \frac{(p - \langle v \rangle)(1 - P_0)}{(1 - p)\langle v \rangle}\left(\frac{(1 - p)\langle v \rangle}{p(1 - \langle v \rangle)}\right)^n, \qquad n \geq 1. \tag{5}$$

Note that this are implicit expressions that have to be solved numerically for general p, since $\langle v \rangle$ depends on P_0 via (2). For $p = 1$, P_0 becomes explicitly

$$P_0(\rho, p = 1) = \frac{2\rho - 1}{\rho} \theta \left(\rho - \frac{1}{2} \right),\tag{6}$$

as one can check easily. $P_0(\rho, p = 1)$ is completely determined by the fact that the first particle of a cluster moves deterministically and all the other particles have smaller hopping probabilities due to the shuffling. For densities $\rho \leq 1/2$ this implies that any state which consists only of clusters of size 1, i.e. separated particles, is stationary. Hence the probability to find a particle directly in front vanishes and we have $P_0 = 0$ (Fig. 2). For densities $\rho > 1/2$ clusters are formed that are separated by exactly one hole, i.e. there are only isolated empty cells in the steady state. It is easy to verify that no pairing of holes can happen, since from the point of view of the holes they jump at least one site backwards and at most to the end of the cluster. As a consequence, $P_n = 0$ for $n \geq 2$. Thus $L - N$ clusters exist, which is then also the number of particles having exactly one hole in front. Thus we obtain $P_1 = (1 - \rho)/\rho$ and $P_0 = 1 - P_1 = (2\rho - 1)/\rho$ in this density regime. These results are exact for $p = 1$ and are reproduced by COMF. Using (3), the flow-density relation is explicitly given by

$$J(\rho, p = 1) = \begin{cases} \rho, & \rho \leq 1/2, \\ \frac{\rho(1-\rho)}{2\rho-1} \left[\exp\left(\frac{2\rho-1}{\rho} \right) - 1 \right], & \rho > 1/2. \end{cases}\tag{7}$$

The fundamental diagram (7) shows a strong asymmetry with respect to $\rho = 1/2$. For densities $\rho \leq 1/2$ each particle can move independently and deterministically, since every particle has at least one hole in front, exactly as in parallel updating [8]. If the density is increased to values greater than $1/2$, clusters of nonvanishing length are formed from which the rightmost particle can move deterministically. The other particles have smaller hopping probabilities. This yields the curvature in the fundamental diagram in the high density regime. In contrast to the parallel and random-sequential dynamics, the shuffled update is not particle-hole symmetric (Fig. 2).

4 Proof of Non-Exactness

In the following it is shown that COMF is not exact for general p. Consider the master equation for the steady-state probability $\mathcal{P}(0^{N-1}, L - N)$, i.e. a cluster of N particles followed by $L - N$ holes:

$$\mathcal{P}(0^{N-1}, M) = \left(\bar{p} + \frac{p^N}{N!} \right) \mathcal{P}(0^{N-1}, M) + \bar{p} \sum_{k=1}^{N-1} \frac{p^k}{k!} \mathcal{P}(0^{N-k-1}, 1, 0^{k-1}, M - 1),\tag{8}$$

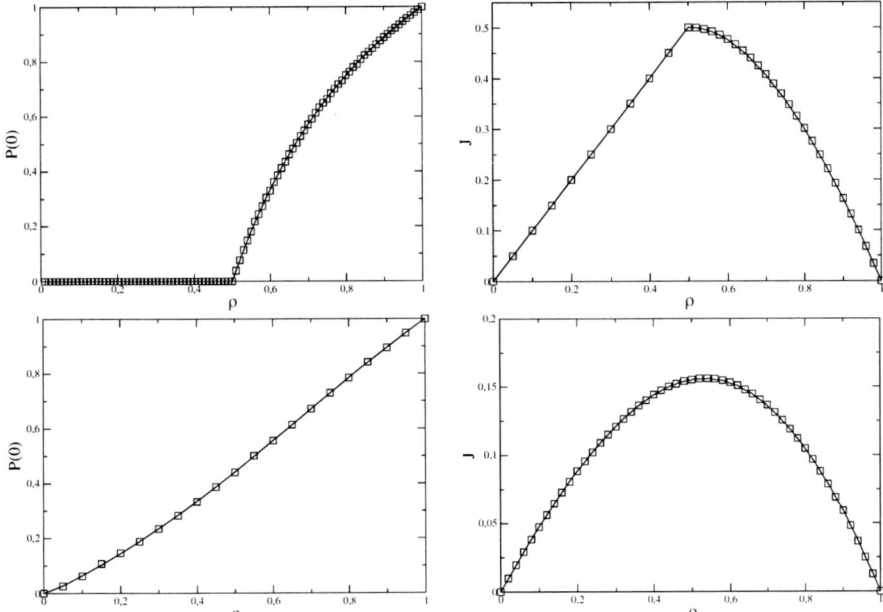

Fig. 2. The probability $P_0(\rho, p)$ (left) and the fundamental diagram $J(\rho, p)$ (right) for $p = 1$ (top) and $p = 0.5$ (bottom). Depicted are the results from COMF (lines) and from computer simulations (squares) for a system consisting of $L = 500$ cells. Note that $P_0(\rho, p = 1)$ is exact.

with $\bar{p} = 1 - p$ and $M := L - N$. Assuming that the steady-state probabilities factorize into probabilities for certain interparticle distances P_n and performing the limit $N \to \infty$ gives $pP_0 P_M = \bar{p} P_1 P_{M-1} (e^p - 1)$. Now, inserting the expression for P_n ($n \geq 1$) from eq. (5) and using (2) leads to the constraint

$$0 = (e^p - 1) \left(\frac{p}{e^{pP_0} - 1} - \frac{1 - P_0}{P_0} \right) - p. \tag{9}$$

This condition is only fulfilled in the limit $P_0 = 1$ (i.e. for $\rho = 1$) and also for $p = 0$. In all other cases, the master equation (8) can not be solved by factorizing probability distributions.

The ASEP with shuffled update can further be mapped onto a generalized zero-range process with parallel update for which a factorization criterion was derived recently [12]. The probability $u_l(m) = p^l/l! - \theta(m - l)p^{l+1}/(l + 1)!$ that exactly l particles leave a cluster of m particles during a timestep [11] can not be written as $u_l(m) = v_l w_{m-l}/\sum_l v_l w_{m-l}$, where v and w are functions that depend only on l and $m - l$ respectively. Therefore the ASEP with shuffled update does also not factorize into single probabilities $\tilde{P}_m$ for the cluster lengths m. This implies that although we know P_n, $n \geq 0$ exactly the fundamental diagram (7) in the case $p = 1$ can not be exact for $\rho > 1/2$ since the flow depends on the particular configuration of clusters.

We just mention that one can show that a truncated model in which only the first two particles of a cluster are allowed to move factorizes into cluster probabilities $\tilde{P}_m$ for $p = 1$ [11].

5 Discussion

The ASEP with periodic boundary conditions and shuffled dynamics was studied. Assuming a factorization of the steady state probability very good approximations for the fundamental diagram were derived. Since the shuffled dynamics is intrinsically stochastic, already for $p = 1$ a nontrivial fundamental diagram (depicted in Fig. 2 (b)) was found. Despite the stochasticity of the update in the regime $\rho \leq 1/2$ all particles move deterministically since each particle is separated by at least one hole to the left and right in the steady state. For higher densities $\rho > 1/2$ also neighboring particles occur which can move with certain probabilities and the flow is increased compared to the purely parallel case. For $p < 1$ the two regimes can no longer be distinguished and the fundamental diagrams become smoother.

It could be proven that the ASEP with shuffled dynamics does not factorize. Therefore the mean-field theory of Sec. 3 which assumes a factorization into interparticle distances is only a good approximation but not exact for $0 < p < 1$. In the case $p = 1$ the exact expression for the probability P_0 that a particle has no hole in front was found. The exact knowledge of this quantity allowed for an exact calculation of the flow for densities $\rho \leq 1/2$, but not for $\rho > 1/2$. This is due to the fact that in this density regime the probability for a particle to move depends on the number of particles in front. So the flow depends on the distribution of clusters which is not known exactly. It could be proven that this distribution also does not factorize. However also in this regime the calculated flow is in impressively good agreement with Monte-Carlo data (see Fig. 2). The fact that the ASEP with shuffled update does not factorize comes as a surprise. So far all update procedures have lead to a factorized steady state. Here the distinguishability of the particles seems to be important whereas in the updates considered previously the particles are basically indistinguishable.

Finally we want to mention an application. The ASEP with shuffled update can be considered as a model for pedestrians moving in one direction in a single lane [5, 6]. This may occur in a corridor which is so narrow that side-by-side motion and overtaking are impossible. To model corridors of arbitrary width and taking into account different step-lengths, in [11, 13] simple generalizations were proposed for which the analytic results derived here already yield good approximations for the fundamental diagrams.

References

1. D. Chowdhury, L. Santen, A. Schadschneider: Phys. Rep. **329**, 199 (2000)
2. K. Nagel and M. Schreckenberg: J. Physique I **2**, 2221 (1992)
3. B. Derrida: Phys. Rep. **301**, 65 (1998)
4. N. Rajewsky, L. Santen, A. Schadschneider, and M. Schreckenberg: J. Stat. Phys. **92**, 151 (1998)
5. A. Keßel, H. Klüpfel, and M. Schreckenberg, in *Pedestrian and Evacuation Dynamics*, edited by M. Schreckenberg and S. D. Sharma (Springer, 2001)
6. H. Klüpfel: *Ph.D. thesis*, Universität Duisburg (2003)
7. A. Schadschneider and M. Schreckenberg: J. Phys. A: Math. Gen. **30**, 69 (1997)
8. A. Schadschneider: Eur. Phys. J. B **10**, 573 (1999)
9. D.A. Smith, R.E. Wilson: these proceedings
10. M. Wölki: *Diploma thesis*, Universität Duisburg-Essen (2005)
11. M. Wölki, A. Schadschneider, and M. Schreckenberg: J. Phys. A: Math. Gen. **39**, 33 (2006)
12. M. R. Evans, S. N. Majumdar, R. K. P. Zia: J. Phys. A: Math. Gen. **37**, 275 (2004)
13. M. Wölki, A. Schadschneider, and M. Schreckenberg, in *Pedestrian and Evacuation Dynamics '05*, edited by M. Schreckenberg (Springer, 2006)

Two-Capacity Flow: Cellular Automata Simulations and Kinematic-Wave Models

Paul Nelson

Department of Computer Science, Texas A&M University, College Station, Texas 77843-3112, pnelson@cs.tamu.edu and
Departments of Nuclear Engineering, of Mathematics and of Civil Engineering, at Texas A&M University

Summary. A simple CA model (CA-184s2ss) is shown to reproduce, at freeway bottlenecks, elements of the observed phenomena of two-capacity flow, including both breakdown and recovery flows, and of wide moving jams. Notwithstanding that both of these phenomena are incompatible with classical kinematic-wave models (KWMs) of traffic flow, sufficiently coarsely time-aggregated CA-184s2ss simulations are shown to be approximated modestly well by the KWM, with a single suitable empirically determined capacity at the bottleneck.

1 Introduction

By "two-capacity" flow we intend observations and related analyses that suggest flows immediately downstream of an enqueued bottleneck can be observed at two distinct values. These distinct values are typically a higher value prior and immediately subsequent to formation of a queue upstream of the bottleneck and a lower value at seemingly random intervals of time subsequent to development of such a queue. This may be regarded as a particular instance of the more general phenomenon of "breakdown" [1], in the sense of a (sudden) drop in speed. (It is more specifically an instance of "spontaneous breakdown.") North American transportation engineers (e.g., [2–4]) have tended to focus on this two-capacity instance of breakdown, particularly at candidate bottlenecks consisting of freeway entrance ramps (merge junctions). Reasons for this focus include:

1. Existence of two-capacity flow at freeway entrance ramps is practically important, not least in that it is the basis for the belief that ramp metering provides a net benefit, as opposed to simply transferring flow from entering to mainline traffic.
2. Two-capacity flow is clearly inconsistent with the classical (lane-aggregated) Lighthill-Whitham-Richards [5, 6] kinematic-wave model (KWM), in that

generally accepted forms of the KWM predict flow immediately down-stream of a point bottleneck [7] will be equal to a single value, the "ca-pacity" of that bottleneck.

3. Existence of two-capacity flow seems to have become widely accepted.

The second item in this list contrasts sharply with observations of sudden speed drops just upstream of a bottleneck, which are readily explained in the KWM context as "shock waves" propagating upstream and forming the tail of a queue having head located at the bottleneck.

Notwithstanding that the two seem fundamentally incompatible, there re-mains the question of just how poorly the classical KWM predicts traffic flow, in the presence of two-capacity flow. The principal objective of this work is to contribute toward an understanding of the possibilities in this regard. More specifically, in Section 4 we demonstrate that a certain (stochastic) cellular automaton (CA), the CA-184s2ss of Section 3, produces some of the elements of two-capacity flow, and then show (Section 5) that nonetheless kinematic-wave simulations of that CA under two-capacity conditions reproduce quite satisfactorily certain aggregate aspects of the CA flow. The following Sec-tion 2 contains a summary of previous observations related to two-capacity flow, and concludes with a description of what we subsequently take as the characteristic signature of two-capacity flow. The final Section 6 contains our conclusions, and suggestions for further related subsequent studies.

Two comments seem in order regarding the philosophy undergirding this work: First, we do not intend the associated CA to be a faithful representation of real traffic flow, but rather to serve as a surrogate for actual traffic flow, in order perhaps to achieve a better understanding of possibilities in the rela-tionships between (inherently stochastic) traffic flow and associated (deter-ministic) macroscopic models. This objective frees us to focus on the simplest possible CA that might reflect some of the key aspects of two-capacity flow. The CA-184s2ss formulated in Section 3 below takes fullest advantage of that freedom. Second, notwithstanding this freedom from reality we make an effort to formulate the key elements of our model, especially the upstream bound-ary condition, in a way that is consistent with some reasonable form of driver behavior. The latter seems crucial to any hope of similarity between the CA and a KWM simulation of it, because solutions of the KWM eventually are strongly boundary driven.

As regards the first of these comments, we note the similarity in objective with a possible use of CA that was suggested by Wolfram [8]: "The derivation of hydrodynamic behavior from microscopic dynamics has never been entirely rigorous. Cellular automata can be considered as providing a simple example in which the necessary assumptions and approximations can be studied in detail." Our use of CA is intended to be exactly along the lines of the second sentence of this quotation, as seems appropriate because the first sentence surely is even more emphatically the case for behavior of vehicular traffic than for hydrodynamics.

Finally, the ultimate conclusion that the KWM can predict some aspects of traffic flow adequately, in the presence of two-capacity flow and notwithstanding the fundamental inconsistency between the two, is somewhat reminiscent of the classical paper of Wigner [9] on "The Unreasonable Effectiveness of Mathematics in the Natural Sciences." More specifically, it is surely an instance of the second reason advanced by Wigner that the "regularities" comprising the "laws of inanimate nature" are "surprising": "the regularity which we are discussing (here the predictions of the KWM) is independent of so many conditions which could have an effect on it (e.g., the presence or absence of two-capacity flow)." Indeed the present work, as well as many others in these proceedings, provides evidence that inroads are well underway toward advancing these surprising regularities into "laws of animate nature." This is precisely as Wigner hoped and perhaps expected, as expressed in the closing paragraph of the cited work.

2 Summary of Observations

Observations of two-capacity flow have been reported by Edie and Foote [10], by Agyemang-Duah and Hall [11], and by Cassidy and Bertini [3], while other studies [12–15] have reported lack of conclusive evidence supporting a reduction in flow subsequent to formation of a queue. From an empirical study Banks [16] (cf. also Banks [17, 18]) indicates that "the hypothesis that flow decreases when it breaks down is confirmed, provided the hypothesis applies to individual lanes," but "when averaged across all lanes...there was no significant change," and concludes that alleged two-capacity flow "is unlikely to provide a basis for metering...."

This variety of conclusions most likely reflects the fact that when differences between the levels of flow at a bottleneck, before and after queue formation, are reported their magnitude tends to be only about 5%. Given such relatively small alleged differences, it is difficult to know whether the different conclusions reached in different studies are attributable to differences in methodologies for obtaining and analyzing data, or to actual differences in traffic behavior at distinct sites, or possibly some combination of these factors. This inherent difficulty in determining existence of the two-capacity phenomenon has been discussed in particularly cogent fashion by Persaud and Hurdle [15]. In view of this difficulty, it is perhaps not surprising that, as suggested in the preceding paragraph, it took some three to four decades to advance beyond the status observed by Wattleworth [18] in 1963: "The question of whether or not the flow downstream of a freeway bottleneck decreases when congestion sets in is currently the subject of much discussion in engineering circles. Research findings support both the yes and no answers to this question. Several studies...suggested that perhaps the question did not have a simple yes or no answer."

Nonetheless, we do seem now to have advanced beyond this uncertainty, in that there appears to be an emerging consensus that two-capacity flow exists. Indeed the discussion now seems to have evolved toward a focus upon how best to incorporate two-capacity flow into standard engineering analyses. See [19], and many more-or-less recent references cited therein). We believe this emerging consensus owes much to the careful data analysis of Cassidy and Bertini [3]. The remainder of this section is therefore primarily devoted to a summary of that work, culminating in a description of the characteristics that we subsequently take as characterizing two-capacity flow.

The "freeway sites used" in [3] are described as "segments of the Queen Elizabeth Way (QEW) and the Gardiner Expressway" that "have been featured in previous studies of capacity." The fundamental idea of the analysis is to plot "scaled flow" $N(x,t) - q_0 t$ versus time t, with "$N(x,t) =$ the cumulative number of vehicles to pass (detector) location x by time t, measured from the passage of some reference vehicle" and "q_0 is defined as the background flow and" t "is the elapsed time from the passage of some reference vehicle." The basic idea is that the rescaling provided by subtracting the "background cumulative count" $q_0 t$ "promotes the visual identification of changing flows." Figure 1 shows an instance of the results from this procedure. The average flow over a time interval is equal to the background count (6100 vph in the present case) plus the slope of the corresponding secant line. The queue at the bottleneck is determined to originate at time 6:18:30. For approximately 12 minutes the corresponding flow downstream approximates 7000 vph, but after that it "breaks down" to a lower value of approximately 6090 vph. After a further elapsed time of approximately seven minutes the flow "recovers" to a higher value of approximately 6890 vph. This pattern of alternating periods of "break down" and "recovery," with differences on the order of 8%, continues until the queue at the bottleneck disappears around 7:54:00 hours.

Figure 2 displays directly a plot of these alternating mean flows, as visually identified from Fig. 1. In the following we take this pattern of alternating flows downstream of a bottleneck as the characteristic signature of two-capacity flow. In the following section we introduce a simple CA model designed to reproduce this signature of two-capacity flow.

"Two-regime" flow refers to the phenomenon of observations of extensive density-flow data scatter at sufficiently high densities [20], and an "inverted lambda" tail in the density-flow scatter plot [21]. The phenomena of two-capacity flow and two-regime flow phenomena are somewhat similar, and sometimes are linked together, but should be treated as distinct. For instance, Hall and Agyemang-Duah [22] effectively note that two-regime flow must be sought where traffic can be in queue, which is to say upstream of what Daganzo [23] terms an "active bottleneck," while two-capacity flow must be observed if at all immediately downstream of an active bottleneck (where "immediately" means absent intervening ramps). Both are prima facie inconsistent with existence of a fundamental diagram, as is implicit to the KWM. We refer the interested reader to [24] for a study of the two-regime phenomenon that has philosophical similarities to the study of the two-capacity phenomenon presented here.

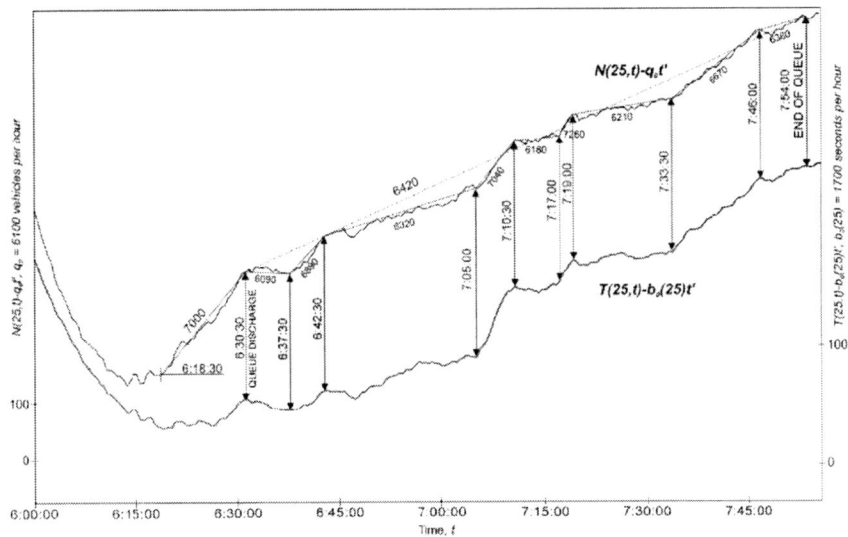

Fig. 1. A typical plot of scaled flow, from Cassidy and Bertini [3].

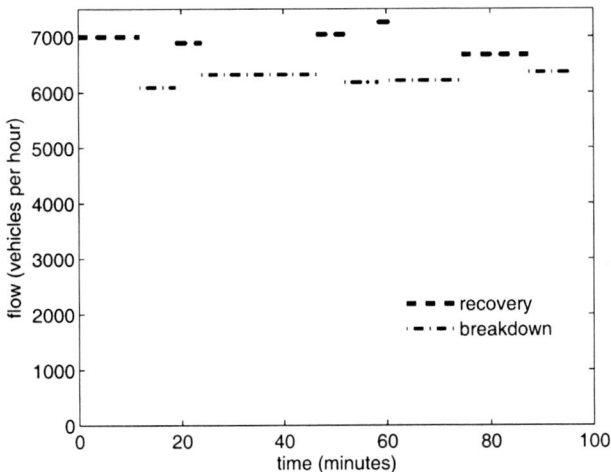

Fig. 2. The characteristic signature of two-capacity flow, as seen downstream of a bottleneck during presence of a queue upstream of that bottleneck.

3 CA-184s2ss

The base model is the instance $v_{\max} = 1$ of the "slow-to-start" class of CA models [25], which is identical to the earlier class of CA-184cc (cc = "cruise control") [26] in the present case. In more detail, a vehicle has speed $v = 1$ during a time step if, and only if, either:

a) It has a (spatial) headway of at least one empty site at the beginning of that time step and its speed was one during the preceding time step; or

b) it has a spatial headway of at least two empty sites at the beginning of that time step.

A bottleneck is then implemented in this base model as a site (cell) at which a Nagel-Schreckenberg "randomization" step occurs, provided the spatial headway of the vehicle at that site is exactly two. That is, if the spatial headway of the vehicles at the bottleneck is exactly two, then a speed determined as one according to the rules of the preceding paragraph is reduced to zero, with prescribed probability p_b. It is common to refer to p_b as "braking probability," but here we prefer the term "probability of hesitation" to reflect the fact that a speed that would "normally" be one can stay at zero either because of a spontaneous braking from a previous speed of one, or because of an unnecessary "hesitation" in accelerating from speed zero to speed one.

In order to complete the specification of an implementation of CA- 184s2ss (s2ss = "slow-to-start stochastic") it is necessary to specify the boundary and initial conditions. In this work the downstream (exiting) boundary condition (BC) is taken as the "infinite supply" condition that the exiting (most downstream) cell is not the bottleneck, and the subsequent two cells are always empty; equivalently, the most downstream vehicle always has speed one, unless it is at the bottleneck and momentarily reduced to zero by the "randomization."

The upstream (entrant) BC is implemented as arrival (with speed one) of a vehicle at the entrant (upstream) cell with probability $p_u = D/(1-D)$, at the second and subsequent steps following the immediately preceding arrival. Here D is some given demand, $0 \leq D \leq 1/2$, which may be regarded as a control variable (along with p_b). It is further assumed that any unmet demand (i.e., vehicles that arrive, but are prevented from entering by the position of their lead vehicle) is shunted to an alternate route. The alternative is to maintain an entrant queue, in some fashion [24].

Throughout the present work the initial conditions correspond to an initially empty section of roadway. For some purposes this requires permitting some time to lapse, in order to reach a steady state.

4 Simulations with CA-184s2ss

Except as explicitly indicated otherwise, the results presented in this section were obtained using an implementation of CA-184s2ss, as described in the preceding section, with 278 cells, labeled from 0 to 277, the bottleneck located at cell 251, probability of hesitation at the bottleneck = .005, and upstream demand = 0.5. Effects from varying some of these parameters will be discussed in subsequent sections.

Flows reported in this section were obtained as the (arithmetic) mean outflow, through the exit cell 277, obtained as the arithmetic mean per time step over some number (*aggregation_time*) of time steps. The value of *aggregation_time*, along with the number of time steps per sample path of the stochastic process CA-184s2ss, varied between sample paths, as will be specified in conjunction with each instance. Note that the times plotted are in number of aggregation periods, not number of time steps.

Figure 3 displays illustrative results, as obtained for four different sample paths, with *aggregation_time* = 45 time steps and 9000 time steps per sample path. The results, subsequent to a rather rapid initial transient, suggest a bistable system, in which the "breakdown" and recovery regimes are both metastable, but roughly equally likely. Further, the flows during "breakdown" and recovery are respectively 1/3 and 1/2 (the latter modulo a bit of combinatorial fluctuation), which are precisely the two likely candidates to be considered capacities for CA-184s2s [26]. However, the durations of each instance of either of these regimes appear to be considerably random. In many respects these results are similar to the Cassidy-Bertini observations of two-capacity flow in actual traffic, as summarized in Fig. 2.

Note that the aggregation time employed in Fig. 3 translates into a maximum of 22 or 23 vehicles per aggregation period, which is a rather small number from the perspective of minimizing effects of fluctuations. With this in mind, Fig. 4 displays the corresponding results for exactly the same sample paths as Fig. 3, but now with an aggregation time of 450 time steps (a maximum of approximately 225 vehicles per aggregation period). The results, past the now nearly negligible initial transient, have somewhat the appearance of an approximation to a constant value of approximately 0.4 (vehicles per time step) but still with a substantial "stochastic" fluctuation. Note that this putative constant value is approximately the arithmetic mean of the two approximately equally likely capacities previously noted.

This observation naturally leads us to inquire regarding the effect of an even coarser aggregation of flows. Figure 5 shows the results for a single sample path of 90,000 time steps, partitioned into 20 aggregation periods of 4500 time steps each. The results obviously are even smoother (i.e., more nearly constant ≈ 0.4) than those of Fig. 4, but even at this coarse level of aggregation there remains a noticeable statistical fluctuation. Our best estimate of the empirical bottleneck capacity, as obtained from an ensemble of five sample paths similar to (and including) that of Fig. 5, is $.393 \pm 003$.

The results just presented and discussed suggest the degree to which two-capacity flow is observed can hinge crucially on the details of the associated data analysis, especially the temporal scale on which flows are aggregated. A similar smoothing of the flows, and consequent diminution of the two-capacity

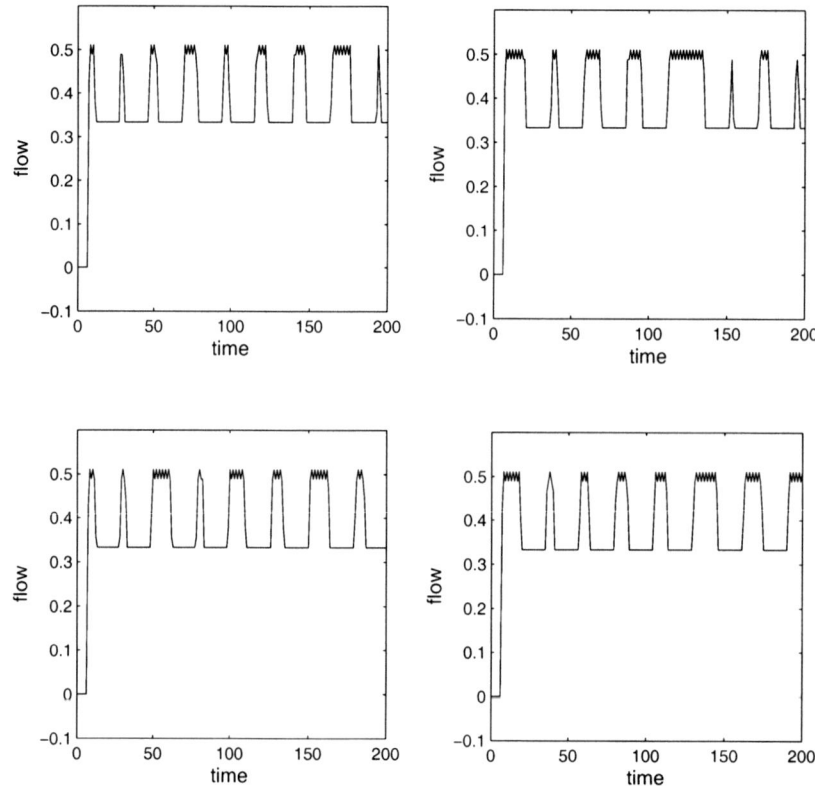

Fig. 3. Flow (vehicles per time step) vs. time (aggregation periods of 45 time steps), for four sample paths of CA-184s2ss, with mean demand $= 1/2$ and probability of hesitation $= .005$.

effect, occurs if one aggregates over an ensemble of sample paths, each of which individually is aggregated on a time scale that displays considerable evidence of two-capacity flow. This is illustrated in Fig. 6, which displays the ensemble-averaged flows over an ensemble of nine sample paths, each of which has a duration of 9000 time steps and is individually aggregated over aggregation periods of *aggregation_time* $= 45$ time steps. See Fig. 7 for the time-dependent flows (also aggregated over periods of 45 time steps) for the individual sample paths.

It seems a reasonable hypothesis that the approximately equal likelihood of flow at each of the smaller and larger "capacities" in the above results stems from the choices of probability of hesitation and length of the roadway section prior to the bottleneck so that there is approximately equal likelihood over time that a (mini)jam stemming from a hesitation either is or is not present on the roadway. This is supported by the results of Fig. 8, which suggest

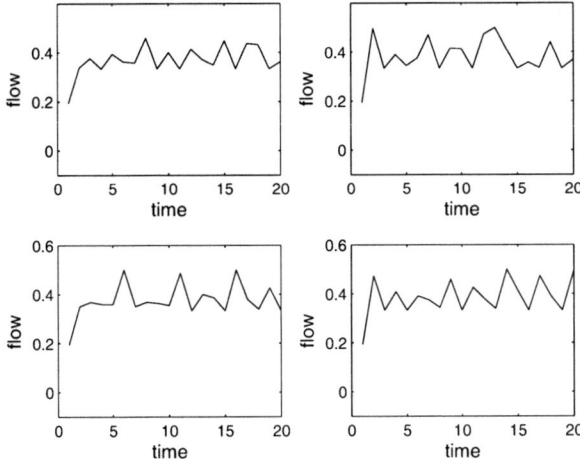

Fig. 4. Flow (vehicles per time step) vs. time (aggregation periods of 450 time steps), for the same four sample paths of CA-184s2ss as in Fig. 3.

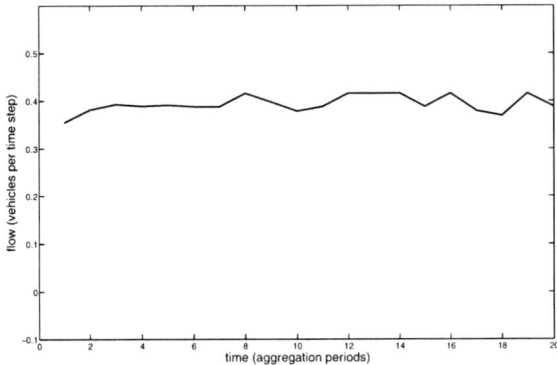

Fig. 5. Flow (vehicles per time step) vs. time (aggregation periods = 4500 time steps) for a sample path of 90,000 time steps, with mean demand = 0.5 and probability of hesitation = .005.

that the higher (lower) capacity flow regime becomes more prevalent (stable) as the probability of hesitation decreases (respectively, increases). Of course any use of that insight to reduce the likelihood of "breakdown" requires some insight into the causes of that hesitation. The fact the breakdown stems from behavior of a very small fraction (0.5%) of the driver-vehicle combinations on the roadway also suggests it may be very difficult to control breakdown. The sensitive dependence upon probability of hesitation, in terms of which regime is predominant, suggests the possibility that details of observations of

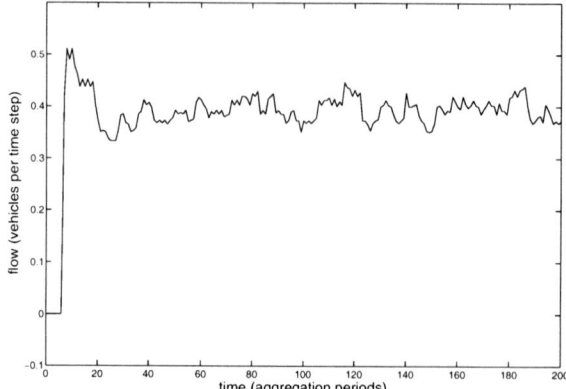

Fig. 6. Flow (vehicles per time step) vs. time (aggregation periods = 45 time steps), averaged over an ensemble of nine sample paths of 9000 time steps each.

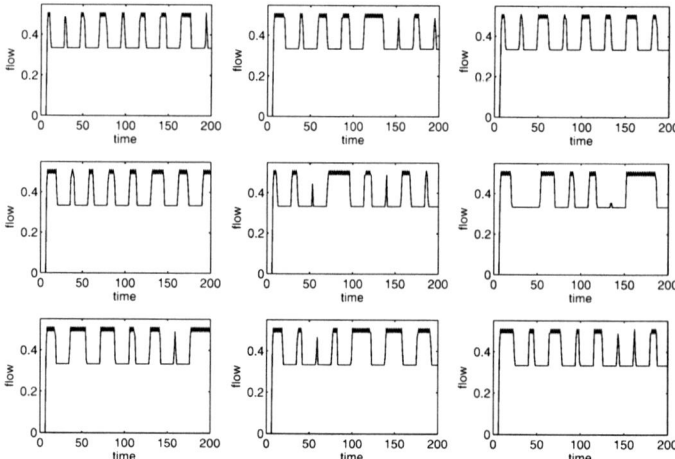

Fig. 7. Flow (vehicles per time step) vs. time (aggregation periods of 45 time steps each) for the nine sample paths of the ensemble of Fig. 6.

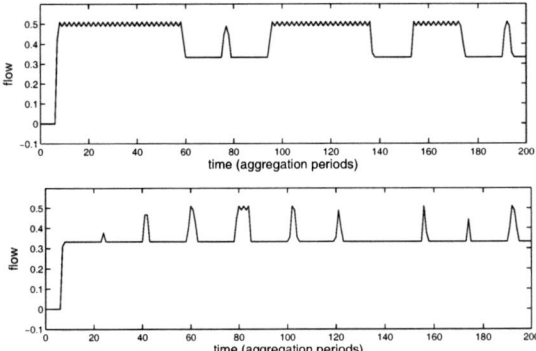

Fig. 8. Flow (vehicles per time step) vs. time (aggregation periods of 45 time steps each) for typical sample paths of 9000 time steps each, with hesitation probabilities $p_b = .00125$ (top) and .02 (bottom).

two-capacity flow could be very dependent upon the composition of the traffic stream in the location and at the time that the observations are taken.

5 Corresponding Kinematic-Wave Results

It is clear that neither the KWM, nor any other deterministic macroscopic model, can possibly approximate well all sample paths corresponding to two-capacity flow as seen at fine levels of temporal aggregation (e.g., Figs. 3, 7 and 8). However, there remains the possibility of acceptable KWM approximations to more coarsely time-aggregated representations of two-capacity flow (e.g., Figs. 4 and 5), or to ensemble aggregates (Fig. 6). The objective of this section is to explore the extent to which this possibility is realized.

Any version of a KWM requires an associated fundamental diagram (density/flow relation) as one of its two essential ingredients, the other being conservation of vehicles. This requirement poses immediate difficulties for CA-184s2s, owing to the well-known fact [26] that it has multiple steady-state flows corresponding to some densities (e.g., 1/2 vehicles per cell). For the moment let us ignore this, to suppose rather that CA-184s2s has a classical FD of the sort shown in Fig. 9, and work out the consequences of that supposition for the time-dependence of the flow immediately downstream of the bottleneck in CA-184s2ss.

In Fig. 9 q_b denotes the capacity of the bottleneck, and the remaining notation is more-or-less obvious. We assume $q_{max} \leq 1/2$, as seems likely from what is well-known about CA-184s2s. Following Nelson and Kumar [7], the corresponding KWM solution of the base scenario of the preceding section then is as follows. At $t = 0$ a shock forms that connects an upstream region having

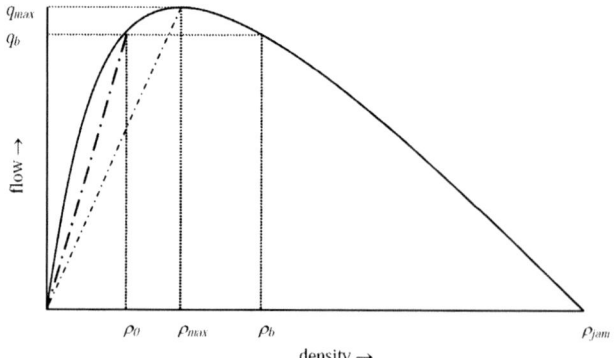

Fig. 9. A hypothetical FD for CA-184s2ss

$(\rho, q) = (\rho_{\max}, q_{\max})$ to a downstream region with $(\rho, q) = (0, 0)$, and according to the shock condition subsequently propagates downstream with speed equal to the slope of the light dashpot line in Fig. 9. This shock subsequently reaches the bottleneck at a time determined by this speed of propagation and the distance from entrant boundary to the bottleneck. At the bottleneck a portion of the incident shock is reflected back upstream, and a portion is transmitted downstream. Although the reflected portion may be of interest for other purposes, only the transmitted portion affects flow downstream of the bottleneck, which is the flow that is relevant to the two-capacity effect. That transmitted portion consists of a shock connecting an upstream region having $(\rho, q) = (\rho_0, q_b)$ to a downstream region with $(\rho, q) = (0, 0)$, and subsequently propagating downstream with speed equal to the slope of the heavy dashpot line in Fig. 9.

It follows that the flows observed downstream of the bottleneck take the following simple form. Prior to some arrival time, which is dependent upon the details of the FD for CA-184s2s, the flow is zero. Subsequent to the arrival time it is equal to the capacity of the bottleneck. If this capacity could, for the base case $p_b = .05$ of the preceding section, be taken as some value in the vicinity of 0.4, then we could proclaim vindication for the KWM and move on. But what is the justification for this value of bottleneck capacity? Further, what is the justification for the assumption that $q_{\max}$ (the CA-184s2s capacity) is greater than this empirical bottleneck capacity, especially given that the capacity corresponding to the vast majority of the initial configurations on a ring road is well-known to be $1/3$?

One answer to these questions, which is very much in the tradition of transportation engineering, is that capacities of both roadways and bottlenecks should be determined on the basis of observational data. If we recall that CA-184s2ss is here supposed to comprise reality, then it is entirely consistent with that empirically grounded philosophy to hold that data such as that in Figs. 4-6 firmly indicate that the capacity of our bottleneck has a value somewhere in the vicinity of $q \approx 0.4$, and that of the roadway upstream of that bottleneck is some value that is not precisely determined by these data, except that it is greater than the empirical bottleneck capacity. This approach absolutely would lead to KWM results (for the outflow) that are quite consistent with those (highly aggregated) empirical results, modulo a typically acceptable amount of "statistical fluctuation."

Of course it is impossible to ignore that our "empirical" results stemmed from a very simple CA model. For this model it seems likely the prosaic argument of the preceding paragraph could be replaced by something more elegant. One commonly used source of FDs for models is simulations on a ring road. An FD from ring-road simulations would certainly be appropriate for use in conjunction with a KWM, because the corresponding closed boundary conditions comprise a computationally accessible alternative to the "long road" envisioned in the title of the seminal Lighthill-Whitham paper [5]. However, we have been unsuccessful in obtaining, from ring-road simulations, a FD displaying the properties delineated above. We believe this is because the "alternate path" BC we implement at the upstream boundary acts, in concert with a nonzero probability of hesitation, to produce phenomena that are qualitatively different from those seen in flow on a ring road.

More specifically, we believe the relevant CA-184s2s capacity is the larger value of $1/2$ rather than the smaller value of $1/3$, because the regularity of the upstream boundary flow (for $D = 1/2$ a vehicle arrives at the entrant boundary exactly every second time step) drives flows in CA-184s2ss, in the region of densities extending from $1/3$ to $1/2$, toward those resulting in the basic CA-184s2s from what Benjamin, Johnson and Hui [25] term as "a certain small sub-set of initial configurations (that) will result in a steady-state region with no traffic queues." However, once one implements a nonzero probability of hesitation within CA-184s2s, then queues (clusters, jams) necessarily form, somewhat randomly, depending upon the details of the implementation of hesitation. For a ring road, the certainty of a slow start that is taken in the present model, and densities no less than $1/3$, there is no possibility for such a cluster to dissipate once formed, because inflow is always at least equal to the outflow of $1/3$. Somewhat similarly, for the present upstream alternate path BCs, a cluster formed at the bottleneck will not only persist, but it also will propagate upstream and grow in size, because the inflow of $1/2 >$ outflow of $1/3$, similarly to the wide jams of Kerner [27]. However, once such a "moving jam" reaches the upstream boundary it will begin to dissipate, because now the inflow of 0 (all vehicles arriving at upstream boundary are shunted to the alternate path) is less than the outflow, which remains at $1/2$ until the jam completely dissipates.

To complete the picture, once the jam at the upstream boundary does disappear, the inflow at that boundary returns to $1/2 =$ the larger of the "two capacities" for CA-184s2s. The boundary between this region of flow $1/2$ and the downstream region of flow 1/3 propagates downstream, and eventually reaches the bottleneck. The flow at and downstream of the bottleneck then returns to 1/2, which is to say a period of recovery is entered. This endures until yet another vehicle hesitates at the bottleneck. For the particular parameters in the preceding section it is apparent that the periods of recovery and of breakdown (existence of a jam) are, in the mean, of approximately equal duration. The vehicle trajectories plotted in Fig. 10 illustrate graphically these cycles of breakdown, followed by eventually successful attempts at recovery, ultimately followed by another instance of breakdown.

More specifically, the white, darker gray, black and lighter gray regions in Fig. 10 represent respectively – in more-or-less standard terms of transportation engineering – spatiotemporal regions of vacuum (no vehicles), free flow, jam and queue discharge (acceleration wave). To essentially rephrase the discussion of the preceding two paragraphs, the initial vacuum transitions to a free flow region, as vehicles entering the roadway arrive. The free flow region transitions to a jam when some vehicle hesitates at the bottleneck, which initially occurs at approximately the 500th time step. This jam widens as it propa-

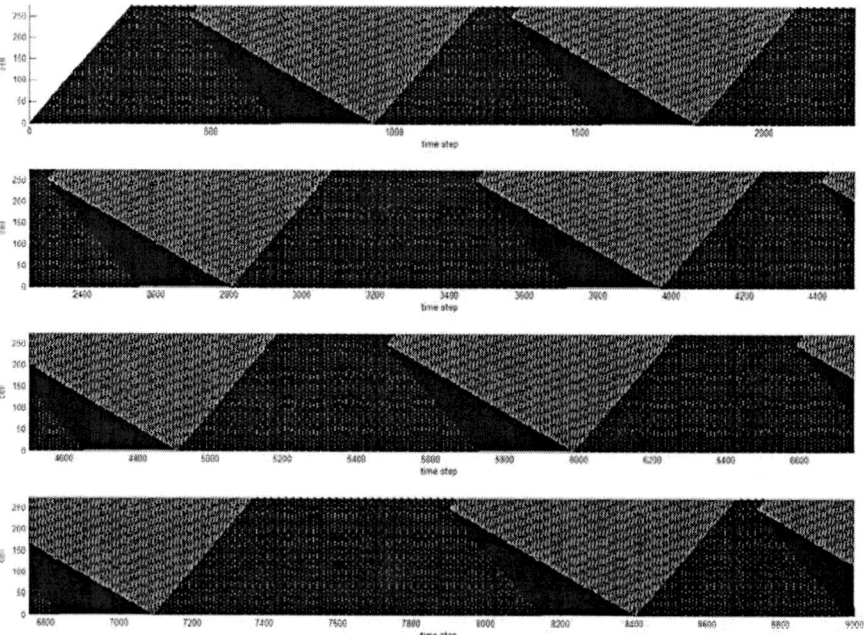

Fig. 10. Vehicle trajectory plot for the initial sample path in the ensemble of Figs. 6 and 7. Vertical axes are $x =$ cell index, $0 \leq x \leq 277$. Horizontal axes are $t =$ time, in steps, for respectively (from top to bottom), $0 \leq t \leq 2250$, $2250 \leq t \leq 4500$, $4500 \leq t \leq 6750$, and $6750 \leq t \leq 9000$.

gates upstream, but eventually begins to narrow when its upstream boundary reaches the entrant section, and ultimately disappears when its downstream boundary reaches the entrant section. As that downstream boundary passes a given location it transitions into a queue (jam) discharge region. When the downstream boundary of the jam reaches the entrant section, there emerges from that boundary a new free flow region, and the cycle repeats. That is, downstream of that boundary there is a queue discharge region that transitions at the upstream entrant section into the downstream boundary of a free flow region, which propagates downstream until it passes the bottleneck, and endures until the next occurrence of the cycle free flow → jam → queue discharge → free flow is initiated by the next hesitation at the bottleneck.

An alternative to ring-road simulations, as a possible more elegant source of a suitable FD, would be an analytic approach via cluster expansions. Under the above circumstances one would expect the desired FD not to be well-approximated by the one developed in [25] via a 2-cluster expansion. It might well be possible to obtain a suitable analytic approximation by similar methods, but we shall not pursue that here.

Finally, note that the free-flow, jam and queue-discharge regions mentioned above correspond to regions of symmetry under minimal translations of length two, one and three cells, respectively. These regions therefore have some claim to be considered as distinct "phases," under the somewhat fundamental definition of a phase transition as a break in symmetry. Of course the extreme regularity and clarity displayed in these transitions is a direct consequence of the regularity of arrival of vehicles at the entrant section, which in turn stems from taking the entrant demand as $1/2$. We have shown that for this value of demand the KWM with the empirically determined bottleneck capacity of ≈ 0.4 predicts flow sufficiently well for purposes such that highly temporally aggregated means suffice. However, for the KWM to be truly useful, with such empirically determined bottleneck capacities, it must perform adequately over the full range of possible entrant demands. We consider this issue in the following section.

6 Demands $< 1/2$

The titular situation opens the door to qualitatively new phenomena, relative to the preceding section, because now there is the possibility of jams disappearing, prior to propagating to the upstream entrant section, through merger with "gaps" in the upstream free flow. The smaller the upstream demand is, the more likely such "cancellation" of jams. We therefore describe in this section the results of some CA experiments with $D < 1/2$ and varying, and comparisons of the results with corresponding predictions of the KWM. Any clues these results might provide toward a basis for the bottleneck capacity of ≈ 0.4 that is suggested empirically by the results of the two preceding sections also are certainly of interest.

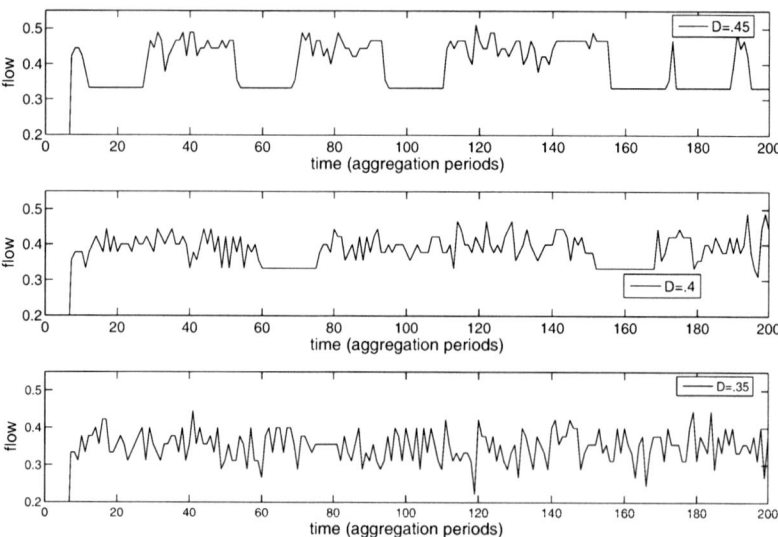

Fig. 11. Flow (vehicles per time step) downstream of the bottleneck vs. time (aggregation periods of 45 time steps), for four sample paths of CA-184s2ss, with various mean demands as specified in the legend and probability of hesitation = .005.

Figures 11, 12 and 13 show results of CA simulations with entrant demands .45, .4 and .35, using the same aggregation levels as respectively Figs. 3–5. At the finest of these aggregation levels (45 time steps), and the largest demand (.45), the only significant difference from the results of Fig. 3 is a notable amount of stochastic noise, presumably from the stochastic nature of arrivals at the upstream boundary, during periods that higher flows (free flow) prevails upstream of the bottleneck. However, the lower flows (queue discharge) remain constant at the lower capacity of 1/3. When the demand decreases to .4 the level of demand-driven high-frequency noise in the higher capacity flow increases even further, and the mean value associated with this demand-driven free flow decreases, but the lower queue-discharge flow remains more-or-less stable. When the demand decreases to .35 the high-frequency noise dominates at all times. At this level of aggregation it is difficult to discern any central tendency of the flows.

When the aggregation period is increased to 450 time steps (Fig. 12) there does emerge something of a central tendency toward mean values; however, some superimposed stochastic effects still appear, especially for the larger values of demand. For even larger aggregation periods (Fig. 13) these are significantly smoothed. The mean flows, downstream of the bottleneck and past the first 2000 time steps, for the three cases of Fig. 13 are .395 ($D = .45$), .380 ($D = .4$) and .354 ($D = .35$). These can be compared to the respective values of .393, .393 and .350 that would be predicted by the KWM, with use of the

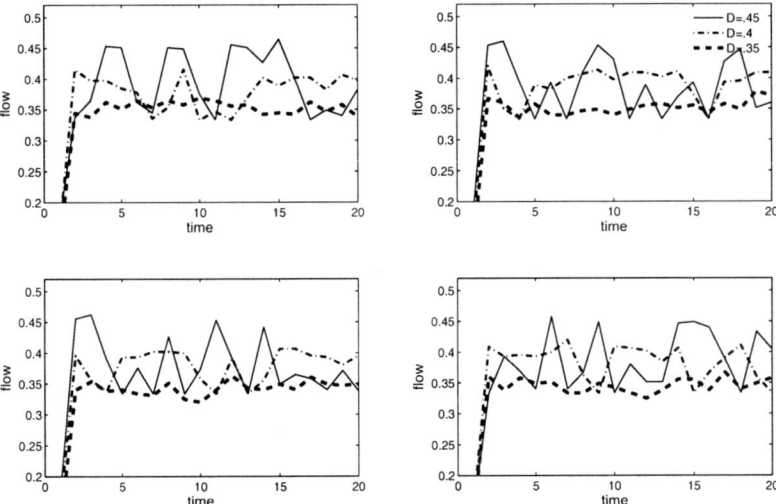

Fig. 12. Flow (vehicles per time step) vs. time (aggregation periods), for four sample paths of CA-184s2ss, now with aggregation periods of 450 time steps.

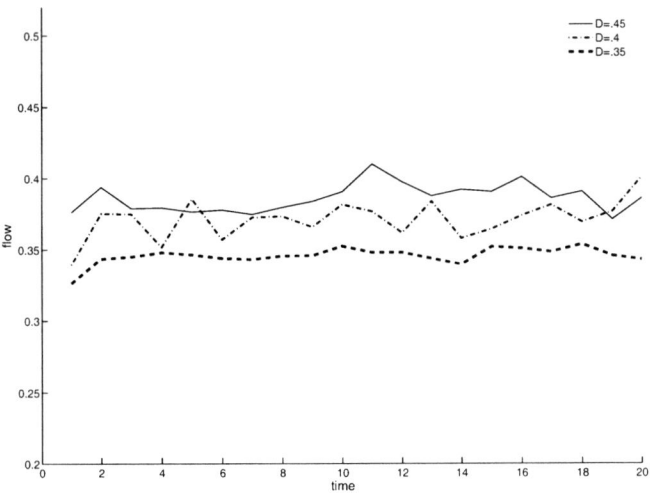

Fig. 13. Flow (vehicles per time step) vs. time (aggregation periods = 4500 time steps) for a sample path of 90,000 time steps, with varying mean demands, as indicated, and probability of hesitation = .005.

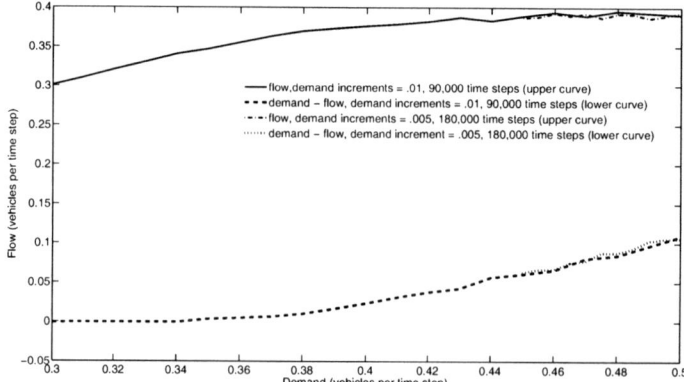

Fig. 14. Demand/flow curves, for CA-184s2ss, with probability of hesitation = .005, and other parameters as specified in the text.

bottleneck capacity of .393 that was determined by the "empirical" observations of Section 4. This is a maximum error of 3.5%, which is comparable to, but smaller than, both the difference between breakdown and recovery flows and the typical accuracy of microsimulation models [28].

Some workers have recently reported evidence of phase transitions in the form of discontinuities of either the first or second kind in demand/flow diagrams. The upper solid curve in Fig. 14, which was obtained by incrementing demand from .3 to .5 in steps of .01 and by aggregating exiting flow from a sample path of 90,000 time steps over the last 88,000 steps, does not discernibly display any such discontinuities. The matching solid lower curve is a plot of the difference between demand and exiting flow, and therefore an approximation[1] to the unmet demand. The two curves together illustrate that unmet demand increases rapidly as demand increases beyond the lower capacity of 1/3, and the flow curve suggests that there remains significant stochastic variation at the higher demand levels, even for this relatively large number of time steps. In order to explore the latter even further, the dashpot curves in Fig. 14 were obtained analogously, except with demand increments from .45 to .5 in steps of .005 and sample paths of 180,000 time steps, with exiting vehicles during the first 2000 time steps neglected in order to achieve an initial steady state. Again there are no apparent discontinuities that would suggest a phase change. However, over the entire range of demands considered there is a distinct gradual change in the slope, and increasing fluctuation in the observed flow, as the demand increases.

---

[1] In order to make it exact it would be necessary to adjust for vehicles in the roadway at beginning and end of the 88,000 steps over which the exiting flow is aggregated.

Figure 15 provides a picture of this evolution of flows, in the form of flow vs. time diagrams similar to Figs. 3, 7, 8 and 11, but for demands varying from .34 to .5, in increments of .02. As demand decreases from the value of 1/2 that was used in the preceding two sections, the first notable effect is high-frequency fluctuations in the flow during periods of high flow (the free-flow value of 1/2). These stem from fluctuations in the demand, and their magnitude increases as demand decreases. In the meantime the queue-discharge flows at the exit initially remain stable at 1/3; however, eventually ($D \approx .42$) one sees fluctuations, driven by the stochastic entrant demand, that have minima even smaller than 1/3. As demand decreases past this point the relative portion of the time during which queue-discharge = 1/3 holds decreases, although some queue discharge remains even at $D = .34$. The decreasing occurrence of queue discharge at the exit stems from the increasing tendency for jams emanating from the bottleneck to be "extinguished" by the increasingly frequent and larger gaps in the preceding region stemming from the entrant boundary, as anticipated at the beginning of this section.

This effect can be seen graphically in the expanded vehicle trajectories of Fig. 16; cf. especially the jams initially forming at $t = 4259$ and 4434. This expanded view also demonstrates that the regions preceding the jams are comprised of alternating gaps (small vacuum regions) and platoons (moving vehicles separated by spatial headways = 1, as in the free-flow regions of the preceding sections). While one conceivably could argue that this represents a "phase" distinct from the vacuum, free-flow, jam and queue-discharge phases previously identified, at the microscopic level it seems a more basic view of this regime is as a fine-scale mixture of the vacuum (empty) and free-flow phases. In any even, when incipient jams are extinguished by gaps in such regions, this tends to happen early in the formation of the jams, when they are relatively small in size, and therefore they tend to be difficult to ascertain visually within more highly aggregated vehicle-trajectory plots such as Fig. 10.

7 Conclusions

The cellular automata CA-184s2ss has been shown to replicate elements of both two-capacity flow and wide moving jams at freeway bottlenecks, as observed in actual traffic (cf. respectively [3] and [27]). Notwithstanding that neither of these phenomena is replicated by the KWM, sufficiently coarsely time-aggregated flows downstream of the (CA-184s2ss simulated) bottleneck yield an empirical capacity. Use of this empirically determined bottleneck within the KWM replicates simulations reasonably well (a few percent), as compared either to coarse time aggregations or ensemble averages of fine time aggregations.

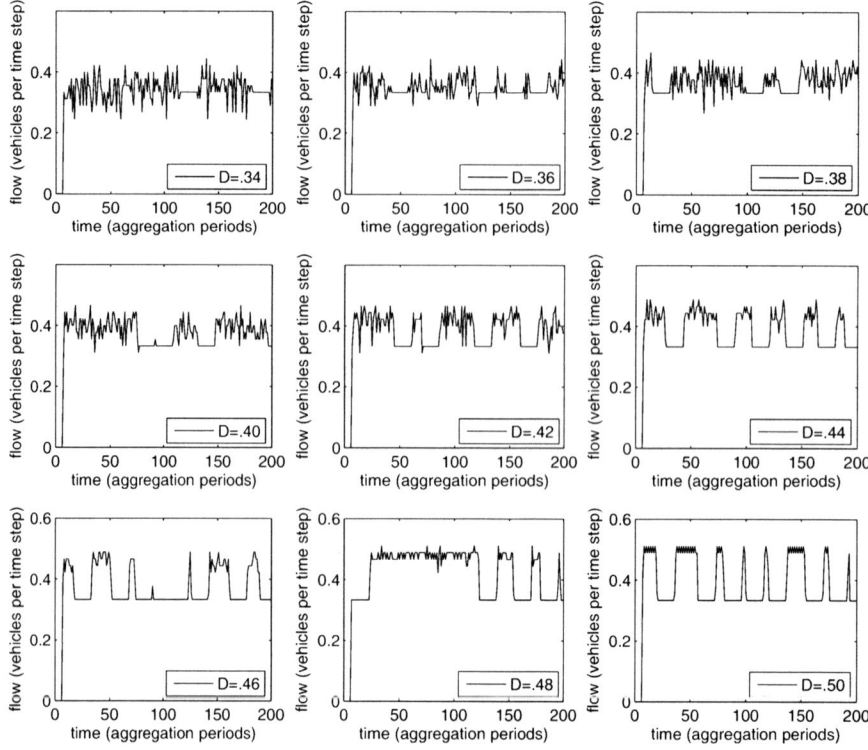

Fig. 15. Flow (vehicles/time step) vs. time (aggregation periods of 45 time steps each) for typical sample paths of 9000 time steps each, with hesitation probability $p_b = .005$, varying demands, and other CA-184s2ss parameters as described in the text.

It would be interesting to make similar comparisons, with the time aggregations carried out over short intervals and with the bottleneck capacity employed in the KWM taken as a stochastic variable, as suggested in [2] and [19].

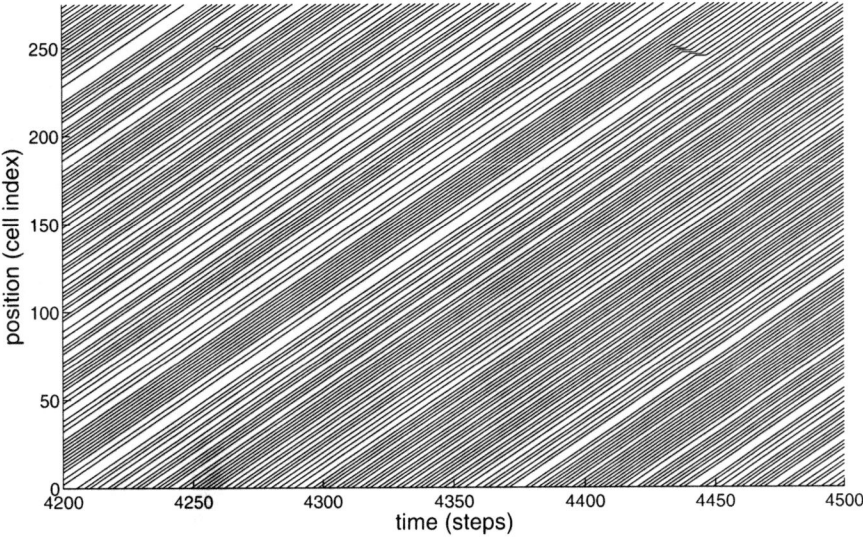

Fig. 16. Expanded view of vehicle trajectories for the sample path of Fig. 15 corresponding to $D = .36$. Note the incipient jams formed at the bottleneck at time steps 4259 and 4434, but extinguished by gaps arriving shortly after their formation.

References

1. R. Kühne, R. Mahnke, in Transportation and Traffic Theory, Proceedings 16th Int. Symp. Transportation and Traffic Theory, edited by H. Mahmassani (Elsevier, Amsterdam, 2005), p. 229.
2. L. Elefteriadou, R. P. Roess, W. R. McShane, Transp. Res. Rec. 1484, 80 (1995).
3. M. J. Cassidy, R. L. Bertini, Transp. Res. B, 33, 25 (1999).
4. J. L. Evans, L. Elefteriadou and N. Gautam, Trans. Res. B, 35, 237 (2001).
5. M. J. Lighthill, G. B. Whitham, Proc. Roy. Soc. London, A229, 317 (1955).
6. P. I. Richards, Operations Research, 4, 42 (1956).
7. P. Nelson, N. Kumar, "Point-constriction, interface and boundary conditions for the kinematic-wave model," in Procs. 83rd Annual Meeting Transportation Research Board, Wash. D. C., Jan. 11-15, 2004 (published in CD format).
8. S. Wolfram, J. Stat. Phys. 45, 471 (1986).
9. E. P. Wigner, Comms. Pure Appl. Math., 13, 1 (1960).
10. L. C. Edie, R. S. Foote, Proceedings of the Highway Research Board, 37, 492 (1960).
11. K. Agyemang-Duah, F. L. Hall, in Procs. Int. Symp. Highway Capacity and Level of Service, edited by U. Brannolte, (A. A. Balkema Press, Rotterdam, 1991), p. 1.
12. L. Newman, Highway Research Record, 167, 14 (1963).
13. B. N. Persaud, Study of a freeway bottleneck to explore some unresolved traffic flow issues, Ph.D. Dissertation, University of Toronto, Canada (1986).

14. F. L. Hall, L. M. Hall, Transp. Res. Rec. 1287, 108 (1990).
15. B. N. Persaud, V. F. Hurdle, in Procs. Int. Symp. Highway Capacity and Level of Service, edited by U. Brannolte (A. A. Balkema Press, Rotterdam, 1991), p. 289.
16. J. H. Banks, Transp. Res. Rec. 1320, 83 (1991).
17. J. H. Banks, Transp. Res. Rec. 1287, 20 (1990); J. H. Banks, Transp. Res. Rec. 1320, 234 (1991).
18. J. A. Wattleworth, Traffic Engineering, 34 15 (1963).
19. W. Brilon, J. Geistefeldt and M. Regler, in Transportation and Traffic Theory, Proceedings 16th Int. Symp. Transportation and Traffic Theory, edited by H. Mahmassani (Elsevier, Amsterdam, 2005), p. 125.
20. J. S. Drake, J. L. Schofer, A. D. May, Highway Research Record, 154, 53 (1967).
21. M. M. Koshi, M., M. Iwasaki, I. Ohkura, in Transportation and Traffic Theory: Proceedings of the 8th ISTTT, edited by V. F. Hurdle, E. Hauer and G. N. Steuart, (University of Toronto Press, Toronto, 1981) p. 403.
22. F. L. Hall, K. Agyemang-Duah, Transportation Research Record, 1320, 91 (1991).
23. C. F. Daganzo, Fundamentals of Transportation and Traffic Operations, Pergamon, Oxford (1997).
24. P. Nelson, "On Two-regime Flow, Fundamental Diagrams and Kinematic-wave Theory," Transportation Science, conditionally accepted.
25. S. C. Benjamin, N. F. Johnson, P. M. Hui, J. Phys. A, 29, 3119 (1996).
26. K. Nagel, M. Paczuski, Phys. Rev. E, 51, 2909 (1995).
27. B. S. Kerner, The Physics of Traffic, Springer, Berlin, 2004.
28. E. Brockfeld, R. D. Kühne, P. Wagner, "Calibration and Validation of Microscopic Traffic Flow Models," in Procs. 83rd Annual Meeting Transportation Research Board, Wash. D. C., Jan. 11-15, 2004 (published in CD format).

Mechanical Restriction Versus Human Overreaction: Accident Avoidance and Two-Lane Traffic Simulations

Andreas Pottmeier[1], Christian Thiemann[1], Andreas Schadschneider[2], and Michael Schreckenberg[1]

[1] Physik von Transport und Verkehr, Universität Duisburg-Essen, 47057 Duisburg, Germany
[2] Institut für Theoretische Physik, Universität zu Köln, 50937 Köln, Germany

Summary. Lee *et.al.* [1] have proposed a cellular automaton model that emphasizes the conflict between human overreaction and limited mechanical capabilities as the origin of congested traffic states. The limited acceleration and deceleration capabilities lead to a rather different approach to realize realistic traffic modeling. But the original model lacks the robustness and usability for more complicated and flexible simulations. In order to allow an extension of the model to two-lane traffic a modification of the original single-lane model is presented that ensures the absence of any collisions in the model dynamics.

1 Introduction

H. K. Lee *et al.* [1, 2] introduced an advanced cellular automaton model for single-lane traffic that allows to reproduce the different forms of synchronized traffic, the pinch effect [3–5] and also short time-headways in free and synchronized flow [6]. In the model, realistic flow properties are a consequence of moderate driving. It takes into account finite acceleration and deceleration properties and also the attitude of the drivers.

Due to the finite deceleration capabilities, the model is not intrinsically accident-free. Usually collisions are avoided by introducing a strict hardcore repulsion between the individual cars. However, this typically leads to processes with a very large deceleration. In the presence of limited deceleration capabilities crashes have to be avoided by choosing the dynamics appropriately. In the original version of the model [1], accidents could occur if the initial state is not chosen carefully. This makes the extension of the model to more complex scenarios like two-lane traffic difficult. Therefore an extension of the model that increases its robustness against accidents is required.

So in parallel to the development of the rule-set for a secure lane change the original one-lane model is adapted. Here the calculation scheme of the parameter γ — which describes the attitude of the driver — has to be changed. The characteristics of the accidents observed in simulations showed that a delayed change from the optimistic to the pessimistic state is the source of the accidents in the original model. The vehicle may react too late to changes in the neighborhood of the leading vehicles, especially of the second-nearest car ahead.

2 Model Definition

First we recapitulate the rule set of the one-lane model. The core is an inequality that defines a velocity c_n^{t+1} which is considered to be safe by the driver:

$$x_n^t + \Delta + \overbrace{\sum_{i=0}^{\tau_f(c_n^{t+1})} (c_n^{t+1} - Di)}^{\tau_f(c_n^{t+1})} \leq x_{n+1}^t + \overbrace{\sum_{i=1}^{\tau_l(v_{n+1}^t)} (v_{n+1}^t - Di)}^{\tau_l(v_{n+1}^t)}. \tag{1}$$

x_n^t and v_n^t are position and velocity of vehicle n, respectively, and Δ represents the minimum gap between the vehicles and is at least the length L of the leading vehicle. Each summation in (1) denotes a deceleration cascade with maximum braking capability D. As long as both $\tau_f(v)$ and $\tau_l(v)$ are set to v/D and $\Delta = L$, the deceleration would end in a bumper-to-bumper configuration. But this is weakened if the human factor is introduced. Note that c_n^{t+1} is not uniquely determined by (1) and the upper limit is used.

In order to model the different behavior of drivers one distinguishes between optimistic and pessimistic driving. The former controls the behavior in free flow. Vehicles drive unhindered and drivers accept "unsafe" gaps, i.e. gaps smaller than those allowing them to react to an emergency braking of the leading vehicle. Short time-headways below 1 sec are possible [6]. The latter governs the driving at high densities. Interactions between the cars are strong and braking is likely. The vehicles drive pessimistic and remain aloof. This leads to the following definition of γ:

$$\gamma_n^t = \begin{cases} 0 & \text{for } v_n^t \leq v_{n+1}^t \leq v_{n+2}^t \quad \text{or} \quad v_{n+2}^t \geq v_{\text{fast}} \\ 1 & \text{otherwise}, \end{cases} \tag{2}$$

where the parameter v_{fast} is slightly smaller than the maximum velocity v_{max}. The upper limits $\tau_f(v)$ and $\tau_l(v)$ of the summation (1) as well as the minimum gap between the cars are determined by

$$\Delta = L + \gamma_n^t \max\{0, \min\{g_{\text{add}}, v_n^t - g_{\text{add}}\}\},$$
$$\tau_f(v) = \gamma_n^t v/D + (1 - \gamma_n^t) \max\{0, \min\{v/D, t_{\text{safe}}\} - 1\}, \tag{3}$$
$$\tau_l(v) = \gamma_n^t v/D + (1 - \gamma_n^t) \min\{v/D, t_{\text{safe}}\}.$$

Here D denotes the deceleration capability of the vehicles and t_{safe} the maximum number of time steps a vehicle observes its own safety when driving optimistically. The remaining update is as follows:

1. $p = \max\{p_d, p_0 - v_n^t(p_0 - p_d)/v_{slow}\}$
2. $\tilde{c}_n^{t+1} = \max\{c_n^{t+1} \mid c_n^{t+1} \text{ satisfies Eq. (1)-(3)}\}$
3. $\tilde{v}_n^{t+1} = \max\{0, v_n^t + a, \max\{0, v_n^t - D, \tilde{c}_n^{t+1}\}\}$
4. $v_n^{t+1} = \max\{0, v_n^t - D, \tilde{v}_n^{t+1} - \eta\}, \quad \text{with } \eta = 1 \text{ if } \text{rand()} < p \text{ or } 0 \text{ else}$
5. $x_n^{t+1} = x_n^t + v_n^{t+1}$

For a detailed description of the update rules, see [1, 2].

As mentioned above the key to a safe one-lane model rests on a more careful determination of γ. Simulations show that two different scenarios lead to a dangerous configuration [7]: (i) if all three vehicles involved in the calculation of γ drive optimistically with the same velocity; (ii) if the velocity difference between v_{n+1} and v_{n+2} is too high. Nevertheless, the frequency of accidents is very low. In the free flow region we could not detect any accidents, at higher densities (e.g. $\rho = 60$ veh/km we found a maximum of the order of 10^8 time-steps in which a vehicle has one accident.

That means critical situations emerge at the transition between optimistic and pessimistic driving as the vehicle reacts with an offset of one time step. Due to the nature of the collisions observed in the original model we changed the definition of γ_n^t. To prevent dangerous situations of the first type the second inequality in $v_n^t \leq v_{n+1}^t \leq v_{n+2}^t$ is strengthened to $v_{n+1}^t < v_{n+2}^t$. Accidents due to a large velocity difference are in general eliminated by a upper limit of the difference between v_n and v_{n+1}.

Additionally it was needed to add a stronger interaction between each vehicle n and $n + 2$. This is done by introducing a brake-light b_n. It denotes whether the vehicle has reduced its velocity because of its surrounding, but not because of dawdling (i.e. the randomization):

$$b_n^t = \begin{cases} 1 & \text{for } \tilde{v}_n^{t+1} < v_n^t \\ 0 & \text{otherwise.} \end{cases} \tag{4}$$

Note, that the brake-light has not the same role as in [8]. Here it provides a way to communicate the presence of a hindrance and therefore a possible change of the optimistic state to the following cars. So each vehicle is able to sense a critical situation early enough.

The parameter γ is now determined considering also the state of the brake-light of the $(n + 2)$th vehicle:

$$\gamma_n^t = \begin{cases} 0 & \text{for } v_n^t \leq v_{n+1}^t < v_{n+2}^t \\ & \text{or } \left(v_{n+2}^t \geq v_{fast} \wedge v_n^t - v_{n+1}^t \leq D \wedge b_{n+2}^t = 0 \right) \\ 1 & \text{otherwise.} \end{cases} \tag{5}$$

The remaining update is unchanged.

3 Results

The changes in the definition of the model dynamics hardly influence the macroscopic and microscopic results [7]. The dynamics of the model is still

capable to reproduce the traffic patterns observed empirically as well as the
short time-headways in the free flow phase and *synchronized traffic*. Fig. 1
shows examples for space-time plots of a system with an on-ramp. The left
part shows synchronized traffic at an on-ramp in which compact jams emerge
because of the so-called *pinch effect* [3, 4]. The right plot in Fig. 1 shows localized synchronized traffic. Note that widening and moving synchronized patterns are reproduced as well. The insertion algorithm follows the one described
in [1]. The following model parameters, which are motivated by empirical facts
and already utilized in [1], are used in the simulations: $a = 1$, $D = 2$, $l = 5$,
$v_{\text{fast}} = 19$, $t_{\min} = 3$, $g_{\text{add}} = 4$, $p_0 = 0.32$, $p_{\text{d}} = 0.11$, $v_{\text{slow}} = 5$, $v_{\max} = 20$. The
length of one cell is chosen to be $\Delta x = 1.5\,\text{m}$ and one time-step to $\Delta t = 1\,\text{sec}$.

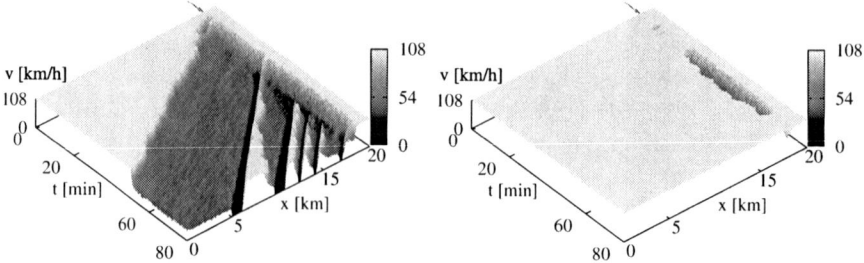

Fig. 1. Impact of an on-ramp: General pattern at $(q_{\text{main}}, q_{\text{ramp}}) = (0.55, 0.16)$ (left),
and localized synchronized pattern at $(0.38, 0.17)$ (right)

Furthermore, we are optimistic to prove rigorously the absence of accidents
in the modified model [7]. In order to estimate the influence of the changes to
the original model made here we have determined how often the new rules are
applied and lead to a different result. In the free flow region the probability of
changed γ_n (see Eq. (5)) compared to the definition in [1] is about 0.004 per
car and second. That means that the influence of the changed rules is small.

4 Two-Lane Traffic

In addition to the modifications that lead to the absence of accidents, simulating realistic scenarios requires the extension to a two-lane model. Here it
is especially important to keep in mind the missing hardcore repulsion. The
lane changes have to take into account safety in the sense of not interfering
with cars on the other lane, but also avoiding accidents in the process of lane
changing. In this contribution we consider only symmetric lane changes.

In the two-lane model each time-step is separated into two sub-steps [9]. In
the first step for each vehicle it is decided whether it will change lane in
the current time-step. In the second step a normal one-lane model time-step
update is applied to each of the two lanes.

The basic idea behind the two-lane model is using the condition for $\tilde{c}_n^{t+1}$
to determine the safety of a possible lane change. In each time-step it is

checked whether the vehicle would be able to drive with the safe velocity if it changes lane. It is also checked whether the follower on the destination lane would be able to drive safely. The parameter $\beta \in \{0, \ldots, D\}$ controls how much the follower may be constrained. For $\beta = 0$ only smooth lane changes are allowed where the follower is not forced to brake at all, while at $\beta = D$ even decelerations with maximal braking capability are acceptable. These conditions constitute the *security criterion* and determine whether a vehicle *can* change lanes without obstructing vehicles on the destination lane or even provoking a dangerous situation.

A vehicle *wants* to change lane only if the *mobility criterion* is satisfied. For the symmetric two-lane model this means that the vehicle can drive faster on the destination lane than on its current lane. This is determined by calculating $\tilde{v}_n^{t+1}$ on the current lane and then virtually changing the vehicle to the other lane, calculating $\tilde{v}_n^{t+1}$ again and comparing these two.

According to [10] a lane change takes $t_{lc} = 3$ seconds in time. Therefore the security criterion must hold for at least three time-steps until a positive mobility criterion can trigger the vehicle to actually change the lane. For each vehicle a new variable ϑ_n^t is introduced that is initially $\vartheta_n^0 = 0$ and acts as a counter for time-steps in which the security criterion is valid.

To formally describe the update rules of the model, $l_n^t \in \{0, 1\}$ denotes the lane used by vehicle n at time t. $\mathcal{F}_l(n)$ denotes the follower and $\mathcal{L}_l(n)$ the leader of the vehicle n on lane l. Some values are calculated for a vehicle virtually changing the lane. In this case, a second subscript is used to specify the lane. Thus $\tilde{v}_{n,l}^{t+1}$ means "$\tilde{v}_n^{t+1}$, if the vehicle would be on lane l".

The update rules of the two-lane model are:

1. $\vartheta_n^{t+1} = \vartheta_n^t + 1$ if all of the following conditions hold, or 0 otherwise:
   - $x_n^t - x_{\mathcal{F}_{1-l}(n)}^t > L + g_{\text{safe}}$ and $x_{\mathcal{L}_{1-l}(n)}^t - x_n^t > L + g_{\text{safe}}$
   - $v_n^t - D \leq \tilde{c}_{n,1-l_n^t}^{t+1}$
   - $v_{\mathcal{F}_{1-l_n^t}(n)}^t - \beta \leq \tilde{c}_{\mathcal{F}_{1-l_n^t}(n)}^{t+1}$
2. $l_n^{t+1} = 1 - l_n^t$ if $\tilde{v}_{n,1-l_n^t}^{t+1} > \tilde{v}_{n,l_n^t}^{t+1}$ and $\vartheta_n^{t+1} \geq t_{lc}$, or l_n^t otherwise.
3. $\vartheta_n^{t+1} = 0$ if $l_n^{t+1} \neq l_n^t$.

Here g_{safe} denotes an optional security gap which becomes important at higher densities where the gap between the vehicles approaches 0.

An important result of this two-lane traffic model that distinguishes this approach from former ones (see e.g. [9, 11] for an overview) is shown in Fig. 2. In the low-density region the number of lane changes N_{lc} rises until an occupation of about 0.08. Than it decreases until an occupation *occ* of about 0.15 when the system runs into the synchronized state. Here N_{lc} nearly reaches 0. This means in the presence of synchronized traffic on both lanes the vehicles do not change anymore. This effect does not depend on t_{lc} and shows the strong synchronization [12] between the two lanes, as lane-changing has no advantage.

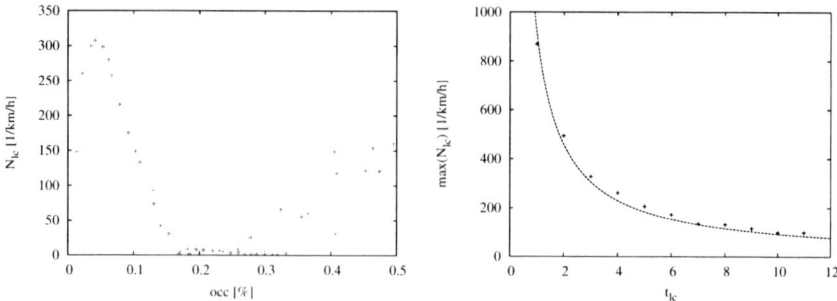

Fig. 2. The number of lane changes N_{lc} per hour and kilometer (left) depending on the occupation occ of the road. The lane change rate disappears in the presence of synchronized traffic. The maximum decreases exponentially with respect to t_{lc}.

The number of lane changes increases again at higher occupations $occ > 0.35$. Here we do have phase-separated traffic and cars reaching the congested area tend to change the lane if the upstream end of the congestion is starting earlier on its lane than on the other. Note that t_{lc} is important for the number of lane changes N_{lc}. The lane change rate decreases exponentially with t_{lc}.

References

1. H. K. Lee, R. Barlovic, M. Schreckenberg, and D. Kim: Phys. Rev. Lett. **92**, 23702 (2004)
2. H. K. Lee, R. Barlovic, M. Schreckenberg, and D. Kim: In *Traffic and Granular Flow '03*, edited by S. P. Hoogendoorn, S. Luding, P. H. L. Bovy, M. Schreckenberg, D. E Wolf. Springer Berlin, (2005)
3. B. S. Kerner: Phys. Rev. Lett **81**, 3797 (1998)
4. B. S. Kerner: Phys. Rev. E **65**, 046138 (2002)
5. B. S. Kerner and S. L. Klenov: J. Phys. A: Math. Gen. **35**, L31 (2002); B. S. Kerner, S. L. Klenov, and D. E. Wolf: J. Phys. A: Math. Gen. **35**, 9971 (2002)
6. W. Knospe, L. Santen, A. Schadschneider, and M. Schreckenberg: Phys. Rev. E **65**, 056133 (2002)
7. C. Thiemann, A. Pottmeier, A. Schadschneider, M. Schreckenberg: in preparation
8. W. Knospe, L. Santen, A. Schadschneider, and M. Schreckenberg: J. Phys. A**33**, L477 (2000)
9. D. Chowdhury, L. Santen, A. Schadschneider: Phys. Rep. **329**, 199 (2000)
10. U. Sparmann: 'Spurwechselvorgänge auf zweispurigen BAB-Richtungsfahrbahnen'. In: Forschung Straßenbau und Straßenverkehrstechnik Heft 263 (Bundesministerium für Verkehr, Bonn-Bad Godesberg, 1978)
11. P. Wagner, K. Nagel, and D. E. Wolf: Physica A **234**, 687 (1997)
12. M. Koshi, M. Iwasaki, and I. Ohkura: In *Proc. 8th International Symposium on Transportation and Traffic Theory* edited by: V. F. Hurdle, E. Hauert, and G. N. Stewart. University of Toronto Press, Toronto, pp. 403–426, (1983)

Ramp Effects in Asymmetric Simple Exclusion Processes

Ding-wei Huang

Department of Physics, Chung Yuan Christian University, Chung-li, Taiwan

Summary. We present analytical results of ramp effects in asymmetric simple exclusion processes. Both on-ramp and off-ramp are included in between the two open boundaries. Exact phase diagrams are obtained analytically in the full parameter space. We find that the order of the two ramps is crucial. When the on-ramp is placed after the off-ramp along the traffic direction, there are only four distinct phases since free flow will not follow a congestion. When the on-ramp is placed before the off-ramp, we observe a new phase. The bottleneck emerges as the flow in between the two ramps saturates to its maximum. Applications to a traffic rotary are discussed.

1 Introduction

Recently, traffic related problems have attracted much attention from physicists [1, 2]. Not only are the problems highly relevant for our modern life, they also provide excellent examples for the phenomena of boundary-induced phase transitions [3, 4]. Traffic flow is basically a one-dimensional phenomenon. With naive intuition, congestions result whenever the in-flow is larger than the out-flow. On the other hand, if the in-flow is less than the out-flow, the vehicles might move freely. However, such impressions are only partially correct. Free flow and congestion can be steady phases on the roadway, instead of transient situations in the naive explanation. Such steady phases present in a system driven far away from equilibrium, where a steady current is maintained asymptotically. The most basic model to capture such a feature is the asymmetric simple exclusion process [5, 6]. The model has been studied thoroughly in a simple configuration of one roadway with two open ends and no ramp. As nontrivial boundaries, ramps can be expected to influence the traffic strongly [7–9]. In this work, we introduce ramps along the roadway and study their influence on the traffic. Exact phase diagrams are obtained. The full parameter space can be completely classified. As a step towards more complex networks, the flow around a traffic rotary is also analyzed.

2 Model

We study the phase diagrams of asymmetric simple exclusion processes with open boundaries and ramps. The system configurations are shown in Fig. 1. A simple roadway is represented by a one-dimensional lattice. Each site can be accommodated by one particle only. At each time step, every particle hops forward to the next site as long as that site was empty in the previous time step. The dynamics in the bulk is fully deterministic. There are four non-trivial sites: particles can be injected from the first site (left end) and from the site designated as the on-ramp; particles can also be removed from the last site (right end) and from the site designated as the off-ramp. Although particles move deterministically along the main road, their injection and removal from these four special sites is stochastic. The injection rates from the left end and the on-ramp are denoted by α_0 and α_1, respectively; the removal rates from the right end and the off-ramp are denoted by β_0 and β_1, respectively. In our previous study [10], we have obtained the analytical phase diagrams in the cases of a single ramp. Now, we consider the case of two ramps, first with the off-ramp placed before the on-ramp as shown in Fig. 1(a). As the two ramps divide the roadway into three homogeneous parts, we can replace the ramp flow by effective boundaries. The flow through the off-ramp β_1 can be represented by a removal rate β' for the first part and an injection rate α' for the second part of the roadway; the flow through the on-ramp α_1 can also be represented by a removal rate β'' for the second part and an injection rate α'' for the third part of the roadway. The regime of the (FFF) phase can be obtained by imposing the constraints $\alpha_0 < \beta'$, $\alpha' < \beta''$, and $\alpha'' < \beta_0$. The effective rates can be solved by balancing the flow across the ramp. The analytical expressions are as follows:

$$\alpha' = \frac{\alpha_0(1 - \beta_1)}{1 + \alpha_0\beta_1} \quad \text{and} \quad \beta' = 1 \ ; \tag{1}$$

$$\alpha'' = \alpha' + \alpha_1(1 + \alpha') \quad \text{and} \quad \beta'' = \frac{\alpha'(1 + \alpha_1)}{\alpha' + \alpha_1(1 + \alpha')} \ . \tag{2}$$

It is interesting to note that the crucial condition is the free flow on the third part of the roadway, i.e., $\alpha'' < \beta_0$. The regimes for other phases can also be obtained similarly. We summarize the results as

$$\text{(FFF)} \quad \alpha_1(1 + \alpha_0) - \beta_1(1 + \beta_0)\alpha_0 < \beta_0 - \alpha_0 \ ; \tag{3}$$

$$\text{(FFJ)} \quad \alpha_1(1 + \alpha_0) - \beta_1(1 + \beta_0)\alpha_0 > \beta_0 - \alpha_0 \ ,$$
$$\alpha_1(1 + \alpha_0)\beta_0 - \beta_1(1 + \beta_0)\alpha_0 < \beta_0 - \alpha_0 \ ; \tag{4}$$

$$\text{(FJJ)} \quad \alpha_1(1 + \alpha_0)\beta_0 - \beta_1(1 + \beta_0) < \beta_0 - \alpha_0 \ ,$$
$$\alpha_1(1 + \alpha_0)\beta_0 - \beta_1(1 + \beta_0)\alpha_0 > \beta_0 - \alpha_0 \ ; \tag{5}$$

$$\text{(JJJ)} \quad \alpha_1(1 + \alpha_0)\beta_0 - \beta_1(1 + \beta_0) > \beta_0 - \alpha_0 \ . \tag{6}$$

We note that the four dimensional parameter space $(\alpha_0, \beta_0, \alpha_1, \beta_1)$ can be completely classified into these four distinct phases. With naive intuition, the two ramps divide the roadway into three sections, and each section can be either free or jam. Thus one would expect eight different phases. However, there are four phases missing: (JFF), (JJF), (JFJ), and (FJF). Along the traffic direction, free flow will not follow a congestion. Basically, the traffic jams emerge as α_0 and/or α_1 increases and resolve as β_0 and/or β_1 increases. The numerical simulations can be exactly reproduced as shown in Fig. 2.

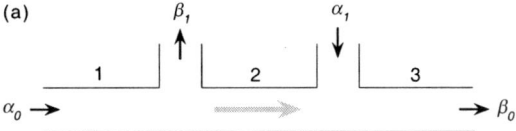

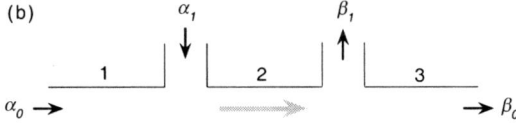

Fig. 1. System configurations: (a) on-ramp α_1 placed after off-ramp β_1; (b) on-ramp α_1 placed before off-ramp β_1. The gray arrow shows the direction of particle hopping. The three parts of the roadway are labelled by the number.

3 Maximum Flow

Next, we switch the order of the two ramps. The on-ramp is now placed before the off-ramp as shown in Fig. 1(b). The results of last section should be revised accordingly. We summarize the results for the four different phases:

(FFF) $\alpha_1(1+\alpha_0) - \beta_1(1+\beta_0)\alpha_0 - \alpha_1\beta_1(1+\alpha_0)(1+\beta_0) < \beta_0 - \alpha_0$,

$\alpha_1(1+\alpha_0) < 1 - \alpha_0$;   (7)

(FFJ) $\alpha_1(1+\alpha_0) - \beta_1(1+\beta_0)\alpha_0 - \alpha_1\beta_1(1+\alpha_0)(1+\beta_0) > \beta_0 - \alpha_0$,

$\alpha_1(1+\alpha_0) - \beta_1(1+\beta_0) < \beta_0 - \alpha_0$;   (8)

(FJJ) $\alpha_1(1+\alpha_0)\beta_0 - \beta_1(1+\beta_0) + \alpha_1\beta_1(1+\alpha_0)(1+\beta_0) < \beta_0 - \alpha_0$,

$\alpha_1(1+\alpha_0) - \beta_1(1+\beta_0) > \beta_0 - \alpha_0$;   (9)

(JJJ) $\alpha_1(1+\alpha_0)\beta_0 - \beta_1(1+\beta_0) + \alpha_1\beta_1(1+\alpha_0)(1+\beta_0) > \beta_0 - \alpha_0$,

$\beta_1(1+\beta_0) < 1 - \beta_0$.   (10)

It is interesting to notice that the above four phases do not completely classify the parameter space $(\alpha_0, \beta_0, \alpha_1, \beta_1)$. In fact, one more distinct phase can be observed. The traffic flow saturates to the maximum value on the second part of the roadway, while the congestion remains on the first part of the roadway and the free flow is maintained on the third part of the roadway. The phase regime can be obtained as follows:

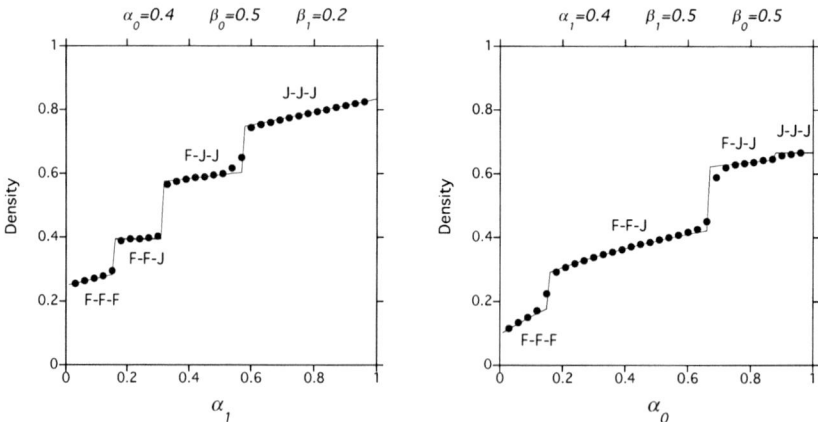

Fig. 2. Bulk density of the system shown in Fig. 1(a) (with three parameters fixed).

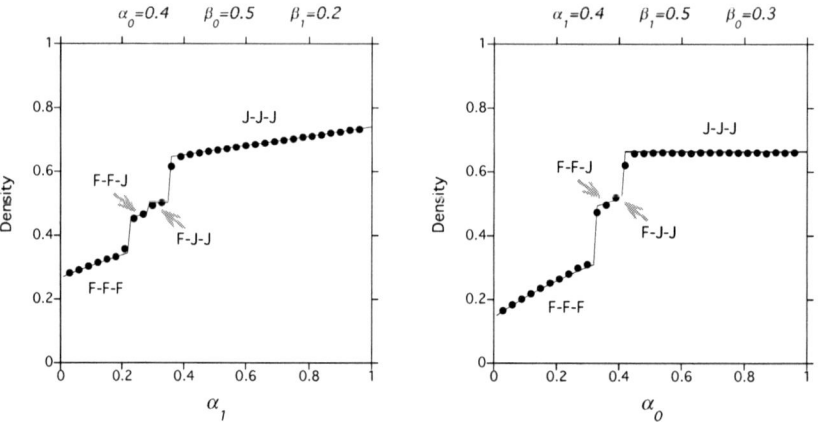

Fig. 3. Bulk density of the system shown in Fig. 1(b) (with three parameters fixed).

$$\text{(JMF)} \quad \alpha_1(1 + \alpha_0) > 1 - \alpha_0 \ ,$$
$$\beta_1(1 + \beta_0) > 1 - \beta_0 \ , \tag{11}$$

where free flow (F), jams (J), and maximum flow (M) can be observed in different parts of the roadway. With these five different phases, the parameter space $(\alpha_0, \beta_0, \alpha_1, \beta_1)$ can then be completely classified. In last section (the off-ramp placed before the on-ramp), the maximum flow can only be observed along the phase boundaries for the extreme conditions $\alpha_0 = 1$ and/or $\beta_0 = 1$. When the off-ramp is placed after the on-ramp, the maximum flow can be observed in an extended regime and becomes a distinct phase. Basically, congestions develop as α_1 increases and the free flow restores as β_1 increase. However, when α_1 and β_1 are larger than certain criteria, the traffic flow would saturate in the middle section of the roadway and a new phase appears. Similarly, the new phase emerges whenever α_0 and β_0 are larger than certain criteria. We emphasize again that the above analytical results are exact. The numerical simulations can be correctly reproduced, see Fig. 3.

4 Traffic Rotary

When periodic boundary conditions are imposed, the system configuration shown in Fig. 1 becomes a traffic rotary. With two on-ramps and two off-ramps, the rotary shown in Fig. 4 can be taken as an alternative to a conventional crossroad. Traffic from west to east is prescribed by α_1 and β_1; traffic from south to north is prescribed by α_2 and β_2. Beside the free flow (FFFF) and congestion (JJJJ), the maximum flow can only be expected in part 1 of the rotary. In such a situation, part 2 will be free flow and part 4 will be congested. Part 3 can be either free or jam. Thus we should have four different phases:

$$\text{(FFFF)} \quad \alpha_1 + \alpha_2 + \alpha_1\alpha_2 < \beta_1 + \beta_2 - \beta_1\beta_2 \; ; \tag{12}$$

$$\text{(JJJJ)} \quad \beta_1 + \beta_2 + \beta_1\beta_2 < \alpha_1 + \alpha_2 - \alpha_1\alpha_2 \; ; \tag{13}$$

$$\text{(MFFJ)} \quad \alpha_1 + \alpha_2 - \alpha_1\alpha_2 < \beta_1 + \beta_2 - \beta_1\beta_2 < \alpha_1 + \alpha_2 + \alpha_1\alpha_2 \; ; \tag{14}$$

$$\text{(MFJJ)} \quad \beta_1 + \beta_2 - \beta_1\beta_2 < \alpha_1 + \alpha_2 - \alpha_1\alpha_2 < \beta_1 + \beta_2 + \beta_1\beta_2 \; . \tag{15}$$

Again, the numerical simulations can be reproduced, see Fig. 5.

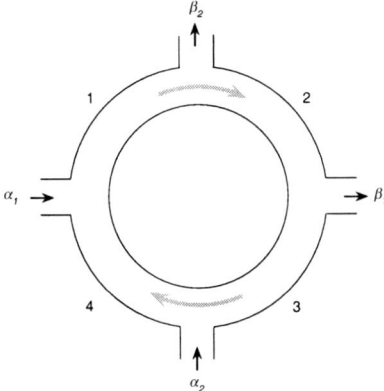

Fig. 4. System configuration of a traffic rotary. Particles move clockwise as shown by the gray arrows. The rotary is then divided into four parts labeled by numbers.

5 Concluding Remarks

We demonstrate that the bulk properties on a roadway are totally controlled by the stochastic ramp-flow through the boundaries. To classify the traffic

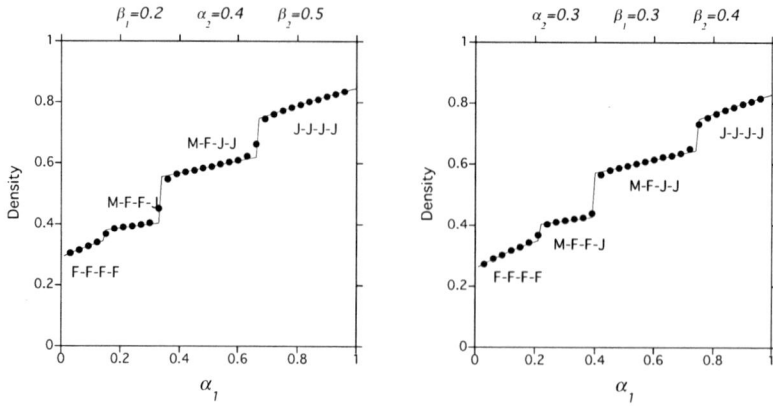

Fig. 5. Bulk density of the system shown in Fig. 4 (with three parameters fixed).

conditions, the roadway can be divided into various parts joined by the ramps. On each part, the traffic is homogeneous and can be characterized as free flow, congestion, or maximum flow. A complete classification in the parameter space is achieved. Exact phase diagrams are obtained analytically. Basically traffic jams emerge as the on-ramp flow increases. Free flow is restored as the off-ramp flow increases. In between these two phases, various kinds of inhomogeneity can develop among different parts of the roadway. Along the traffic direction free flow will not follow a congestion directly. To constitute a bottleneck, the maximum flow must appear in between the downstream free-flow and the upstream congestion, which also implies that the on-ramp must be placed before the off-ramp along the traffic direction. In the simple model studied, the present results are exact. These transparent results might be useful to analyze real traffic networks.

References

1. D. Chowdhury, L. Santen, and A. Schadschneider, Phys. Rep. **329**, 199 (2000).
2. R. Mahnke, J. Kaupužs, and I. Lubashevsky, Phys. Rep. **408**, 1 (2005).
3. J. Krug, Phys. Rev. Lett. **67**, 1882 (1991).
4. C. Appert and L. Santen, Phys. Rev. Lett. **86**, 2498 (2001).
5. B. Derrida and M. R. Evans, in *Nonequilibrium Statistical Mechanics in One Dimension*, V. Privman (Ed.), Cambridge University Press, Cambridge, UK, 1997.
6. B. Derrida, Phys. Rep. **301**, 65 (1998).
7. B. S. Kerner and H. Rehborn, Phys. Rev. Lett. **79**, 4030 (1997).
8. D. Helbing and M. Treiber, Phys. Rev. Lett. **81**, 3042 (1998).
9. H. Y. Lee, H. W. Lee, and D. Kim, Phys. Rev. Lett. **81**, 1130 (1998).
10. D. W. Huang, Phys. Rev. **E72**, 016102 (2005).

Linking Cellular Automata and Optimal-Velocity Models Through Wave Selections at Bottlenecks

Peter Berg and Justin Findlay

Faculty of Science, University of Ontario Institute of Technology,
2000 Simcoe Street N, Oshawa, ON, L1H 7K4, Canada

Summary. A bottleneck simulation of road traffic on a loop, using the cellular automata Nagel-Schreckenberg model (with $p = 0$), reveals three types of stationary wave solutions. They consist of i) two shock waves at the bottleneck boundaries, ii) one shock wave at the boundary and one on the "open" road and iii) the trivial solution, i.e. homogeneous, uniform flow. These solutions are selected dynamically from a range of stationary wave solutions, similar in fashion to the wave selection in a bottleneck simulation of the optimal-velocity model. This is yet another indication that CA and OV models share certain underlying dynamics, although the former are discrete in space and time while the latter are continuous.

Cellular automata (CA) models have been widely used to simulate traffic flow on highways and road networks [4, 8], in particular the Nagel-Schreckenberg model [6, 7]. Together with car-following (CF) and continuum models, they represent the three most popular classes of traffic models.

Analytical work by Berg *et al.* [2] and Lee *et al.* [5] has established a link between car-following models based on ordinary differential equations, and continuum models based on partial differential equations. While an analytical link between CA models and either CF or continuum models is still missing (mean-field theory aside), the dynamics of all three classes exhibit many common features such as sub-critical bifurcations, limit cycles and pattern formation [3, 4].

Bottlenecks are the major cause for highway congestion and, therefore, have been studied in some detail [3, 4, 9]. In this paper, a wave selection analysis of a bottleneck simulation reveals a fundamental link between the dynamics of CA models and optimal-velocity (OV) models, which belong to the class of CF models.

1 Cellular Automata Bottleneck Simulation

For traffic on a loop (periodic boundary conditions) of length L, a bottleneck of length L_B is located at $0 \leq x \leq L_B$. The system is simulated using the

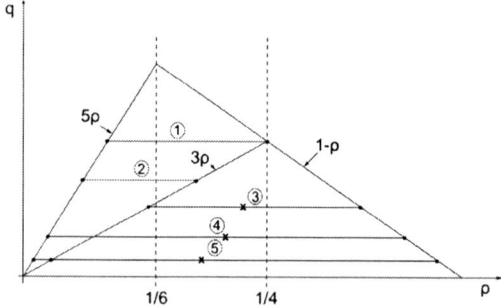

Fig. 1. Stationary shock waves visualized in the fundamental diagram: From six possible wave connections (the trivial uniform flow solution in Fig. 3c is not shown here) only three emerge as dynamical solutions, displayed in Figs. 3a-3c.

Nagel-Schreckenberg (NS) CA model [6] for vanishing randomness ($p = 0$) with a reduction in top speed from $v^{max} = 5$ on the "open" road to $v_B^{max} = 3$ in the bottleneck. All other model parameters remain the same. Initially, N cars are randomly distributed along the road and the system is updated up to $t = 10^6$ time steps. We set $L_B = 200$ and $L = 1000$.

Note that the wave selection on a loop is fundamentally different from wave selection on an "open" road with different conditions at the upstream and downstream boundary, respectively. Also, we set $p = 0$ in order to avoid the jam formation in the NS model, which would interfere with the stationary wave patterns.

2 Wave Selection in the Fundamental Diagram

We will use the fundamental diagram (FD), i.e. flux versus density, to interpret our numerical results. It is shown for both the bottleneck (q_B: flux, ρ: density) and the open road (q_o)

$$q_B = \begin{cases} 3\rho & ; 0 \leq \rho \leq 1/4, \\ 1 - \rho ; 1/4 < \rho \leq 1 \end{cases} \quad , \quad q_o = \begin{cases} 5\rho & ; 0 \leq \rho \leq 1/6, \\ 1 - \rho ; 1/6 < \rho \leq 1 \end{cases} \quad (1)$$

in Fig. 1. Note that they merge into the same function for $\rho \geq 1/4$.

Generally speaking, we could expect as many as six stationary wave solutions for the bottleneck simulation as $t \rightarrow \infty$. In the FD, five of them are visualized as chords with zero gradient due to vanishing wave speed. They connect plateaus between the bottleneck and the open road (case 2, 3 and 4) in case of one stationary shock wave at each bottleneck boundary. However, they can also entail a plateau connection on the open road as in cases 1 and 5. These five stationary wave patterns are shown in Fig. 2 in terms of density distribution along the loop. In addition, there is the trivial wave solution of homogeneous

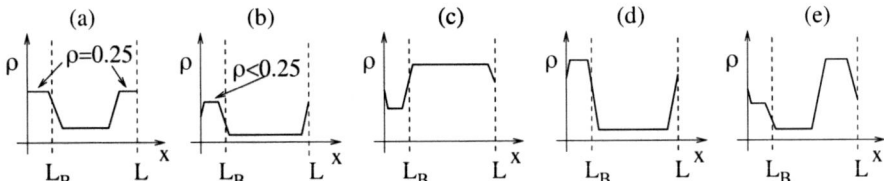

Fig. 2. Stationary wave patterns of Fig. 1: While case (a) and (b) occur in the simulations (cases (1) and (2) in Fig. 1, respectively), as shown in Figs. 3a and 3b, cases (c), (d), and (e) do not emerge. This is a very close analogy to the wave pattern selection of a bottleneck simulation with the optimal-velocity model (see Fig. 4) [10].

uniform flow. In principle, we could think of further wave patterns but we will restrict the analysis to the simplest cases featured here.

We found in our simulations that only three wave patterns are selected from this range of possible wave solutions. They consist of the following (ρ: average density on the loop):

- Case 1: $0.17 \leq \rho < 0.25$
  Stationary wave pattern that connects two plateaus by one shock at the downstream boundary of the bottleneck and one classical (Lax) shock on the open road (see Figs. 2a and 3a): The resulting bottleneck headway (distance between cars) is exactly at $d_B = 4$. On the open road we find the headway to be near $d_n = 7$, or exactly $d_o = 20/3$ on average.
- Case 2: $0 \leq \rho < 0.17$
  Stationary wave pattern that connects two plateaus by shocks at the upstream and downstream boundary of the bottleneck, respectively (see Fig. 2b and 3b): In the bottleneck $d_B > 4$ and on the open road $d_o > 20/3$. Note that it takes a very long time for the system to reach steady-state due to the small interaction of cars on the open road. Hence, Fig. 3b should be considered as a transient, quasi-steady state.
- Case 3: $\rho > 0.25$
  Trivial flow solution, i.e. homogeneous, uniform flow: Unless the average headway is close to an integer number, as is the case in Fig. 3c, the individual headways d_n oscillate around the average headway d. However, this is an effect solely due to the discretization of space, and the flow solution can still be considered uniform.

Three open questions remain:

1. Why is there a transition between the structures at $\rho = 0.17$ and $\rho = 0.25$?
2. What determines the location of the shock on the open road in case 1?
3. Why do we not observe the other wave patterns?

We will now briefly elaborate on all three questions.

The two headways in Fig. 3b are determined by the conservation of cars and by imposing zero wave speed (zero gradient of the chord in the FD). This can

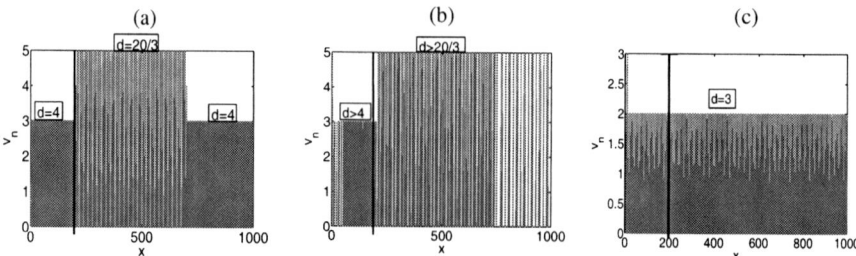

Fig. 3. Stationary wave patterns of bottleneck simulation: Vehicle speed v_n vs. position x; d: average headway. (a) $0.17 < \rho < 0.25$ (here: $\rho = 0.20$): Two shocks emerge, one at the downstream bottleneck boundary and one classical (Lax) shock on the open road. (b) Stationary wave pattern of bottleneck simulation with average density $\rho < 0.17$ (here: $\rho = 0.142$): Two shocks emerge, one at each bottleneck boundary. (c) Trivial flow solution for bottleneck simulation with average density $\rho > 0.25$ (here: $\rho = 0.333$).

be written as two equations with two unknowns, the bottleneck headway d_B and the open road headway d_o. Neglecting finite size effects, conservation of cars reads

$$\frac{L - L_B}{d_o} + \frac{L_B}{d_B} = N. \tag{2}$$

We find for the wave speed criterion (equal fluxes q_B and q_o in both road segments)

$$q_B = q_o \quad \Rightarrow \quad 3\rho_B = 5\rho_o \quad \Rightarrow \quad \frac{3}{d_B} = \frac{5}{d_o}, \tag{3}$$

where ρ_B and ρ_o denote the bottleneck and open road density, respectively. The system (2)-(3) can be solved for d_B. It yields

$$d_B = \frac{\frac{3}{5}L + \frac{2}{5}L_B}{N}. \tag{4}$$

In Fig. 3b, we have $L = 1000$, $L_B = 200$, $N = 142$ and, hence, $d_B = 4.79$. This equals $\rho_B = 1/d_B = 0.21$, which coincides with the numerical value. The value for d_o follows correspondingly.

The maximum amount of vehicles that the wave structure in Fig. 3b can support, however, is reached when $d_B = 4.0$ and, determined by zero wave speed, $d_o = 20/3$. For $L_B = 200$, this corresponds to an average density of

$$\rho = 0.2\frac{1}{4} + 0.8\frac{1}{20/3} = 0.17, \tag{5}$$

which is coincidentally close to $q = 1/6$, the maximum of q_o. If we increase ρ beyond this, the wave pattern in Fig. 3a is triggered, with a bottleneck headway exactly at $d_B = 4$ and $d_o = 20/3$ for the other plateau value. This is shown by chord 1 in Fig. 1. The length of the second plateau L_p is now determined by the conservation of cars alone. Setting $\rho = 0.2$ in Fig. 3a, we write

$$\rho = \frac{1}{4}(1 - L_p/L) + \frac{1}{20/3}(L_p/L) \quad \Rightarrow \quad L_p = 500, \tag{6}$$

which is very close to the numerical value of $L \approx 510$. Note that finite size effects impose limits on the accuracy of estimates. For $\rho \geq 0.25$, this wave pattern must vanish and we are left with the trivial flow solution of Fig. 3c.

3 Link to Optimal-Velocity Model

While the above analysis explains the transition between the three wave patterns, which are observed in the numerical simulations, it does not answer why the remaining wave patterns in Fig. 1 do not occur.

A similar bottleneck simulation with the (linearly stable) OV model [1] has been carried out by Ward *et al.* [10]. Here, the bottleneck was simulated by a reduction factor in the optimal velocity function. Again, three wave patterns emerge from the numerical results, as indicated by the chords in the fundamental digram in Fig. 4.

These wave patterns can partly be explained by kinematic wave theory (characteristics crossing boundaries between bottleneck and "open" road) and a phase-plane analysis of stationary travelling waves in the corresponding continuum model [2, 5, 10]. For example, case 2 of the OV model is a connection between two saddle points. A full explanation of the wave structures might require a higher-order continuum model though. For case 1 of the CA model, however, we cannot define a characteristic wave speed at the maximum of the bottleneck flux ($\rho = 1/4$) and the method of characteristics cannot be applied to explain the emerging wave pattern.

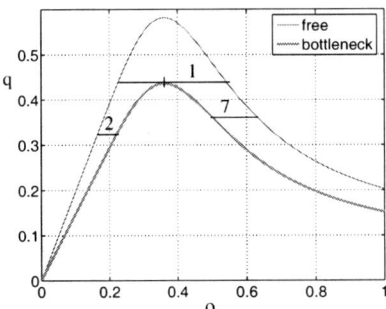

Fig. 4. Fundamental diagram and wave selection of a bottleneck simulation with the (linearly stable) OV model [10]. Three stationary wave patterns emerge. Cases (1) and (2) of the OV simulation correspond to cases (1) and (2) of the CA simulation (Fig. 1). Case (7) corresponds to the trivial flow solution of Fig. 3c due the identical fundamental diagrams of the bottleneck and the open road for $\rho \geq 0.25$ in Fig. 1.

Nevertheless, there is a one-to-one correspondence between the three solutions of the OV and the CA model:

1. OV model, case 1 $\leftrightarrow$ CA model, case 1
2. OV model, case 2 $\leftrightarrow$ CA model, case 2
3. OV model, case 7 $\leftrightarrow$ CA model, trivial solution (Fig. 3c)

Note that the two fundamental diagrams for the CA model in Fig. 1 merge into one another for $\rho \geq 0.25$ (maximum of q_B lies on q_o-curve, chord 7 merges into one point). This also implies that, for case 1 of the OV model, we actually find three plateaus similar to Fig. 2e, with the bottleneck flux again at its maximum.

Kinematic wave theory alone cannot fully explain the selection of stationary shocks in the bottleneck simulation, since one always finds at least one shock with two diverging characteristics. On the other side, stationary wave patterns can be expected at the boundaries due to a non-smooth change in model parameters. Therefore, it is the dynamics of the model which determine the actual observable patterns, and these resemble each other in the OV and CA model simulations.

4 Conclusion and Future Work

In this paper, the dynamics of a bottleneck simulation exhibit a link between cellular automata and optimal-velocity traffic models. This is surprising in some sense since it connects a model, which is discrete in space and time, to a model, which is continuous in space and time.

Future work will focus on a **CA** model with two different fundamental diagrams for the bottleneck and the open road, respectively, and how this compares to the bottleneck simulations of the OV model. This can be achieved by choosing two different "dawdling" probabilities $0 < p_o < p_B < 1$. However, this entails the formation of jams, which overlap with the stationary wave patterns.

References

1. M. Bando, K. Hasebe, A. Nakayama et al.: Phys. Rev. E **51**, 1035 (1995)
2. P. Berg, A. Mason, A. Woods: Phys. Rev. E **61**, 1056 (2000)
3. D. Helbing: *Verkehrsdynamik* (Springer, Heidelberg 1997)
4. B.S. Kerner: *The Physics of Traffic* (Springer, Heidelberg 2004)
5. H.K. Lee, H.W. Lee, D. Kim: Phys. Rev. E **64**, 056126 (2001)
6. K. Nagel, M. Schreckenberg: J. Phys. I France **2**, 2221 (1992)
7. A. Schadschneider, M. Schreckenberg: J. Phys. A: Math. Gen. **26**, L679 (1993)
8. A. Schadschneider: Physica A **285**, 101 (2000)
9. Y. Sugiyama, A. Nakayama et al.: 'Observation, Theory and Experiment for Freeway Traffic as Physics of Many-Body System'. In: *Traffic and Granular Flow '03*, ed. by S.P. Hoogendoorn, S. Luding, P.H.L. Bovy, M. Schreckenberg, D.E. Wolf, (Springer, Heidelberg 2005) pp.45–58
10. J. Ward, R.E. Wilson, P. Berg: these proceedings.

Linking Synchronized Flow and Kinematic Waves

Jorge A. Laval

Laboratoire Ingénierie Circulation Transport LICIT (INRETS/ENTPE), France

Summary. This paper shows that including the effects of lane-changing activity in kinematic wave theory reveals the physical mechanisms and reproduces the main empirical features that motivated Kerner's three-phase theory. This is shown using a hybrid representation of traffic flow where lane-changing vehicles are treated as discrete particles with realistic accelerations embedded in a continuous multilane kinematic wave stream. We show that this parsimonious four-parameter model reproduces the three phases identified by Kerner, including phase transitions and jam formation. We conclude that synchronized flow and wide-moving jams differ only in their lane-changing spatiotemporal patterns, but obey the same conservation laws and boundary conditions. Freeway segments with one, two and three junctions are analyzed.

1 Introduction

Kerner's three-phase theory [1] was introduced for explaining complex traffic features, such as the capacity drop [2–6], hysteresis [7], stop-and-go waves [8–13] and other complex traffic patterns [14–16]. It has been the subject of intense debate in recent years [17]. In particular, there is no consensus on (i) whether or not the so-called synchronized flow should be considered as a separate phase, and (ii) whether traffic jams arise spontaneously or are caused by bottlenecks. This paper shows that lane-changing activity is at the core of the matter, and helps to provide an answer to these important questions.

Recently, a multilane hybrid (MH) model [18] that requires only four observable parameter has been shown to explain most of the above-mentioned traffic complexities. These parameters are the triangular fundamental diagram (ie, free-flow speed, u, wave speed, w and jam density, κ), and a behavioral parameter, τ, in time units, which could be roughly interpreted as the time to complete a lane change maneuver.

The MH model is based on the effects of "disruptive lane-changing maneuvers"; i.e., a lane-changing vehicle acts as a moving bottleneck on its target lane while it accelerates to the speed prevailing on that lane. The ensuing disruption creates voids in the target traffic stream and triggers other lane changes. These lane changes, in turn, create other voids. And voids reduce capacity! It turns out that this simple physical principle explains the capacity drop on bottlenecks caused by lane-drops, moving obstructions and merge bottlenecks; see [18, 19].

This paper shows that the MH model reproduces the main features of three-phase theory. To this end, Sec. 2 describes the input data for the MH model; Sections 3 to 6 present simulations of the scatter in the fundamental diagram, the outflow from wide-moving jams, the "catch effect" and the spontaneous emergence of jams, respectively. Finally, a brief discussion is included in Sec. 7.

2 Input Data

All the numerical experiments in this paper assumed the following parameter values: $\tau = 4$ sec, $u = 112.7$ km/h (70 mph), $w = -22.5$ km/h (-14 mph) and $\kappa = 139.7$ veh/(km*lane) (225 veh/mile*lane)). We used a time-step of $\Delta t = 0.6$ sec, but the results are independent of Δt.

We have assumed that all the lane-changing particles correspond to cars with a maximum acceleration given by $a = a_0(1 - v/v_{max})$, where v is the car's current speed, while v_{max} and a_0 are the car's maximum speed and acceleration at zero speed, respectively. We chose a car with average performance features, such that $v_{max} = 123.8$ km/h (76.9 mph) and $a_0 = 3.4$ m/s^2 (11.17 ft/s^2).

3 Scatter in Empirical Fundamental Diagrams

Empirical traffic data gathered from loop-detectors exhibit a wide scatter in the congested branch of the fundamental diagram. In fact, synchronized flow was introduced for reproducing this scatter.

Our results indicate that the scatter is a combination of two factors: (1) a disruptive lane change creates congestion upstream and free-flow downstream; (2) different lanes may be in different regimes at a given point in time; eg, left lane in free-flow and right lane in congestion. In both cases the aggregate

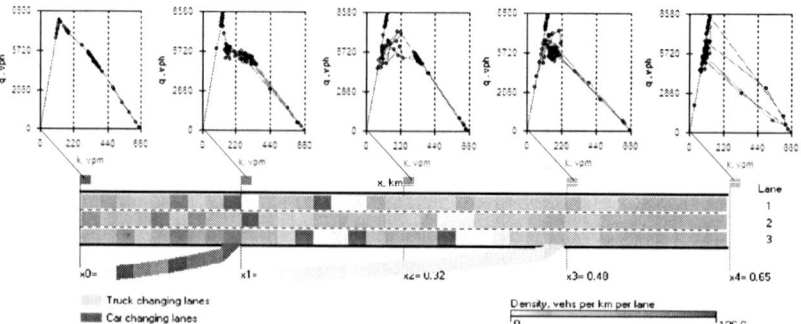

Fig. 1. Fundamental diagrams collected at locations $x_0, \ldots x_4$.

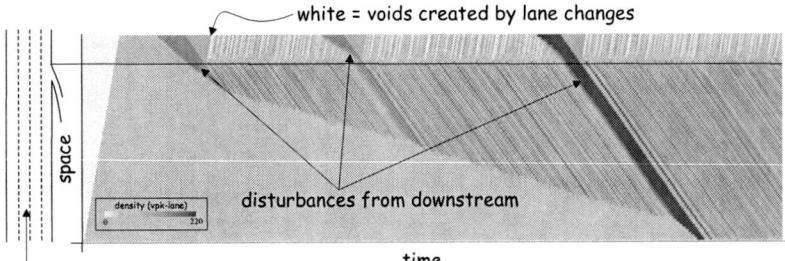

Fig. 2. Time-space density map showing the simulation of the "catch effect".

state will fall inside the fundamental diagram. The precise location of this point in the fundamental diagram depends on the proportion of time spent in each state.

To illustrate this, Fig. 1 presents the simulated fundamental diagrams collected at 5 evenly-spaced locations $(x_0, \ldots x_4)$ on a three-lane freeway segment with two on-ramps. The simulation consisted in varying on-ramp demand rates and introducing exogenous moving jams, in order to create a wide variety of traffic conditions on all detectors. The aggregation interval was 30 seconds. It can be seen an important scatter at $(x_1, \ldots x_4)$. We conclude that the free-flow state created by a disruptive lane change (void) and the congested state upstream of it produces aggregate states inside the fundamental diagram and may explain the scatter in empirical observations.

4 The "Catch Effect"

It has been observed that when a disturbance coming from upstream propagates through an initially uncongested on-ramp, it induces a bottleneck that can last for long periods of time.

The time-space density map resulting from the simulation of the "catch effect" is shown in Fig. 2. Three disturbances from downstream were exogenously introduced in the simulation. Notice that after the passage of the first disturbance through the on-ramp, disruptive lane-changing maneuvers appear. This can be seen as white areas in the figure, which represent the voids in flow that a disruptive maneuver produces in traffic stream. As a consequence, the bottleneck discharge rate decreases, which explains the capacity drop and why congestion gets caught at the on-ramp and does not vanish.

5 Outflow from Wide-Moving Jams

Empirical evidence indicates that the inflow to a wide-moving jam, q_{in}, is consistently higher than the outflow from the jam, q_{out}; ie,

$$q_{in} > q_{out}. \tag{1}$$

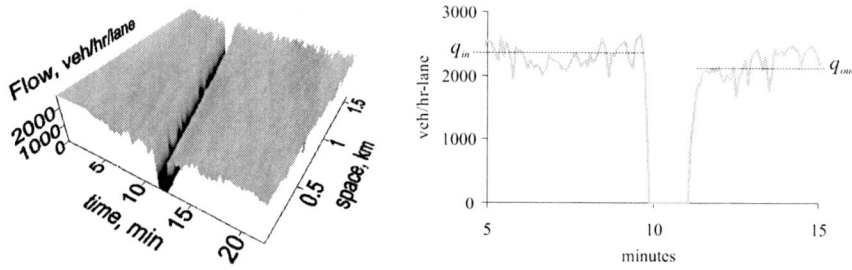

Fig. 3. Simulation results of a wide-moving jam propagating through a three-lane freeway: spaciotemporal pattern of the mean flow across lanes (left); lane-average flow vs time at location $x = 0.8$ km (right).

To the author's acknowledge, existing traffic flow models capture this effect by means of additional exogenous parameters; ie, this phenomenon is imposed to the models rather than being a consequence of the underlying theory. In the MH model, however, a lower outflow is obtained naturally because of lane-changing maneuvers taking place near the downstream front of the jam. The simulation results of a wide-moving jam propagating through a three-lane freeway is shown in Fig. 3. The left part of the figure shows the spaciotemporal pattern of the mean flow across lanes, where it is clear that the jam propagates with constant speed on both ends, as observed empirically. The right part of the figure shows a cross-section of this surface at location $x = 0.8$ km, where it is evident how condition (1) is satisfied.

6 Stop-and-Go at on-Ramp Bottlenecks

Complex traffic patterns have been observed at on-ramp bottlenecks, most notably stop-and-go waves and the spontaneous emergence of wide-moving jams. The following experiments show how these complex features arise naturally in the proposed theory. Fig. 4 shows the propagation of disturbances across three on-ramps. Notice that wide-moving jams propagate at a constant speed and tend to get wider as they propagate through the on-ramps. This is in qualitative agreement with the observations in [1].

Fig. 5 shows the emergence stop-and-go waves on a freeway segment with a single on-ramp. Freeway demand was held constant at 90% of its capacity. Notice how the level of on-ramp demands determines both the frequency and magnitude of stop-and-go waves.

7 Discussion

This paper has showed that the effects of lane-changing activity near bottlenecks may be the main cause of traffic instabilities. Our results suggest the

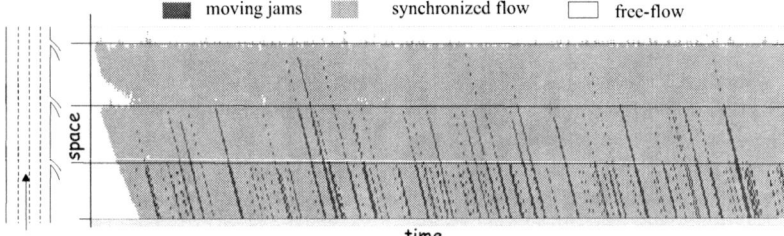

Fig. 4. Propagation of disturbances across three on-ramps.

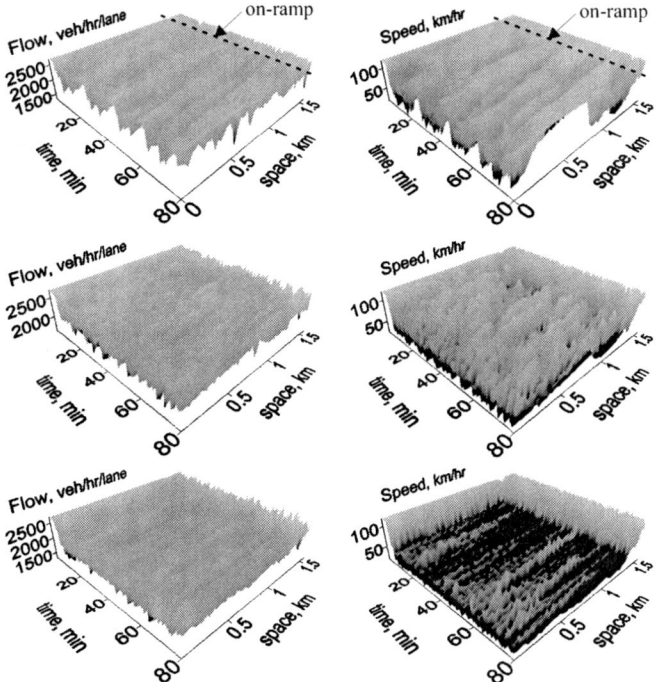

Fig. 5. Stop-and-go at on-ramp bottlenecks for low on-ramp demand (top, 400 veh/h), medium on-ramp demand (middle, 700 veh/h) and high on-ramp demand (bottom, 1100 veh/h). On-ramp at $x=1.2$ km.

following physical explanation to traffic instabilities at merge bottlenecks: on-ramp queues determine the speed at which traffic enters the freeway. This is approximately the same speed at which shoulder-lane vehicles will start their lane-changing maneuvers. Therefore, the lower the speed of entering traffic the greater the voids created by lane changes from the shoulder lane and the greater the losses in capacity. It also follows that the stop-and-go waves observed at on-ramp bottlenecks may be caused by changes in traffic conditions at the on-ramp.

References

1. B S Kerner. *The physics of traffic*. Springer, 2004.
2. HS Mika, JB Kreer, and LS Yuan. Dual mode behavior of freeway traffic. *High. Res. Rec.* **279**: 1–13, 1969.
3. K Agyemang-Duah and FL Hall. Some issues regarding the numerical value of freeway capacity. In U.Brannolte, editor, *International Symposium on Highway Capacity*, pages 1–15, Balkema, Rotterdam, 1991.
4. FL Hall and K Agyemang-Duah. Freeway capacity drop and the definition of capacity. *Transportation Research Record, TRB* **1320**:91–98, 1991.
5. B S Kerner and H Rehborn. Experimental features and characteristics of traffic jams. *Phys. Rev. E* **53**: R1297–R1300, 1996.
6. B Persaud, S Yagar, and R Brownlee. Exploration of the breakdown phenomenon in freeway traffic. *Transportation Research Record, TRB* **1634**: 64–69, 1998.
7. J Treiterer and JA Myers. The hysteresis phenomenon in traffic flow. In D. J. Buckley, editor, *6th Int. Symp. on Transportation and Traffic Theory*, pages 13–38, A.H. and A.W. Reed, London, 1974.
8. DC Gazis, R Herman, and G Weiss. Density oscillations between lanes of a multi-lane highway. *Operations Research* **10**: 658–667, 1962.
9. G F Newell. Theories of instability in dense highway traffic. *J. Opns. Res. Japan* **1**(5): 9–54, 1962.
10. K Smilowitz, C Daganzo, J Cassidy, and R Bertini. Some observations of highway traffic in long queues. *Trans. Res. Rec.* **1678**: 225–233, 1999.
11. MJ Cassidy and M Mauch. An observed traffic pattern in long freeway queues. *Trans. Res. A* **2**(35): 143–156, 2001.
12. J M Del Castillo. Propagation of perturbations in dense traffic flow: a model and its implications. *Trans. Res. B* **2**(35): 367–390, 2001.
13. M Mauch and MJ Cassidy. Freeway traffic oscillations: Observations and predictions. In M.A.P. Taylor, editor, *15th Int. Symp. on Transportation and Traffic Theory*, Pergamon-Elsevier, Oxford, U.K., 2002.
14. B S Kerner and H Rehborn. Experimental properties of phase transitions in traffic flow. *Phys. Rev. Lett.* **79**: 4030–4033, 1997.
15. B S Kerner and H Rehborn. Theory of congeste traffic flow: self-organization without bottlenecks. In A. Ceder, editor, *14th Int. Symp. on Transportation and Traffic Theory*, pages 147–177, Pergamon, New York, N.Y., 1999.
16. B S Kerner. Complexity of synchronized flow and related problems for basic assumptions of traffic flow theories. In H. M. Zhang, editor, *Networks and Spatial Economics*, pages 35–76. Kluwer Academic Publishers, Boston, USA, 2001.
17. CF Daganzo, M Cassidy, and R Bertini. Possible explanations of phase transitions in highway traffic. *Trans. Res. A* **5**(33): 365–379, 1999.
18. JA Laval and CF Daganzo. Lane-changing in traffic streams. *Trans. Res. B (In Press)*, 2005.
19. JA Laval, M Cassidy, and CF Daganzo. Impacts of lane changes at merge bottlenecks: A theory and strategies to maximize capacity. these proceedings.

Probabilistic Description of Traffic Breakdown

Reinhard Mahnke[1] and Reinhart Kühne[2]

[1] Rostock University, Institute of Physics, D–18051 Rostock, Germany;
   reinhard.mahnke@uni-rostock.de
[2] German Aerospace Center, Institute of Transportation Research,
   D–12489 Berlin, Germany; reinhart.kuehne@dlr.de

Summary. Traffic breakdowns are described by a balance equation that models the dynamics of jam formation by the following two contributions. There are discharge rate depending on the length of the congestion and an adhesion rate mainly depending on the traffic volume of the considered road section. With this balance equation it is feasible to calculate the dynamics of traffic pattern formation especially the first passage time for a transition from free flow condition to congested traffic including the influence of the parameters affecting the discharge and adhesion rates. As a simple approximation we consider constant attachment rate as well as constant detachment rate.

Starting with the probability density and furtheron with the cumulative probability for breakdowns the change in the incident duration distribution is calculated and qualitatively given. The paper concludes with recommendations for a comprehensive operation improvement and provides necessary steps for a long lasting stabilization of traffic for a given vehicular flow time series pattern.

1 Introduction

In the probabilistic description a traffic breakdown is defined as a car cluster formation process. For this we consider a model of traffic flow on a freeway section and study the spontaneous formation of a jam regarded as a large car cluster arising on the road. The cluster is specified by its size n, the number of aggregated cars (see Fig. 1). Its internal parameters, namely, the headway distance and, consequently, the speed of cars in the cluster are treated as fixed values independent of the cluster size n.

We note that in the model under consideration there can be only one cluster on the road. The free flow phase is specified also by the corresponding headway distance that, however, depends strictly speaking on the car cluster size n. When a vehicular cluster arises on the road its further growth is due to the attachment of the free cars to its upstream boundary, whereas the cars located near its downstream boundary accelerate to leave it, which decreases the cluster size. These processes are treated as random changes of the cluster size n by ± 1 (see Fig. 2) and the cluster evolution is described in terms of time variations of the probability function $P(n,t)$ for the cluster to be of size n at time t. Then following [1–4] we write the balance equation called one–step Master equation governing the cluster evolution

$$\frac{\partial P(n,t)}{\partial t} = w_+(n-1)P(n-1,t) + w_-(n+1)P(n+1,t)$$
$$- [w_+(n)P(n,t) + w_-(n)P(n,t)] . \tag{1}$$

Here we have the simplest case when the attachment rate $w_+(n)$ to the cluster takes into account strictly speaking the net time gap for a freely moving car to move up to the cluster. If net time gap and gross time gap are taken equal then this rate is just the traffic flow q

$$w_+ = q \tag{2}$$

which is constant. The rate of the cars escaping from the cluster at its down-stream front also is given as a constant

$$w_-(n) = \frac{1}{\tau} . \tag{3}$$

The value τ can be interpreted as the characteristic time needed for the first car in the cluster to leave it and to go out from its downstream boundary at a distance about the headway distance in the current free flow state [5, 6]. An example of such a cluster trajectory is shown in Fig. 3. Due to the problem under consideration, we introduce boundary conditions for the trajectory $n(t)$. It is naturally to define $n = 0$ as the reflecting boundary and cluster size $n(t)$ is always $n(t) \geq 0$. On the other hand, $n = n_{esc}$ is the absorbing or escape cluster size and the breakdown phenomenon appears when $n(t) = n_{esc}$ [5–7].

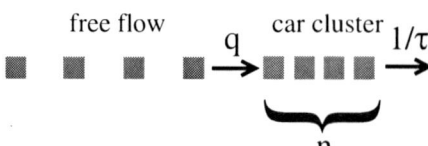

Fig. 1. Definition of car cluster formation with attachment rate q (inflow) and detachment rate $1/\tau$ (outflow).

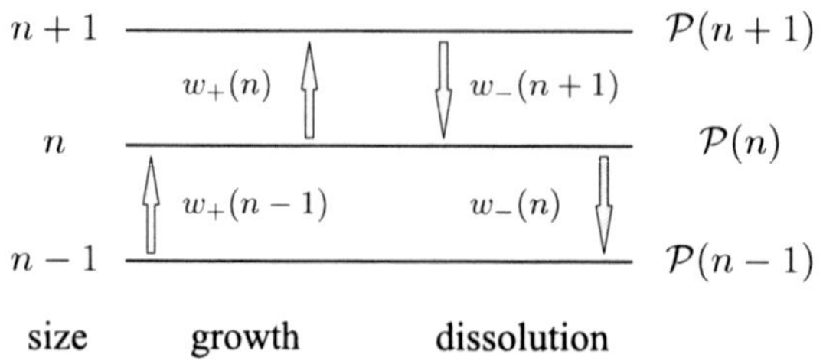

$$\begin{array}{lccc}
n+1 & \rule{3cm}{0.4pt} & & P(n+1) \\
 & w_+(n) \uparrow \quad & \downarrow w_-(n+1) & \\
n & \rule{3cm}{0.4pt} & & P(n) \\
 & \uparrow w_+(n-1) & w_-(n) \downarrow & \\
n-1 & \rule{3cm}{0.4pt} & & P(n-1) \\
\text{size} & \text{growth} & \text{dissolution} &
\end{array}$$

Fig. 2. Schematic illustration of cluster evolution by one–step Master equation.

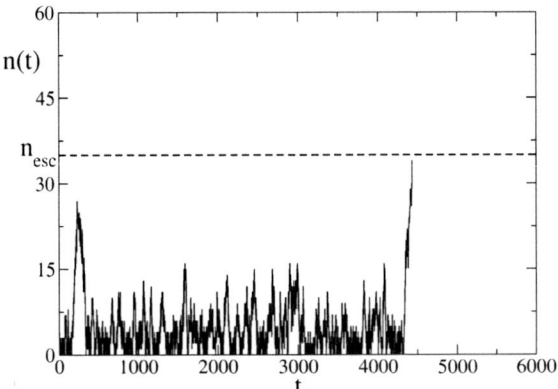

Fig. 3. Example of stochastic trajectory $n(t)$ with initial condition $n(t = 0) = 0$, reflecting boundary at $n = 0$ veh. and absorbing boundary at $n_{esc} = 35$ veh.

2 Drift-Diffusion Approximation

In order to apply well developed techniques of escape theory [3, 4] to the analysis of the traffic breakdown probability we approximate the discrete balance equation (1) by the corresponding Fokker–Planck or drift–diffusion equation because in the case under consideration the kinetic coefficients $w_+(n)$, $w_-(n)$, first, vary smoothly on scales about unity and, second, are approximately equal to each other. The following transformations allows us to find an appropriate approximation.

1. The Kramers–Moyal expansion

$$\frac{\partial p(x,t)}{\partial t} = \sum_{n=1}^{\infty} \frac{(-1)^n}{n!} \frac{\partial^n}{\partial x^n} \left[\alpha_n(x,t) p(x,t) \right] \tag{4}$$

for the probability distribution function $p(x,t)$ in a continuum approximation

$$p(x,\, t)\, dx \approx P(n,\, t) \tag{5}$$

with the expansion coefficients given by the moments of the transition rates $w(x', x, t)$ from state x to state x', i. e.,

$$\alpha_n(x,t) = \int_{-\infty}^{\infty} (x' - x)^n w(x', x, t)\, dx' . \tag{6}$$

2. Discrete variable n is transformed to a new continuous one

$$x = \frac{n}{n_{ecs}} . \tag{7}$$

3. Initial condition $n_0 = 0$ means no jam at the beginning

$$x_0 = \frac{n_0}{n_{ecs}} = 0 . \tag{8}$$

4. Reflecting boundary $n = 0$ is transformed to left value $a = 0$.
5. Absorbing boundary $n = n_{esc}$ is transformed to rigth value $b = 1$.
6. Diffusion coefficient D is given as an independent parameter.
7. Drift coefficient v is calculated from the rates difference

$$v = w_+(n) - w_-(n) = q - 1/\tau . \tag{9}$$

We can immediately consider the influence of the drift value v with the help of the potential $U(x) = -vx$ (see Fig. 4) which provides driving force.

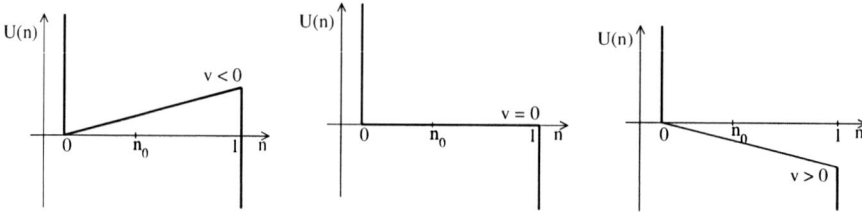

Fig. 4. The linear potential $U(n)$ for three different scenarios according to the different values of the drift parameter v.

As a result we obtain the well-known drift–diffusion equation with initial and boundary conditions

$$\frac{\partial p(x,t)}{\partial t} = -v\frac{\partial p(x,t)}{\partial x} + D\frac{\partial^2 p(x,t)}{\partial x^2} . \tag{10}$$

The probability density $p(x,t)$ satisfies initial condition (delta–function)

$$p(x, t = 0) = \delta(x - x_0) \tag{11}$$

and boundary conditions, i. e. there is a reflecting boundary at $x = a = 0$ where the flux j vanishes

$$j(x = 0, T) = v\, p(x = 0, T) - D\left.\frac{\partial p(x,T)}{\partial x}\right|_{x=0} = 0 \tag{12}$$

and an absorbing boundary at $x = b = 1$ where the probability density is zero

$$p(x = 0, t) = 0 . \tag{13}$$

The method how to get a solution of this initial–boundary–value–problem is shown in [8] where the details of calculation are given. Here we would like

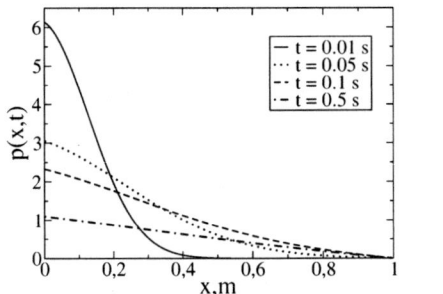
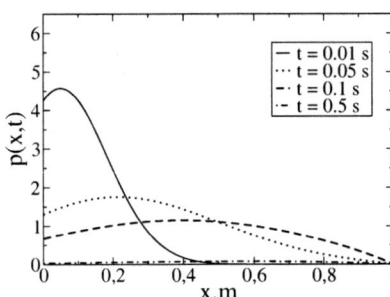

Fig. 5. The solution of Fokker–Planck equation (10) – (13) for two different values of drift parameter v ($v = -1$ m/s (left plot) and $v = 3$ m/s (right plot)) and for fixed diffusion term $D = 1$ m^2/s.

to present the final results. According to the fact, that drift parameter v can take any real values we should separate three different cases (see Fig. 5)

$$\frac{v}{D} > -2 \ : \ p(x,t) = 2 e^{\frac{v}{2D} x}$$

$$\times \sum_{m=0}^{\infty} \frac{e^{-[\tilde{k}_m^2 + (v/2D)^2] Dt}}{1 + \frac{v/2D}{\tilde{k}_m^2 + (v/2D)^2}} \sin\left[\tilde{k}_m\right] \sin\left[\tilde{k}_m \left(1 - x\right)\right] \ ; \quad (14)$$

$$\frac{v}{D} = -2 \ : \ p(x,t) = e^{-x} \left(3 e^{-Dt} \left(1 - x\right) \right.$$

$$\left. + 2 \sum_{m=1}^{\infty} \frac{e^{-[\tilde{k}_m^2 + 1] Dt}}{1 - \frac{1}{\tilde{k}_m^2 + 1}} \sin\left[\tilde{k}_m\right] \sin\left[\tilde{k}_m \left(1 - x\right)\right] \right) \ ; \quad (15)$$

$$\frac{v}{D} < -2 \ : \ p(x,t) = 2 e^{\frac{v}{2D} x} \left(- \frac{e^{-[-z_0^2 + (v/2D)^2] Dt}}{1 + \frac{v/2D}{(-z_0)^2 + (v/2D)^2}} \sinh\left[z_0\right] \sinh\left[z_0 \left(1 - x\right)\right] \right.$$

$$\left. + \sum_{m=1}^{\infty} \frac{e^{-[\tilde{k}_m^2 + (v/2D)^2] Dt}}{1 + \frac{v/2D}{\tilde{k}_m^2 + (v/2D)^2}} \sin\left[\tilde{k}_m\right] \sin\left[\tilde{k}_m \left(1 - x\right)\right] \right) \quad (16)$$

where the values $\tilde{k}_m$ and z_0 are solutions of transcendental equations (see Fig. 6 and Fig. 7)

$$\tan \tilde{k}_m = -\frac{2D}{v} \tilde{k}_m \quad (17)$$

$$\tanh \tilde{z}_0 = -\frac{2D}{v} \tilde{z}_0 \ . \quad (18)$$

We would like to mention that the smallest or ground–state wave vector $\tilde{k}_0$ vanishes when v/D tends to -2 from above, and no continuation of this

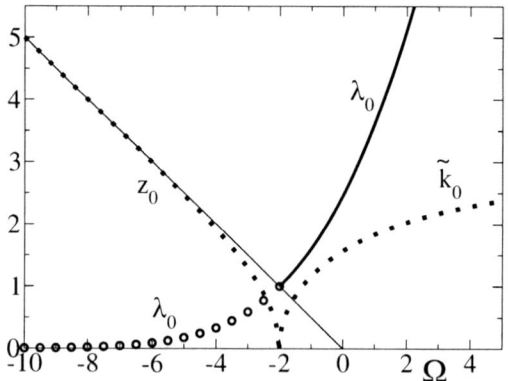

Fig. 6. The wave number $\tilde{k}_0$ and its imaginary part z_0 for $\Omega = v/D \geq -2$ and $\Omega = v/D \leq -2$, respectively, and the eigenvalue $\lambda_0 = \tilde{k}_0^2 + (v/2D)^2$ ($\Omega = v/D \geq -2$) and $\lambda_0 = -z_0^2 + (v/2D)^2$ ($\Omega = v/D \leq -2$) for the ground state $m = 0$. The thin straight line shows the approximation $z_0 \approx -\Omega = v/D$ valid for large negative $\Omega = v/D$.

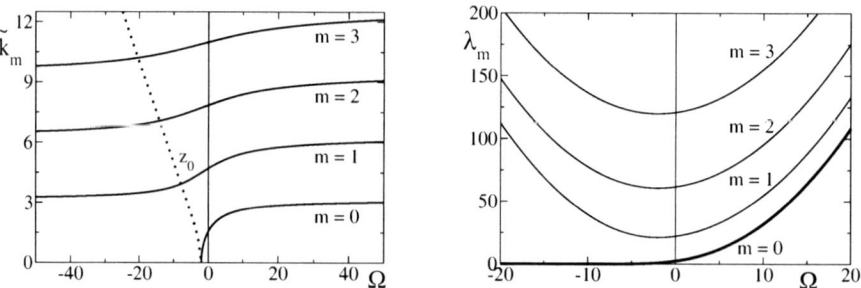

Fig. 7. The parameter $\Omega = v/D$ dependence of wave numbers $\tilde{k}_m$ and eigenvalues λ_m for ground state $m = 0$ and excited states $m = 1, 2, 3$.

solution exists on the real axis for $v/D < -2$. A purely imaginary solution $\tilde{k}_0 = iz_0$ appears instead, where z_0 is real, see Fig. 6.

3 Breakdown Probability Density

In terms of probabilistic modelling a traffic breakdown is an event when the system's state which started at time $t = 0$ with $x = 0$ (free flow) reaches for the first time $x = b = 1$ where the escape value $b = 1$ is a given cluster size regarded as overcritical. Following Risken [4] the distribution function of the first passage times T is given by the probability distribution flux through the absorbing boundary at $x = b = 1$ out of the considered domain $0 \leq x \leq 1$. We calculate this first passage time distribution from continuity equation (conservation law) for our system with reflecting boundary at $x = 0$ and absorbing boundary at $x = 1$, which gives us the relation

$$\frac{d}{dt} \int_0^1 p(x, t)dx + \mathcal{P}(t, x = 1) = 0 \tag{19}$$

for breakdown probability flux $\mathcal{P}(t, b)$ at $x = 1$ (congestion of size $n = n_{esc}$) with $x_0 = 0$ (initial free flow traffic situation). The first passage time probability density depends on drift and diffusion values in the following way

$$\frac{v}{D} > -2 : \; \mathcal{P}(t, x = 1) = 2e^{\frac{v}{2D}} \sum_{m=0}^{\infty} \frac{e^{-\left(\tilde{k}_m^2 + (v/2D)^2\right)Dt}}{1 + \frac{v/2D}{\tilde{k}_m^2 + (v/2D)^2}} \tilde{k}_m \sin\left[\tilde{k}_m\right] ; \tag{20}$$

$$\frac{v}{D} = -2 : \; \mathcal{P}(t, x = 1) = e^{-1}\left(3\,D\,e^{-Dt}\right.$$

$$+ 2D \sum_{m=1}^{\infty} \frac{e^{-\left(\tilde{k}_m^2 + 1\right)T}}{1 - \frac{1}{\tilde{k}_m^2 + 1}} \tilde{k}_m \sin\left[\tilde{k}_m\right]\Bigg) ; \tag{21}$$

$$\frac{v}{D} < -2 : \; \mathcal{P}(t, x = 1) = 2\,D e^{\frac{v}{2D}}\left(\frac{-e^{-\left(-\tilde{z}_0^2 + (v/2D)^2\right)Dt}}{1 + \frac{v/2D}{-\tilde{z}_0^2 + (v/2D)^2}} \tilde{z}_0 \sinh\left[\tilde{z}_0\right]\right.$$

$$+ \sum_{m=1}^{\infty} \frac{e^{-\left(\tilde{k}_m^2 + (v/2D)^2\right)Dt}}{1 + \frac{v/2D}{\tilde{k}_m^2 + (v/2D)^2}} \tilde{k}_m \sin\left[\tilde{k}_m\right]\Bigg) . \tag{22}$$

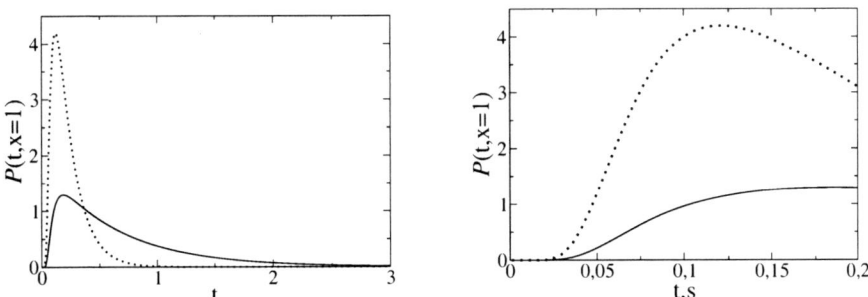

Fig. 8. The first passage time probability density $\mathcal{P}(t, x = 1)$ for two different values of drift parameter v ($v = -1$ m/s (smooth curve) and $v = 3$ m/s (dotted curve)) at a fixed diffusion term $D = 1$ m^2/s. The right plot shows the time lag in detail.

4 Breakdown Probability

A quantity of practical interest is the probability $W\left(v, D, t = t_{obs}\right)$ that break-down takes place within a given time interval $t \in \left[0, t_{obs}\right]$. It is obtained by integrating the breakdown probability density as follows

$$W\left(v, D, t = t_{obs}\right) = \int\limits_{0}^{t_{obs}} \mathcal{P}(t, x = 1)\, dt \ . \tag{23}$$

The result of integration in three different cases where v/D is larger, equal, or smaller than -2 reads

$$\frac{v}{D} > -2 \ : \ W\left(v, D, t_{obs}\right) = 2\, e^{\frac{v}{2D}}$$

$$\times \sum_{m=0}^{\infty} \frac{1 - e^{-\left(\tilde{k}_m^2 + (v/2D)^2\right)D\, t_{obs}}}{\tilde{k}_m^2 + (v/2D)^2 + v/2D}\, \tilde{k}_m \sin\left[\tilde{k}_m\right] \ ; \tag{24}$$

$$\frac{v}{D} = -2 \ : \ W\left(v, D, t_{obs}\right) = e^{-1}\left(3\left(1 - e^{-D\, t_{obs}}\right)\right.$$

$$\left. + 2 \sum_{m=1}^{\infty} \frac{1 - e^{-\left(\tilde{k}_m^2 + 1\right)D\, t_{obs}}}{\tilde{k}_m} \sin\left[\tilde{k}_m\right]\right) \ ; \tag{25}$$

$$\frac{v}{D} < -2 \ : \ W\left(v, D, t_{obs}\right) = 2e^{\frac{v}{2D}}\left(-\frac{1 - e^{-\left(-\tilde{z}_0^2 + (v/2D)^2\right)D\, t_{obs}}}{-\tilde{z}_0^2 + (v/2D)^2 + v/2D}\, \tilde{z}_0 \sinh\left[\tilde{z}_0\right]\right.$$

$$\left. + \sum_{m=1}^{\infty} \frac{1 - e^{-\left(\tilde{k}_m^2 + (v/2D)^2\right)D\, t_{obs}}}{\tilde{k}_m^2 + (v/2D)^2 + v/2D}\, \tilde{k}_m \sin\left[\tilde{k}_m\right]\right) \tag{26}$$

5 Weibull Distribution as Fit Function

The shape of $W(v, D, t = t_{obs})$ in any case reminds to the stochastic distributions used in reliability assessment and support the ideas of Regler [9] and others [10–12]. These authors use Weibull distributions as fitting curves with enough parameters to match a broad variety of cummulative distribution functions. The Weibull distribution is calculated from

$$W(q) = 1 - \exp\left[-\left(\frac{q}{\beta}\right)^{\alpha}\right] \tag{27}$$

where α and β are parameters of the distribution. We have fitted our calculation for the cumulative (breackdown) probability by Weibull distribution (Fig. 9). The following transformations relating the parameters in our equations to the physical observables have been used

$$x = l_{\text{eff}}\, n\,,\tag{28}$$

$$v = \left(q - \frac{1}{\tau}\right) l_{\text{eff}}\,,\tag{29}$$

$$D = \frac{1}{2}\left(q + \frac{1}{\tau}\right) l_{\text{eff}}^2\,,\tag{30}$$

where l_{eff} is the effective length of a car. Here q is the vehicular flow and τ is the characteristic reaction (relaxation) time constant, as introduced earlier.

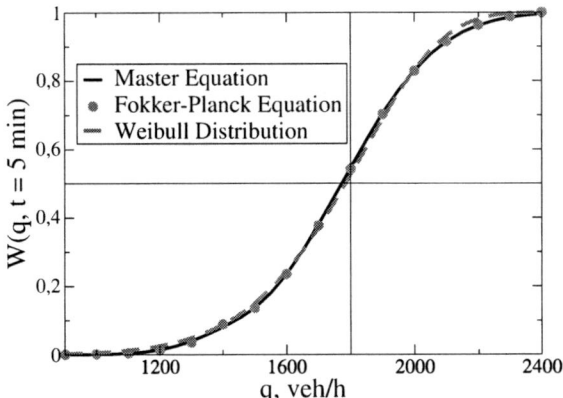

Fig. 9. Cumulative breakdown probability $W(q, t = t_{obs})$. The parameters of calculation (Fokker–Planck equation) and simulation (Master equation) are escape cluster size $n_{esc} = 19$ veh., effective length of car $l_{\text{eff}} = 7$ m, relaxation time $\tau = 2$ s, absorbing boundary $x_{esc} = 133$ m, observation time $t_{obs} = 5$ min. For comparison Weibull distribution is shown with parameter values $\alpha = 8.3$ and $\beta = 1865$.

6 Conclusion

A method how to calculate the traffic breakdown from a physical point of view have been discussed and developed. A brief summary of calculation results in a continuum drift–diffusion approximation is presented and the calculated breakdown probability is compared to the Weibull distribution. In the following one needs to compare the results with real empirical data to conclude with recommendations for a comprehensive operation improvement and provide necessary steps for a long lasting stabilization of traffic for a given vehicular flow time series pattern.

References

1. R. Mahnke, N. Pieret: Stochastic Master–Equation Approach to Aggregation in Freeway Traffic, Phys. Rev. E **56** (1997) 2666.
2. R. Mahnke, J. Kaupužs: Stochastic Theory of Freeway Traffic, Phys. Rev. E **59** (1999) 117.
3. C. W. Gardiner: *Handbook of Stochastic Methods*, Springer, Berlin, Heidelberg, 2002.
4. H. Risken: *The Fokker–Planck Equation*, Springer, Berlin, Heidelberg, 1996.
5. R. Kühne, R. Mahnke: Controlling Traffic Breakdowns, Presentation at 16th International Symposium on Transportation and Traffic Theory (ISTTT), University of Maryland, Washington, 2005.
6. R. Mahnke, J. Kaupužs and I. Lubashevsky: Probabilistic Description of Traffic Flow, Phys. Rep. **408** (2005) 1–130
7. R. Kühne, R. Mahnke, I. Lubashevsky, J. Kaupuvzs: Probabilistic Description of Traffic Breakdowns, Phys. Rev. E **65** (2002) 066125.
8. J. Hinkel: How to Calculate Traffic Breakdown Probability?, Contribution to TGF2005, this volume, Springer, 2006.
9. M. Regler: Verkehrsablauf und Kapazität auf Autobahnen (Traffic Flow and Capacity on Freeways), Dissertation, Universität Bochum, Januar 2004.
10. M. Lorenz, L. Elfteriadou: A Probabilistic Approach to Defining Freeway Capacity and Breakdown. In: Brilon, W. (ed.) 4th Int. Symp. on Highway Capacity, June 27 - July 1, 2000, Hawaii, Transp. Res. E Circular EDCO/8.
11. Kühne, R.: Application of Probabilistic Traffic Pattern Analysis to Capacity, Siam Conference, San Diego, July 9 – 19, 2001.
12. Brilon, W.; Zurlinden, H.: Überlastungswahrscheinlichkeiten und Verkehrsleistung als Bemessungskriterium for Straßenverkehrsanlagen. Forschung Straßenbau und Straßenverkehrstechnik, Heft 870, Bundesministerium für Verkehr, Bau- und Wohnungswesen, Abt. Straßenbau, Straßenverkehr Bonn (Hrsg.), August 2003.

How to Calculate Traffic Breakdown Probability?

Julia Hinkel

Institut für Physik, Universität Rostock, D–18051 Rostock, Germany
`julia.hinkel@uni-rostock.de`

Summary. We would like to calculate the traffic breakdown probability distribution which is related to a first-order phase transition from free flow to congested flow. Intuitively we introduce the notion of breakdown probability density as a function of time to reach some significant large car cluster size (first passage time problem). The calculations are based on an initial–boundary–value Fokker–Planck equation including balance condition of the probability flux.

1 Drift-Diffusion Problem with Absorbing and Reflecting Boundary

Let us consider the initial–boundary–value–problem (shown schematically in Fig. 1) with constant diffusion coefficient D and constant drift coefficient v. Our task is to calculate the probability density $p(x, t)$ to find the system in state x (exactly in the interval $[x; x + dx]$) at time t. The dynamics of $p(x, t)$ is given by the drift–diffusion–equation (see [1, 2]) as well as initial and boundary conditions

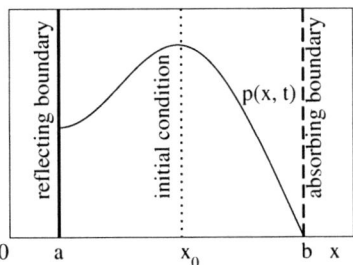

Fig. 1. Schematic picture of the boundary–value problem showing the probability density $p(x , t)$ in the interval $a \leq x \leq b$.

$$\frac{\partial p(x,t)}{\partial t} = -v\frac{\partial p(x,t)}{\partial x} + D\frac{\partial^2 p(x,t)}{\partial x^2} \ . \tag{1}$$

The solution of (1) satisfies the (delta–function) initial condition

$$p(x, t = 0) = \delta(x - x_0) \tag{2}$$

and boundary conditions, i. e. at $x = a$ is a reflecting boundary (no flux j at left border)

$$j(x = a, t) = vp(x = a, t) - D\left.\frac{\partial p(x,t)}{\partial x}\right|_{x=a} = 0 \tag{3}$$

and at $x = b$ is an absorbing boundary

$$p(x = b, t) = 0 \ . \tag{4}$$

2 Dimensionless Drift-Diffusion Equation

It is convenient to formulate the drift–diffusion problem in dimensionless variables. For this purpose we define a new variable $0 \le y \le 1$ instead of $a \le x \le b$ by $y = \dfrac{x - a}{b - a}$ and a new time $T = \dfrac{D}{(b-a)^2} t$ and one dimensionless control parameter (scaled drift v which may have positive, zero, or negative values)

$$\Omega = \frac{v}{D}(b - a) \tag{5}$$

and a new probability density by $P(y, T)dy = p(x, t)dx$ and therefore $P(y, T) = (b-a)p(x, t)$. As a result, the equations (1) to (4) can be rewritten as follows:

$$\frac{\partial P(y,T)}{\partial T} = -\Omega\frac{\partial P(y,T)}{\partial y} + \frac{\partial^2 P(y,T)}{\partial y^2} \tag{6}$$

with the initial condition (delta–function)

$$P(y, T = 0) = \delta(y - y_0) \tag{7}$$

and boundary conditions, i. e. reflecting boundary at $y = 0$

$$J(y = 0, T) = \Omega P(y = 0, T) - \left.\frac{\partial P(y,T)}{\partial y}\right|_{y=0} = 0 \tag{8}$$

and absorbing boundary at $y = 1$

$$P(y = 1, T) = 0 \ . \tag{9}$$

3 Solution in Terms of Orthogonal Eigenfunctions

To find the solution of the well–defined drift–diffusion problem, first we take the dimensionless form (6) – (9) and use a transformation to a new function Q by

$$Q(y,T) = e^{-\frac{\Omega}{2}y}P(y,T) . \tag{10}$$

It results in a dynamics without first derivative called reduced Fokker–Planck–equation

$$\frac{\partial Q(y,T)}{\partial T} = -\frac{\Omega^2}{4}Q(y,T) + \frac{\partial^2 Q(y,T)}{\partial y^2} . \tag{11}$$

According to (10) the initial condition is transformed to

$$Q(y,T=0) = e^{-\frac{\Omega}{2}y_0}P(y,T=0) , \tag{12}$$

whereas the reflecting boundary condition at $y = 0$ becomes $J(y = 0, T) = \frac{\Omega}{2}Q(y = 0, T) - \frac{\partial Q(y,T)}{\partial y}\Big|_{y=0} = 0$. The absorbing boundary condition at $y = 1$ now reads $Q(y = 1, T) = 0$. The solution of reduced equation (11) can be found by the method of separation of variables (see [3]). A separation ansatz is $Q(y,T) = \chi(T)\psi(y)$. Hence,

$$Q(y,T) = \sum_{m=0}^{\infty} C_m e^{-\lambda_m T}\psi_m(y) . \tag{13}$$

and by using the initial condition (12) and the transformation (10) we obtain the solution $P(y,T)$ of the Fokker–Planck–equation (9) as (see Fig. 2)

$$P(y,T) = e^{\frac{\Omega}{2}(y-y_0)}\sum_{m=0}^{\infty} e^{-\lambda_m T}\psi_m(y_0)\psi_m(y) \tag{14}$$

with eigenfunction of ground state ($m = 0$)

$$\psi_0(y) = \begin{cases} \sqrt{\dfrac{2}{1 + \frac{\Omega}{2}\frac{1}{k_0^2 + \Omega^2/4}}}\ \sin\left[\tilde{k}_0(1-y)\right] , & \Omega > -2 \\[2ex] \sqrt{3}\,(1-y) , & \Omega = -2 \\[2ex] \sqrt{\dfrac{2}{1 + \frac{\Omega}{2}\frac{1}{-z_0^2 + \Omega^2/4}}}\ \sinh\left[z_0(1-y)\right] , & \Omega < -2 \end{cases} \tag{15}$$

and all other eigenfunctions

$$\psi_m(y) = \sqrt{\dfrac{2}{1 + \frac{\Omega}{2}\frac{1}{k_m^2 + \Omega^2/4}}}\ \sin\left[\tilde{k}_m(1-y)\right] \quad m = 1, 2, \dots . \tag{16}$$

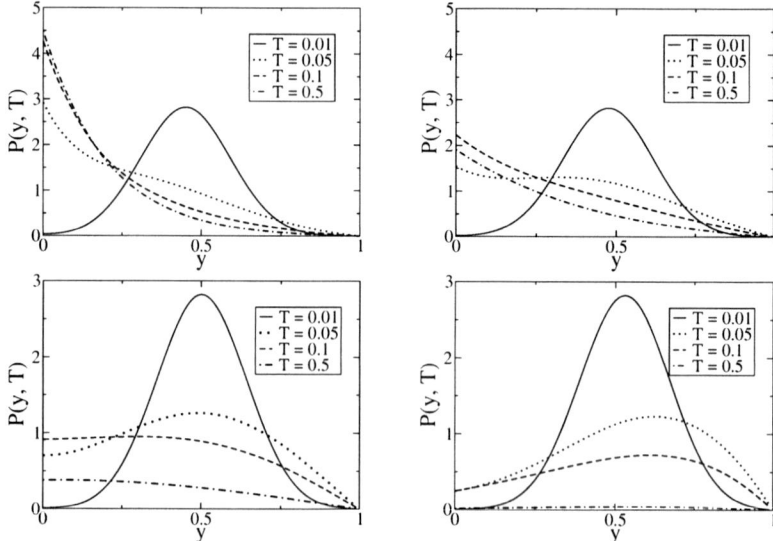

Fig. 2. The solution of drift–diffusion Fokker–Planck–equation with initial condition $y_0 = 0.5$ for different values of the control parameter Ω, i. e. $\Omega = -5$ (top left), $\Omega = -2.5$ (top right), $\Omega = 0.1$ (bottom left), $\Omega = 3$ (bottom right).

The eigenvalue of ground state ($m = 0$) is given by

$$\lambda_0 = \begin{cases} \tilde{k}_0^2 + \Omega^2/4, & \Omega > -2 \\ 1, & \Omega = -2 \\ -\tilde{z}_0^2 + \Omega^2/4, & \Omega < -2 \end{cases} \tag{17}$$

and all others eigenvalues are $\lambda_m = \tilde{k}_m^2 + \Omega^2/4$ for $m = 1, 2, \ldots$, where the wave numbers are calculated from transcendental equation

$$\tilde{k}_0 : \tan \tilde{k}_0 = -\frac{2}{\Omega} \tilde{k}_0 \qquad \Omega > -2 \tag{18}$$

$$z_0 : \tanh z_0 = -\frac{2}{\Omega} z_0 \qquad \Omega < -2 \tag{19}$$

$$\tilde{k}_m : \tan \tilde{k}_m = -\frac{2}{\Omega} \tilde{k}_m \qquad m = 1, 2, \ldots \ . \tag{20}$$

4 First Passage Time Probability Density

The balance equation in our open system (see [2]) has the form

$$\frac{\partial}{\partial T} \int_0^1 P(y,T) dy + \overline{\mathcal{P}}(T, y = 1) = 0.$$ (21)

The quantity $\overline{\mathcal{P}}(T, y = 1) dT$ is the probability that the absorbing boundary at $y = 1$ is reached for the first time within the time interval $[T, T + dT]$ (since it is forbidden to return and then reach it once again). Hence, $\overline{\mathcal{P}}(T, y = 1)$ is the first passage time probability density (see [2]) sometimes called breakdown probability density (see Fig. 3). As before at Sect. 3, we should separate three different cases

$\Omega > -2 :$ $\overline{\mathcal{P}}(T, y = 1) = 2e^{\frac{\Omega}{2}(1 - y_0)}$

$$\times \sum_{m=0}^{\infty} \frac{e^{-\left(\tilde{k}_m^2 + \Omega^2/4\right)T}}{1 + \frac{\Omega}{2} \frac{1}{\tilde{k}_m^2 + \Omega^2/4}} \tilde{k}_m \sin\left[\tilde{k}_m(1 - y_0)\right];$$ (22)

$\Omega = -2 :$ $\overline{\mathcal{P}}(T, y = 1) = e^{-(1 - y_0)} \left[3(1 - y_0) e^{-T} \right.$

$$\left. + 2 \sum_{m=1}^{\infty} \frac{e^{-\left(\tilde{k}_m^2 + 1\right)T}}{1 - \frac{1}{\tilde{k}_m^2 + 1}} \tilde{k}_m \sin\left[\tilde{k}_m(1 - y_0)\right] \right];$$ (23)

$\Omega < -2 :$ $\overline{\mathcal{P}}(T, y = 1) = 2e^{\frac{\Omega}{2}(1 - y_0)} \left[-\frac{e^{-\left(-\tilde{z}_0^2 + \Omega^2/4\right)T}}{1 + \frac{\Omega}{2} \frac{1}{-\tilde{z}_0^2 + \Omega^2/4}} \tilde{z}_0 \sinh\left[\tilde{z}_0(1 - y_0)\right] \right.$

$$\left. + \sum_{m=1}^{\infty} \frac{e^{-\left(\tilde{k}_m^2 + \Omega^2/4\right)T}}{1 + \frac{\Omega}{2} \frac{1}{\tilde{k}_m^2 + \Omega^2/4}} \tilde{k}_m \sin\left[\tilde{k}_m(1 - y_0)\right] \right].$$ (24)

5 Cumulative Breakdown Probability

The probability that the absorbing boundary $y = 1$ is reached within certain observation time interval $0 \le T \le T_{obs}$ is given by the cumulative (breakdown) probability (see [4])

$$W(\Omega, T = T_{obs}) = \int_0^{T_{obs}} \overline{\mathcal{P}}(T, y = 1) dT.$$ (25)

Finally, we would like to show the results for the cumulative (breakdown)

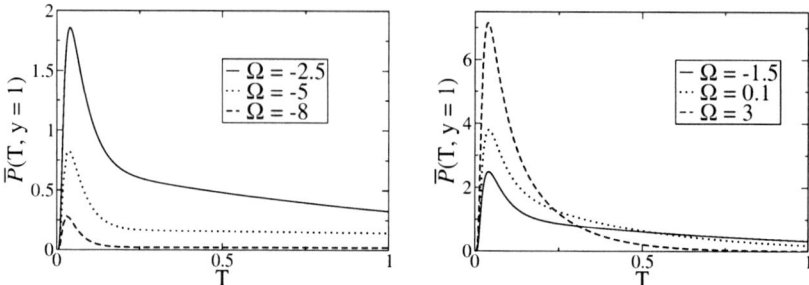

Fig. 3. The first passage time probability density distribution $\overline{\mathcal{P}}(T, y = 1)$ for $\Omega < -2$ (left) and $\Omega > -2$ (right).

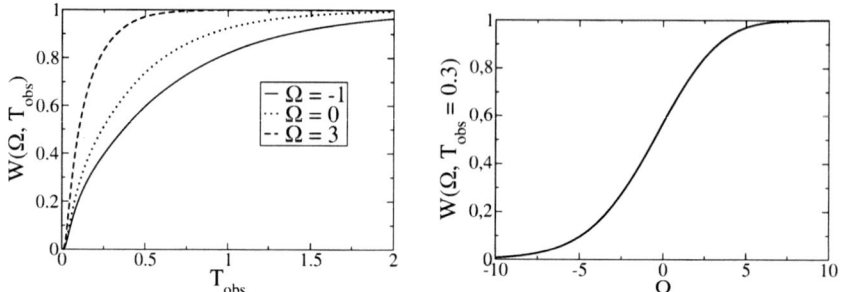

Fig. 4. The cumulative probability $W(\Omega, T_{obs})$ as function of observation time T_{obs} (with fixed Ω, left) and as function of parameter Ω (with fixed $T_{obs} = 0.3$, right).

probability (25) with respect to different values of the control parameter Ω. The solution (see Fig. 4) reads

$$\Omega > -2 : \ W(\Omega, T_{obs}) = 2\, e^{\frac{\Omega}{2}(1-y_0)}$$
$$\times \sum_{m=0}^{\infty} \frac{1 - e^{-(\tilde{k}_m^2 + \Omega^2/4)T_{obs}}}{\tilde{k}_m^2 + \Omega^2/4 + \Omega/2}\, \tilde{k}_m \sin\left[\tilde{k}_m(1 - y_0)\right] (26)$$

$$\Omega = -2 : \ W(\Omega, T_{obs}) = e^{-(1-y_0)}\left[3\left(1 - e^{-T_{obs}}\right)(1 - y_0)\right.$$
$$\left. + 2\sum_{m=1}^{\infty} \frac{1 - e^{-(\tilde{k}_m^2 + 1)T_{obs}}}{\tilde{k}_m} \sin\left[\tilde{k}_m(1 - y_0)\right]\right]; \qquad (27)$$

$$\Omega < -2 : \ W(\Omega, T_{obs}) = 2e^{\frac{\Omega}{2}(1-y_0)}\left[-\frac{1 - e^{-(-z_0^2 + \Omega^2/4)T_{obs}}}{-z_0^2 + \Omega^2/4 + \Omega/2} z_0 \sinh\left[\tilde{z}_0(1 - y_0)\right]\right.$$
$$\left. + \sum_{m=1}^{\infty} \frac{1 - e^{-(\tilde{k}_m^2 + \Omega^2/4)T_{obs}}}{\tilde{k}_m^2 + \Omega^2/4 + \Omega/2} \tilde{k}_m \sin\left[\tilde{k}_m(1 - y_0)\right]\right]. \qquad (28)$$

These mathematical results are applied to traffic theory for comparison with empirical observation (see [4, 5]).

Acknowledgements

I am indebted to R. Mahnke (Rostock), R. Kühne (Berlin), J. Kaupužs (Riga) and E. Shchukin (Rostock) for the collaboration and participations in discussions.

References

1. H. Risken: The Fokker-Planck Equation. Method of Solution and Applications, Springer, Berlin, 1996
2. C. W. Gardiner: Handbook of Stochastic Methods for Physics, Chemistry and the Natural Science, Springer, Berlin, 2004
3. S. Flügge: Practical Quantum Mechanics, Springer, Berlin, 1999
4. R. Kühne, R. Mahnke: International Symposium on Transportation and Traffic Theory, Washington, 2005
5. R. Mahnke, R. Kühne: Probabilistic Description of Traffic Breakdown, these proceedings

Models for Highway Traffic and Their Connections to Thermodynamics

Hans Weber[1], Reinhard Mahnke[2], Jevgenijs Kaupužs[3], and Anders Strömberg[4]

[1] Luleå University of Technology, Department of Physics, SE–97187 Luleå, Sweden
   `Hans.Weber@ltu.se`
[2] Rostock University, Institute of Physics, D – 18051 Rostock, Germany
   `reinhard.mahnke@uni-rostock.de`
[3] Institute of Mathematics and Computer Science, University of Latvia, LV–1459
   Riga, Latvia; `kaupuzs@latnet.lv`
[4] Luleå University of Technology, Department of Physics, SE–97187 Luleå, Sweden
   `anesot-0@student.luth.se`

Summary. Models for highway traffic are studied by numerical simulations. Of special interest is the spontaneous formation of traffic jams. In a thermodynamic system the traffic jam would correspond to the dense phase (liquid) and the free flowing traffic would correspond to the gas phase. Both phases depending on the density of cars can be present at the same time. A model for a single lane circular road has been studied. The model is called the optimal velocity model (OVM) and was developed by Bando, Sugiyama, et al. We propose here a reformulation of the OVM into a description in terms of potential energy functions forming a kind of Hamiltonian for the system. This will however not be a globally defined Hamiltonian but a locally defined one as it is a dynamical model. The model defined by this Hamiltonian will be suitable for Monte Carlo simulations.

1 Bando Model

We report a suggested reformulation of the Bando Model [1, 2], to a model including a kind of thermodynamics. The Bando model is a deterministic model for traffic flow. We restrict the work to a one-dimensional single lane circular road shown in Fig. 1.

Velocity of car i is denoted by v_i and position by x_i, (in a dimensionless formulation the set of equations (1) for velocity u and position y (right)). The Bando model is defined by the following set of equations

$$
\begin{cases}
\frac{d}{dt}v_i = \frac{1}{\tau}\left(v_{\mathrm{opt}}(\Delta x_i) - v_i\right) & \frac{d}{d\hat{t}}u_i = (u_{\mathrm{opt}}(\Delta y_i) - y_i) \\
\frac{d}{dt}x_i = v_i & \frac{d}{dt}y_i = \frac{1}{b}u_i \\
v_{\mathrm{opt}}(\Delta x_i) = v_{\max}\frac{(\Delta x)^2}{D^2+(\Delta x)^2} & u_{\mathrm{opt}}(\Delta y_i) = \frac{(\Delta y)^2}{1+(\Delta y)^2} \\
& b = \frac{D}{v_{\max}\tau}
\end{cases}
\tag{1}
$$

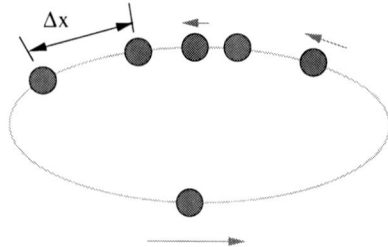

Fig. 1. One-dimensional circular road with periodic boundary conditions. The cars are represented by blue filled circles, their velocities are marked by red arrows and a headway distance is marked by Δx.

The control parameters are the maximal velocity v_{max}, the time scale τ and the interaction distance D. The optimal velocity $v_{opt}(\Delta x)$ is a function of headway (bumper-to-bumper) distance $\Delta x_i = x_{i+1} - x_i$. The average density of cars is $c = N/L$.

The acceleration of the cars is given by $a_i = \frac{dv_i}{dt}$ and is given by the Bando model (Eq. (1)). This acceleration corresponds to a force on the car according to Newton's second law $F = ma = \frac{dv}{dt}$. Finally we get the potential energy V_{pot} associated with the force F, as we know the displacement it acts over. This potential energy can be used to reformulate the Bando model into a Hamiltonian for the system of cars. It should be noted however that the Hamiltonian obtained is not a true Hamiltonian as it is only locally defined. But at least we can still use it to perform Monte Carlo simulations.

Below we show some earlier numerical results for the Bando model [3], where the equations where integrated with a 4th order Runge–Kutta method. Under certain conditions the traffic separates into two phases, a dense (= jam) and a dilute (= free flow) one. Very much like in a liquid–gas transition, we can use the difference in densities as the order parameter. The results in Fig. 2 show coexisting dense and dilute phases for certain values of b and c, where the cars move in either a jam with $u = u_{min}$ or in a free flow with $u = u_{max}$. The corresponding phase diagram can be seen in Fig. 3.

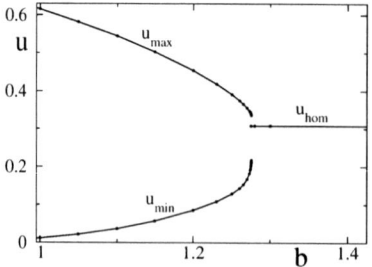 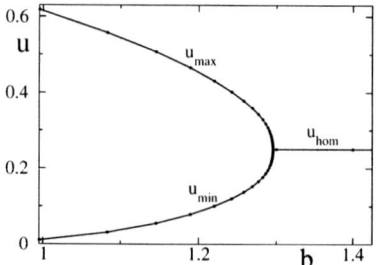

Fig. 2. The subcritical bifurcation diagram at $c = 1.5$ (left) and the critical bifurcation diagram at $c = \sqrt{3}$ (right) for a system of $N = 60$ cars.

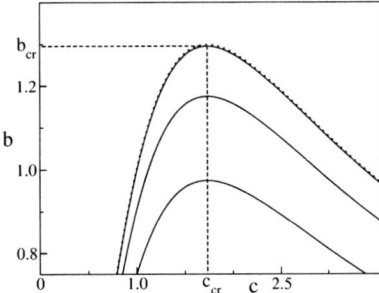

Fig. 3. Phase diagram of the Bando Model. The critical value for b is $b_{cr} = 3\sqrt{3}/4$.

2 Monte Carlo Simulations

Many complex problems (with many degrees of freedom) such as magnetic systems, gases, super conductors, atoms, nuclear decay, telephone switchboards, etc. can be analysed by Monte Carlo (MC) simulations. We will take as an example the 2D Ising model. This is a model for magnetic systems and also other ones with two states of the configuration variables like a binary alloy. In Fig. 4 a graphical representation of the two-dimensional Ising model on a square lattice is shown.

The Ising model in two dimensions has an exact solution, the famous Onsager solution [4] and in principle one should not need to perform Monte Carlo simulations on it. But it has become very famous and it has reached a position within condensed matter physics similar to the Bohr atomic model for hydrogen.

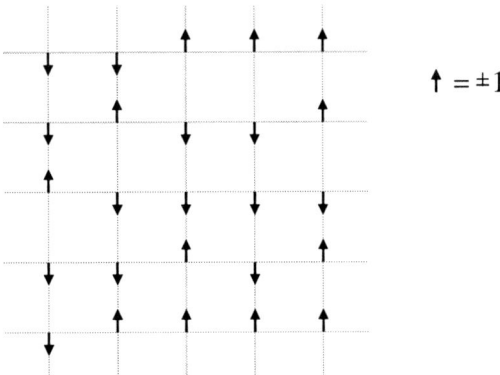

Fig. 4. The two-dimensional Ising model on a square lattice. At each lattice site there is a configuration variable $S_i = \pm 1$ shown here as an arrow.

Every state of the Ising model has an energy according to the Hamiltonian of the system

$$H = - \sum_{\langle i,j \rangle} S_i \cdot S_j \ .$$

The thermodynamic properties are given by the partition function $Z = \sum_l e^{-H_l/k_B T}$. From Z we can calculate "any" thermodynamic property of the system. Most (nearly all) systems are however too complicated to be solved analytically in a closed form and we have to revert to Monte Carlo simulations (see [5]) in order to learn more about the model in question.

Now we will describe how to do Monte–Carlo simulations in practice. For a configuration of spins S_i, the Metropolis procedure is:

1. Generate a new state by changing one of the spins $S_j \rightarrow S_j + \Delta S_j$ of the Ising model.
2. Calculate the energy difference ΔE.
3. Accept the new state if $\Delta E < 0$, else if $\Delta E > 0$, accept the new state if $r < e^{-\Delta E/k_B T}$ where r is a random number $r \in [0,1]$, otherwise keep the old value.
4. Go to step 1.

Monte Carlo usually is used for equilibrium properties, but can be used for dynamics as well. There are other Monte Carlo procedures as well, as the heat-bath method [5].

3 Example of Driven System

It is not directly clear that one should be able to get any sensible results from Monte Carlo simulations for a driven system. A driven system is not in equilibrium and thereby we are no assured that equilibrium methods should apply. But we will show an example of how it can be done. One of the authors has performed a Monte Carlo simulation of a current–voltage (IV) characteristics for a superconducting (SC) film. In a SC film there are vortex pairs (see Fig. 5) which are thermally excited. Vortices interact logarithmically $V(r) = \ln(r)$ and hence system is a 2D Coulomb gas. A Monte Carlo move consists of adding $\pm$–pairs (charge neutral) at random position and random orientation. From MC dynamics we obtain IV current – voltage characteristics. Due to the electric field, Lorentz force gives different energies to the created pairs depending on their orientation (see Fig. 6). The energy contribution due to the Lorentz force introduces a local part into the Hamiltonian. Hence, there is no global Hamiltonian! Non-linear IV characteristics of the form

$$V \propto I^a$$

(with $a = 3$ at the critical temperature $T = T_c$) obtained from experiments and Monte Carlo simulations well coincide with each other (see Fig. 6).

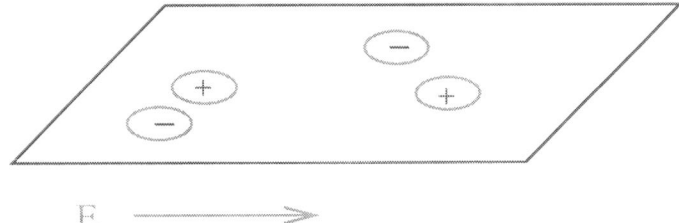

Fig. 5. A two-dimensional superconductor with 2 vortex pairs in it. The electrical field is in the direction of E.

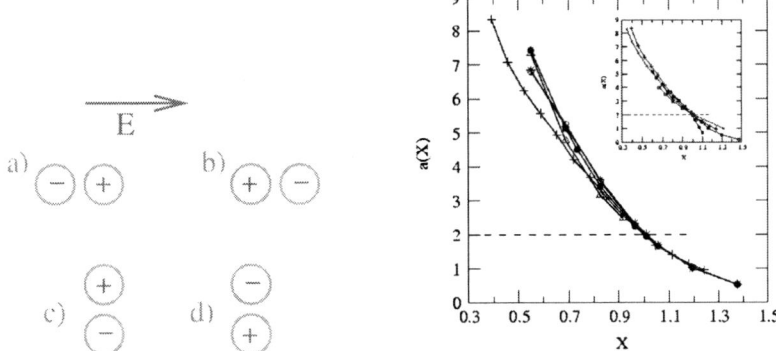

Fig. 6. Four possible orientations of a vortex pair are shown (left) with respect to the driving electrical field E. The energy of a particular pair depends on its orientation. Configurations a) and b) have different energy whereas c) and d) are not effected by the presence of the field E. The curves (right figure) show Monte Carlo simulation results together with experimental ones (taken from [6]). The different curves fall on top of each other suggesting that we can recover the experimental facts with a simple driven model. Here X is the reduced temperature T/T_c.

4 Traffic Flow

Now we describe the Metropolis procedure for the cars. The basic idea is simple: we reformulate the Bando model in terms of locally defined potential energies.

1. For a car i make a random change in velocity $\Delta v \in [-\Delta v_{max}, \Delta v_{max}]$.
2. The force F acting on a car is known from the Bando model (Eq. (1)).
3. Calculate the change in energy ΔE (potential + kinetic) due to the proposed change in velocity Δv of the car.
4. Use Metropolis algorithm to determine if the change Δv is accepted.
5. Move the car with either its new or old velocity in time step Δt.

There is an extra parameter in the problem — the ratio m/T where m is the mass of the car and T is the temperature. Note however that T is not a real

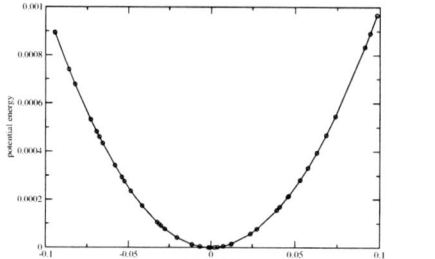 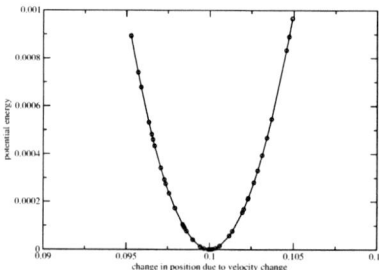

Fig. 7. The potential energy for a single car in a system with $N = 3$ and $L = 30$. The left figure shows the potential energy of a car at fixed position as a function of its velocity. The right figure shows the potential energy of a car with fixed velocity as a function of position.

temperature, it is a meassure of the strength of fluctuations. The potential energy V_{pot} is also unusual as it is a function of both velocity v_i and position x_i.

We have made test runs with a preliminary program for the Bando model defined in terms of potential and kinetic energies. These preliminary runs have been restricted to only take the potential energy into account, as this corresponds closely to the Bando model. In Fig. 7 we show the potential energy of a single car in a system consisting of only 3 cars. These three cars are set to the homogeneous solution v(equidistant) $= 1.000$ and $c = \Delta x = 10.0$. In Fig. 7 the functional form of the potential is that of a parabola with its minimum at the homogeneous solution.

References

1. M. Bando, K. Hasebe, A. Nakayama, A. Shibata, Y. Sugiyama: Japan J. Indust. and Appl. Math. **11**, 203 (1994); Phys. Rev. E **51**, 1035 (1995).
2. M. Bando, K. Hasebe, K. Nakanishi, A. Nakayama, A. Shibata, Y. Sugiyama: J. Phys. I France **5**, 1389 (1995).
3. J. Kaupužs, H. Weber, J. Tolmacheva, R. Mahnke: Applications to Traffic Breakdown on Highways, ECMI (Riga 2002). In: *Progress in Industrial Mathematics* at ECMI 2002 (Eds.: A. Buikis, R. Ciegis, A. D. Fitt), pp. 133-138, Springer-Verlag, Berlin (2004).
4. L. Onsager: Phys. Rev. **65**, 117-149 (1944).
5. M. E. J. Newman and G. T. Barkema: *Monte Carlo Methods in Statistical Physics*, (Clarendon Press, Oxford 1999) pp 45–85.
6. H. Weber, M. Wallin and H. J. Jensen: Phys. Rev. B **53**, 8566 (1996).

Variance-Driven Traffic Dynamics

Martin Treiber[1], Arne Kesting[1], and Dirk Helbing[1,2]

[1] Technische Universität Dresden, Institute for Transport & Economics,
Andreas-Schubert-Strasse 23, D-01062 Dresden, Germany
[2] Collegium Budapest – Institute for Advanced Study,
Szentháromság u. 2, H-1014 Budapest, Hungary

Summary. We investigate the adaptation of the time headways in car-following models as a function of the local velocity variance, which is a measure of the inhomogeneity of traffic flows. We apply our *meta-model* to several car-following models and simulate traffic breakdowns in open systems with an on-ramp bottleneck. Single-vehicle data generated by 'virtual detectors' show a semi-quantitative agreement with microscopic data from the Dutch freeway A9. This includes the observed distributions of the net time headways and times-to-collisions for free and congested traffic, and the velocity variance as a function of traffic density. Macroscopic properties such as the observed wide scattering of flow-density data are reproduced as well, even for deterministic simulations. We explain these phenomena by a self-organized variance-driven process that leads to the spontaneous formation and decay of long-lived platoons.

1 Introduction

One of the open questions of traffic dynamics is a microscopic understanding of the observed wide variation of the time headways which is closely related to the wide scattering of flow-density data in the congested traffic regime (see, e.g., Refs. [1–3] for an overview). Moreover, the most probable value of the time headway in congested traffic is larger by a factor of about 2 compared to free traffic, see Fig. 1(a).

With the increasing availability of single-vehicle data, further statistical properties of traffic became the subject of investigation such as the velocity variance as a function of the traffic density [4], or the distribution of the times-to-collision (TTC), which is surprisingly invariant with respect to density changes (compared to distance, time gap, or velocity distributions), see Fig. 1(b).

In this contribution, we propose a variance-driven adaptation mechanism (VDT mechanism), according to which drivers increase their time gaps T when the local traffic dynamics is unstable or largely varying.

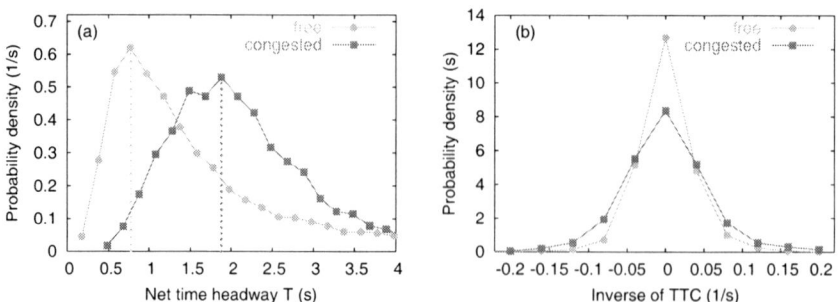

Fig. 1. Empirical statistical properties of cars following any kind of vehicle obtained from single-vehicle data from the left lane of the Dutch freeway A9 from Haarlem to Amsterdam. (a) Net time headway; (b) Inverse times-to-collision, for two traffic situations: The data set for 'free traffic' includes all single-vehicle data where the one-minute average of velocities was above 20 m/s, and the traffic flow above 1000 vehicles/h. 'Congested traffic' includes all data where the one-minute average of the velocities was below 15 m/s.

2 Variance-Driven Time Headways (VDT)

We will formulate the VDT mechanism in terms of a *meta-model* applicable to any car-following model containing the 'safe' time headway or a related parameter such as the desired (equilibrium) distance. Some examples are the optimal-velocity model (OVM) [5], the velocity-difference model (VDIFF) [6], the intelligent-driver model (IDM) [7], or the Gipps model [8].

We assume that smooth traffic flow allows for lower values of the time headway than disturbed traffic flow, i.e., the actual time headway

$$T = \alpha_T T_0 = \min\left(\alpha_T \sup max, 1 + \gamma V_n\right). \tag{1}$$

is increased in nonperturbed traffic with respect to the minimum time headway T_0 by a factor $\alpha_T \geq 1$. Furthermore, we characterize disturbed traffic flow (such as stop-and-go traffic) by the *local variation coefficient*

$$V_n = \frac{\sqrt{\theta_n}}{\bar{v}_n}, \tag{2}$$

where the local velocity average $\bar{v}_n$ and velocity variance θ_n, are calculated from the own velocity v_i and the velocities of the $(n-1)$ predecessors $(j-i)$:

$$\bar{v}_n = \frac{1}{n}\sum_{j=0}^{n-1} v_{j-i}, \qquad \theta_n = \frac{1}{n-1}\sum_{j=0}^{n-1}(v_{j-i} - \bar{v}_n)^2. \tag{3}$$

The VDT has three parameters, namely the number n of vehicles to determine the local velocity variance, the maximum multiplication factor $\alpha_T \sup max$

by which the time headway is increased compared to perfectly smooth traffic, and the sensitivity γ. The factor $\alpha_T \sup max$ can be estimated from empirical time-headway distributions as the ratio of the most probable time headways for congested and free traffic, respectively, while the sensitivity γ is calibrated to the empirical variation coefficient $V_n = V_n(\rho)$ as a function of density (Fig. 2). Throughout this contribution, we will assume the values $n = 5$, $\alpha_T \sup max = 2.2$, and $\gamma = 4$.

Notice that, in the special case $n = 2$, the local variation coefficient for vehicle i is given by $V_2^{(i)} = \sqrt{2}|v_i - v_{i-1}|/(v_i + v_{i-1})$, i.e., the VDT mechanism adds a contribution to the underlying car-following model which is proportional to the velocity difference to the immediate predecessor. For $n > 2$, the VDT includes some anticipation beyond this vehicle which is expected to be an essential ingredient for human driving [9].

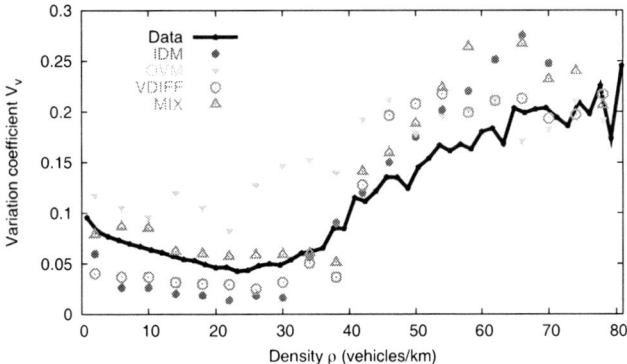

Fig. 2. Velocity variation coefficient $\sqrt{\theta}/\langle v_i \rangle$ as a function of overall traffic density from single-vehicle data of the Dutch freeway ("Data"), and from a virtual detector 4 km upstream of the on-ramp, when the VDT is simulated with various underlying models.

2.1 Acceleration Noise

Since the VDT is essentially based on fluctuations of the velocity, it is to be expected that purely deterministic underlying models yield unrealistic results due to the lack of an initial source triggering fluctuations. For simplicity, we will just add a white (independent and δ-correlated) noise term [10] to the deterministic car-following acceleration $a \sup (det)_i$ according to

$$\dot{v}_i = a \sup (det)_i(t) + \sqrt{Q}\,\xi_i(t).$$ (4)

Here, Q denotes the fluctuation strength (we will assume $Q = 0.1 \mathrm{m}^2/\mathrm{s}^3$ for all simulations unless stated otherwise), and the white noise $\xi_i(t)$ is assumed to be unbiased and δ-correlated:

$$\langle \xi_i \rangle = 0, \quad \langle \xi_i(t)\xi_j(t') \rangle = Q\delta_{ij}\delta(t - t'). \tag{5}$$

The Kronecker symbol δ_{ij} is 1, if $i = j$ and zero otherwise, while the Dirac function $\delta(t)$ is defined by $\int_{-\infty}^{\infty} \delta(t')\,dt' = 1$ and $\delta(t) = 0$ for $t \neq 0$.

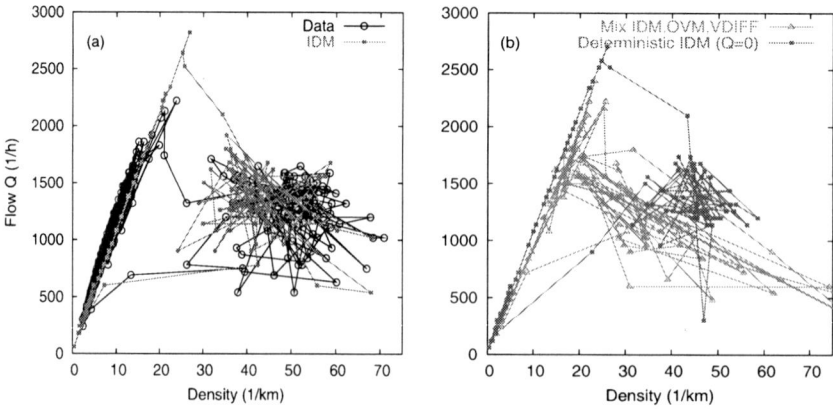

Fig. 3. Empirical and simulated flow-density data obtained from aggregated data (aggregation interval: 60 s) from real and virtual detectors. The empirical curve ("Data") is obtained by aggregating single-vehicle data on the Dutch freeway A9.

3 Simulations of a Traffic Scenario

In the following, we will apply the VDT to three car-following models, namely the intelligent-driver model (IDM) [7], the optimal-velocity model (OVM) [5], and the velocity-difference model (VDIFF) [6], which augments the OVM by a term proportional to the velocity difference. We will also simulate hetero-geneous traffic (referred to as MIX) consisting of a mixture of 1/3 IDM, 1/3 OVM, and 1/3 VDIFF. For each model, we assume 80 % 'cars', and 20 % 'trucks' which differ only in the desired velocity (35 m/s and 25 m/s, respectively). For the model equations and parameters, we refer to Ref. [11].

We have simulated a single-lane road section of total length 15 km with an on-ramp of merging length $L_{\mathrm{rmp}} = 200$ m located at $x_{\mathrm{rmp}} = 12$ km, from which a constant flow of 400 vehicles/h merges to the main road [11]. Instead of explicitly modelling on-ramp lane changes, we have inserted the ramp vehicles centrally into the largest gap within the merging region. At the merging, the velocity was 60% of that of the respective front vehicle.

We have started the simulation with free traffic and increased the traffic de-mand at the in-flowing boundary linearly from 300 vehicles/h at $t = 0$ s to 3000 vehicles/h at $t = 2400$ s. Afterwards, we decreased the inflow linearly to 300 vehicles/h until $t = 4800$ s. In case the inflow exceeded capacity, we delayed the insertion of new vehicles at the upstream boundary. To enable a direct comparison with detector data, we implemented a 'virtual detector' 4 km upstream of the on-ramp recording the passage time, velocity, and type of each vehicle.

Figure 2 compares the local variation coefficient, Eq. (2) with $n = 5$, as obtained from empirical single-vehicle data, with that obtained from the virtual detector when simulating the VDT with the three mentioned models and with the model mix. Both the decrease with density for low densities and the distinct increase at $\rho \approx 35$ veh./km are reproduced. Notice that this dependence on the density is an emergent result and not contained in the model assumptions.

Figure 3 (a) shows that the fundamental diagram obtained from simulations of the VDT with the IDM agrees, in a statistical sense, quantitatively with empirical observations. Remarkably, wide scattering is even observed in a completely deterministic simulation (Fig. 3 (b)).

Figure 4 shows simulated statistical single-vehicle properties. Both the wide time-gap distribution and the shift of the maximum with traffic density are reproduced, although the agreement is not quantitative (cf. Fig. 1). The wider distributions obtained for the model mix suggest that heterogeneity in the population of driver-vehicle units plays an essential role. Finally, the distribution of times-to-collision agrees nearly quantitatively with the data for the OVM and the model mix, but not for the IDM. In conclusion, the model mix can reproduce, at least semiquantitatively, the fundamental diagram, the variance function, and the distributions of time headways and times-to-collision.

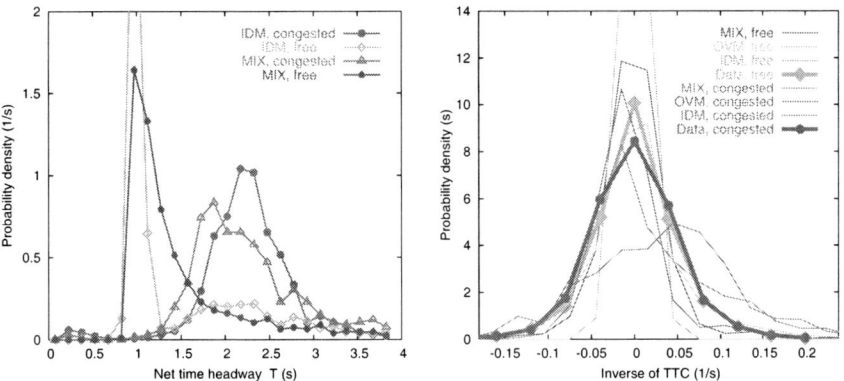

Fig. 4. (a) Distribution of the net time headways of cars following any kind of vehicle (cars or trucks); (b) distribution of the inverse $(v_i - v_{i-1})/s_i$ of the times-to-collision obtained from 'virtual detectors' of VDT simulations with different underlying models.

4 Discussion

In the variance-driven time headway (VDT) model put forward in this paper, the desired safety time headway increases with the local velocity variance. Therefore, when traffic becomes unstable, larger time gaps are held in order to keep up a reasonable level of safety, which is reflected in the nearly unchanged times-to-colision. This causes, on the one hand, a capacity drop. On the other hand, time headways are expecetd to be small in platoons consisting of vehicles driving with similar velocities. Therefore, platoons are rather long-lived and time gaps are much larger between them. This explains (at least, partially) the large scattering of flow-density data [3] in synchronized traffic flow [2].
We note that an understanding of the effects of the velocity variance is crucial for devising measures to avoid traffic breakdowns: The VDT feedback mechanism is triggered most likely near sources of sustained velocity variations, for example in the merging, diverging, or weaving zones near freeway intersections. Particularly, it is essential to avoid merging and diverging maneuvers at high velocity differences, e.g., by increasing the length of the acceleration lane at on-ramps and off-ramps. The simulations illustrate this point: When simulating on-ramp vehicles merging with the velocity of the nearby main-road vehicles (as compared to 60% of this velocity), we have observed a delayed traffic breakdown and sometimes even no breakdown at all for the same traffic demand.
Finally, the distinct increase of the time headways after traffic breakdown allows for vehicle-based options to increase the traffic performance and stability by means of adaptive cruise control systems [12].

Acknowledgements

The authors would like to thank for partial financial support by the Volkswagen AG within the BMBF project INVENT.

References

1. D. Helbing, "Traffic and related self-driven many-particle systems," Rev. Mod. Phys. **73**, 1067–1141 (2001).
2. B. S. Kerner, *The Physics of Traffic* (Springer, Heidelberg, 2004).
3. K. Nishinari, M. Treiber, and D. Helbing, "Interpreting the wide scattering of synchronized traffic data by time gap statistics," Phys. Rev. E **68**, 067101 (2003).
4. M. Treiber, A. Hennecke, and D. Helbing, "Derivation, properties, and simulation of a gas-kinetic-based, non-local traffic model," Phys. Rev. E **59**, 239–253 (1999).
5. M. Bando, K. Hasebe, A. Nakayama, A. Shibata, and Y. Sugiyama, "Dynamical model of traffic congestion and numerical simulation," Phys. Rev. E **51**, 1035–1042 (1995).

6. R. Jiang, Q. Wu, and Z. Zhu, "Full velocity difference model for a car-following theory," Phys. Rev. E **64**, 017101 (2001).
7. M. Treiber, A. Hennecke, and D. Helbing, "Congested traffic states in empirical observations and microscopic simulations," Phys. Rev. E **62**, 1805–1824 (2000).
8. P. G. Gipps, "A behavioural car-following model for computer simulation," Transp. Res. B **15**, 105–111 (1981).
9. M. Treiber, A. Kesting, and D. Helbing, "Delays, inaccuracies and anticipation in microscopic traffic models," Physica A **359**, 729–746 (2006).
10. D. Helbing and M. Treiber, "Fokker-Planck equation approach to vehicle statistics," (2003), `cond-mat/0307219`.
11. M. Treiber, A. Kesting, and D. Helbing, "Understanding widely scattered traffic flows, the capacity drop, platoons, and times-to-collision as effects of variance-driven time gaps," preprint `physics/0508222` (2005).
12. A. Kesting, M. Treiber, M. Schönhof, F. Kranke, and D. Helbing, "'Jam-avoiding' adaptive cruise control (ACC) and its impact on traffic dynamics", these proceedings.

Stability of Steady State Solutions in Balanced Vehicular Traffic

Florian Siebel and Wolfram Mauser

Department of Earth and Environmental Sciences, University of Munich,
Luisenstraße 37, D-80333 Munich, Germany

Summary. Recently we proposed an extension to the traffic model of Aw, Rascle and Greenberg. The extended traffic model can be written as a hyperbolic system of balance laws and numerically reproduces the reverse λ shape of the fundamental diagram of traffic flow. In the current work we analyze the steady state solutions of the new model and their stability properties. In addition to the equilibrium flow curve the trivial steady state solutions form two additional branches in the flow-density diagram. We show that the characteristic structure excludes parts of these branches resulting in the reverse λ shape of the flow-density relation. The upper branch is metastable against the formation of synchronized flow for intermediate densities and unstable for high densities, whereas the lower branch is unstable for intermediate densities and metastable for high densities. Moreover, the model reproduces the characteristic properties of wide moving jam formation and propagation.

1 Balanced Vehicular Traffic: The BVT Model

The BVT model (balanced vehicular traffic model, see [5, 6]) generalizes the traffic model of Aw, Rascle and Greenberg [1, 2] by prescribing a more general source term to the momentum equation. The evolution of traffic density ρ and velocity v is described by the following hyperbolic system of balance laws

$$\frac{\partial \rho}{\partial t} + \frac{\partial (\rho v)}{\partial x} = 0 , \tag{1a}$$

$$\frac{\partial (\rho(v - u(\rho)))}{\partial t} + \frac{\partial (\rho v(v - u(\rho)))}{\partial x} = \beta(\rho, v)\rho(u(\rho) - v) , \tag{1b}$$

where $u(\rho)$ is the equilibrium velocity and $\beta(\rho, v)$ is the effective relaxation coefficient (see Fig. 1). In the traffic model of Aw, Rascle and Greenberg, $\beta(\rho, v) = 1/T$, where $T = const$ is the *relaxation time*. In moving observer coordinates, the momentum equation reads $d(v - u)/dt = -\beta(v - u)$. The characteristic speeds of the BVT model are $\lambda_1 = v + \rho u'(\rho) \leq v$, which is related to shocks and rarefaction waves, and $\lambda_2 = v$, which is related to contact

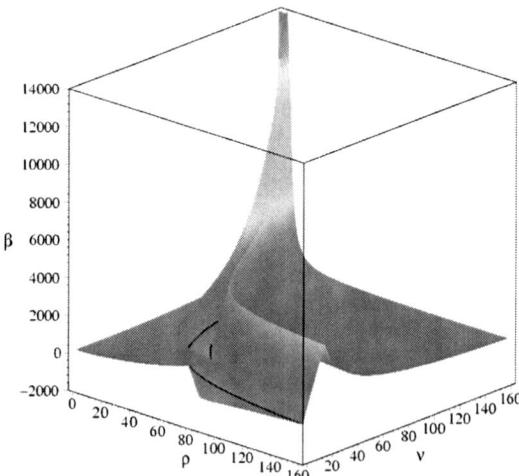

Fig. 1. The effective relaxation coefficient $\beta(\rho, v)$, for units $[\rho] = 1/\text{km/lane}$, $[v] = \text{km/h}$ and $[\beta] = 1/\text{h}$.

discontinuities. The appearance of negative $\beta(\rho, v) < 0$ can be motivated by the incorporation of a finite *reaction time* τ in the momentum equation (see [5] for more details).

2 Steady State Solutions

In order to obtain a better understanding of the BVT model we study the steady state solutions of the system (1a)-(1b). For steady state solutions there is a linear coordinate transformation $(t, x) \rightarrow (\Theta, z)$ such that all time derivatives with respect to Θ vanish. As a consequence steady state solutions lie on straight lines in the fundamental diagram of traffic flow, i.e. $\rho v = q + \rho w$, where $q, w = \text{const}$. They further fulfill

$$(\lambda_1 - w)\frac{dv}{dz} = \beta(\rho, v)(u(\rho) - v) . \tag{2}$$

2.1 Trivial Steady State Solutions

The trivial steady state solutions $(dv/dz = 0$, see Fig. 2) are

- $v = u(\rho)$: the equilibrium velocity curve,
- $\beta(\rho, v) = 0$, lower branch: the jam line,
- $\beta(\rho, v) = 0$, upper branch: the high-flow branch.

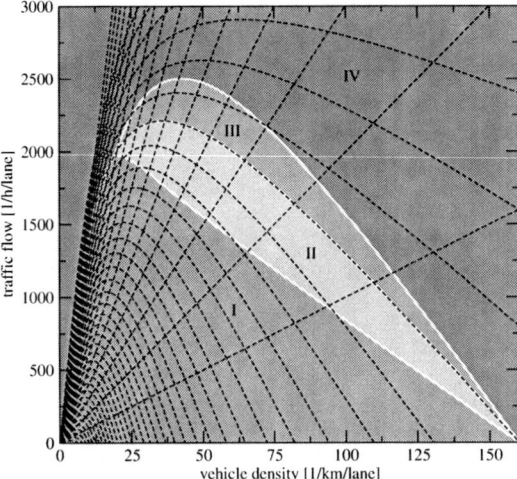

Fig. 2. Trivial steady state solutions and the characteristic structure of the BVT model. The trivial steady state solutions are the equilibrium flow-curve (borders between regions I and IV and regions II and III), the jam line (border between regions I and II), and the high-flow branch (border between regions III and IV). The dashed lines correspond to characteristic curves with gradient λ_1 and λ_2 respectively.

2.2 Non-Trivial Steady State Solutions

The non-trivial steady state solutions (i.e. $dv/dz \neq 0$) are monotonous solutions linking the trivial steady state branches mentioned before. They cover a wide range of states in the fundamental diagram of traffic flow, in particular regions II and III of Fig. 2.

3 Stability of Trivial Steady State Solutions

Motivated by [4] we study the stability properties of the trivial steady state solutions.

3.1 Linear Stability

The finite propagation speeds restrict the stability of steady state solutions to regions where the characteristic cone covers the steady state solution, i.e. $\lambda_1 \leq w \leq \lambda_2$, where w is given by

$$w = \frac{d(\rho v)}{d\rho} \ . \tag{3}$$

These intuitive results can be confirmed by a formal linear stability analysis (see [6]).

3.2 Nonlinear Stability

Numerical simulations [6] show that the linearly stable branches of non-equilibrium steady states are only metastable (in the sense of [3]) , i.e. for sufficiently
large perturbations the solutions depart from the steady states. We summarize
the overall stability properties in Fig. 3.

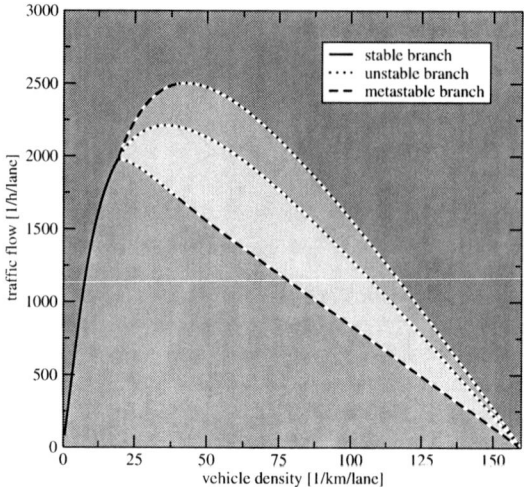

Fig. 3. Stability properties of the trivial steady state solutions in the BVT model.

4 Identification of Free Flow, Wide Moving Jams and Synchronized Flow

We try to relate the traffic states of the BVT model to the three phases of
traffic flow according to [3].

4.1 Free Flow

We interpret the stable branch of equilibrium flow and the metastable section
of the high-flow branch as free flow. Hence, according to Fig. 3, free flow is
metastable for intermediate densities.

4.2 Wide Moving Jams

Wide moving jams can be understood as spatially extended (quasi) steady
state solutions at the jam line. With a quasi steady state we describe a traffic
state for which the constants of steady states (see Sec. 2) can vary slightly.
Using this definition, wide moving jams have the following characteristics [6]

- The outflow from wide moving jams is nearly constant and far below the maximum free flow.
- The propagation speed of the downstream front of wide moving jams is nearly constant, wide moving jams travel upstream with a velocity of about 15 km/h.
- Wide moving jams travel through bottlenecks.
- Wide moving jams are not formed spontaneously from free flow.

4.3 Synchronized Flow

All other traffic states in the congested regime, including (non-trivial) steady states and narrow jams make up synchronized flow. Thus, states of synchronized flow cover a wide range in the flow-density diagram. In numerical simulations, the precise location of data points of synchronized flow in the fundamental diagram strongly depends on the initial and boundary data.

5 Traffic Flow at a Bottleneck

We exemplarily model a 7 km long section of a two lane highway with a bottleneck between 5 and 6 km, prescribing periodic boundary conditions. We use constant initial data of free flow in the metastable regime, $\rho = 75$ [1/km]. At the bottleneck, we modify the velocity according to $v \to v + (u(\rho) - v - 0.1 \ km/h)|\sin(\pi x)|$. Figure 4 shows the simulation results for the density and the velocity.
In particular, we observe that

- synchronized flow forms at the bottleneck and further upstream.
- a wide moving jam forms which travels through the bottleneck.
- narrow moving jams can merge and can be swallowed by the wide moving jam.
- narrow moving jams can be caught by the bottleneck.

6 Conclusion

Continuum models of traffic flow using an equilibrium flow-density curve have been criticized for the one-dimensionality of steady states. The BVT model [5, 6], which is a generalization of the model of Aw, Rascle and Greenberg [1, 2], still uses an equilibrium flow-density curve. However, this flow-density curve does not describe traffic states in the congested regime directly. As a consequence of the model, steady states cover a wide range in the congested regime of the fundamental diagram. The equilibrium flow-density curve is still very important, as it determines the characteristic and stability properties of the model. The upper branch of the trivial non-equilibrium steady state

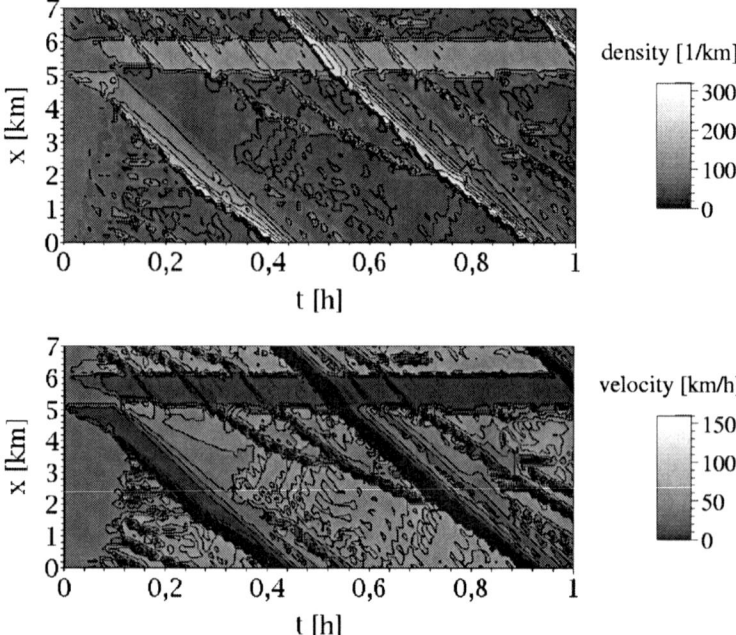

Fig. 4. Formation and propagation of wide moving jams. The upper panel shows the evolution of the density, whereas the lower panel shows the evolution of the velocity. At the bottleneck located between 5 and 6 km, synchronized flow forms, which finally leads to a wide moving jam. This wide moving jam moves with a velocity of about -15 km/h (i.e. upstream) and swallows moving jams during this propagation. It further travels through the bottleneck.

solutions is metastable against the formation of synchronized flow for intermediate densities and unstable for high densities. The model can thus explain the metastability of free flow for intermediate densities. It further reproduces the main characteristics of wide moving jams.

References

1. A. Aw, M. Rascle: SIAM J. Appl. Math. **60**, 916 (2000)
2. J.M. Greenberg: SIAM J. Appl. Math. **62**, 729 (2001)
3. B.S. Kerner: *The Physics of Traffic.* (Springer, Berlin 2004)
4. H.K. Lee, H.-W. Lee, D. Kim: Phys. Rev. E **69**, 016118 (2004)
5. F. Siebel, W. Mauser: SIAM J. Appl. Math. **66**, 1150 (2006)
6. F. Siebel, W. Mauser: Phys. Rev. E **73**, 066108 (2006)

Wave Selection Problems in the Presence of a Bottleneck

Jonathan Ward[1], R. Eddie Wilson[1], and Peter Berg[2]

[1] University of Bristol, Department of Engineering Mathematics, Bristol BS8 1TR
[2] UOIT, Faculty of Science, Oshawa, ON, L1H 7K4, Canada

Summary. The Optimal–Velocity (OV) model is posed on an inhomogeneous ring–road and the consequent spatial traffic patterns are described and analysed. Parameters are chosen throughout for which all uniform flows are linearly stable, and a simple model for a bottleneck is used in which the OV function is scaled down on a subsection of the road. The large-time behaviour of this system is stationary and it is shown that there are three types of macroscopic traffic pattern, each consisting of plateaus joined together by sharp fronts. These patterns solve simple flow and density balances, which in some cases have non-unique solutions. It is shown how the theory of characteristics for the classical Lighthill-Whitham PDE model may be used to explain qualitatively which solutions the OV model selects. However, fine details of the OV model solution structure may only be explained by higher order PDE modelling.

1 Introduction

The aim of this paper is to understand the steady state wave profiles that emerge in car-following models in the presence of spatial inhomogeneity. We simulate traffic with the Optimal Velocity model [1], posed on a ring-road that is made inhomogeneous by adding a simple model for a bottleneck, in which the Optimal Velocity function is reduced by a constant factor for some portion of the road. The surprise in this paper is that such a simple model set-up can display non-trivial solution structure.

The modelling of traffic flow can be understood at two distinct levels: microscopically, whereby each vehicle is considered individually, and macroscopically whereby traffic is considered as a continuous fluid. The simple discrete model that we consider here develops stable stationary patterns as $t \to \infty$, which can be understood by drawing parallels with continuum models.

The paper is set out as follows. In Sect. 2, we describe the OV model set-up that we use for the remainder of this paper, including precise details of how the spatial inhomogeneity is applied. Then in Sect. 3, we outline our numerical simulation and coarse-graining procedure, and we show results of three numerical experiments with qualitatively different solution structure as $t \to \infty$ (see Fig. 1). Sections 4 and 5 analyse these experiments using classical kinematic wave theory, firstly by analysing simple flow and density balances and then by using characteristic arguments to explain the wave selection principles. We calculate explicitly a phase diagram describing where the different solution types occur. Finally in Sect. 6, we conclude and indicate the success and failures of higher order continuum models in explaining the fine details of the solution structure.

2 Problem Set-Up

We consider the traffic patterns formed by a large number N of identical vehicles driving on a unidirectional single-lane ring-road of length L. Overtaking is not considered. Vehicles move in continuous space x and time t, and their displacements and velocities are labelled $x_n(t)$ and $v_n(t)$ respectively. We suppose that the direction of motion is in increasing x, and moreover that vehicles are labelled $n = 1, 2, \ldots, N$ in the downstream direction. For the vehicles' equations of motion, we adopt the well-known Optimal Velocity (OV) car-following model [1] for which

$$\dot{x}_n = v_n, \tag{1}$$
$$\dot{v}_n = \alpha \left\{ V(h_n; x) - v_n \right\}. \tag{2}$$

Here dot denotes differentiation with respect to time, and the rate constant $\alpha > 0$ is known as the sensitivity. The variable $h_n := x_{n+1} - x_n$ gives the headway, or gap to the vehicle in front, and loosely speaking the OV model describes the relaxation of traffic to a safe speed which is defined in terms of this gap. Note that under open boundary conditions one would need to prescribe the trajectory of the lead vehicle N, but on the ring-road we assume merely that it follows vehicle 1, so that $h_N = L + x_1 - x_N$.

The novelty in this paper is that we use an inhomogeneous OV function which takes the form

$$V(h_n; x) := \begin{cases} r_B V(h_n), & 0 \le x \bmod L < \hat{L}L, \\ V(h_n), & \hat{L}L \le x \bmod L < L, \end{cases} \tag{3}$$

and which is thus scaled down by a reduction factor $0 < r_B < 1$ for a proportion $0 < \hat{L} < 1$ of the ring-road under consideration. (Note that for sake of brevity, the vehicles' displacements $x_n(t)$ are set-up as monotone increasing and unbounded, although henceforth, we interpret all displacements modulo L.)

In (3), V with a single argument denotes a spatially independent OV function, and for concreteness, we adopt the standard [1] S-shape

$$V(h) = \tanh(h - 2) + \tanh(2). \tag{4}$$

However, qualitatively similar results should be recovered by any V for which 1. $V(0) = 0$, 2. $V' \geq 0$ and 3. $V(h) \to V_{\max}$ as $h \to \infty$. The detailed structure of V is not important because throughout we choose $\alpha \geq 2\max V'$, so that all uniform flows are linearly stable. Consequently, the patterns that we observe are forced only by the spatial inhomogeneity and not by spontaneous flow breakdown effects.

3 Numerical Procedure and Simulation Results

We now supplement equations (1–3) with the uniformly spaced initial data

$$x_n = nh_* \qquad \text{and} \qquad v_n = V(h^*) \quad \text{for } n = 1, 2, \ldots, N. \tag{5}$$

Here $h^* := L/N$ is the mean spacing. Note that for the limiting (no bottleneck) cases where either $r_B = 1$ or $\hat{L} = 0$, (5) gives a uniform flow solution of (1–3) in which $x_n = nh_* + tV(h^*)$. However, in general we should expect the bottleneck to redistribute traffic. In order to investigate the resulting patterns, we solve the initial value problem (1–5) numerically using a standard fixed step fourth-order Runge-Kutta solver.

After some experimentation with the solver, we conclude that the traffic always settles down to a stationary profile as $t \to \infty$, although the transient processes can sometimes be very long. Here stationarity means that suitably defined macroscopic density and velocity variables become steady, although they are non-trivially dependent on space x, and consequently vehicles' motions are in fact periodic as $t \to \infty$, since as they drive around the ring-road, they move repeatedly through the spatial pattern and experience traffic jams, free-flowing regimes etc. Note however that if we chose smaller values of sensitivity α than presented here, so as to force the linear instability of a range uniform flows, then the macroscopic variables could also be non-trivially time-dependent as $t \to \infty$.

Taking into account the above discussion, the results that we display shortly show stationary macroscopic density profiles $\rho(x)$ rather than individual vehicle trajectories. The simplest way to relate microscopic and macroscopic variables is via $\rho(x_n, t) = 1/h_n(t)$, although it is well-known [2, 3] that this relationship holds exactly only for entirely homogeneous situations. Therefore we use a coarse-grained [2] density

$$\rho(x,t) = \int_L dx' dt' \phi(x - x', t') \sum_n \delta(x_n(t') - x'), \tag{6}$$

with

$$\phi(x,t) = \frac{1}{2\pi\sigma^2} \exp(-x^2/2\sigma^2)\delta(t), \tag{7}$$

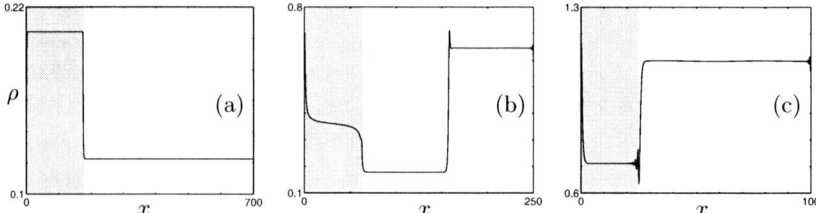

Fig. 1. Stationary $t \to \infty$ coarse-grained density profiles $\rho(x)$. The portions of solution profiles within the bottleneck are indicated by shading. (a) Light traffic $h_* = 7.0$ ($\rho_* = 0.142857$), see Example 1; (b) Medium traffic $h_* = 2.5$ ($\rho_* = 0.4$), see Example 3; (c) Heavy traffic $h_* = 1.0$ ($\rho_* = 1.0$), see Example 2.

which is thus obtained from distributional point density by convolving with a Gaussian test function whose characteristic length scale σ is chosen large enough so as to smooth out individual vehicles but small enough so as to retain macroscopic features. A macroscopic flow variable $q(x, t)$ may be obtained in a similar way by coarse-graining the discrete velocity v_n, and then a coarse-grained velocity is given by $v(x, t) := q(x, t)/\rho(x, t)$. Note that since we are usually seeking a steady density, there are computational short-cuts and the cheapest procedure is to calculate $\rho(x)$ by coarse-graining in time the numerical trajectory of a single vehicle as it drives once around the ring-road. We now give three examples of the eventual stationary profiles $\rho(x)$ which show how the structure changes as the mean headway h^* is varied. To simplify matters, all other parameters are held fixed as follows: $N = 100$ vehicles, bottleneck reduction factor $r_B = 0.6$, bottleneck nondimensionalised length $\hat{L} = 0.25$ and sensitivity $\alpha = 2.0$. Later we consider how the qualitative solution structures change as functions of the three problem parameters $\rho_* := 1/h_*$, r_B and $\hat{L}$.

Example 1. We take $h_* = 7.0$ which corresponds to light traffic (large h_*, small ρ_*). See Fig. 1(a). The $t \to \infty$ steady density profile $\rho(x)$ adopts a two-plateau form, with an almost constant density ρ_B attained in the bottleneck and a lower (almost constant) density ρ_1 on the remainder of the loop. At each end of the bottleneck, the two density plateaus are joined by sharp, almost shock-like fronts.

Example 2. We take $h_* = 1.0$ which corresponds to heavy traffic. See Fig. 1(c). In a similar fashion to Example 1, $\rho(x)$ adopts a two-plateau form. However this time the bottleneck density ρ_B is less than the density ρ_1 on the unconstrained part of the loop. Like Example 1, there are also sharp, shock-like fronts at each end of the bottleneck, although here they have a more complicated oscillatory structure.

Example 3. We now take $h_* = 2.5$ which may be regarded as an intermediate case. See Fig. 1(b). In contrast to the two previous examples, $\rho(x)$ now has a three-plateau form. The density as before adopts an almost constant (but slightly S-shaped) profile $\rho \simeq \rho_B$ within the bottleneck, with fronts at each end. Downstream of the bottleneck is a low density ρ_1 region, whereas upstream is a high density ρ_2 region, which may be thought of as a queue waiting to enter the bottleneck. There is thus an extra internal shock-like front in the unconstrained part of the loop, where the fast traffic that has come out of the bottleneck rejoins the queue to enter it. Unlike the other fronts we have encountered so far, that joining ρ_1 and ρ_2 is not locked on a discontinuity in the model; nevertheless, it is stationary.

Further simulation may be used to show how the Fig. 1 profiles are related to each other. If one starts with the Fig. 1(b) structure (Example 3) and decreases the mean headway (increases the mean density), then the queue upstream of the bottleneck grows in length until it reaches the downstream boundary of the bottleneck, and swamps the entire unconstrained part of the loop. At this point, the internal shock vanishes and the Fig. 1(c) structure is recovered. Conversely, if one starts with Fig. 1(b) and decreases the mean density, the queue upstream of the bottleneck shortens until it vanishes altogether. At that point, the internal shock is absorbed into the upstream boundary of the bottleneck and the Fig. 1(a) structure is recovered.

4 Density and Flow Balances

We now begin an explanation of the structures seen in Section 3. Later we derive a phase diagram which predicts when each will occur. Since the observed structures resemble constant density plateaus separated by classical shocks, we attempt an explanation based on kinematic wave theory [6]. To this end, we introduce the fundamental (flow) diagram $Q(\rho) = \rho \hat{V}(\rho)$ where $\hat{V}(\rho) = V(1/\rho)$ is the continuum counterpart to the discrete OV function V. As is well-known, Q is usually a unimodal function. With choice (4), Q attains its maximum value $Q_{max} \simeq 0.58$ at $\rho_{max} \simeq 0.36$. In the bottleneck, the fundamental diagram Q is scaled by r_B.

Firstly we consider the two-plateau structures of Figs. 1(a) and (c). Since the fronts are sharp, negligibly few vehicles are contained within them at any one time. We may therefore approximate the density $\rho(x)$ with a piecewise-constant profile consisting of ρ_B within the bottleneck and ρ_1 in the unconstrained part of the loop. It thus follows that

$$\hat{L}\rho_B + (1 - \hat{L})\rho_1 = \rho_*, \tag{8}$$

$$Q(\rho_1) = r_B Q(\rho_B), \tag{9}$$

which describe respectively the conservation of vehicles and a flow balance (the latter is necessary since the observed profiles are stationary). Equations (8), (9) are thus a pair of simultaneous equations to solve for ρ_1 and ρ_B, where the remaining parameters ρ_*, r_B and $\hat{L}$ are prescribed.

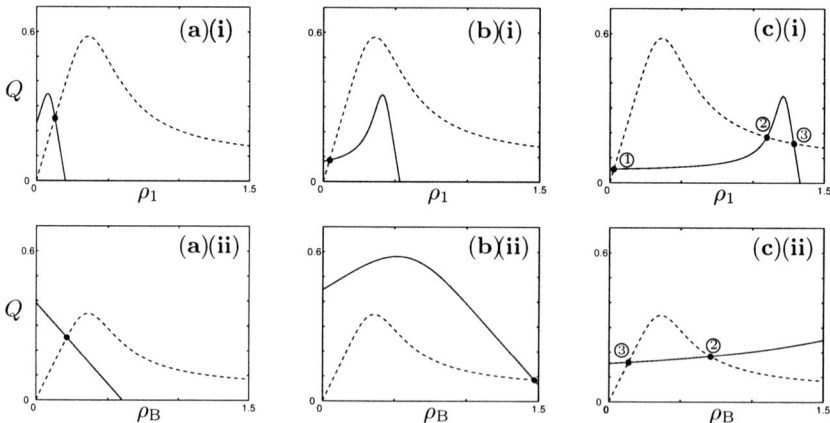

Fig. 2. Solution structure of (8), (9). Panels (a-c) correspond directly to panels (a-c) in Fig. 1. Top row (i) indicates solutions of (10) and bottom row (ii) of the equivalent equation (11). The extra numbering in panels (c)(i,ii) allows the (ρ_1, ρ_B) solution pairs to be identified.

We must therefore examine the (ρ_1, ρ_B) solution structure of (8), (9) and this is achieved via Fig. 2. To see this, note that ρ_B may be eliminated from (8), (9) to give

$$Q(\rho_1) = r_B Q \left[\frac{\rho_* - (1 - \hat{L})\rho_1}{\hat{L}} \right], \qquad (10)$$

and the left and right hand sides of this equation are plotted against ρ_1 in Figs. 2(a-c)(i). Note alternatively that ρ_1 can be eliminated from (8), (9) to give

$$r_B Q(\rho_B) = Q \left[\frac{\rho_* - \hat{L}\rho_B}{1 - \hat{L}} \right], \qquad (11)$$

and as a cross-check, the left and right hand sides of this equation are plotted against ρ_B in Figs. 2(a-c)(ii). Further, parameters have been chosen so that the panels (a-c) correspond directly to panels (a-c) in Fig. 1. Firstly, the light traffic diagrams Figs. 2(a)(i,ii) indicate a unique (ρ_1, ρ_B) solution pair and it may be shown that this is indeed corresponds to values obtained in Example 1.

However, in the heavy traffic diagrams Figs. 2(c)(i,ii), there are clearly three (ρ_1, ρ_B) solution pairs: what determines which pair is selected in the corresponding Example 2? Finally, in the intermediate case of Figs. 2(b)(i,ii), there is a unique (ρ_1, ρ_B) solution pair, however, the corresponding Example 3 selects instead a three-plateau structure. It now remains to identify extra principles which explain the solution selection in cases (b) (Example 3) and (c) (Example 2).

5 Wave Selection Via Characteristics

We now use characteristic arguments from kinematic wave theory [6, Chap. 2] to explain the observed wave selection behaviour. We focus initially on Example 2 (heavy traffic), see Figs. 1(c) and Figs. 2(c)(i,ii), and then later we consider the three-plateau case.

We recall that in kinematic wave theory, characteristics are lines (or line segments) in the (x, t) plane on which density is conserved. Further, it is well-known that the local velocity of a characteristic with density ρ is given by $Q'(\rho)$. Consequently, characteristics propagate downstream in light traffic and upstream in heavy traffic. When characteristics converge, one obtains a classical shock, whereas when they diverge, one obtains a (non-stationary) expansion fan.

Figure 3 develops a characteristic analysis of the (ρ_1, ρ_B) solution pairs found in Fig. 2(c)(i,ii). The key point to note is that the solution pairs numbered 1 and 3 straddle $\rho = \rho_{max}$ at which both the unconstrained $Q(\rho)$ and bottleneck $r_B Q(\rho)$ fundamental diagrams attain their maxima. These solution pairs can be disregarded, because the consequent density profiles would involve patterns of characteristics with both positive and negative slopes. This means that at either the upstream or downstream boundary of the bottleneck, there would necessarily be a non-stationary expansion fan which would not agree with the $t \to \infty$ stationary results.

In contrast, solution pair 2 is non-straddling and involves only characteristics with negative slopes, see Fig. 3 panel (iv). In this sketch, neither the upstream or downstream boundary of the bottleneck has a classical (compressive) shock. Rather, at each boundary the characteristics cross through the shock which is forced solely by the model discontinuity at that point. It may be shown that this solution agrees with that found by discrete simulation in Section 3 and moreover that it is a proper solution of the Lighthill-Whitham-Richards model in that it may be reached via the solution of the initial value problem [8].

We now turn our attention to the three-plateau case (Example 3, Fig. 1(b)), for which it may be shown that the analysis of Section 4 predicts a straddling, and hence invalid solution pair (ρ_1, ρ_B). The resolution is thus to approximate the density $\rho(x)$ by a piecewise-constant profile with three components: ρ_B (density in bottleneck) and ρ_1, ρ_2 (densities in unconstrained part of loop). The density and flow balances thus yield respectively

$$\hat{L}\rho_B + \beta(1 - \hat{L})\rho_1 + (1 - \beta)(1 - \hat{L})\rho_2 = \rho_*, \tag{12}$$

$$Q(\rho_1) = Q(\rho_2) = r_B Q(\rho_B), \tag{13}$$

where $0 < \beta < 1$ parametrises the internal shock position separating ρ_1 and ρ_2. We thus have three equations, but four unknowns, namely β, ρ_1, ρ_2 and ρ_B, and we require extra information to fix a unique solution. By studying characteristic diagrams, it becomes clear that a solution without diverging characteristics (and hence non-stationary expansion fans) is only possible if

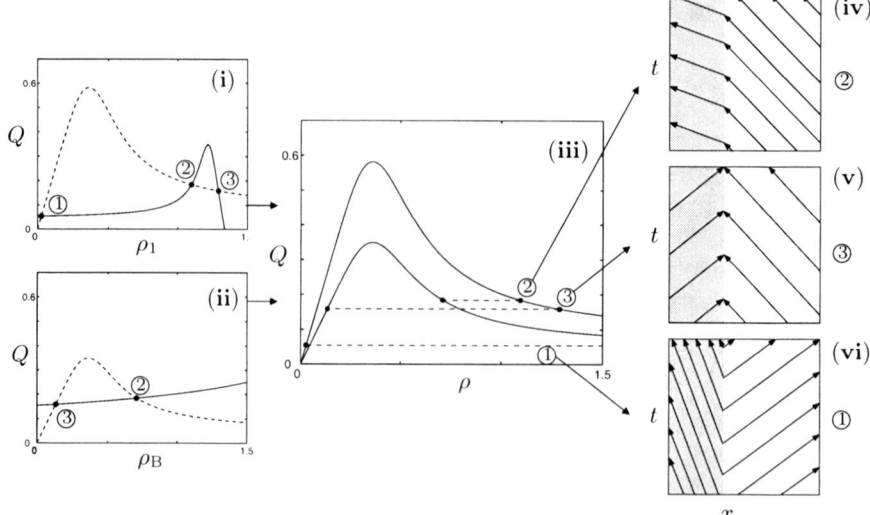

Fig. 3. Characteristic analysis for the (ρ_1, ρ_B) solution pairs from Figs. 2(c)(i,ii). Panel (iii) shows the location of solution pairs, joined by horizontal lines representing flow balance, on the fundamental diagrams Q and $r_B Q$. Characteristic pictures for each of the three root pairs are shown in panels (iv), (v) and (vi): the bottleneck is denoted by shading. Panels (v) and (vi) cannot give stationary profiles since they predict an expansion fan at the up- and down-stream ends of the bottleneck respectively. Hence solution pair 2 from panel (iv) is selected. Note that in panel (iii), this solution pair is non-straddling in the sense that both ρ_1 and ρ_B are the same side of the fundamental diagram maximum.

$$\rho_B = \rho_{\max}, \tag{14}$$

i.e., if the flow inside the bottleneck is maximised. Further, when supplemented by (14), system (12), (13) can be solved uniquely for ρ_1, ρ_2 and β, and it may be shown that this solution agrees with the discrete simulations. The characteristic structure is shown in Fig. 4. In particular, it involves non-standard waves at the up- and down-stream ends of the bottleneck. However it may be shown via the solution of the initial value problem that these are admissible solutions of the Lighthill-Whitham-Richards model [8].

We now turn our attention to the computation of a phase diagram. Since in the three-plateau case we have $\rho_B = \rho_{\max}$, the values of ρ_* where solutions change from two plateau solutions to three plateau solutions can be calculated. At the thresholds, β is either 0 or 1 and $\rho_B = \rho_{\max}$, thus eliminating ρ_1 or ρ_2 in (13) using (12), leaves only

$$r_B Q(\rho_{\max}) = Q \left[\frac{\rho_* - \hat{L}\rho_{\max}}{1 - \hat{L}} \right], \tag{15}$$

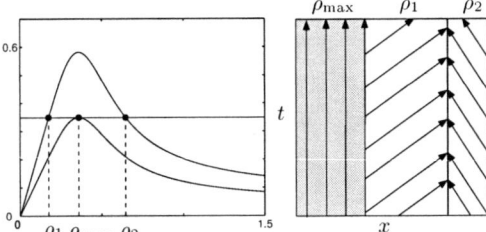

Fig. 4. Characteristic analysis for the three-plateau case: only the configuration shown with flow maximised in the bottleneck avoids expansion fans. Note that the characteristics inside the bottleneck have zero velocity and hence this structure is on the very boundary of becoming an expansion fan. The internal shock between ρ_1 and ρ_2 is classical since at it the characteristics converge.

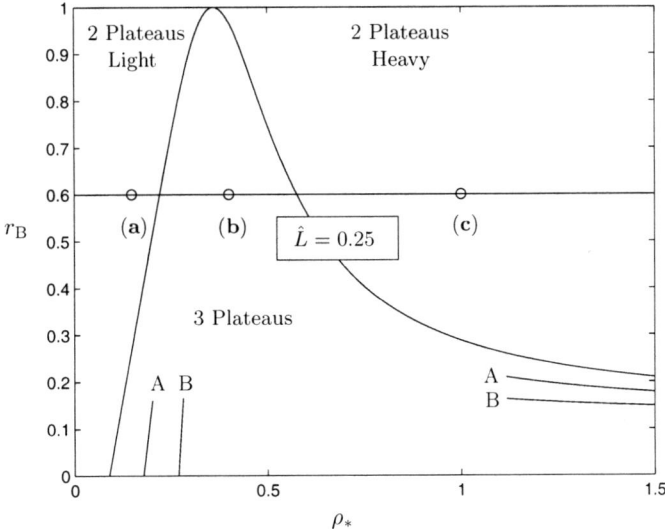

Fig. 5. Phase diagram derived from (15) for bottleneck length $\hat{L} = 0.25$. The points marked (a), (b), and (c) correspond to panels (a), (b) and (c) in Figs. 1 and 2. The line segments denoted A and B indicate to where the phase boundary would move for $\hat{L} = 0.5$ and $\hat{L} = 0.75$ respectively.

as a relation between the problem parameters that holds at the transition, see Fig. 5. In particular, we may partition the (ρ_*, r_B) plane according to whether the three-plateau solution occurs, or according to which type of two-plateau solution occurs, and the boundary in this plane depends on $\hat{L}$ in a manner that we can determine explicitly. In particular, increasing the length of the bottleneck shrinks the domain where the three-plateau solution occurs.

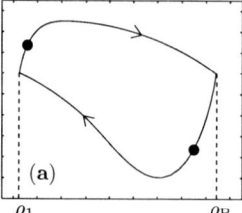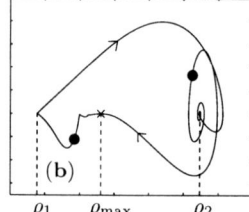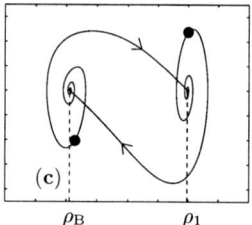

Fig. 6. Numerical (ρ, ρ_x) phase plots corresponding to Figs. 1(a–c). The discs mark the boundaries of the bottleneck. (a) Light traffic: heteroclinic cycle connecting $(\rho_1, 0)$ and $(\rho_B, 0)$. (b) Three plateau case: saddle at $(\rho_1, 0)$; saddle–node at $(\rho_B, 0)$ explaining S–shaped structure; complex fixed point at $(\rho_2, 0)$. (c) Heavy traffic: complex fixed points at $(\rho_1, 0)$ and $(\rho_2, 0)$.

6 Higher Order Modelling

We have shown how first order kinematic wave theory explains the principal qualitative features of the discrete simulation results presented in Fig. 1. This theory however is based on a piecewise-constant ansatz for the density profile $\rho(x)$, and does not explain, for example, the S-shaped profile in the bottleneck in Fig. 1(b), nor does it explain the internal structure of the shocks. To analyse these features, we should resort to higher order PDE approximations of the OV model [2, 3]. Using the work of [7] as motivation, we use finite differences to obtain the spatial derivative of the coarse-grained density so that we may display numerical (ρ, ρ_x) phase portraits: see Figs. 6(a–c).

In each of these phase portraits, trajectories spend most time in the vicinity of fixed points which correspond to the constant density plateaus in Figs. 1(a–c). These fixed points are then linked via rapid transits across the phase portrait which describe the interior structure of the shocks. Note that in Figs. 6(a–c), the bottleneck boundaries are in fact 'mid-shock' and are denoted by small solid discs.

In the light traffic portrait Fig. 6(a), we observe a heteroclinic-cycle connecting saddle-like fixed points, which agrees with the analytical prediction of [7]. However, in the heavy traffic portrait Fig. 6(c), the numerical trajectory crosses itself, and we observe complicated fixed-points which resemble projections of Shilnikov points. The conclusion in this case is that the dynamics cannot be represented in two dimensions. In fact, one may show that the second-order theory [7] predicts a pair of stable node fixed points, and hence cannot produce the required connections. Instead, the solution seems to adopt a sharper profile which brings higher derivatives into play, and which thus permits the required connections in a higher dimensional phase space. In the three-plateau case Fig. 6(b), [7] predicts that the bottleneck fixed point is at saddle-node bifurcation, in the vicinity of which, trajectory behaviour is polynomial in x. This observation can be used to explain the S-shape bottleneck profile in Fig. 1(b).

Thus to sum up, from a simple model for a bottleneck on a loop, we have observed interesting, stationary wave patterns in the OV model as $t \to \infty$. These patterns consist generally of two or three plateaus separated by shock–like structures. We have built an understanding of these patterns using a PDE approach, principally by using first order kinematic wave theory. However, the fine details require higher order modelling with a momentum equation – this is work in progress.

References

1. M. Bando, K. Hasebe et al, Phys Rev. E **51**, 1035 (1995)
2. H.K. Lee, H. W. Lee, D. Kim, Phys. Rev E **64**, 056126 (2001)
3. P. Berg, A. Mason, A.W. Woods, Phys Rev. E **61**, 1056 (2000)
4. M.J. Lighthill, G.B. Whitham, Proc. R. Soc., A**229**, 317 (1955)
5. P.I. Richards, Operations Research **4**, 42-51 (1956)
6. G.B. Whitham: *Linear and Nonlinear Waves* (Wiley, New York, 1974)
7. R.E. Wilson and P. Berg, *Traffic and Granular Flow '01*, (Springer, Berlin, 2003)
8. W.L. Jin and H.M Zhang, Trans. Sci. **37(3)**, 294 (2003)

Impacts of Lane Changes at Merge Bottlenecks: A Theory and Strategies to Maximize Capacity

Jorge Laval[1], Michael Cassidy[2], and Carlos Daganzo[2]

[1] Laboratoire Ingénierie Circulation Transport LICIT (INRETS/ENTPE), France
[2] Department of Civil and Environmental Engineering, University of California, Berkeley

Summary. Recent empirical observations at freeway merge bottlenecks have revealed (i) a drop in the bottleneck discharge rate when queues form upstream, (ii) an increase in lane-changing maneuvers simultaneous with this "capacity drop", and (iii) a reversal of the drop when the ramp is metered.

This paper shows that a simple vehicle lane-changing theory, which has been shown to explain related phenomena at lane-drops and moving bottlenecks, also explains the new phenomena at merges. In this theory, lane-changing vehicles are modeled as discrete particles endowed with realistic accelerations, and are embedded in a multilane stream where each lane obeys the kinematic wave model. This theory is parsimonious: only one of its four parameters has to be calibrated by running the model.

Our simulations show that the theory predicts surprisingly well the cumulative flows at all locations, the vehicle trip times, the number of lane-changing maneuvers, the capacity drop, its recovery upon metering, and the distribution of these measures across lanes and over time. Applications are discussed.

1 Introduction

The kinematic wave (KW) model [1, 2], when applied with a triangular fundamental diagram (KWT) [3], is arguably the simplest means to explain basic traffic features, such as the spatial extent of queues and average vehicle densities within these queues [4–9]. But more complex traffic features, such as the capacity drop [10–14], hysteresis [15], the capacity of moving bottlenecks [16] and stop-and-go waves [5, 17–21] cannot be explained with such a simple model.

Many of these features, however, are explained by a multilane hybrid (MH) theory that combines the KW model with discrete lane changes treated as moving bottlenecks [22]. This paper shows that the theory also explains traffic behavior at merges. Section 2 describes the key concepts of the theory in [22]

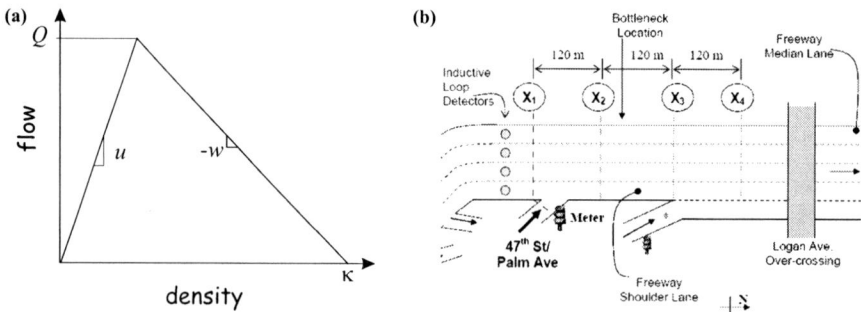

Fig. 1. (a) Triangular fundamental diagrams; (b) site description (taken from [23]).

and the proposed treatment of merges. Section 3 tests the model and section 4 discusses its results.

2 Parameters of the Multilane Hybrid Model with Merges

The multilane hybrid (MH) model in [22] treats each lane as a KWT traffic stream, interrupted by lane-changing vehicles. The KWT model has three parameters that can be measured by direct observation: a "free-flow" speed, u (km/h), a wave speed, w (km/h), and a "jam density", κ (veh/km). One can also define a "capacity", Q (veh/h), related to the previous three parameters by: $Q/u + Q/w = \kappa$; see Fig. 1a. Typical values (used on all our tests) are: $u = 112.7$ km/h (70 mph), $w = 22.5$ km/h (14 mph) and $\kappa = 139.7$ veh/km (225 veh/mile). Therefore, $Q = 2,625$ veh/h.

Lane-changing vehicles are treated in the MH model as discrete particles with variable speeds and accelerations (v, a), subject to upper bounds: $v \leq v_{max}; a \leq a_{max}$. These bounds are parameters of the model, but they can be chosen without running the model, simply by analyzing the vehicle fleet. For all of our tests we chose features of an average car on level terrain: $v_{max} = 123.8$ km/h and $a_{max} = 3.4(1 - v/v_{max})$ m/s^2.

Lane changes are assumed to be triggered by speed differences between adjacent lanes, and drivers' desire for travelling faster. This is modeled by a probability rate for lane changing (probability per unit time), π, which defines the behavior of a driver experiencing a speed deficit, $\Delta v \geq 0$, relative to a neighboring lane. The behavioral relationship is assumed to be of the form $\pi = \Delta v/(u\tau)$, where τ is a behavioral parameter with units of time. This is the only parameter that is estimated by running the model.

Given u, w, κ, τ, and the upstream traffic demand, the model can be simulated in discrete time by inspecting the system at intervals Δt (sec); see [22]. It predicts the flow and accumulations by lane at any desired set of locations and the number of lane changes between every pair of adjacent lanes. As explained in [22], the model reproduces the capacity drop at lane-drops because lane changers with bounded accelerations introduce voids in the traffic stream, which pass through a bottleneck to the detriment of its discharge rate.

The on-ramp was modeled as an additional freeway lane with the same above-mentioned KWT parameters. Since our site did not have an acceleration lane, we neglected its length; i.e., lane changes to/from the freeway can only take place at a point. Given that several on-ramp vehicles may enter the freeway at the same time, an additional condition is imposed at the boundary with the freeway. At this boundary, the on-ramp capacity is expanded by 40 %. Notice that this expanded capacity does not induce on-ramp inflows greater than Q; it merely implements the boundary condition for merges proposed in [25] by increasing the priority for entering traffic to the levels observed in [24].

3 Tests

In this section we show how the MH model replicates the empirical observations at an on-ramp merge in [23]. We describe these observations first.

3.1 Field Measurements

The experiments in [23] are the first to reveal some of the mechanisms that lead to lower bottleneck discharge rate after queues formed upstream. The site is a stretch of northbound Freeway 805 in San Diego, California; see Fig. 1b. The merge formed by the metered on-ramp at 47th St/Palm Ave on-ramp, is a recurrent active bottleneck. The experiments were conducted during ten morning rush periods in summer and fall 2003. Capacity drops and recoveries due to select ramp metering strategies were observed. The rush periods of October 15th and October 21st were selected for analysis in this paper.

Detailed traffic data were manually extracted from videos. The empirical data for October 15th are presented in Figs. 2a and 3a-b, while the data for October 21st correspond to Figs. 4a and 5a-b. These figures display the following time series:

- Oblique queueing diagrams: a transformation of cumulative vehicle count vs time, $N(t)$, measured at the four locations labelled X_1 through X_4 in Fig. 1b; see [26]. Notice that to guarantee flow conservation ramp inflows were added to the counts upstream of the bottleneck.
- Vehicle accumulations: the number of vehicles in the shoulder lane (only) between locations X_1 and X_3, as per the illustration directly to the right of Fig. 3a. These accumulations were sampled from video every 5 sec and the curve presents the averages of these counts over 1-min intervals.

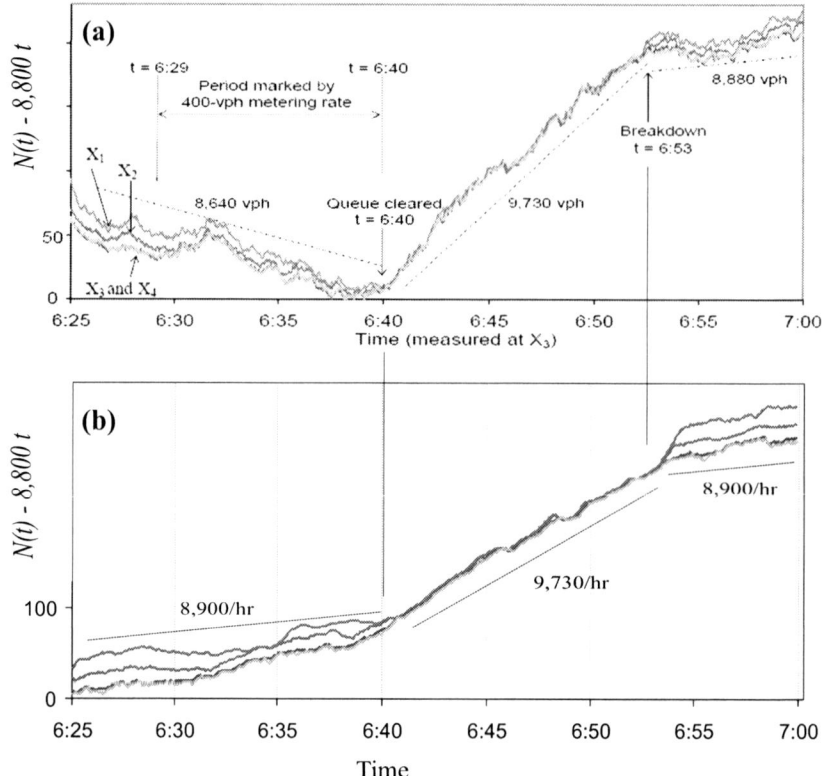

Fig. 2. Oblique queueing diagrams on October 15st: (a) measured (taken from [23]); (b) simulated.

- Oblique cumulative lane changes, $LC(t)$: this plot corresponds to the commutative number of vehicle lane changes between X_1 and X_3 on an oblique coordinate system to amplify the variations in the lane changing activity. Only lane changes leaving the two rightmost lanes were considered; see schematic to the right of Fig. 3b.

Examination of figures 2a, 3a-b, 4a and 5a-b reveals that (i) the capacity drop occurs simultaneously with an increase in lane-changing counts and shoulder lane vehicle accumulation, and that (ii) controlling the ramp-metering rate could mitigate this lane changing and accumulation, so that high merge capacities could be restored.

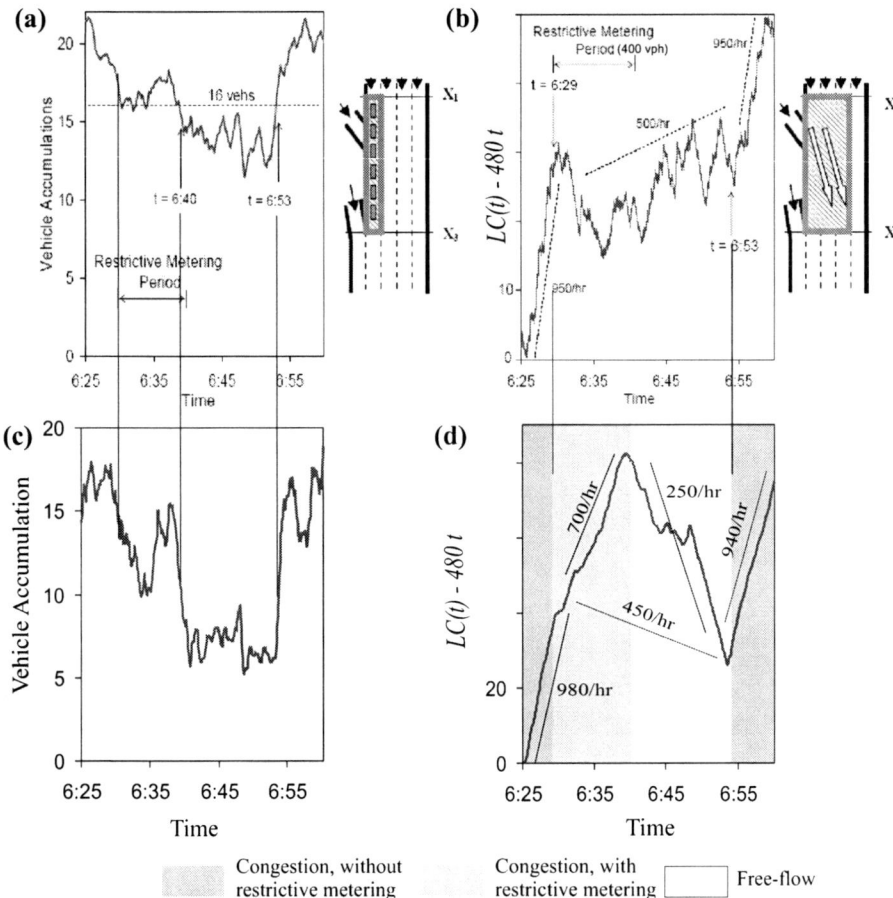

Fig. 3. Time series of accumulations and lane-changing: (a) measured shoulder lane accumulation on October 15st (taken from [23]); (b) measured lane-changing flows on October 15st (taken from [23]); (c) measured shoulder lane accumulation; (d) measured lane-changing flows.

3.2 Simulation Results

The input data for the simulations consisted of the lane-specific traffic demands measured in 30-sec intervals at the loop-detector located upstream of X_1 (see Fig. 1b[3]), and the demand on the on-ramp taken from [23]. The behavioral parameter $\tau = 4$ sec was found to replicate the number of lane

---

[3] Notice that when a queue reaches this detector it no longer measures demand but the bottleneck discharge rate. When this happened we extrapolated the last demand value into future time steps.

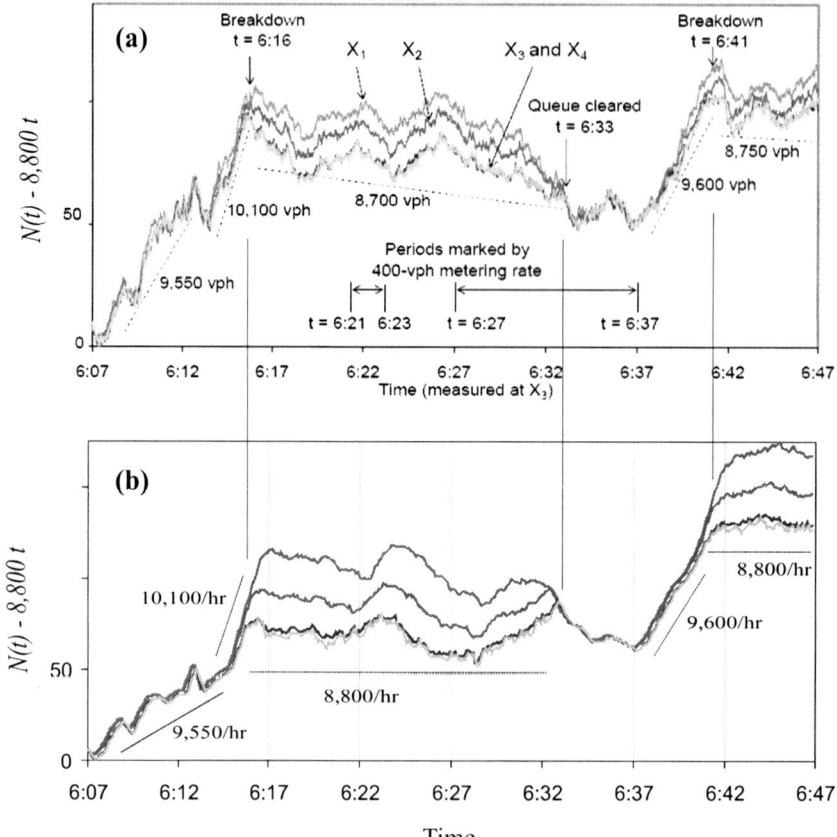

Fig. 4. Oblique queueing diagrams on October 21st: (a) measured (taken from [23]); (b) simulated.

changes during congested periods and was used throughout. We also chose $\Delta t = 0.6$ sec, but any small value for the time increment would work similarly. Simulation results are shown below the corresponding empirical charts on Figs. 2 to 5. Vertical solid lines connecting the empirical and simulated charts have been added to facilitate comparisons.

Examination of Figs. 2 and 4 reveals that the simulation accurately predicts the cumulative count curves at all locations. In all cases, the theory predicts bottleneck activation times to within 30 seconds of the observed times. Predicted bottleneck discharge rates are within 3% of those observed.

Although discrepancies exist between the predicted and observed curves of shoulder-lane accumulations and cumulative lane changes (Figs. 3 and 5), key features of these microscopic measures are reproduced by the theory. In particular, the predicted time series of shoulder lane accumulations exhibit similar

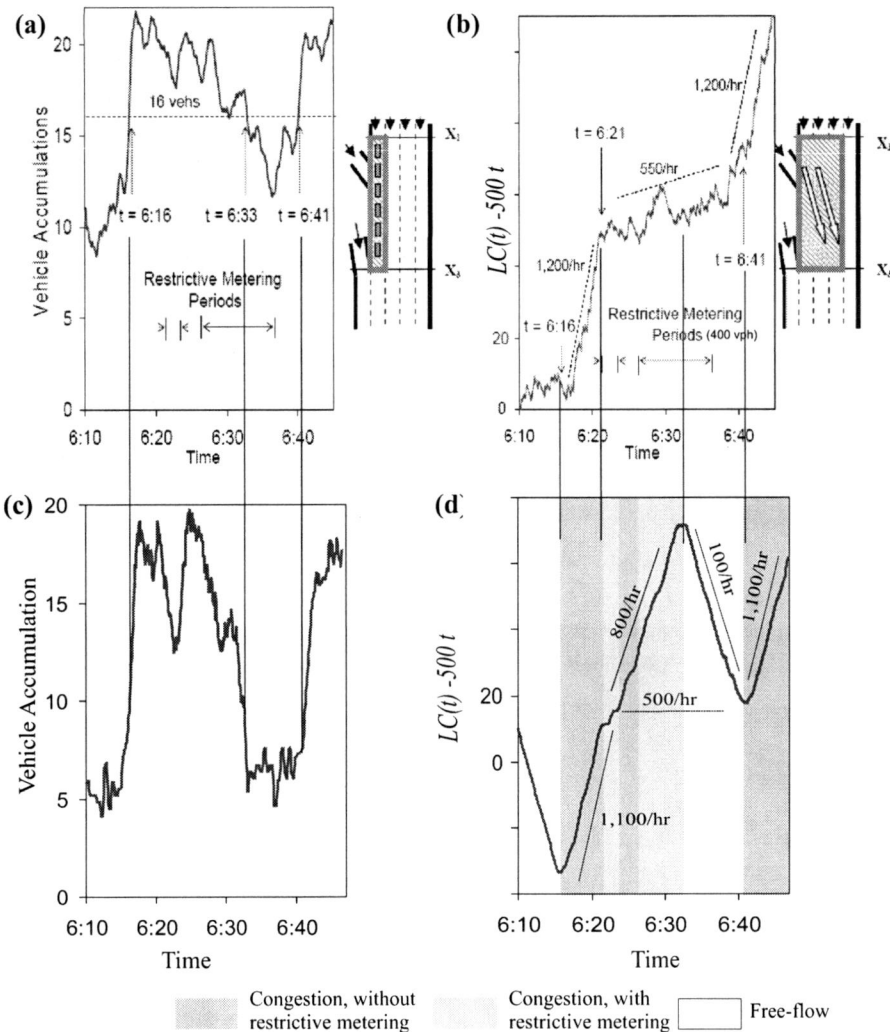

Fig. 5. Time series of accumulations and lane-changing: (a) measured shoulder lane accumulation on October 21st (taken from [23]); (b) measured lane-changing flows on October 21st (taken from [23]); (c) measured shoulder lane accumulation; (d) measured lane-changing flows.

shapes to those observed, though the theory underpredicts the numeric values. Predicted lane-changing maneuvers match those observed during congested times outside of restrictive metering periods (this traffic regime is color-coded dark gray in Figs. 3d and 5d). Prediction errors arise for the other regimes; these being congestion during restrictive metering periods (light gray in the

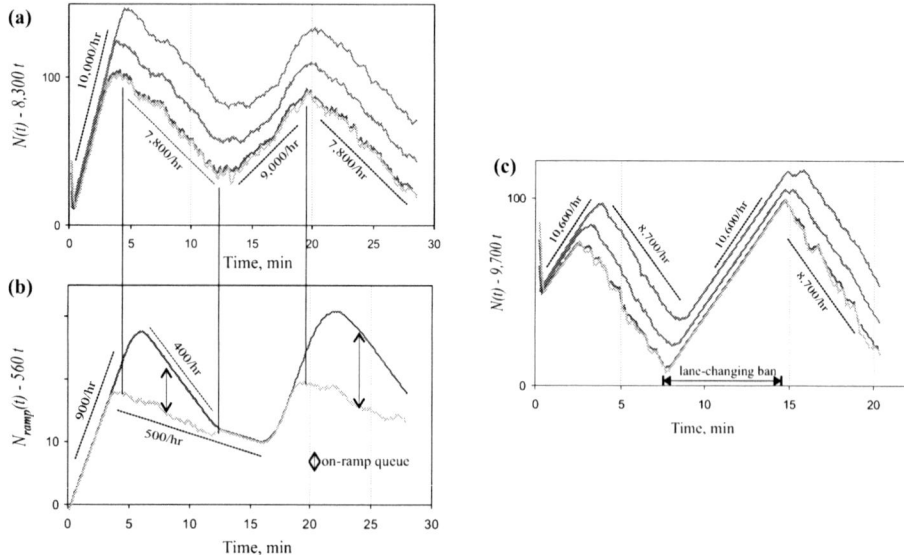

Fig. 6. Simulation of the effects of two control schemes: (a) oblique queueing diagrams on the freeway and (b) on the on-ramp; (c) oblique queueing diagrams on the freeway with a temporary lane-changing ban.

figures) and free-flow conditions (white). Notably, however, the total number of lane-changing maneuvers predicted over these two regimes combined closely match observed numbers. The observed times of regime change are also closely matched by the theory.

4 Discussion

The generally good fit between theory and observation – particularly on the important aggregate measures – is encouraging, given the paucity of model parameters. A perfect match between theory and observation should not be expected at the microscopic level given (i) that observations vary significantly from day to day; and (ii) that our behavioral assumption was the simplest possible. In fact we find it surprising that a single-parameter model could reproduce so much detail.

With this in mind, we now use the model to assess the capacity enhancements generated by two control schemes: one focusing on on-ramp flow, the other on freeway lane-changing maneuvers. Predictions for the first of these solutions are displayed in Fig. 6a-b. These were produced for a constant freeway demand and for on-ramp demands that cause the ramp queue to grow slowly and then recede. Note how three capacities arise: a full capacity of 10,000 veh/h that arises with no queues on the freeway nor on the on-ramp; an intermediate capacity of 9,000 veh/h with a queue only on the freeway; and a low capacity of 7,800 veh/h with queues both on the freeway and the on-ramp. The latter two capacities correspond to roughly 10 and 20% of the full capacity, and are consistent with the range of capacity drops reported in the literature. Notice that both drops occur when the on-ramp input flows are 500 veh/h in both cases. Therefore, the 1,200 veh/h difference in discharge flow from the merge most likely occurs because queued on-ramp vehicles enter the freeway at low speeds. Thus, the model predicts that preventing on-ramp queues is of critical importance. Note too that varying traffic conditions on the on-ramp induce oscillations in the discharge rate and can contribute to oscillations in the freeway queue.

Predictions for the second control scheme are shown in Fig. 6c. These predictions are made for the same site under constant freeway and on-ramp demands, but now a lane-changing ban is imposed on freeway traffic in $t \in [7, 14]$ min. It is clear how full capacity is restored during the period of no lane-changing activity. Notice that full recovery was possible because no queues formed on the on-ramp. Had there been a ramp queue, merge capacity would have been partially recovered, as per our preceding results. These topics are currently under further investigation by the authors.

References

1. M.J. Lighthill and G.B. Whitham. On kinematic waves. I Flow movement in long rivers. II A theory of traffic flow on long crowded roads. *Proc. Roy. Soc.* 229(A), 281–345 (1955).
2. P. I. Richards. Shockwaves on the highway. *Opns. Res.* 4, 42–51 (1956).
3. G. F. Newell. A simplified theory of kinematic waves in highway traffic, I general theory, II queuing at freeway bottlenecks, III multi-destination flows. *Trans. Res. B* 27, 281–313 (1993).
4. M. Cassidy and J. Windover. Methodology for assessing dynamics of freeway traffic flow. *Trans. Res. Rec.* 1484, 73–79 (1995).
5. M. Mauch and M.J. Cassidy. Freeway traffic oscillations: Observations and predictions. In M.A.P. Taylor, editor, *15th Int. Symp. on Transportation and Traffic Theory*, Pergamon-Elsevier, Oxford,U.K. (2002).
6. M. Cassidy and R. Bertini. Some traffic features at freeway bottlenecks. *Trans. Res. B* 1(33), 25–42 (1999).
7. C.F. Daganzo, M. Cassidy, and R. Bertini. Possible explanations of phase transitions in highway traffic. *Trans. Res. A* 5(33), 365–379 (1999).
8. V.F. Hurdle and B. Son. Road test of a freeway model. *Trans. Res. A* 7(34), 537–564 (2000).
9. J.C. Muñoz and C.F. Daganzo. The bottleneck mechanism of a freeway diverge. *Trans. Res. A* 6(36), 483–505 (2002).
10. H.S. Mika, J.B. Kreer, and L.S. Yuan. Dual mode behavior of freeway traffic. *High. Res. Rec.* 279, 1–13 (1969).

11. K. Agyemang-Duah and F.L. Hall. Some issues regarding the numerical value of freeway capacity. In U.Brannolte, editor, *International Symposium on Highway Capacity*, pages 1–15, Balkema, Rotterdam (1991).
12. F.L. Hall and K. Agyemang-Duah. Freeway capacity drop and the definition of capacity. *Transportation Research Record, TRB* 1320, 91–98 (1991).
13. B. S. Kerner and H. Rehborn. Experimental features and characteristics of traffic jams. *Phys. Rev.* E53, R1297–R1300 (1996).
14. B. Persaud, S. Yagar, and R. Brownlee. Exploration of the breakdown phenomenon in freeway traffic. *Transportation Research Record, TRB* 1634, 64–69 (1998).
15. J. Treiterer and J.A. Myers. The hysteresis phenomenon in traffic flow. In D. J. Buckley, editor, *6th Int. Symp. on Transportation and Traffic Theory*, pages 13–38, A.H. and A.W. Reed, London (1974).
16. J.C. Muñoz and C.F. Daganzo. Moving bottlenecks: a theory grounded on experimental observation. In M.A.P. Taylor (Ed.), *15th Int. Symp. on Transportation and Traffic Theory*, pp. 441–462, Pergamon-Elsevier, Oxford, U.K. (2002).
17. D.C. Gazis, R. Herman, and G. Weiss. Density oscillations between lanes of a multi-lane highway. *Operations Research* 10, 658–667 (1962).
18. G. F. Newell. Theories of instability in dense highway traffic. *J. Opns. Res. Japan* 1(5), 9–54 (1962).
19. K. Smilowitz, C. Daganzo, J. Cassidy, and R. Bertini. Some observations of highway traffic in long queues. *Trans. Res. Rec.* 1678, 225–233 (1999).
20. M.J. Cassidy and M. Mauch. An observed traffic pattern in long freeway queues. *Trans. Res. A* 2(35), 143–156 (2001).
21. J. M. Del Castillo. Propagation of perturbations in dense traffic flow: a model and its implications. *Trans. Res. B* 2(35), 367–390 (2001).
22. J.A. Laval and C.F. Daganzo. Lane-changing in traffic streams. *Trans. Res. B* (2005), in press.
23. M. Cassidy and J. Rudjanakanoknad. Increasing capacity of an isolated merge by metering its on-ramp. *Trans. Res. B* 10(39), 896–913 (2005).
24. M. Cassidy and S. Ahn. Driver turn-taking behavior in congested freeway merges. *Transportation Research Record, TRB* (2005), forthcoming.
25. C. F. Daganzo. The nature of freeway gridlock and how to prevent it. In J. B. Lesort, editor, *13th Int. Symp. on Transportation and Traffic Theory*, pages 629–646, Elsevier, New York (1996).
26. E. W. Weisstein. *CRC Concise Encyclopedia of Mathematics, Second Edition.* CRC Press (2002).

Modeling a Bottleneck by the Aw-Rascle Model with Phase Transitions

Paola Goatin

Laboratoire d'Analyse Non linéaire Appliquée et Modélisation, I.S.I.T.V., Université du Sud Toulon - Var, 83162 La Valette du Var Cedex, France. (goatin@univ-tln.fr)

Summary. We describe a new model of vehicular traffic flow with phase transitions. The model is obtained by coupling the classical Lighthill-Whitham-Richards equation with the 2×2 system proposed by Aw and Rascle. We show an application to the modeling of a bottleneck, and compare the results with those obtained using other models.

1 Introduction

One of the first models introduced to describe traffic flow is the well known Lighthill-Whitham [8] and Richards [9] (LWR) model, which reads

$$\partial_t \rho + \partial_x[\rho v(\rho)] = 0, \tag{1}$$

where $\rho \in [0, R]$ is the mean traffic density, and $v(\rho)$, the mean traffic velocity, is a given non-increasing function, non-negative for ρ between 0 and the positive maximal density R, which corresponds to a traffic jam. This simple model expresses conservation of the number of cars, and relies on the assumption that the car speed depends only on the density. Nevertheless, experimental data (Fig. 1, right) suggest that a good traffic flow model should exhibit two qualitative different behaviors:

- for low densities, the flow is *free* and essentially analogous to the LWR model;
- at high densities the flow is *congested* and has one more degree of freedom (it covers a 2-dimensional domain).

The first continuous model showing phase transitions has been introduced by Colombo [3]. In this paper we briefly describe a new traffic flow model with phase transitions obtained combining the Aw-Rascle 2×2 model [2] (in the following referred to as the AR model) with the LWR equation (the model has been studied in detail in [6]). We show an application to the modeling of a bottleneck, and compare the results with the ones obtained using the LWR model and the biphasic model described in [3].

2 Brief Description of the Model

The model under consideration has been introduced in [4, 6]. The LWR equation and the AR model describe the *free* flow and the *congested* phase, respectively. More precisely, the model reads

Free flow: Congested flow:
$(\rho, v) \in \Omega_f$ $(\rho, v) \in \Omega_c$

$$\partial_t \rho + \partial_x[\rho v] = 0 \qquad \begin{cases} \partial_t \rho + \partial_x[\rho v] = 0 \\ \partial_t \big[\rho(v + p(\rho))\big] + \partial_x\big[\rho v(v + p(\rho))\big] = 0 \end{cases}$$

$$v = v_f(\rho) = (1 - \rho/R)\, V \qquad p(\rho) = V_{\text{ref}} \ln(\rho/R)$$

$$\tag{2}$$

where R is the maximal possible car density, V is the maximal speed allowed and V_{ref} a given reference velocity. The sets Ω_f and Ω_c denote the free and the congested phases respectively. In Ω_f there is only one independent variable, the car density ρ. In Ω_c the variables are the car density ρ and the car speed v or, equivalently, the conservative variables ρ and $y := \rho v + \rho p(\rho)$; see [2]. The "pressure" function p plays the role of an *anticipation factor*, taking in account drivers' reactions to the state of traffic in front of them.

It is reasonable to assume that if the initial data are entirely in the free (resp. congested) phase, then the solution will remain in the free (resp. congested) phase for all time. Thus we are led to take Ω_f (resp. Ω_c) to be an invariant set for (2), left (resp., right). The resulting domain is given by

$$\Omega_f = \{(\rho, v) \in [0, R_f] \times [V_f, V] : v = v_f(\rho)\}\,,$$

$$\Omega_c = \{(\rho, v) \in [0, R] \times [0, V_c] : p(r) \le v + p(\rho) \le p(R)\}\,,$$

where $V_f > V_c$ are the threshold speeds, i.e. above V_f the flow is free and below V_c the flow is congested. The parameter $r \in\,]0, R]$ depends on the environmental conditions and determines the width of the congested region. The maximal free-flow density R_f must satisfy $V_f + p(R_f) = p(R)$ (that is $V_f + V_{\text{ref}} \ln(R_f/R) = 0$ with our choice of the pressure). In order to get this condition, we are led to assume $V_{\text{ref}} < V$. It is easy to check that the *capacity drop* in the passage from the free phase to the congested phase [7] is then automatically satisfied. In order to resume, we have the following order relation between the speed parameters:

$$V > V_{\text{ref}} > V_f > V_c.$$

Fig. 1 shows that the shape of the invariant domain is in good agreement with experimental data. A detailed description of the Riemann solver and further analytical results are given in [6].

3 Initial-Boundary Value Problems

From the point of view of traffic flow, it is natural to consider Initial Boundary Value Problems. We start considering the case of a road starting at $x = 0$

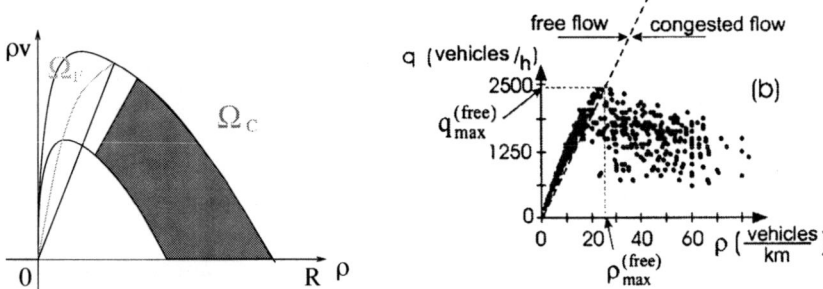

Fig. 1. Left: invariant domain for (2). Right: experimental data, taken from [7].

where the inflow $\tilde{f}$ is regulated. This leads to study the following Riemann problem with boundary

$$\begin{cases} \partial_t \rho + \partial_x [\rho \cdot v_f(\rho)] = 0 & (\rho, y) \in \Omega_f, \; t \geq 0, \; x \geq 0 \\ \begin{cases} \partial_t \rho + \partial_x [\rho \cdot v] = 0 \\ \partial_t y + \partial_x [y \cdot v] = 0 \end{cases} & (\rho, y) \in \Omega_c, \; t \geq 0, \; x \geq 0 \\ (\rho, y)(0, x) = (\bar{\rho}, \bar{y}) & x \geq 0, \\ (\rho v)(t, 0) = \tilde{f} & t \geq 0. \end{cases} \tag{3}$$

We denote the maximum possible flow along the considered road by $F = R_f V_f$.

Proposition 1. *With reference to (3), if*

$$V_{\text{ref}} \geq V \left(1 - \frac{r}{eR}\right), \tag{4}$$

then for all $(\bar{\rho}, \bar{y}) \in \Omega_f \cup \Omega_c$, there exists a threshold $f^{\max} = f^{\max}(\bar{\rho}, \bar{y})$ such that for all $\tilde{f} \in [0, f^{\max}]$ the Riemann problem for (3) admits a solution in the sense of [1, Definition NC]. More precisely, there exists a unique state $(\tilde{\rho}, \tilde{y}) \in \Omega_f \cup \Omega_c$ such that the flow at $(\tilde{\rho}, \tilde{y})$ is $\tilde{f}$ and the standard solution to the Riemann problem (2) with data $(\tilde{\rho}, \tilde{y})$ and $(\bar{\rho}, \bar{y})$ consists only of waves having positive speed.

1. *If $(\bar{\rho}, \bar{y}) \in \Omega_f$, then $f^{\max} = F$ and $(\tilde{\rho}, \tilde{y})$ is in Ω_f. The solution consists of a 2-wave in the free phase.*
2. *If $(\bar{\rho}, \bar{y}) \in \Omega_c$, then there exist a $f^{\min} = f^{\min}(\bar{\rho}, \bar{y})$ such that:*
   a) *If $f^{\min} \leq \tilde{f} \leq f^{\max}$, $(\tilde{\rho}, \tilde{y})$ is the unique intersection between the curve $\rho v(\rho, y) = \tilde{f}$ and the 2-wave through $(\bar{\rho}, \bar{y})$. The solution consists of a simple 2-wave.*
   b) *If $\tilde{f} < f^{\min}$, then $(\tilde{\rho}, \tilde{y})$ is the unique state in Ω_f such that $\tilde{\rho} v_f(\tilde{\rho}) = \tilde{f}$. The solution consists of a phase boundary and a 2-wave.*

Moreover, the Riemann Solver is continuous in L^1_{loc}.

Condition (4) ensures that $\sup_{\Omega_f \cup \Omega_c} \lambda_1 < 0$, hence all waves of the first family are exiting the domain $x \geq 0$, $t \geq 0$ and the problem is *non-characteristic* [1]. In practice, inequality (4) means that, if the maximal speed V is not too high, the anticipation factor, which is proportional to V_{ref}, forces informations to move backward.

The proof is detailed in [5, 6]. Note that, as remarked in [5], the incoming flow $\tilde{f}$ can be slightly greater than the flow $\bar{\rho} v(\bar{\rho})$ present on the road.

Once the Riemann solver is available, well posedness for the Initial-Boundary Value Problem can be proved as in [5], for all initial and boundary data with bounded total variation.

We consider now the somewhat symmetric case of a road whose outflow at $x = 0$ is regulated. At the level of Riemann problem, this can be modeled by

$$\begin{cases} \partial_t \rho + \partial_x \left[\rho \cdot v_f(\rho) \right] = 0 & (\rho, y) \in \Omega_f, \ t \geq 0, \ x \leq 0 \\ \partial_t \rho + \partial_x \left[\rho \cdot v \right] = 0 \\ \partial_t y + \partial_x \left[y \cdot v \right] = 0 & (\rho, y) \in \Omega_c, \ t \geq 0, \ x \leq 0 \\ (\rho, y)(0, x) = (\bar{\rho}, \bar{y}) & x \leq 0, \\ (\rho v)(t, 0) \leq \tilde{f} & t \geq 0. \end{cases} \tag{5}$$

Proposition 2. *With reference to (5), condition (4) implies that for all $(\bar{\rho}, \bar{y}) \in \Omega_f \cup \Omega_c$, and for all possible flows $\tilde{f} \in [0, F]$ the Riemann problem (5) admits a solution in the sense of [1, Definition NC]. More precisely, there exists a unique state $(\tilde{\rho}, \tilde{y}) \in \Omega_f \cup \Omega_c$ such that the flow at $(\tilde{\rho}, \tilde{y})$ is less or equal $\tilde{f}$ and the standard solution to the Riemann problem (2) with data $(\bar{\rho}, \bar{y})$ and $(\tilde{\rho}, \tilde{y})$ consists only of waves having negative speed. If the flow at $(\bar{\rho}, \bar{y})$ is less or equal $\tilde{f}$, then $(\tilde{\rho}, \tilde{y}) = (\bar{\rho}, \bar{y})$, otherwise:*

1. *If $(\bar{\rho}, \bar{y}) \in \Omega_f$, then $(\tilde{\rho}, \tilde{y})$ is in Ω_c. The solution consists of a phase transition eventually followed by a 1-wave in the congested phase.*
2. *If $(\bar{\rho}, \bar{y}) \in \Omega_c$, then the solution consists of a simple 1-wave, $(\tilde{\rho}, \tilde{y})$ being the intersection of the 1-Lax curve through $(\bar{\rho}, \bar{y})$ and the line $\rho v = \tilde{f}$.*

Proof. Let us assume that the flow at $(\bar{\rho}, \bar{y})$ is greater than $\tilde{f}$, and let (ρ_m, v_m) be the intersection point between the 1-Lax curve through $(\bar{\rho}, \bar{y})$ and the line $\rho v = \tilde{f}$, i.e. the solution of the equation $\rho \bar{v} + \rho V_{ref} \ln(\bar{\rho}/\rho) = \tilde{f}$, with $\rho_m > \bar{\rho}$. If $(\rho_m, y_m = y(\rho_m, v_m)) \in \Omega_c$, then $(\tilde{\rho}, \tilde{y}) = (\rho_m, y_m)$. Otherwise, $(\tilde{\rho}, \tilde{y})$ is the intersection point between the 1-Lax curve through $(\bar{\rho}, \bar{y})$ and the line $v = V_c$, i.e. the solution of the equation $\bar{v} + V_{ref} \ln(\bar{\rho}/\rho) = V_c$.

4 An Example: Modeling a Bottleneck

In this section, we compare the model introduced here with the LWR model (1) and the biphasic model introduced by Colombo in [3], which reads:

Free flow: Congested flow:
$(\rho, q) \in \Omega_f$ $(\rho, q) \in \Omega_c$

$$\partial_t \rho + \partial_x [\rho v] = 0 \qquad \begin{cases} \partial_t \rho + \partial_x [\rho v] = 0 \\ \partial_t q + \partial_x [(q - Q)v] = 0 \end{cases} \qquad (6)$$

$$v = v_f(\rho) \qquad\qquad v = v_c(\rho, q) = \left(1 - \frac{\rho}{R}\right)\frac{q}{\rho}.$$

Here, Q is a parameter of the road under consideration and the weighted flow q is a variable originally motivated by the linear momentum in gas dynamics. The two phases are defined by

$$\Omega_f = \{(\rho, q) \in [0, R_f] \times [0, +\infty[\,:\, v_f(\rho) \geq V_f, \, q = \rho \cdot V\},$$
$$\Omega_c = \left\{(\rho, q) \in [0, R] \times [0, +\infty[\,:\, v_c(\rho, q) \leq V_c, \, \frac{q - Q}{\rho} \in \left[\frac{Q_- - Q}{R}, \frac{Q_+ - Q}{R}\right]\right\},$$

where the parameters $Q_- \in]0, Q[$ and $Q_+ \in]Q, +\infty[$ depend on the environmental conditions and determines the width of the congested region.

We consider traffic on a highway described by the interval $[-2, 2]$, in which the number of lanes is reduced from three to two at $x = 0$. This is simulated by setting the maximal density $R = 1$ for $x < 0$, and $R = 2/3$ for $x > 0$. Modeling this problem requires the solution of two Riemann problems with boundary, namely (3) for $x > 0$ and (5) for $x < 0$. In the example shown here, we have chosen initial data on the left-hand side so that the incoming flux is higher than the maximal possible flux in the two-lane region. This causes the traffic congestion shown by Fig. 2, 3.

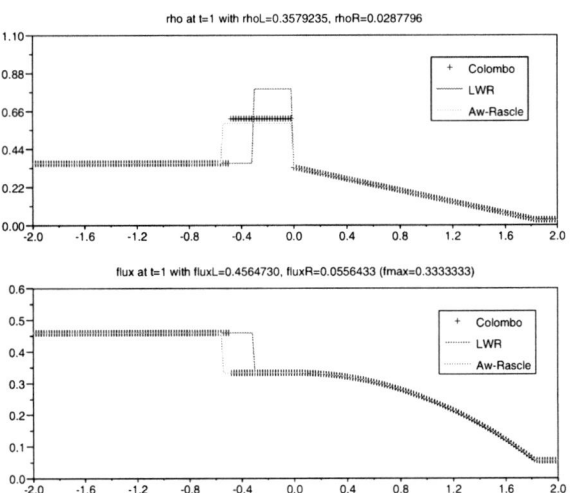

Fig. 2. Bottleneck at $x = 0$ with incoming flux f_l higher than the maximal possible flux at $x > 0$.

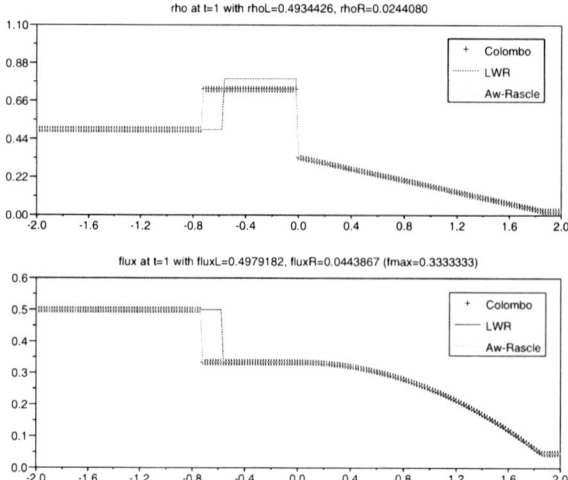

Fig. 3. Bottleneck at $x = 0$ with incoming flux $f_l = F$, the maximal possible incoming flux at $x < 0$.

Initial data u_l for $x < 0$ and u_r for $x > 0$ are taken in the free phase Ω_f. The solutions given by the three models present the same behavior:

- a shock (hiding a phase transition for models (2) and (6)) moving backward in the three lane region, upstream the congested traffic;
- a discontinuity (under-compressive shock) at $x = 0$, corresponding to the bottleneck;
- a rarefaction wave moving forward in the free phase.

In particular, the LWR-Aw-Rascle coupling (2) and the LWR-Colombo coupling (6) are in good agreement, especially if the flux of the incoming traffic is equal to F (Figure 3).

The numerical integrations for Figs. 2, 3 rely on the choices: $R = 1$, $r = 0.47$, $V = 2$, $V_f = 1$, $Q = 0.5$, $Q_- = 0.25$, $V_c = 0.85$, $R_f = 0.5$ for $x < 0$ and $r = 0.41$, $V_f = 1$, $Q = 1/3$, $Q_- = 1/6$, $V_c = 0.85$, $R_f = 1/3$ for $x > 0$; $V_{\text{ref}} = V \frac{1 - R_f/R}{\ln(R/R_f)}$.

References

1. D. Amadori, R.M. Colombo: 'Continuous dependence for 2×2 conservation laws with boundary', J. Differential Equations **138**, 226–266 (1997).
2. A. Aw, M. Rascle: 'Resurrection of "second order" models of traffic flow', SIAM J. Appl. Math. **60**, 916–938 (2000).
3. R.M. Colombo: 'Hyperbolic Phase Transitions in Traffic Flow', SIAM J. Appl. Math. **63**, 708–721 (2002).
4. R.M. Colombo, P. Goatin: 'Traffic Flow Models with Phase Transitions', Flow Turbulence Combust., to appear.
5. R.M. Colombo, P. Goatin, F.S. Priuli: 'Global Well Posedness of a Traffic Flow Model with Phase Transitions', Nonlinear Anal. A, to appear.
6. P. Goatin: 'The Aw-Rascle Traffic Flow Model with Phase Transitions', Math. Comput. Modeling **44**, 287-303 (2006).
7. B.S. Kerner: 'Phase Transitions in Traffic Flow'. In *Traffic and Granular Flow '99*, ed. by D. Helbing, H. Hermann, M. Schreckenberg, D. Wolf (Springer, 2000) pp. 253–283.
8. M.J. Lighthill, G.B. Whitham: 'On Kinematic Waves. II. A Theory of Traffic Flow on Long Crowded Roads', Proc. Roy. Soc. London. Ser. A **229**, 317–345 (1955).
9. P.I. Richards: 'Shock waves on the highway', Operations Res. **4**, 42–51 (1956).

Solvability and Metastability of the Stochastic Optimal Velocity Model

Masahiro Kanai[1], Katsuhiro Nishinari[2], and Tetsuji Tokihiro[1]

[1] University of Tokyo, 3-8-1 Komaba Meguro-ku, Tokyo 153-8914, Japan
[2] University of Tokyo, 7-3-1 Hongo Bunkyo-ku, Tokyo 113-8656, Japan

Summary. In a recent paper (Phys. Rev. E **72**, 035102 (2005)) we have proposed a stochastic optimal velocity model which includes two *exactly solvable* stochastic models. It can be regarded as a stochastic version of the optimal velocity model. We find that the model shows striking metastability (i.e. long-lived metastable states, dynamical phase transition, and sharp spontaneous metastability breaking) as well as solvability. In this work, we present additional explanations of the solvability and metastability, which are helpful in understanding the traffic dynamics of the model.

1 Introduction

For the last several decades traffic dynamics has attracted much attention from physicists and mathematicians, since it is a typical example of non-equilibrium statistical mechanics of *self-driven many-particle systems* [1–3].

There are a lot of models for traffic flow proposed thus far, from various viewpoints, such as macroscopic and microscopic, differential equations and cellular automata, deterministic and probabilistic. Among these traffic models, the *optimal velocity model* (OV) achieves a remarkable success although it is a simple deterministic model [4, 5]. It takes the form

$$\frac{d^2 x_i}{dt^2} = a\left[V(x_{i+1} - x_i) - \frac{dx_i}{dt}\right],\tag{1}$$

where $x_i = x_i(t)$ is the position of i-th vehicle at time t. A function V is called the optimal velocity function, which gives an optimal speed according to the headway $x_{i+1} - x_i$. Note that the parameter a corresponds to the sensitivity of drivers which plays an important role in the stability of traffic flow [4]. Moreover, it is reported that the OV model shows a kind of solvability, i.e., it has an analytic solution of jam flow [6, 7].

In our recent paper [8], we have proposed a *stochastic optimal velocity (SOV) model* which includes two *exactly solvable* stochastic processes. It can be regarded as a stochastic extended version of the OV model. In this work, we present additional explanations of the solvability and metastability of the SOV model.

2 The Stochastic Optimal Velocity Model

2.1 General Scheme

First of all, we explain the general framework of our stochastic CA model for one-lane traffic. The roadway, being divided into cells, is regarded as a one-dimensional array of L sites, and each site contains one vehicle at most. Let M_i^t be a stochastic variable which denotes the number of sites at which the i-th vehicle moves at time t, and $w_i^t(m)$ be the probability of $M_i^t = m$ $(m = 0, 1, 2, \ldots)$. Then, we assume as a principle of motion that the probability $w_i^{t+1}(m)$ depends on the probability distribution $w_i^t(0), w_i^t(1), \ldots$, and the positions of vehicles $x_1^t, x_2^t, \ldots, x_N^t$ at the adjacent time. The updating procedure is as follows:

- Calculate the next intention w_i^{t+1} $(i = 1, 2, \ldots, N)$ from the present intentions $w_i^t(0), w_i^t(1), \ldots$ and positions $x_1^t, x_2^t, \ldots, x_N^t$:

$$w_i^{t+1}(m) = f(w_i^t(0), w_i^t(1), \ldots; x_1^t, \ldots, x_N^t; m) \qquad (2)$$

- Determine the number of sites M_i^{t+1} that a vehicle moves (i.e. the velocity) probabilistically according to the intention w_i^{t+1}.
- Each vehicle moves as

$$x_i^{t+1} = x_i^t + \min(\Delta x_i^t, M_i^{t+1}) \quad (\forall i), \qquad (3)$$

where $\Delta x_i^t = x_{i+1}^t - x_i^t - 1$ denotes the headway.

The hard-core exclusion rule is incorporated through the second term of the right hand side of (3).

The probability distribution w_i^t imports the driver's intention and the uncertainty of operation into a traffic model, and there is no physical counterpart of it. We hence call it the *intention* in the sense that the moving vehicles are not driven by a kind of external force field but by themselves.

2.2 The SOV Model

In what follows, we assume $w_i^t(m) \equiv 0$ for $m \geq 2$. Note that $\sum_{m=0}^{\infty} w_i^t(m) = 1$ by definition. Setting $v_i^t = w_i^t(1)$ we have $w_i^t(0) = 1 - v_i^t$ and for the expectation value $\langle M_i^t \rangle = \sum_{m=0}^{\infty} m w_i^t(m)$ we then obtain $\langle M_i^t \rangle = v_i^t$. From (2) we have

$$w_i^{t+1}(1) = v_i^{t+1} = f(v_i^t; x_1^t, x_2^t, \ldots; 1) \qquad (4a)$$
$$w_i^{t+1}(0) = 1 - v_i^{t+1}, \qquad (4b)$$

and we therefore express the intention by v_i^t instead of w_i^t. As long as vehicles move separately (i.e. $\Delta x > 0$), the positions are updated according to the simple form

$$x_i^{t+1} = \begin{cases} x_i^t + 1 & \text{with probability } v_i^t \\ x_i^t & \text{with probability } 1 - v_i^t, \end{cases} \tag{5}$$

and consequently we have

$$\langle x_i^{t+1} \rangle = \langle x_i^t \rangle + v_i^t \tag{6}$$

in the sense of the expectation value. This equation expresses that the intention v_i^{t+1} can be regarded as the average velocity at time t.

Let us take the evolution equation

$$v_i^{t+1} = (1 - a)v_i^t + aV(\Delta x_i^t), \tag{7}$$

in (2), where a $(0 \leq a \leq 1)$ is a parameter and the function V takes values in $[0, 1]$ so that v_i^t should be within $[0, 1]$. Equation (7) consists of two terms, i.e., a term turning over the intention v_i^t into the next, and an effect of the situation (the headway Δx_i^t). The intrinsic parameter a indicates the sensitivity of vehicles to the traffic situation, and the larger a is, the less time a vehicle takes to change the intention.

A discrete version of the OV model is expressed as

$$x_i(t + \Delta t) - x_i(t) = v_i(t)\Delta t, \tag{8a}$$

$$v_i(t + \Delta t) - v_i(t) = a\Big[V(\Delta x_i(t)) - v_i(t)\Big]\Delta t, \tag{8b}$$

where $\Delta x_i(t) = x_{i+1}(t) - x_i(t)$, and Δt is a time interval. Due to the formal correspondence between (7) and (8b), we call the stochastic CA model defined by (7) the *stochastic optimal velocity model* hereafter.

3 Solvability of the SOV Model

As we described in the preceding paper [8], the SOV model recovers two models of stochastic processes, the asymmetric simple exclusion process (ASEP) and the zero range process (ZRP), when the parameter a takes the values 0 and 1 (both ends of the domain). The ASEP is well known to be exactly solvable, and we can give an explicit formula for the flux:

$$Q(\rho) = \frac{1}{2}\left(1 - \sqrt{1 - 4p\rho(1 - \rho)}\right), \tag{9}$$

where ρ denotes the density of vehicles and p is the probability of a vehicle hopping. The ZRP includes the ASEP as special case and is also exactly solvable in the sense that we can make an exact calculation of the flux. The flux $Q(\rho)$ is known to be calculated from the OV function $V(x)$ through an iteration process [9].

$$Q(\rho) = \rho \sum_{x=0}^{L-N} V(x)p(x), \tag{10}$$

where $p(x)$ gives the probability of a certain headway taking the value of x and is calculated as follows:

$$h(x) := \begin{cases} 1 - V(1) & (x = 0) \\ \frac{1-V(1)}{1-V(x)} \prod_{y=1}^{x} \frac{1-V(y)}{V(y)} & (x > 0) \end{cases}, \tag{11}$$

and from this $h(x)$ we have

$$p(x) := h(x)\frac{Z(L-N-1, N-1)}{Z(L, N)}, \tag{12}$$

where $Z(L, N)$ is iteratively calculated through the recursion formula

$$Z(L, N) = \sum_{x=0}^{L-N} Z(L - x - 1, N - 1)h(x), \tag{13a}$$

$$Z(x, 1) = h(x - 1), \qquad Z(x, x) = h(0), \tag{13b}$$

where L is the number of sites, and N the number of vehicles.

Furthermore, we find that as a approaches 1 the SOV model reduces to ZRP. Figure 1 shows the fundamental diagrams with two values of the sensitivity $a = 0.3$ and 0.8, where we take an ordinary OV function

$$V(\Delta x) = \frac{\tanh(\Delta x - 3/2) + \tanh 3/2}{1 + \tanh 3/2}. \tag{14}$$

In this case, the exact fundamental diagram $(a = 1)$ has a good agreement with the sensitivity $a > 0.6$.

4 Metastability of the SOV Model

In contrast with the case of $a \to 1$, the SOV model does never approach ASEP although $a \to 0$, and shows some novel features instead. Since we already had a detailed discussion about this point in [8], we briefly review that case in this section.

First of all, we point out that the SOV model shows an ASEP-like property and deviates from ASEP as time advances (i.e. the system approaches a stationary state). Fig. 2 shows fundamental diagrams of the SOV model with the OV function (14) and the sensitivity $a = 0.01$. They are plotted at each time stage $t = 10$ and 1000, starting from uniform states (and random states) with

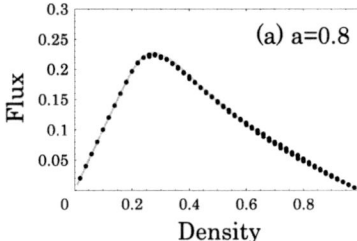 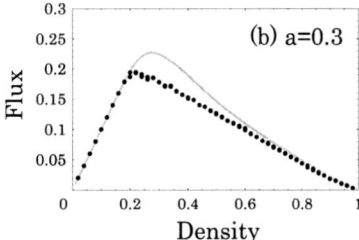

Fig. 1. The fundamental diagram of the SOV model with sensitivity (a) $a = 0.8$ and (b) $a = 0.3$. In both figures, we also draw the exact fundamental diagram (gray line) of ZRP, i.e. the SOV model with $a = 1$, for the sake of comparison.

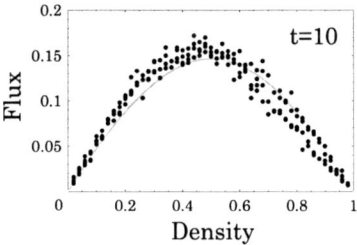 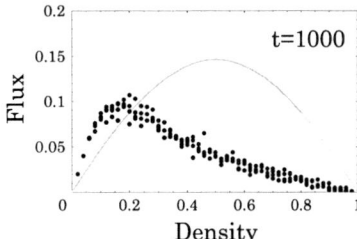

Fig. 2. Fundamental diagrams of the SOV model with the OV function (14) and the sensitivity $a = 0.01$. They are plotted at each time stage $t = 10$ and 1000, starting from uniform states (and random states) with $p (\equiv v_i^0) = 0.5$. The exact curve (gray) of ASEP with probability $p = 0.5$ are included for comparison. The system size is $L = 1000$, and the number of samples is 4 at each density.

$p (\equiv v_i^0) = 0.5$. Then, the exact diagram of ASEP with the probability p is figured out by use of (9). In Fig. 2, we find that a curve similar to the diagram of ASEP appears only for the first few steps ($t = 10$) and then changes the shape rapidly ($t = 1000$). Surprisingly, when the diagram becomes stationary, it allows a discontinuous point and two overlapping stable states around the density $\rho \sim 0.14$ (Fig. 3(a) $t = 10000$).

Figure 3(a) shows the fundamental diagram following those in Fig. reffd2, and (b) shows a closeup of (a) around the discontinuous point. We find that, in the region of density at which the flux changes discontinuously, there appears more than one stable state at the same density. The stable states are categorized by their properties, i.e., the highest-flux states are free flow where vehicles move freely at the maximum velocity, the middle-flux states are congested where vehicles create many small clusters, and the lowest-flux states are jammed where a big jam moves backwards.

Moreover, it is remarkable that before reaching the low-flux stable state the traffic flow stays at higher-flux states for some time. In the present case, we

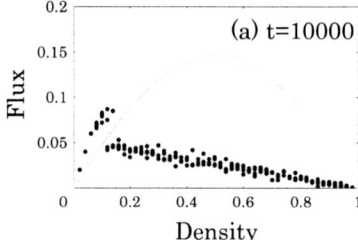 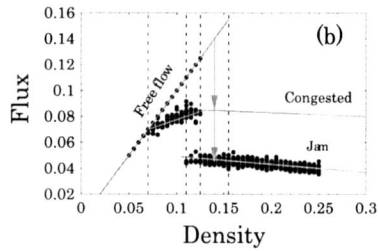

Fig. 3. (a) The fundamental diagram following those in Fig. 2. (b) A closeup of Fig. (a) around the discontinuous point $\rho \simeq 0.14$. They are plotted at each time stage $t = 10000$ with the same condition of Fig. 2.

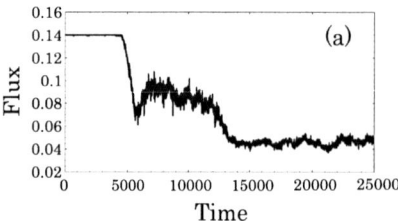 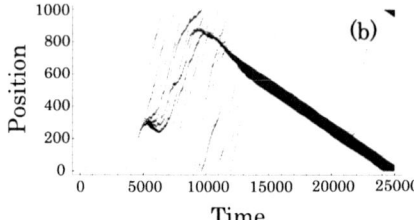

Fig. 4. (a) The time evolution of flux at the density $\rho = 0.14$ starting from the uniform state. We observe two plateaus at the flux $Q = 0.14$ with a lifetime $T \simeq 5000$, and $Q \simeq 0.08$ with $T \simeq 7000$ before reaching the stationary jam state. (b) The corresponding spatio-temporal diagram, where vehicles (black dots) move from bottom up.

observe a *dynamical phase transition* at the density $\rho \simeq 0.14$ as indicated by successive arrows in Fig. 2. Figure 4 shows the flux changing stepwise over time and the spatio-temporal pattern. In these figures, we observe apparent changes of traffic condition from a free flow to a jam through a congested state.

5 Conclusion and Remark

In this work, we exhibit the solvability and metastability of the SOV model. According to the sensitivity parameter, the SOV model changes the macro-scopic property from solvable to metastable. It is noteworthy that although the SOV model includes the ASEP at the sensitivity parameter $a = 0$, it does not show an ASPE-like behavior for any small a. Moreover, the metastability contains a quite rich structure, i.e., some long-lived metastable states appear and break their metastability spontaneously. We stress that in the present model it is not an external field, but an internal (microscopic) mechanism that drastically changes the macroscopic property of states, although that mechanism is not defined explicitly.

Results for the SOV model under open boundary conditions, with step-like OV function [10], and a multi-velocity version of the SOV model will be reported in the nearest future.

References

1. D. Chowdhury, L. Santen and A. Schadschneider, Phys. Rep. **329**, 199 (2000)
2. D. Helbing, Rev. Mod. Phys. **73**, 1067 (2001)
3. T. Nagatani, Rep. Prog. Phys. **65**, 1331 (2002)
4. M. Bando, K. Hasebe, A. Nakayama, A. Shibata and Y. Sugiyama, Phys. Rev. E **51**, 1035 (1995)
5. M. Bando, K. Hasebe, K. Nakanishi, A. Nakayama, A. Shibata and Y. Sugiyama, J. Phys. I France **5**, 1389 (1995)
6. Y. Sugiyama and H. Yamada, Phys. Rev. E **55**, 7749 (1997)
7. K. Nakanishi, K. Itoh and Y. Igarashi, Phys. Rev. E **55**, 6519 (1997)
8. M. Kanai, K. Nishinari and T. Tokihiro, Phys. Rev. E **72**, 035102(R) (2005)
9. O. J. O'Loan, M. R. Evans, and M. E. Cates, Phys. Rev. E **58**, 1404 (1998)
10. M. Kanai, K. Nishinari and T. Tokihiro, J. Phys. A **39**, 2921 (2006).

Modeling of Flows with Power-Law Spectral Densities and Power-Law Distributions of Flow Intensities

Bronislovas Kaulakys, Miglius Alaburda, Vygintas Gontis, Tadas Meskauskas, and Julius Ruseckas

Institute of Theoretical Physics and Astronomy of Vilnius University, A. Gostauto 12, LT-01108 Vilnius, Lithuania

Summary. We present analytical and numerical results of modeling of flows represented as correlated non-Poissonian point process and as Poissonian sequence of pulses of different size. Both models may generate signals with power-law distributions of the intensity of the flow and power-law spectral density. Furthermore, different distributions of the interevent time of the point process and different statistics of the size of pulses may result in $1/f^\beta$ noise with $0.5 \lesssim \beta \lesssim 2$. A combination of the models is applied for modeling Internet traffic.

1 Introduction

Modeling and simulations enable one to understand and explain the observable phenomena and predict new ones. This is true, as well, for mathematical studies and modeling of traffic flow with the aim to get a better understanding of phenomena and avoid some problems of traffic congestion. Traffic phenomena are complex and nonlinear, they show cluster formation, huge fluctuations and long-range dependencies. Almost twenty years ago it was detected from empirical data that fluctuations of a traffic current on a expressway obey a $1/f$ law for low spectral frequencies [1]. Similarly, $1/f$ noise is observable in the flows of granular materials [2, 3].

$1/f$ noise, or $1/f$ fluctuations are usually related with power-law distributions of other statistics of the fluctuating signals, first of all with the power-law decay of autocorrelations and the long-memory processes (see, e.g., the comprehensive bibliography of $1/f$ noise on the website [4], review articles [5, 6] and references in the recent paper [7]). The appearance of clustering and large fluctuations in traffic and granular flows may be a result of synchronization of the ensemble of the nonlinear system subjected to common random external perturbations, which may result in nonchaotic behavior of Brownian-type motions, intermittency and $1/f$ noise [8, 9].

Traffic and granular flows usually may be considered as consisting of discrete identical objects such as vehicles, pedestrians, granules, packets and so on. They may be represented as consisting of pulses or elementary events and further simplified to a point process model [7, 10–12]. Moreover, from the modeling of traffic it was found that $1/f$ noise may be the result of clustering and jumping [10] similar to the point process model of $1/f$ noise [7, 11, 12]. On the other hand, $1/f$ noise may be conditioned by the flow consisting of uncorrelated pulses of variable size with a power-law distribution of pulse durations [13]. In Internet traffic the flow of the signals primarily is composed of power-law distributed file sizes. The files are divided by the network protocol into equal packets [14]. Therefore, the total incoming web traffic is a sequence of packets arising from a large number of requests. Such a flow exhibits $1/f$ fluctuations as well [14, 15].

Long-range correlations and power-law fluctuations of expressway traffic flow have recently been observed on a wide range of time-scales from minutes to months and investigated using the method of detrended fluctuation analysis [16]. There are no explanations why traffic flow exhibits $1/f$ noise behavior in such a large interval of time.

It is the purpose of this paper to present analytical and numerical results for the modeling of flows represented as sequences of different pulses and as a correlated non-Poissonian point process resulting in $1/f$ noise and to apply these results to the modeling of Internet traffic.

2 Signal as a Sequence of Pulses

We will investigate a signal of flow consisting of a sequence of pulses,

$$I(t) = \sum_k A_k(t - t_k).$$
(1)

Here the function $A_k(t - t_k)$ represents the shape of the pulse k having influence on the signal $I(t)$ in the region of time t_k.

2.1 Power Spectral Density

The power spectral density of the signal (1) can be written as

$$S(f) = \lim_{T \to \infty} \left\langle \frac{2}{T} \sum_{k,k'} e^{i\omega(t_k - t_{k'})} \int_{t_i - t_k}^{t_f - t_k} \int_{t_i - t_{k'}}^{t_f - t_{k'}} A_k(u) A_{k'}(u') e^{i\omega(u - u')} du\, du' \right\rangle,$$
(2)

where $\omega = 2\pi f$, $T = t_f - t_i \gg \omega^{-1}$ is the observation time and the brackets $\langle \ldots \rangle$ denote the averaging over realizations of the process. We assume that the pulse shape functions $A_k(u)$ decrease sufficiently fast when $|u| \to \infty$. Since $T \to \infty$, the bounds of the integration in Eq. (2) can be changed to $\pm\infty$.

When the time moments t_k are not correlated with the shape of the pulse A_k, the power spectrum is [2]

$$S(f) = \lim_{T \to \infty} \frac{2}{T} \sum_{k,k'} \left\langle e^{i\omega(t_k - t_{k'})} \right\rangle \left\langle \int_{-\infty}^{+\infty} \int_{-\infty}^{+\infty} A_k(u) A_{k'}(u') e^{i\omega(u-u')} du\, du' \right\rangle.$$

(3)

After introduction of the functions [13]

$$\Psi_{k,k'}(\omega) = \left\langle \int_{-\infty}^{+\infty} A_k(u) e^{i\omega u} du \int_{-\infty}^{+\infty} A_{k'}(u') e^{-i\omega u'} du' \right\rangle$$

(4)

and

$$\chi_{k,k'}(\omega) = \left\langle e^{i\omega(t_k - t_{k'})} \right\rangle$$

(5)

the spectrum can be written as

$$S(f) = \lim_{T \to \infty} \frac{2}{T} \sum_{k,k'} \chi_{k,k'}(\omega) \Psi_{k,k'}(\omega).$$

(6)

2.2 Stationary Process

Equation (6) can be further simplified for the stationary process. Then all averages can depend only on $k - k'$, i.e.,

$$\Psi_{k,k'}(\omega) \equiv \Psi_{k-k'}(\omega)$$

(7)

and

$$\chi_{k,k'}(\omega) \equiv \chi_{k-k'}(\omega).$$

(8)

Equation (6) then reads

$$S(f) = \lim_{T \to \infty} \frac{2}{T} \sum_{k,k'} \chi_{k-k'}(\omega) \Psi_{k-k'}(\omega).$$

(9)

Introducing a new variable $q \equiv k - k'$ and changing the order of summation yields

$$S(f) = \lim_{T \to \infty} \frac{2}{T} \sum_{q=1}^{k_{\max}-k_{\min}} \sum_{k=k_{\min}}^{k_{\max}-q} \chi_q(\omega) \Psi_q(\omega)$$

$$+ \lim_{T \to \infty} \frac{2}{T} \sum_{q=k_{\min}-k_{\max}}^{-1} \sum_{k=k_{\min}-q}^{k_{\max}} \chi_q(\omega) \Psi_q(\omega) + \lim_{T \to \infty} \frac{2}{T} \sum_{k=k_{\min}}^{k_{\max}} \Psi_0(\omega). \quad (10)$$

Here $k_{\min}$ and $k_{\max}$ are minimal and maximal values of the index k in the interval of observation T. Eq. (10) may be simplified to the structure

$$S(f) = 2\bar{\nu}\Psi_0(\omega) + \lim_{T \to \infty} 4 \sum_{q=1}^{N} \left(\bar{\nu} - \frac{q}{T}\right) \operatorname{Re} \chi_q(\omega)\Psi_q(\omega) \tag{11}$$

where $\bar{\nu}$ is the mean number of pulses per unit time and $N = k_{\max} - k_{\min}$ is the number of pulses in the time interval T.

If the sum $\frac{1}{T}\sum_{q=1}^{N} q \operatorname{Re} \chi_q(\omega)\Psi_q(\omega) \to 0$ when $T \to \infty$, then the second term in the sum vanishes and the spectrum is

$$S(f) = 2\bar{\nu}\Psi_0(\omega) + 4\bar{\nu} \sum_{q=1}^{\infty} \operatorname{Re} \chi_q(\omega)\Psi_q(\omega) = 2\bar{\nu} \sum_{q=-\infty}^{\infty} \chi_q(\omega)\Psi_q(\omega). \tag{12}$$

2.3 Fixed Shape Pulses

When the shape of the pulses is fixed (k-independent) then the function $\Psi_{k,k'}(\omega)$ does not depend on k and k' and, therefore, $\Psi_{k,k'}(\omega) = \Psi_{0,0}(\omega)$. Then equation (6) yields the power spectrum

$$S(f) = \Psi_{0,0}(\omega) \lim_{T \to \infty} \frac{2}{T} \sum_{k,k'} \chi_{k,k'}(\omega) \equiv \Psi_{0,0}(\omega)S_\delta(\omega). \tag{13}$$

Eq. (13) represents the spectrum of the process as a composition of the spectrum of one pulse,

$$\Psi_{0,0} = \left| \int_{-\infty}^{+\infty} A_k(t)e^{i\omega t} dt \right|^2, \tag{14}$$

and the power density spectrum $S_\delta(\omega)$ of the point process

$$I_\delta(t) = a \sum_{k} \delta(t - t_k) \tag{15}$$

with the area of the pulse $a = 1$.

3 Stochastic Point Processes

The shapes of the pulses mainly influence the high frequency power spectral density, i.e., at $\omega \geq 1/\Delta t_p$, with Δt_p being the characteristic pulse length. Therefore the power spectral density at low frequencies for not very long pulses is mainly conditioned by the correlations between the transit times t_k, i.e., the signal may be approximated by the point process.

The point process model of $1/f^\beta$ noise has been proposed [11, 12], generalized [7], analysed and used for financial systems [17]. It has been shown that when the average interpulse, interevent, interarrival, recurrence or waiting times $\tau_k = t_{k+1} - t_k$ of the signal diffuse in some interval, the power spectrum of such process may exhibit the power-law dependence, $S_\delta(f) \sim 1/f^\beta$, with $0.5 \lesssim \beta \lesssim 2$. The distribution density of the signal (15) intensity defined as $I = 1/\tau_k$ may be of the power-law, $P(I) \sim I^{-\lambda}$, with $2 \leqslant \lambda \leqslant 4$, as well. The exponents β and λ are depending on the manner of diffusion-like motion of the interevent time τ_k and, e.g., for the multiplicative process are interrelated [7, 17]. For the pure multiplicative process [7]

$$\beta = 1 + \alpha, \quad \lambda = 3 + \alpha, \tag{16}$$

where α is the exponent of the power-law distribution, $P_k(\tau_k) \sim \tau_k^\alpha$, of the interevent time. In general, for relatively slow fluctuations of τ_k, the distribution density of the flow I,

$$P(I) \sim P_k(I^{-1})I^{-3}, \tag{17}$$

is mostly conditioned by the multiplier I^{-3}. Since the point process model has recently [7, 17] been analysed rather properly we will not repeat the analysis here and present only some new illustrations.

Figure 1 demonstrates that for essentially different distributions of τ_k, the power spectra and distribution densities of the point processes are similar.

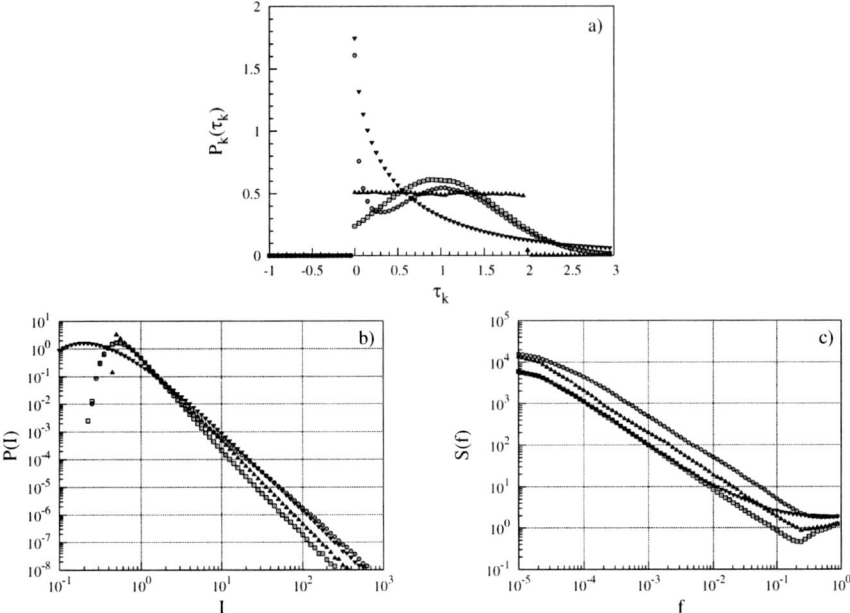

Fig. 1. Distribution densities of the interevent time τ_k, (a), of the flow $I(t)$, (b), and of the power spectra $S(f)$, (c), for different point processes with slow diffusion-like motion of the average interevent time. Different symbols correspond to different types of generation of the interevent sequences.

Further we proceed to the flow consisting of the pulses of different durations and application of this approach for modeling of the Internet traffic.

4 Flow Consisting of Pulses of Variable Duration

When the occurrence times t_k of the pulses are uncorrelated and distributed according to a Poisson process, the power spectrum of the random pulse train is given by Carlson's theorem

$$S(f) = 2\bar{\nu} \left\langle |F_k(\omega)|^2 \right\rangle, \tag{18}$$

where

$$F_k(\omega) = \int_{-\infty}^{+\infty} A_k(t) e^{i\omega t} dt \tag{19}$$

is the Fourier transform of the pulse A_k. Suppose that the random parameters of the pulses are the duration and the area (integral) of the pulse. We can take the form of the pulses as

$$A_k(t - t_k) = T_k^\rho A \left(\frac{t - t_k}{T_k} \right), \tag{20}$$

where T_k is the characteristic duration of the pulse. The value of the exponent $\rho = 0$ corresponds to the fixed height but different durations, the telegraph-like pulses, whereas $\rho = -1$ corresponds to constant area pulses but of different heights and durations, and so on.

For the power-law distribution of the pulse durations,

$$P(T_k) = \begin{cases} \frac{\delta+1}{T_{\max}^{\delta+1} - T_{\min}^{\delta+1}} T_k^\delta, & T_{\min} \leq T_k \leq T_{\max}, \\ 0, & \text{otherwise}, \end{cases} \tag{21}$$

from Eqs. (18) and (19) we have the spectrum

$$S(f) = \frac{2\bar{\nu}(\delta+1)}{(T_{\max}^{\delta+1} - T_{\min}^{\delta+1})\omega^{\delta+2\rho+3}} \int_{\omega T_{\min}}^{\omega T_{\max}} |F(u)|^2 u^{\delta+2\rho+2} du. \tag{22}$$

For $\tau_{\max}^{-1} \ll \omega \ll \tau_{\min}^{-1}$ when $\delta > -1$ the expression (22) may be approximated as

$$S(f) \approx \frac{2\bar{\nu}(\delta+1)}{(T_{\max}^{\delta+1} - T_{\min}^{\delta+1})\omega^{\delta+2\rho+3}} \int_0^\infty |F(u)|^2 u^{\delta+2\rho+2} du. \tag{23}$$

Therefore, the random pulses with the appropriate distribution of the pulse duration (and area) may generate signals with the power-law distribution of the spectrum with different slopes. So, the pure $1/f$ noise generates, e.g., the fixed area ($\rho = -1$) with the uniform distribution of the durations ($\delta = 0$) sequences of pulses, the fixed height ($\rho = 0$) with the uniform distribution of the inverse durations $\gamma = T_k^{-1}$ and all other sequences of random pulses satisfying the condition $\delta + 2\rho = -2$.

In such a case we have from Eq. (23)

$$S(f) \sim \frac{(\delta + 1)\bar{\nu}}{(T_{max}^{\delta+1} - T_{min}^{\delta+1})f}. \tag{24}$$

5 Internet Traffic

In this Section we will apply the results of Section 4 for modeling Internet traffic. The incoming traffic consists of a sequence of packets, which are the result of the division of the requested files by the network protocol (TCP). The maximum size of a packet is 1500 bytes. Therefore, the information signal is as in the point process (15) with pulse area $a = 1500$ bytes. Further, we will analyse the flow of the packets and will measure the intensity of the flow in packets per second. In such a system of units in Eq. (15) we should put $a = 1$.

We exploit the empirical observation [14, 18] that the distribution of the file sizes x may be described by the positive Cauchy distribution

$$P(x) = \frac{2}{\pi} \frac{s}{s + x^2} \tag{25}$$

with the empirical parameter $s = 4100$ bytes. This distribution asymptotically exhibits the Pareto distribution and follows Zipf's law $P(X > x) \sim 1/x$. The files are divided into packets of a maximum size of 1500 bytes or less by the network protocol. In Internet traffic the packets spread into the Poissonian sequence with average inter-packet time τ_p (see Fig. 2). The total incoming flow of the packets to the server consists of packets arising from the Poissonian request of the files with average interarrival time of files τ_f.

The files are requested from different servers located at different distance. This results in the distribution of the average inter-packet time τ_p in some interval. For reproduction of the empirical distribution of the interpacket time τ_k we assume the uniform distribution of $\lg \tau_k$ in some interval $[\tau_{k,min}, \tau_{k,max}]$, similarly to the McWhorter model of $1/f$ noise [7]. As a result, the presented model reproduces sufficiently well the observable non-Poissonian distribution of the arrival interpacket times and the power spectral density, as well (see Fig. 3).

6 Conclusion

In the paper it was shown that processes exhibiting $1/f$ noise and power-law distribution of the intensity may be generated starting from the signals as

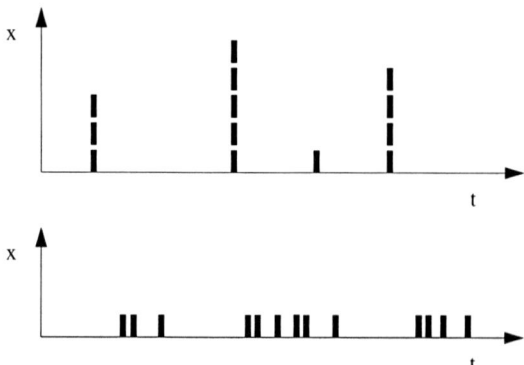

Fig. 2. Division of the requested files into equal size packets with some inter-packet time.

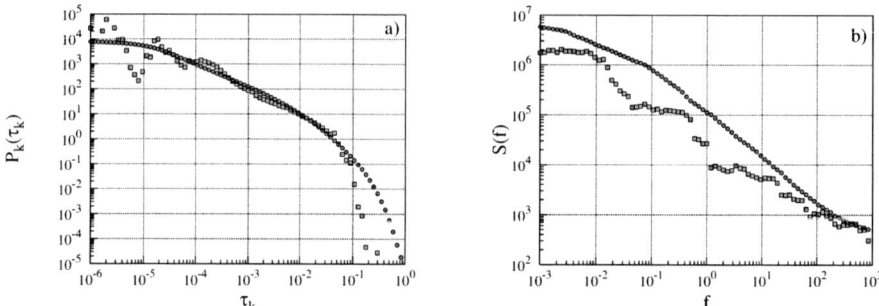

Fig. 3. Distribution densities of (a) the interpacket time τ_k, and (b) the power spectra, for the simulated point process (open circles ○) and empirical data (open squares □). The used parameters are as in the empirical data [14, 18], $\tau_f = 0.101s$, $\tau_{k,min} = 11.6\mu s$ and $\tau_{k,max} = 1000\, \tau_{k,min}$.

sequences of constant area pulses with correlated appearance times as well as of different size Poissonian pulses. Combination of both approaches enables the modeling of signals in Internet traffic.

Acknowledgements

Support by the Lithuanian State Science and Studies Foundation is acknowledged.

References

1. T. Musha, H. Higuchi: Jpn. J. Appl. Phys. **15**, 1271 (1976);
2. K. L. Schick, A. A. Verveen: Nature **251**, 599 (1974).
3. G. Peng, H. J. Herrman: Phys. Rev. E **51**, 1745 (1995).
4. W. Li: www.nslij-genetics.org/wli/1fnoise.
5. M. B. Weismann: Rev. Mod. Phys. **60**, 537 (1988).
6. H. Wong: Microelectron. Reliab. **43**, 585 (2003).
7. B. Kaulakys, V. Gontis, M. Alaburda: Phys. Rev. E **71**, 051105 (2005).
8. B. Kaulakys, G. Vektaris: Phys. Rev. E **52**, 2091 (1995).
9. B. Kaulakys, F. Ivanauskas, T. Meskauskas: Int. J. Bifurcation and Chaos **9**, 533 (1999).
10. X. Zhang, G. Hu: Phys. Rev. E **52**, 4664 (1995).
11. B. Kaulakys, T. Meskauskas: Phys. Rev. E **58**, 7013 (1998).
12. B. Kaulakys: Phys. Lett. A **257**, 37 (1999).
13. J. Ruseckas, B. Kaulakys, M. Alaburda: Lith. J. Phys **43**, 223 (2003).
14. A. J. Field, U. Harder, P. G. Harrison: IEE Proc.-Commun. **151**, 355 (2004).
15. V. Gontis, B. Kaulakys, J. Ruseckas: AIP Conf. Proceed. **776**, 144 (2005).
16. S. Tadaki et al: these proceedings and private communication.
17. V. Gontis, B. Kaulakys: Physica A **343**, 505 (2004); **344**, 128 (2004).
18. A. J. Field, U. Harder, P. G. Harrison:
    http://www.doc.ic.ac.uk/~uh/QUAINT/data/.

Relationship Between Non-Markovian- and Drift-Fokker-Planck Equation

Knud Zabrocki[1], Svetlana Tatur[2], Steffen Trimper[1], and Reinhard Mahnke[3]

[1] Fachbereich Physik, Martin-Luther-Universität Halle-Wittenberg, D–06099 Halle/Saale, Germany; trimper@physik.uni-halle.de
[2] Dept. of Physics, Ural State University, 620083 Ekaterinburg, Russia; svetlana.tatur@mail.ru
[3] Institut für Physik, Universität Rostock, D–18051 Rostock, Germany; reinhard.mahnke@uni-rostock.de

Summary. Based on the stochastic description of transport phenomena the relationship between a non–Markovian evolution equation and the Fokker–Planck equation with drift is investigated. Memory is included by direct coupling between initial and current values of probability density. We present the result for three different initial distributions.

1 The Non-Markovian Fokker-Planck Equation

In media with a spatial–temporal accumulation process transport phenomena should be described by a stochastic approach [1] based on probabilities. The time evolution of the probability density could depend on the history of the sample to which it belongs, i. e. the changing rate of the probability should be influenced by the changing rate in the past and so the evolution equation of the probability has to be supplemented by memory terms. A recent overview is given in [2]. Here the modification we proposed is to replace the conventional Fokker-Planck equation by [3]

$$\partial_t p(\mathbf{r}, t) = \mathcal{M}(\mathbf{r}, t; p, \nabla p) + \int\limits_0^t dt' \int\limits_{-\infty}^\infty d^d r' \mathcal{K}(\mathbf{r} - \mathbf{r}', t - t'; p, \nabla p)\, \mathcal{L}(\mathbf{r}', t'; p, \nabla p).$$

(1)

This equation is of convolution type and consists of two competing parts standing for processes on different timescales. The first part manifested by the operator $\mathcal{M}$ characterizes the instantaneous and local process, whereas the second part with the operators $\mathcal{K}$ and $\mathcal{L}$ represent the delayed processes, the memory. In general all the operators may be non-linear in $p(\mathbf{r}, t)$ and $\nabla p(\mathbf{r}, t)$. The specification of them has to be according to the physical situation, which one deals with. One main feature of the quantity $p(\mathbf{r}, t)$ is, that it is conserved

$$\frac{dP(t)}{dt} = \frac{d}{dt} \int_{-\infty}^{\infty} d^d r \, p(\mathbf{r}, t) = 0 . \tag{2}$$

To preserve p the instantaneous term $\mathcal{M}$ has to be related to a (probability) current, e. g., $\mathcal{M} \propto \nabla \cdot \mathbf{j}$. For an arbitrary polynomial kernel $\hat{K}(z)$, where

$$\hat{K}(z) = \int d^d r \, \mathcal{K}(\mathbf{r}, z) \tag{3}$$

with the Laplace-transform $\mathcal{K}(\mathbf{r}, z) = \int_0^{\infty} dt \, e^{-zt} \, K(\mathbf{r}, t)$ the conservation law (2) is not fulfilled in general. A possible choice where it is fulfilled is $\mathcal{L} \equiv -\partial_t p(\mathbf{r}, t)$. For a detailed discussion, see [3].

2 Diffusion with Time Independent Memory Kernel

Let us consider the evolution equation

$$\partial_t p(\mathbf{r}, t) = D \nabla^2 p(\mathbf{r}, t) - \int_0^t dt' \int_{-\infty}^{\infty} d^d r' K(\mathbf{r} - \mathbf{r}', t - t') \, \partial_{t'} p(\mathbf{r}', t') \tag{4}$$

as a special case of (1). This Fokker-Planck equation relates $p(\mathbf{r}, t)$ to $p(\mathbf{r}, t')$ with $0 < t' < t$ unlike to a conventional one, where the evolution is only dependent on the probability at present time. Moreover (4) offers a coupling between $\partial_t p(\mathbf{r}, t)$ and $\partial_{t'} p(\mathbf{r}, t')$. The mixing of time scales leads to a substantial modification of the long time limit. As one of the simplest choices we took a strictly spatial local, but time independent kernel

$$K(\mathbf{r}, t) = \mu \, \delta(\mathbf{r}) , \tag{5}$$

where the parameter $\mu > 0$ characterizes the strength of the memory. By this choice the spatial and temporal variables are decoupled. Inserting the kernel in (4) one gets

$$\partial_t p(\mathbf{r}, t) = D \nabla^2 p(\mathbf{r}, t) - \mu \, [p(\mathbf{r}, t) - p_0(\mathbf{r})] \quad \text{with} \quad p_0(\mathbf{r}) \equiv p(\mathbf{r}, t = 0) . \tag{6}$$

The time independence of the kernel means that all times t' $(0 < t' < t)$ in the past have the same weight and so there is a very strong memory with a direct coupling of the instantaneous value to the initial value. The memory induced feedback to the initial value appears as a driving force. Without this coupling one can interpret the equation as a description of a particle, which performs a diffusive motion, where the probability density $p(\mathbf{r}, t)$ decays on a time scale μ^{-1}. As (6) is a linear equation and so the solution of it can be found analytically for arbitrary initial conditions

$$p(\mathbf{r}, t) = e^{-\mu t} \int_{-\infty}^{\infty} d^d r' \, p_0(\mathbf{r}') \left[G(\mathbf{r} - \mathbf{r}', t) + \mu \int_0^t dt' \, G(\mathbf{r} - \mathbf{r}', t - t') e^{\mu t'} \right],$$

$$\tag{7}$$

where $G(\mathbf{r}, t)$ is the Green's function of the conventional diffusion equation. From the general solution, some properties could be followed easily such as if the initial distribution is non-negative $p_0(\mathbf{r})$, so the $p(\mathbf{r}, t)$ does provided $\mu > 0$. The second moment $s(t)$ could be calculated

$$s(t) \equiv \int_{-\infty}^{\infty} \mathbf{r}^2 \, p(\mathbf{r}, t) \, d^d r = \frac{2 \, d \, D \, (1 - e^{-\mu t})}{\mu} \int_{-\infty}^{\infty} p_0(\mathbf{r}) \, d^d r + \int_{-\infty}^{\infty} \mathbf{r}^2 \, p_0(\mathbf{r}) \, d^d r \, .$$

$$\tag{8}$$

Notice that for the limit of vanishing memory $\mu \to 0$ one can easily verify that the last equation shows conventional diffusive behavior. The selection of the initial distribution is the essential point in our model and so three example are given to illustrate the solution of (6). Without lack of generality we concentrate our calculation on the one-dimensional case. It can be generalized to higher dimensions.

Delta-Starting Distribution

The first (more academic) example is the delta–starting distribution $p_0(x) = p_0 \, \delta(x)$. Substituting this in (7) the following solution is calculated

$$p(x, t) = \frac{p_0}{\sqrt{4 \pi D t}} e^{-\left(\mu t + \frac{x^2}{4 D t}\right)} + \frac{p_0 \, \kappa}{4} \left[f_+(x; D, \mu) + f_-(x; D, \mu) \right] \tag{9}$$

$$f_\pm = e^{\pm \kappa x} \left[\operatorname{erf}\left(\frac{\pm x}{\sqrt{4 D t}} + \sqrt{\mu t} \right) - \operatorname{sgn}(\pm x) \right], \tag{10}$$

where $\operatorname{erf}(x)$ is the error function and $\kappa = \sqrt{\mu / D}$. The first part is the solution of the homogeneous equation showing temporal decay with time constant μ^{-1}. In the long time limit the system shows a non-trivial stationary solution

$$\lim_{t \to \infty} p(x, t) \equiv p_s(x) = \frac{p_0 \, \kappa}{2} e^{-\kappa |x|} \, . \tag{11}$$

Such a stationary solution is due to the permanent coupling to the initial distribution and the greater $\mu > 0$ the stronger is this effect and more pronounced are the deviations from the pure diffusive behavior ($\mu = 0$).

Gaussian and Exponential Initial Conditions

For an arbitrary initial condition $p_0(\mathbf{r})$ the stationary solution can be directly calculated by solving the differential equation $\nabla^2 p_s(\mathbf{r}) = \kappa^2 \left[p_s(\mathbf{r}) - p_0(\mathbf{r})\right]$ with $\kappa^2 = \mu/D$. It results in

$$p_s(\mathbf{r}) = \frac{\kappa^{\frac{d+2}{2}}}{(2\pi)^{\frac{d}{2}}} \int d^d r' \, \frac{p_0(\mathbf{r}')}{|\mathbf{r} - \mathbf{r}'|^{\frac{d-2}{2}}} \, K_{\frac{d-2}{2}}\left(\kappa \, |\mathbf{r} - \mathbf{r}'|\right) \,, \tag{12}$$

where K_ν is a modified Bessel function, which could be expressed by standard functions for odd dimensions, i. e. $(d = 1, 3)$, and for even dimensions offers a logarithmic behavior [4]. In case of the Gaussian distribution $p_0(x) = p_0 \, e^{-\lambda x^2}$, the integration of (12) leads to

$$p_s(x) = \frac{p_0 \, \kappa}{4} \sqrt{\frac{\pi}{\lambda}} \, e^{\beta^2} \left[g_+(x; \beta, \lambda, \kappa) + g_-(x; \beta, \lambda, \kappa)\right] \tag{13}$$

with

$$g_\pm(x; \beta, \lambda, \kappa) = e^{\pm \kappa x} \, \mathrm{erfc}\left(\beta \pm x\sqrt{\lambda}\right) \quad \text{and} \quad \beta = \sqrt{\frac{\mu}{4\lambda D}} = \frac{\kappa}{2\sqrt{\lambda}}. \tag{14}$$

For exponential starting distribution $p_0(x) = p_0 \, e^{-\lambda |x|}$ the calculation shows

$$p_s(|x|) = \begin{cases} p_0 \, \frac{\lambda \kappa^2}{\kappa^2 - \lambda^2} \left[\frac{e^{-\lambda |x|}}{\lambda} - \frac{e^{-\kappa |x|}}{\kappa}\right] & \text{for} \quad \lambda \neq \kappa \\[2ex] p_0 \, \frac{1 + \kappa |x|}{2} \, e^{-\kappa |x|} & \text{for} \quad \lambda = \kappa \end{cases} \tag{15}$$

3 Relationship to Fokker-Planck Equation with Drift Term

The conventional form of the Fokker-Planck equation, where an external force is considered, has in the one-dimensional case the following form

$$\frac{\partial p(x, t)}{\partial t} = D \frac{\partial^2 p(x, t)}{\partial x^2} - \frac{\partial}{\partial x} \left[f(x) \, p(x, t)\right] \,, \tag{16}$$

where D is the diffusion coefficient, supposed to be constant (independent of x) here and $f(x)$ is the drift coefficient or the force, for which $f(x) = -dU(x)/dx$ with $U(x)$ as corresponding potential. On the one hand, the diffusion coefficient D measures the intensity of the noise and represents the stochastic part of motion, whereas the drift coefficient $f(x)$ corresponds to the force experienced by the system and so it describes the deterministic part of motion. In this subsection we calculate the force $f(x)$ which corresponds to the potential $U(x)$ for the different starting distributions $p_0(x)$, in such a way that both Fokker-Planck equations (6) and (16) are equivalent. To do this the deterministic parts of both equations are compared in the long time limit, in the stationary state. The formal solution is found by integration

$$f(x) = \frac{\mu \int\limits_0^x [p_s(\xi) - p_0(\xi)]\, d\xi}{p_s(x)} + \frac{C}{p_s(x)}\,, \qquad (17)$$

where C is an integration constant, which one could set to zero. To show the equivalence mathematically, one has to do this comparison in the following way. First take an arbitrary function $h(x)$ with bounded support, then integrate the product of h and the deterministic part of (6) resp. (16) over the complete real line, and finally compare the results of these integrations. If both integrations are equal, then the functions are equal.

4 Results

Finally we present the results for three different initial distributions, depicted in Fig. 1, the corresponding stationary solutions is shown in Fig. 2, the drift term and the corresponding potential in Fig. 3 and Fig. 4, respectively. For the delta-like starting distribution the drift term can be calculated to

$$f(x) = -\sqrt{\mu D}\,\mathrm{sign}(x) \qquad (18)$$

and so one can verify the following potential

$$U(x) = \sqrt{\mu D}\,|x|\ . \qquad (19)$$

In case of a Gaussian initial distribution an analogue calculation leads to

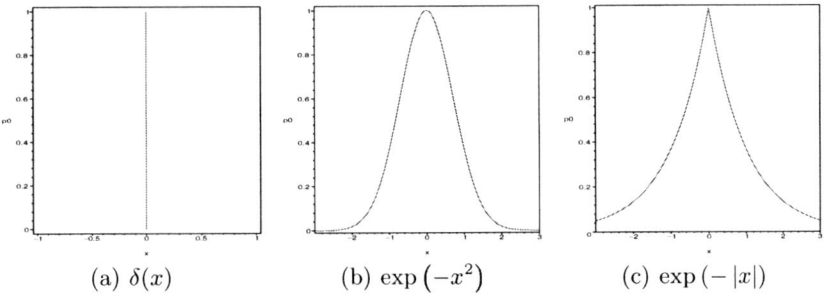

(a) $\delta(x)$ (b) $\exp\left(-x^2\right)$ (c) $\exp\left(-|x|\right)$

Fig. 1. Starting distribution $p_0(x)$.

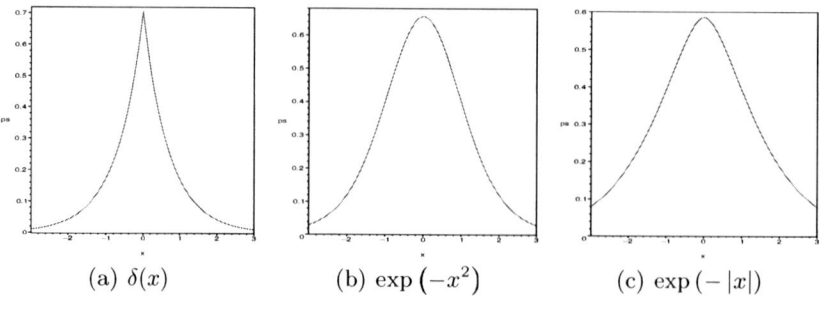

(a) $\delta(x)$ (b) $\exp\left(-x^2\right)$ (c) $\exp\left(-|x|\right)$

Fig. 2. Stationary state $p_s(x)$.

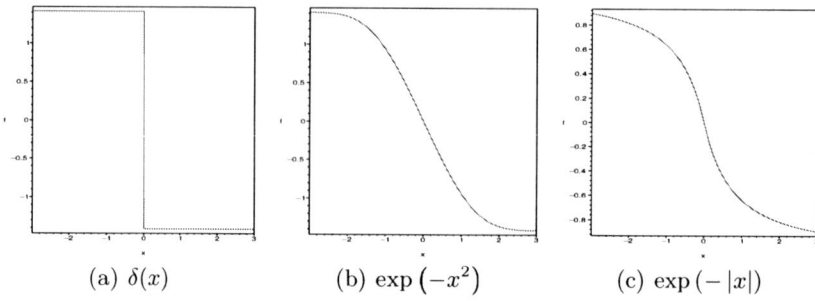

(a) $\delta(x)$ (b) $\exp\left(-x^2\right)$ (c) $\exp\left(-|x|\right)$

Fig. 3. Drift coefficient $f(x)$.

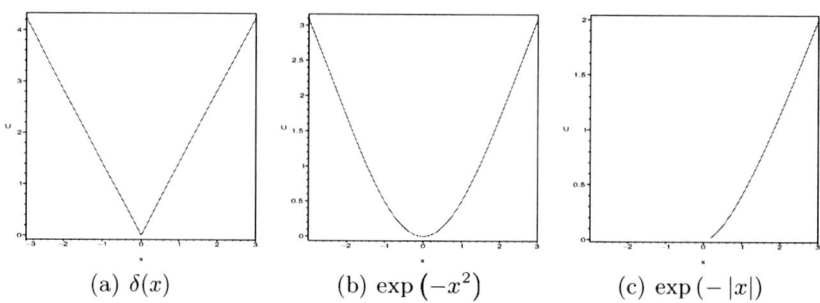

(a) $\delta(x)$ (b) $\exp\left(-x^2\right)$ (c) $\exp\left(-|x|\right)$

Fig. 4. Potential $U(x)$.

$$f(x) = \sqrt{\mu D} \, \frac{g_+(x; \beta, \lambda, \kappa) - g_-(x; \beta, \lambda, \kappa)}{g_+(x; \beta, \lambda, \kappa) + g_-(x; \beta, \lambda, \kappa)} \, , \tag{20}$$

whereas an exponential starting distributions yields to the drift term

$$f(x) = \sqrt{\mu D} \, \mathrm{sign}(x) \, \lambda \, \frac{e^{-\kappa |x|} - e^{-\lambda |x|}}{\kappa \, e^{-\lambda |x|} - \lambda \, e^{-\kappa |x|}} \, . \tag{21}$$

The underlying potentials can be obtain after an integration. The results are shown in Fig. 4. A generalization of the results for higher dimensions and a more general discussion will be published elsewhere [5].

Acknowledgements

The authors (S. T. and K. Z.) acknowledge support by the DFG (SFB 418) as well as by DAAD (S. Tatur).

References

1. R. Mahnke, J. Kaupužs and I. Lubashevsky: Probabilistic description of traffic flow, Phys. Rep., **408**, 1–130, 2005.
2. T. D. Frank, *Nonlinear Fokker-Planck Equations*, Springer-Verlag, Berlin, 2005.
3. S. Trimper and K. Zabrocki: Phys. Lett. A, **331**, 423–431, 2004.
4. M. Abramowitz and I.A. Stegun, *Handbook of Mathematical Functions* Dover Pub. New York, 1972.
5. K. Zabrocki, R. Mahnke, and S. Trimper, in preparation.

Part VI

Traffic Flow: Empirical Results
and Applications

Accidents in Platoons of Vehicles

Cécile Appert-Rolland[1] and Ludger Santen[2]

[1] Université de Paris-Sud, Laboratoire de Physique Théorique, Bâtiment 210, F-91405 Orsay Cedex, France - appert@th.u-psud.fr
[2] Universität des Saarlandes, Fachrichtung Theoretische Physik, D-66041 Saarbrücken, Germany - santen@lusi.uni-sb.de

Summary. In dense vehicular traffic cars often drive at close distances, they form clusters or platoons. Within these platoons, time headways are observed, which are often even shorter than the reaction time of the drivers - a situation which is potentially dangerous.
Here we propose a simple dynamical model for a platoon undergoing emergency braking, which takes into account the individual variations of the reaction time and braking capacities. We apply the model to real platoons, i.e. platoons which have been identified in large sets of single-vehicle traffic data. We use our results in order to compare the impact of different possible regulations (speed limit, minimum headway).

1 Introduction

One of the remarkable empirical features of highway traffic is the fact that vehicles frequently undergo the security distance. Although the recommended security distance amounts to 1.8 sec in Germany and the observed time headway distribution has in general a maximum around 1. sec, one frequently observes time headways far below one second [1]. These extremely short time headways have important consequences concerning the structure, performance and security of vehicular traffic. In this work we focus on the security aspects of dense highway traffic. The search for causes of accidents in vehicular traffic has attracted broad scientific interest. Studies on this subject include the identification of dangerous situations in model generated configurations. Although many interesting results have been obtained using this model-based approach it is not obvious how these results depend on the particular model. Therefore we try to minimize the modeling part throughout our analysis by taking empirical data of highway traffic in combination with a simple braking model. The usage of empirical data implies that we consider properly all relevant correlations between time-headways and velocities. This procedure should lead to more significant results than obtained in previous studies. As we are interested in the risk related to the structure of traffic flows we analyze a typical situation where a chain reaction of brakings leads to an accident. This will happen in platoons, i.e. clusters of vehicles driving at short distances. Let us consider a platoon including N vehicles ($i = 0 \ldots N - 1$). Each vehicle has a velocity v_i, a reaction time τ_i and a braking capacity a_i^*. The time-headway t_i^h is defined here as the temporal distance between the rear end of the preceeding vehicle $i - 1$ and the front end of vehicle i. Now we assume that one car in this platoon is braking with deceleration a_0. If time headways are small enough, the following cars have to brake as well in order to avoid collision. More precisely, if $t_i^h < \tau_i$, vehicle i has to brake harder than vehicle $i - 1$. If this is true for several vehicles in a row it finally may happen that the required deceleration exceeds the braking capacity of the car. In this case a collision occurs.

Apart from this mechanism it is possible that a driver didn't even start to brake before colliding with the preceeding car if his reaction time is too long (or the time headway too short). As a result, the accident probability is increasing with the position in the platoon. This well-known effect has already been studied in ideal platoons e.g. having constant initial time headways, in particular to study the impact of various cruise control devices [2]. In order to improve the realism of the approach some authors have introduced probability distributions for the reaction times or braking capacities [3]. Although this approach recognizes the variability of time headways, speeds and reaction times it disregards the fact that e.g. velocity-velocity or velocity-time headway correlations are non-negligible in dense vehicular traffic. Correlations can be included by directly identifying the platoons in the empirical data sets (see section 3 for details).

In the scenario that we consider, cars undergo emergency braking. We thus have to introduce a dynamical model for such emergency braking, which we keep as simple as possible: Braking of car i triggers the braking of the following car $i + 1$, which starts to brake after a reaction time τ_{i+1}. Then the car slows down with the weakest constant deceleration sufficient to avoid accidents, until it stops. This choice of a constant deceleration law is consistent with deceleration records for emergency braking on tracks (see e.g. [4]), although other empirical studies indicate that unexpected braking on real roads with non professional drivers is rather described by a two step process [5]. However, so far there exists only a small number of empirical studies on this subject, such that it is difficult to single out one of the two possibilities. Therefore we have chosen the constant deceleration law for simplicity.

The event that triggers the chain of emergency brakings is the sudden braking of an initiating car with a constant deceleration a_0. We chose a modest value for a_0 (between 3 to $5m/s^2$), well below the average maximum braking capacity. The choice of the model implies that we have to assign two parameters to each vehicle, which are not included in the data set: the reaction time of the drivers and the maximum braking capacities. These are taken from probability distributions, which rely on the outcome of e.g. car-following experiments. There is a consensus to take a log-normal distribution for the reaction times. The parameters of this distribution differ depending whether the driver expects the considered event or not. In a platoon, drivers expect to have to adjust their velocity to the preceeding cars very often, and thus they react quite rapidly if a weak acceleration is required. However, they do not expect an emergency braking, and it is a well known fact that, due to their fear of rear-end collision with the following car, drivers hesitate to brake too hard. Therefore we used a distribution of reaction times which corresponds to unexpected events. The log-normal distribution for the reaction times has to be cut off beyond a certain value, that we took equal to 2s following [4]. Still, some authors [6] claim that some much longer reaction times may be observed in real traffic, and the effect of these will be considered in future work. It is however reasonable to assume that in platoons, where people expect to have to adapt their speed all the time, very long reaction times would be exceptional.

In this paper, we took for the braking capacity distribution a Gaussian centered around $7m/s^2$, and truncated below 6 and above $8m/s^2$, except when stated otherwise.

We stress the fact that the braking capacity and reaction time distribution are the only model parameters that can not be directly taken from the single vehicle data. Contrary the structure of the platoons is directly accessible and does not depend on model parameters of any kind. In order to obtain representative values of the accident probabilities by means of numerical simulations we have considered 500 to 1000 realizations of the probability distributions.

2 Criterion for Accidents

A first estimate for the number of accidents in a platoon can be obtained simply by comparing the final positions of the cars after braking. Obviously an accident must have occurred if the order of cars is exchanged compared to the initial positions. The corresponding criterion on deceleration requirements reads

$$\frac{1}{a_i} \leq \left(\frac{v_{i-1}^0}{v_i^0}\right)^2 \frac{1}{a_{i-1}} + \frac{2(t_i^h - \tau_i)}{v_i^0} \ . \tag{1}$$

However, some of the accidents may not be identified by means of the final positions. Figure 1 shows such an example, where the trajectories of two cars intersect, although the final positions do not indicate the occurrence of an accident. In order to consider such kinds of accidents as well, one has to introduce an alternative criterion in a certain range of parameters [7]

$$\frac{1}{a_i} \leq \frac{2(v_{i-1}^0 - v_i^0)\tau_i + 2v_i^0 t_i^h - a_{i-1}\tau_i^2}{2a_{i-1}v_i^0 t_i^h + (v_{i-1}^0 - v_i^0)^2} \ . \tag{2}$$

In our study on, which relies on a large set of real data, we found that about 15% of the collisions, i.e. a non-negligible fraction, are identified by means of (2).

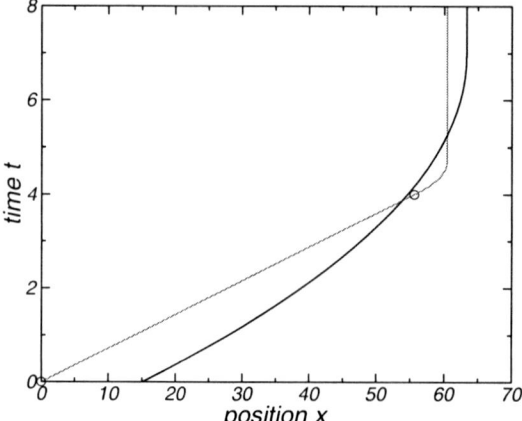

Fig. 1. Trajectories of two successive cars. The circle indicates the end of the reaction time for the 2nd driver. When two cars are very close, but the second car has a deceleration capacity much stronger than the first car, it is possible to have an intersection between the trajectories that would not be detected if one looks only on the final positions.

3 Description of the Data Set

The data have been collected between March, 30, 2000 and May, 16, 2000 on the German highway A3 between the junction Duisburg-Wedau and the highway-intersection Kreuz-Breitscheid. The traffic stream characteristics at this location have been established by magnetic loops, one for each lane. The chosen location is well apart from on- and off-ramps or intersections, such that the data set should represent the bulk properties of real vehicular traffic. It also important to note that there is no speed limit applied at this section of the highway. The whole data set comprises measurements of about 780000 cars. By means of the detection devices it is possible to measure the passing time (up to a precision of $1/100$ sec), the speed of a vehicle, and the occupation rate of the loop. These direct measurements can be used in order to calculate the length of a vehicle, spatial and temporal distances between two cars and various other quantities of interest. The lower bound for the velocity measurement is 10 km/h, i.e., velocities of slower vehicles are not measured. The relation between direct measurements, and the spatial quantities as e.g.the distance headways are based on the assumption that the vehicles pass the detector at a constant and representative speed. This assumption is not valid for cars in a jam, where the detected speeds are much higher than the average speeds. As mentioned above the single-vehicle data allow for the determination of the time-headway t^h and the distance-headway gap of the n-th vehicle via

$$t^h(n) = t_n - t_{n-1} - \frac{l_{n-1}}{v_{n-1}} \tag{3}$$

and

$$gap(n) = v_n(t_n - t_{n-1}) - l_{n-1} \tag{4}$$

where we assume that v_n and v_{n-1} are constant. t_n denotes the time the n-th vehicle passes the detector, l_n and v_n its length and velocity.

3.1 t^h as a Function of the Velocity

It is an obvious fact that the accident probability of a car largely depends on its distance to the vehicle in front. While the spatial headway strongly depends on the speed of the cars, it has been argued that the temporal headways are rather insensitive to the speed of the cars. Here we want to evaluate the velocity dependence of the time headway distribution. We have plotted the distribution for the t^h of all vehicles having a velocity between V and $V + \Delta V$ where $\Delta V = 30$ km/h. Now, we take the value of t^h for which the distribution is maximum (t_{max}^h), and plot it as a function of V. We choose the maximum t_{max}^h instead of the average value, because the average value is strongly influenced by the long tail for large t^h's - which is not of interest, as it reflects only the average flux. We rather expect the velocity to have an impact on the distribution at short headways. Our observation is that t_{max}^h is almost constant for all velocities greater than 80 km/h. Below this value, t_{max}^h increases as the velocity V decreases. Drivers are impressed by driving too close (regarding spatial distance) to the preceeding car, while they are less sensitive to temporal distances.

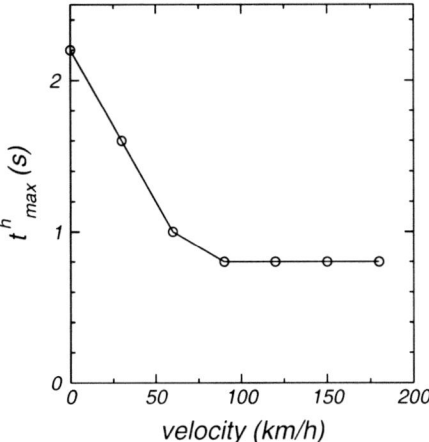

Fig. 2. t_{max}^h as a function of the velocity.

3.2 Identification of the Platoons

We define a platoon as a set of cars within which all time headways are less than a certain threshold T_{max}, while the time headways in front and behind the platoon are larger than T_{max}. This crude definition does not aim at characterizing a precise dynamical structure - some more specialized definitions of platoons have been given, which take into account for example the speeds of the cars, etc. Here, the extraction of the platoons from the data could rather be viewed as a pre-filtering of the data, on which the results should not depend.

4 Impact of Various Security Measures

As the probability that a car brakes suddenly is not known, the absolute number of accidents we identify does not have any meaning. The relevant quantity is the relative number of accidents, when one compares two situations. Here, we use the result for the original data as a reference state. Then, we modify the data in order to mimic various security measures. We compute the ratio of the number of accidents with a given security measure to the number of accidents in the reference state (see figure 3).

In a previous work [7] we have shown that our results depend only weakly on the choice for T_{max}. Thus we present here only the results for $T_{max} = 3$ and $7s$.

Now we study the dependence of the results with respect to the amplitude of the stimulus, i.e. the amplitude a_0 of the first braking car. The effects of the security measures (which are described below) are quite similar when the amplitude a_0 takes the values 3, 4, or $5m/s^2$.

The first possible security measure is the application of a speed limit, which we introduced in the following way: if the average velocity of a platoon $\bar{v}$ is above the speed limit v_{max}, all the velocities within the platoon are rescaled by the factor $v_{max}/\bar{v}$. In this way the relative speed dispersion is kept within the platoon. We also keep the time-headways between cars when applying the speed limit. This is justified by fig. 2, though it is possible that real speed limitations would have an impact on time-headways. But it is not obvious a priori in which direction the effect would be: one the one hand, more cars would drive at about the same speed, and thus there may be more competition between them. On the other hand, less drivers attempt to overtake as they have to respect the speed limit. Therefore the number of aggressive drivers should be reduced. A study by [1] shows that the fraction of drivers exceeding the speed limit ($130km/h$) is the same if you consider the subsets of vehicles with a time headway less than $0.5s$ or $1.0s$. This is in favor of a constant time headway for all velocities, at least for velocities above a certain threshold (around $80km/h$ in our case).

Fig. 3 shows that, counterintuitively, a speed limit increases the number of accidents. This is a consequence of the non-linearity of the braking trajectories, as explained in [7]. Of course, for most other accident scenarii, an increase of the speed would be a serious drawback.

The next points in figure 3 refer to a modification of the braking capacity distribution. As expected, an increase (decrease) of the width of the distribution increases (decreases) the number of accidents.

The first significant improvement on the number of accidents is obtained when the average of the braking capacity distribution is increased from 7 to $9m/s^2$. The effect is however limited, and in practice, it would be rather impossible to improve the braking capacities of *all* cars without increasing the width of the distribution.

The suppression of short time headways is by far the most efficient measure. If all time headways smaller than $1.8s$ are replaced by time headways equal to 1.8, almost all collisions are suppressed. Interestingly, even the suppression of time headways below $1s$ only reduces the number of accidents drastically - more than half of the accidents are avoided.

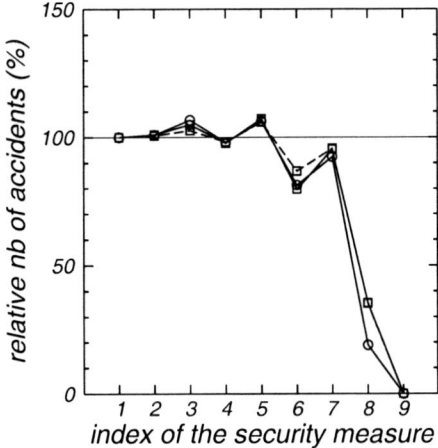

Fig. 3. Ratio of the number of accidents for various security measures. The 'x'-axis refers to these measures with the following correspondence: (1) is the reference state - by definition it is 100%. (2) Speed limitation $v_{max} = 130km/h$. (3) $v_{max} = 110km/h$. (4) Braking capacities $a^* \in [6.8, 7.2](m/s^2)$. (5) $a^* \in [5, 9](m/s^2)$. (6) $a^* \in [8, 10](m/s^2)$. (7) Time headway minimum $t^h \geq 0.5$. (8) $t^h \geq 1.0$. (9) $t^h \geq 1.8$. Symbols circles, squares, x correspond respectively to an amplitude $a_0 = 3$, 4, and $5m/s^2$. Solid (dashed) lines indicate that $T_{max} = 3s$ $(7s)$.

5 Perspectives

A further improvement of the model should address the possibility that a vehicle reacts not only to the vehicle just in front, but also to other preceeding vehicles. This was already addressed in [8, 9] and recent work by S. Hoogendoorn [10] indicates such an influence of more than one preceeding car. When a car-following model is considered, it is easy to give a different weight - or sensibility - to the interaction depending on which pair of vehicles is considered. The results by S. Hoogendoorn indicate that if α_0 is the sensibility for nearest neighbors, then a sensibility of the order of $\alpha_0/2$ should typically be taken for next nearest neighbors.

In our case, which considers emergency braking, such a weighting is not possible. Either one brakes or one does not brake. Our proposal would be that, though the action of braking would still follow the signal of the immediately preceeding car, preceeding cars could have an influence on the reaction time distribution. I.e. if the driver observes that the 2nd car in front is braking, he prepares himself to react more rapidly.

One could also take advantage of the fact that in our data, the type of the vehicles is known (cars, trucks...). One could thus incorporate the fact that for example, a car behind a truck can only react on the vehicle directly in front, while other configurations allow for a larger visibility.

6 Conclusion

We have shown that if something unexpected occurs on the road, as many drivers drive with a short headway, it is likely that they won't be able to avoid a collision. This suggests three possible levels of action to minimize the number of accidents:

- Decrease the number of unexpected events: this could be obtained for example by regularizing the flow, inciting the drivers to have more homogeneous velocities, etc.
- Suppress short time-headways: the equipment of cars with automatic distance control could be helpful.
- Improve reaction times: cars could be equipped with alert devices, etc.

Acknowledgements

LS acknowledges support by the Deutsche Forschungsgemeinschaft under Grant No. SA864/2-2. The authors are grateful to the Landesbetrieb Straßenbau NRW for providing the empirical data.

References

1. M. Aron, M.-B. Biecheler, and J.-F. Peytavin. Sécurité routière - temps inter-véhiculaires et vitesse. quels enjeux de sécurité sur l'autoroute? *Recherche Transports Sécurité*, 64:3–17, 1999.

2. M. Brackstone, M. McDonald, and B. Sultan. A collision model for the assessment of the safety benefits of avcss. *Proc. of the 6th ITS World Congress, Toronto, Canada.*, Nov., 1999.

3. J. Carbaugh, D.N. Godbole, and R. Sengupta. Safety and capacity analysis of automated and manual highway systems. *Transportation Research Part C*, 6:69–99, 1998.

4. Rodger J. Koppa. Human factors. In *Traffic Flow Theory*, pages 3–1, 2000.

5. Des progrès pour la sécurité - l'automobile citoyenne. *Les dossiers du CCFA*.

6. R.J. Kiefer, M.T. Cassar, C.A. Flanagan, C.J. Jerome, and M.D. Palmer. Surprise braking trials, time-to-collision judgments and "first look" maneuvers under realistic rear-end crash scenarios. In DC Contract DTFH61-01-X-00014, Washington, editor, *Performed by Crash Avoidance Metrics Partnership (CAMP)*, August 2005.

7. C. Appert and L. Santen. Accidents in dense vehicular traffic. *preprint*, 2005.

8. H. Lenz, C.K. Wagner, and R. Sollacher. Multi-anticipative car-following model. *The European Physical Journal B - Condensed Matter*, 7:331–335, 1999.

9. M. Treiber, A. Kesting, and D. Helbing. Delays, inaccuracies and anticipation in microscopic traffic models. *Physica A*, 360:71–88, 2006.

10. S.P. Hoogendoorn, S. Ossen, and M. Schreuder. Multi-anticipative car-following behavior: and empirical analysis. *preprint*, 2005.

Jam-Avoiding Adaptive Cruise Control (ACC) and its Impact on Traffic Dynamics

Arne Kesting[1], Martin Treiber[1], Martin Schönhof[1], Florian Kranke[2], and Dirk Helbing[1,3]

[1] Technische Universität Dresden, Institute for Transport & Economics,
   Andreas-Schubert-Strasse 23, D-01062 Dresden, Germany
[2] Volkswagen AG, Postfach 011/1895, D-38436 Wolfsburg, Germany
[3] Collegium Budapest – Institute for Advanced Study,
   Szentháromság u. 2, H-1014 Budapest, Hungary

Summary. Adaptive-Cruise Control (ACC) automatically accelerates or decelerates a vehicle to maintain a selected time gap, to reach a desired velocity, or to prevent a rear-end collision. To this end, the ACC sensors detect and track the vehicle ahead for measuring the actual distance and speed difference. Together with the own velocity, these input variables are exactly the same as in car-following models. The focus of this contribution is: What will be the impact of a spreading of ACC systems on the traffic dynamics? Do automated driving strategies have the potential to improve the capacity and stability of traffic flow or will they necessarily increase the heterogeneity and instability? How does the result depend on the ACC equipment level?

We discuss microscopic modeling aspects for human and automated (ACC) driving. By means of microscopic traffic simulations, we study how a variable percentage of ACC-equipped vehicles influences the stability of traffic flow, the maximum flow under free traffic conditions until traffic breaks down, and the dynamic capacity of congested traffic. Furthermore, we compare different percentages of ACC with respect to travel times in a specific congestion scenario. Remarkably, we find that already a small amount of ACC equipped cars and, hence, a marginally increased free and dynamic capacity, leads to a drastic reduction of traffic congestion.

1 Introduction

Traffic congestion is a severe problem on European freeways. According to a study of the European Commission [1], its impact amounts to 0.5% of the gross national product and will increase even up to 1% in the year 2010. Since building new infrastructure is no longer an appropriate option in most (Western) countries, there are many approaches towards a more effective road usage and a more 'intelligent' way of increasing the capacity of the road network. Examples of advanced traffic control systems are, e.g., 'intelligent' speed limits, adaptive ramp metering, or dynamic routing. These examples are based on a centralized traffic management, which controls the operation and the response to a given traffic situation. In this contribution, we focus on a local strategy based on autonomous vehicles, which are equipped with adaptive cruise control (ACC) systems. The motivation is that a jam-avoiding driving strategy of these automated vehicles might also help to increase the road capacity and thus decrease traffic congestion. Moreover, ACC systems become commercially available to an increasing number of vehicle types.

An ACC system is able to detect and to track the vehicle ahead, measuring the actual distance and speed difference. Together with the own speed, these input data allow the system to calculate the required acceleration or deceleration to maintain a selected time headway, to reach a desired velocity, or to prevent a rear-end collision. It should be emphasized that ACC systems control the longitudinal driving task. Merging, lane changing or gap-creation for other vehicles still needs the intervention of the driver. ACC systems promise a gain in comfort and safety in applicable driving situations, but they are not yet applied in congested traffic conditions. The next generation of ACC will successfully extend the application range to all speed ranges and most traffic situations on freeways including stop-and-go traffic. This leads to the question: In which way does a growing market penetration of ACC-equipped vehicles influence the capacity and stability of traffic flow? Although there is considerable research on this topic [2], there is even no clarity up to now about the sign of the effect. Some investigations predict a positive effect [3, 4], while others are more pessimistic [5, 6].

The contribution is organized as follows: We start with a discussion of modeling issues concerning the description of human vs. automated driving and pinpoint the differences between ACC-driven vehicles and human drivers. In Sec. 3, we will model three ACC driving styles which are explicitly designed to increase the dynamic capacity and traffic stability by varying the individual driving behavior. Since the impact on the traffic dynamics could solely be answered by means of traffic simulations, in Sec. 4 we perform a simulation study of mixed freeway traffic with a variable percentage of ACC vehicles. In Sec. 5, we conclude with a discussion of our results.

2 Modeling Human and Automated (ACC) Driving Behavior

Most microscopic traffic models describe the acceleration and deceleration of each individual 'driver-vehicle unit' as a function of the distance and velocity difference to the vehicle in front and the own velocity [7, 8]. Some of these car-following models have been successful in reproducing the characteristic features of macroscopic traffic phenomena such as traffic breakdowns, the scattering in the fundamental diagram, traffic instabilities, and the propagation of stop-and-go waves or other patterns of congested traffic. While these collective phenomena can be described by macroscopic, fluid-dynamic traffic models as well [9], microscopic models are more appropriate to cope with the heterogeneity of mixed traffic, e.g., by representing individual driver-vehicle units by different parameter sets or even by different models.

Remarkably, the input quantities of car-following models are exactly those of an ACC system. As in microscopic models, the ACC controller unit calculates the acceleration with a negligible response time. Therefore, one might state that car-following models describe ACC systems more accurately than human drivers despite of their intention to reproduce the traffic dynamics of human driving behavior.

Thus the question arises, how to take into account the *human* aspects of driving for a realistic description of the traffic dynamics. The nature of human driving is apparently more complex. First of all, the *finite reaction time* of humans results in a delayed response to the traffic situation. Furthermore, human drivers have to cope with imperfect estimation capabilities resulting in *perception errors* and *limited attention spans*. These destabilizing influences alone would lead to a more unsafe driving and a high number of accidents if the reaction time reached the order of the time headway. But in day-to-day situations the contrary is observed: In dense (not yet congested) traffic, the modal value of the time headway distribution on German or Dutch freeways (i.e., the value where it reaches its maximum) is around 0.9 s [10–12], which is of the same order of typical reaction times [13]. Moreover, single-vehicle data for German freeways [10] indicate that some drivers even drive at headways as low as 0.3 s, which is below the reaction time by a factor of at least 2-3 even for a very attentive driver. For principal reasons, therefore, safe driving is not possible in this case when considering only one vehicle in front.

This suggests that human drivers achieve additional stability and safety by scanning the traffic situation *several vehicles ahead* and by *anticipating* future traffic situations. The question is, how this behavior affects the overall driving behavior and performance with respect to ACC-like driving mimicked by car-following models. Do the stabilizing effects (such as anticipation) or the destabilizing effects (such as reaction times and estimation errors) dominate, or do they effectively cancel out each other? The *human driver model* (HDM) [14] extends the car-following modeling approach by explicitly taking into account reaction times, perception errors, spatial anticipation (more than one vehicle ahead) and temporal anticipation (extrapolating the future traffic situation). It turns out that the destabilizing effects of reaction times and estimation errors can be compensated for by spatial and temporal anticipation [14]. One obtains essentially the same longitudinal dynamics, which explains the good performance of the simpler, ACC-like car-following models. Thus, for the sake of simplicity, we model both automated ACC-driving and human driving with the same microscopic traffic model, but differentiate the driving strategies by different parameter sets.

3 Jam-Avoiding ACC Driving Strategies

As discussed in the previous section, both human drivers and ACC-controlled vehicles are effectively described by the car-following model approach. Here, we will use the *intelligent driver model* (IDM) [15], according to which the acceleration of each vehicle α is a continuous function of the velocity v_α, the net distance gap s_α, and the velocity difference (approaching rate) Δv_α to the leading vehicle:

$$\dot{v}_\alpha = a \left[1 - \left(\frac{v_\alpha}{v_0} \right)^4 - \left(\frac{s^*(v_\alpha, \Delta v_\alpha)}{s_\alpha} \right)^2 \right]. \tag{1}$$

The deceleration term depends on the ratio between the effective 'desired minimum gap'

$$s^*(v, \Delta v) = s_0 + vT + \frac{v \Delta v}{2\sqrt{ab}} \tag{2}$$

and the actual gap s_α. The minimum distance s_0 in congested traffic is significant for low velocities only. The dominating term in stationary traffic is vT, which corresponds to following the leading vehicle with a constant safe time headway T. The last term is only active in non-stationary traffic and implements an accident-free, 'intelligent' driving behavior including a braking strategy that, in nearly all situations, limits braking decelerations to the 'comfortable deceleration' b. The IDM guarantees crash-free driving. The parameters for the simulations are given in Table 1.

In order to design a jam-avoiding behavior for the ACC vehicles, we modify the ACC model parameters. The (average) time headway has a direct relation to the maximum (static) road capacity: Neglecting the length of vehicles leads to the approximative relationship $Q \approx 1/T$ between the flow Q and the headway T (cf. Eq. (3) below). The crucial parameter controlling the capacity is, therefore, the safe time headway which is an explicit parameter of the IDM. Moreover, the system performance is not only determined by the time headway distribution, but also depends on the *stability* of traffic flow. An ACC driving behavior aiming at increasing the traffic performance should, therefore, additionally consider a driving strategy which is able to stabilize the traffic flow, e.g. by a faster dynamic adaptation to the new traffic situation. The stability is mainly affected by the IDM parameters 'maximum acceleration' and 'desired deceleration', see [15].

In the following, we will investigate the potentials of three different parameter sets for jam-avoiding driving behavior, varying the IDM parameters T, a and b. In order to refer to the values given in Table 1, we express the parameter changes by simple multipliers. For example, $\lambda_a = 2$ represents an increased ACC parameter $a' = \lambda_a a$, where a is the value listed in Table 1.

(1) The reduction of the time headway T by a factor $\lambda_T = 2/3$ has a positive impact on the capacity. The other model parameters of Table 1 remain unchanged, i.e., in particular, $\lambda_a = 1$, $\lambda_b = 1$.

(2) Besides setting $\lambda_T = 2/3$, we increase the desired acceleration by choosing $\lambda_a = 2$. The faster acceleration towards the desired velocity increases the traffic stability.

(3) The additional reduction of the desired deceleration by $\lambda_b = 1/2$ corresponds to a more cautious and more anticipative driving style. This behavior also increases the stability.

Model Parameter	Value
Desired velocity v_0	120 km/h
Save time headway T	1.5 s
Maximum acceleration a	1.0 m/s^2
Desired deceleration b	2.0 m/s^2
Jam distance s_0	2 m

Table 1. Model parameters of the *intelligent driver model* (IDM) used in our simulations. The vehicle length is 5 m. In order to model jam-avoiding ACC strategies, we modify the safe time headway parameter T, the 'maximum acceleration' a and the 'desired deceleration' b by multipliers λ_T, λ_a, and λ_b, respectively.

4 Microscopic Simulations of Mixed Traffic

Let us now investigate the impact of ACC vehicles which are designed to enhance the capacity and stability of traffic flows. We will simulate mixed traffic consisting of human and automated (ACC) longitudinal control with a variable percentage of ACC vehicles.

Our simulation is carried out for a single-lane road with an on-ramp serving as bottleneck and with open boundary conditions. To keep matters simple, we replace an explicit modeling of the merging of ramp vehicles to the main road by inserting ramp vehicles centrally into the largest gap within a 300 m long ramp section. In order to generate a sufficient velocity perturbation in the merge area, the speed of the accelerating on-ramp vehicles at the time of insertion is assumed to be 50% of the velocity of the respective front vehicle. Moreover, we neglect trucks and multi-lane effects. While these aspects are relevant in real traffic, they do not change the picture qualitatively. Nevertheless, the induction of a second driver-vehicle type, e.g., ACC vehicles, always has the potential to reduce the traffic performance by an increased level of heterogeneity. We have compared the simulation results with Gaussian distributed model parameters, but found no qualitative difference for this single-lane scenario.

4.1 Spatiotemporal Dynamics and Travel Time

Let us now demonstrate that already a moderate increase in the dynamic capacity obtained by a small percentage of jam-avoiding ACC vehicles may have a significant effect on the system performance.

We have simulated idealized rush-hour conditions by linearly increasing the inflow at the upstream boundary over a period of 2 hours from 1200 vehicles/h to 1600 vehicles/h. Afterwards, we have linearly decreased the traffic volume to 1000 vehicles/h until $t = 5$ h. Moreover, we have assumed a constant ramp flow of 280 vehicles/h. Since the maximum overall flow of 1880 vehicles/h exceeds the road capacity, a traffic breakdown is provoked at the bottleneck. We have used the IDM parameters from Table 1 and parameter set (3) for ACC vehicles, i.e., $\lambda_T = 2/3, \lambda_a = 2, \lambda_b = 1/2$.

Figure 1 shows the spatiotemporal dynamics of the traffic density for 0% and 10% ACC vehicles. The increased capacity obtained by the induced ACC vehicles leads to a strong reduction of the traffic jam already for a small percentage of ACC vehicles. For 30% ACC vehicles, the traffic jam disappears completely.

An increased percentage of jam-avoiding ACC vehicles has a strong effect on the travel time: Figure 2 shows the actual and cumulated travel times for various ACC percentages. At the peak of congestion ($t = 3.2$ h), the travel time for individual drivers is nearly triple that of the uncongested situation ($t < 1$ h). Already 10% ACC vehicles reduce the maximum travel time delay of individual drivers by about 30% (Fig. 2(a)), and the cumulated time delay (which can be associated with the economic cost of this jam) by 50% (Fig. 2(b)). Several factors contribute to this enhanced system performance. First, an increased ACC percentage leads to a delay of the traffic breakdown. Second, the ACC vehicles reduce the maximum queue length significantly. Third, the jam dissolves earlier. These effects, which are responsible for the drastic increase in the system performance already for a small proportion of jam-avoiding ACC vehicles, will be investigated in the following.

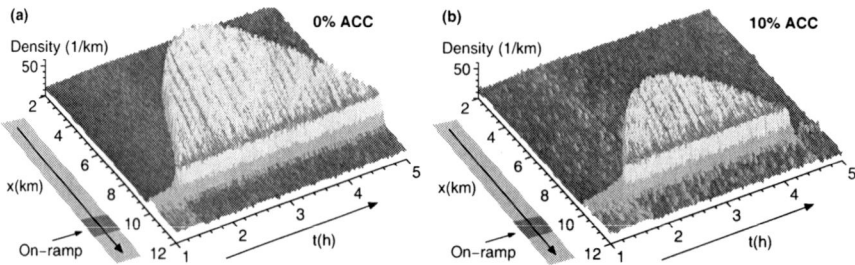

Fig. 1. Spatiotemporal dynamics of the traffic density (a) without ACC vehicles and (b) with 10% ACC vehicles (parameter set (3)). Already a small increase in the road capacity induced by a small percentage of jam-avoiding ACC vehicles leads to a significant reduction of traffic congestion (light high-density area).

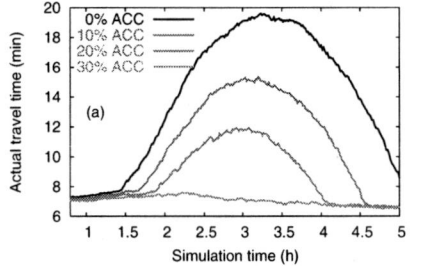 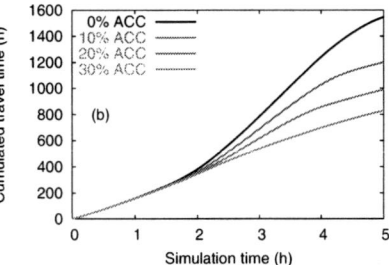

Fig. 2. Time series for (a) the actual and (b) the cumulated travel times for simulation runs with different percentages of ACC vehicles. The traffic breakdown leads to a significant prolongation of travel time. A proportion of 30% ACC vehicles can completely prevent the traffic breakdown.

4.2 Maximum Capacity in Free Traffic

The *static road capacity* $Q_{\max}^{\text{theo}}$, which corresponds to the maximum of the flow-density diagram, is mainly determined by the average time headway T. However, the theoretical capacity depends also on the 'effective' length $l_{\text{eff}} = l_{\text{veh}} + s_0$ of a driver-vehicle unit and is given by

$$Q_{\max}^{\text{theo}} = \frac{1}{T} \left(1 - \frac{l_{\text{eff}}}{v_0 T + l_{\text{eff}}} \right). \tag{3}$$

The maximum capacity $Q_{\max}^{\text{free}}$ before traffic breaks down (which is a *dynamic* quantity), however, is typically lower than $Q_{\max}^{\text{theo}}$, since it depends on the traffic stability as well. Therefore, we have analyzed the 'maximum free capacity' resulting from the traffic dynamics as a function of the average time headway T and the percentage of ACC vehicles. Our related simulation runs start with a low upstream inflow and linearly increase the inflow with a rate of $\dot{Q}_{\text{in}} = 800$ vehicles/h². We have checked other progression rates as well, but found a marginal difference only.

For determining the traffic breakdown, we have used 'virtual detectors' located 1 km upstream and downstream of the on-ramp location. In analogy to the real-world double-loop detectors, 'virtual detectors' count the passing vehicles, measure the velocities, and aggregate the data within a time interval of one minute. For each simulation run we have recorded the maximum flow before traffic has broken down (single dots in Fig. 3(a)). Due to the complexity of the simulation and the 1-min data aggregation, $Q_{\max}^{\text{free}}$ varies stochastically. We have, therefore, averaged the data with a linear regression using a Gaussian weight of width $\sigma = 0.2$, and plotted the expectation value and the standard deviation.

Figure 3(a) shows the maximum free capacity as a function of the ACC percentage for the three different parameter sets representing different ACC driving styles. $Q_{\max}^{\text{free}}$ increases approximately linearly with increasing percentage of ACC vehicles. The parameter a mainly increases the traffic stability, which leads to a delayed traffic breakdown and, thus, to higher values of $Q_{\max}^{\mathrm{free}}$. Remarkably, the values are nearly identical with those for heterogenous traffic consisting of driver-vehicle units with Gaussian distributed parameters.

In Fig. 3(b) the most important parameter, the time headway T, is varied for a homogeneous ensemble of 100% ACC vehicles. Obviously, $Q_{\max}^{\mathrm{free}}$ decreases with increasing T. Furthermore, the dynamic quantity $Q_{\max}^{\mathrm{free}}$ remains always lower than the theoretical capacity $Q_{\max}^{\mathrm{theo}}$ given by Eq. (3), which is only reached for perfectly stable traffic. The three parameter sets show the influence of the IDM parameters a and b: The acceleration a has a strong impact on traffic stability, while the stabilizing influence of b is smaller. Finally, as the difference between $Q_{\max}^{\mathrm{theo}}$ and the dynamic maximum free capacity $Q_{\max}^{\mathrm{free}}$ increases for lower values of T, one finds that a smaller T reduces stability as well.

In order the assess the potentials of various driving styles, we have evaluated an approximate relationship as a function of the ACC equipment level α_{ACC}. The relative gain γ in system performance is given by

$$\gamma \approx [0.95(1 - \lambda_T) + 0.07\lambda_a + 0.08(1 - \lambda_b)]\,\alpha_{\mathrm{ACC}}. \qquad (4)$$

Thus, λ_T is the most crucial parameter, while λ_b has hardly any influence. For example, lowering the time headway by $\lambda_T = 0.7$ with $\alpha_{\mathrm{ACC}} = 1$ results in a maximum gain of $\gamma \approx 30\%$.

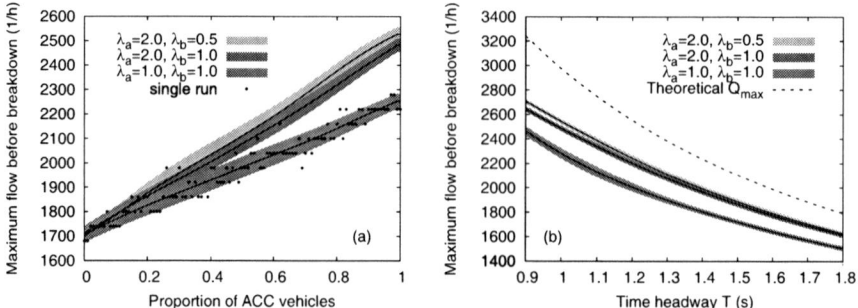

Fig. 3. Maximum free capacity as a function of (a) the percentage of ACC vehicles, and (b) the time headway T for 100% ACC vehicles. We have simulated three different parameter sets for ACC vehicles with $\lambda_T = 2/3$ and varying values of λ_a and λ_b (see main text). Dots indicate results of single simulation runs, while the solid lines correspond to averages over several simulations and the associated bands to plus/minus one standard deviation.

4.3 Dynamic Capacity After a Traffic Breakdown

Let us now investigate the system dynamics after a traffic breakdown. The crucial quantity is the *dynamic capacity*, i.e., the downstream outflow from a traffic congestion Q_{out} [16]. The difference between the free capacity Q_{max}^{free} and Q_{out} is denoted as *capacity drop* with typical values between 5% and 30%. We have used the same simulation setup as in the previous section. After a traffic breakdown was provoked by an increasing inflow, we have averaged over the 1-min flow data of the 'virtual detector' 1 km downstream of the bottleneck. We have identified the congested traffic state by filtering out for velocities smaller than 50 km/h at a cross-section 1 km upstream of the bottleneck. Again, we have averaged over multiple simulation runs by applying a Gaussian-weighted linear regression.

Figure 4(a) shows the dynamic capacity for a variable percentage of ACC vehicles for the three different parameter sets specified before. Interestingly, the capacity increase is not linear as in Fig. 3(a). Above approximately 50% ACC vehicles, the dynamic capacity increases faster than for lower percentages. We explain this behavior with an 'obstruction effect': the faster accelerating ACC vehicles are hindered by the slower accelerating drivers. In fact, the slowest vehicle type determines the dynamic capacity, which could be called a 'weakest link effect'. In conclusion, distributed model parameters have a quantitative effect on the outflow from congested traffic (it is lower than for homogeneous traffic with averaged parameters), while such an effect is not observed for the free-flow capacity!

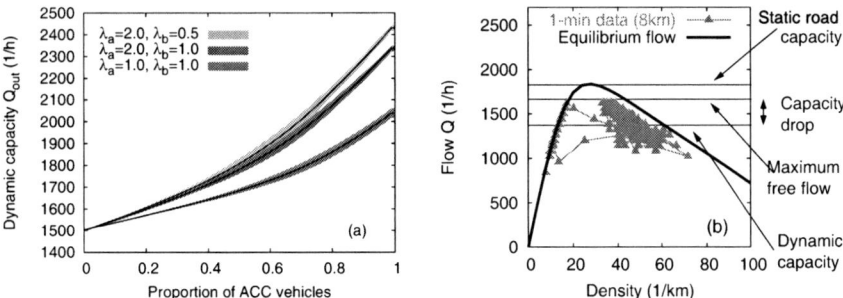

Fig. 4. (a) Dynamic capacity as a function of the percentage of ACC vehicles. The curves represent three different parameter sets corresponding to different ACC driving strategies. The results from multiple simulation runs are averaged using a linear regression with a Gaussian weight of width $\sigma = 0.2$. (b) Flow-density data for the traffic breakdown determined from a 'virtual' detector 2 km upstream of the bottleneck without ACC vehicles. The equilibrium flow-density curve of identical vehicles corresponds to the parameter set given in Table 1.

5 Discussion

Adaptive cruise control (ACC) systems are already available on the market. The next generations of ACC systems will extend their range of applicability to all speeds, and it is assumed that their spreading will grow in the future. In this contribution, by means of microscopic traffic simulations we have investigated the impact that an automated longitudinal driving control of ACC systems based on the intelligent driver model (IDM) is expected to have on the traffic dynamics.

ACC systems are closely related to car-following models as their reaction is restricted to a leading vehicle. Moreover, we have explained why such a car-following approach also captures the main aspects of longitudinal driver behavior so well. We, therefore, expect that both ACC systems and human driver behavior will mix consistently in future traffic flows although the driving operation is fundamentally different.

The equipment level of ACC systems provides an interesting option to enhance the traffic performance by automated driving strategies. In order to analyze the potentials, we have studied ACC driving styles, which are explicitly designed to increase the capacity and stability of traffic flows. We have varied the percentage of ACC vehicles and found that already a small proportion of ACC vehicles, which implies a marginally increased free and dynamic capacity, leads to a drastic reduction of traffic congestion. Furthermore, we have shown that, capacity and stability do have similar importance for the traffic dynamics.

We have assumed that the ACC systems have a more jam-avoiding driving style than the human drivers. One might additionally take into account inefficient human behavior when traffic gets denser and the time headway increases with increasing local velocity variance [12, 17]. In this case, a constant time headway policy for automated driving is expected to improve the system performance even more.

Up to now, ACC systems are only optimized for the user's driving comfort and safety. In fact, present ACC systems may have a negative influence on the system performance when their percentage becomes large. The design of ACC strategies, which also consider their impact on traffic dynamics, will be crucial for the next ACC generations.

Furthermore, we propose to implement an 'intelligent' ACC strategy that adapts the ACC driving style *dynamically* to the overall traffic situation. For example, in dense, but not yet congested traffic, a jam-avoiding parameter set could help to delay or suppress traffic breakdowns as shown in our simulations, while in free traffic a parameter set mimicking natural driver behavior may be applied instead. The respective 'traffic state' could be autonomously detected by the vehicles using the history of their sensor data in combination with digital maps. Moreover, inter-vehicle communication could contribute information about the traffic situation in the neigborhood, e.g., by detecting the downstream front of a traffic jam [18].

Acknowledgements

The authors would like to thank Hans-Jürgen Stauss, and Klaus Rieck for the excellent collaboration and the Volkswagen AG for partial financial support within the BMBF project INVENT.

References

1. "European Commission (Energy & Transport), White Paper European transport policy for 2010: time to decide,", COM (2001) 370 final.
2. M. Minderhoud, *Supported Driving: Impacts on Motorway Traffic Flow* (Delft University Press, Delft, 1999).
3. M. Treiber and D. Helbing, "Microsimulations of freeway traffic including control measures", Automatisierungstechnik **49**, 478–484 (2001).
4. L. Davis, "Effect of adaptive cruise control systems on traffic flow", Phys. Rev. E **69**, 066110 (2004).
5. G. Marsden, M. McDonald, and M. Brackstone, "Towards an understanding of adaptive cruise control", Transportation Research C **9**, 33–51 (2001).
6. B. S. Kerner, *The Physics of Traffic* (Springer, Heidelberg, 2004).
7. D. Helbing, "Traffic and related self-driven many-particle systems", Review of Modern Physics **73**, 1067–1141 (2001).
8. K. Nagel, P. Wagner, and R. Woesler, "Still flowing: old and new approaches for traffic flow modeling", Operations Research **51**, 681–710 (2003).
9. M. Treiber, A. Hennecke, and D. Helbing, "Derivation, properties, and simulation of a gas-kinetic-based, non-local traffic model", Phys. Rev. E **59**, 239–253 (1999).
10. W. Knospe, L. Santen, A. Schadschneider, and M. Schreckenberg, "Single-vehicle data of highway traffic: Microscopic description of traffic phases", Phys. Rev. E **65**, 056133 (2002).
11. B. Tilch and D. Helbing, "Evaluation of single vehicle data in dependence of the vehicle-type, lane, and site", in *Traffic and Granular Flow '99*, D. Helbing, H. Herrmann, M. Schreckenberg, and D. Wolf, eds., (Springer, Berlin, 2000), pp. 333–338.
12. M. Treiber, A. Kesting, and D. Helbing, "Understanding widely scattered traffic flows, the capacity drop, and platoons as effects of variance-driven time gaps", Phys. Rev. E 74, 016123 (2006)
13. M. Green, "'How long does it take to stop?' Methodological analysis of driver perception-brake Times", Transportation Human Factors **2**, 195–216 (2000).
14. M. Treiber, A. Kesting, and D. Helbing, "Delays, inaccuracies and anticipation in microscopic traffic models", Physica A **359**, 729–746 (2006).
15. M. Treiber, A. Hennecke, and D. Helbing, "Congested traffic states in empirical observations and microscopic simulations", Physical Review E **62**, 1805–1824 (2000).
16. C. Daganzo, M. Cassidy, and R. Bertini, "Possible explanations of phase transitions in highway traffic", Transportation Research B **33**, 365–379 (1999).
17. M. Treiber, A. Kesting, and D. Helbing, "Variance-driven traffic dynamics and statistical aspects of single-vehicle data", in this volume.
18. M. Schönhof, A. Kesting, M. Treiber, and D. Helbing, "Inter-Vehicle Communication on highways: Statistical properties of information propagation", in this volume.

Inter-Vehicle Communication on Freeways: Statistical Properties of Information Propagation in Ad-Hoc Networks

Martin Schönhof[1], Arne Kesting[1], Martin Treiber[1], and Dirk Helbing[1,2]

[1] Technische Universität Dresden, Institute for Transport & Economics,
Andreas-Schubert-Straße 23, D-01062 Dresden, Germany
[2] Collegium Budapest – Institute for Advanced Study, Szentháromság u. 2,
H-1014 Budapest, Hungary

Summary. The function of adaptive cruise control (ACC) systems can be enhanced by information flows between equipped cars, i.e., by upstream transmission of messages about the current traffic situation. Message transport within one driving direction is obviously rather restricted for small percentages of equipped cars due to the limited broadcast range. Thus, we consider vehicles in the opposite driving direction as possible relay stations. Analytical results based on a Poisson approximation, which are in accordance with empirical traffic data, show the efficiency and velocity of information propagation based on transversal message hopping. The obtained propability distributions of the transmission times are compared with numerical results of microscopic traffic simulations. By simulating the formation of a typical traffic jam, we show how information about distant bottlenecks and jam fronts reaches upstream equipped cars, which then can optimize their driving strategies.

1 Introduction

Inter-vehicle communication (IVC) is widely regarded as a powerful concept for the transmission of traffic-related information. In contrast to the common communication channels, which operate with a centralized broadcast concept via radio or mobile-phone services, IVC is designed as a local service based on ad-hoc networks. Vehicles equipped with a short-range radio device, broadcast messages which are received by all other equipped cars within the limited broadcast range. The message transmission is not controlled by a central station, and, therefore, no further infrastucture is needed. Supported by the technological progress and the falling prices for corresponding hardware, the market for short-range communication devices is growing, and wireless local-area networks (WLAN) spread more and more.

In this contribution we will focus on the propagation of information via IVC equipped vehicles. Since IVC will start with a small equipment level, it is crucial to investigate the functionality and the statistical properties of the message hopping processes. Fast and reliable information spreading is a necessary precondition for a successful implementation of this technology. The traffic information of interest can be generated by the IVC equipped cars themselves, if each car reports about the traffic conditions it currently faces. This results in a completely decentralized, autonomous traffic surveillance and information system. While the information must be transported over distances of about 1 km in upstream direction, the broadcast range is only of the order of 250 m. We, therefore, also consider equipped vehicles in the opposite driving direction as transmitter cars.

Apart from the single drivers the whole traffic system may benefit from IVC as well [2]. *Adaptive cruise control* (ACC) automates the braking and accelerating of a car. While the objectives of the currently available ACC systems are to enhance the comfort and safety of driving, there has been no focus on their effect on the capacity of the freeway, except for the general positive effects of avoiding accidents. Transmission of traffic information via IVC could help ACC systems to recognize the traffic situation faster and more reliably. Moreover it could help ACC systems to increase road capacity by allowing it to reduce the time headway just when it is about to leave the downstream front of a traffic jam.

Our contribution is organized as follows: After a discussion of message transport strategies for freeways and their statistics (Section 2), we will present in Section 3 a simulation scenario, where information about a traffic jam is transported upstream by cars of the other driving direction. Afterwards, we will summarize our contribution and give a short outlook.

2 Statistics of Message Transport on Freeways

2.1 Message Transport Strategies

In the context of freeway traffic, messages normally have to travel upstream in order to be valuable for their receivers. In general, there are two strategies, how a message can be transported upstream via IVC (or mixtures of both): Either the message hops from an IVC car to a subsequent IVC car within the same driving direction – which will be called *longitudinal hopping*, or the message hops to an IVC car of the other driving direction which takes the message upstream and delivers it back to cars of the original driving direction. The second mechanism, where vehicles of the opposite direction act as relay stations, will be referred to as *transversal hopping* (cf. Fig. 1).

2.2 Spatial Distribution of Equipped Vehicles

If the market penetration is low, the encounter of an IVC equipped vehicle with another one is seldom. In good approximation, the positions of the IVC

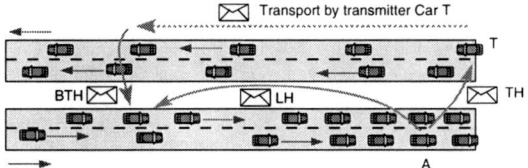

Fig. 1. Transport of a traffic information message on a freeway: When car "A" enters a traffic jam, it broadcasts a related message. This is received by a subsequent car via longitudinal hopping ("LH") and by an equipped transmitter car "T" of the other driving direction via transversal hopping ("TH"). The message can travel with the transmitter "T" upstream, until it is delivered back to the original driving direction by back transversal hopping ("BTH"). In the main text, we will discuss which message passing mechanism is more efficient.

cars therefore can be assumed independent of each other, even for high traffic densities. With the additional assumption of a constant overall traffic density ρ on all lanes of the analyzed driving direction, and for a given percentage (equipment level, market penetration) α of IVC vehicles, it follows that the number of IVC vehicles on a given road section is Poisson distributed. Thus, the headways Δs between consecutive equipped vehicles are distributed exponentially:

$$f_{\Delta s}(x) = \lambda e^{-\lambda x} \quad \text{with} \quad \lambda = \alpha \rho. \tag{1}$$

This assumption is very well supported by empirical data, cf. Fig. 2. Evaluating the data of single cars passing a freeway cross section, it is possible to obtain the distribution of distances between IVC equipped vehicles for scenarios of different equipment levels. Even for a single lane, this distance is exponentially distributed for small equipment levels. However, above a level of 20%, the form of the distribution gets more and more similar to the Erlang/Pearson III distribution of headways [5].

2.3 Longitudinal Message Hopping

Longitudinal hopping is only possible, if there is an upstream receiver in the broadcast range of the sending car. For message transport over a certain distance, there has to be a closed chain of IVC cars: Every single distance between two subsequent IVC equipped cars must be smaller than the broadcast range for a certain time span. This is very unlikely for a low equipment level. The following example presents a more detailed analysis.

For a given maximum broadcast range $r_{\max}$, the probability of finding an upstream receiver for longitudinal hopping is given by

$$P(\Delta s < r_{\max}) = \int_0^{r_{\max}} f_{\Delta s}(x)dx = 1 - e^{-\lambda r_{\max}}. \tag{2}$$

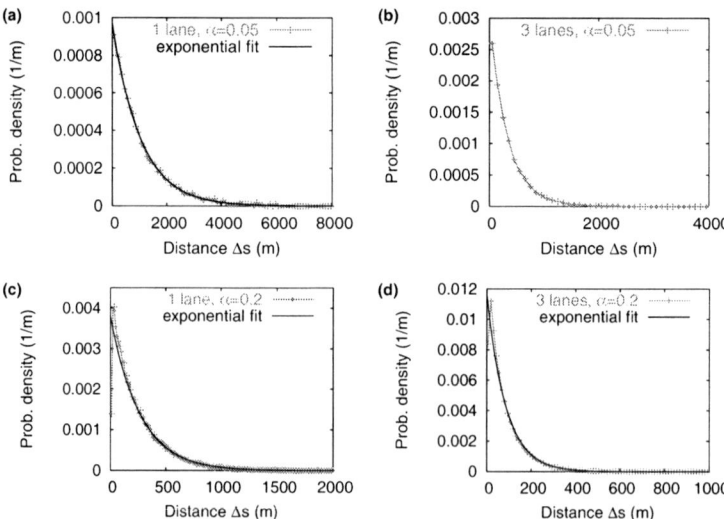

Fig. 2. Probability density of distances between IVC equipped vehicles based on single vehicle data for the freeway I-880. Each car entering the upstream boundary of the investigated freeway stretch have, with probability α randomly and independently, been chosen to be an 'equipped' car. The resulting fraction of α chosen cars corresponds to an IVC market penetration of α. Using the time headways Δt between consecutive equipped vehicles, we have obtained the distance Δs for every equipped car i via $\Delta s_i = \Delta t_i V_{i-1}$, where the equipped car $i-1$ is the predecessor of car i, and V_{i-1} its velocity. The single vehicle data were recorded in 1993 at cross section 6 (29300 feet distance from Mariana) of freeway I-880, Hayward, California, in direction north [13]. Data of congested or light traffic (velocity < 60 km or flow < 1000/h/lane) have been omitted. Only the right lane has been taken into account in **(a)** and **(c)**. In **(b)** and **(d)**, the three rightmost lanes from altogether five lanes have been considered.

Considering an overall density of $\rho = 30$ veh/km on two lanes, $\alpha = 0.05$, and $r_{max} = 250$ m, we obtain a probability of 31% for a message hop. If we require that the information should be available at least $r_u = 1000$ m upstream of a detected traffic event, the information has to hop at least four times. Because of the statistical independence of the hopping processes, the probability for n successful hops is given by

$$P_n = \left(1 - e^{-\lambda r_{max}}\right)^n . \tag{3}$$

That is, the probability for four successful hops is only $(0.31)^4 \approx 1\%$. Note that this is an upper limit for the transmission probability, as not every hop will bridge exactly 250 m. Thus, normally more than 4 hops will be necessay, which further reduces the transmission probability.

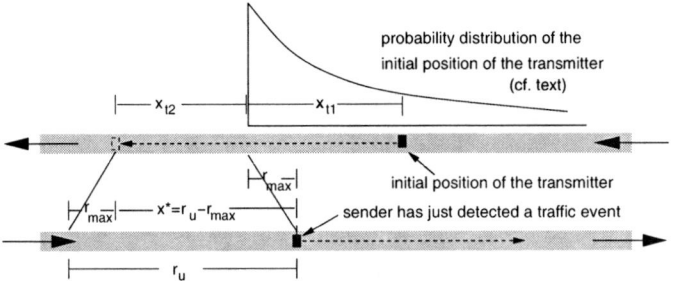

Fig. 3. Initial spatial configuration and labelling of the distances: The sender has just detected an event and broadcasts a corresponding message. The first encountered equipped car of the other direction, the transmitter, may be downstream or upstream (left) of the sender, but in the latter case within the broadcast range r_{max} (left cutoff of probability distribution). If the transmitter is out of the broadcast range (for large x_{t1}), the message will not be received immediately. The time, when the message is picked up by the transmitter does not directly affect the time t which is needed to deliver the message the distance r_u upstream of the initial sender position. However, both times depend, of course, on x_{t1}.

2.4 Transversal Message Hopping

With longitudinal hopping, a message either reaches its "destination" at once or never. Via transversal hopping, a message reaches *always* the destination point $r_u = 1000$ m upstream of the position where it has been generated. The message is available at this point as soon as the first encountered equipped car of the other direction, the *transmitter*, has moved a distance $x^* = r_u - r_{max}$ upstream from the place of message generation. The remaining distance can be bridged via wireless communication (cf. Fig. 3). The time t, when this is completed, depends on the initial position of the transmitter at the time the message is generated and on its velocity v_{tr}. The initial distance of the transmitter from the "retransmission point" x^* consists of two parts, x_{t1} and x_{t2} (cf. Fig. 3). Thus, we obtain

$$t = \frac{x_{t1} + x_{t2}}{v_{tr}}. \tag{4}$$

x_{t2} is given by

$$x_{t2} = r_u - 2r_{max} \tag{5}$$

(cf. Fig. 3), while the stochastic quantity x_{t1} is determined by the gap distribution between two IVC cars, i.e., its probability density is given by

$$f_{x_{t1}}(x) = f_{\Delta s}(x) = \lambda e^{-\lambda x} \Theta(x). \tag{6}$$

Here, the Theta-function $\Theta(x)$ is 1 for positive arguments x, and zero, otherwise.

Let us now calculate the cumulative distribution $P(t < \tau)$ of arrival times t. According to Eq. (4), the message arrives at a time $t < \tau$, if $x_{t1} < \tau v_{tr} - x_{t2}$. Therefore, the probability that the information is succesfully transmitted until time τ can be calculated as

$$P(t < \tau) = P(x_{t1} < \tau v_{tr} - x_{t2}) \tag{7}$$

$$= \int_0^{\tau v_{tr} - x_{t2}} f_{x_{t1}}(x)\, dx \tag{8}$$

$$= \Theta\left(\tau - \frac{r_u - 2r_{max}}{v_{tr}}\right)\left(1 - e^{-\lambda(2r_{max} + v_{tr}\tau - r_u)}\right) \tag{9}$$

Because $f_{x_{t1}}(x) = 0$ for $x < 0$ (see Eq. (6)), the probability distribution is zero if, in the case of a small value of τ, the upper bound of the integral in Eq. (8) becomes negative. This results in the Theta function $\Theta(\tau v_{tr} - x_{t2}) = \Theta\left(\tau - \frac{r_u - 2r_{max}}{v_{tr}}\right)$ in Eq. (9). Since the probability of a transmission before the time $\frac{r_u - 2r_{max}}{v_{tr}} = \frac{x_{t2}}{v_{tr}}$ is zero, this is the minimal possible transmission time. It occurs, when the transmitter only needs to pass the distance x_{t2}, i.e., if it is initially as far as possible upstream (corresponding to maximum of the distribution in Fig. 3).

In Figure 4, the information transport within the same driving direction is compared to the information transport via a transmitter of the opposite driving direction. In the first case, the message is instantaneously available a certain distance r_u upstream of a recognized traffic event (if we neglect the broadcasting time). However, because of the low equipment rate, the transmission succeeds only with a very small probability that does not change in time. Either the information reaches the destination more or less at once, or never. In the case of transversal hopping, the message needs at least 18 seconds, but after 36 seconds, the message is available with a probability of 50%. An 36-seconds old information 1000 m ahead of the event is still very valuable: For example, in 36 seconds a possible disturbance of the traffic flow may travel (with a characteristic speed of ≈ 15 km/h) 150 m upstream. Hence, for the receiver of this information, there are 850 m left to react to the traffic event (e.g. stop-and-go wave).

2.5 Microscopic Simulation of Inter-Vehicle Communication

In order to test these analytical results, we have carried out a multi-lane traffic simulation of a 10 km freeway stretch with two independent driving directions and altogether four lanes. We have used the *intelligent driver model* (IDM) [12] complemented by a lane changing algorithm [11] (see Fig. 5 below). The parameters have been selected as in Ref. [2], whereas the desired velocities have been chosen Gaussian distributed with an rms value of 18 km/h around $v_0 = 120$ km/h. We have used open boundary conditions with a constant inflow at the upstream boundary of $Q = 1240/h/lane$.

The microscopic simulation approach allows for a detailed modeling of the message broadcast and receipt mechanisms of IVC equipped vehicles (colored vehicles in Fig. 5). To obtain the statistics of message propagation, the equipped vehicles have generated a "dummy" message while crossing the position $x = 5$ km. In Fig. 6, the results of the simulation are compared to the analytical results based on the Poisson approximation (cf. Sec. 2.2). The percentage of vehicles equipped with the IVC device has been varied and the traffic density measured by 'virtual' detectors as in Ref. [2]. The results show a very good agreement with our analytical calculations (Eq. 9).

3 Application: Upstream Transport of Traffic-Related Information Via Transversal Hopping

Let us now demonstrate the message propagation mechanism with a microscopic traffic simulation. We have simulated the two driving directions of an altogether four-lane freeway. In one driving direction, we have triggered a wide moving cluster (also called a "Moving Localized Cluster" [3, 8]), while traffic was freely flowing in the other driving direction (see Fig. 5). Two types of messages have been generated: (i) If the velocity of a vehicle equipped with

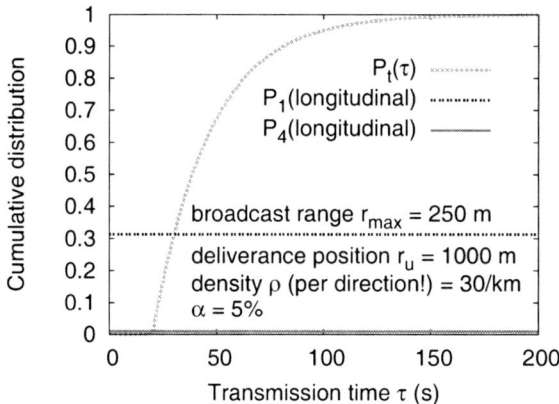

Fig. 4. Probability distribution of the time interval between the generation of a message and its availability 1 km upstream of the event for the two broadcast strategies. If only cars of the same driving direction are used for message transmission, at least 4 successful "hops" are necessary. The transmission probability is, therefore, $P_4 = P_1^4 = 0.01$ or less (cf. text). When transmitter cars of the opposite driving direction are used, the message needs at least 18 seconds, but after 36 seconds, the message is available with a probablilty of about 50%. The velocity of the transmitters has been assumed to be $v_{tr} = 100$ km/h. The minimal time for the message transfer is $\frac{r_u - 2r_{max}}{v_{tr}} = 18$ s.

traffic
jam

IVC
equipped
vehicles

Fig. 5. Screenshot of the traffic scenario discussed in Sec. 3. The microscopic simulation approach allows one to combine traffic dynamics with the microscopic mechanisms of broadcasting and receiving messages via inter-vehicle communication (IVC). The colored cars are equipped with the functionality of generating, sending and receiving information. In the driving direction towards the reader, a stop-and-go wave propagates through the system. The equipped vehicles in the opposite driving direction are used as transmitter cars enabling a "transversal" message hopping. This process allows for a fast information propagation in upstream direction.

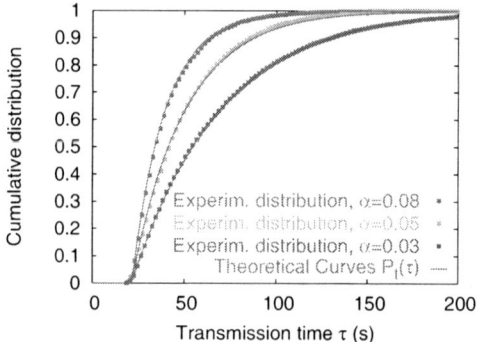

Fig. 6. Message transport via transmitter cars in the opposite driving direction: Comparison of Eq. (9) (solid lines) with the simulated distribution of the time intervals τ until a message is available 1000 m upstream of a traffic event (symbols). The assumed IVC parameters were the broadcast range $r_{max} = 250$ m and the minimal delivery range $r_u = 1000$ m. Applying the vehicle parameters in Ref. [2] and choosing an inflow of $Q = 1240$ veh/h/lane, we have a transmitter velocity of $v_{tr} = 85$ km/h and an overall density of $\rho = 29$ veh/km in each direction. The simulations have been carried out with equipment rates of $\alpha = 3\%$, $\alpha = 5\%$, and $\alpha = 8\%$.

an IVC device dropped below 30 km/h, the car started to broadcast the message "start of traffic jam" with the time and position of its detection. (ii) If the velocity exceeded the velocity 30 km/h, the message "end of traffic jam" was being broadcasted.

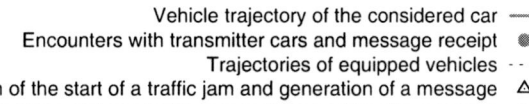

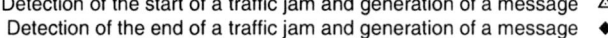

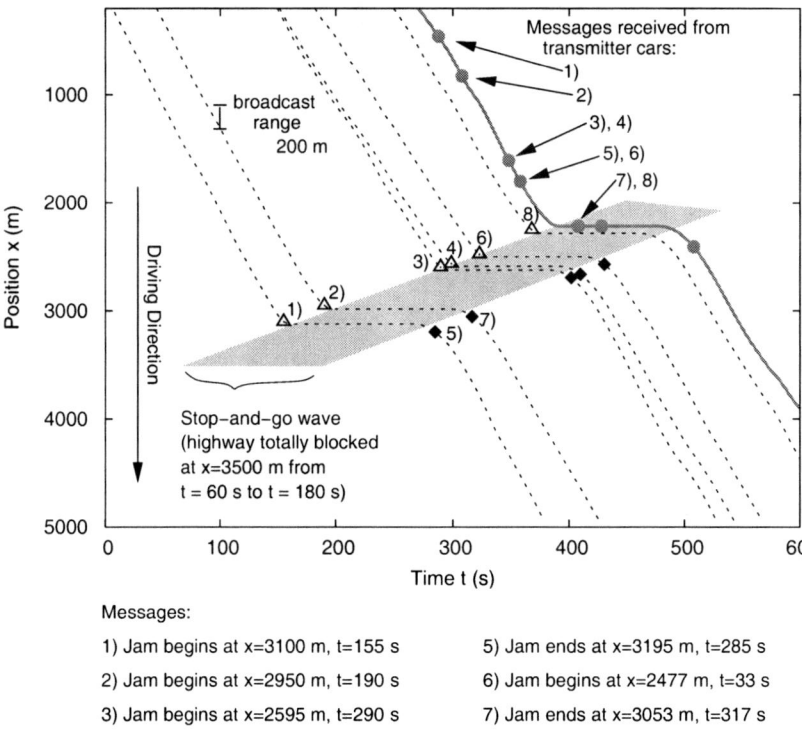

Messages:

1) Jam begins at x=3100 m, t=155 s	5) Jam ends at x=3195 m, t=285 s
2) Jam begins at x=2950 m, t=190 s	6) Jam begins at x=2477 m, t=33 s
3) Jam begins at x=2595 m, t=290 s	7) Jam ends at x=3053 m, t=317 s
4) Jam begins at x=2563 m, t=299 s	8) Jam begins at x=2250 m, t=368 s

Fig. 7. Spatiotemporal diagram of a traffic simulation, for which the trajectories of vehicles equipped with inter-vehicle communication (IVC) devices are displayed by dashed lines. The equipment level is 3%. While the cars encounter a propagating stop-and-go wave, they start to broadcast messages about the begin and the position of the stop-wave and the following start-wave as labeled by numbers in the diagram. Since the broadcast range of 200 m does not allow for a reliable message propagation only in the driving direction (see scale in the diagram), the messages are transported by equipped (transmitter) cars of the opposite driving direction (trajectories not shown). Finally, the receipt of the propagating messages is marked for a specific vehicle (solid trajectory). This considered car gets the information about the position of the traffic jam, and, additionally, the expected travel time, for the first time 2 km upstream. The reliability of the information increases by the receipt of additional messages, which confirm and update the reconstruction of the expected traffic situation for the individual driver.

The spatiotemporal traffic dynamics and the processes of sending and receiving messages are shown in Fig. 7. Due to the low equipment rate of $\alpha = 3\%$, the equipped vehicles have an average distance to each other that exceeds the broadcast range of the IVC device. An upstream message propagation only within one driving direction would, therefore, lead to a fast breakdown of the information chain (see Fig. 7) as stated in Sec. 2.3. Thus, we have used IVC-equipped vehicles as transmitters in the other driving direction. Fig. 7 numbers the generated messages and shows their delivery to a specific vehicle. Remarkably, the considered vehicle gets the first information about the traffic congestion already 2 km before encountering the stop-wave. The information is confirmed and updated by subsequent messages provided by other vehicles. The up-to-date information about the expected traffic situation could be used to warn drivers or to set-up a strategically operating adaptive cruise control (ACC) system [2].

4 Summary and Outlook

The market penetration of adaptive cruise control (ACC) is steadily growing. By means of inter-vehicle communication (IVC), the performance of these systems can be increased by accurate and up-to-date messages about the traffic situation ahead.

For receiving and transmitting up-to-date information on a short timescale, it is promising to use an entirely decentralized system like an ad-hoc-network of vehicles equipped with inter-vehicle communication technology – especially if these equipped cars on the road also gather the traffic information that is transmitted.

A problem of such a short-range communication system is that it may not work properly for a low equipment rate. In this contribution, we have, therefore, presented a communication strategy for inter-vehicle communication that operates well for low equipment rates by using cars on the opposite driving direction as relay stations. For example, even for an equipment rate of 5% only, a traffic-information message will be passed 1 km upstream with a probability of 50% within 36 seconds. The simulations of Fig. 7 showed that even lower equipment rates enable effective communication in realistic situations. A further step is to develop and implement traffic-state dependent strategies for ACC [2] that react to IVC information in a situation-specific way.

Acknowledgements

The authors would like to thank Hans-Jürgen Stauss, and Klaus Rieck for the excellent collaboration and the Volkswagen AG for partial financial support within the BMBF project INVENT.

References

1. L. Davis: Phys. Rev. E **69**, 066110 (2004)
2. A. Kesting, M. Treiber, M. Schönhof, F. Kranke, D. Helbing: ' 'Jam-avoiding' adaptive cruise control (ACC) and its impact on traffic dynamics.', these proceedings
3. B.S. Kerner, H. Rehborn: Phys. Rev. E **53**, 1297-1300 (1996)
4. D. Helbing: Rev. Mod. Phys. **73**, 1067–1141 (2001)
5. M. Krbalek, D. Helbing: Physica A **333**, 370-378 (2004)
6. L. Neubert, L. Santen, A. Schadschneider, M. Schreckenberg: Phys. Rev. E **60**, 6480-6490 (1999)
7. B. Rao, P. Varaiya: Transp. Res. Rec. **1408**, 35–43 (1993)
8. M. Schönhof, D. Helbing: 'Empirical features of congested traffic states and their implications for traffic modeling', Transp. Sci., submitted
9. B. Tilch, D. Helbing: 'Evaluation of single vehicle data in dependence of the vehicle-type, lane, and site'. In: *Traffic and Granular Flow '99*, ed. by D. Helbing, H. Herrmann, M. Schreckenberg, D. Wolf (Springer, Berlin 2000) pp. 333–338
10. M. Treiber, D. Helbing: Automatisierungstechnik **49**, 478–484 (2001)
11. M. Treiber, D. Helbing: 'Realistische Mikrosimulation von Strassenverkehr mit einem einfachen Modell'. In: *16. Symposium Simulationstechnik at Rostock, Germany, September 2002*, ed. by D. Tavangarian, R. Grützner, pp. 514–520 (2002)
12. M. Treiber, A. Hennecke, D. Helbing: Phys. Rev. E **62**, 1805–1824 (2000)
13. `http://www.clearingstelle-verkehr.de`, Deutsches Zentrum für Luft- und Raumfahrt e.V., Berlin

Effects of Advanced Traveller Information Systems on Agents' Behaviour in a Traffic Scenario

Thorsten Chmura[1,2], Johannes Kaiser[1], Thomas Pitz[1,2], Mark Blumberg[1], and Marco Brück[1]

[1] Laboratory for Experimental Economics, University of Bonn, 53113 Bonn, Germany
[2] Shanghai Jiao Tong University, Antai School of Managment, Shanghai 200052, People's Republic of China

Summary. A genetic algorithm approach is used to study the behaviour of agents in a simulation of a daily route choice. There are two roads to choose and we show that there is a welfare enhancing effect of an Advanced Traveller Information System (ATIS) in comparison to the standard case without an ATIS. In the first case it is remarkable that not all agents follow the recommendation of the ATIS and the equilibrium distribution is only approximately attained.

1 Introduction

With an increasing amount of traffic world wide (NSTC (1999)), congestion is a daily routine for many travellers and commuters and it is unlikely that this situation will change for the better for many traffic systems in the near future (BVBW (2001)). To solve the problem it becomes more important to supply Intelligent Transportation Systems (ITS), which lead to a better usage of available traffic networks. Kwan and Golledge (1998) emphasise that Advanced Traveller Information Systems (ATIS) are a key factor for the implementation of successful ITS.

This paper examines how individual drivers react on information from an ATIS, in a situation of daily home-to-work-route choice. Mahmassani et al. (1997) distinguish three possible reactions of work commuters on peak period congestion: First, variations of the departure time, second, changes in frequency, purpose, and duration of intervening stops and third, selecting a different route. These results are based on an empirical study in Austin, Texas. Another study presented by Stern et al. (1998) investigates commuters' behaviour in the Netherlands. In agreement with Mahmassani's result, the most common reaction is a variation in the working time, which results in a variation of trip starting times, and a variation of route to/from work. Other observable reactions are increasing home working activities, increased usage of public transportation, job changes and changes of home locations.

In Section 2 we describe a simple model of route choice under influence by an ATIS, in Section 3 a genetic algorithm to simulate agents' behaviour is described and in Section 4 we present results of some simulations of agents' behaviour within our model.

2 Action Models

To study the effects of an ATIS, we use two models to capture the effects of the genetic algorithm. First, we set up a basic model of agent behaviour without an ATIS and second we extend this model by an ATIS.

2.1 Basic Model

The basic model consists of agents who have two route options to drive to work. So they have to decide between two roads, which differ in their transport capacity. In the following they will be called S_big and S_small. There is no outside option, neither short term orientated as using public transport systems or not going to work, nor long term orientated as moving nearer to working place or getting another job. The resulting action tree is shown in figure 1.

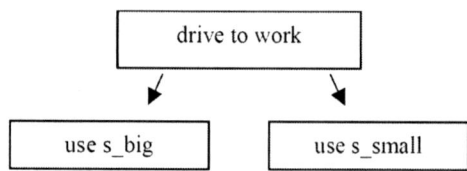

Fig. 1. Basic model action tree.

The agents decide in a random order, the numbers of agents driving on S_big respective S_small are call n_big and n_small. The resulting agent's utility is equal to the use under perfect circumstances - no other drivers on the road - minus the time lost by other drivers jamming the road. S_big enables a greater utility under perfect circumstances and is less affected by higher utilisation. Altogether this model is very similar to the experiment described by Selten et al. (2005).

2.2 Basic Model Extended by ATIS

Now we introduce an ATIS in form of a traffic radio in our model. After a specific number of agents have decided on which route to take, a route recommendation is given. This recommendation consists just of the information which route would actually yield to a higher utilisation. As the interval of recommendations gets smaller, the quality of the ATIS should improve. The decision tree is then modified to give reaction options to the recommendation. First the agent has to decide whether he believes in the recommendation or not - in the latter case he will decide in the same way as in the basic model. If he believes in the recommendation, on the one hand he could directly follow this recommendation or on the other hand respond in the contrarian way. Figure 2 shows the corresponding action tree.

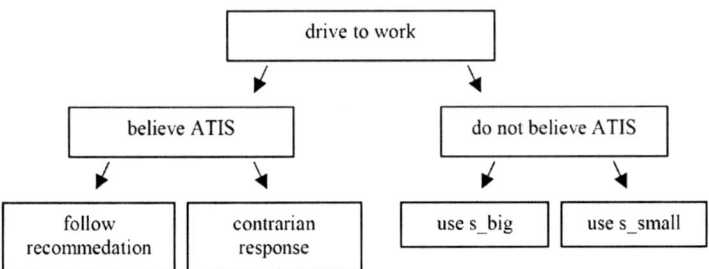

Fig. 2. Extended model action tree.

3 The Genetic Algorithm

To get dynamics into the simulation, a genetic algorithm is used to change the agents' action attributes. Each attribute has a binary code which is called 'gene' in the following. A set of genes makes up a gene pool. Following Pitz (2005) a genetic action tree $G(T)$ is defined as:

1. T is an action tree.
2. For each agent, action type and attribute $C(i, a)$ is the gene pool (a set of bits).
3. For each action type, there exists a decision degree, which is a probability for choosing that action type. There might be a condition Δ if the action type (a special node in the action tree) is disabled. For instance, the agent might be willing to buy something, but is out of money.
4. For each gene $c(i, a)$ out of $C(i, a)$ there exists a fitness function $f(c(i, a))$, which is determined contingent of the outcome of the action.
5. $d(c(i, a))$ is the semantic of $c(i, a)$, which describes the contingency of the action in respect to the gene in the simulation.

If the agent is not obliged to a certain action, it will follow the action tree from the root to the leafs. Every attribute of each action type with its coded value $c(i, a)$ will be filled using a uniform distribution. A violation of the condition

Δ_{H1} or Δ_{H2} will cancel the sub-nodes $H1$ or $H2$ respectively[3]. This is the way we have chosen to force the agent to perform a certain action, if necessary. In the case that there are two choices left, the gene pool is used to determine the agent's action. A $c(i,a)$ is taken out of the gene pool $C(i,a)$ (again by a uniformly distributed probability) and if $d(c(i,dh1)) < dh1$, then the left node is chosen and vice versa.

Reaching the root of the decision tree, every action attribute has its coded value $c(i,a)$ and each of them is evaluated by the fitness function $f(c(i,a))$ dependent on the outcome when the agent has run the action tree. Since every gene now has a certain value, the action is specified and can be carried out. Each action changes the environment and this will be the evaluation basis for the fitness f of the agent's action attributes.

Three principles are used by the genetic algorithm: mutation, selection and cross over. A mutation is created by a mutated copy of a gene which is changed at 'selected' random places. With a probability p (which is in our case anti-proportionally related to its fitness) the agent's gene will be replaced by this mutation. The better its fitness, the lower is the probability of changing the gene. When all the genes of each agent are assembled to one large gene pool, 5% of this gene pool is randomly mixed by a cross over, that is, one set of genes is taken to another location and the former set is overwritten by the replaced set.

We used a general framework developed in a seminar at the Laboratory for Experimental Economics, University of Bonn, intended for use with genetic action trees. Instead of setting the fitness immediately after each agent's choice, the fitness of the decision genes is calculated after the activation of all agents for each genetic cycle. In our model the fitness of every agent's decision gene is the sum of utilities of this agent over a genetic cycle.

4 Results

We now combine our agents' action models with the genetic algorithm and make some behaviour simulations. We begin with some simulation calibrations, then we go on to present some results for the basic model and finally we introduce the ATIS in two different quality levels. All results we present in this section come from simulation runs of the genetic algorithm. If something is said to be significant, it is in the context of a t-test at the 5%-level.

4.1 Calibration

Each run of the simulation covers 200 rounds with 100 agents. Every 10 rounds the genetic algorithm is called and the mutation is done - the genetic cycle. The other parameters of the framework are set to their default values, too.

---

[3] $H1$ denotes the "left" node, $H2$ the "right" node.

Just to mention the basic setup, every decision degree is set to 50 per cent. The implemented utility function for an agent using S_big is:

$$U\_big = 3.50 - 0.02 \cdot n\_big. \tag{1}$$

For the other agents, who are driving on S_small, the utility function is:

$$U\_small = 3.00 - 0.03 \cdot n\_small. \tag{2}$$

For this utility functions an equilibrium and a first best solution could be determined. In equilibrium 70 agents should use S_big and 30 agents should use S_small. In this agents-to-road distribution no agent could increase his own utility by deviating. The equilibrium utility sum of all agents would be 210.00. For the first best solution 67 agents should use S_big and 33 agents should use S_small, so the social welfare optimal maximised utility sum of 211.05 could be reached.

4.2 Basic Model

Beginning with the simplest situation, the decision between S_small and S_big, the agents show an interesting behaviour, see figure 3. The equilibrium number of agents on S_big should be 70. Notice that in this equilibrium the realised utilities of agents using S_big respective S_small are exact equal, but no agent could increase his own utility by deviating.
Several runs of the simulation show an average utilisation of S_big of 64, which is lower than the equilibrium. The resulting mean utility sum is about 207.73, which is significantly lower than the equilibrium utility sum of 210.00, and has a variance of 0.066613. This variance is due to a huge fluctuation on the roads within a run of the simulation, sometimes even less than 50 agents are on S_big, but the following analysis will only consider the averages of several runs.

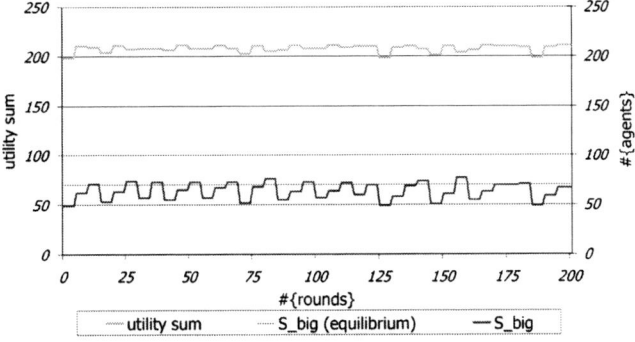

Fig. 3. Simulation results, basic model.

The explanation for this finding lies in the functioning of the genetic algorithm. Even if the equilibrium distribution of agents on the roads is reached, there is always some cross-over-mutation which destabilises the equilibrium in the long term. Additionally, it is unlikely that this equilibrium is reached even in the short term. There is always a mutation of the agents on the relatively full road and relatively low utility, even in the case of being close to the equilibrium on S_big. Either they win or lose. The losers will always run through the genetic algorithm with a hundred percent chance.

Consider the case when there is less than the equilibrium number of 70 agents on S_big. They will enjoy a higher utility and therefore their fitness function will be higher than the ones of the agents on the relatively crowded S_small. A smaller portion of the agents (the number on S_small) will mutate and, despite the fact of some cross-over mutation, will change to users of S_big - not many. As highly likely result, more than 70 agents will use S_big now. In the case of a larger than equilibrium number of users of S_big, they will get the lower utility and therefore will mutate. There is a great number of agents possibly switching back to S_small and so there is again a relatively large group of agents on S_small.

Empirical evidence of such a back switching behaviour is given by a study of Tacken and de Boer (1989, 1991). An improvement in the beltway around Amsterdam created a change in people's driving behaviours and 'traditional' congestion places become less used which results in less congestion. After this happened, people began to switch back to 'traditional' behaviours again. In our simulation similar results are obtained using an exponential utility function instead of our specified linear function, which is why we stick to the simpler linear function.

4.3 Basic Model Extended by ATIS

Now we introduce the traffic radio. Since there is almost no significant difference in results between the quality of the traffic radio and the results, we pick two arbitrary values for the traffic radio update interval: 20 and 1, the latter one is the reference case for a perfect traffic radio. Starting with the former case, the variance of the agents' utility decreases and the mean utility increases significantly, a typical simulation run is presented in figure 4.

The mean utility sum is now about 209.81 with a variance of 0.008558. Compared to the simulations without ATIS the mean utility sum has increased and the variance decreased. This happened because the number of agents on the road is now much closer to the equilibrium distribution in every round. Now, between 40 and 60 per cent will consider the traffic radio in their decision - either by direct or contrarian response and the ratio between direct and contrarian response is a little above 1/2, see figure 5. In the case of a perfect traffic radio where the update interval is 1 the mean utility sum goes on increasing to 210.09 and the variance decreases to 0.001651. However, no serious change in agents' behaviour is observable.

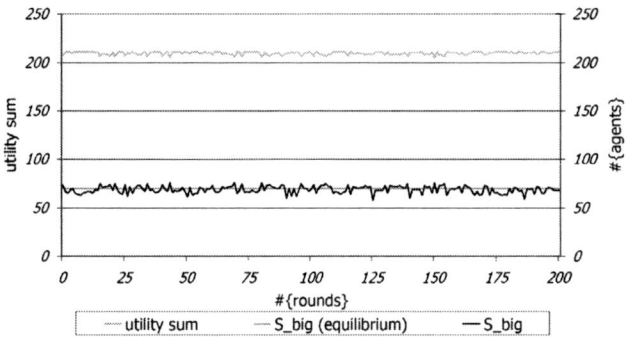

Fig. 4. Route choice, model with ATIS, interval = 20.

Finally, the average utility increases in both cases in comparison to the case without traffic radio, but the agents' utility still varies a lot within a simulation.

So why does not every agent follow the traffic radio directly, because this would lead the agents to the equilibrium? It's again the genetic algorithm in combination with the model's implementation. When the genes force the agent's behaviour to the region of the equilibrium, much of the outcome is determined by chance. For some there is no need to consider the traffic radio if there is a group large enough who does - their concern will create a sort of "public good", they compensate for the agents who ignore the traffic radio and even for the agents who choose the contrarian response, because these agents are outnumbered by the factor 2, as mentioned above. This leads to the quite unstable outcome, for instance, there should not be any contrarian responses in the case of a perfect traffic radio. A simulation with disabled contrarian

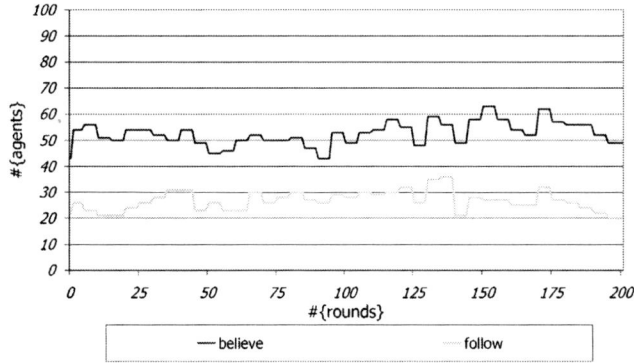

Fig. 5. Reaction to ATIS, model with ATIS, interval = 20.

response node is welfare enhancing for average utility and its variance will be significantly better[4]. A simple[5] OLS regression supports our explanation, see figure 6:

$$AVG\_U_i = \beta_0 + \beta_1 * AVG\_n\_Believe_i + \beta_2 * AVG\_n\_dont\ Follow_i + \epsilon_i \qquad (3)$$

Model 1: OLS estimates using the 39 observations 1-39

Dependent variable: AVG_U

VARIABLE	COEFFICIENT	STD.DEV.	T-STAT	2Prob(t > |T|)
0) const	83,1481	0,664550	125,119	< 0,00001 ***
11) n_Believe	0,0167860	0,00800095	2,098	0,042987 **
10) n_dont_Follow	-0,0405304	0,0127177	-3,187	0,002970 ***

unadjusted R² = 0,22322

F-statistic (2, 36) = 5,17259 (p-value = 0,0106)

Fig. 6. Model 1: OLS estimates using the 39 observations 1-39.

For interpretation: the more agents minded the traffic radio in a simulation run, the higher the average utility, as shown by the positive coefficient of "Believe". The average utility decreases, the higher the number of agents who respond in a contrarian way.

In the end one finds that the genetic algorithm works fine to embrace the equilibrium. But because of the special circumstances mentioned above it still leaves room for some marginal opportunity to improve welfare by an ATIS, here the traffic radio.

5 Conclusions

This paper examines the effect of ATIS on commuter behaviour. Our simulation shows that introduction of ATIS in form of traffic radio leads to significantly higher utility with lower variance. It is interesting that our simulations have shown that not every agent will follow the recommendation directly to achieve this effect. Another - slightly surprising - result is that there are still agents who respond in a contrarian way. Further research should focus on how to set incentives for drivers to believe and follow ATIS. As every driver who does not follow the recommendation has an external effect on the others' utilities, it should be tried to make them for this effect reliable. A Vickrey-Clarke-Groves mechanism (Vickrey (1961), Clarke (1971), Groves (1973)) could be

---

[4] Further results are available from the authors on request.

[5] Since each simulation run is a random draw of the same data generating process, OLS should be applicable. Simulation parameters are: traffic radio update interval 20 and the static linear utility function.

used for that purpose, but a problem may occur, as not every individual's behaviour could be observed. Another direction for further research could be using of genetic algorithm on more complex travel behaviour models, like Hivert's (1997) SATCHMO framework or like the Prism-Constrained Activity-Travel Simulator described by Kitamura et al. (1996).

Appendix: Specifications

Parameters, Variables and Constants of the Decision Tree

S_big	constant for big road
S_small	constant for small road
S_neutral	constant for no road
n_activated	number of agents, already activated in actual round
n_big, n_small	number of agents on the roads
E_interval	interval of actualisation of route recommendation (for no recommendations set E_interval> number of agents)
Recommendation	route recommendation, possible values: S_big, S_small, S_neutral

Decision Tree

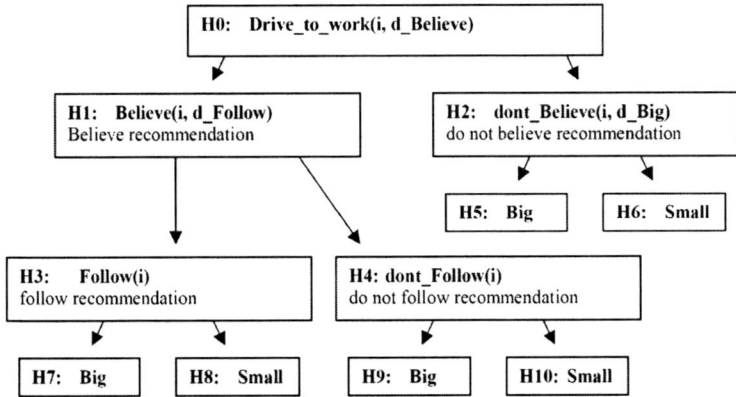

Fig. 7. Decision Tree.

References

1. BVBW, Bundesministerium für Verkehr, Bau- und Wohnungswesen (2001): "Auswirkungen neuer Informations- und Kommunikationstechniken auf Verkehrsaufkommen und innovative Arbeitspltze im Verkehrsbereich", www.bmvbw.de (German).

2. Clarke, E. (1971): "Multipart Pricing of Public Goods" Public Choice, 2, 19-33. Groves, T. (1973): "Incentives in Teams" Econometrica, 41, 617-631.

3. Hivert, L. (1997): "SATCHMO: a knowledge-based system for mode-choice modelling" in: Understanding Travel Behaviour in an Era of Change (P. R. Stopher and M. Lee-Gosselin, eds.) pp. 351-379, Pergamon, Oxford.

4. Kitamura, R.; Fujii, S.; Otsuka, Y. (1996): "An analysis of induced travel demand using a production model system of daily activity and travel which incorporates time-space constraints" Paper presented at the 5th world congress of the Regional Science Association International. Tokyo.

5. Kwan, M.-P.; Golledge, R. G.; Speigle, J. (1998): "Information Representation for Driver Decisions Support Systems" in: Theoretical Foundations of Travel Choice Modeling (Gärling, T; Laitila, T.; Westin, K.; eds.) pp. 281-303, Pergamon.

6. Mahmassani, H. S.; Gregory S.; Caplice, C. G. (1997): "Daily Variation of Trip Chaining, Scheduling, and Path Selection Behaviour of Work Commuters" in: Understanding Travel Behaviour in an Era of Change (P. R. Stopher and M. Lee-Gosselin, eds.) pp. 351-379, Pergamon, Oxford.

7. NTSC, National Science and Technology Council (1999): "National Transportation Science and Technology Strategy" Committee on Technology, Subcommittee on Transportation Research and Development, http://www.volpe.dot.gov/infosrc/strtplns/nstc/strategy99/index.html.

8. Pitz, T., Chmura, T. (2005): "Genetic Action Trees - A New Concept for Social and Economic Simulation", Computational Economics from Economics, Working Paper, Archive EconWPA number 0507002.

9. Selten, R.; Schreckenberg, M.; Pitz, T.; Chmura, T.; Kube, S. (2005): "Commuters Route Choice-Behaviours" Games and Economic Behaviour (resubmitted).

10. Stern, E. (1998): "Travel Choice in Congestions: Modeling and Research Needs" in: Theoretical Foundations of Travel Choice Modeling (Gärling, T; Laitila, T.; Westin, K.; eds.) pp. 173-199, Pergamon.

11. Tacken, M.; de Boer, E. (1989): "Flexitime and the spread of traffic peak hour. An analysis of conditions and behaviour" unpublished manuscript, OSPA, Delft University of Technology, Delft, The Netherlands (in Dutch).

12. Tacken, M.; de Boer, E. (1991): "Change in spread of travel and working times due to opening of the Amsterdam Orbital Motorway" unpublished manuscript, OSPA, Delft University of Technology, Delft, The Netherlands (in Dutch).

13. Vickrey, W. (1961): "Counterspeculation, Auctions, and Competitive Sealed Tenders" Journal of Finance, 16, 8-37.

Chaotic Traffic Flows on Two Crossroads Caused by Real-Time Traffic Information

Minoru Fukui[1], Katsuhiro Nishinari[2], Yasushi Yokoya[3], and Yoshihiro Ishibashi[4]

[1] Nakanihon Automotive College, Sakahogi-cho, Gifu-ken 505-0077, Japan
(fukui3@cc.nagoya-u.ac.jp)
[2] Department of Aeronautics and Astronautics,Faculty of Engineering, the University of Tokyo, Tokyo, 113-8656, Japan
(tknishi@mail.ecc.u-tokyo.ac.jp)
[3] Japan Automobile Research Institute, Karima, Tsukuba, Ibaraki,305-0822, Japan (yyokoya@jari.or.jp)
[4] School of Business, Aichi Shukutoku University, Nagakute-cho, Aichi-ken 480-1197, Japan (yishi@asu.aasa.ac.jp)

Summary. A cellular automaton traffic model of cars on two single-lane roads crossing at one point is studied. At the crossing real-time traffic information is displayed and drivers determine the driving direction based on the information about the trip-time. The traffic flow and density of the cars on the roads oscillate between free flow and jam states, and their time-behavior is chaotic. The real-time traffic information lures the driver to a road with shorter trip-time and causes too much concentration of the traffic by the nonlinearity of the velocity-density relation.

1 Introduction

Nowadays, the automobile is necessary for the usual civil life. However, an automobile mass society gave birth to social problems such as traffic jams, traffic accidents and air pollution. To solve these problems, the automobile itself and the complex system of automobiles and roads are improved by the electric and information technology. Then safety and energy efficiency of the automobile are improved and environmental pollution is being reduced. Real-time traffic information is collected and supplied to improve traffic flow and the efficient usage of the road network. Sequences of traffic lights are effective in optimizating the flow on urban streets. Traffic information, however, occasionally induces instabilities of the traffic flow or an abnormal concentration of cars on the road network. The timing of collection and supply of traffic information essentially influences the dynamics of the traffic flow. Effects of announcing global traffic information have been studied in a two-route traffic model introducing two types drivers: dynamic and static ones [1,2]. Optimum operation of a set of traffic lights was studied in two urban streets with an intersection [3] and in a two-dimensional urban street network [4]. Yokoya [5] has investigated traffic flows in two single-lane roads intersecting at one point, where each driver decides which route he should take by the information constantly provided by a road-side facility. He found an abnormal concentration of cars induced by the provided information.

In the present paper, traffic flows are investigated in a CA model of a road system where two single-lane roads intersect at one point. A traffic-sign board supplies drivers with real-time traffic information at the intersection and the driver decides the direction of motion based on this information. The update rule of the driver is different from Yokoya's rule at the intersection. Traffic flow fluctuates complicatedly in time about an equilibrium flow. The dynamics of the traffic flow is analyzed from the point of view of chaos.

2 A Model with Traffic Information

The road system consists of two one-dimensional (one-lane and one-way) roads, which cross at one point. There is no traffic signal but a board displaying real-time traffic information at the crossing. In cellular automaton (CA) traffic models, the road is expressed by a chain of cells, with cyclic boundary condition, and the cell at the crossing is shared by the two roads (Fig. 1). The car moves forward following the rule-184 CA [6]. On two roads, named X-road and Y-road, the cars are allowed only to go ahead in $+X$-axis direction or $+Y$-axis direction, respectively. The update is carried out simultaneously for both roads. This means that there is no traffic signal at the crossing. In order to avoid collision of cars entering simultaneously into the crossing from both roads, one car is selected with equal probability and allowed to move. Traffic information is displayed at the crossing and updated every time-step. The driver at the crossing gets information about the trip-times on both roads and chooses the road with the shorter trip-time. If the trip-times are equal for both roads, the driver does not change the direction of motion. The trip-time is calculated by the reciprocal of the mean velocity of all cars on each road. The present model is not deterministic, because it contains a stochastic process in the approach of the cars into the crossing.

We simulate traffic flow of both roads (X-road and Y-road) in the model described above. As a first simulation, the road length L of both roads is set at 500 cells and an equal number of 180 cars ($N_x = N_y = 180$) are put randomly on both roads. Then the cars advance obeying the rules described above. The initial density $d = N/L$ is 0.36. The sum of the number of cars on both roads is conserved. The mean velocity V of cars is calculated by the ratio of the number of the movable cars to the total number of cars on the respective road. Time-series of variations of the density and the velocity on both roads are shown in Fig. 2. Though the densities d_x and d_y of cars on both roads initially are equal, one gradually increases whereas the other decreases. After they reach a maximum or minimum, respectively, they turn back towards the other extreme.

Thus they continue to oscillate in antiphase between different maxima and minima in a complicated way. The velocities V_x and V_y of the cars show a more complicated behaviour (Fig. 2(right)). They vary between the value 1, corresponding to the free flow, and small values corresponding to jams. When the velocity V_x for the X-road surpasses V_y for Y-road, just then the density d_x on the X-road is at the minimum and begins to increase, i.e. all cars crowd into X-road. When d_x catches up with d_y after some time-steps, V_x is still larger than V_y. As a result, the d_x goes on increasing to the maximum. Thus the d_x continues to oscillate between different maxima and minima. The situation that almost all cars concentrate in one road with the other road becoming empty repeats itself.

Fig. 3(left) shows a velocity-density $(V-d)$ diagram made from the time-series data in Fig. 2. The straight-line part of $V_x = 1$ corresponds to free flow on the X-road and the curve $(1/d_x - 1)$ for $d_x > 1/2$ corresponds to the jamming state. When a point (V_x, d_x) is sited on the line of the free flow, it approaches a critical point of $(d_x = 1/2, V_x = 1)$ on the line as time passes and then goes down on the curve of the jam. After the velocity decreases to a minimum point where almost all cars are on the X-road, it turns back through another route with lower velocity and returns to free flow again. Thus it travels around clockwise on the different trajectories in the shape like a frame of glasses, which are never the same trajectories.

A flow-density diagram $(F-d)$ calculated from the same time-series data is given in Fig. 3(right). The flow F_x on the X-road travels around clockwise on the different trajectories through the free flow state, the jam state and the mixed states of free flow region and jam region. The flow F_y travels around on the flow-density space following the flow F_x trajectory with the conservation relation $N_x + N_y = 2dL$, though it never coincides with the F_x trajectory. Fig. 4 gives a spatiotemporal pattern of cars advancing on the X-road and the Y-road, which corresponds to one turn in the flow-density diagram. A

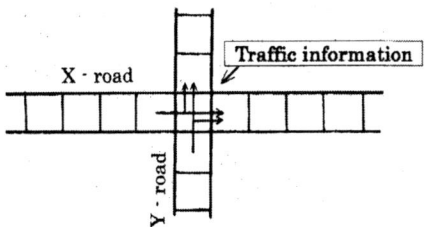

Fig. 1. Crossroad system. Cars on X-road and Y-road are bound for east and north, respectively. The roads are under the periodic boundary condition. Traffic information is displayed on the board at the intersection. The driver chooses the road with the shorter trip-time.

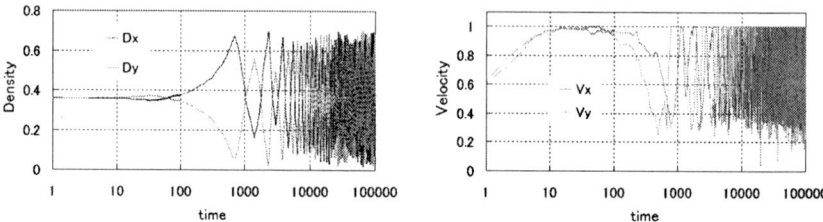

Fig. 2. Time variations of (left) car densities and (right) velocities on the X- and Y-roads for mean density $d = 0.36$.

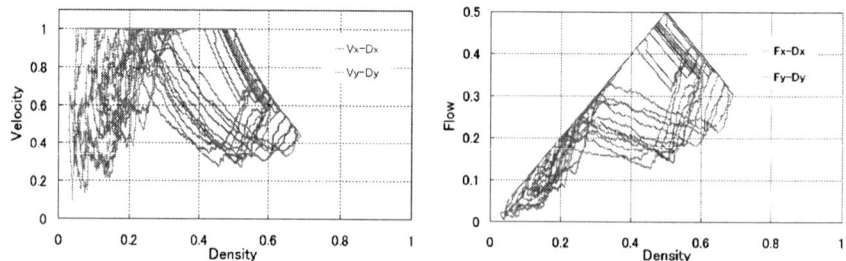

Fig. 3. Trajectory in (left) velocity-density and (right) in flow-density space for $d = 0.36$.

dot indicates a car in the figure. The light color region shows the free flow configuration ...10101... of cars, where 1 indicates a car. The dark color region shows the jam of the configuration such as ...011110....

We perform the similar simulations for the car densities of $d = 0.5$ and $d = 0.62$. The car densities oscillate complicatedly in the similar way. The flow travels around clockwise on the different trajectories in the flow-density space. The oscillation of the car density, i.e. the overcrowding of cars occurs in the density range of $0.25 < d < 1$.

3 Character of the Traffic Flow

We examine the character of these fluctuating traffic flows in the viewpoint of chaos. We find an attractor of the density d_x in the m-dimensional phase space and the correlation dimension of the fluctuating d_x. To embed the time-series data of d_x in the m-dimensional phase space, a set of points $(d_x(t), d_x(t + \delta), d_x(t + 2\delta), ..., d_x(t + (m - 1)\delta)$ is conventionally made from the $d_x(t)$ data. In the present system, $d_x(t)$ oscillates roughly periodically. The mean period depends linearly on the length L of the road and is numerically expressed as $T = 2.87L$ for $d = 0.36$. Then we put $\delta = 250$, which is usually set to

Fig. 4. A spatiotemporal pattern of cars advancing on the X-road (left) and Y-road (right), respectively. Cars move to the right and time increases from top to bottom. The number of cells is $L = 500$ and the intersection is located at $L = 100$ from the right edge.

some fractions of the mean period T in order to draw round trajectories. The dimension m of the space is estimated from the correlation dimension calculated by the Grassberger-Procaccia algorithm (GP method) [7]. In the GP method, the following correlation integral $C^m(\epsilon)$ is calculated,

$$C^m(\epsilon) = \lim_{N \to \infty} \frac{1}{N^2} \sum_{i,j=1}^{N} H(\epsilon - |r(i) - r(j)|), \tag{1}$$

where $H(t)$ is the Heaviside function, $r(i)$ is the coordinate of the i-th point in the m-dimensional space and $|r(i) - r(j)|$ indicates the distance between the i-th and the j-th points. The correlation integrals for the points of d_x are calculated in various m-dimensional spaces. They are assumed to be scaled as $C^m(\epsilon) \propto \epsilon^{\nu(m)}$. The correlation exponent $\nu(m)$ is given from the gradient of a linear part in the graph of $C^m(\epsilon)$ vs. ϵ expressed on log-log scale (Fig. 5(left)). Thus $\nu(m)$ is calculated for various m, shown in Fig. 5(b). The correlation dimension D_2 is estimated from the infinite m limit as $D_2 = 2.4$ (a fractal value) for $d = 0.36$.

We put the embed dimension m to 3 from the correlation dimension D_2 and embed the time series data of d_x in the 3-dimensional phase space. The strange attractors for d_x at $d = 0.36$ are plotted in the 3-dimensional phase space, shown in Fig. 6(left). They look like the shape of two-dimensional crushed springs. Next, we calculate the largest Lyapunov exponent of the attractors. The largest Lyapunov exponent is estimated from the following parameter $\lambda(t, \tau)$ [8]

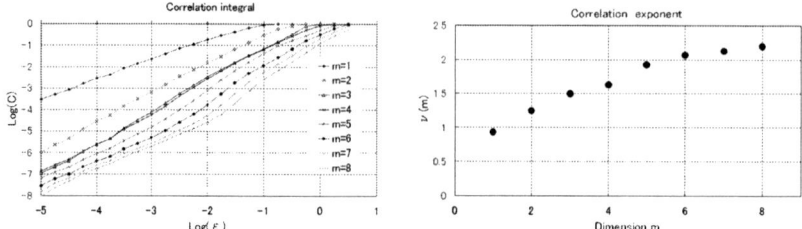

Fig. 5. (left) Correlation integrals $C^m(\epsilon)$ for $m = 1, \ldots, 8$ and (right) correlation exponent $\nu(m)$ for various embed dimensions m for $d = 0.36$.

$$\Lambda_i(t, \tau) = \frac{|r_i(t + \tau) - r_i'(t + \tau)|}{|r_i(t) - r_i'(t)|}, \qquad \lambda(t, \tau) = \frac{1}{N\tau} \sum_{i=1}^{N} \log \Lambda_i(t, \tau), \qquad (2)$$

where $r_i(t)$ is the coordinate of the i-th point and $r_i'(t)$ that of the neighbor point nearest i-th point at $t = 0$. $\log \Lambda_i(t, \tau)$ is averaged for all possible points of r_i. Fig. 6(right) shows $\lambda(t, \tau)$ for d_x for $d = 0.36$ calculated on the various values of t and τ. We estimate the largest Lyapunov exponent to be $+8 \cdot 10^{-3}$ from the plateau on the curve in this figure. We apply this method also to the d_x in the density $d = 0.5$ and 0.62 and list the results $\lambda(t, \tau)$:

$$\lambda(t, \tau) = \begin{cases} 8 \cdot 10^{-3} & \text{for } d = 0.36, \\ 5 \cdot 10^{-3} & \text{for } d = 0.5, \\ 6 \cdot 10^{-3} & \text{for } d = 0.62. \end{cases} \qquad (3)$$

In these densities, the largest Lyapunov exponents are all positive. It means that the density varies chaotically in the traffic flow on the crossing road with the real time traffic information. The chaotic property originates from

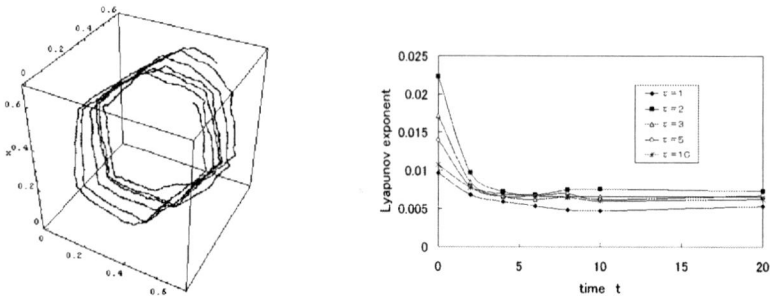

Fig. 6. (left) Strange attractor of the density d_x for $d = 0.36$. (right) the largest Lyapunov exponent.

the stochastic process included in the approach of car into the intersection from both roads. The real-time traffic information lures the driver to a road with shorter trip-time and causes too much concentration of the traffic. In the present traffic model, all drivers are sensitive to the real-time information. It does not improve the mean total flow of both roads.

References

1. J. Wahle, A.L.C. Bazzan, F. Klügl, M. Schreckenberg. Physica A **287**, 669 (2000)
2. K. Lee, P.M. Hui, B.H. Wang, N.F. Johnson. J. Phys. Soc. Jpn. **70**, 3507 (2001)
3. E.M. Fouladvand, Z. Sadjdi, R. Shaebani. J. Phys. A: Math. Gen. **37**, 561 (2004)
4. R. Barlovic, E. Brockfeld, A. Schadschneider and M. Schreckenberg. Phys. Rev. E **64**, 056132 (2001)
5. Y. Yokoya. Phys. Rev. E **69**, 016121 (2004)
6. S. Wolfram. Rev. Mod. Phys. **55**, 601 (1983)
7. P. Grassberger and I. Procaccia. Physica D **9**, 189 (1983).
8. S. Sato, M. Sano and Y. Sawada. Prog. Theor. Phys. **77**, 1 (1987)

A Vehicle Detection and Tracking Approach Using Probe Vehicle LIDAR Data

Bin Gao[1] and Benjamin Coifman[1,2]

[1] Department of Electrical and Computer Engineering, The Ohio State University Columbus, Ohio, 43210, USA

[2] Department of Civil and Environmental Engineering and Geodetic Science, The Ohio State University Columbus, Ohio, 43210, USA

Summary. Detection, identification and tracking of multiple moving targets have important applications in transportation and vehicle control areas. In this paper we present our approach to detect, recognize and track the vehicles within the detection region of a moving probe vehicle, based on the data collected by multiple sensors, including LIDAR and GPS. This paper develops a methodology to group the LIDAR measurements into targets, classify the targets as vehicles or fixed objects, and track the vehicular targets within lanes using a Kalman-filter. One important feature of this approach is that we track all of the observations in world coordinates, allowing us to average over many samples and ideally many runs to differentiate between the fixed objects (road boundaries) and moving objects (vehicles).

1 Introduction

Vehicle detection and tracking is important in various traffic analysis applications including traffic flow measurement, driver behavior analysis and many vehicle control applications, e.g., [1], emergency braking systems and autonomous driving systems, e.g., [2]. LIght Detection And Ranging (LIDAR) is one tool used to detect the location of the nearest obstacle within a specific angular range and it is starting to be deployed on probe vehicles to detect other vehicles, pedestrians, road boundaries or other objects in traffic. Most traffic studies employ detectors at a fixed location or do not collect information on multiple vehicles. In contrast, a probe vehicle equipped with LIDAR can collect information on vehicles over a range of both space and time. Most LIDAR sensors have the advantage of large detection region and ability to sense the outline of the objects, but approaches have to be designed to associate the observations from a given vehicle both in space and time.

This association problem is typically broken into the following three steps, clustering, classification and tracking, as follows. Clustering is a process to group the LIDAR measurements in a given frame from the same object together, classification sorts the clusters into several different classes, and tracking is a process to associate clusters representing a given vehicle from frame to frame. Several approaches have been designed to cluster the LIDAR range data [3,4], recognize different types of objects from the clustered data [3,5,6,7] and track the objects [8,4,9].

It remains difficult to accurately differentiate between vehicles and non-vehicle objects (mostly road boundaries) in the classification process. Examples of separating the road boundaries from vehicles are presented in [3,5,6], but the methods still need to be improved to cope with partial occlusion problems, detection errors and noise. Borrowing the idea of background subtraction from the image processing field, this paper presents a new method to accurately distinguishing between road boundaries and vehicles in the LIDAR data stream by constructing a density image of data points in a world coordinate system. The density image is rich in information about the road boundaries and shows advantage in retrieving the location of the road boundaries, particularly when the probe vehicle makes multiple runs past the same location. The boundary detection method, together with a clustering process and a Kalman-filter based tracking algorithm, renders an effective approach to detect and track the vehicles using the LIDAR range data.

2 Methodology

In this section, an approach is developed to cluster the LIDAR range data based on their spatial distributions, classify the clusters as vehicle objects or non-vehicle objects, and then track the vehicle objects from frame to frame. The LIDAR sensor used in this study has an angular range of $180°$, an angular resolution of $\Delta\alpha = 0.5°$, a range limit of 80 m, a ranging resolution of 0.01 m, and a scanning rate of 3Hz. Fig. 1a shows the vehicle coordinate system (x, y), with the origin at the probe vehicle position and the y axis along the heading direction of the probe vehicle, superimposed on a world coordinate system (X, Y), whose origin is at some fixed location in the world, the Y axis pointing north and X axis pointing east. These two coordinate systems will be used in the following analysis.

2.1 Clustering

The clustering process divides the data points into m mutually exclusive subsets, $C_1, \ldots, C_m$, each subset is a cluster representing the points from a same object. The clustering process is carried out in the following two steps.

Step 1: group the data points based on their proximity. In this step, the grouping rules are based on the grouping rule presented in [3] and two types of constraints: velocity-spacing constraints and lane width constraints.

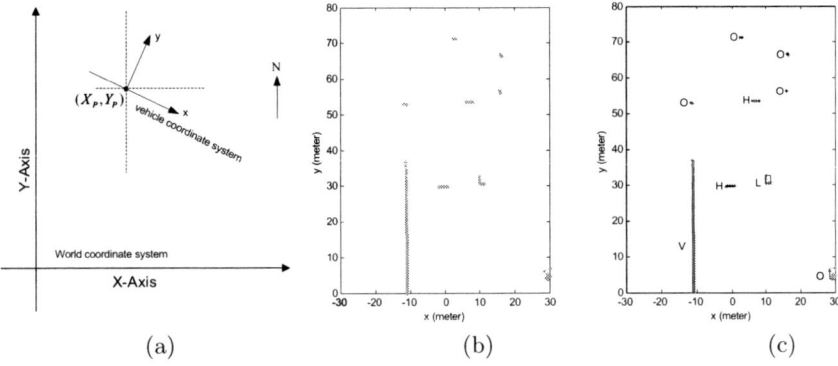

(a)   (b)   (c)

Fig. 1. (a) Vehicle and world coordinate systems, (b) one frame of LIDAR data, (c) the resulting clusters, each surrounded by a rectangle, and the shape category is shown on the left side of each cluster.

Step 2: each cluster resulting from step 1 is classified into one of four shape categories for future use in classification step. The four shape categories are:

    (1) "V" shape: the cluster fits a vertical line segment.
    (2) "H" shape: the cluster fits a horizontal line segment.
    (3) "L" shape: the cluster fits two perpendicular line segments.
    (4) "O" shape: other than above three cases.

Fig. 1c shows the result of the clustering process on one frame of LIDAR data, in which each cluster is surrounded by a rectangle and the shape category is shown on the left side of each cluster.

2.2 Classification

The classification process tries to classify the clusters into two classes: vehicles and non-vehicle objects. The classification process is developed in three stages: density image construction, road boundary detection and final classification.

(1) Density Matrix Construction

By using supplementary devices including GPS and a gyroscope, the location and heading of the probe vehicle can be measured so the LIDAR data points measured relative to the vehicle's coordinates can be converted into a world coordinate system. At which point the density of LIDAR data are tallied in world coordinates at 1 m resolution using data from multiple runs. Because the vehicles are moving objects while the road boundaries are static, after several runs the roadway will be covered with a relatively low density of observations resulting from vehicles observations while the road boundaries will be much denser because they are always observed in the same location.

If the probe vehicle regularly passes through a congested region the density matrix may still contain high densities in the roadway, making difficult the task of differentiating between vehicles and fixed objects. To compensate this effect we look for evidence of vehicle passages, each cell in the matrix is decreased by 10 each time a probe vehicle trajectory passes through that location, and it is decreased by 3 each time an "L" shape cluster is observed at those coordinates since this shape usually corresponds to vehicles. Fig. 2a shows the density image generated from 1 run only (no compensation), the darker the cell the denser the number of observations. Fig. 2b-c show respectively the density image generated from 11 runs no compensation, and the same image after the compensation process (all negative values are forced to zero in this figure).

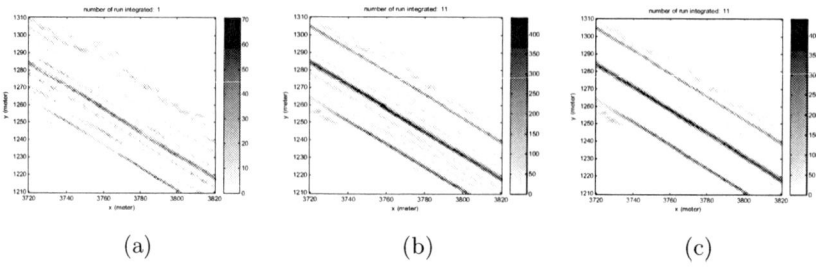

(a)   (b)   (c)

Fig. 2. The density matrix (a) from 1 run; (b) from 11 runs without compensation; (c) from the same 11 runs with compensation (any negative values are forced to zero in this figure for presentation only).

Comparing Fig. 2a and b, as the number of runs used in density image increases, the contrast between the road boundaries and the area on the road becomes larger. Comparing Fig. 2b and c, it can be seen that the compensation process further improves the contrast between road boundaries and the locations on the road so that they are separable.

(2) Road Boundary Detection

The road boundaries can be detected based on the assumption of a simplified road model in the vehicle coordinate system: the section of the road has left and right boundaries at some yet to be determined $x = w_1 < 0$ and $x = w_2 > 0$, respectively, an orientation along the y axis and zero curvature. All of the objects between the left and the right boundaries are assumed to be vehicles, and all other objects are considered non-vehicle objects. The road boundary detection process consists of following 3 steps.

Step 1: For each frame of LIDAR data, build a boundary-voting function that describes how likely a point on the x axis is beyond a boundary. Projecting the density matrix into vehicle coordinates, the density along the x axis is determined between $-M$ and M (where the exact value of M should be large enough to capture the entire road and is set empirically to 50 m for this study), e.g., Fig. 3a. A simple pulse detection method then used to find the first peak on either side of the vehicle in this curve. Then a boundary-voting function, which is piecewise constant, is built based on the density function and the pulses detected to denote: likely road, unsure, or likely off-road, e.g., Fig. 3b. This boundary-voting function shows that the points beyond the detected peaks are likely to be beyond the boundaries of the road, any points between the peaks are probably road, and the points with a negative density are very likely to be roadway. The boundary-voting function generated in this step in the i-th sample (or frame) is denoted f_i.

Step 2: Build boundary-voting functions based on the shape of the clusters and the estimated tracking positions (as defined in Sec. 2.3 below) in each frame. The process of deriving the shape and position based voting functions is as follows. First for the i-th frame and N subsequent frames (where N is large enough to extend beyond the range of the LIDAR) project the future probe vehicle positions into the vehicle coordinate system of this frame. For each cluster in the i-th frame, find the minimum absolute distance between the mean position of the points in the cluster and the coordinates of the probe vehicle's future trajectory along the road, x^*. A boundary-voting function is built for each cluster based on its shape and x^*, as shown in Fig. 4 for $x^* > 0$. The previous classification is superseded if the cluster is too large (width plus length exceeds 33.5 m) and instead is labeled "non-vehicle" shaped clusters. Finally, a boundary-voting function is generated for every estimated position of a tracked vehicle as well. When $x^* < 0$ the functions are simply reflected over the y axis. The resulting boundary-voting functions from Fig. 4 are denoted g_k, and they are used to modify the existing boundary-voting functions $f_{k-T}, \ldots, f_{k+T}$, in the adjacent frames (T is an integer constant, set to 5 in this study).

Step 3: Calculate the final boundary-voting function as a weighted sum of all the boundary-voting functions built in steps 1 and 2. For frame k corresponding to the given g_k, adjust the f_k in the adjacent frames by the following weighting:

$$f_{k+t} = f_{k+t} + g_k \frac{T+1-|t|}{T+1}, \qquad t \in \{-T, \ldots, T\}. \qquad (1)$$

Step 4: After completing steps 1-3 for all clusters that could influence the current frame, the boundary positions are calculated for that frame. The left boundary position is defined as $w_1^i = \max\{x | x < 0, f_i(x) > 0\}$, and the right boundary position is defined as $w_2^i = \min\{x | x > 0, f_i(x) > 0\}$.

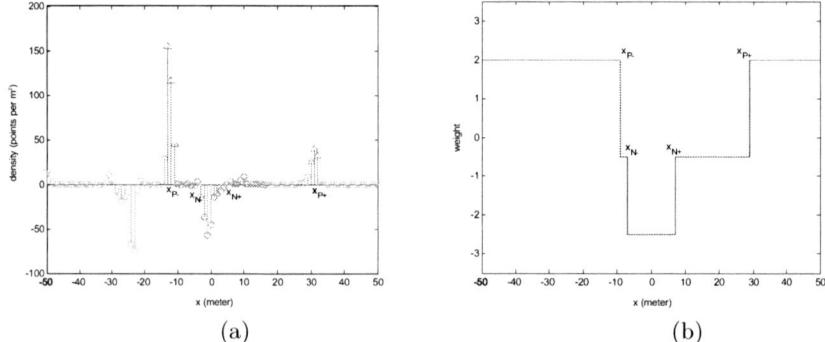

(a)   (b)

Fig. 3. An example of pulse detection and the boundary-voting function: (a) the stem plot of the density function as a function of x, with the pulse detected shown in triangle markers; (b) the boundary-voting function corresponding to the density function in (a), i.e., the resulting f_i curve.

(3) Final Classification

Finally, each cluster in each frame is classified as vehicle or non-vehicle objects. A given cluster is considered a vehicle if its coordinates fall on the road, $w_1^k < x^* < w_2^k$, otherwise it is classified as a non-vehicle object. Fig. 5 shows the final boundary-voting function, the boundary locations, (w_1, w_2), and the classification results of this frame of data.

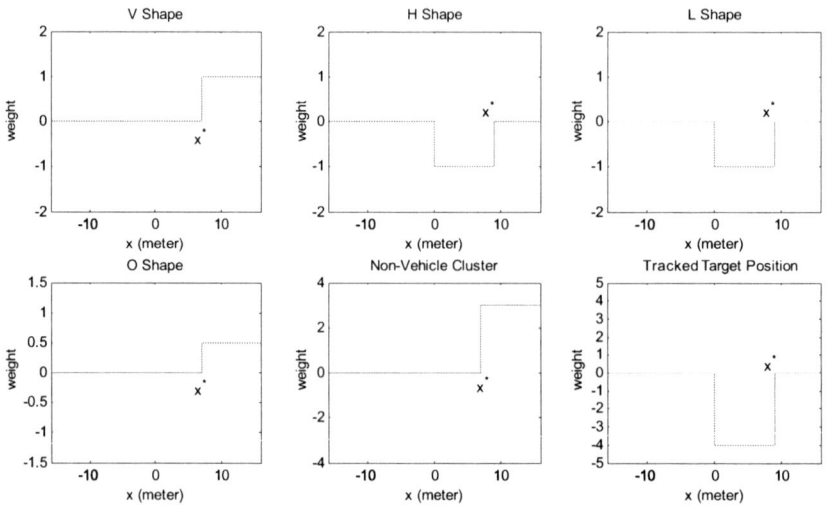

Fig. 4. The boundary-voting function of objects with $x > 0$ for the four shapes, the "non-vehicle" clusters and the estimated tracking positions. The curves are reflected over the y axis for objects with $x < 0$.

2.3 Tracking

The tracking process tries to associate the objects that are classified as vehicles from frame to frame, keeping track of each vehicle individually. The tracking algorithm is built based on a one-dimensional two-state Kalman-filter. For each frame of LIDAR data the tracking process contains two sub-processes, estimation and matching, as follows.

(1) Estimation

Based on the historical trajectories of the targets from the previous frames, their estimated positions in the current frame can be calculated using a Kalman-filter (the detailed process of Kalman-filtering based tracking algorithm can be found in [9]). This work tracks the clusters in vehicle coordinates using independent filters for horizontal and vertical movements.

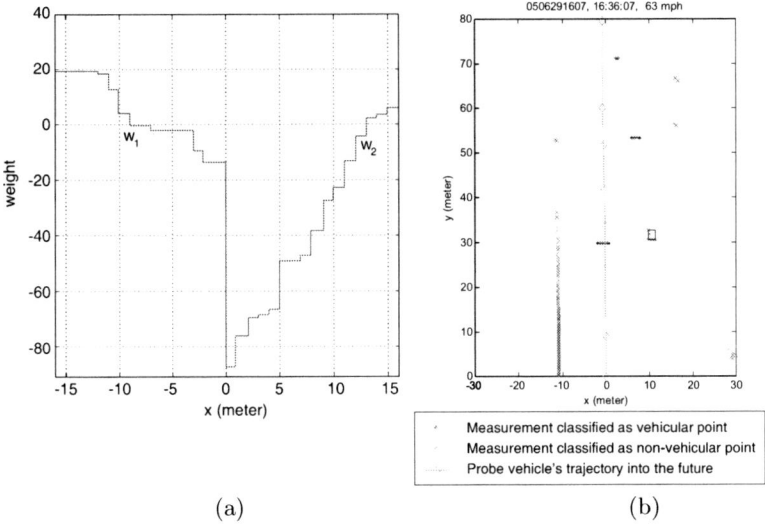

(a)   (b)

Fig. 5. (a) The boundary-voting function and the boundary locations; (b) the classification result, the points surrounded by rectangles are distinct clusters classified as vehicles.

(2) Matching

The matching process matches each estimated position to either a measured position or a newly initialized target. The matching process uses the following two strategies:

Strategy 1: lane-matching strategy: first distribute the measured positions (MP) and the estimated positions (MP) laterally across the road into a series of lane regions, where a given cluster is considered to be in the k-th lane provided it satisfies, $k \cdot 3.6\text{m} - 1.8\text{m} \leq x^* < k \cdot 3.6\text{m} + 1.8\text{m}$. Then the EPs and MPs are matched in order within each lane. A weighted matching cost is calculated for each pair of matched EP and MP as $c = a(\tilde{x}_i - x_j)^2 + b(\tilde{y}_i - y_j)^2$, where $a = 9$ and $b = 1$ were established empirically to penalize discrepancies in the x direction greater than the y direction. If any pair has a cost $c > c_{\max}$, the maximum tolerable matching cost, all of the matching in the corresponding lane is cancelled and the matching will be determined by strategy 2.

Strategy 2: Without distributing the EPs and MPs into lane regions, find the lowest possible cost of matching an unmatched EP to an unmatched MP; if the lowest cost found is below $c_{\max}$, match the corresponding EP and MP, remove them from the unmatched set, and find the lowest remaining matching cost and repeat until the lowest matching cost is above $c_{\max}$, match the rest unmatched MPs (if any) to newly initialized targets.

After the matching is complete for a frame of LIDAR data, the Kalman-filter parameters are updated for the estimation step of the next frame [9]. Fig. 6 shows the tracking results from two consecutive frames. The tracking numbers attached to the vehicle clusters show the association between the detected vehicles in the two frames.

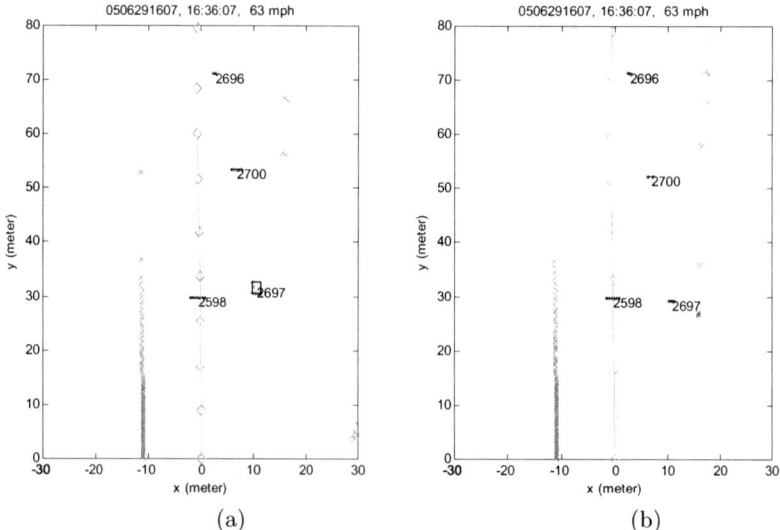

Fig. 6. Tracking results from two consecutive frames: the numbers attached to the vehicle objects are the tracking numbers describing the association between the two frames; (a) is the earlier frame, and (b) is the later frame.

3 Results

The clustering, classification and tracking processes are applied to probe vehicle LIDAR data collected on a pre-defined route in Columbus, Ohio, USA. Some results are verified manually against video from a camera mounted on the probe vehicle.

The overall rate of error in the results are low, and the results show that the approach proves to be reasonable and effective in most of the frames. However, the quality of the results varies due to traffic conditions, more errors occurred in congested traffic than under light traffic conditions. Examples of the three typical errors found in the verified samples are shown in Fig. 7 and discussed below.

Error type 1: Over-segmenting of one vehicle. In the clustering process one vehicle may be separated into more than one cluster due to partial occlusions. As shown in Fig. 7a, clusters numbered 5558 and 5572 are actually from the same vehicle. The reason of this type of error is partial occlusion of the vehicle and a low distance threshold setting for the measurement angle in step 2 of the clustering process.

Error type 2: Failure to recognize an irregular-shaped vehicle. As shown in Fig. 7b, on the left side of the future probe vehicle trajectory, some vehicle features are incorrectly classified as road boundaries. From the video images, the object on the left is found to be a semi-truck. Sometimes trucks appear not to be a regular "H" or "L" shaped objects, as most vehicles are expected to be, and this shape problem increases the value of the boundary-voting function, resulting in classification of the truck as non-vehicle object.

Error type 3: Incorrect classification due to GPS errors. As shown in Fig. 7c, the future probe vehicle trajectory comes very close to a section of the road boundary. The sudden shift evident in the figure is infeasible and arises due to incorrect positions given by the GPS receiver. In this case the incorrect GPS positions lead to an inaccurate future probe vehicle trajectory and then result in incorrect classification results for the clusters.

4 Conclusions

The clustering, classification and tracking methods presented herein have several advantages. First of all, in the classification process, the road boundary detection method is effective in separating the vehicles from the road boundaries. Its performance only improves when the number of runs on the same route is increased. Secondly, the road boundary detection method is robust to most noise except GPS position errors. Finally, the clustering, classification and tracking processes give a comprehensive analysis and reasonable interpretation of the LIDAR range data.

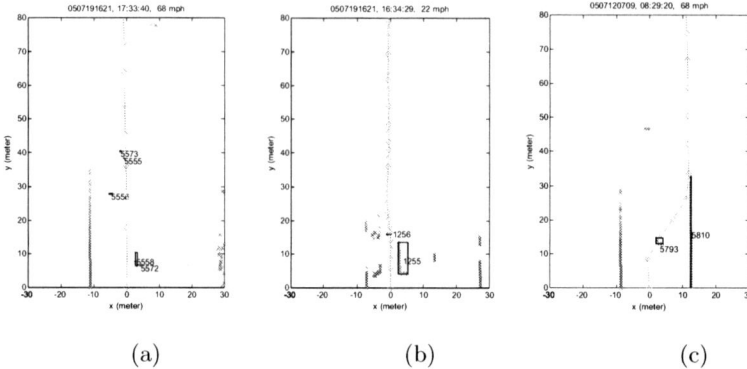

Fig. 7. Three typical errors found in the result verification process: (a) an example of over-segmentation; (b) an example of irregular shaped objects; (c) an example of GPS position errors.

The main limitation of the designed approach is that it is so sensitive to GPS errors, which can result in wrong position or orientation of the probe vehicle and finally incorrect classification results. The problem can be addressed with inertial navigation tools that will allow for dead-reckoning. The approach also needs to be improved to solve the over-segmentation problem and cope with irregular shaped vehicles that occasionally appear in the LIDAR data. The former can be addressed in part by looking at the evolution of the groups over time, identifying cases where two targets merge or split off from a single target. On-going work is progressing on the following improvements: (1) automatic detection of GPS measurement errors when the probe vehicle trajectory overlaps with or becomes very close to the road boundaries, (2) combining the clustering and tracking processes together to solve the over-segmenting problem, (3) utilize multiple hypothesis approaches to decrease the error rate [10], (4) Better verification methods to allow for the processing of much more data than is feasible with manual verification.

Acknowledgements

This material is based upon work supported in part by the National Science Foundation under Grant No. 0133278.

References

1. M. Sergi, C. Shankwitz, M. Donath: LIDAR-based vehicle tracking for a virtual mirror. Intelligent Vehicles Symposium, 2003. Proceedings of the IEEE, Page 333-338 (2003).
2. A. Ewald, V. Willhoeft: Laser Scanners for Obstacle Detection in Automotive Applications. Proceedings of the IEEE Intelligent Vehicles Symposium. USA (2000).
3. J. Sparbert, K.C. Dietmayer, D. Sweller: Lane Detection and Street Type Classification using Laser Range Images. Proceedings of ITSC 2001, IEEE 4th International Conference on Intelligent Transport Systems, Oakland (2001).
4. A. Mendes, Luis Conde Bento and U. Nunes: Multi-target Detection and Tracking with a Laserscanner. 2004 IEEE Intelligent Vehicles Symposium, Italy (2004).
5. W.S. Wijesoma, K.R.S. Kodagoda, and Arjuna P. Balasuriya: Road-Boundary Detection and Tracking Using Ladar Sensing. IEEE Transactions on Robotics and Automation, Vol. 20, No. 3 (2004).
6. R. Mazl, L. Preucil: Building a 2D environment map from laser range-finder data. Intelligent Vehicles Symposium 2000. Proceedings of the IEEE, p. 290-295 (2000).
7. K.Ch. Fuerstenberg, K.C.J. Dietmayer, V. Willhoeft: Pedestrian recognition in urban traffic using a vehicle based multilayer laserscanner. Intelligent Vehicle Symposium, 2002, IEEE. Volume 1, Page 31-35 (2002).
8. D. Sweller, K. Fuerstenberg, K. Dietmayer: Vehicle and Object Models for Robust Tracking in Traffic Scenes using Laser Range Images. Proceedings of ITSC 2003, IEEE 5th International Conference on Intelligent Transportation Systems, Singapore (2002).
9. Eli Brookner: Tracking and Kalman Filtering Made Easy. (John Wiley, 1998).
10. D. Streller and K.Dietmayer: Object Tracking and Classification Using a Multiple Hypothesis Approach. 2004 IEEE Intelligent Vehicles Symposium, p. 808 (2004).

Multi-Anticipative Car-Following Behavior: An Empirical Analysis

Serge P. Hoogendoorn[1], Saskia Ossen[1], and Marco Schreuder[2]

[1] Transport & Planning Department, Delft University of Technology, The Netherlands
[2] Traffic Research Centre, Netherlands Ministry of Transport, The Netherlands

Summary. Using vehicle trajectories for a motorway collected from a helicopter, and a newly developed approach to parameter identification, this paper provides new empirical evidence of multi-anticipative car-following by estimating the driver-specific parameters of the different multi-anticipative car-following models. We investigate the nature of the multi-leader stimuli, providing insight into the number of vehicles ahead to which drivers react and the kind of stimuli drivers respond to. The paper also shows that there is large inter-driver variability in multi-leader driving behavior. The implications of our research findings for microscopic modeling are discussed as well.

1 Introduction

Several researchers have suggested that driving behavior cannot be described adequately by only considering the vehicle directly in front. Drivers anticipate on traffic conditions further downstream by considering also the pre-predecessor or second-leader. The notion of multi-anticipative behavior reaches back to the late sixties, when the well known car-following model of [1] was extended by [2]. More recently, [3] extended the model of [4] to include multiple vehicle interactions, showing how the reaction to multiple vehicles stabilizes the dynamic behavior of the model, while retaining the fundamental macroscopic properties of the traffic flow. Moreover, the multi-anticipative car-following model is able to describe synchronized traffic flow conditions. In [5], a similar view is taken as the Ideal Driver Model (IDM) is generalized with multi-vehicle interaction behavior.

The aim of our investigation is to provide empirical evidence for multi-anticipative car-following behavior, while at the same time providing insight into the type of stimuli to which a driver reacts. Furthermore, the availability of microscopic data gives us the opportunity to quantify the inter-driver differences in the car-following parameters, which turn out to be very important to correctly describe multi-anticipative car-following behavior.

2 Multi-Anticipative Car-Following Models

In this paper the models of [2] and [3] are taken as a starting point to investigate multi-leader car-following behavior from empirical trajectory data.

2.1 Bexelius Model Family

A straightforward model incorporating multi-anticipatory behavior is the model of Bexelius, first proposed in [2]:

$$\dot{v}_i(t) = \sum_{j=1}^{m} \alpha_j \left(v_{i-j}(t-\tau) - v_i(t-\tau) \right) = \sum_{j=1}^{m} \alpha_j \Delta v_i^{(j)}(t-\tau) . \tag{1}$$

In (1), α_j for $j = 1, ..., m$ describes the sensitivity with respect to the relative speed $\Delta v_i^{(j)}$ of vehicle $i - j$ (the j−th vehicle ahead); τ denotes the reaction time. The Bexelius model provides a simple description of multi-anticipative car following behavior, enabling mathematical analysis of for instance platoon stability. The model has several drawbacks. For instance, the additive form may not always correctly capture multi-anticipative behavior. A driver may respond to the second (or third, or fourth) leader when the relative speed with respect to that specific leader is large. This is why we propose (and later, cross-compare) the following, simple modifications of (1) (Bexelius type 2):

$$\dot{v}_i(t) = \tilde{\alpha} \min \left\{ \Delta v_i^{(1)}(t-\tau), \Delta v_i^{(2)}(t-\tau), ..., \Delta v_i^{(m)}(t-\tau) \right\} \tag{2}$$

and (Bexelius type 3):

$$\dot{v}_i(t) = \min \left\{ \tilde{\alpha}_1 \Delta v_i^{(1)}(t-\tau), \tilde{\alpha}_2 \Delta v_i^{(2)}(t-\tau), ..., \tilde{\alpha}_m \Delta v_i^{(m)}(t-\tau) \right\} . \tag{3}$$

The fact that drivers only respond to relative speeds and not to, for instance, distances may arguably not yield a realistic description car-following behavior. The Helly model (and its generalizations) discussed next aims to remedy that by including a distance dependent factor as well.

2.2 Generalized Helly Models

The linear model of Helly [6] is given by the following relation:

$$\dot{v}_i(t) = \alpha_1 \Delta v_i^{(1)}(t-\tau) + \beta_1 \left(\Delta x_i^{(1)}(t-\tau) - S_i^{(1)} \right) \tag{4}$$

where the desired distance is defined by a simple linear function of the driving speed

$$S_i^{(1)} = s_i + T v_i . \tag{5}$$

In Eq. (4), $\Delta x_i^{(1)}$ denotes the distance between vehicle i and the vehicle directly ahead; $S_i^{(1)}$ denotes the desired distance, which is assumed to be a function of the speed $v_i(t)$; β_1 is the sensitivity with respect to the difference between the current distance $\Delta x_i^{(1)}$ and the desired distance $S_i^{(1)}$.

To generalize the Helly model with multi-anticipatory behavior, we can for instance include the speed relative to the second leader, or we can include the distance with respect to the second leader. We propose the following Generalized Helly (GH) model:

$$\dot{v}_i(t) = \sum_{j=1}^{m_1} \alpha_j \Delta v_i^{(j)}(t - \tau) + \sum_{j=1}^{m_2} \beta_j \left(\Delta x_i^{(j)}(t - \tau) - S_i^{(j)} \right) \tag{6}$$

where $j = 1, \ldots, m$ denotes the leaders to which a driver responds. In (6), m_1 and m_2 denote the number of leaders with respect to whose relative speed or deviation from the desired distance a driver response. We will refer to the model using the notation GH-m_1-m_2 in the remainder of the paper. To keep the number of parameters limited, we propose the following expression for the desired distance $S_i^{(j)}$:

$$S_i^{(j)} = S_0 + jTv_i . \tag{7}$$

2.3 Model of Lenz

A recent approach to multi-anticipatory car-following modeling is due to [3] (which will be referred to as the *Lenz model* in the remainder):

$$\dot{v}_i(t) = \sum_{j=1}^{m} \kappa_j \left[V \left(\frac{\Delta x_i^{(j)}(t)}{j} \right) - v_i(t) \right] \tag{8}$$

where $V(\Delta x)$ is a equilibrium speed function describing the speed of the follower in relation to the distances to the vehicles ahead; the parameters κ_j denote the sensitivity to the jth leader. Please note the direct relation with the fundamental diagram describing the macroscopic properties of traffic flow. Hoogendoorn and Ossen [7] showed poor average model performance compared to the other car-following models. To improve performance, we propose to include a true reaction time τ as follows:

$$\dot{v}_i(t) = \sum_{j=1}^{m} \kappa_j \left[V \left(\frac{\Delta x_i^{(j)}(t - \tau)}{j} \right) - v_i(t - \tau) \right] . \tag{9}$$

For the remainder of the paper, the following specification for the equilibrium speed V is used:

$$V(\Delta x) = v_0 \left[\left\{ 1 + \exp \left(\frac{1000}{\gamma \Delta x} - \frac{10}{2.1} \right) \right\}^{-1} - 5.34 \cdot 10^{-9} \right] , \tag{10}$$

where v_0 (the free speed) and γ are parameters to be estimated from the data; see [3] for details.

3 Parameter Identification Approach

The driver-specific parameters of the considered car-following models are estimated using microscopic traffic data collected for individual drivers. The vehicle trajectories used are collected by an airborne data collection system, and describe the positions x_i of all vehicles in the observed traffic stream at fixed time instants t_k. For the data considered here, positions are available each 0.1 s, so $t_k = t_0 + 0.1k$. By using each trajectory separately, we in fact find the car-following parameters for one specific driver.

3.1 Generalized Form of Car-Following Models

In discretized form all models considered here can be expressed as follows:

$$v_i(t_{k+1}) = v_i(t_k) + h \cdot a_i(t_k|\theta) \tag{11}$$
$$= v_i(t_k) + h \cdot a_i(\mathbf{y}(t_k), \mathbf{y}(t_k - \tau)|\theta)$$
$$= f(h, \mathbf{y}(t_k), \mathbf{y}(t_k - \tau)|\theta)$$

In Eq. (11), θ denotes the set of parameters describing the car-following behavior, such as the reaction time, the sensitivity, etc.; $h > 0$ denotes the time-step used for discretization of the model. The vector $\mathbf{y}(t_k)$ denotes the state that is relevant for driver i at time instant t_k. This state includes all stimuli that are present in a specific model. We assume that the relation between the speed data and the predicted speed is as follows:

$$v_i^{obs}(t_{k+1}) = v_i(t_{k+1}) + \varepsilon(t_k) = f(h, \mathbf{y}(t_k), \mathbf{y}(t_k - \tau)|\theta) + \varepsilon(t_k). \tag{12}$$

The error term $\varepsilon(t_k)$ is introduced to reflect errors in the modeling, similar to the error term used in multivariate linear regression. Note that the error terms $\varepsilon(t)$ are generally serially correlated, which will be handled later in the section. For now, let us assume that the error term is normally distributed with mean zero and standard deviation σ.

3.2 Maximum Likelihood Estimation

Since we can generally observe (either directly or indirectly) the state $y(t_k)$ from our available data, we can use Eq. (12) to determine a prediction for the speed. According to the model, the difference between the prediction and the observation follows the normal distribution with mean 0 and standard deviation σ. Assuming that the errors are uncorrelated, the probability of a set of observations $k = 1, ..., n$ can be determined from Eq. (12), yielding the log-likelihood $\tilde{L}$ of the sample:

$$\tilde{L}(\theta, \sigma^2) = -\frac{n}{2} \ln \left(2\pi\sigma^2\right) - \frac{1}{2\sigma^2} \sum_{k=1}^{n} \left(v_i^{obs}(t_{k+1}) - f(h, \mathbf{y}(t_k), \mathbf{y}(t_k - \tau)|\theta)\right)^2.$$

$$\tag{13}$$

Maximum-Likelihood (ML) estimation involves finding the parameters that maximize the log-likelihood. A necessary condition for the optimum allows determination of the standard deviation:

$$\frac{\partial \tilde{L}(\theta, \sigma^2)}{\partial \sigma^2} = 0 \implies \hat{\sigma}^2 = \frac{1}{n} \sum_{k=1}^{n} \left(v_i^{obs}(t_{k+1}) - f(h, \mathbf{y}(t_k), \mathbf{y}(t_k - \tau)|\theta) \right)^2 .$$

(14)

The ML estimate for the variance is given by the MSE of the predictions and the observations. For the remaining parameters, the ML estimates can be determined by numerical optimization.

Using the approach proposed in [8], serial correlation is dealt with by transforming the non-linear regression mode. The estimates will be the same as in case of the non-transformed model, and only the error term will be different, enabling the correct statistical analysis of the model. The likelihood-ratio test is used to cross-compare two different car-following models. The likelihood-ratio test accounts for the number of parameters thereby enabling correctly comparing simple and complex models.

Despite the fact that the presented approach is very generic, in the remainder it will be applied only to relatively simple models with at most six parameters. The main reason for this is the fact that the trajectory data only contains sufficient 'information' to identify a relatively small number of parameters.

4 Trajectory Data Used

The vehicle trajectory data used here was collected using a new data collection approach [9] using an air-borne observation platform (a helicopter). Using dedicated image processing software, vehicles are detected and tracked. This yields trajectory data covering approximately 500 m of motorway roadway stretch; the spatial resolution is smaller than 40 cm, while the temporal resolution is 0.1 s. Two datasets have been used in the parameter identification, both pertaining to the afternoon rush-hour. One of these was collected during the afternoon peak hour at the three-lane A15 motorway to the South of the Dutch city of Rotterdam (referred to as the Waalhaven site). During the entire period in which data were collected, congestion was quite heavy (average speeds of 7 m/s). The other dataset (the Everdingen site) was collected at the A2 motorway near the Dutch city of Utrecht and is characterized by stop-and-go traffic conditions. From the collected data, all the relevant variables (positions, distances, speeds, relative speeds, etc.) can be determined.

5 Overall Estimation Results

The average performance of a particular model is expressed in terms of the average log-likelihood value (i.e. averaged over all driver-specific estimations). For each individual driver we establish which of the model performs best in

Model	m	Everdingen			Waalhaven		
		$\tilde{L}$	% ref	# best	$\tilde{L}$	% ref	# best
CHM (ref)	1	-550.6	0.0%	2	-1098	0.0%	3
Bexelius (GH-2-0)	2	-495.3	10.0%	5	-1024	6.8%	4
Bexelius (GH-3-0)	3	-467.3	15.1%	4	-980.1	10.8%	1
GH-1-1	2	-453.2	17.7%	8	-892.9	18.7%	8
GH-2-1	2	-408.4	25.8%	12	-832.5	24.2%	24
GH-3-1	1	-397.8	27.8%	18	-809.9	26.3%	23
GH-1-2	2	-438.4	20.4%	7	-862.5	21.5%	4
GH-1-3	3	-426.2	22.6%	8	-838.6	23.6%	7
GH-2-2	2	-394.9	28.3%	8	-790.7	28.0%	17
GH-m_1-m_2				72			91
Bexelius type 2	2	-558.9	-1.5%	4	-1628	-48.2%	1
Bexelius type 3	2	-548.1	0.5%	16	-1613	-46.9%	0
Lenz	2	-467.5	15.1%	52	-854	22.2%	45
Total				144			228

Table 1. Overview of estimation performance for the considered models. Note that the differences in the values of the log-likelihood between the two sites (Everdingen and Waalhaven) are primarily caused by the average number of sample points of which a single observation consists.

log-likelihood terms. Best implies that the considered model passes the log-likelihood test for the optimal parameter settings when compared to any of the other models, including a zero-acceleration reference model. This kind of analysis into inter-driver variability is discussed in detail in [10].

5.1 Model Performance Comparison

Tab. 1 shows an overview of the performance of the considered models, indicating the number of leaders m, the average log-likelihood $\tilde{L}$, the relative improvement of the log-likelihood compared to the CHM model, and the number of times a specific model performed best.

The results depicted in Tab. 1 provide valuable insights into nature of multi-anticipative of car-following behavior. Let us start by concluding that including multiple leaders yields a much better description of car-following behavior, as can be seen from the average values of the log-likelihood of the multi-leader models compared to the single-leader models. From a behavioral point of view, this result provides empirical evidence that drivers do not only react on their direct leader, but anticipate further downstream, reacting on the behavior of the second and even the third leader. The extent in which this occurs will be considered in the next section.

We see that the Generalized Helly models with multiple leaders on average have the best log-likelihood values. In particular, the GH-2-2 model has an average log-likelihood value of -394.9 and -790.7 for the Everdingen and Waal-haven datasets respectively. The average improvements over the reference CHM model are respectively equal to 28.3% and 28.0%. In comparison, the Lenz model shows improvements of only 15.1% and 22.2% (compared to the CHM model) respectively for the two sites. Note that also the performance of the GH-3-1 model is also very good. This implies that including a third leader will further improve the description of car-following behavior.

From the respective model performances, we can also conclude that including the relative speed differences with respect to the second and third leader (GH-2-1 and GH-3-1 models) yields a larger improvement than including the multi-leader differences between the current and the desired distance (i.e. the GH-1-2 and GH-1-3 models). It appears that regarding the second, third leader, drivers are more susceptible to relative speeds than to distances (although from a human factors viewpoint, the opposite might be expected).

Note finally that, although the percentage for which a specific model of the Generalized Helly model family performs best is small (e.g. the GH-2-2 model is only optimal for predicting driving behavior in 28.3% and 12.4% of all cases for the Everdingen and Waalhaven sites), we see optimal performance in 50.0% and 66.4% of the cases, for the respective measurement sites if we look at the entire family of Generalized Helly models. This leads to the conclusion that the Generalized Helly models are well suited to describe multi-leader driving behavior. Nevertheless, the percentage of drivers for which the Lenz model yields an optimal description is considerable (for 24.1% and 32.8% of all drivers respectively for the two data sets). This provides evidence for the hypothesis that multi-anticipative behavior cannot be captured correctly by one single model type. Rather, different modeling paradigms – i.e. the Generalized Helly models and Lenz type models – are required. This is in line with other findings regarding variability in driving behavior [10]. That said, the GH model family turns out to provide on average the best description of car-following behavior.

6 Analysis of Generalized Helly Models

This section provides an analysis of the Generalized Helly models GH-3-1 and GH-2-2. These models have been chosen due to their relative good performance.

6.1 Parameter Distribution of GH-3-1 Models

Let us start by considering the GH-3-1 model. We are interested in the correlations between the parameters of the model. Tab. 2 shows the results of the analysis which was established by considering all drivers for which the GH-3-1 model performed best (in 39% of all cases). Note also that the table shows the average results for both datasets (Everdingen and Waalhaven).

	τ	α_1	β_1	α_2	α_3
average	1.213	0.289	0.060	0.065	0.072
std. dev.	0.281	0.307	0.083	0.113	0.133
correlation matrix					
	τ	α_1	β_1	α_2	α_3
τ	1	-0.300	-0.255	-0.138	-0.027
α_1		1	-0.056	-0.161	0.012
β_1			1	0.179	-0.100
α_2				1	-0.159
α_3					1

Table 2. Estimation results for GH-3-1 model. The table shows the average parameter estimates, the standard deviation and the inter-driver parameter correlations.

Note that the estimate for the reaction time τ is of the correct order, given the findings reported in literature. The table shows that drivers are on average most sensitive to the behavior of first leader, as is to be expected. The sensitivities with respect to the second and third leader are approximately of the same size. The sensitivities are approximately 25% of the sensitivity to the first leader. We can thus again conclude that multi-anticipative behavior plays a significant role when describing car-following behavior. It is also interesting to see that the sensitivity with respect to the second and third leader is not declining. Note also that the standard deviations are very high, in particular regarding the sensitivities to the relative speed. This implies that the inter-driver differences in the response to the first, second and third leader are large. In part, this is explained by the fact that the vehicle composition is heterogeneous (person-cars, trucks, etc.). Also note that looking at the *inter-driver correlations*, we can conclude that these are relatively small (all are less than 0.35).

Interestingly, if we compare the sum of the sensitivities α_i, for $i = 1, 2, 3$ to the sensitivity α_1 in the original Helly model (GH-1-1), we see that these values are both approximately equal to 0.4 s^{-1}. In a way, the sensitivity is spread out over the different leaders. This holds approximately for all models of the Generalized Helly family.

6.2 Parameter Distribution of GH-2-2 Models

Tab. 3 shows the results of the analysis. Note that these results only pertain to the drivers for which the GH-2-2 model outperformed the other two models (which occurred in 25% of all cases).

From Tab. 3 the average values of the reaction time and the sensitivities become apparent. Although the sensitivities describing the response to the direct leader are larger than those describing the response to the second leader, we can again conclude that the latter is considerable. This holds in particular

	τ	α_1	β_1	α_2	α_3
average	1.231	0.278	0.052	0.107	0.021
std. dev.	0.371	0.272	0.079	0.107	0.039
correlation matrix					
	τ	α_1	β_1	α_2	β_2
τ	1	-0.338	-0.254	0.084	-0.257
α_1		1	0.131	-0.200	-0.131
β_1			1	0.112	0.363
α_2				1	-0.066
β_2					1

Table 3. Estimation results for GH-2-2 model. The table shows the average parameter estimates, the standard deviation and the inter-driver parameter correlations.

	τ	κ_1	κ_2	v_0	γ
average	1.035	0.196	0.150	32.255	7.093
std. dev.	0.397	0.205	0.206	7.323	7.124
correlation matrix					
	τ	κ_1	κ_2	v_0	γ
τ	1	-0.143	-0.143	0.061	-0.002
κ_1		1	0.290	-0.086	0.457
κ_2			1	-0.236	0.090
v_0				1	-0.229
γ					1

Table 4. Estimation results for Lenz model (with two leaders). The table shows the average parameter estimates, the standard deviation and the inter-driver parameter correlations.

for the parameter α_1, which is only two times as large as the parameter α_2, meaning that the multi-leader behavior is indeed considerable. Also note that the standard deviations in the parameter values are relatively large, in particular for the parameters β_1 and β_2. The correlations between the parameter values are small.

7 Parameter Distribution of the Lenz Model

Let us now consider the Lenz model. Tab. 4 shows an overview of the average parameter estimates, the standard deviations, and the correlation between them, determined for all drivers for which the Lenz model performed better than the two other models (in 36% of the cases). In particular note the average value of the reaction time of 1.04 s, which was neglected in the original work of Lenz et al. [3].
The table shows that the sensitivity to the first leader is higher than the sensitivity to the second, although the differences are not particularly large.

696   Serge P. Hoogendoorn et al.

The results also show that the estimate for the free speed is in line with our expectations (recall that the data was collected from two motorways, which have a maximum speed of 33.3 m/s). The correlations between the parameters are not particularly high: all are less than 0.5. The variances are however quite substantial, again indicating large differences between the drivers.

8 Conclusions and Recommendations

In this paper we have applied a new maximum likelihood estimation approach to identify driver-specific parameters of multi-anticipative car-following models using vehicle trajectory data. The approach allows for statistical analysis of the model estimates. For instance, the standard error of the parameter estimates can be determined, as well as the correlation of the estimates. Also, we can easily test whether a specific model outperforms the other models using the likelihood-ratio test. The estimation approach has been applied to a number of existing as well as new car-following models that somehow include multi-anticipative behavior. The estimation results show that incorporating multi-anticipative behavior substantially improves the extent in which the models can explain driver behavior. In a several number of cases it turns out that the best performing models include three leaders. Drivers appear to be more responsive to the relative speed than to the difference between the desired distance and the actual distance with respect to the second and third leader.

Not all multi-anticipative behavior can be described by one single modeling paradigm. Rather, it turns out that different modeling approaches are needed to correctly explain driving behavior for specific drivers. Specifically, generalized Helly models and the model of Lenz both show good performance, but generally for different drivers. This is in line with other findings regarding variability in driving behavior [10]. Based on inter-driver variances in the parameter estimates, it turns out that differences between drivers who are described by the same model are also large. From this we can conclude that the extent in which drivers react to the second and the third leader can vary substantially between drivers. Inter-driver parameter correlations are generally small.

These findings have important implications to the current microscopic simulation practice. For one, most of the commercial simulation models include only the first leader, while in fact multiple leaders are to be considered to correctly describe driving behavior. Furthermore, differences in driving behavior are generally described – if at all – by considering a limited number of homogeneous groups of drivers which differ only in the parameter values describing their behavior. Our analyses have shown that on top of different parameter values, different models are needed to correctly describe driver heterogeneity. Such inter-driver differences are studied in detail in [10].

Future research is aimed at identifying more involved multi-anticipative car-following models, including more leaders, but also including higher model complexity. This however requires modifications to the parameter estimation procedure proposed here, in such a way that it would be possible to include data from multiple drivers in order to have sufficient data available to identify the multitude of model parameters. Another approach would be to collect more data for a single driver. This requires improvements to the current data collection system. Current research is guided in both directions.

Acknowledgements

The research described in this paper is in part funded by the Traffic Research Center AVV of the Dutch Ministry of Transportation, Public Works and Water Management. The research is part of the research programme "Tracing Congestion Dynamics – with Innovative Traffic Data to a better Theory", sponsored by the Dutch Foundation of Scientific Research MaGW-NWO.

References

1. D. Gazis, R. Herman, and R. Rothery. Nonlinear follow-the-leader models of traffic flow. Operation Research 4(9), 545–567 (1961)
2. S. Bexelius. An extended model for car-following. Transp. Res. 2(1), 13–21 (1968)
3. H. Lenz, C. Wagner, and R. Sollacher. Multi-anticipative car-following model. Eur. Phys. J. B7, 331–335 (1999)
4. M. Bando, K. Hasebe, A. Nakayama, A. Shibata, and Y. Sugiyama. Dynamical model of traffic congestion and numerical simulation. Phys. Rev. E51, 1035 (1995)
5. M. Treiber, A. Kesting, and D. Helbing. Delays, inaccuracies and anticipation in microscopic traffic models. Physica A360, 71 (2006)
6. W. Helly. Simulation of bottlenecks in single lane traffic flow. In *International Symposium on the Theory of Traffic Flow*, pp. 207–238, Elsevier, New York (1959).
7. S. P. Hoogendoorn and S. Ossen. Parameter estimation and analysis of car-following models. In H. Mahmassani, editor, *16th International Symposium on Traffic and Transportation Theory, Maryland.*, Elsevier, Amsterdam (2005)
8. S. P. Hoogendoorn and S. Ossen. Empirical analysis of two-leader car-following behaviour. European Journal of Transportation and Infrastructure Research, to appear (2005)
9. S. Hoogendoorn. Microscopic traffic data collection by remote sensing. Transportation Research Records 1855, 121–128 (2002)
10. S. Ossen, S. P. Hoogendoorn, and B. Gorte. Inter-driver differences in car-following: A vehicle trajectory based study. In *Transportation Research Board Annual Meeting*, Washington D.C., Transportation Research Board (2006)

Statistical Analysis of Floating-Car Data: An Empirical Study

M. Ebrahim Fouladvand[1,2] and Amir H. Darooneh[1]

[1] Department of Physics, Zanjan University, P.O. Box 45196-313, Zanjan, Iran
[2] Institute for Studies in Theoretical Physics & Mathematics (IPM), Department of Nano-Science, P.O. Box 19395-5531, Tehran, Iran.

Summary. We present results of a statistical analysis of empirical floating-car data. Our investigations are based on analyzing the time series of four basic quantities namely velocity, velocity difference, spatial gap and the acceleration associated to some instrumented cars. We try to identify the moving phases of the instrumented vehicle according to the statistical properties of its velocity time series. Moreover, by exploring the two-point joint probabilities, we propose a new approach for modelling vehicular dynamics based on the floating car data.

1 Introduction

Empirical observations on the spatio-temporal structure of traffic flow have revealed inherent complexities both on microscopic and macroscopic levels [1–12]. Quite recently significant progress has been made towards the thorough understanding of traffic flow dynamics by introducing several improved models [13–19]. Inevitably, in order to compare the microscopic single-vehicle predictions of each model to reality, one has to know the empirical behaviour of typical cars in different traffic states. So far the empirical data were mainly gathered via induction loops installed at fixed locations of the road. Although one obtains useful information about the flow, this scheme is inadequate to provide the necessary information about the long time behaviour of individual cars. In order to get insight into the real-life driving behaviour of individual drivers, one needs a time record of individual cars. Principally, this type of data can be obtained by instrumentation of a car with lidar/radar detectors. These detectors can simultaneously measure the velocity and the acceleration of the instrumented car, the velocity of its leader and the spatial gap to its leader. This floating-car data can be used to test the validity or the development of more sophisticated models of vehicular movement.

It is our major objective in this paper to report on a detailed statistical analysis of empirical floating car time series of four basic quantities i.e., velocity, velocity difference, spatial gap and acceleration/deceleration. On the account of this analysis, we try to classify the driving states of the floating car. This can give useful information about the traffic state in the environment of the floating car. Besides, we try to introduce a new approach for treating the car-car interaction on more realistic grounds.

2 Time Series Analysis of Floating-Car Data

If we look at the velocity record of a typical vehicle during a finite time interval, we would certainly realize that this velocity, $v(t)$, is a seemingly erratic and fluctuating function of time and its statistical properties depend on the global traffic congestion around the vehicle. Similar arguments correspond to the other single-vehicle quantities such as the spatial gap, or headway as is often called, to the leader vehicle $g(t)$, the vehicle acceleration $a(t)$ and the velocity difference $\Delta v(t) = v_l(t) - v(t)$ to its leader vehicle ($v_l(t)$ denotes the leader's velocity). It is our objective in this paper to introduce some characteristic aggregate statistical functions which give us a better insight into the stochastic aspects of traffic flow and enables us to establish a more realistic modelling framework of the driving rules and strategies. We shall now focus on the velocity of a particular vehicle say i. In empirical measurements, time is measured in discrete multiples of τ and the position of each vehicle is recorded as the multiple of a space grid denoted by δx. The time and space discretisation induces a discretisation for the velocity denoted by δv which is given by $\delta v = \frac{\delta x}{\tau}$. Regarding this fact, the integer-valued velocity ranges from 0 to $v_{max} = n_{max}\delta v$. By this notion, the velocity time series gives rises to the integer-valued velocity distribution function denoted by $P^i(v; \delta v, T)$. It is the relative frequency of the integer velocity v of the i-th vehicle during the period $[n_1\tau, n_2\tau]$.

3 Empirical Results

In this section we obtain some of the distribution functions in the above sections. We recall that most of the present data in the literature has been gathered through loop detectors at various points of the road [11, 12, 21–26]. We do not intend to discuss these types of data. The readers can refer to review articles and related papers in the field [4, 5, 9, 10, 12, 21–23, 25, 26]. There are basic differences between floating-car data and those obtained from fixed loop detectors. Each of these types of data give their own useful information. Specifically, fixed detectors measure the local properties of traffic flow, namely flow, occupancy, average velocity etc, at certain locations of the road. However, they can not give us illustrative information about the individual cars behaviour unless lots of detectors are installed which seems infeasible. On the other hand, to gain significant insight into the vehicular dynamics, it is salient to analyse the car-car interaction. Fixed detectors are inadequate and unable to provide enough information for such vital analysis. Therefore, having floating-car data seems unavoidable [27–29]. Recently there has been an increasing attempt to gather floating vehicle data mainly in order to calibrate the parameters needed for the modelling of vehicular dynamics in the framework of car-following models [20, 31–34]. The data we have analysed have been gathered from some equipped cars on German highways [30]. They contain time series of v, Δv, g and a. These four quantities have been recorded at 0.1 s intervals. The leader velocity is measured with radar while the follower velocity is measured by Lidar technique. The number data in each figure is ten times the duration of measurement. The precision of acceleration is 0.125 m/s^2.

Let us begin by showing the time series of v, Δv, g and a. The following sets of figures exhibit the time series of above-mentioned quantities for different driving situations. We have analysed the statistical properties of these time series by taking direct time averaging. Based on their statistical properties, four relatively different driving states have been identified. We call them fast (F), relatively fast (RF), slow (S) and very slow (VS) states. Fig. 1 considers the fast driving state.

Generally speaking, the relative deviations are small. We have evaluated the temporal auto-correlation of v, Δv and g. All of them are weakly correlated over time scales up to 10 s. and anti correlated for τ greater than 10 s. As can be seen from the graphs (and confirmed by mathematics) there is strong anti-correlation between velocity and the velocity difference to the leader up to 10 s. Between velocity and the gap, One observes a weak short correlation up to 3 s and a strong anti correlation between 4 s and 20 s. Between g and Δv we observe a rather strong correlation up to 20 s.

Next (Fig. 2) we consider a relatively fast driving state. The driving behaviour can be inferred by looking at the velocity time series. In comparison to the fast driving time series, one observes that fluctuations are enhanced. The average velocity of the car has reduced to 28 m/s. The range of velocity is wider and includes 22 to 32 m/s. The velocity standard deviation has notably increased to 2.56 m/s. Consequently the velocity relative deviation has sharply increased. Concerning the velocity difference, both the average, and its standard deviation have increased. For the gap, both standard and relative deviations have increased. Auto-correlation analysis shows that the velocity is correlated up to 18 s while Δv and g are more correlated (up to 30 s). v and g are correlated up to 6 s and uncorrelated after 6 s. Similar arguments apply to the case of v and Δv. Concerning the velocity difference and the gap, they are weakly correlated up to 5 s and then become uncorrelated ($\tau > 5$ s).

Fig. 3 exhibits the floating car behaviour in a slower driving state. As observed, the average velocity is further reduced to 19 m/s. This may correspond to moving in a higher congested environment. Compared to Fig. 2, velocity standard and relative deviations are notably reduced. However the velocity difference turns out to be more erratic since its standard deviation has increased in comparison to Fig. 2. Furthermore, the gap fluctuations are suppressed. Besides the value of the average velocity, a distinctive feature of

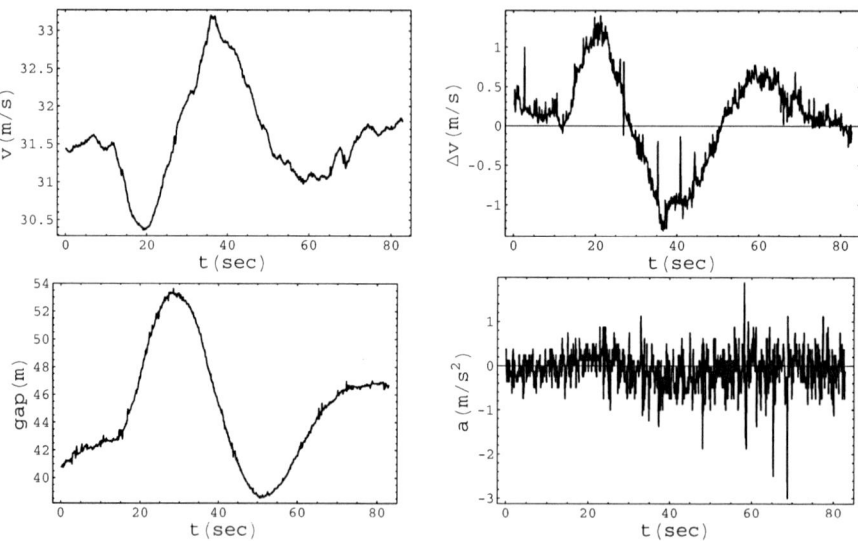

Fig. 1. Single-vehicle time series: velocity, velocity difference, gap and acceleration for an 80-second drive in fast driving state. Statistical properties of the time series are as follows: $\bar{v} = 31.6$ m/s, $\sigma_v = 0.64$ m/s; $\overline{\Delta v} = 0.1$ m/s, $\sigma_{\Delta v} = 0.57$ m/s; $\bar{g} = 44.9$ m, $\sigma_g = 4.15$ m.

the time series of Fig. 2 is the reduction of the velocity standard deviation. This may be related to a high degree of synchronisation of the vehicle's velocity to its leader's velocity. Since the driving interval is not long enough, the auto- and cross-correlations do not give rise to meaningful results.

The next set of figures (Fig. 4) exhibits the floating car behaviour in a very much slow driving situation. Although the average velocity is very small, the velocity standard deviation is relatively very large leading to a large velocity relative deviation. In comparison to Fig. 3, the velocity standard deviation has a considerable larger value. Concerning velocity difference, the average value is nearly zero but its standard deviation is larger than the value in Fig. 3. Finally, while the average gap has decreased to 8 m, the gap standard deviation is relatively high. The very slow state has the highest relative deviation of gaps among the times series discussed so far. v and g are correlated up to 20 s. In sharp contrast, Δv is correlated only over short time scales up to $3 - 4s$. Concerning the cross-correlations, there is correlation between v and g up to 100 s. v and Δv are nearly uncorrelated. Between Δv and g one observes a fluctuating cross-correlation between negative and positive values.

We have summarized the statistical properties of the above time series denoted by fast (F), relatively fast (RF), slow (S) and very slow (VS) in Table 1.

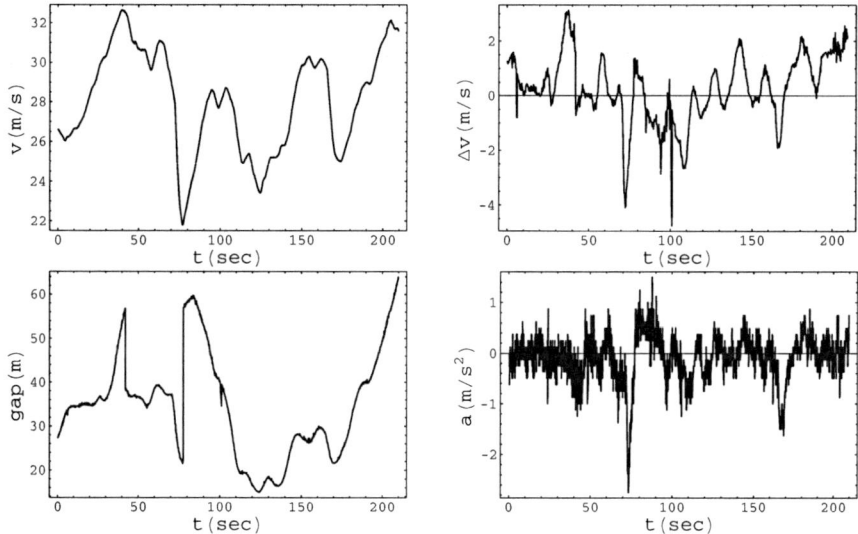

Fig. 2. Single-vehicle time series: velocity, velocity difference, gap and acceleration for a 200-second drive in a relatively fast driving state. The statistical properties are as follows: $\bar{v} = 28.1$ m/s, $\sigma_v = 2.56$ m/s; $\overline{\Delta v} = 0.27$ m/s, $\sigma_{\Delta v} = 1.21$ m/s; $\bar{g} = 34.5$ m, $\sigma_g = 11.7$ m.

traffic state →	F	RF	S	VS
$\bar{v}$ (m/s)	31.6	28	19	3.7
σ_v (m/s)	0.64	2.56	0.28	0.8
$\overline{\Delta v}$ (m/s)	0.1	0.27	−0.13	0.06
$\sigma_{\Delta v}$ (m/s)	0.57	1.21	0.27	0.8
$\bar{g}$ (m)	45	34	29	8
σ_g (m/s)	4	11	1.5	4

Table 1: Statistical properties of different driving states.

3.1 Two-Point Distribution Functions

Although one-point distribution functions give us useful information on quantification of driving behaviours, many important features lie beyond the one-point functions and one has to consider higher joint distributions. In this section we present some two-point functions obtained from the empirical data and will discuss their importance for a successful modelling of vehicular dynamic at the microscopic level. Here we show two basic two-point distribution functions, namely $P_2(v, g), P_2(\Delta v, g)$ for the traffic states discussed so far. The grid values are the same as for the one-point functions, i.e., velocity grid = 1 m/s, velocity difference grid = 0.25 m/s, gap grid = 1 m and acceleration/deceleration grid = 0.1 m/s^2.

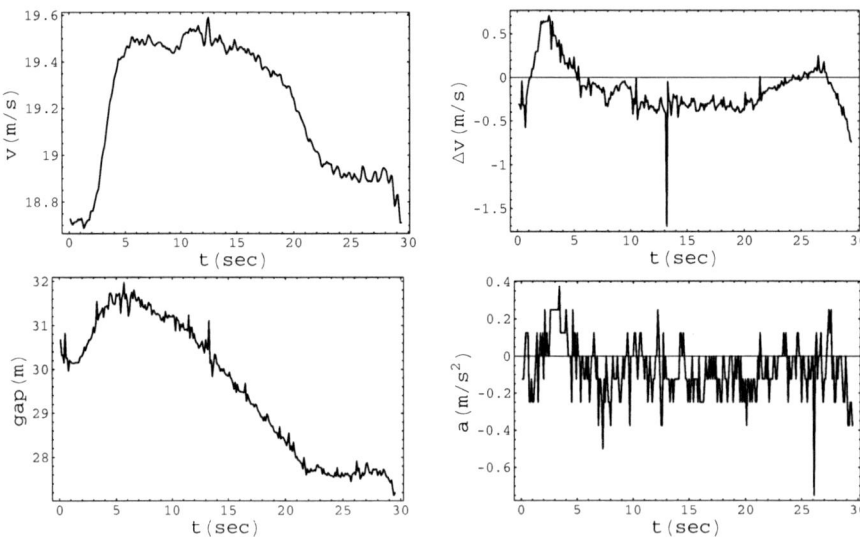

Fig. 3. Single-vehicle time series for velocity, velocity difference, gap and acceleration for an 30-second drive in a slow driving state. The statistical properties are as
follows: $\bar{v} = 19.2$ m/s, $\sigma_v = 0.28$ m/s; $\overline{\Delta v} = -0.13$ m/s, $\sigma_{\Delta v} = 0.27$ m/s ; $\bar{g} = 29$ m,
$\sigma_g = 1.5$ m.

We shall now investigate, in some detail, the characteristics of these distributions. First, let us discuss $P_2(v, g)$. From this distribution we get useful
information which relates the velocity to the gap value. According to the free
flow graph, the relative frequencies are scattered in a 2D area. If in the two-
dimensional $v - g$ plane we mark those grids having large amplitudes in the
$P_2(v, g)$, then we can obtain insight on how the gap and velocity are dependent on each other. The same arguments can be applied to $P_2(v, \Delta v)$ and
$P_2(\Delta v, g)$.
In the RF driving state, one observes that the degree of dependence between
velocity and gap has increased at more points. This can be verified by close
examination of the diagram. In fact, we notice the number of relatively large
columns are increased which consequently gives rise to more marked points
in the 2D $g - v$ plane. This implies the optimal velocity assumption can be
justified (although not precisely). The optimal velocity curve can be obtained
by fitting through the points in $v - g$ plane at which $P_2(v, g)$ has notable
values. We have not plotted the two-point functions in the S state due to
insufficient number of data points. Finally in the VS state, the dependence
between v and g substantially reduces. This is manifested by looking at the
distribution itself and noting the small number of grids having notable value
of $P_2(v, g)$. This may limit the validity of the optimal velocity assumption in

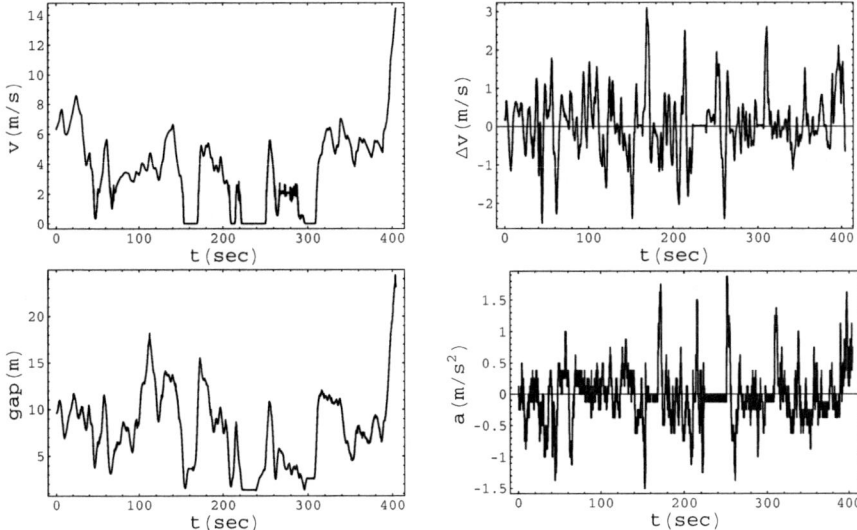

Fig. 4. Single-vehicle time series for velocity, velocity difference, gap and acceleration for an 400-second drive in a very slow driving state. The statistical properties are as follows: $\bar{v} = 3.7$ m/s, $\sigma_v = 2.5$ m/s; $\overline{\Delta v} = 0.06$ m/s, $\sigma_{\Delta v} = 0.8$ m/s; $\bar{g} = 7.8$ m, $\sigma_g = 4.1$ m.

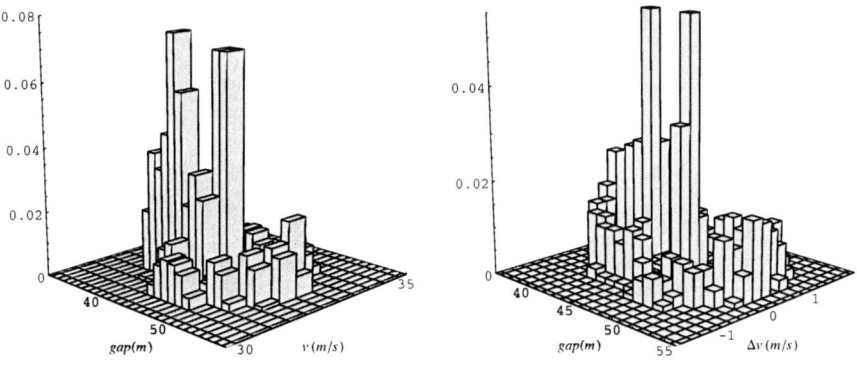

Fig. 5. Two-point functions $P_2(v, g)$ and $P_2(\Delta v, g)$ for a fast (F) driving state.

this driving state. Nevertheless, we remark the confirmation of this conclusion needs analysis of further data.

Next, we discuss the characteristics of $P_2(\Delta v, g)$ in different states. As can be seen by examining the high value grids, in F and RF states, Δv and g are dependent to each at certain points in the 2D $g - \Delta v$ plane. This is suppressed in the VS state. Concerning $P_2(v, \Delta v)$, the diagrams (not shown here) tell us that in the fast state, v and Δv are dependent only in a limited curve-like

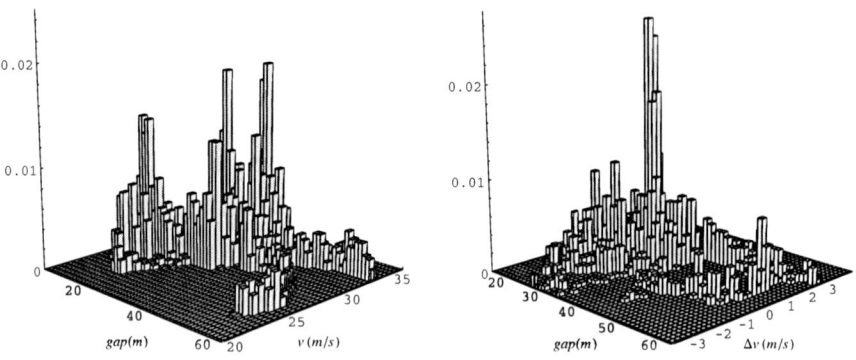

Fig. 6. Two-point functions $P_2(v, g)$ and $P_2(\Delta v, g)$ for a relatively fast (RF) driving state.

region of the $v - \Delta v$ plane whereas, in RF state, the dependence region appears as 2D area in the $v - \Delta v$ plane. By contrast, in the VS state, the dependence region shrinks and appears in a more restricted region of the $v - \Delta v$ plane.

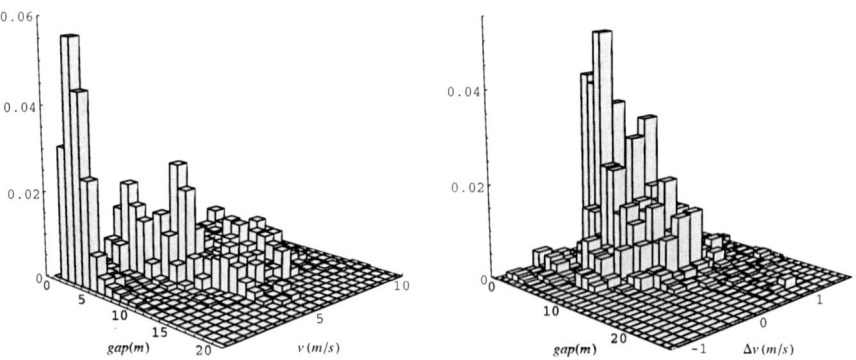

Fig. 7. Two-point functions $P_2(v, g)$ and $P_2(\Delta v, g)$ for a very slow (VS) driving state.

4 Summary and Concluding Remarks

In this paper we have analysed the floating-car data taken from instrumented vehicles. Our findings suggest the existence of four different driving states classified as fast, relatively fast, slow and very slow state. Generally speaking, our analysis demonstrates that the degree of validity of the optimal velocity assumption depends on the driving state. This gives a rather important hint for the improvement of the car-following approach [33]. Knowing the distribution functions, allows us to develop a general framework for modeling of vehicular dynamics. As explained, the diverse types of driver reactions to stimuli received from the traffic ahead of them gives rise to heterogeneous driving strategies. The manifestation of these strategies is reflected in the non-trivial joint distribution functions of driving quantities. This suggests that if we can measure these joint functions in different traffic situations i.e., free, synchronized and congested, then one can make use of them in order to establish a realistic choice of driving strategies by the appropriate designation of the acceleration a in terms of v, g and Δv. Let us clarify this point. Apparently we know that the car's acceleration a is related to its velocity v, its gap g and the velocity difference Δv but we do not know the quantitative relationship. The subtle point is that this relationship is not a functional form in which a is assumed to be a function of v, g and Δv. The empirical data confirms the existence a multitude of acceleration values for fixed values of v, g and Δv. Therefore we are talking about the probability of having an acceleration value provided the velocity, gap and the velocity difference have values v, g and Δv, respectively. The following procedure gives us this conditional probability on a numerical basis. First we evaluate the four- and three-point functions $P_4(a, g, v, \Delta v)$ and $P_3(g, v, \Delta v)$. Then we proceed by finding the conditional probability that the car's acceleration is a given the velocity, gap and the velocity difference have the values v, g and Δv respectively. This conditional probability is obtained as follows:

$$P(a|g, v, \Delta v) = \frac{P_4(a, g, v, \Delta v)}{P_3(g, v, \Delta v)} \tag{1}$$

Moreover, for those car-following models which use a functional dependence of the acceleration in terms of v, g and Δv, the above conditional form of the dependence of a in terms of v, g and Δv can be exploited to derive a functional form by finding the average value of the acceleration with respect to the above probability distribution as follows:

$$a(g, v, \Delta v) = \sum_a aP(a|g, v, \Delta v) \tag{2}$$

Finally, it must be mentioned that our data is related to only a few instrumented cars. In order to draw decisive and exhaustive conclusions, one has to obtain sufficiently large data-sets from a variety of drivers. Analysis of future floating-cars data will shed more light upon the problem.

Acknowledgements

We are deeply indebted to the *Institute of Transport Research (IVF)* at German Aerospace centre (DLR) and in particular Peter Wagner for providing us with the empirical data. The data have been gathered by the Robert Bosch GmbH and are available on the IVF web site [30].

References

1. B. Kerner, *Physics of Traffic*, Springer, 2004.
2. A.D. May, *Traffic Flow Fundamentals*, Prentice Hall (1990).
3. C.F Daganzo, *Transportation and Traffic Theory*, Elsevier (1993).
4. D. Chowdhury, L. Santen and A. Schadschneider, Phys. Rep. **329**, 199 (2000).
5. D. Helbing, Rev. Mod. Phys. **73**, 1067 (2001).
6. B.S. Kerner, Networks and Spatial Economics **1**, 35 (2001).
7. H.J. Herrmann, D. Helbing, M. Schreckenberg, and D.E. Wolf (eds.), *Traffic and Granular flow* (Springer, Berlin, 2000).
8. M. Fukui, Y. Sugiyama, M. Schreckenberg, and D.E. Wolf (eds.) *Traffic and Granular flow* (Springer, Tokyo, 2002).
9. B. S. Kerner, *Phys. Rev. E* **65**, 046138 (2002).
10. I. Lubashevsky, R. Mahnke, P. Wagner, and S. Kalenkov, Phys. Rev. E **66**, 016117 (2002).
11. L. Neubert, L. Santen, A. Schadschneider, and M. Schreckenberg, Phys. Rev. E **60**, 6480 (1999).
12. W. Knospe, L. Santen, A. Schadschneider, and M. Schreckenberg, Phys. Rev. E **65**, 056133 (2002).
13. D. Helbing and M. Schreckenberg, Phys. Rev. E **59**, R2505 (1999).
14. M. Treiber, A. Hennecke, and D. Helbing, Phys. Rev. E **62**, 1805 (2000).
15. W. Knospe, L. Santen A. Schadschneider, and and M. Schreckenberg, J. Phys. A: Math. Gen. **33**, L477 (2000).
16. B. S. Kerner and S.L Klenov, J. Phys. A: Math. Gen. **35**, L31 (2002).
17. B. S. Kerner, S.L Klenov, and D. E Wolf, J. Phys. A: Math. Gen. **35**, 9971 (2002).
18. R. Jiang and Q.S. Wu, J. Phys. A: Math. Gen. **36**, 381 (2003).
19. K. Nagel, P. Wagner, and R. Woesler, Oper. Res. **51**, 681 (2003).
20. P. Wagner, arXive cond-mat/0411066.
21. B. Kerner and H. Rehborn, Phys. Rev. E **53**, R1297 (1996); Phys. Rev. E **53**, R4275 (1996).
22. B. Kerner and H. Rehborn, Phys. Rev. Lett. **79**, 4030 (1997);
    B. Kerner, Phys. Rev. Lett. **81**, 3797 (1998).
23. M. Treiber, A. Henneke, and D. Helbing, Phys. Rev. E **62**, 1805 (2000).
24. B. Kerner, Phys. Rev. E **65**, 046138 (2002).
25. B. Kerner and S.L. Klenov, Phys. Rev. E **68**, 036130 (2003).
26. W. Knospe, L. Santen, A. Schadschneider, and M. Schreckenberg, Phys. Rev. E **70**, 016115 (2004).
27. E.P. Todosiev and L.C. Barabosa, Traffic Engineering **34**, 17 (1963).
28. M. Brackstone, B. Sultan, and M. McDonald, Transp. Res. F **5**, 329 (2002).
29. P. Hidas and P. Wagner, *German Aerospace Centre (DLR)* preprint 2004.
30. The data have been gathered by the *Robert Bosch GmbH* on the highway A 8 along the route between Stuttgart and Karlsruhe in 1995. The data are available on the web-site: *http://www.clearingstelle-verkehr.de/cs/verkehrsdaten*
31. P. Hidas, Traffic Enginering + Control **39**(5), 300 (1998).
32. G.S. Gurusinghe, T. Nakatsuji, Y. Azuta, P. Ranjitkar, and Y. Tanaboriboon, *Transportation. Res. Board conference 2003*, paper no TRB2003-004137 (2003).
33. P. Wagner and Ihor Lubashevsky, Arxive cond-mat/0311192.
34. J. H. Banks, Transportation Research B **37**, 539 (2003).

Scale-Free Features
in the Observed Traffic Flow

Sin-ichi Tadaki[1], Macoto Kikuchi[2], Akihiro Nakayama[3], Katsuhiro Nishinari[4], Akihiro Shibata[5], Yuki Sugiyama[6], and Satoshi Yukawa[7]

[1] Computer and Network Center, Saga University, Saga 840-8502, Japan
[2] Cybermedia Center, Osaka University, Toyonaka 560-0043, Japan
[3] Department of Physics and Earth Sciences, University of Ryukyus ,Okinawa 903-0213, Japan
[4] Department of Aeronautics and Astronautics, University of Tokyo, Bunkyo 113-8656, Japan
[5] Computing Research Center, KEK, Tsukuba 305-0801, Japan
[6] Graduate School of Information Science, Nagoya University, Nagoya 464-8601, Japan
[7] Department of Applied Physics, University of Tokyo, Bunkyo 113-8656, Japan

Summary. Scale-free features in traffic flow are discussed based on observed data. The observed traffic flow is known to contain some periodic motions reflecting our social activities. By employing the DFA (Detrended Fluctuation Analysis) method, the traffic flow can be shown to exhibit daily periodic oscillations with scale-free fluctuations.

1 Introduction

Understanding properties of traffic flow on expressways has improved mainly based on mathematical models and their computer simulations since early the 1990's [1, 2]. Many interesting features have been studied from the viewpoints of nonequilibrium statistical physics, pattern formation and transportation phenomena. Since the early data analysis by Kerner and Rehborn [3], we have been aware of the proper complex behavior of the traffic flow observed on real expressways. The complexity appearing in the fundamental diagram (density-flow relation) has been discussed in the context of *synchronized flow* [4].

The observed temporal behavior of the expressway traffic is a complex mixture of various time-scales from the dynamical behavior of cars and external sources including human social activities. The characteristic response time of cars, for instance, is of the order of one second. The phase transition between the free-flow and congestion states takes of the order of 10 minutes. The daily periodic behavior of the congestion seems to reflect human social activities.

The existence of power-law fluctuations in traffic flow has been discussed previously. Since the pioneering observation by Musha and Higuchi [5], however, a limited number of observational studies have been reported on power-law fluctuations in expressway traffic flow [6–8]. The observed power-law fluctuation in these studies is limited to time-scales shorter than a few hours.

Power-law fluctuations in traffic flow have also been discussed based on simulation results [9–11]. Granular flow in vertical pipes, which has been thought to be a related phenomenon, also shows $f^{-4/3}$ fluctuations [12]. These results will correspond to the observed power-law fluctuations on short time-scales. We need a filter to extract the proper dynamical fluctuations from raw data. One of methods for subtracting trends from non-stationary raw sequential data is the *detrended fluctuation analysis* (DFA) method [16, 17]. In this report, by employing the DFA, we study the long-range correlation hidden in the raw data of the traffic flow.

2 Detrended Fluctuation Analysis

The detrended fluctuation analysis (DFA) has been invented first for analyzing the long-range correlation in DNA (Deoxyribonucleic Acid) sequences [16, 17]. The method has been applied to various non-stationary time series.

The method is described as follows. First the *profile* $y(t)$ of the raw temporal data $\{u(t)\}$ $(0 \le t < T)$ is defined:

$$y(t) = \sum_{i=0}^{t} [u(i) - \langle u \rangle], \tag{1}$$

where $\langle u \rangle = T^{-1} \sum_{t=0}^{T-1} u(t)$ is the temporal average value of the raw data $\{u(t)\}$.

The entire time sequence of the profile $y(t)$ of length T is divided into T/l non-overlapping segments of length l. The *local trend* $\tilde{y}_n(t)$ in the n-th segment is defined by fitting the raw profile $y(t)$ in the segment. Here we employ the *first order DFA* method, where the linear least square method is used to fit the profile.

The *detrended profile* $y_l(t)$ is defined as the deviation of the original profile $y(t)$ from the local trend $\tilde{y}_n(t)$:

$$y_l(t) = y(t) - \tilde{y}_n(t), \quad \text{if } nl \le t < (n+1)l. \tag{2}$$

The variance of the detrended sequence is defined as the mean-square of the detrended profile:

$$F^2(l) = \frac{1}{T} \sum_{t=0}^{T-1} y_l^2(t). \tag{3}$$

By analyzing the dependence of the variance $F(l)$ on the segment length l, we find the long-range correlations in non-stationary time sequences. If the variance $F(l))$ obeys a power law

$$F(l) \sim l^{\alpha}, \tag{4}$$

the power spectrum $P(k)$ of the time sequence $u(t)$ behaves as

$$P(k) \sim k^{-\beta}, \quad \beta = 2\alpha - 1. \tag{5}$$

3 Data and Results

In this report we analyze the observed traffic flow data provided by the Japan Highway Public Corporation [18]. The data was obtained at the 468 km point near Seta East IC (Interchange) on the Meishin Expressway connecting Kobe to Nagoya. There were two lanes bound for Kobe (West) in 1999, when the observation was performed. The time sequence of the flow for 5 minutes on Aug. 11th, 1999, is shown in Figs. 1. Traffic congestion in the morning and the evening is observed, which can be seen in the behavior of the average speed. No significant indication for congestion, however, can be recognized in the time-sequence of the flow.

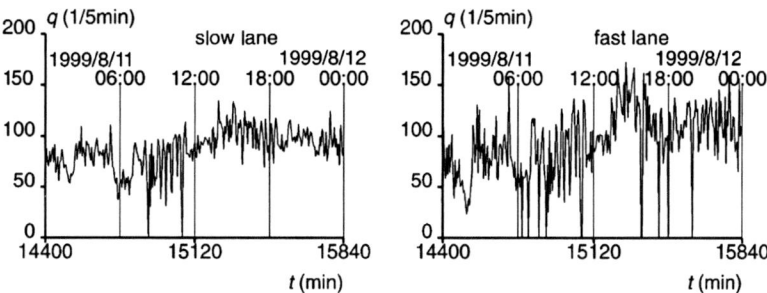

Fig. 1. The time sequence of the traffic flow observed on Aug. 11th, 1999 for the fast lane (right) and the slow lane (left).

The power spectrum of the monthly traffic flow data shows a power-law spectrum on long time-scales. The power spectrum contains the largest peak corresponding to the daily periodic motion. It is difficult to see the power-law behavior in the short time-scale region, which is observed by Musha-Higuchi [5]. The Detrended Fluctuation Analysis (DFA) is applied to the one year data of the traffic flow observed there in 1999. The profile $y(t)$ is constructed from the traffic flow $q(t)$:

$$y(t) = \sum_{i=0}^{t} [q(i) - \langle q \rangle]. \tag{6}$$

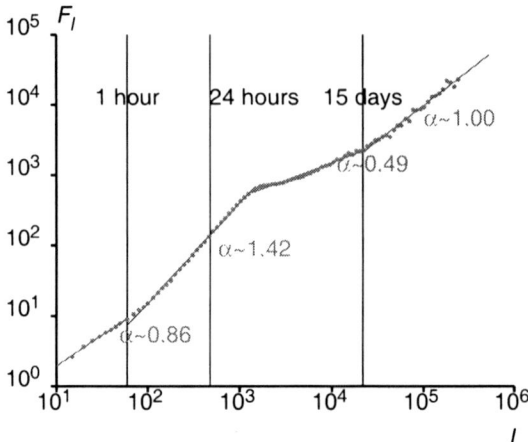

Fig. 2. The dependence of the variance $F(l)$ on the segment size l for the traffic flow data of 1999.

The dependence of the variance $F(l)$ on the segment size l is shown in Fig. 2. There are three crossover time-scales: 1 hour, 24 hours and 15 days. Both in the short time-scale less than 1 hour and in the long time-scale greater than 15 days, the exponent shows $\alpha \sim 1$ which corresponds $1/f$ fluctuation [19]. The feature shown in Fig. 2 is typical: a power-law fluctuation with a periodic trend, which has been studied with artificial time sequences by Hu et al. [20]. The central crossover point indicates the period of the trend. The exponent in both outer regions shows the exponent of the fluctuation. Namely, the traffic flow seems to be a periodic time sequence of one-day period with power-law fluctuations.

4 Modified Traffic Flow Without Daily Trend

Here we confirm that the observed traffic flow consists of a periodic time sequence with power-law fluctuations. The time sequence contains a daily trend as shown in Fig. 2. Hence we define the daily trend of the traffic flow $\tilde{q}_{\text{daily}}(\tau)$ as follows:

$$\tilde{q}_{\text{daily}}(\tau) = \frac{1}{D} \sum_{d=0}^{D-1} q(d \times T_{\text{day}} + \tau), \tag{7}$$

where $0 < \tau < T_{\text{day}} = 24 \cdot 60$ minutes and D is the number of days in a year ($D = 365$ for the year 1999). Then the traffic flow data $q(t)$ is replaced with the modified traffic flow $q'(t)$:

$$q'(t) = q(t) - \tilde{q}_{\text{daily}}(t \bmod T_{\text{day}}). \tag{8}$$

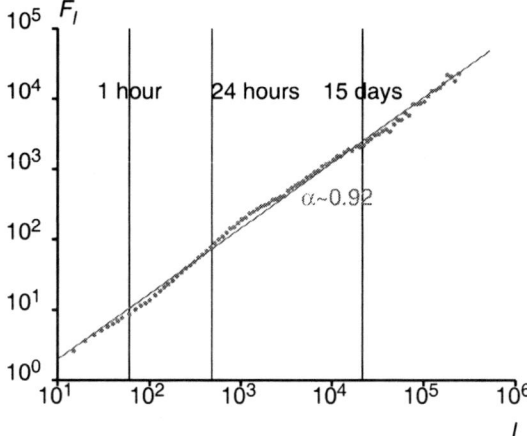

Fig. 3. The dependence of the variance $F(l)$ on the segment size l for the modified traffic flow data $q'(t)$ of 1999.

The DFA method is applied to the modified traffic flow $q'(t)$ (Fig. 3). The weak daily trend remains. The variance $F(l)$, however, can be fitted by $F(l) \sim l^\alpha$ with $\alpha \sim 0.92$. Namely the power-law fluctuation is certainly observed in the wide range from the short time-scale of the order of minutes to the long time-scale of the order of months.

The power-law fluctuation discussed here is a proper feature of traffic flow. It is independent of whether the flow contains a congestion or not. We analyze other observational data, which was taken at the 133.11 km point near Taki IC of Ise Expressway. The traffic at this point flows smoothly without congestion. Moreover the road has only one lane at the observational point. The DFA method is also applied to the modified traffic flow without daily trend. The observed features, however, are almost the same as those in Fig. 3.

5 Summary and Discussion

The time sequence of traffic flow is analyzed and the existence of power-law fluctuations is discussed. Musha and Higuchi have found power-law fluctuations on time-scales shorter than a few hours. A simple power spectrum analysis shows the power-law fluctuations on a time-scale longer than a day. We employ the detrended fluctuation analysis (DFA) method to analyze the long-range correlation in the observed traffic flow data. The result shows that there are three crossover time-scales: one hour, 1 day and 15 days. The feature is typical for a periodic sequence with power-law fluctuations.

To confirm that the traffic flow is a mixture of a daily periodic motion with a power-law fluctuation, modified traffic flow data is defined by extracting the averaged daily trend from the raw data. The modified traffic flow data shows power-law fluctuations on a wide range of time-scales by the DFA method.

We employed the first-order DFA method, in which local trends are fitted with linear equations. The DFA method can be extended to n-th order by using n-th order polynomials to construct local trends. We applied higher-order DFA methods up to fourth order and obtained qualitatively the same results as with the first order DFA method.

The power-law fluctuation observed in simulations will correspond to the observed fluctuations in the short time-scale region. Namely the power-law fluctuation in the short time-scale comes from the microscopic dynamical behavior. A small fluctuation of the car motion in the headway distance or speed will be relaxed according to the microscopic dynamical laws. The density fluctuation will propagate upstream. The propagation of the high density fluctuation is extinguished in the low density region. This is the origin of the power-law behavior on the short time-scale.

There is no obvious explanation why the power-law fluctuation extends to time-scales longer than a few months and with the same exponent. It is difficult to think that the microscopic dynamics generates such a long range power-law fluctuation. Some macroscopic dynamical laws seem to have the same properties as the microscopic traffic dynamics.

Acknowledgements

The authors thank the Japan Highway Public Corporation for providing us the observation data. A part of this work is financially supported by Grant-in-aid No. 15607014 from Ministry of Education, Science, Sports and Culture, Japan.

References

1. M. Fukui, Y. Sugiyama, M. Schreckenberg, and D. E. Wolf (Eds.): *Traffic and Granular Flow '01* (Springer, Berlin, 2003).
2. S. P. Hoogendoorn, S. Luding, P. H. L. Bovy, M. Schreckenberg, and D. E. Wolf (Eds.): *Traffic and Granular Flow '03* (Springer, Berlin, 2005).
3. B. S. Kerner and H. Rehborn: Phys. Rev. E**53**, R1297 (1996).
4. B. S. Kerner and H. Rehborn: Phys. Rev. E**53**, R4275 (1996).
5. T. Musha and H. Higuchi: Jpn. J. Appl. Phys. **15**, 1271 (1976).
6. P. Wagner and J. Peinke: Z. Naturforsch. **52a**, 600 (1997).
7. L. Neubert, L. Santen, A. Schadschneider, and M. Schreckenberg: Phys. Rev. E**60**, 6480 (1999).
8. K. Nishinari and M. Hayashi (Eds.): *Traffic Statistics in Tomei Express Way* (The Mathematical Society of Traffic Flow, Japan, 1999).
9. S. Yukawa and M. Kikuchi: J. Phys. Soc, Jpn. **65**, 916 (1996).
10. S. Tadaki, M. Kikuchi, Y. Sugiyama, and S. Yukawa: J. Phys. Soc, Jpn. **67**, 2270 (1998).

11. S. Tadaki, M. Kikuchi, Y. Sugiyama, and S. Yukawa: J. Phys. Soc, Jpn. **68**, 3110 (1999).
12. O. Moriyama, N. Kuroiwa, M. Matsushita, and H. Hayakawa: Phys. Rev. Lett. **80**, 2833 (1998).
13. S. Tadaki, K. Nishinari, M. Kikuchi, Y. Sugiyama, and S. Yukawa: J. Phys. Soc, Jpn. **71**, 2326 (2002).
14. S. Tadaki, K. Nishinari, M. Kikuchi, Y. Sugiyama, and S. Yukawa: Physica A**315**, 156 (2002).
15. M. Kikuchi, A. Nakayama, K. Nishinari, Y. Sugiyama, S. Tadaki, and S. Yukawa: in *Traffic and Granular Flow '01* (Springer, Berlin, 2003) p. 257.
16. C.-K. Peng, S. V. Buldyrev, S. Havlin, and M. Simons, H. E. Stanley and A. L. Goldberger: Phys. Rev. E**49**, 1685 (1994).
17. C.-K. Peng, S. Havlin, H. E. Stanley, and A. L. Goldberger: Chaos **5**, 82 (1995).
18. The Japan Highway Public Corporation: `http://www.jhnet.go.jp`.
19. S. Tadaki, M. Kikuchi, A. Nakayama, K. Nishinari, A. Shibata, Y. Sugiyama, and S. Yukawa: in *Traffic and Granular Flow '03* (Springer, Berlin, 2005) p. 59.
20. K. Hu, P. Ch. Ivanov, Z. Chen, P. Carpena, and H. E. Stanley: Phys. Rev. E**64**, 011114 (2001).

States of Traffic Flow in the Deep Lefortovo Tunnel (Moscow): Empirical Data

Ihor Lubashevsky[1], Cyril Garnisov[2], Reinhard Mahnke[3], Boris Lifshits[2], and Mikhail Pechersky[2]

[1] A. M. Prokhorov General Physics Institute of Russian Academy of Sciences, Vavilov str., 38, Moscow, 119311 Russia (`ialub@fpl.gpi.ru`)
[2] Research and Project Institute for City Public Transport, Sadovo-Samotechnay, 1, Moscow, 103473 Russia (`mpechersk@tochka.ru`)
[3] Universität Rostock, Institut für Physik, D–18051 Rostock, Germany (`reinhard.mahnke@uni-rostock.de`)

Summary. The paper analyses traffic flow data collected in the Lefortovo tunnel (Moscow) in 2004/05. First, it shows the presence of cooperative traffic dynamics in this tunnel and, second, studies the phase portrait of the vehicle ensemble in the velocity-density plane. In particular, the regions of regular and stochastic dynamics are found and the presence of dynamical traps is demonstrated.

1 Traffic Flow in Long Tunnels

Traffic flow dynamics in long highway tunnels has been studied individually since the middle of the last century (see, e.g., Refs. [1, 2]). Interest in this problem is due to several reasons. The first, and maybe main one, is safety. Jam formation in long tunnels is rather dangerous and detecting the critical states of vehicle flow leading to jams is of prime importance for the tunnel operation. However, the tunnel traffic in its own right is also an attractive object for studying the basic properties of vehicle ensembles on highways because, on one hand, the individual car motion is more controllable inside tunnels with respect to velocity limits and lane changing. On the other hand, long tunnels typically are well equipped for monitoring the car motion practically continuously along them, which provides a unique opportunity to receive a detailed information about the spatio-temporal structures of traffic flow.

By this paper we start the analysis of the basic properties exhibited by congested tunnel traffic based on empirical data collected during the last time in several new deep long tunnels located on the 3rd circular highway of Moscow. Here preliminary results for the Lefortovo tunnel (Fig. 1) are presented. It comprises two branches where the upper one is a deep linear three lane tunnel with a length of about 3 km. Exactly in this branch the presented data were

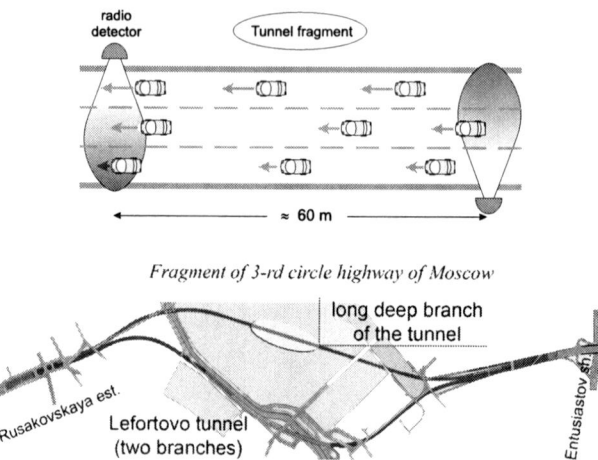

Fig. 1. Lefortovo tunnel structure.

collected. The tunnel is equipped with a dense system of stationary radiodetetors distributed uniformly along it chequerwise at spacing of 60 m. Because of the technical features of the detector traffic flow on the left and right lanes is measured at a spacing of 120 m whereas on the middle lane the spatial resolution is 60 m. The data were averaged over 30 s.

Each detector measures three characteristics of the vehicle ensembles; the flow rate q, the car velocity v, and the occupancy k for three lanes individually. The occupancy is analog to the vehicle density and is defined as the total relative time during which vehicles were visible in the view region of a given detector within the averaging interval. It is measured in percent.

2 Observed Cooperative Motion of Vehicle Ensemble

This section demonstrates that the observed traffic flow indeed exhibits cooperative dynamics when the vehicle density becomes high enough. To do this figure 2 (upper frames) depicts the phase planes $\{k, v\}$ and $\{k, q\}$ with the distribution of the traffic flow states fixed by all the detectors on 31.05.2004. These phase planes were divided into cells of size about $1\% \times 2$ km/h and $1\% \times 0.02$ car/s, respectively. The number of states is measured with frequency $1/30$ s^{-1} and those falling in a chosen cell were counted, giving the corresponding distributions. These distributions are represented here in some relative units in form of level contours. The left side of each window matches the free flow states as is clearly seen in the right window, where the darkened region visualizes an upper fragment of the flow-density relation of the free car motion. However the obtained distributions even for the free flow are widely scattered which seems to be due to the essential heterogeneity of the

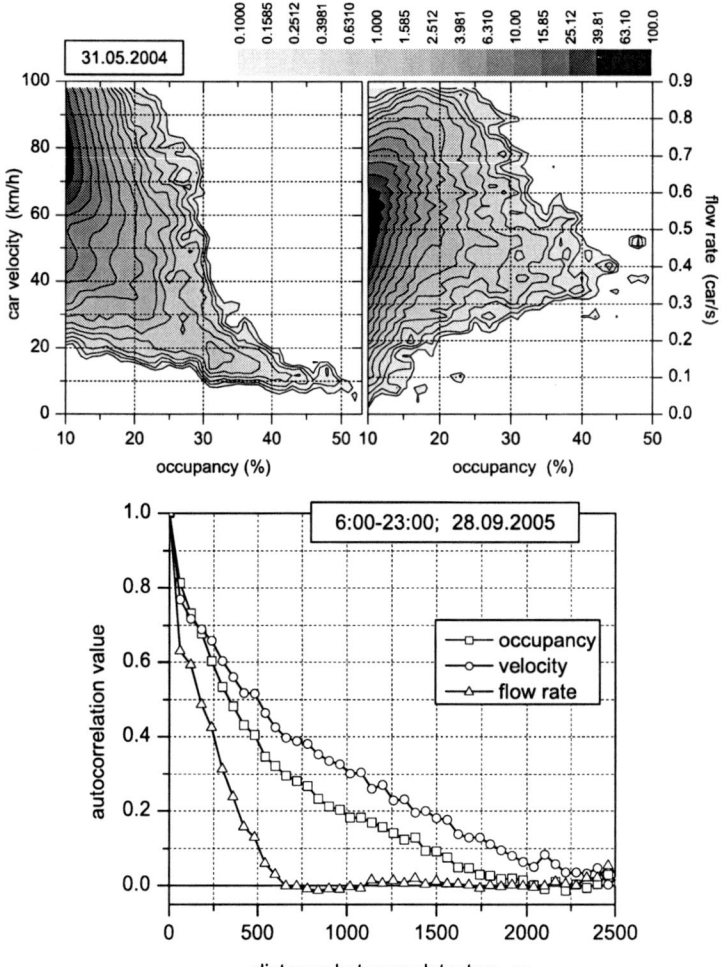

Fig. 2. Fundamental diagrams (upper frames) and autocorrelations in the occupancy, car velocity, and flow rate measured by differing detectors vs. the distance between them (lower frame).

free flow with respect to the headway distances. The middle parts of these windows visualize another mode of traffic flow corresponding to the so-called widely scattered states or the synchronized vehicle motion (for a review see Refs. [3, 4]). In fact here the distribution levels cover rather wide regions and do not follow each other so frequently as in the left part. Exactly this mode is usually related to the cooperative vehicle motion.

Figure 2 (lower frame) exhibits the spatial autocorrelations in the occupancy, car velocity, and flow rate measured by different detectors at the middle lane on 28.09.2005 when congested traffic was dominant. In agreement with the single-vehicle data [5] the congested vehicle motion is characterized by essential correlations especially in the car velocity. The flow rate measurements are correlated substantially only within several neighboring detectors (on scales about several hundred meters) whereas the velocity measurements as well as the occupancy ones are correlated at half of the tunnel length, i.e. at scales about one kilometer.

3 Pattern of Vehicle Ensemble Dynamics in the Phase Space

The characteristics of the vehicle ensemble dynamics in the phase space $\{k, v\}$ were studied in the following way, replicating actually the technique of Ref. [6] used in a similar analysis. The plane $\{k, v\}$ is divided into cells $\{\mathcal{C}\}$ of size $2.5\% \times 2.5$ km/h. Let at time t the traffic flow measurements of a given detector fall in a cell $\mathcal{C}_i$ and in the averaging time $dt = 30$ s the next measurements of the same detector are located in a cell $\mathcal{C}_j$. Then the vector $d\mathbf{r} := \{dk_t, dv_t\}$ such that $dk_t = k_j - k_i$ and $dv_t = v_j - v_i$ describes the system motion on the phase plane at the given point $\mathbf{r}_i := \{k_i, v_i\}$ at time t. These vectors were calculated using the data collected on 28.09.2005 by all the detectors. Averaging the vectors found gives the drift field $\mathbf{V}_m(\mathbf{r}) = \langle d\mathbf{r} \rangle / dt$ and the intensity $D(\mathbf{r})$ of an effective random force determined as

$$Ddt = \sqrt{\langle |d\mathbf{r}| \rangle^2 - \langle d\mathbf{r} \rangle^2}.$$

Figure 3 exhibits these fields. The upper window depicts the ratio $\eta :=$ $D/|\mathbf{V}|_m$, namely, its variations from 0 up to 3.5. The white region comprises the cells where no measurements were obtained. The hatched domain matches the ratio $\eta > 3.5$, where the vehicle ensemble dynamics can be regarded as purely random. The region between them contains several levels of the variation of the ratio η. The level $\eta = 1.0$ is singled out in Fig. 3. For smaller values of η the dynamics of the vehicle ensemble becomes practically regular. The lower window of Fig. 3 shows the drift field $\mathbf{V}_m(\mathbf{r})$. Since its intensity changes essentially at different parts of the plane $\{k, v\}$ two frames are used to visualize it. In the left frame the drift field is zoomed in by three times relative to the right one. Let us consider them individually. The system dynamics in the right frame is rather regular and the fie;d $\mathbf{V}_m(\mathbf{r})$ corresponds to the irrelievable drift of vehicle ensemble to smaller velocities and higher densities. In other words, it is some visualization of the jam formation. In fact one or two jams were obvserved on that day. It should be noted that the transition region separating the left frame pattern being rather chaotic and the given one is relatively thin. It is located at $k = 35\%$ and has a thickness less then

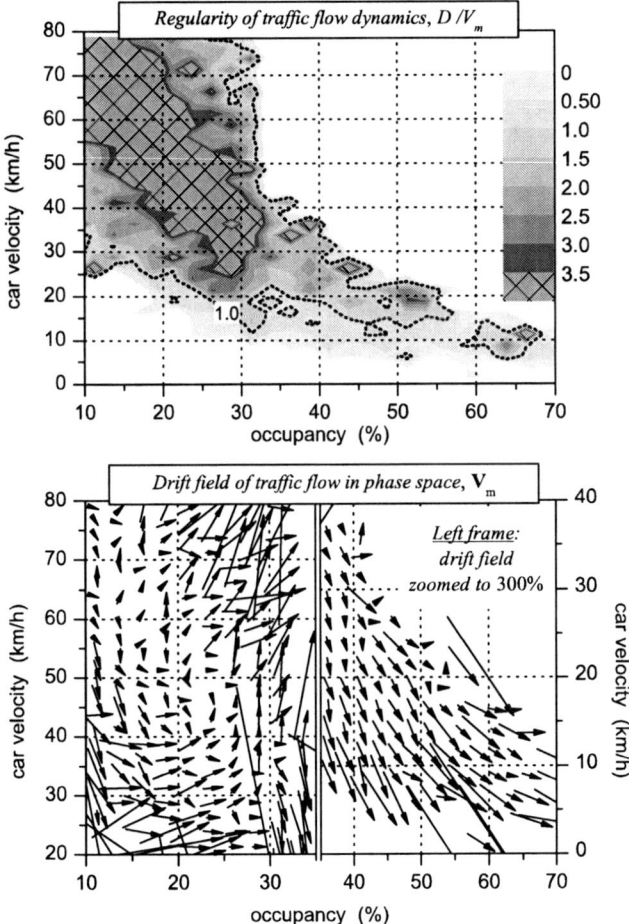

Fig. 3. The upper window visualizes the distribution of the ratio between the random and regular components of the effective forces. The lower window depicts the regular drift field.

5%. So the observed jam formation seems to proceed via some breakdown in the cooperative vehicle motion, which is an agreement with other data [4].

The pattern shown in the left frame matches the upper one in structure. Inside a neighborhood Q_0 of the decreasing frame diagonal the traffic dynamics is practically pure chaotic, at least, the found values of $\mathbf{V}_m(\mathbf{r})$ are relatively small and their directions do not form any regular pattern. As it must, outside this domain the field $\mathbf{V}_m(\mathbf{r})$ becomes more regular and the obtained data enable us to estimate its characteristic direction. Unexpectedly, it turns out that the field $\mathbf{V}_m(\mathbf{r})$ crossing this neighborhood does not change its direction for backward one as it should be if the domain Q_0 has contained a zero set of the regular field $\mathbf{V}_m(\mathbf{r})$. Such behavior of a dynamical system can be explained using the notion of dynamical traps predicting also the existence of a long-lived state multitude as a consequence of some nonequilibrium phase transitions caused by the human bounded rationality [7–9].

4 Conclusions

The paper presents a preliminary analysis of traffic flow data collected in the Lefortovo tunnel located on the 3rd circular highway of Moscow or, more precisely, in its upper linear branch being a deep three lane tunnel of length 3 km. Radiodetectors for vehicle motion are distributed chequerwise at practically uniform spacing of 60 m. The measured data are averaged over 30 s.

It is shown that the observed congested traffic in fact exhibits cooperative phenomena in vehicle motion, namely, there is a region of widely scattered states on the fundamental diagrams which is related typically to the appearance of synchronized traffic. Besides, the spatial autocorrelations in the occupancy, vehicle velocity, and flow rate measured by different detectors are found to be essential. Especially it concerns the correlations in the velocity and occupancy, their correlation length takes values of about 1 km. The occupancy data are correlated on substantially shorter scales about 200–300 m.

The phase portrait of the vehicle ensemble dynamics on the occupancy-velocity plane is also studied. It is demonstrated that there are two substantially different regions on it. One matches the cooperative vehicle motion and encloses some kernel where the dynamics is purely chaotic. It is essential that the found regular drift outside this region does not change the direction when crossing it. The latter feature is some prompt to apply the concept of dynamical traps for describing phase transition in congested traffic. The other part of the phase plane corresponds to the irreversible stage of jam formation. The two regions are separated by a rather narrow transition layer located at $k = 35\%$, which demonstrates that the observed jams originated inside a congested traffic via some breakdown.

Acknowledgements

This paper was supported in part by DFG Project 436 RUS 17/122/04, RFBR Grants 05-01-00723, 05-07-90248, and Moscow Grant 1.1.258.

References

1. H.C. Chin and A.D. May: Examination of the Speed-Flow Relationship at the Caldecott Tunnel. In: *Transportation Research Record* **1320**, Transportation Reserch Board (NRC, Washington, DC, 1991), pp. 1–15.
2. R.W. Rothery: Car following models. In: *Traffic Flow Theory* Transportation Research Board, Special Report **165**, ed. by N. Gartner, C.J. Messer, and A.K. Rathi (1992), Chap. 4.
3. D. Helbing: Rev. Mod. Phys. **73**, 1067 (2001).
4. B.S. Kerner: *Physics of traffic flow* (Springer, Berlin, 2004).
5. L. Neubert, L. Santen, A. Schadschneider, and M. Schreckenberg: Phys. Rev. E **60**, 6480 (1999).
6. R. Friedrich, S. Kriso, J. Peinke, P. Wagner: Phys. Lett. A **299**, 287 (2002).
7. I. Lubashevsky, R. Mahnke, P. Wagner, and S. Kalenkov: Phys. Rev. E **66**, 016117 (2002)
8. I. Lubashevsky, M. Hajimahmoodzadeh, A. Katsnelson, and P. Wagner: Eur. Phys. J. B **36**, 115 (2003).
9. I. Lubashevsky, R. Mahnke, M. Hajimahmoodzadeh, and A. Katsnelson: Eur. Phys. J. B **44**, 63 (2005).

List of Contributors

Miglius Alaburda
Institute of Theoretical Physics and
Astronomy
Vilnius University
A. Gostauto 12
LT-01108 Vilnius
Lithuania
miglius@itpa.lt

Bruno Andreotti
ESPCI-Universités Paris 6 and 7
Laboratoire de Physique et Mécanique
des Milieux Hétérogénes
UMR7636
10, rue Vauquelin
75005 Paris
France
andreotti@pmmh.espci.fr

Cécile Appert-Rolland
Université de Paris-Sud
Laboratoire de Physique Théorique
Bâtiment 210
F-91405 Orsay Cedex
France
Cecile.Appert-Rolland@th.u-psud.fr

Igor S. Aranson
Materials Science Division
Argonne National Laboratory
9700 South Cass Avenue
Argonne, IL 60439
USA
aronson@anl.gov

André Audoly
L.P.M.C - Université de Nice
Parc Valrose
06108 Nice Cedex 2
France
audoly@unice.fr

Marcel Ausloos
Supratecs
University of Liège
Liège B-4000
Belgium
Marcel.Ausloos@ulg.ac.be

Arnaud Banos
SET Laboratory
University of Pau
France
arnaud.banos@univ-pau.fr

Andrea Baldassarri
Dipartimento di Fisica
Università "La Sapienza"
P.le A. Moro 2
00185 Roma
Italy
andrea.baldassarri@roma1.infn.it

Peter Berg
Faculty of Science
University of Ontario Institute of
Technology
2000 Simcoe Street N
Oshawa, ON, L1H 7K4
Canada
Peter.Berg@uoit.ca

Mark Blumberg
Laboratory for Experimental Economics
University of Bonn
53113 Bonn
Germany
mark@blumberg-online.de

Maik Boltes
Central Institute for Applied Mathematics
Forschungszentrum Jülich GmbH
52425 Jülich
Germany
m.boltes@fz-juelich.de

Georges Bossis
L.P.M.C - Université de Nice
Parc Valrose
06108 Nice Cedex 2
France
bossis@unice.fr

André Brahic
Université Paris VII Denis Diderot
A.I.M., C.E.A. Saclay
France
Member of the Cassini spacecraft
Imaging Team
brahic@cea.fr

Marco Brück
Laboratory for Experimental Economics
University of Bonn
53113 Bonn
Germany
mbrueck@gmx.net

Alexander P. Buslaev
Moscow State Automobile and Road
Technical University

64, Leningradsky pr.
Moscow
Russia
busl@math.madi.ru

Michael Cassidy
Department of Civil and Environmental
Engineering
University of California
Berkeley
cassidy@ce.berkeley.edu

Christophe Chalons
Université Paris 7 - Denis Diderot &
Laboratoire J.-L
Lions
U.M.R. 7598
Université Pierre et Marie Curie
Boîte courrier 187
75252 Paris Cedex 05
France
chalons@math.jussieu.fr

Thorsten Chmura
Laboratory for Experimental Economics
University of Bonn
53113 Bonn
Germany
    and
Shanghai Jiao Tong University
Antai School of Managment
Shanghai 200052
People's Republic of China
chmura@uni-bonn.de

Debashish Chowdhury
Department of Physics
Indian Institute of Technology
Kanpur 208016
India
debch@iitk.ac.in

Massimo Pica Ciamarra
Dip.to di Scienze Fisiche
Università di Napoli "Federico II"
INFM-Coherentia, INFN and AMRA
Napoli
Italy
picaciamarra@na.infn.it

Eric Clement
ESPCI-Universités Paris 6 and 7
Laboratoire de Physique et Mécanique
des Milieux Hétérogénes
UMR7636
10, rue Vauquelin
75005 Paris
France
erc@ccr.jussieu.fr

Benjamin Coifman
Department of Electrical and Computer
Engineering
The Ohio State University
Columbus, Ohio, 43210
USA
    and
Department of Civil and Environmental
Engineering and Geodetic Science
The Ohio State University
Columbus, Ohio, 43210
USA
coifman.1@osu.edu

Antonio Coniglio
Dip.to di Scienze Fisiche
Università di Napoli "Federico II"
INFM-Coherentia, INFN and AMRA
Napoli
Italy
coniglio@na.infn.it

Winnie Daamen
Delft University of Technology
Faculty of Civil Engineering and
Geosciences - Transport & Planning
Stevinweg 1
2628 CN Delft
The Netherlands
w.daamen@tudelft.nl

Carlos Daganzo
Department of Civil and Environmental
Engineering
University of California
Berkeley
daganzo@ce.berkeley.edu

Blanche Dalloz-Dubrujeaud
IUSTI
Université de Provence
5 rue Enrico Fermi
F-13453 Marseille Cedex 13
France
blanche.dalloz@polytech.univ-mrs.fr

Fergal Dalton
Istituto dei Sistemi Complessi - CNR
via del Fosso del Cavaliere 100
00133 Roma
Italy
fergal.dalton@isc.cnr.it

Amir H. Darooneh
Department of Physics
Zanjan University
P.O. Box 45196-313
Zanjan
Iran

Manuel Díez-Minguito
Institute 'Carlos I' for Theoretical and
Computational Physics
    and
Departamento de Electromagnetismo y
Física de la Materia
Universidad de Granada
E-18071 Granada
Spain
mdiez@onsager.ugr.es

Martin R. Evans
SUPA, School of Physics,
The University of Edinburgh
Mayfield Road
Edinburgh EH9 3JZ
Scotland
martin@ph.ed.ac.uk

Justin Findlay
Faculty of Science
University of Ontario Institute of
Technology
2000 Simcoe Street N
Oshawa, ON, L1H 7K4
Canada
justinfindlay@gmail.com

Sebastian Fischer
Physik Department
TU München
85748 Garching
Germany
and
Hahn–Meitner Institut
Glienicker Straße 100
14109 Berlin
Germany
sfischer@ph.tum.de

M. Ebrahim Fouladvand
Department of Physics
Zanjan University
P.O. Box 45196-313
Zanjan
Iran
and
Institute for Studies in Theoretical
Physics & Mathematics (IPM)
Department of Nano-Science
P.O. Box 19395-5531
Tehran
Iran
foolad@iasbs.ac.ir

Erwin Frey
Arnold Sommerfeld Center and CeNS
Department for Physics
Ludwig-Maximilians-Universität
München
Theresienstrasse 37
D-80333 München
Germany
frey@lmu.de

Minoru Fukui
Nakanihon Automotive College
Sakahogi-cho
Gifu-ken 505-0077
Japan
fukui3@cc.nagoya-u.ac.jp

Adam Gadomski
University of Technology and Agriculture
Bydgoszcz PL85-796
Poland
agad@atr.bydgoszcz.pl

Bin Gao
Department of Electrical and Computer
Engineering
The Ohio State University
Columbus, Ohio, 43210
USA
gao.82@osu.edu

Mauro Garavello
Dipartimento di Matematica e
Applicazioni
Università di Milano Bicocca
Via R. Cozzi 53 - Edificio U5
20125 Milano
Italy
mauro.garavello@unimib.it

Cyril Garnisov
Research and Project Institute for City
Public Transport
Sadovo-Samotechnay 1
Moscow, 103473
Russia

Pedro L. Garrido
Institute 'Carlos I' for Theoretical and
Computational Physics
and
Departamento de Electromagnetismo y
Física de la Materia
Universidad de Granada
E-18071 Granada
Spain
plgarrido@onsager.ugr.es

Paola Goatin
Laboratoire d'Analyse Non linéaire
Appliquée et Modélisation
I.S.I.T.V.
Université du Sud Toulon - Var
83162 La Valette du Var Cedex
France
goatin@univ-tln.fr

Abhimanyu Godara
Dept. of Civil Eng., Institute of
Technology, Banaras Hindu
University, 221005 Varanasi, India
abhimanyu_godara@rediffmail.com

Vygintas Gontis
Institute of Theoretical Physics and
Astronomy
Vilnius University
A. Gostauto 12
LT-01108 Vilnius
Lithuania
gontis@itpa.lt

Andreas Götzendorfer
Experimentalphysik V
Universität Bayreuth
D–95440 Bayreuth
Germany
andreas.goetzendorfer@
uni-bayreuth.de

Céline Goujon
IUSTI
Université de Provence
5 rue Enrico Fermi
F-13453 Marseille Cedex 13
France
    and
Instituto de Investigaciones en
Materiales
Universidad Nacional Autónoma de
México
Apdo. Postal 70-360
Cd. Universitaria
México D.F. 04510
México
celine.goujon@polytech.univ-mrs.fr

Sean Gourley
Physics Department
Clarendon Laboratory
Parks Road
Oxford, OX1 3PU
U.K.
s.gourley1@physics.ox.ac.uk

Yan Grasselli
EAI Tech - CERAM
rue A. Einstein
BP 085
06902 Sophia Antipolis Cedex
France
yan.grasselli@cote-azur.cci.fr

Philip Greulich
Institut für Theoretische Physik
Universität zu Köln
D-50937 Köln
Germany
pg@thp.uni-koeln.de

Rafał Grochowski
Mechanische Verfahrenstechnik
Universität Dortmund
D–44227 Dortmund
Germany
rafal.grochowski@bci.uni-dortmund.de

Rosemary J. Harris
Institut für Festkörperforschung
Forschungszentrum Jülich
D–52425 Jülich
Germany
r.harris@fz-juelich.de

Dirk Helbing
Institute for Economics and Traffic
Technische Universität Dresden
Andreas-Schubert-Str. 23
D-01062 Dresden
Germany
    and
Collegium Budapest – Institute for
Advanced Study
Szentháromság u. 2
H-1014 Budapest
Hungary
helbing1@vwi.tu-dresden.de

Hans J. Herrmann
Departamento de Física
Universidade Federal do Ceará
Campus do Pici
60451-970, CE
Brazil
hans@icp.uni-stuttgart.de

Julia Hinkel
Institut für Physik
Universität Rostock
D–18051 Rostock
Germany
julia.hinkel@uni-rostock.de

Hauke Hinsch
Arnold Sommerfeld Center and CeNS
Department for Physics
Ludwig-Maximilians-Universität
München
Theresienstrasse 37
D-80333 München
Germany
hauke.hinsch@physik.lmu.de

Andreas Hoffmann
Westfälische Wilhelms-Universität
Münster
Institut für Theoretische Physik
Wilhelm-Klemm-Str. 9
D-48149 Münster
Germany
andreash@uni-muenster.de

Serge P. Hoogendoorn
Delft University of Technology
Faculty of Civil Engineering and
Geosciences – Transport & Planning
Stevinweg 1
2628 CN
Delft
The Netherlands
s.p.hoogendoorn@tudelft.nl

Mao-Bin Hu
School of Engineering Science
University of Science and Technology of
China
Hefei 230026
P.R.China
humaobin@ustc.edu.cn

Ding-wei Huang
Department of Physics
Chung Yuan Christian University
Chung-li
Taiwan
dwhuang@phys.cycu.edu.tw

Yoshihiro Ishibashi
School of Business
Aichi Shukutoku University
Nagakute-cho
Aichi-ken 480-1197

Japan
yishi@asu.aasa.ac.jp

Rui Jiang
Institute for Economics and Traffic
Technische Universität Dresden
Andreas-Schubert-Str. 23
D-01062 Dresden
Germany
  and
University of Science and Technology of
China
Hefei 230026
P.R.China
rjiang@ustc.edu.cn

Alexander John
Universität zu Köln
Institut für Theoretische Physik
50937 Köln
Germany
aj@thp.uni-koeln.de

Neil F. Johnson
Physics Department
Clarendon Laboratory
Parks Road
Oxford, OX1 3PU
U.K.
n.johnson@physics.ox.ac.uk

Johannes Kaiser
Laboratory for Experimental Economics
University of Bonn
53113 Bonn
Germany
johannes.kaiser@uni-bonn.de

Masahiro Kanai
University of Tokyo
3-8-1 Komaba Meguro-ku
Tokyo 153-8914
Japan
kanai@ms.u-tokyo.ac.jp

Yuko Kanayama
Department of Aeronautics and
Astronautics
Faculty of Engineering

University of Tokyo
Hongo, Bunkyo-ku
Tokyo 113-8656
Japan
yu_kanayama@hotmail.com

Bronislovas Kaulakys
Institute of Theoretical Physics and
Astronomy
Vilnius University
A. Gostauto 12
LT-01108 Vilnius
Lithuania
kaulakys@itpa.lt

Jevgenijs Kaupužs
Institute of Mathematics and Computer
Science
University of Latvia
LV–1459 Riga
Latvia
kaupuzs@latnet.lv

Arne Kesting
Institute for Economics and Traffic
Technische Universität Dresden
Andreas-Schubert-Str. 23
D-01062 Dresden
Germany
kesting@vwi.tu-dresden.de

Jan Kierfeld
Max Planck Institute of Colloids and
Interfaces
Science Park Golm
14424 Potsdam
Germany
Jan.Kierfeld@mpikg.mpg.de

Macoto Kikuchi
Cybermedia Center
Osaka University
Toyonaka 560-0043
Japan
kikuchi@cmc.osaka-u.ac.jp

Wolfram Klingsch
Institute for Building Material
Technology and Fire Safety Science

University of Wuppertal
Pauluskirchstrasse 7
42285 Wuppertal
Germany
klingsch@uni-wuppertal.de

Stefan Klumpp
Max-Planck-Institut für Kolloid- und
Grenzflächenforschung
Wissenschaftspark Golm
14424 Potsdam
Germany
klumpp@mpikg-golm.mpg.de

Hubert Klüpfel
TraffGo HT GmbH
Falkstraße 73–77
47057 Duisburg
Germany
kluepfel@traffgo-ht.com

Victor L. Knoop
Delft University of Technology
Faculty of Civil Engineering and
Geosciences – Transport & Planning
Stevinweg 1
2628 CN
Delft
The Netherlands
v.l.knoop@citg.tudelft.nl

Henning Arendt Knudsen
Dept. of Physics
University of Oslo
P.O.Box 1048 Blindern
NO-0316 Oslo
Norway
arendt@phys.ntnu.no

Xiang-Zhao Kong
School of Engineering Science
University of Science and Technology of
China
Hefei 230026
P.R.China
xzkong@mail.ustc.edu.cn

Roger Kouyos
Theoretical Biology
Eidgenössische Technische Hochschule
Zürich
Universitätsstrasse 16
CH-8092 Zürich
Switzerland
roger.kouyos@env.ethz.ch

Pavel Kraikivski
Max Planck Institute of Colloids and
Interfaces
Science Park Golm
14424 Potsdam
Germany
Pavel.Kraikivski@mpikg.mpg.de

Klaus Kroy
Institut für Theoretische Physik
Universität Leipzig
Augustusplatz 10/11
04109 Leipzig
Germany
Klaus.Kroy@itp.uni-leipzig.de

Florian Kranke
Volkswagen AG
Postfach 011/1895
D-38436 Wolfsburg
Germany
florian.kranke@volkswagen.de

Christof A. Krülle
Experimentalphysik V
Universität Bayreuth
D–95440 Bayreuth
Germany
christof.kruelle@uni-bayreuth.de

Natalia Kruszewska
University of Technology and Agricul-
ture
Bydgoszcz PL85-796
Poland
nkruszewska@atr.bydgoszcz.pl

Reinhart Kühne
German Aerospace Center
Institute of Transportation Research
D–12489 Berlin
Germany;
reinhart.kuehne@dlr.de

Torsten Kühne
Max Planck Institute of Colloids and
Interfaces
Science Park Golm
14424 Potsdam
Germany
Torsten.Kuehne@mpikg.mpg.de

Ambarish Kunwar
Department of Physics
Indian Institute of Technology
Kanpur 208016
India
ambarish@iitk.ac.in

Sylvain Lassarre
GARIG
Institut National De Recherche Sur Les
Transports Et Leur Securite
Champs Sur Marne, France
lassarre@inrets.fr

Jorge A. Laval
Laboratoire Ingénierie Circulation
Transport LICIT
(INRETS/ENTPE)
France
jlaval@gmail.com

Boris Lifshits
Research and Project Institute for City
Public Transport
Sadovo-Samotechnay 1
Moscow, 103473
Russia
atp@mgtnip.ru

Mao Lin
School of Automobile and Traffic
Engineering
Jiangsu University
Zhenjiang
Jiangsu
maolin19821126@163.com

Stefan J. Linz
Westfälische Wilhelms-Universität
Münster
Institut für Theoretische Physik
Wilhelm-Klemm-Str. 9
D-48149 Münster
Germany
slinz@uni-muenster.de

Reinhard Lipowsky
Max Planck Institute of Colloids and
Interfaces
Science Park Golm
14424 Potsdam
Germany
lipowsky@mpikg.mpg.de

Thomas Lippert
Central Institute for Applied Mathematics
Forschungszentrum Jülich GmbH
52425 Jülich
Germany
th.lippert@fz-juelich.de

Ihor Lubashevsky
A.M. Prokhorov General Physics
Institute of The Russian Academy of
Sciences
Vavilov str. 38
Moscow, 119311
Russia
ialub@fpl.gpi.ru

Reinhard Mahnke
Institut für Physik
Universität Rostock
D–18051 Rostock
Germany
reinhard.mahnke@uni-rostock.de

Florent Malloggi
ESPCI-Universités Paris 6 and 7
Laboratoire de Physique et Mécanique
des Milieux Hétérogénes
UMR7636
10, rue Vauquelin
75005 Paris
France
malloggi@ccr.jussieu.fr

Yousuke Matsuo
Laboratory of Physics
College of Science and Technology
Nihon University
Chiba 274-8501
Japan
matsuo@phys.ge.cst.nihon-u.ac.jp

Joaquín Marro
Institute 'Carlos I' for Theoretical and
Computational Physics
    and
Departamento de Electromagnetismo y
Física de la Materia
Universidad de Granada
E-18071 Granada
Spain
jmarro@ugr.es

Wolfram Mauser
Department of Earth and Environmen-
tal Sciences
University of Munich
Luisenstraße 37
D-80333 Munich
Germany
w.mauser@iggf.geo.uni-muenchen.de

Tadas Meskauskas
Institute of Theoretical Physics and
Astronomy
Vilnius University
A. Gostauto 12
LT-01108 Vilnius
Lithuania
meska@ktl.mii.lt

Melanie J. I. Müller
Max-Planck-Institut für Kolloid- und
Grenzflächenforschung
Wissenschaftspark Golm
14424 Potsdam
Germany
mmueller@mpikg.mpg.de

José Daniel Muñoz
Departamento de Fisica
Universidad Nacional de Colombia
Cra 30 # 45- 03
Bogotá, Colombia

Akio Nakahara
Laboratory of Physics
College of Science and Technology
Nihon University
Chiba 274-8501
Japan
nakahara@phys.ge.cst.nihon-u.ac.jp

Akihiro Nakayama
Department of Physics and Earth
Sciences
University of Ryukyus
Okinawa 903-0213
Japan
nakayama@sci.u-ryukyu.ac.jp

Alireza Namazi
Universität zu Köln
Institut für Theoretische Physik
50937 Köln
Germany
an@thp.uni-koeln.de

Paul Nelson
Department of Computer Science
Texas A&M University
College Station
Texas 77843-3112
pnelson@cs.tamu.edu

Mario Nicodemi
Dip.to di Scienze Fisiche
Università di Napoli "Federico II"
INFM-Coherentia, INFN and AMRA
Napoli
Italy
nicodem@na.infn.it

Bernard Nienhuis
Institute for Theoretical Physics
Universiteit van Amsterdam
Valckenierstraat 65
1018 XE Amsterdam
The Netherlands
nienhuis@science.uva.nl

Katsuhiro Nishinari
Department of Aeronautics and
Astronautics

Faculty of Engineering
University of Tokyo
Hongo
Bunkyo-ku
Tokyo 113-8656
Japan
tknishi@mail.ecc.u-tokyo.ac.jp

Yasushi Okada
Department of Cell Biology and
Anatomy
Graduate School of Medicine
University of Tokyo
Hongo
Bunkyo-ku
Tokyo 113-0033
Japan
yokada@m.u-tokyo.ac.jp

Luis Eduardo Olmos
Departamento de Fisica
Universidad Nacional de Colombia
Cra 30 # 45- 03
Bogotá, Colombia
leolmoss@unal.edu.co

Saskia Ossen
Transport & Planning Department
Delft University of Technology
The Netherlands
s.j.l.ossen@tudelft.nl

Srdjan Ostojic
Institute for Theoretical Physics
Universiteit van Amsterdam
Valckenierstraat 65
1018 XE Amsterdam
The Netherlands
srdjan@science.uva.nl

Mikhail Pechersky
Research and Project Institute for City
Public Transport
Sadovo-Samotechnay 1
Moscow, 103473
Russia
mperchersk@tochka.ru

Zhang Peng
School of Automobile and Traffic
Engineering
Jiangsu University
Zhenjiang
Jiangsu
zhangpeng547@126.com

Alberto Petri
Istituto dei Sistemi Complessi - CNR
via del Fosso del Cavaliere 100
00133 Roma
Italy
alberto.petri@isc.cnr.it

Benedetto Piccoli
Istituto per le Applicazioni del Calcolo
"M. Picone"
Viale del Policlinico 137
00161 - Roma
Italy
piccoli@iac.rm.cnr.it

Luciano Pietronero
Dipartimento di Fisica
Università "La Sapienza"
P.le A. Moro 2
00185 Roma
Italy
   and
Istituto dei Sistemi Complessi - CNR
Via dei Taurini 19
00185 Roma
Italy
luciano.pietronero@roma1.infn.it

Thomas Pitz
Laboratory for Experimental Economics
University of Bonn
53113 Bonn
Germany
   and
Shanghai Jiao Tong University
Antai School of Managment
Shanghai 200052
People's Republic of China
tpitz@uni-bonn.de

Giorgio Pontuale
Istituto dei Sistemi Complessi - CNR
via del Fosso del Cavaliere 100
00133 Roma
Italy
giorgio.pontuale@isc.cnr.it

Thorsten Pöschel
Center for Musculoskeletal Surgery
Charité, University Medicine Berlin
Free and Humboldt-University of Berlin
Augustenburger Platz 1
D-13353 Berlin
Germany
thorsten.poeschel@charite.de

Andreas Pottmeier
Physik von Transport und Verkehr
Universität Duisburg-Essen
47057 Duisburg
Germany
pottmeier@traf1.uni-duisburg.de

Ingo Rehberg
Experimentalphysik V
Universität Bayreuth
D–95440 Bayreuth
Germany
ingo.rehberg@uni-bayreuth.de

Mustapha Rouijaa
Experimentalphysik V
Universität Bayreuth
D–95440 Bayreuth
Germany
musrouij@web.de

Julius Ruseckas
Institute of Theoretical Physics and
Astronomy
Vilnius University
A. Gostauto 12
LT-01108 Vilnius
Lithuania
julius@itpa.lt

Ludger Santen
Universität des Saarlandes
Fachrichtung Theoretische Physik

D-66041 Saarbrücken
Germany
santen@lusi.uni-sb.de

Alessandro Sarracino
Dip.to di Scienze Fisiche
Università di Napoli "Federico II"
INFM-Coherentia, INFN and AMRA
Napoli
Italy
sarracino@na.infn.it

Andreas Schadschneider
Universität zu Köln
Institut für Theoretische Physik
50937 Köln
Germany
as@thp.uni-koeln.de

Martin Schönhof
Institute for Transport & Economics
Technische Universität Dresden
Andreas-Schubert-Straße 23
D-01062 Dresden
Germany
martin@vwitme011.vkw.tu-dresden.de

Michael Schreckenberg
Universität Duisburg-Essen
Campus Duisburg
Physik von Transport und Verkehr
47057 Duisburg
Germany
schreckenberg@traf1.uni-duisburg.de

Marco Schreuder
Traffic Research Centre
Netherlands Ministry of Transport
The Netherlands
M.Schreuder@avv.rws.minvenw.nl

Armin Seyfried
Central Institute for Applied Mathematics
Forschungszentrum Jülich GmbH
52425 Jülich
Germany
a.seyfried@fz-juelich.de

Akihiro Shibata
Computing Research Center
KEK
Tsukuba 305-0801
Japan
ashibata@post.kek.jp

Florian Siebel
Department of Earth and Environmental Sciences
University of Munich
Luisenstraße 37
D-80333 Munich
Germany
f.siebel@iggf.geo.uni-muenchen.de

David A. Smith
Bristol Centre for Applied Nonlinear Mathematics
Department of Engineering Mathematics
University of Bristol
Queen's Building
University Walk
Bristol BS8 1TR
United Kingdom
D.A.Smith@bristol.ac.uk

Bernhard Steffen
Central Institute for Applied Mathematics
Forschungszentrum Jülich GmbH
52425 Jülich
Germany
b.steffen@fz-juelich.de

Anders Strömberg
Luleå University of Technology
Department of Physics
SE–97187 Luleå
Sweden
anesot-0@student.luth.se

Yuki Sugiyama
Graduate School of Information Science
Nagoya University
Nagoya 464-8601
Japan
sugiyama@phys.cs.is.nagoya-u.ac.jp

Sin-ichi Tadaki
Computer and Network Center
Saga University
Saga 840-8502
Japan
tadaki@cc.saga-u.ac.jp

Jakub Tadych
University of Technology and Agriculture
Bydgoszcz PL85-796
Poland
tajak@co.bydgoszcz.pl

Alexander G. Tatashev
Moscow State Automobile and Road
Technical University
64, Leningradsky pr.
Moscow
Russia
omm@vmat.madi.ru

Svetlana Tatur
Dept. of Physics
Ural State University
620083 Ekaterinburg
Russia
svetlana.tatur@mail.ru

Christian Thiemann
Physik von Transport und Verkehr
Universität Duisburg-Essen
47057 Duisburg
Germany
mail@christian-thiemann.de

Kai-Uwe Thiessenhusen
Deutsches Zentrum für Luft- und
Raumfahrt
Institut für Verkehrsforschung
Rutherfordstr. 2
12489 Berlin
Germany
Kai-Uwe.Thiessenhusen@dlr.de

Nathalie Thomas
Instituto de Investigaciones en
Materiales
Universidad Nacional Autónoma de
México
Apdo. Postal 70-360
Cd. Universitaria
México D.F. 04510
México
nathalie.thomas@polytech.univ-mrs.fr

Tetsuji Tokihiro
University of Tokyo
3-8-1 Komaba Meguro-ku
Tokyo 153-8914
Japan
toki@ms.u-tokyo.ac.jp

Martin Treiber
Institute for Economics and Traffic
Technische Universität Dresden
Andreas-Schubert-Str. 23
D-01062 Dresden
Germany
martin@mtreiber.de

Steffen Trimper
Fachbereich Physik
Martin-Luther-Universität Halle-
Wittenberg
D–06099 Halle/Saale
Germany
trimper@physik.uni-halle.de

Peter Wagner
Deutsches Zentrum für Luft- und
Raumfahrt
Institut für Verkehrsforschung
Rutherfordstr. 12
12489 Berlin
Germany
Peter.Wagner@dlr.de

Peter Walzel
Mechanische Verfahrenstechnik
Universität Dortmund
D–44227 Dortmund
Germany
p.walzel@bci.uni-dortmund.de

Jonathan Ward
University of Bristol
Department of Engineering Mathematics
Bristol BS8 1TR
Jon.Ward@bristol.ac.uk

Hans Weber
Luleå University of Technology
Department of Physics
SE–97187 Luleå
Sweden
Hans.Weber@ltu.se

Jochen H. Werth
University of Duisburg-Essen
Campus Duisburg
Lotharstr. 1
D-47048 Duisburg
Germany
werth@comphys.uni-duisburg.de

R. Eddie Wilson
Bristol Centre for Applied Nonlinear
Mathematics
Department of Engineering Mathematics
University of Bristol
Queen's Building
University Walk
Bristol BS8 1TR
United Kingdom
RE.Wilson@bristol.ac.uk

Dietrich E. Wolf
University of Duisburg-Essen
Campus Duisburg
Theoretische Physik
Lotharstr. 1
47048 Duisburg
Germany
wolf@comphys.uni-duisburg.de

Marko Wölki
Universität Duisburg-Essen
Campus Duisburg
Physik von Transport und Verkehr
47057 Duisburg
Germany
woelki@traf1.uni-duisburg.de

Qing-Song Wu
School of Engineering Science
University of Science and Technology of
China
Hefei 230026
P.R.China
qswu@ustc.edu.cn

Yong-Hong Wu
Department of Mathematics and
Statistics
Curtin University of Technology
Perth WA6845
Australia
Y.Wu@curtin.edu.au

Eiji Yamada
Graduate School of Information Science
Nagoya University
Nagoya 464-8601
Japan

Marina V. Yashina
Moscow State Automobile and Road
Technical University
64, Leningradsky pr.
Moscow
Russia
yamv@math.madi.ru

Yasushi Yokoya
Japan Automobile Research Institute
Karima
Tsukuba
Ibaraki,305-0822
Japan
yyokoya@jari.or.jp

Satoshi Yukawa
Department of Applied Physics
University of Tokyo
Bunkyo 113-8656
Japan
yuk@bopper.t.u-tokyo.ac.jp

Chang Yulin
School of Automobile and Traffic
Engineering
Jiangsu University
Zhenjiang
Jiangsu
P.R. China
ylchang@ujs.edu.cn

Knud Zabrocki
Fachbereich Physik
Martin-Luther-Universität Halle-
Wittenberg
D–06099 Halle/Saale
Germany
zabrocki@physik.uni-halle.de

Stefano Zapperi
Dipartimento di Fisica
Università "La Sapienza"
P.le A. Moro 2
00185 Roma
Italy

and
Istituto dei Sistemi Complessi - CNR
Via dei Taurini 19
00185 Roma
Italy
stefano.zapperi@roma1.infn.it

Gong Zhen
School of Automobile and Traffic
Engineering
Jiangsu University
Zhenjiang
Jiangsu

Henk J. van Zuylen
Delft University of Technology
Faculty of Civil Engineering and
Geosciences – Transport & Planning
Stevinweg 1
2628 CN
Delft
The Netherlands
h.j.vanzuylen@citg.tudelft.nl

Printed in the United States
152493LV00007B/1/P